Igneous and Metamorphic Petrology

INTERNATIONAL SERIES IN THE EARTH SCIENCES

ROBERT R. SHROCK, *Consulting Editor*

DE SITTER · Structural Geology

EWING, JARDETZKY, AND PRESS · Elastic Waves in Layered Media

GRIM · Clay Mineralogy

HEINRICH · Mineralogy and Geology of Radioactive Raw Materials

HEISKANEN AND VENING MEINESZ · The Earth and Its Gravity Field

HOWELL · Introduction to Geophysics

JACOBS, RUSSELL, AND WILSON · Physics and Geology

OFFICER · Introduction to the Theory of Sound Transmission

SHROCK AND TWENHOFEL · Principles of Invertebrate Paleontology

TURNER AND VERHOOGEN · Igneous and Metamorphic Petrology

IGNEOUS
AND METAMORPHIC
PETROLOGY

Francis J. Turner
John Verhoogen

DEPARTMENT OF GEOLOGY
UNIVERSITY OF CALIFORNIA, BERKELEY

SECOND EDITION

McGRAW-HILL BOOK COMPANY, INC.
1960   New York   Toronto   London

Preface

This book is intended for the use of advanced students, research workers, and teachers in the field of petrology. Its aim is to present a unified general impression of the origin and evolution of rocks that are generally believed to have crystallized, or to have been profoundly modified, at high temperatures and at pressures such as prevail from the earth's surface to a depth of perhaps 20 km. This impression must necessarily be marred and blurred by omissions and misinterpretations on the part of the authors, for it is based on data (of field association, mineralogy, chemistry, and fabric of the rocks themselves) too voluminous to be grasped and adequately handled by one or even two students. Moreover, its limitations reflect the prejudices and necessarily restricted geological experience of two individuals.

The science of petrology is underlain by chemical and physical foundations. Many of the most important advances in petrology during the past half century are the result of laboratory experiments of a physical or a chemical nature, and especially of investigation of phase relationships in various chemical systems (particularly in silicate systems) under controlled conditions. Modern petrological literature abounds in references to the concepts of equilibrium and stability, to the kinetics of chemical reaction, to mechanisms of diffusion of matter in and between various media, and to allied topics in the realm of physical chemistry. We have therefore devoted considerable space, especially in Chaps. 2, 14, and 17, to consideration of just such fundamental principles, treated mainly from the thermodynamic standpoint. The reader is advised to refer to this material whenever the physical and chemical implications of petrological data or of laboratory experiment are discussed in the text. We believe that the quantitative physicochemical approach is justified on the ground that discussion of most problems of petrogenesis—and direction of research along the most profitable lines—is possible only in the light of some understanding of the complex behavior of multicomponent systems under a wide range of physical conditions.

Wherever possible throughout the text reference is made to the increasingly voluminous experimental observations related to petrologically significant systems. Experimental and geological observations may lead to divergent or even irreconcilable conclusions. In this conflict lies one of the most stimulating aspects of modern petrology—the demand for evaluation of the relative merits of experimental and geological data. When, as occurs all too commonly, we find ourselves—at least temporarily —at an impasse, we prefer to admit the dilemma, rather than to resolve it by rejecting outright either the accumulated fruit of geological experience or the results of well-conducted laboratory experiments. In this dilemma the geologist must appreciate the inherent difficulties of experimentation in silicate systems, where reactions commonly are slow and equilibrium difficult to establish. He can appreciate such difficulties only by recourse to chemistry and thermodynamics.

No attempt has been made to compile comprehensive bibliographies. Footnote references are given, for the most part, to illustrative papers written in the English language and published within the last two or three decades. Attention is thus focused on modern work easily accessible to English-speaking readers; other works, written in foreign languages, could equally well have been cited. Many of the papers selected include extensive references to the great contributions of classic writers in the field of petrology, including works written in languages other than English.

In conclusion we wish particularly to acknowledge our gratitude to the Geological Society of America for permission to reproduce, in the first edition, lengthy verbatim excerpts and many figures from the Society's *Memoir 30* (Mineralogical and Structural Evolution of Metamorphic Rocks, F. J. Turner, 1948). Some of this material, including Figs. 96–104, 107–117, has been retained in the present edition, but those portions relating to metamorphic facies (Chaps. 19 and 20, Figs. 66–94) have been revised to conform to *Memoir 73* of the Geological Society of America (Metamorphic Reactions and Metamorphic Facies, W. S. Fyfe, F. J. Turner, J. Verhoogen, 1958). In fact Chapters 18 to 24, dealing with metamorphism, are based largely on *Memoirs 30* and *73*.

Francis J. Turner
John Verhoogen

Contents

CHAPTER 1

Introduction

THE PROBLEM OF CLASSIFICATION OF ROCKS

Definition and classification of different kinds of rocks are essential in furnishing a language by which petrological concepts may be conveyed and observational data concerning rocks may be intelligibly recorded. However, there is no general agreement as to the most satisfactory basis of classification or as to the extent to which precision of definition is desirable and practicable. In consequence, a number of radically different systems of classification are in current use, and details of classification are the same in no two standard textbooks.

Teaching of formal courses in petrology and field geology demands some system of classification in which common broad types of rocks may be identified and described in terms of characters that may be observed in the hand specimen and in the field. Such characters are those relating to field occurrence, mineral composition, and fabric (structure or texture). Since many rock-forming minerals commonly occur in grains of microscopic size, some of the criteria of classification must be subject to confirmation with the microscope. Other requirements of a satisfactory general classification are simplicity, lack of unnecessary ambiguity, and some degree of genetic significance.

Classification must be simple enough for teacher and student to grasp and retain its essentials, without continual reference to petrological dictionaries or textbooks. Ambiguity of nomenclature is undesirable. It is unfortunate that the use of such common names as *gabbro* and *diorite* should be complicated by the preference of individual petrographers for definitions based alternatively on the feldspar or on the dark minerals present in the rocks in question. But vagueness of definition arising from natural transition from one class of rock to another is neither avoidable nor undesirable. Any classification which distinguishes sharply between

1

granite and syenite or between andesite and basalt is misleading in that it fails to take into account existing transitional relationships between these classes of rocks.

Since rocks are end products of recurring evolutionary processes that are at least partially understood, most petrologists favor a genetic basis for classification. The processes of origin of some common rocks can be observed directly (*e.g.*, basaltic lava flows or the quartz sands of desert dunes). For other rocks (*e.g.*, sandstone, limestone, diabase) the mode of origin may be deduced with confidence from mineralogical and structural analogies with rocks whose origin is susceptible to direct observation. But there are yet other common rocks (*e.g.*, granites and granodiorites) that may originate in several ways, none of which imprints upon the rock any infallible criterion of origin. Purely genetic classification of rocks is therefore impracticable. Rather each class and type of rock must be defined in terms of recognizable diagnostic characters (notably mineral composition and fabric) so chosen that they bring out as far as possible the mode of origin and any orderly relation that may exist between one rock and another. A similar problem has been solved satisfactorily by zoologists, who employ a genetic classification of the animal kingdom based on observable morphological characters.

The chemical and physical processes that are responsible for the origin of rocks may be grouped broadly as follows:

1. Crystallization of minerals and solidification of glass from siliceous melts at high temperatures (*igneous* processes)
2. Recrystallization and mutual reaction of minerals in solid rocks at high temperatures without intervention of a silicate-melt phase (*metamorphic* processes)
3. Exchange of ions between minerals of solid rocks and migrating fluid phases (usually aqueous gases or solutions) over a wide range of temperatures and pressures (*metasomatic* processes)
4. *Sedimentary* processes, in the broad sense, namely,
   a. Weathering of preexisting rocks
   b. Transport of weathering products by such media as wind and moving water
   c. Deposition of suspended material from air or water, solution and precipitation of minerals by water, and secretion of dissolved materials through organic agencies

The problem of classification of rocks would be simplified, at least in so far as the major divisions are concerned, if the processes involved in the origin of any one common kind of rock were invariably confined to one or other of these four categories. It would be convenient, too, if there were ubiquitous and infallible criteria of igneous, metamorphic, metasomatic, and sedimentary origin. Neither of these conditions holds good, however, although they are to some extent implied by the adoption of any precisely defined genetic classification of rocks.

In this book the conventional threefold classification of rocks into igneous, metamorphic, and sedimentary groups will be followed, because it is commonly possible to tell from field and petrographic characters that either igneous or metamorphic or sedimentary processes have played at least the dominant role in the origin of most rocks. But the diagnostic characters of some are much more sharply defined and obvious than those of others. Moreover there are still important classes of rocks that combine the features of two of the main groups; such for example are the migmatites, in part igneous and in part metamorphic; such, too, are the pyroclastic rocks, composed largely of igneous fragments deposited as sediment from suspension in air or water. It is generally recognized, too, that in depth, igneous and metamorphic phenomena merge imperceptibly, so that it may be impossible to determine whether a given mass of granitic rock is of truly igneous or combined igneous, metamorphic, and metasomatic origin.

Boundaries between all divisions in the rock classification here followed are regarded as vague and transitional. For convenience of student and teacher, synoptic schemes of classification will be given; but these do not represent in true perspective the mutual relationship of all rock types. Precise definition generally will be avoided. Natural classes of rocks lend themselves to description by referring to typical illustrative examples, rather than to exact limitation in terms of precise mineralogical or chemical data. For the purposes of this book a simple classification is sufficient. Highly complex schemes such as that of A. Johannsen [1] or the norm classification,[2] both including only igneous rocks, serve a useful purpose as reference catalogues for recording and grouping mineral combinations, variations in fabric, and chemical analyses. However, they do not furnish a satisfactory medium for petrological discussion.

IGNEOUS AND METAMORPHIC ROCKS AS CHEMICAL SYSTEMS

Rocks which have originated largely through the operation of igneous or metamorphic processes consist of assemblages of minerals—crystalline phases—which have been held at high temperatures, in mutual contact, for periods of time which are long compared with the duration of most simple laboratory experiments. In most, if not all, cases the temporary presence of a fluid phase—e.g., a silicate melt or an aqueous intergranular

[1] A. Johannsen, A Descriptive Petrography of the Igneous Rocks, vol. I, pp. 140–158, University of Chicago Press, Chicago, 1931.

[2] J. P. Iddings, Igneous Rocks, vol. I, pp. 348–393, Wiley, New York, 1909.

fluid—has provided a medium for mutual reaction between the associated minerals as they crystallized.   The mineral assemblages of both igneous and metamorphic rocks have thus had some opportunity to attain a state of mutual tolerance or equilibrium at high temperature.   This is not so with most clastic sedimentary rocks, the constituent fragments and grains of which have been brought together almost fortuitously from widely different sources at temperatures too low to allow mutual reaction.

Some of the most widely accepted classifications of igneous and metamorphic rocks depend upon an assumption that rocks of this kind have indeed reached a state approximating to internal equilibrium at the high temperatures at which they crystallized, and have not been substantially affected by subsequent cooling to ordinary temperatures.   The constant recurrence of a few relatively simple mineral assemblages in common igneous and metamorphic rocks supports, though it does not prove, the validity of this interpretation.   Some understanding of the nature of equilibrium and stability, especially as applied to multicomponent silicate systems, is therefore an essential background for discussion of the problems of igneous and metamorphic petrogenesis.   It is also important to realize that the same general principles governing stability and equilibrium in silicate systems apply equally to metamorphic and to igneous petrogenesis.   The main difference concerns the temporary fluid phase—a silicate melt present in abundance during igneous crystallization, an aqueous pore fluid present in relatively small quantities in metamorphism.   Some of the principles governing equilibrium and related phenomena are discussed in the chapter which now follows.

CHAPTER 2

Principles of Chemical Equilibrium Applied to Rocks

INTRODUCTORY STATEMENT

The object of petrology is to study changes that occur spontaneously in rock masses: liquid magmas which solidify, solid rocks which melt partially or totally, sediments which undergo chemical or physical transformations. Petrology is thus concerned essentially with a flow of particles, ions, atoms, molecules, or whole crystals, moving into, through, or out of rocks, changing within any rock from one phase to another, or changing position within the same phase. The physical chemist who wishes to analyze such phenomena, and bring order and logic to them, may attempt to do so by means of experiments aimed at reproducing in the laboratory the conditions which prevail in nature. The natural systems with which he is concerned are, however, of such complexity and the conditions under which they evolve are, in some aspects, so imperfectly understood that laboratory experiments are not likely to provide a clear and complete picture of what really happens in the crust of the earth. The tremendous amount of experimental work that has been done on the synthesis of hydrothermal minerals, for instance, has still not produced a clear picture of how such minerals actually form. We must then fall back on more powerful tools, on some more general laws; and we turn instinctively to thermodynamics, which is a science of the possible and the impossible, of what may happen and what may not. These all-embracing laws govern all natural transactions, and there is no doubt that even rock masses must abide by them. Our first purpose is therefore to present these laws in a convenient form.

Thermodynamics starts from two postulates, known as the first and second "laws," for which there is no proof except that they have, so far, been found valid in every instance. From this starting point, and from a few additional physical hypotheses, the whole structure of this science is derived by a process of mathematical deduction. Thermodynamics is thus essentially mathematical, although the mathematics involved are relatively simple. Now since our purpose is mainly to deal with real physical systems, the nonmathematical reader may wonder why the mathematics are necessary at all; and it is quite true that most results of thermodynamic reasoning could be obtained purely from physical and chemical measurements. This procedure would have, however, two rather serious defects. In the first place, it would not allow us to predict what is likely to happen in a system which has not been investigated experimentally—and most of our petrological systems have not. In the second place, if we had to express nonmathematically our results in a form as general, and as rigorous, as a simple mathematical formula, we would have to use pages to say what can be expressed in a line by means of an equation. In the long run it takes less trouble to master the mathematics of the situation than to do without them.

There is no doubt, therefore, that the serious student of petrology who wishes to interpret correctly new data and explore successfully new fields will find ample reward for the efforts he makes at mastering the thermodynamic tools. But there is also no doubt that a complete command of the mathematical language of physical chemistry is not essential to an understanding of the more descriptive aspect of petrology. We have attempted therefore to summarize, rather than to explain, certain important physical principles, the exact significance of which cannot, however, be fully appreciated without reference to fuller treatments such as are given in standard texts in thermodynamics. We must thus warn the reader against misconceptions which may arise from too literal an interpretation of the following pages.

CONDITIONS FOR EQUILIBRIUM. THE CHEMICAL POTENTIAL

General Relations.[1] *Internal Energy.* The fundamental concepts to be introduced first are the internal energy and entropy of a system.[2] The

---

[1] The reader should refer for more details and for rigorous proof of many statements in this chapter to classic texts such as G. N. Lewis and M. Randall, *Thermodynamics and the Free Energy of Chemical Substances,* McGraw-Hill, New York, 1923; E. A. Guggenheim, *Modern Thermodynamics by the Method of J. W. Gibbs,* Methuen, London, 1933; H. S. Taylor, *Treatise on Physical Chemistry,* Macmillan, New York, 1931; I. Prigogine and R. Defay, *Treatise on Thermodynamics, Based on the Methods of Gibbs and De Donder,* vol. 1, Chemical Thermodynamics, Longmans, Green, New York, 1954; G. W. Morey, The Application of thermodynamics to heterogeneous equilibria, *Jour. Franklin Institute,* vol. 194, pp. 425–484, 1928. See also P. Lafitte, *Introduction à l'étude des roches métamorphiques et des gites métallifères,* Masson, Paris, 1957.

[2] The word *system* is used here in its thermodynamic sense to describe any particular portion of the universe in which we are interested, as, for instance, a galaxy, or a

change in the internal energy is defined as the total energy the system gains or loses in the course of any process. This energy may be supplied to, or taken from, the system in any form: heat, mechanical work, radiation, etc. According to the first law of thermodynamics, all systems which are in the same state have the same internal energy, irrespective of the manner in which the system was brought into that state. The change in internal energy dE in the course of any transformation depends only on the initial and final states of the system, not on the particular transformation through which the system has been brought from the one to the other.

The internal energy E is therefore defined formally as a function of the state of the system, with the property that, in any infinitesimal process taking place without any alteration of composition, internal energy increases by an amount which is equal to the heat dq absorbed by the system plus the work dw done on the system.

$$dE = dq + dw \qquad (2\text{-}1)$$

If the system at a pressure P is in mechanical equilibrium with its surroundings, the work dw corresponding to a change in volume dV is

$$dw = -P \, dV \qquad (2\text{-}2)$$

so that

$$dE = dq - P \, dV \qquad (2\text{-}3)$$

Reversible Processes. A reaction is said to proceed reversibly when the system is balanced in such a manner that an infinitesimal change in the conditions will cause the reaction to proceed in the opposite direction. When ice melts at $0°C.$ and atmospheric pressure, the reaction is reversible in so far as an infinitesimal drop in temperature will cause the water to freeze again; the same reaction proceeding at $10°C.$ is not reversible because a very small drop in temperature will not make the water freeze. The reaction

$$CaCO_3 + SiO_2 \rightleftarrows CaSiO_3 + CO_2$$

is reversible at $270°C.$ and when the pressure of CO_2 is 1 bar, because any small increase or decrease in the pressure of CO_2 will start the reaction running from right to left or from left to right respectively.

A ball running on a plane surface would be proceeding reversibly, according to this definition, if any infinitesimal change in the slope could

certain amount of gas enclosed in a container, or a given assemblage of crystals, or a small portion inside a crystal. The word *system* is used also by petrologists in a somewhat different and broader sense, as in the expression "the system albite-anorthite." In the latter connotation, *system* means all possible thermodynamic systems consisting of mixtures of albite and anorthite, either solid or liquid or both, and in any proportion. Wherever there is possibility of confusion between these two meanings, we shall use the expression *limited system* for the thermodynamic system as defined above.

make it run in the opposite direction.   This can be true only if the slope is infinitesimally small to begin with, in which case the velocity of the ball will also be very small.   A reaction is thus reversible only when the conditions differ infinitesimally from the equilibrium conditions, and the velocity of a reversible reaction is infinitesimally small.   All reactions proceeding at a finite rate are therefore irreversible, and occur only when the system has been somewhat removed from the equilibrium conditions. The reversible reaction is no more than an idealized, though very useful, concept.

Entropy.   This fundamental difference between reversible and irreversible processes is the basis for introducing the concept of entropy. Entropy is defined by its two most important properties, *viz.*: (1) in any reversible process the change in entropy of the system is measured by the heat received by the system divided by its absolute temperature; (2) in any spontaneous irreversible process the change in entropy is greater than this amount.   The second law of thermodynamics states that this entropy S, like internal energy, is a function of the state of the system, being independent of the manner in which the system was brought to that state; and that for any change in the course of which the heat absorbed is dq,

$$
\begin{aligned}
dS &> dq/T &&\text{for any spontaneous irreversible process} \\
dS &= dq/T &&\text{for a reversible process}
\end{aligned}
\right\} \quad (2\text{--}4)
$$

For instance, when 1 mole (60 gm.) of cristobalite melts reversibly at 1713°C., or 1986°K., the heat absorbed (heat of fusion) is 2035 cal.   The change in entropy is therefore

$$
\Delta S_{\text{melting}} = {}^{2035}\!/_{1986} = 1.02 \qquad \text{cal./deg. mole}
$$

Combining (2–3) and (2–4), we obtain, for any reversible process,

$$
dE = T\, dS - P\, dV \tag{2--5}
$$

We note that for a system which is thermally insulated $dq - 0$; hence $dS > 0$ for any irreversible process, that is, for any actual process occurring with finite velocity.   Since the universe as a whole is by definition an isolated system, its entropy must thus continuously increase.   This is often given as a statement of the second law.

The practical importance of entropy in problems of chemical equilibrium arises from the fact that, as appears already in Eq. (2–5), entropy bears a somewhat similar relation to temperature as volume does to pressure.   Just as pressure influences chemical reactions through volume changes connected with the reactions, temperature affects chemical reactions through entropy changes connected with these reactions.   The entropy of a pure substance is a positive quantity which may be calculated at any temperature T if the specific heat c of this substance is known at all temperatures between zero (absolute) and T; for the specific heat (at constant pressure) is defined as the ratio of the supplied heat dq to the corresponding change in temperature dT (that is, $c = dq/dT$) we have, by (2–4),

$$dS = \frac{dq}{T} = \frac{c\,dT}{T} = c\,d\ln T$$

and

$$S - S_0 = \int_0^T c\,d\ln T$$

S_0 being the entropy of the substance at the absolute zero (which is zero for perfect crystals; see below).

Physical Significance of Entropy. The concept of energy is familiar enough. A system contains energy by virtue of (1) its capacity to do mechanical or electrical work; (2) the amount of radiation it contains; (3) the amount of heat it contains. (Note that this last term is not as easy to define exactly as might at first appear, for heat is sometimes defined in terms of temperature, while temperature in turn is usually defined in terms of heat.) For present purposes we may perhaps think of the internal energy of a body as consisting of two terms: the potential energy determined by the forces acting between its constituent particles, and the kinetic energy of motion of these particles. The first term is the more important one in determining the energy of solids, while the latter, which predominates in determining the energy of gases, corresponds roughly to the usual concept of heat.

The concept of entropy, on the other hand, is not generally familiar, and attempts to represent entropy in concrete terms are liable to be misleading. It is often said, for instance, that TS represents the "part of the energy which is bound as heat," a statement which is generally not correct. Perhaps the most easily grasped interpretation of entropy is that which is given by statistical mechanics. Statistical mechanics is the branch of physical chemistry which attempts to predict the behavior of a system by considering the dynamic behavior of an individual particle (as deduced from the laws of mechanics), and applying the results statistically to cover the average behavior of all particles in the system. In this type of treatment a function is introduced which turns out to have all the properties of entropy stated in the second law, but which, in addition, is related to the "probability" or "spread" of the system, that is, to the number of ways in which it is possible to combine properties of individual particles in order to reproduce the observed, macroscopic, properties of the system. The significance of this quantity may perhaps be understood by considering two examples. Consider first a gas, the molecules of which are moving within a container of given volume. This gas may be described in terms of certain macroscopic properties such as temperature or pressure; this pressure, in turn, may be interpreted in terms of collisions of molecules striking against the walls of the container. We know from elementary kinetic theory that all particles do not have exactly the same velocity; some move faster than others. But it is irrelevant, for the purpose of determining the pressure, whether molecule a has velocity v_1 and molecule b velocity v_2 or vice versa. There are many ways in which velocities may be assigned to individual molecules consistently with the observed pressure. To use the language of statistical mechanics we say that there is a large number of possible microscopic states, or complexions of the system, each complexion consisting of a certain assignment of velocities and coordinates to individual particles. If W is the "probability" of the system, $i.e.$, the number of such possible complexions, the entropy per mole is defined as

$$S = k \ln W$$

where k is a constant (Boltzmann constant) equal to the usual gas constant R divided by the number of particles in 1 mole ($k = 1.38 \times 10^{-16}$ erg/deg.). The entropy thus increases as the logarithm of the probability.

There is no formal definition of "spread" in terms of W, but the significance of the word may be understood from the fact that the entropy of a gas turns out to be directly proportional to the standard deviation, or "spread," of the curve representing the frequency distribution of velocities among individual molecules.

Turn now to the opposite extreme represented by a perfect crystal, without flaws of any kind, at the absolute zero of temperature. In such a state, particles have no motion, all velocities are equal and have zero value, and the coordinates of each particle are rigidly defined in terms of lattice constants. Hence there is no spread, either with regard to velocity or to position, and consequently there is only one possible microscopic state, or complexion, of the system. Hence

$$S = k \ln W = k \ln 1 = 0$$

and the entropy of a perfect crystal at the absolute zero is zero. If heat is supplied to the crystal and its temperature raised, the particles are set in motion and the entropy increases, corresponding to an increasing spread in position and oscillatory motions. At the melting point, when the particles leave their appointed positions in the crystal lattice to take up random translational motion in the liquid, the entropy increases still further. The entropy of a liquid is thus larger than that of the corresponding crystal at the same temperature.

Space will not permit us to consider the formal steps by which it is shown that the entropy defined in terms of W is really the same as entropy defined by (2–4); but it is appropriate to note that the statement of the second law (that the entropy increases in every irreversible process in an isolated system) means that an isolated system tends, as a result of any irreversible process, to a more probable, or more disordered, state with a greater spread.   There are many ways in which disorder may be increased even in the absence of thermal motion or heat.   For instance, black and white balls in separate boxes form a more ordered system than a random mixture of the two; mixing of the balls therefore increases the entropy. Similarly, two gases will mix spontaneously because of the corresponding increase in disorder or entropy; and the process of mixing is irreversible because, once mixed, the two kinds of molecules do not tend to segregate in different portions of the container (or, at least, there is an overwhelming probability that they will not do so).

It is necessary to emphasize the word "isolated" in the above statement of the second law, for it is clear from Eq. (2–4) that entropy may decrease if dq is negative, i.e., if enough heat is withdrawn from the system.   For instance, our two gases may be separated again, with decrease in entropy, if they are cooled to the condensation point of one of them.   Two liquids may be separated also by supplying heat, as in distillation.   What the second law means in such cases is that although the entropy of any system may actually decrease, the entropy of the rest of the universe will increase by a larger amount, as a result of the operations by which entropy has been locally reduced.   The exact computation of this entropy increase would, of course, be extremely difficult; it would be necessary, for instance, to determine the entropy changes connected with the cerebral work of the chemist who has designed the distillation plant.

Our principal objective is to determine in which direction, under given physical conditions, a reaction will run spontaneously.   The second law states that it will run in the direction appropriate for increasing disorder, but the first law says that energy changes within the system must balance the energy received from, or lost to, the surroundings.   Suppose, for instance, that increasing disorder requires addition of energy, as when the particles of a crystal are made to oscillate more rapidly; if no energy is supplied from outside, increase in entropy cannot take place, in agreement with the experimental fact that the temperature of a thermally insulated crystal does not increase spontaneously.   Or suppose that increase in volume is a necessary requirement for a more random arrangement of particles; then if the system is under external pressure P, increase in entropy or randomness cannot occur if the system does not have a supply of energy from which to draw the required amount of work $P\,dV$ to be done against the external pressure.   Thus the possibility for occurrence of a reaction will depend on (1) change in entropy; (2) supply or expenditure of heat involved in (1); (3) the work that has to be done by, or against, all external constraints.

To take all three factors into account simultaneously and automatically, a number of functions have been introduced, mainly as a matter of convenience. Take, for instance, the quantity H (enthalpy) defined formally as

$$H = E + PV \qquad (2\text{-}6)$$

Remembering the rules for differentiation of a product, we have, for any infinitesimal transformation,

$$dH = dE + P\,dV + V\,dP$$

If the transformation occurs at constant pressure ($dP = 0$) we have, by (2-3),

$$dH = dq$$

so that the change in enthalpy in any process at constant pressure is measured by the heat received (H, for this reason, is also known as the "heat content"). If the reaction occurs at constant entropy ($dS = 0$) but reversibly, $dq = 0$ and

$$dH = 0$$

If the reaction occurs at constant entropy but irreversibly, $dq < 0$ and

$$dH < 0$$

Thus a reaction runs spontaneously, at constant pressure and entropy, in the direction for which H decreases.

Consider now the function F (elsewhere commonly designated A) which is the Helmholtz free energy and is defined as

$$F = E - TS \qquad (2\text{-}7)$$

Then, for any reversible process at constant temperature,

$$dF = dE - T\,dS - S\,dT = dE - T\,dS = dw$$

by (2-1) and (2-4). Hence the change in F is measured by the work done in a reversible isothermal reaction (F, for this reason, is also known as the "maximum work"). Now if a reaction runs reversibly at constant temperature and volume

$$dF = dE - T\,dS = dq - P\,dV - T\,dS = dq - T\,dS = 0$$

and, if the reaction is irreversible,

$$dF < 0$$

Hence a reaction runs spontaneously, in a system held at constant temperature and volume, in the direction in which F decreases.

Thirdly, consider the function

$$G = E + PV - TS \qquad (2\text{-}8)$$

which is known as the Gibbs free energy (commonly designated F when the Helmholtz free energy is noted A). For any reaction at constant T and P we have

$$dG = dE - T\,dS + P\,dV$$

If the reaction is reversible,

$$dE = T\,dS - P\,dV$$

and

$$dG = 0$$

If the reaction is spontaneous and irreversible,

$$dE < T\,dS - P\,dV$$

by (2-3) and (2-4), and

$$dG < 0$$

Hence the Gibbs free energy decreases in any reaction proceeding spontaneously at constant temperature and pressure.

Note finally that for any reversible reaction at constant V and S

$$dE = 0$$

and, if the reaction is irreversible,

$$dE < 0 \qquad (2\text{-}9)$$

These four quantities E, H, F, G, which have the fundamental common property of determining the direction in which a reaction will run under certain experimental conditions, are known as "thermodynamic potentials." The two potentials which we shall find most useful are F and G; they have, as has been shown, the important properties that for any spontaneous reaction at P, T constant

$$dG < 0 \qquad (2\text{-}10)$$

whereas for any spontaneous reaction at V, T constant

$$dF < 0 \qquad (2\text{-}11)$$

Remembering that "reversible reaction" is synonymous with "equilibrium," we have the corresponding equilibrium conditions:

At constant P and T:

$$dG = 0$$

At constant V and T:

$$dF = 0$$

It cannot be overemphasized at this stage that the quantity to be used as a measure of the driving force or, as we say, of the "affinity" of a reaction depends on the variables used, that is, on the experimental conditions under which the reaction is proceeding. For instance, the affinity of a reaction running at constant V, T is measured by changes in F; that of a reaction at constant P and T is measured by changes in G.

Open Systems. We have considered so far only systems in which the amount of each component is held constant. If the number n_i of moles of any component i varies in the course of the transformation, a certain amount of i being added to, or subtracted from, the system, it is reasonable to suppose that the internal energy E will vary proportionally to the amount of each component. As we know already by (2–5) that E is a function of S and V, we write now

$$E \equiv E(S,V,n_i)$$

meaning that the internal energy of a system is a function of its entropy, its volume, and the number of moles of each component.

We now apply the principle of superposition of effects; that is, we state that the change in E, resulting from simultaneous changes in volume, entropy, and composition, is the same as would result if we proceeded by steps, changing only one factor at a time while all other factors are held constant, and adding all successive steps. We write this in the following manner

$$dE = \left(\frac{\partial E}{\partial V}\right)_{S,n_i} dV + \left(\frac{\partial E}{\partial S}\right)_{V,n_i} dS + \sum_i \left(\frac{\partial E}{\partial n_i}\right)_{S,V,n_j} dn_i \qquad (2\text{–}12)$$

where the sum $\sum_i$ means that we consider successively changes in all components, and n_j designates the number of moles of all components except i. The symbol ∂/∂ stands, as usual, for partial differentiation, that is, differentiation which is carried out with respect to one variable, all the other variables being held constant. The variables which are held constant in each step are indicated by subscripts. Thermodynamicists are very particular about these subscripts and insist that they should never be omitted, as they serve to describe the physical conditions under which the experiment is being carried out. It is important to realize that the same experiment carried out under different conditions leads to different results. For instance, if we change the pressure on a perfect gas and measure the corresponding change in volume, we find different results, depending on whether the experiment is carried out at constant temperature or at constant entropy ("adiabatically"). Similarly, we shall find that the behavior of a given magma may be changed appreciably by changing the conditions under which it is held.

Obviously, Eq. (2–12) should reduce to (2–5) when $dn_i = 0$.   Hence, we must have

$$\left(\frac{\partial E}{\partial V}\right)_{S,n_i} = -P \qquad \left(\frac{\partial E}{\partial S}\right)_{V,n_i} = T$$

which may serve as a thermodynamic definition of T and P.

If we define the chemical potential μ_i of component i by the relation

$$\mu_i \equiv \left(\frac{\partial E}{\partial n_i}\right)_{S,V,n_j} \tag{2–13}$$

Eq. (2–12) becomes

$$dE = T\,dS - P\,dV + \sum_i \mu_i\,dn_i \tag{2–14}$$

Note that S, V, n_i are extensive properties, that is, properties which depend on the quantity of the system, in contradistinction to P, T, μ_i, which do not depend on the quantity of the system.   Let us now increase the quantity of the system without altering the conditions, that is to say, let us integrate (2–14) for constant values of P, T, μ_i.   Then

$$E = TS - PV + \sum_i \mu_i n_i$$

and, by comparison with (2–8),

$$G = \sum_i \mu_i n_i \tag{2–15}$$

that is, the free energy of a system is equal to the sum of the chemical potentials of each of its components multiplied respectively by the number of moles of each component in the system.

For a system of one component containing 1 mole

$$G = \mu \tag{2–16}$$

The chemical potential of a pure component is thus equal to its free energy per mole.

By writing relations similar to (2–12) for the other functions H, F, G, one obtains the following relations which are similar to (2–13) and which may all serve to define the chemical potential:

$$\mu_i = \left(\frac{\partial E}{\partial n_i}\right)_{S,V,n_j} = \left(\frac{\partial H}{\partial n_i}\right)_{S,P,n_j} = \left(\frac{\partial F}{\partial n_i}\right)_{V,T,n_j} = \left(\frac{\partial G}{\partial n_i}\right)_{P,T,n_j} \tag{2–17}$$

The last of these identities is perhaps the one that is most commonly used to define μ.

Fundamental Properties of the Chemical Potential.   *Equilibrium Conditions.*   For any spontaneous process occurring at constant P and T, we have

$$dG < 0 \tag{2–10}$$

In particular, if we consider the transfer, at constant pressure and temperature, of dn_i moles of component i from phase α to phase β, the free energy variation is

$$dG = - \left(\frac{\partial G^\alpha}{\partial n_i}\right)_{P,T,n_j} dn_i + \left(\frac{\partial G^\beta}{\partial n_i}\right)_{P,T,n_j} dn_i$$

or, by (2–17),

$$dG = - \mu_i^\alpha \, dn_i + \mu_i^\beta \, dn_i \qquad (2\text{–}18)$$

If the transfer occurs spontaneously, we must have, because of (2–10),

$$\mu_i^\alpha > \mu_i^\beta \qquad (2\text{–}19)$$

A *necessary condition for i to migrate spontaneously from phase α to phase β is that its chemical potential should be greater in phase α than in phase β.* For instance, water will vaporize only if its chemical potential is greater in the liquid than in the vapor. If no transfer occurs, *i.e.*, if the two phases are in equilibrium with respect to i, we must have

$$\mu_i^\alpha = \mu_i^\beta$$

The general conditions for equilibrium in a system of c components 1, 2, . . . , c distributed in φ phases α, β, . . . are thus simply

$$\left.\begin{aligned}
\mu_1^\alpha = \mu_1^\beta = \cdots = \mu_1^\varphi \\
\cdots \\
\mu_c^\alpha = \mu_c^\beta = \cdots = \mu_c^\varphi
\end{aligned}\right\} \qquad (2\text{–}20)$$

which mean that, at equilibrium, the chemical potential of each component of the system is the same in all phases in which it is present.

Similarly, let us write the equation corresponding to a reaction between chemical species a, b, p, q, . . . in the form

$$v_a a + v_b b + \cdots = v_p p + v_q q + \cdots$$

where v_a, v_b . . . are the stoichiometric coefficients of the reaction. This reaction will run spontaneously from left to right at constant pressure and temperature according to (2–10) only if the free energy decreases, that is, if the free energy of the reactants is greater than that of the products. Now the free energy of the reactants, by (2–15), is equal to the sum of the products obtained by multiplying the chemical potential of each component by the stoichiometric coefficient of the corresponding species; and the same is true for the products of the reaction. Hence a necessary condition that the reaction may run from left to right is that

$$v_a \mu_a + v_b \mu_b + \cdots > v_p \mu_p + v_q \mu_q + \cdots \qquad (2\text{–}21)$$

and when equilibrium is reached, the two sides of this equation become equal. It has already been noted that a condition such as (2–10) applies only to reactions at constant P and T; at constant V and T the condition for a spontaneous reaction is $dF < 0$. Equation (2–21), on the contrary, is always valid; but it must be remembered that when a reaction runs at constant volume, the pressure usually changes as the reaction proceeds, so that the chemical potentials which appear on both sides of (2–21) are not computed at the same pressure.

If a system initially in equilibrium is to remain in equilibrium when certain conditions (say, temperature and pressure) are changed, these changes must affect the chemical potential of each component by equal amounts in all phases in which it is present. It is by expressing such conditions mathematically that most of the useful relations are obtained. The first step is therefore to find how the chemical potential changes as a function of P and T; and this requires first a definition of partial molar quantities.

Partial Molar Quantities. When salt is dissolved in water it may be observed that the volume of the solution is not equal to the sum of the volumes of pure water and solid salt. Such changes in volume on solution result from the interaction of the molecules of the solute with those of the solvent, which may be pulled farther apart or, on the contrary, closer together. The partial molar volume of salt in water is then defined as the change in volume of a large amount of solution when a small amount of salt is added to it at constant temperature and pressure. A partial molar volume may be positive, negative, or zero.

Similarly, solution of a solute in a solvent may be accompanied by certain thermal effects; heat may be given off, or absorbed from the surroundings. Such heat exchanges involve a change in entropy. There are also certain changes in entropy which are not connected with any heat effect, but which arise from the irreversibility of the process of mixing: salt, once dissolved at constant pressure and temperature, cannot be recovered by an infinitesimal change in the conditions. Thus the partial molar entropy of salt in water is defined as the change in entropy of a large amount of solution when a small amount of salt is added to it at constant temperature and pressure. Similar definitions may be given for any other partial molar quantity (*e.g.*, partial molar energy). One usually denotes the partial molar volume or the partial molar entropy of component i of a solution by such symbols as $\bar{v}_i$, $\bar{s}_i$, and we write

$$\bar{v}_i = \left(\frac{\partial V}{\partial n_i}\right)_{P,T,n_j} \qquad \bar{s}_i = \left(\frac{\partial S}{\partial n_i}\right)_{P,T,n_j}$$

We note that the partial molar volume and partial molar entropy of a pure component (that is, of a phase consisting of a single pure substance) are identical with the corresponding molar quantities, which are always positive. The last identity in Eq. (2–17) shows also that the chemical potential of any component i of a solution is equal to its partial molar free energy. Lewis and Randall,[3] for instance, use the expression "partial molar free energy" rather than "chemical potential."

The quantities $\bar{v}_i$, $\bar{s}_i$ are extremely important in practical problems because the change in chemical potential of component i when the pressure or temperature is changed is found, as will be shown presently, to be equal respectively to $\bar{v}_i$ times the change in pressure, or to $\bar{s}_i$ times the change in temperature, the latter product being taken with a minus sign. In other words, if $\bar{v}_i$ is positive, the chemical potential of i increases with increasing pressure. If $\bar{s}_i$ is positive, the chemical potential decreases with increasing temperature, and vice versa. The chemical potential of a component of a solution always increases when the concentration of this constituent increases;[4] it may increase, decrease, or be unaffected by any change in the concentration of some other component.

Molar Fraction. It is convenient to express the composition of a phase by the "molar fractions" of its components. If a phase α contains n_1 moles of component 1, n_2 moles of component 2, n_c moles of component c, the molar fraction N_i^α of component i in this phase α is, by definition,

$$N_i^\alpha = \frac{n_i}{n_1 + n_2 + \cdots + n_c} = \frac{n_i}{\sum_i n_i} \qquad (2\text{–}22)$$

Obviously, we must have

$$\sum_i N_i^\alpha = 1 \qquad (2\text{–}23)$$

meaning that the sum of the molar fractions of all components of a phase is always equal to 1.

It may then be shown that if $\bar{v}_1$, $\bar{v}_2$, . . . are the partial volumes of components 1, 2, . . . , the total volume of V^α of the phase α is

$$V^\alpha = \sum_i n_i \bar{v}_i^\alpha \qquad (2\text{–}24)$$

The "mean molar volume" v^α, or volume of the quantity of the phase that contains altogether 1 mole, is

$$v^\alpha = \frac{V^\alpha}{\sum_i n_i^\alpha} = \sum_i N_i^\alpha \bar{v}_i^\alpha \qquad (2\text{–}25)$$

Similar relations obtain for all other extensive properties, *e.g.*,

$$s^\alpha = \sum_i N_i^\alpha \bar{s}_i^\alpha$$

[3] Lewis and Randall, *op. cit.*, p. 203, 1923.
[4] Guggenheim, *op. cit.*, p. 24, 1933.

when s^α is the entropy of the quantity of the phase that contains altogether 1 mole.

Gibbs-Duhem Relation.   By differentiation of (2–8) we obtain, for any
phase α,

$$dG^\alpha = dE^\alpha - T\,dS^\alpha - S^\alpha\,dT + P\,dV^\alpha + V^\alpha\,dP$$

where dE^α is given by (2–14).   By differentiation of (2–15) we have also

$$dG^\alpha = \sum_i \mu_i^\alpha\,dn_i^\alpha + \sum_i n_i^\alpha\,d\mu_i^\alpha$$

so that finally

$$-V^\alpha\,dP + S^\alpha\,dT + \sum_i n_i^\alpha\,d\mu_i^\alpha = 0 \tag{2–26}$$

Dividing throughout by $\sum_i n_i^\alpha$, (2–26) becomes

$$-v^\alpha\,dP + s^\alpha\,dT + \sum_i N_i^\alpha\,d\mu_i^\alpha = 0 \tag{2–27}$$

a very important relation, known as the Gibbs-Duhem equation, which
shows how the chemical potentials in any phase α must change when
the pressure and temperature change by amounts dP and dT, respectively.

If the system consists of several phases (*e.g.*, solid, liquid, gas) an equation of type (2–27) may be written for each of these.   If, in addition, the
φ phases forming the system are in equilibrium with respect to all components of the system, the variation of the chemical potential of any
component must be the same in all phases in which it is present, so that
we have φ equations involving the $(c+2)$ variables dT, dP, $d\mu_1$, . . . ,
$d\mu_c$.   Several applications of this relation will be made later (see, for
instance, the derivation of the Clausius-Clapeyron relation on page 22, and
the determination of the vapor pressure in a two-component, three-phase
system, page 409).

If phase α contains only one component, and consists therefore of a pure
substance, we have, for this substance,

$$-v\,dP + s\,dT + d\mu = 0 \tag{2–28}$$

In particular, for any change in pressure at constant temperature ($dT = 0$),

$$d\mu = v\,dP$$

and similarly, at constant pressure ($dP = 0$),

$$d\mu = -s\,dT$$

which we write

$$\left(\frac{\partial\mu}{\partial P}\right)_T = v \qquad \left(\frac{\partial\mu}{\partial T}\right)_P = -s \tag{2–29}$$

More generally, the variation with pressure or temperature of the chemical potential of any component i of a phase α of constant composition may be written as

$$\left(\frac{\partial \mu_i^\alpha}{\partial P}\right)_{T,n_i,n_j} = \bar{v}_i^\alpha \qquad \left(\frac{\partial \mu_i^\alpha}{\partial T}\right)_{P,n_i,n_j} = -\bar{s}_i^\alpha \qquad (2\text{-}30)$$

The most general form for the variation of the chemical potential, including changes in composition, is

$$d\mu_i = \left(\frac{\partial \mu_i}{\partial P}\right)_{T,n_i,n_j} dP + \left(\frac{\partial \mu_i}{\partial T}\right)_{P,n_i,n_j} dT + \sum_i \left(\frac{\partial \mu_i}{\partial n_i}\right)_{P,T,n_j} dn_i$$

$$= \bar{v}_i\, dP - \bar{s}_i\, dT + \sum_i \left(\frac{\partial \mu_i}{\partial n_i}\right)_{P,T,n_j} dn_i \qquad (2\text{-}31)$$

an expression which we shall use, for instance, to derive the change in solubility of a solid with pressure or temperature (see page 24).

Explicit Forms of the Chemical Potential. The chemical potential of a pure perfect gas may be written under the form

$$\mu = \mu^0 + RT \ln P \qquad (2\text{-}32)$$

where R is the gas constant (1.986 cal./deg. mole = 83.15 bars cm.^{3}/deg. mole), and μ^0 is a function of temperature only. Indeed, by differentiation of (2-32) at constant T, we obtain

$$\left(\frac{\partial \mu}{\partial P}\right)_T = \frac{RT}{P}$$

By (2-29) we have also

$$\left(\frac{\partial \mu}{\partial P}\right)_T = v$$

so that

$$v = \frac{RT}{P}$$

which is the usual form of the perfect-gas law.

Suppose now that we have a gaseous phase, which obeys the perfect-gas law, and which is a mixture of several components. The "partial vapor pressure" p_i of component i is then defined as

$$p_i = PN_i \qquad (2\text{-}33)$$

when P is the total pressure, and the chemical potential of i in this gas mixture is written as

$$\mu_i = \mu_i^0 + RT \ln p_i \qquad (2\text{-}34)$$

Since $\sum_i N_i = 1$, we have

$$\sum_i p_i = P$$

The "partial vapor pressure" of any component i of a liquid or solid solution is defined as the partial vapor pressure of this component in the saturated vapor in equilibrium with this phase. Since at equilibrium the chemical potential of any component is the same in all phases in which it is present, the partial vapor pressure of any component in the gas phase may be taken as a measure of its chemical potential in the solid or liquid phases in equilibrium with the gas.

If the gas phase does not behave as a perfect gas, we write instead of (2-34),

$$\mu_i = \mu_i^0 + RT \ln p_i^* \tag{2-35}$$

where p_i^* is the "fugacity." Since it can be shown experimentally that all gases become perfect at sufficiently low pressure, we may write p_i^* in the form

$$p_i^* = p_i f_i \tag{2-36}$$

where f_i is the "activity coefficient" of i, which becomes 1 at sufficiently low pressure. Combining (2-33), (2-36), and (2-35), we write

$$\mu_i = \mu_i^0 + RT \ln PN_i f_i \tag{2-37}$$

where f_i depends on pressure and temperature, and also on composition if P is large.

For a condensed phase (solid or liquid solution), we write similarly

$$\mu_i = \mu_i^0 + RT \ln N_i f_i \tag{2-38}$$

where μ_i^0 is now a function of pressure and temperature, and f_i is the activity coefficient. The quantity $N_i f_i$ is also known as the *activity*. If $f_i = 1$, the solution is said to be *ideal,* and

$$\mu_i = \mu_i^0 + RT \ln N_i \tag{2-38a}$$

where μ_i^0 is the chemical potential of pure i ($N_i = 1$) under the same conditions. All solutions become ideal when sufficiently dilute; some solutions remain ideal for all concentrations. The liquid solution albite-anorthite at high temperature is a good example of a solution which remains ideal for all compositions.

To sum up, the chemical potential usually may be written as a function of pressure, temperature, and composition, the relation between these quantities being of the same general form for gaseous mixtures, or liquids, or solid solutions.

Two-phase Systems. We are now in a position to apply the preceding relations to a number of simple systems. We shall begin by considering various equilibria involving two phases only.

1. Suppose we have a system consisting of a pure component i in two

phases 1 and 2, for instance, water and ice.   For each of these phases we may write an equation of type (2–27), *viz.*,

$$-v^1\, dP + s^1\, dT + d\mu_i^1 = 0$$
$$-v^2\, dP + s^2\, dT + d\mu_i^2 = 0$$

At equilibrium $\mu_i^1 = \mu_i^2$. Furthermore, if equilibrium is to be maintained between these two phases for any change of pressure and temperature, we must have $d\mu_i^1 = d\mu_i^2$.   Hence

$$\frac{dP}{dT} = \frac{s^1 - s^2}{v^1 - v^2} \tag{2–39}$$

If phase 1 is a pure liquid and phase 2 the corresponding solid, Eq. (2–39) gives the variation of the equilibrium temperature, *i.e.*, of the melting point, as a function of pressure.   If Δh is the heat of melting, we have, putting $v^1 - v^2 = \Delta v$, and since $\Delta s = s^1 - s^2 = \Delta h/T$,

$$\frac{dP}{dT} = \frac{\Delta h}{T\, \Delta v} \tag{2–40}$$

a relation which is known as the Clausius-Clapeyron equation, and which may be applied to any two-phase system of one component.   It gives, for instance, the variation with pressure of the boiling point of a liquid (Δh is then of course the heat of vaporization), or the variation with temperature of the pressure at which a polymorphic transition between two solid phases may occur (see page 468).   Many applications of this simple formula may be made in systems of petrological significance.

2. Consider now the equilibrium between a pure solid and its vapor in the case where the pressure on the solid may be varied independently from that on the vapor.   For any change of pressure at constant temperature we have, by (2–28), for the solid phase

$$d\mu = v^s\, dP$$

If the solid is to remain in equilibrium with its own vapor, which we suppose to be a perfect gas, the chemical potential in the vapor phase must change by an equivalent amount.   Hence, from (2–32) if p is the vapor pressure at the (constant) temperature of the experiment

$$d\mu = v^s\, dP = RT\, d\ln p$$

or

$$d\ln p = \frac{v^s}{RT}\, dP \tag{2–41}$$

Let p_0 be the vapor pressure when the pressure acting on the solid is P_0. Then, integrating from P_0 to P_1, and assuming v^s to be constant,

$$p_1 = p_0 \exp\left[\frac{v^s}{RT} (P_1 - P_0) \right] \tag{2–42}$$

a relation giving the vapor pressure of a solid as a function of the pressure on the solid.   The following are the values of p_1/p_0 at $0°C$. for various minerals and for $P_0 = 1$ bar, $P_1 = 1,000$ bars:

Orthoclase	115
Anorthite	99.5
Albite	84.7
Leucite	49.4
Hypersthene	5.5
Olivine	4.9
Corundum	3.0
Quartz	2.7

This table shows that, neglecting the compressibilities of the solid phases (*i.e.*, assuming that v^s is constant in the range P_0 to P_1), the application of a pressure of 1,000 bars to solid orthoclase will increase 115-fold its vapor pressure; the effect on orthoclase being roughly 40 times greater than that on quartz.   Since this effect is proportional to e^{P_1}, it increases very rapidly when P_1 increases.[5]

Similarly, consider how pressure applied to the solid phases only would affect the dissocation of calcite $CaCO_3$.   At equilibrium

$$CaCO_3 = CaO + CO_2$$

and the chemical potential (free energy per mole) of calcite (component 1) is equal to the sum of the chemical potentials of CaO (component 2) and of CO_2 (component 3).   If the pressure on the two solid phases is increased by an amount dP at constant temperature, maintenance of equilibrium requires that

$$d\mu_1 = d\mu_2 + d\mu_3$$

or

$$v_1\, dP = v_2\, dP + RT\, d \ln p^*$$

where p^* is the fugacity of CO_2.   Integrating from P_0 to P_1, and assuming that v_1 and v_2 do not change appreciably, we obtain

$$p_1^* = p_0^* \exp\left[(v_1 - v_2)(P_1 - P_0)/RT\right]$$

Let $P_1 - P_0 = 1,000$ bars, $T = 600°C$., $v_1 = 37$ cm.^{3}, $v_2 = 25.5$ cm.^{3} Then

$$p_1^* = 1.37\, p_0^*$$

If the activity coefficient of CO_2 remained unchanged, as if, for instance, CO_2 behaved as a perfect gas, the dissociation pressure of CO_2 would be

---

[5] Prof. J. W. Winchester has called the writers' attention to the fact that if the compound, when it vaporizes, dissociates into n different species, the exponent in (2-42) must be divided by n, *i.e.*,

$$p_1 = p_0 \exp\left[v^s(P_1 - P_0)/nRT\right]$$

increased by 37 per cent. However, at 600°, the activity coefficient of CO_2 varies from 1.025 at 25 bars to 1.36 at 1,000 bars; [6] the dissociation pressure would hardly change at all in the pressure range considered.

A simple manner of increasing the pressure on the solid phases is to add to the system solid-vapor a gas which does not react chemically with either the solid or its vapor. This addition would increase the total pressure in the system, and increase therefore the chemical potential of the solid; but it would not affect the chemical potential of the substance of the solid in the vapor phase. For, although the pressure in the vapor phase would increase, the molar fraction of this substance in the vapor would decrease by a corresponding amount. The only way in which equilibrium could be restored after addition of the inert gas would be by evaporation of a certain amount of the solid. We shall return to this point in Chap. 14 when discussing the "solubility" of a solid in a gaseous phase.

3. Let us consider now a pure solid 1 in equilibrium with its solution in a solvent 2. We now change the pressure and temperature. The composition of the pure solid will, of course, remain unchanged, but the number of moles of component 1 in the solution will presumably vary by an amount dn_1^2. It may be shown that in any phase, for any component i, there is always a relation of the type

$$\left(\frac{\partial \mu_i}{\partial n_i}\right)_{P,T} dn_i = \left(\frac{\partial \mu_i}{\partial N_i}\right)_{P,T} dN_i \tag{2-43}$$

so that, by (2-31), we have, in the solid phase,

$$d\mu = v^1 dP - s^1 dT \tag{2-44}$$

and in the solution

$$d\mu_1^2 = \bar{v}_1^2 dP - \bar{s}_1^2 dT + \left(\frac{\partial \mu_1^2}{\partial N_1^2}\right)_{P,T} dN_1^2 \tag{2-45}$$

and the condition for maintenance of equilibrium is obtained by equating (2-44) and (2-45). For instance, if the temperature is changed at constant pressure ($dP = 0$),

$$\frac{dN_1^2}{dT} = \frac{\bar{s}_1^2 - s^1}{(\partial \mu_1^2/\partial N_1^2)_{P,T}} \tag{2-46}$$

and for a change of pressure at constant temperature ($dT = 0$),

$$\frac{dN_1^2}{dP} = -\frac{\bar{v}_1^2 - v^1}{(\partial \mu_1^2/\partial N_1^2)_{P,T}} \tag{2-47}$$

These two equations give respectively the change in solubility of component 1 with temperature and pressure. Since $(\partial \mu_i/\partial N_i)_{P,T}$ is always

[6] A. J. Majumdar and R. Roy, Fugacities and free energies of CO_2 at high pressures and temperatures, Geochim. et Cosmochim. Acta, vol. 10, pp. 311–315, 1956.

positive in a stable phase, the solubility will increase with temperature if the partial molar entropy of the solute in the solution is greater than the molar entropy of the pure solid. $\bar{s}_1^2 - s^1$ may be either positive or negative so that the solubility of certain salts will increase with increasing temperature, while the solubility of other salts may decrease.

For an ideal solution, we have, by (2–38), setting $f_i = 1$,

$$\left(\frac{\partial \mu_i}{\partial N_i}\right)_{P,T} = \frac{RT}{N_i} \tag{2-48}$$

so that (2–46) becomes

$$\frac{dN_1^2}{dT} = \frac{N_1^2 \,\Delta h}{RT^2} \tag{2-49}$$

where Δh is the heat of solution per mole. The value of Δh may be determined experimentally, and it is usually found to depend on the temperature and on the concentration of the solution.

The examples given above are intended to show how the general relations developed in this chapter may be applied to particular instances; a great many other applications of these same principles can be made, but they would serve no useful purpose at this stage. The essential point to remember is that any equilibrium problem may be studied simply by writing an equation expressing the rule that for any given set of conditions the chemical potential of any component is the same in all phases in which it is present; we then solve the corresponding equations for the desired variables.

Phase Rule. *General Statement.* The object of the phase rule is to state how many things we must know about a system in order to predict all its other properties and characteristics. The answer will depend in the first place on what properties and characteristics we want to predict. In some cases we may wish to know only the values of the intensive variables which characterize the system, while in other cases we may wish to know also something about the extensive variables such as the mass and volume of each phase. We must state also what factors we are allowed to neglect. For instance, it turns out in many cases of practical importance that the chemical potentials depend on, and therefore the conditions for equilibrium involve, only the pressure, the temperature, and the concentration of the various constituents in the various phases. In other cases, however, factors such as the curvature of the interfaces separating various phases, or position in a gravitational field or in an electrical field, become important in determining the states of equilibrium of a system, and must therefore be taken into account. It follows that there must be several phase rules, each of which is applicable to a particular set of conditions. An example of a phase rule applicable to a system in which certain interfaces are curved will be found on page 463.

If we now consider more particularly the usual case where the chemical potentials depend only on pressure, temperature, and molar fractions, and if we do not care for any information concerning the extensive or "capacity" characteristics of the system (mass, volume, etc.), the well-known answer to our question is that, in a system of c components distributed in φ phases, all intensive variables—P, T, and molar fractions of all c components in all φ phases—may be determined whenever the values of $(c + 2 - \varphi)$ of these variables are given. This quantity $(c + 2 - \varphi)$ is the "variance" w of the system. A system with a negative variance cannot be in equilibrium; for variance 0 there is only one pressure and one temperature at which the system may consist of the given number of phases, the composition of each of these being determined by these particular values of P and T. For variance 1, we may assign arbitrarily the pressure or the temperature or the molar fraction of any one component in any one phase, and still find values of the other variables such that the system will be in equilibrium with the number of phases indicated; and so on.

This phase rule is obtained simply by counting the number of unknowns and the number of relations to determine them. Let us consider, for instance, a system of c components in φ phases. Equilibrium requires that the chemical potential of each component should be the same in all phases in which it is present, so that we have $c(\varphi - 1)$ equations of type (2–20), stating that the system is in equilibrium. Now the chemical potentials depend on pressure, on temperature, and on the molar fraction of each component in every phase; there are thus altogether $2 + c\varphi$ variables P, T, N_1^1, . . . , N_c^φ. All these molar fractions are of course not mutually independent, for we know that for each phase $\sum_i N_i = 1$; so that we have φ additional relations of this type. All told, there are $c(\varphi - 1) + \varphi$ relations between $c\varphi + 2$ variables. Elementary algebra tells us that in such a case if we assign arbitrary values to $w = c\varphi + 2 - [c(\varphi - 1) + \varphi] = c + 2 - \varphi$ variables, we shall be able to determine all the other variables from these arbitrary values.

This quantity w is the "variance" of the system, and the relation

$$w = c + 2 - \varphi \tag{2–50}$$

expresses the Gibbs phase rule. It states that the state of any system is defined, as far as the intensive variables P, T, N_1^1, . . . , N_c^φ are concerned, by w of these quantities.

If any component j is not present in any phase γ, there is no equilibrium condition involving μ_j^γ and therefore we have one equation less; but we have the additional relation $N_j^\gamma = 0$, and the variance therefore remains unchanged.

If a solid phase has a fixed composition, the molar fractions are determined by the stoichiometric formula; for instance, in pure albite $NaAlSi_3O_8$, the molar fraction of Na_2O is $\frac{1}{8}$, that of Al_2O_3 is $\frac{1}{8}$, and that of SiO_2 is $\frac{3}{4}$. For such a pure phase there are no compositional variables, and the total number of variables of the system is correspondingly reduced. This does not alter the variance, because the number of equilibrium conditions is also reduced. We cannot indeed equate the chemical potential of, say, Al_2O_3 in a melt in equilibrium with albite with that of Al_2O_3 in albite, for the latter has no meaning. This is easily seen by referring to Eq. (2–17), which defines the chemical potential of a component in terms of the change in free energy of a phase when the number of moles of that component is changed, the number of moles of all other components of the phase remaining constant; if the phase has a fixed composition, this cannot be done.

Attention should be called again to the significance of the number 2 that appears in the phase rule. This number is determined on the assumption that the intensive variables required to describe the state of the system consist only of concentrations, one pressure, and one temperature. This assumption may not necessarily be valid; for instance, systems may be mechanically constructed so that two or more parts of the system are at different pressures. This increases the number of variables and also the variance. If calcium carbonate is in equilibrium with CaO and CO_2, all phases being at the same pressure, the system has two components and three phases and is therefore univariant, meaning that temperature automatically determines the pressure; but if the gas phase CO_2 and solid phases ($CaCO_3$ and CaO) are at different pressures, the variance is 2, meaning that we may choose both the temperature and the pressure acting on the solid phases and still be able to find a gas pressure at which the system will be in equilibrium (see page 23).

Duhem's Theorem.[7] An important special case is that of a closed system. A system is said to be closed if no masses are added to or subtracted from it in the course of the transformation under consideration. Hence the mass of any component at any time is equal to the mass initially present plus or minus the quantity of this component which has formed or has disappeared in the course of the transformation. For instance, if a system consisting of liquid water and water vapor is closed, the amount of liquid water present at any time is the amount initially present, plus the amount of vapor which may have condensed, minus the amount of water which may have evaporated. Stating that a system is closed amounts thus to writing c conditions, c being the number of components of the system.

The basic assumption of the phase rule, as stated above, is that the state of a system is sufficiently described by the values of the intensive variables

[7] Prigogine and Defay, *op. cit.*, p. 188, 1954.

P, T, etc.   In contradistinction to this, we now consider that the state of the system is completely defined only if all variables, intensive and extensive, are known.   Actually, all the extensive variables may be determined if we know the mass of each phase and the intensive variables; for instance, the volume of phase α is equal to the mass of this phase times the specific volume, which is an intensive variable.   Thus the number of variables required to define completely the state of a closed system is equal to the number of variables required to define its intensive properties plus φ, φ being the number of phases.   We have then c additional equations of constancy of mass expressing that the system is closed, φ additional variables, and the variance becomes

$$w = c + 2 - \varphi - c + \varphi = 2 \qquad (2\text{-}51)$$

a remarkable relation, due to Duhem, which states that the state of a closed system is completely defined by the values of two independent variables only.   The initial masses of each component must, of course, be given.   For instance, given a closed system consisting of 12 components with known initial masses and of seven phases, the values of two variables, intensive or extensive, will be sufficient to determine all the extensive and intensive properties of the system.

We must, however, be careful in choosing our independent variables that we do not conflict with the phase rule for intensive variables (2-50).   For if the system is invariant in the sense of Eq. (2-50), we cannot choose arbitrarily values for any of the intensive variables, and the two variables required to describe completely the state of the system must necessarily be extensive ones.   For instance, if we consider the invariant equilibrium between ice, water, and water vapor, P and T are determined from the conditions for equilibrium.   Thus to determine the mass and volume of each phase and the total volume of the system we must be given two extensive variables, such as the mass of the ice and volume of the vapor phase.   The mass of this phase may then be found, since the specific volume of water vapor depends on pressure and temperature only.   The mass of the liquid phase is then found by difference, the total mass of water initially present in the system being stated.   On the other hand, if $w > 2$ in Eq. (2-50), we may choose P and T as independent variables, and by Duhem's theorem, the states of equilibrium of the system will be determined completely from these two quantities.   To every set of values of P and T will correspond a set of the other intensive and extensive variables, each set of values describing a different state of equilibrum of the system.   These various states of equilibrium may differ with regard to intensive or extensive properties of the various phases.

It follows that, when rock masses are subjected to conditions of varying temperature and pressure, either the composition or the relative masses of the various phases must change in response to changes in P and T, irrespective of the number of components or of the number of phases. If neither the composition nor the total volume nor the mass of each phase changes when P and T are varied, it may be safely concluded that the system is not, or was not, in equilibrium, or that the system is not closed, *i.e.*, that matter must have entered or left the system.

The importance of Duhem's theorem lies in the fact that if any geological system (a mass of rock or magma) may ever be considered to be closed, *i.e.*, if it does not exchange matter with the surroundings, then the state of this system with known initial masses of its components may be determined at any time from the values of two variables only, *e.g.*, temperature and volume, or the masses of any two phases, etc. To be sure, it would not be an easy task for us to predict what this state will be, for this would require a complete knowledge of the manner in which the chemical potential of each component in each phase changes as a function of the physical variables (P, T, V, etc.) and of composition, and this information we usually do not have. Duhem's theorem is nevertheless very useful in a qualitative way.

Components. Whereas in some cases the meaning of the word *component* is clear, this is not always so. If a system contains water, and water only, there is only one component, which is water. But if the system consists of a solution of water and albite ($NaAlSi_3O_8$), it may be asked whether the components are H_2O-Na-Al-Si-O, or H_2O-Na_2O-Al_2O_3-SiO_2, or some other combination of molecules more complex than any of the preceding ones. Clearly, the choice of components should be such that the chemical composition of each and every phase of the system can be stated in terms of these components. This could be accomplished by taking as components all the s chemical species present in the system; but this number s is not necessarily equal to the number c that enters (2–51), for some of these species may be capable of reacting and each such reaction introduces a new equilibrium condition that reduces the variance by 1. For instance, in a gas phase in which CO, CO_2, and O_2 are present, $s = 3$ but there is a possible reaction

$$CO + \tfrac{1}{2}O_2 = CO_2$$

which introduces an equilibrium condition of type (2–21) stating that the sum of the free energies of the reactants equals the sum of the free energies of the product. As the variance is defined as the difference between the number of variables and the number of conditions that these variables must satisfy, the phase rule becomes

$$w = s - r + 2 - \varphi$$

where r is the number of distinct chemical reactions that occur in the system. The number of independent components c is then $c = s - r$. For instance, in a system consisting of brucite $Mg(OH)_2$, periclase MgO, and water, $s = 3$, but $r = 1$ because of the reaction

$$MgO + H_2O = Mg(OH)_2$$

and $c = 2$. This is also the number of components we should have obtained directly by noting that the composition of each and every phase can be described in terms of the two components MgO and H_2O.

No ambiguity regarding variance can arise from the choice of components if one is careful to remember the definition of variance: number of variables minus number of relations between them.[8]   For example, a system consisting of albite and water could be indifferently described in terms of the two components H_2O and $NaAlSi_3O_8$ or in terms of the four components H_2O-Al_2O_3-Na_2O-SiO_2.   The latter choice introduces additional compositional variables which are, however, automatically determined by the stoichiometric formula of albite.   The obvious choice is to pick the smallest possible number of components, namely, two in the case just mentioned.   Two components, however, would be inadequate if the solid phase were to dissolve nonstoichiometrically in water, for the composition of the solution could not be expressed in terms of water and $NaAlSi_3O_8$ only.

In a system of given composition the number of chemical species may change with changing physical conditions, e.g., temperature.   For instance, at high temperature water partially dissociates

$$H_2O = H_2 + \tfrac{1}{2}O_2$$

There would thus be three constituents instead of one, and only one chemical reaction between them: $c = s - r = 2$.   Actually, this does not change the variance if neither hydrogen nor oxygen reacts with any of the other phases present, as we also have the additional relation that the molar fraction of hydrogen is twice that of oxygen.   This is not true, of course, if some of the phases are susceptible to oxidation, in which case oxygen must be regarded as an additional component.

In most cases, petrologists find it convenient to choose simple oxides (SiO_2, CaO, Al_2O_3, etc.) as components.   They also commonly use mineral names (e.g., albite) to designate components. This usage, of course, is incorrect, for the name "albite" refers exclusively to a solid phase with distinct physical and crystallographic properties which are not present when the substance is, say, dissolved in a melt.   Because it is very inconvenient to talk of "the component $NaAlSi_3O_8$," mineral names are commonly used in a loose sense.

[8] For an excellent discussion of the phase rule, see J. Zernicke, *Chemical Phase Theory*, pp. 5–16, Kluwer, Deventer, 1957.

Goldschmidt's Mineralogical Phase Rule. It was first recognized by Goldschmidt that as pressure and temperature vary from point to point within the earth, any rock having appreciable extension in space must necessarily have been subjected to a range of temperatures and pressures. If metamorphism could occur with zero variance ($\varphi = c + 2$), the chances that the exact temperature and pressure corresponding to the invariant point would be realized and maintained under natural conditions would be very remote indeed. This same limitation applies even to systems with one degree of freedom ($\varphi = c + 1$) where, for example, over a range of pressure (depth) the temperature would have to be maintained everywhere at the value appropriate to the local pressure. Thus for stable assemblages to form in large masses in other than exceptional conditions, the variance must be at least equal to 2, *i.e.*,

$$\varphi \leqq c$$

This may be stated as Goldschmidt's mineralogical phase rule: The *maximum* number of crystalline phases that can coexist in rocks in stable equilibrium is equal to the number of components. If a fluid phase existed in the pores of the rock at the time of reconstitution, this maximum number would be $c - 1$, water (or whatever forms the fluid phase) being counted as a component.

Korzhinsky's "Mobile" and "Inert" Components.[9] In a closed system, which does not exchange matter with its surroundings, the final amount of any component is equal to, and determined by, its initial amount. Systems may exist, however, in which some components cannot be exchanged with the surroundings, but others can. For instance, a system could be enclosed in a membrane permeable only to some of its components. Korzhinsky calls "inert" those components the amount of which is fixed by the initial amount present in the system, while "mobile" components can be exchanged with the surroundings; the final amount of these mobile components is thus unrelated to their initial amount and depends only on their activity in the surrounding medium. If the activity of such a mobile component is constant—as when a crystal is in contact with an infinite amount of fluid of constant composition, pressure, and temperature—the activity of the mobile component in the reaction product must be equal to this constant value. This introduces an additional equilibrium condition (or removes one variable), and the variance is thereby decreased by one unit. If, for instance, a crystal of orthoclase is in contact with a solution in which the activity of sodium is constant, a mixed K-Na feldspar will form in which the Na/K ratio will be determined by the activity of Na in

---

[9] D. S. Korzhinsky, Mobility and inertness of components in metasomatism, *Acad. sci. U.S.S.R. Bull., Sér. géol.*, 1, pp. 35–60, 1936; Phase rule and geochemical mobility of elements, *Internat. Geol. Cong. Repts., 18th session*, pt. II, pp. 50–57, London, 1950.

the solution, and only one solid phase will form, although the system has two components. Similarly, if MgO reacts with water at a constant chemical potential such that the reaction

$$MgO + H_2O = Mg(OH)_2$$

runs from left to right, only one solid phase will form although two phases could coexist in this two-component system. Thus Korzhinsky writes the mineralogical phase rule under the form [10]

$$\varphi = c - c_m = c_i$$

where c_m is the number of mobile components with fixed activities that are independent of the original composition of the system, and c_i is the number of inert components. Thus c_m is equal to the difference between the total number of components and the number of phases. For instance, Korzhinsky notes that only three phases (e.g., diopside-phlogopite-scapolite, or scapolite-phlogopite-spinel, or diopside-scapolite-orthoclase) ever coexist in certain metasomatic Archean complexes of Siberia, although the mineral composition suggests a much larger number of components (MgO, CaO, SiO_2, Al_2O_3, K_2O, Na_2O, H_2O, CO_2, SO_3, O_2, FeO); he concludes that only three of these were inert. Korzhinsky [11] considers that water and carbon dioxide behave as absolutely mobile components in metamorphic processes; thus the number of simultaneously stable minerals does not depend on the presence or absence of water or CO_2. Similarly, the final H_2O or CO_2 content of the rocks does not depend on their initial content.

Whether Korzhinsky's phase rule is generally applicable to geological processes remains to be seen. The fundamental assumption is that the activity of the mobile components is fixed, independently of what happens in the system. This assumption would hold, for instance, if a rock were permeated by an infinite amount of fluid of constant P, T, and concentration; but it would not hold true if the rock were permeated by a limited amount of solution, the composition of which would change by reaction with the rock. If a limited amount of water reacts with periclase, brucite will form and coexist in equilibrium with the remaining periclase, the activity of the water dropping to the level consistent with equilibrium between these two phases; the activity of the water is therefore not independent of reactions occurring in the system.

[10] See also J. B. Thompson, The thermodynamic basic for the mineral facies concept, Am. Jour. Sci., vol. 253, pp. 65–103, 1955.

[11] D. S. Korzhinsky, Dependence of mineral stability on depth, Mém. Soc. Russe Minéral., ser. 2, vol. 66, pp. 385–396, 1937.

STABILITY

Definition of Stability. Having thus dealt briefly with the conditions for equilibrium, we turn now to a consideration of what is required to disturb this equilibrium, or, in other words, to a study of the stability of a system. This matter is of great importance to petrologists since they are concerned primarily with transformations which occur in rock masses. If these rock masses were in equilibrium to begin with, and if this equilibrium were such that it could not be disturbed by changes in conditions such as are encountered in the earth's crust, these rock masses would never undergo any transformation, and the subject of petrology would be cut very short.

This study will fall logically into two steps. In the first place, we assume that all the "physical" variables such as pressure, temperature, volume, etc., are kept constant. Thus the only possible changes are "chemical," *i.e.*, changes in composition. Then in the second step we shall consider what happens when these physical variables are made to change, *e.g.*, what happens when a system previously in equilibrium is heated or subjected to a greater pressure. This second step will lead to a general theorem known as the moderation theorem, or, under a less general form, to Le Châtelier's principle. In the meantime, however, we shall thus be concerned essentially with the behavior of a system in which only the "chemical" variables, such as composition, are allowed to vary.

A system is defined as stable if any disturbance in its chemical composition tends to disappear spontaneously in the course of time. The word *chemical* must be understood here in a rather broad sense, for we include among possible "chemical reactions" such changes as passage of a component from one phase to another. Liquid water and ice would be considered in this respect as two different chemical species. Thus, for instance, a system consisting of liquid water and ice in a definite proportion will be said to be stable if any change in this proportion tends spontaneously to disappear in the course of time, the system thus returning to its original composition. If, on the contrary, changing a small quantity of ice into water were to start a reaction which caused further melting, the equilibrium water-ice would not be stable. A system in which a disturbance tends to grow is said to be unstable or metastable, according to the minimum size of the initial disturbance that will keep on growing. A system is unstable if an infinitesimally small disturbance grows; it is metastable if only certain disturbances of finite size will grow. To put this in the language that was originally used by Gibbs, a metastable phase is stable with respect to all "adjacent" phases, that is, to all phases which differ in their properties only infinitesimally from the original one; but it is unstable with respect to at least one phase with properties differing finitely from it. A "stable" phase is stable with respect to all phases, adjacent or not; an unstable phase is not even stable with respect to adjacent phases.

A classical example of a metastable phase is an "undercooled" liquid, *i.e.,* a liquid below its melting point, which is stable with respect to all other liquids differing only infinitesimally from it, but which is unstable with respect to a certain solid phase which has a different entropy, different volume, etc. If we disturb the undercooled liquid by adding to it "germs" of this solid phase, these germs will grow in the course of time. If the liquid were originally above its melting point these germs would disappear by melting and the liquid would accordingly be stable. An example of an unstable phase is that of a solution of two partially miscible liquids in their unmixing range, for any very small change in the conditions will cause the solution to split into two liquids of different composition.

Fluctuations. The difference between metastable and unstable phases may perhaps be grasped best by considering the "fluctuations" that are likely to occur in a homogeneous phase. To understand what is meant by these fluctuations, let us consider first the case of a gas. We know from elementary kinetic theory that the molecules in a gas are in constant motion. If we could observe with a sufficiently powerful microscope what happens in any small portion of the gas, we would see a constant stream of molecules entering and leaving this small volume, and although the average number of molecules present in this volume would remain constant when observed over a long period, the actual number present at any instant would fluctuate around this mean value. Now since the density of a gas is determined by the number of particles per unit volume, it follows that the density of a very small portion of the gas is not a constant, but varies with time, and from place to place in the gas. In the same sense, we know that all molecules in a gas do not travel about with the same velocity. If u is the average velocity, a certain number of molecules may have velocity $2u$, or $u/2$; a smaller number have velocities $3u$, or $u/3$; and so on. Nor is the velocity of any one molecule constant in time: each molecule is constantly being accelerated or slowed down by collision with other molecules. It follows that the average velocity of the particles present at any time in a small portion of the gas is not constant; and since this average velocity is related to the temperature, we may say that the temperature of a very small portion of a gas fluctuates in time and space, just as the density; and so do all other properties of the gas— energy, entropy, etc.

Such local and transient oscillations of the properties of a small portion of a homogeneous phase around the average value for the phase as a whole are referred to as fluctuations, and they constitute a very general phenomenon. The magnitude of these fluctuations is usually small, except under certain special circumstances; at the critical point of a gas, for instance, the density fluctuations become so marked as to be noticeable experimentally (opalescence) and to lead finally to a splitting of the gas phase into two distinct phases of unequal density.

The result of such fluctuations is to form constantly within any homogeneous phase very small masses differing infinitesimally from the average, and forming thus what we have called adjacent phases. Now if the phase in which these fluctuations occur is stable or metastable, its free energy is less than that of any adjacent phase, and these adjacent phases tend spontaneously to destroy themselves. But if the initial phase is unstable, certain fluctuations may lead to the formation of a small mass which has lower free energy and which will therefore tend to grow at the expense of the original phase. The essential difference between metastable and unstable states is that in the latter infinitesimal fluctuations may lead to the formation of a small portion of the phase with less free energy, which will therefore tend to grow; whereas in the metastable state much larger fluctuations would be required. Such large fluctuations are, however, exceedingly rare, and the chances that a nonadjacent phase might form are therefore very small. Metastable states may thus persist over great lengths of time.

Conditions for Stability. Let us now attempt to write down more specifically the conditions for stability. We consider a system in equilibrium and "disturb" this equilibrium by changing the "chemical" variables; for instance, to a system consisting of water we add a small amount of ice. Presumably the system so disturbed will no longer be in equilibrium, and something will happen: either the ice will melt, or the water will freeze. It is important to notice, however, that if something happens at a finite rate, the reaction is not a reversible one, since reactions are strictly reversible only if they occur under infinitesimally small driving forces, and hence with infinitesimal velocity. Strictly speaking, then, we have according to the second theorem (see Eq. 2–4)

$$T \, dS - dq > 0$$

Let us put

$$T \, dS - dq = dq' \tag{2–52}$$

where dq' is a positive quantity. Let us now define the chemical state of a closed system by a variable ξ; a change in ξ indicates the extent to which a certain reaction has taken place. For instance, if we consider the reaction by which water is changed to ice, if dn_1 is the number of moles of water freezing in a given time, and $dn_1 = - dn_2$ is the number of moles of ice forming in the same time, we write

$$d\xi = dn_1 = - dn_2$$

Then the velocity v of the reaction we define as

$$v = \frac{d\xi}{dt} \qquad (2\text{-}53)$$

where t is time. This velocity of freezing is thus the number of moles of water frozen in a given interval of time.

Following De Donder,[12] we define the affinity A of an irreversible reaction by the relation

$$A = \frac{dq'}{d\xi}$$

Hence

$$dq' = A \, d\xi = Av \, dt$$

or

$$\frac{dq'}{dt} = Av$$

and, since dq' and dt are always positive,

$$Av > 0 \qquad (2\text{-}54)$$

except for a reversible process for which $A = 0$ and $v = 0$.

We can now make the following formal definitions: if A and v are both simultaneously equal to zero, the system is said to be in *true equilibrium.* If $A \neq 0$, but $v = 0$, the system is said to be in *false,* or *metastable, equilibrium.* A metastable state is thus characterized by an affinity which is not zero, but by a zero velocity of reaction. This velocity is zero simply because the probability of a fluctuation sufficiently large to start the reaction is vanishingly small (see page 34).

Let us now return to our definition of stability. We have said that a system is stable for constant values of the physical variables if any change in the chemical variable ξ results in a reaction tending to bring the system back to its original state. In other words, if a system initially in the state (x_0, y_0, ξ_0)* is brought to a state $(x_0, y_0, \xi_0 + \delta\xi)$, it is said to be stable if the velocity v of the reaction which shall result from this disturbance is such that

$$v \, \delta\xi < 0 \qquad (2\text{-}55)$$

Consider as an example a liquid and its vapor in equilibrium when there is a certain proportion of liquid to vapor in a closed container of given volume. We now change this proportion by vaporizing a further amount of liquid. If this disturbance causes a process of condensation, that is,

[12] T. De Donder and P. Van Rysselberghe, *Thermodynamic Theory of Affinity,* p. 19, Stanford University Press, Stanford, Calif., 1936.

* x and y stand here for any pair of physical variables, for instance P and T, or V and T.

if the reaction runs from vapor to liquid, the system will return spontaneously to its original state. It will then be said to be stable. An undercooled liquid, on the other hand, is not stable, because the addition of a small amount of the solid phase causes a further amount of the solid to form, that is, the reaction runs in the same direction as the disturbance.

If the system is initially in equilibrium, its affinity $A_0 = 0$. In the disturbed state, the affinity becomes

$$A = A_0 + \delta A = \delta A = \left(\frac{\partial A}{\partial \xi}\right)_{x,y} \delta \xi$$

But we always have, by (2–54),

$$Av > 0$$

Hence we must have

$$\left(\frac{\partial A}{\partial \xi}\right)_{x,y} v \, \delta \xi > 0$$

and (2–55) may be satisfied only if

$$\left(\frac{\partial A}{\partial \xi}\right)_{x,y} < 0 \tag{2–56}$$

The condition that a given state be one of equilibrium is thus

$$A = 0 \tag{2–57}$$

and the condition that this equilibrium be stable is

$$\left(\frac{\partial A}{\partial \xi}\right)_{x,y} < 0 \tag{2–58}$$

We must emphasize this distinction between condition for equilibrium and condition for stability. An equilibrium may be unstable. The reader is reminded here of the stone which stands motionless on the top of a hill, but which may be caused to roll down the slope by the slightest provocation; the stone is thus in equilibrium, but the equilibrium is unstable. The same stone resting on the floor of the valley is in stable equilibrum. A stone resting in a small depression halfway up the hill is in metastable equilibrium for, although it is stable with respect to all immediately adjacent positions, it may be dislodged by a sufficiently large thrust and sent tumbling down to the bottom of the valley.

We must remember also that a condition for stability such as (2–58) refers to one particular reaction we have in mind, e.g., crystallization, or vaporization, etc. There is no general criterion to cover stability with respect to any conceivable reaction; each possible reaction has to be considered separately. A given phase may at the same time be stable with respect to one reaction (for example, unmixing of two liquids) and unstable with respect to another reaction (crystallization). Quartz at 900°C. may be unstable with respect to the transformation to tridymite, and yet stable with respect to decomposition by some particular chemical reagent. The word *stability* is to be used only in conjunction with some-thing else—stability with respect to some definite process. When it is said, for short, that a mineral is stable, the statement implies that the reader should know with respect to what; and when we say, rather loosely, that "quartz is a stable mineral at room temperature" we mean that it is stable with respect to reactions that are likely to occur in nature at room temperature; yet quartz at this temperature is unstable with respect to attack by hydrofluoric acid.

The subscripts x and y which appear in (2–58) carry an important im-plication: the condition for stability will depend on the physical conditions of the experiment. This is due to the fact that the affinity itself (see page 13) depends on the conditions of the experiment, that is, the vari-ables which are being held constant in this particular experiment. For instance, as we shall see presently, a system might be stable with respect to the possibility of a certain reaction occurring at constant V and T, and yet be unstable with respect to this same reaction occurring at constant P and T.

Equation (2–58) may be put into a number of different forms, depend-ing on our particular choice of the variables x and y. Indeed, returning to the definition of A and to Eq. (2–52), we see by (2–3) that for any spontaneous process in a closed system

$$dE = T \, dS - P \, dV - A \, d\xi$$

which reduces to (2–5) only if $A = 0$, that is, if the system is in equilibrium and the reaction reversible. Then from our definitions of H, F, G we have also

$$dH = T \, dS + V \, dP - A \, d\xi$$
$$dF = -S \, dT + P \, dV - A \, d\xi$$
$$dG = -S \, dT + V \, dP - A \, d\xi$$

hence

$$-A = \left(\frac{\partial E}{\partial \xi}\right)_{V,S} = \left(\frac{\partial H}{\partial \xi}\right)_{S,P} = \left(\frac{\partial F}{\partial \xi}\right)_{V,T} = \left(\frac{\partial G}{\partial \xi}\right)_{P,T} \tag{2–59}$$

The last of these relations, for instance, indicates that the affinity of a reaction is a measure of the rate at which the free energy changes as the reaction proceeds at constant T and P. Equation (2–58) then becomes, for the variables P and T,

$$\left(\frac{\partial A}{\partial \xi}\right)_{P,T} = -\left(\frac{\partial^2 G}{\partial \xi^2}\right)_{P,T} < 0$$

whereas the equilibrium condition is

$$A = -\left(\frac{\partial G}{\partial \xi}\right)_{P,T} = 0$$

Together, these two relations indicate that the condition that a system be in a state of stable equilibrium at constant P and T is that the free energy should be minimum. Similarly, the condition for stability at constant V and T is that the Helmholtz free energy F be minimum. For constant S and V, E must be minimum, whereas H should be minimum for constant S and P. These requirements are all contained in the much briefer statement of Eq. (2–58) that a system is stable with respect to a given reaction if the affinity of this reaction decreases as the reaction proceeds.

The affinity A may be written under a still different form, since a chemical reaction involving the reactants a, b, . . . and the products p, q, . . . may be written under the form

$$-v_a a - v_b b - \cdots + v_p p + v_q q = 0$$

in which we give a minus sign to the stoichiometric coefficients of the reactants (those which disappear as the reaction proceeds) and a plus sign to the products. The variable ξ is then defined as

$$d\xi = \frac{dn_a}{-v_a} = \frac{dn_b}{-v_b} = \cdots = \frac{dn_p}{v_p} = \frac{dn_q}{v_q}$$

We have then [13]

$$\left(\frac{\partial G}{\partial \xi}\right)_{P,T} = \sum_i v_i \left(\frac{\partial G}{\partial n_i}\right)_{P,T,n_j}$$

so that, by (2–17) and (2–59),

$$A = -\sum_i v_i \mu_i \qquad (2\text{–}60)$$

and the condition that the reaction may proceed is

$$\sum_i v_i \mu_i < 0$$

in agreement with (2–21).

The exact significance of the affinity A should be kept in mind. By definition, A is linked essentially to dq', which depends on the irreversibility of a reaction. For a reversible reaction, dq' and A are both zero, meaning that the system is at all stages in a state of equilibrium. A reaction may proceed with finite velocity only if the system is not in equilibrium, in which case the reaction must be irreversible; and the irreversibility is measured, at P, T constant, by the change in free energy. The reversible

[13] De Donder and Van Rysselberghe, *op. cit.*, p. 32, 1936.

process, which is the one commonly considered in textbooks, is merely an ideal process. It is because actual processes such as occur in rocks may be irreversible that the unraveling of equilibrium and stability conditions in rocks proves such a formidable task.

Pure Phases. To illustrate the effect of physical conditions on stability relations, let us consider the case of a reaction involving only pure substances, that is, one which does not involve solutions or gaseous mixtures. An example of such a reaction is the dissociation of calcium carbonate

$$CaCO_3 \rightarrow CaO + CO_2$$

which involves three phases (solid $CaCO_3$, solid CaO, gas CO_2) each of which consists of one pure component.

The affinity of this reaction is, by (2–60),

$$A = \mu_{CaCO_3} - \mu_{CaO} - \mu_{CO_2}$$

Since the composition of each of these phases remains unchanged as the reaction proceeds (although the mass of each phase obviously changes), the chemical potentials depend on pressure and temperature only; and if the reaction is made to run at constant pressure and temperature, the chemical potentials remain constant at all times, independently of the degree to which the reaction has proceeded. Hence if the affinity was positive initially (as it should be if the reaction is to run from left to right), it remains so until the reaction has run to completion, and can never become zero. Furthermore, since A is independent of ξ, $(\partial A/\partial \xi)_{P,T}$ cannot be negative, as required by (2–58). Hence there can be no state of stable equilibrium, at P and T constant, in which all three phases are simultaneously present.

The situation becomes different if the reaction is made to run at constant volume in a closed container; for as the reaction proceeds the amount of CO_2 increases, and if the total volume is constant, the pressure in the container must increase, and the chemical potentials will therefore vary as a function of time. Now the effect of pressure on the chemical potential of a pure component is proportional to its molar volume [see Eq. (2–30)], and as the molar volume of CO_2 is much greater than that of CaO or $CaCO_3$, the chemical potential of CO_2 increases correspondingly more. The affinity thus decreases gradually as more CO_2 is formed, and it may eventually become zero. At this stage the reaction would stop, the three components remaining now in equilibrium.

Such considerations may lead to interesting geologic conclusions. If two minerals A and B begin to react to produce C and D

$$A + B \rightarrow C + D$$

the reaction, once started, must run to completion if the minerals are under constant pressure and temperature, but a stable state of equilibrium may be reached, with all four phases present, at constant volume and temperature. Finding all four minerals together in the same rock does not mean that a stable state was not reached; it may mean simply that the volume of the system was held constant.

Whether rocks should be considered to be at constant pressure or at constant volume depends apparently on the mechanical properties of the enclosing masses. Masses held between strong competent layers which will not yield under excess pressure are presumably at constant volume (layers of shale between quartzites). On the other hand, hard nodules in a soft, plastic mass capable of considerable yield would probably be under conditions of constant pressure.

Moderation Theorem. We now proceed to the second step in our investigation, that is, to consider what happens when the physical conditions (*e.g.*, temperature, pressure) are altered. We consider a given reaction with given affinity, this affinity being, as we know, the sum of the chemical potentials of the reactants minus the corresponding sum for the products, the potential of each component being multiplied by its stoichiometric coefficient in the reaction. If the affinity is positive, the reaction proceeds spontaneously from left to right, *i.e.*, from reactants to products; its velocity is then said to be positive. If, on the contrary, the affinity is negative, the reaction proceeds spontaneously from right to left, and its velocity is said to be negative. If reactants and products are in equilibrium, the affinity is zero, and the velocity is also zero. In other words, the affinity A and the velocity always have the same sign, and if $A = 0$, $v = 0$ also. (The case $A \neq 0$, $v = 0$ corresponds, as we have said, to metastable equilibrium.)

Now since the chemical potential of any substance depends on pressure and temperature, so does the affinity of any reaction; and if this affinity is zero at any temperature and pressure, it is usually not zero at different temperatures and pressures. Let δA be the change in affinity produced by "disturbances," *i.e.*, by departures from the equilibrium condition. The affinity being now δA instead of zero, a reaction will, in general, begin to proceed at a certain rate v. It may then be shown from (2–54) that if the system was initially in equilibrium we have necessarily $\delta A \, v > 0$, that is, δA and v always have the same sign.

This fundamental relation may be put under a number of different forms, depending on the nature of the disturbance. For instance, if a system originally in equilibrium is disturbed by a rise in temperature, the pressure being constant, a reaction will start which will proceed in a direction such as to be endothermic; and if this reaction is made to run at constant pressure and entropy ("adiabatically"), it will result in a change in temperature in the system which is of opposite sign to the change which started the reaction. In other words, the temperature effect of the reaction is to moderate the temperature change that started it. In the same manner, if the pressure is raised above its equilibrium value, a reaction starts which will result in a decrease in volume of the system; and if this reaction is made to run at constant volume, it will cause the pressure in the system to vary in the opposite direction to that of the pressure change which started it; there is then a moderation of pressure. This is what is usually stated as Le Châtelier's principle; but it must be pointed out that this principle, as usually given, is subject to certain exceptions.[14] For instance, if the volume of a system is increased beyond its equilibrium value, a reaction starts which, if it proceeded at constant pressure and temperature, would cause the volume of the system to increase. Thus, if liquid water and vapor are in equilibrium, an increase in volume causes more water to vaporize. This vaporization, proceeding at constant pressure and temperature, would result also in an increase in volume. In this case, the change that occurs in the system by changing one of the factors of the equilibrium (volume) is not such as, if it occurred alone, would cause a variation of opposite sign of the factors considered, as stated in Le Châtelier's principle.

KINETICS

General Statement. We have said that a necessary condition for any reaction to proceed is that its affinity should be positive; but this condition is not sufficient to ensure that the reaction will proceed with a finite velocity. Many examples are known of reactions which, although thermodynamically possible, proceed so slowly that, in practice, they may be considered as not occurring at all (metastable states). Between these reactions with zero velocity, and other reactions which proceed almost instantaneously, there is a complete gradation; and a rather remarkable fact is that the actual velocity of a reaction is usually not directly proportional to its affinity, at least when this affinity is large, *i.e.*, under conditions far removed from equilibrium. Certain reactions with large affinity proceed very slowly, while others proceed rapidly in spite of a rather small free energy change. Thermodynamics, while predicting the sign of the velocity, can tell us nothing about its magnitude.

This matter, however, is of great importance, particularly to petrologists. The reader is presumably already familiar with the effects of rates of crystallization, as compared with rates of cooling, on the texture of igneous rocks. Glassy rocks, for instance, form by rapid cooling. Reactions which, for lack of time, have not proceeded to completion are a possible cause of magmatic differentiation. The study of many metamorphic processes would be impossible if all reactions proceeded instantaneously,

[14] Prigogine and Defay, *op. cit.*, pp. 266–269, 1954.

for rock masses would adjust themselves with very great speed to changing conditions and would retain no clues, except possibly textural ones, as to their previous history.

Consider rocks, containing a typical mineral A, that are believed to represent a high-grade (high temperature) product of metamorphism of pelitic rocks. It is possible (hypothesis a) that A survives in the present low-temperature environment because the reaction by which it formed when the temperature was rising failed to proceed with sufficient velocity in the opposite direction when the temperature was subsequently lowered; A would be metastable under surface conditions. This explanation is the usual one. It is also conceivable (hypothesis b) that A is stable under surface conditions, but does not form in sediments because the necessary reactions are too slow at low temperature. Finally (hypothesis c), it is possible that A is not stable even at high temperature, but crystallized at these high temperatures because the velocity of the reaction leading to its formation is greater than that of the reaction leading to the formation of the truly stable form. In hypothesis c, A would be metastable from the time it formed, a condition not unknown among minerals (see Ostwald's step rule, page 480). The matter can be settled only when thermodynamic relations in systems containing A have been determined experimentally.

As a further example, consider the potash feldspar, $KAlSi_3O_8$, which occurs in two polymorphous forms that are thought to differ with respect to the degree of ordering of Al and Si ions in their respective lattices. The ordered form, microcline, is stable at temperatures below a few hundred degrees; the disordered form, sanidine, is stable at high temperatures. Yet the feldspar of low-temperature hydrothermal veins (adularia) has the sanidine structure, and must have been metastable from the moment of formation. Its appearance at low temperatures reflects a higher velocity for the reaction leading to the disordered structure than for the reaction leading to an ordered arrangement of Al and Si ions. In volcanic rocks sanidine crystallizes at high temperature as the truly stable polymorph. Its subsequent survival at low temperatures is due to rapid quenching to temperatures so low that the velocity of inversion is zero.

The common mineral assemblages of metamorphic rocks are generally regarded as "stable" over specific ranges of temperature and pressure. This is the view tentatively held by the authors in interpreting metamorphic facies (Chap. 18). But it must be remembered that any such assemblage may perhaps have been metastable from the moment it formed, and some assemblages certainly are metastable under existing physical conditions to which they are subject as laboratory specimens. The kinetics of possible reactions, as well as thermodynamic relations, influences the changes that occur in any petrological system with changing temperature and pressure throughout its geological history.   The same holds true for systems experimentally investigated in the laboratory.   A study of reaction rates is therefore essential to petrology.

The problem is a formidable one.   Even in simple cases the prediction of a reaction velocity is no small task, and geological reactions are not simple ones.   Petrologists are concerned mostly with heterogeneous processes, *i.e.*, processes involving several phases, such as solid and liquid (crystallization, solution), or different solid phases, or solids and gases, etc.   Now a process occurring in such heterogeneous systems cannot be simple; for it probably occurs in a number of steps, each of which takes place with a different velocity.   For instance, any process occurring at an interface will depend (1) on the rate at which the reaction itself proceeds and (2) on the rate at which reactants are carried to, and products carried away from, this interface.   A crystal cannot dissolve faster than its particles, torn loose from the lattice, may be carried away into the solution. There is usually also a problem of conduction of heat involved, for a solid can melt no faster than the necessary heat is being supplied to it; nor can a gas condense at an appreciable rate if the heat of condensation is not carried away sufficiently rapidly.

It is clear that the over-all rate of a process which involves several successive steps will be determined by the rate of the slowest of these; and in many processes of geological importance there is little doubt that the rate-determining factor is the velocity of diffusion of reactants and products in the various phases involved.   A short study of the process of diffusion will therefore be in order.

Diffusion.[15]   Diffusion is the spontaneous process which tends to maintain a constant concentration everywhere in a homogeneous phase.   Since the chemical potential of any component of a homogeneous phase at constant T and P increases with its concentration, and since matter tends spontaneously to flow from a high to a low potential, any component will tend to migrate from a high concentration to a lower one, and equilibrium may be reached within a phase only where the concentration of each component is uniform throughout this phase.   The force F_{ix} acting on any particle i in the direction x is

$$F_{ix} = -\frac{1}{N_A} \frac{\partial \mu_i}{\partial x} \tag{2-61}$$

N_A (the Avogadro number 6.02×10^{23}) being introduced here to reduce the force per mole $-\dfrac{\partial \mu_i}{\partial x}$ to its value per particle.   Now let us suppose a

---

[15] An excellent account of diffusion processes, particularly in solids, is to be found in W. Jost, *Diffusion in Solids, Liquids, Gases*, Academic Press, New York, 1952; in the following account his treatment and notations are adopted.   See also R. M. Barrer, *Diffusion in and through Solids*, Cambridge University Press, London, 1941.

frictional resistance to the movement of particle i which is proportional to its velocity; then in the steady state the velocity v_{ix} will be such that this frictional resistance shall be equal to F_{ix}. Let B_i be the "coefficient of mobility" of the particle i, that is, the velocity it assumes under unit force; then

$$v_{ix} = B_i F_{ix} = - \frac{B_i}{N_A} \frac{\partial \mu_i}{\partial x} \qquad (2\text{--}62)$$

and the rate of flow of i, that is, the number S_{ix} of moles of i flowing per unit time across a unit surface normal to x, will be

$$S_{ix} = c_i v_{ix} = - \frac{c_i B_i}{N_A} \frac{\partial \mu_i}{\partial x} \qquad (2\text{--}63)$$

c_i being the number of moles of i per unit volume.

Now for an ideal solution, since the concentration is proportional to the molar fraction

$$\frac{\partial \mu_i}{\partial x} = \frac{RT}{c_i} \frac{\partial c_i}{\partial x}$$

and

$$S_{ix} = - D_i \frac{\partial c_i}{\partial x} \qquad (2\text{--}64)$$

where $D_i = RTB_i/N_A$ is the "diffusion coefficient" of i. Equation (2–64) expresses what is known as *Fick's law*. It states that the rate of flow in any direction per unit surface and per unit time is proportional to the concentration gradient in that direction. The diffusion coefficient has the dimensions of the square of a distance divided by time. The rate of flow S_{ix} must of course be expressed in terms of the same units as c; if c is given in moles per cubic centimeter, S_{ix} refers to the number of moles crossing a unit surface per unit time. If c is expressed as a number of particles per unit volume, S_{ix} gives a number of particles flowing across a unit surface per unit time.

There is a useful and simple relation giving $\bar{x}$, the average distance reached by a particle after time t from a plane boundary. This average distance is the sum of all the distances traveled by all particles divided by the total number of particles and is found to be

$$\bar{x}^2 = 2Dt \qquad (2\text{--}65)$$

The diffusion coefficient in solids is usually a very small quantity, of the order of 10^{-5} to 10^{-20} cm.^{2}/sec. For instance, the diffusion coefficient of carbon in iron at 925°C. is 1.15×10^{-7} cm.^{2}/sec., so that after 1 year (3.15×10^7 sec.) the average distance through which particles of carbon have moved from the boundary between pure carbon and pure iron would be

$$\bar{x}^2 = 2 \times 1.15 \times 10^{-7} \times 3.15 \times 10^7 \text{ cm.}^2$$

or

$$\bar{x} = 2.69 \text{ cm.}$$

The diffusion coefficient of silver in gold at 100°C. is of the order of 10^{-20}, so that in the course of the whole of geologic time (5×10^9 years), the average penetration of silver in gold would be less than a millimeter.

Since the diffusion rate depends on the frictional resistance offered to the motion of the particles, we should expect some relation between diffusion coefficients in solution and viscosity of the solution. The velocity of a particle of radius r moving under unit force in a medium of viscosity η is, by Stokes' law,

$$B = \frac{1}{6\pi\eta r} \qquad (2\text{-}66)$$

so that

$$D = \frac{RT}{6\pi\eta r N_A} \qquad (2\text{-}67)$$

a relation which is known as the Stokes-Einstein law, and which has been verified approximately in many cases. It shows that diffusion coefficients should decrease with increasing viscosity of the solution; and we might expect liquids which are very viscous at their melting point to crystallize more slowly than more fluid ones. This is in fact the circumstance to which Morey [16] assigns the very existence of glasses, crystallization being slow because of the slowness with which particles are supplied to the faces of the growing crystals.

Bowen [17] has made a study of the mutual diffusion of liquid diopside and plagioclase at a temperature of about 1500°C. He finds that the diffusion coefficient depends on the concentration, being less in the more viscous mixtures. The diffusion coefficient of diopside in plagioclase of composition Ab_2An_1 is 1.73×10^{-7} cm.²/sec.; that of diopside in plagioclase Ab_1An_2 is 2.31×10^{-6} cm.²/sec. These diffusion coefficients are "average" values, since diopside itself exerts a notable influence on the viscosity of the liquid into which it diffuses.

For liquids of low viscosity, the possibility of turbulent motion favors mixing; the rate of diffusion may still be expressed by a relation of type (2-64), where D is now the "coefficient of eddy diffusion," which is many times greater than the ordinary diffusion coefficient. This is the reason why stirring is sometimes an efficient means of increasing the velocity of certain reactions in solutions.

[16] G. W. Morey, *Properties of Glass*, p. 34, Reinhold, New York, 1938.
[17] N. L. Bowen, Diffusion in silicate melts, *Jour. Geology*, vol. 29, pp. 295–317, 1921.

Velocity of Formation of New Phases.[18] This matter of rates of crystallization is not, however, as simple as might appear from the above statement on the viscosity of glass-forming materials. The rate of crystallization depends indeed not only on the velocity with which molecules, ions, or atoms diffuse toward the boundaries of the growing crystals, but also on the rate at which nuclei, or embryonic crystals, form in the first place. This problem of nucleation is a very complex one. The fundamental reason for the appearance of nuclei of a new phase in a homogeneous phase is the existence of fluctuations, that is, transient local deviations from the normal state. If the initial phase is stable, these transient deviations disappear spontaneously in the course of a very short time; but if the phase is unstable, they may start to grow, forming embryos of the new phase. If the initial phase is metastable, only fluctuations of finite magnitude will grow. These large fluctuations are very infrequent, their frequency decreasing exponentially as their magnitude increases.

A new phase never forms at exactly the temperature and pressure at which it is in equilibrium with the parent phase; there must be a certain amount of overstepping (*Überschreitung*) into the region of metastability or instability of the parent phase. The reason for this is that the affinity of the transformation at the equilibrium point is zero, by definition of equilibrium. But if the affinity is zero, so is necessarily the velocity of the transformation.

If we consider now more particularly a crystal forming from a solution or melt, a certain amount of undercooling is required before the crystal may start to grow. The amount of undercooling required to start crystallization depends on a large number of factors. One of these is the surface tension, or surface energy, at the contact crystal melt. As we shall point out later, a certain amount of energy is required to form a new surface, and this energy is to be obtained from the energy released by crystallization, which increases as the difference between the chemical potentials of liquid and solid increases, that is, as we move away from the equilibrium point. The probability of formation of a nucleus decreases exponentially with increasing surface tension. Then, as will also be explained, the difference in chemical potential between solid and liquid depends on the size of the solid grains; a very small grain has a higher chemical potential than a larger one. It follows that for a slight degree of undercooling, only large nuclei are stable, and the probability of a large fluctuation is, as we know, small. Thus, only very few nuclei are present to begin with. This number increases with increasing undercooling, but when the undercooling becomes great, that is, when the temperature drops noticeably, the rate of diffusion in the melt decreases rapidly. Therefore the rate of

[18] For further information see M. Volmer, *Kinetik der Phasenbildung*, T. Steinkopf, Leipzig, 1939 (particularly pp. 61–65).

growth of each nucleus is now small and crystallization is slow, in spite of the large number of nuclei present at that stage.   The matter is further complicated by the problem of disposing sufficiently rapidly of the heat released by crystallization; crystallization can indeed proceed no faster than the heat of crystallization can be carried away.   Because of the large number of contributing factors, rates of crystallization depend very much on the nature and past history of the system, and no general prediction can be made.

Rates of crystallization in relation to degree of undercooling affect markedly the texture, or fabric, of igneous rocks (see Chap. 3).   A rock consisting of a few large grains is believed to have crystallized under conditions of mild undercooling; the number of nuclei is small, but the rate of growth of each nucleus is large, the rate of diffusion being maintained at a relatively high level by the relatively high temperature.   A rock consisting of a large number of small grains, on the contrary, is believed to have solidified under a higher degree of undercooling.   Glasses form when the undercooling is so great that the velocity of crystallization is practically zero.   Different textures may of course result from the same degree of undercooling in different materials.   The degree of undercooling under which a magma solidifies depends essentially on the rate at which heat is carried away from the magma, and this depends on such factors as average temperature of the surroundings, thermal conductivity of these surroundings, and ratio of the outer surface of the body to its total volume.

Winkler [19] has studied experimentally the effects of undercooling and of velocity of cooling upon the mean grain size of nepheline separating from an artificial melt.   He shows how observed variations in grain size of pyroxene and plagioclase across the width of basic dikes might be used to deduce the temperatures at which the minerals in question crystallized, on the rather dubious assumption that the data determined for nepheline apply equally well to pyroxene and plagioclase.

Temperature Coefficient of Reaction Rates.   Perhaps the best insight into the mechanics of rate processes (diffusion, chemical reactions, changes of phases, etc.) is gained from a study of the variations of these rates as a function of temperature.   Reaction rates increase exponentially with increasing temperature; the effect of pressure is, in comparison, very small. Reaction rates are also found to depend on certain "activation" energies or, to use more modern language, on the "height of certain potential barriers"; the higher the barrier, the slower the rate.   Most reaction rates may be expressed in the form

$$k = Ae^{-E/RT} \tag{2-68}$$

[19] H. Winkler, Crystallization of basaltic magma as recorded by variation in crystal-size in dikes, *Mineralog. Mag.*, vol. 28, pp. 557–574, 1949.

where E is the "energy of activation" or "heat of activation"; it is the energy (per mole) that a particle must acquire before it can enter into the reaction. The quantity $e^{-E/RT}$ is, accordingly, the fraction of all particles which possess this energy at a given time, and the factor A, which is a "frequency factor," represents the number of these particles which enter into the reaction per unit time. Eyring [20] and coworkers suggest that in any process there is an intermediate, critical "activated" state through which the particles must go in order to complete the reaction, so that the rate of the reaction depends essentially on the number of particles which are at any time in this activated state. Eyring assumes further that there is a state of thermodynamic equilibrium between activated and ordinary particles, so that the number of activated particles is determined by the difference in free energy between the ordinary and activated states. This free energy of activation may be computed, or at least there is a theoretical possibility of computing it, by use of partition functions, from a knowledge of size, mass, moments of inertia, and modes of vibration of the particles; a knowledge of the potential energy associated with molecules in their different possible configurations is also required. The theory has met with considerable success in accounting for reaction rates in a number of different processes, including chemical reaction, diffusion, rates of viscous and plastic flow, etc.

This theory leads to a number of interesting conclusions. In the first place, the rate of all processes for which the free energy of activation is zero is the same, being a constant times the temperature. At the absolute zero of temperature, all reaction rates indistinctively are zero. On the other hand, since the rate is determined by the free energy of activation rather than by the corresponding energy, the same reactions should proceed faster, all other factors remaining the same, when it runs in the direction for which the entropy of activation is greatest, and this will usually also be the direction in which the entropy of the system itself increases; vaporization should thus be faster than condensation, melting faster than crystallization, or the transformation of gray to white tin faster than the reverse process. It is impossible to predict what the magnitude of this effect would be, and it may be quite small in some cases. It might be possible, however, to explain in this manner why certain metamorphic reactions appear to run faster in the direction resulting from an increase in temperature than in the direction resulting from a decrease in temperature. This point will be returned to in Chap. 17.

[20] S. Glasstone, K. J. Laidler, and H. Eyring, *The Theory of Rate Processes,* McGraw-Hill, New York, 1941.

CHAPTER 3

Characteristics and Classification
of Igneous Rocks

MAGMAS

Nature of Magmas. Igneous rocks are formed by solidification of hot mobile rock material termed *magma*. Magma is commonly defined as molten rock matter. In magmas poured from volcanic vents, which are termed *lavas*, a complex liquid silicate phase is invariably conspicuous and in most cases greatly predominant. Nevertheless the flowing magma typically contains suspended crystals and bubbles of gas, and to refer to such a physically complex mixture of several phases as "molten rock matter" is unwarranted simplification. Moreover there is good reason to believe that many bodies of igneous rock, consolidated before reaching the earth's surface, were formed from mobile intrusive rock material that was only partially liquid at the time of emplacement. It is doubtful, for example, if some intrusive granites were ever entirely or even largely liquid. Yet such rocks are conventionally classified as igneous by virtue of their intrusive relation to surrounding rocks, and there are no obvious infallible criteria to indicate whether or not the parent intrusive material was partially crystalline. So we must either greatly restrict the category of undoubtedly igneous rocks or broaden our concept of magma. The latter course seems preferable. Accordingly, the term *magma* is used in this book to include all naturally occurring mobile rock matter that consists in noteworthy part of a liquid phase having the composition of silicate melt. This would exclude such materials as pure sulfide or carbonate melts, for which special terms, such as *sulfide magma,* could be used if necessary.

From the point of view of the physical chemist a magma should be regarded as a multicomponent system consisting of a liquid phase, or "melt,"

50

and of a number of solid phases such as suspended crystals of olivine, pyroxene, plagioclase, etc.   A gas phase may also be present under certain conditions.[1]   The liquid phase is essentially a mutual solution of all components.   This solution is probably very different from the familiar aqueous solution of simply ionized salts in which cations such as Na^+ and Ca^{++} and anions such as $(SO_4)^{--}$ and Cl^- are present abundantly.   Silicates are by far the most important components of igneous rocks, and chemically equivalent material makes up the great bulk of the liquid phase of a magma.   The physical state of these silicates in the melt is not known with certainty.   Strongly coherent structures made of $[SiO_4]$ groups, similar to those found in many silicate minerals, probably occur in the melt phase of a magma accompanied by free cations such as Na^+, Ca^{++}, and Mg^{++}. These linked silicate structures in the liquid phase may be envisaged as consisting of $[SiO_4]$ and $[AlO_4]$ tetrahedra appropriately connected to build composite groups whose composition approximates to $[SiO_2]$, $[AlSiO_4]$, $[AlSi_2O_6]$, $[AlSi_3O_8]$, etc.   These will henceforth be referred to as *ionic groups*.   The degree of association, or complexity, of these groups is presumably strongly dependent on the temperature; it depends also on magmatic composition, as it appears that addition of relatively small amounts of (OH) or F causes a breakdown of the larger units, decreasing thereby the viscosity of the melt.

Judging from the composition of gases erupted at volcanoes, the gas phase consists dominantly of water with minor amounts of CO_2, HCl, HF, SO_2, H_2BO_3, etc.   At all depths greater than a few hundred meters, water is above its critical pressure (see page 405).   It is therefore immaterial whether we consider this fluid phase as a "gas" or as a "liquid"; the essential point is that it may be brought *continuously* (by increasing the pressure, or decreasing the temperature, or both) from the state of a thin vapor, such as rises from a solidifying lava flow, to that of a "fluid" with a density of the same order as that of a liquid under ordinary conditions. For example, the density of water "vapor" at 400°C. and 2,000 bars is 0.75 gm./cc.*

Cooling Behavior of Magmas.   A magma, cooling through a given temperature interval, is affected by physical and chemical reactions which, according to the moderation theorem (see page 41), must run in such a direction as to be exothermic: *e.g.*, condensation of gas, crystallization from a liquid, chemical reactions giving off heat.   Some of these reactions may, at first sight, seem anomalous; for instance vaporization, if accompanied by simultaneous crystallization, may take place with release of

[1] See J. Verhoogen, Thermodynamics of a magmatic gas phase, *California Univ., Dept. Geol. Sci. Bull.*, vol. 28, pp. 91–136, 1949 (particularly pp. 99–101).

* *Handbook of Physical Constants* (*Geol. Soc. America Special Paper 36*, 1942), p. 211.

heat.[2] Even a simple mixture of albite and water, cooling under constant external pressure, may pass through a complicated sequence of condensation, crystallization, boiling, resorption of early-formed crystals, recondensation, and so forth.[3]

If the magma may be considered as a closed system, *i.e.*, if it does not exchange matter with its surroundings, the composition and relative amounts of the various phases present at any temperature can be found from the bulk composition of the system and the value of one other variable, intensive or extensive (*e.g.*, the total volume of the system; see Duhem's theorem, page 27). We should thus expect magmas of different composition to differ somewhat in their sequence of crystallization, even if the physical conditions are identical. Conversely, the same magma evolving under different physical conditions must necessarily react differently. The sequence of events in a magma cooling under constant external pressure is not the same as in a magma cooling at constant volume. Since the physical and mechanical conditions which prevail in magmatic surroundings are not clearly understood and probably are highly variable, it is difficult to predict from theoretical or experimental grounds any generally applicable course of behavior of magmas undergoing crystallization.

There is even greater latitude of behavior if the magma does not behave as a closed system—and there is little doubt that, indeed, most magmas are not closed. In such a magma, since the number of components is very large, the number of intensive variables required to define the state of the system in terms of the phase rule [Eq. (2–50)] is also very large, except perhaps at the latest stages of crystallization when the number of phases is also large. Thus our knowledge of the physical conditions prevailing in a magma is usually insufficient to allow us to make any precise prediction of its behavior. Failure to realize that a magma is in general a multivariant system has at times led to much discussion which, because it refers to hypothetical conditions not even approximately described, is of no petrological significance. There is generally no way of saying that "the magma now will do this" or "the magma now must do that"; all we can say, at most, is that the magma might, or again might not, behave in a specified manner. In fact, the only concrete evidence of how a given magma has behaved is usually that afforded by the chemistry, mineralogy, and fabric of the rock in question, if these features are capable of accurate interpretation.

The problem is further complicated by possible uncertainty whether a magma was in internal equilibrium and in equilibrium with its surroundings. It may even be suggested that, in some respects, the cooling of a

[2] For an elementary discussion of this phenomenon of "second boiling point," see Chap. 14.

[3] See Chap. 14.

magma is not a reversible process, as melting a rock does not usually reproduce, in reversed order, the successive phenomena of crystallization of the corresponding magma. Irreversibility could arise, for instance, if the magma were not in thermal equilibrium, *i.e.*, if the temperature at any given moment were not uniform throughout the melt. The fact that many magmas intruded at different times and at different places seem to have behaved in the same general way is no absolute proof that their evolution is an equilibrium process, for irreversible processes are also known to obey certain laws. For example, they tend to run in such a manner that the rate of production of entropy shall be minimum,[4] and they may obey a generalized form of the moderation theorem. It cannot be doubted, because of the analogy between the behavior of certain natural magmas and that of artificial melts, that equilibrium is approximated in many magmas, but this is perhaps not necessarily a universal condition.

Difficult as the prediction of the behavior of a magma may be, we can nevertheless make a distinction between magmas which cool at depth and magmas, such as lava flows, which cool on the surface. This difference is due in part to changes in equilibrium dependent on the difference in pressure in these two contrasted environments, and in part to differences in the kinetics of the cooling process. At the surface, cooling will be comparatively rapid, with the result that crystallization need not occur at all, the magma solidifying instead to a glassy metastable state. Or where crystallization does occur, certain reactions may fail to run to completion; olivine, for instance, may be transformed only partially into pyroxene, the corresponding mineral phase stable at lower temperature in presence of excess silica. It is to be noted that the rate of cooling depends not only on depth, but also on the size and shape of the intrusive body; small bodies with comparatively large surface for a given volume (*e.g.*, tabular masses) cool more rapidly than larger bodies of near-spherical shape. In fact, the rate of cooling seems much the same, whether the body cools at a depth of 100 m. or of 1,000 m. Typical features of rapidly cooled masses may thus be found in thin tabular bodies intruded at considerable depths, but may be absent from thick bodies injected near the surface.

An important difference between magmas which cool at or near the surface and magmas which cool at great depths is correlated with the behavior of the volatile components, mainly water. The solubility of water in silicate melts appears to increase, within certain limits, with increasing pressure, the molar volume of water vapor being much larger at low pressure than the partial molar volume of the water in the melt. Magmas which reach the surface may thus lose the greater part of their volatile content. These volatile components play an important role in two respects. In the first place, they have comparatively low molecular weights,

<hr/>

[4] I. Prigogine, *Thermodynamique des phénomènes irréversibles*, Dunod, Paris, 1947.

and their molar fraction in the melt is large compared to their concentration in weight per cent; for instance, the molar fraction of water in a 6 per cent solution of water in albite is close to $\frac{1}{2}$. Small amounts of water therefore change noticeably the chemical potentials of the other components of the melt [see Eq. (2–38)], leading notably to a large depression of the melting point of the various silicates involved.   A rapid release of pressure may thus be equivalent, with respect to crystallization, to a rapid chilling.   On the other hand, it appears that such components as H_2O, F, and Cl decrease considerably the viscosity of silicate melts, a fact which Buerger [5] attributes to the breaking down of Si—O—Si bridges when O is replaced by (OH) or F.

The viscosity of a melt may be taken to illustrate how the physical properties of a magma, and hence its behavior, may depend on composition and environment.   The viscosity of silicate melts is known to decrease very rapidly with increasing temperature.   It probably increases, at constant temperature, with increasing pressure.   It is also markedly dependent on the silica content of the melt, being much larger for silica-rich magmas than for silica-poor ones.   This viscosity is also affected, as we have said, by the volatile content, although there are no experimental determinations of the magnitude of this effect.   It follows that it is almost impossible to predict what the viscosity of a natural magma would be; and there is little ground for assuming, for instance, that a granitic magma at depth should be more or less viscous than a basaltic magma closer to the surface, but at the same temperature.   Striking local variations in viscosity are sometimes observed in apparently uniform lavas erupted simultaneously at the same volcano.

Magmatic Temperatures.   Temperatures measured in flowing lavas are mostly in the range 900°C. to 1100°C.[6]   Most measurements refer to lavas of basaltic, basanitic, or andesitic composition, and most of the higher values refer to basaltic lava.   Zies [7] records the temperature of largely crystalline "hornblende-andesite" lava extruded from the Guatemalan

[5] M. J. Buerger, The structural nature of the mineralizing action of fluorine and hydroxyl, *Am. Mineralogist*, vol. 33, pp. 744–747, 1948.

[6] T. Minakami, The explosive activities of volcano Asama in 1935, *Earthquake Research Inst. Bull.*, vol. 13, pt. 3, p. 634, 1935; J. Verhoogen, New data on volcanic gases: the 1938 eruption of Nyamlagira, *Am. Jour. Sci.*, vol. 237, p. 664, 1939; E. G. Zies, Temperature measurements at Parícutin volcano, *Am. Geophys. Union Trans.*, vol. 27, pp. 178–180, 1946; T. A. Jaggar, Origin and development of craters, *Geol. Soc. America Mem.* 21, p. 150, 1947; T. Minakami, Report on the volcanic activities in Japan during 1939–47, *Bull. volcanologique*, sér. 2, tome 10, p. 57, 1950; T. Ishikawa, New eruption of Usu volcano, Hokkaido, Japan, during 1943–1945, *Jour. Faculty Sci. Hokkaido Univ.*, ser. 4, vol. 7, no. 3, p. 252, 1950.

[7] E. G. Zies, Temperatures of volcanoes, fumaroles and hot springs, *Temperature: Its Measurement and Control in Science and Industry*, p. 377, Rheinhold, New York, 1941.

volcano Santiaguito as 725°C.   Extreme temperatures of 1150° to 1350°C. have been measured in gas-charged lava of blowing cupolas in Hawaii and elsewhere.[8]   Within the crust, magmas doubtless remain partly or even largely liquid at temperatures much lower than those of actively flowing lavas at the surface.   Green hornblende and biotite are common minerals of the more siliceous igneous rocks, and their textural relation to associated minerals and glass indicates that they may crystallize while much of the magma is still liquid.   In contact with air, green hornblende is converted to oxyhornblende at 750°C., whereas some igneous biotites decompose at 850°C.   Muscovite, an essential mineral of many granites, cannot crystallize much above 700°C., even at water pressures of several thousand bars.[9]   Experimental investigation [10] of crystallization phenomena in hydrous feldspathic melts has shown that melts approximating granite in composition can exist at high water pressures compatible with plutonic conditions at temperatures below 700°C.   From the data of this kind that were available to him, Larsen [11] concluded that within the earth's crust the temperature of basaltic magma is normally less than 1000°C. (probably 800° to 900°C.), and the temperature of more siliceous magmas is 600° to 700°C.   A more probable range of intracrustal magmatic temperatures compatible with modern experimental and petrographic data is 700° to 1100°C.   The lower temperatures in this range would refer to water-saturated granitic magmas, the higher temperatures to pyroxene-andesite and basalt.

VOLCANIC AND PLUTONIC ROCKS

Igneous rocks have already been defined as rocks formed by consolidation of magma.   Such a definition can be considered satisfactory only if clear criteria of magmatic origin can be recognized in the majority of igneous rocks.

Magmas erupted at the surface from volcanic vents are termed *lavas*. They solidify as volcanic rocks whose strictly igneous origin is beyond doubt.   Magmas must also exist locally within the crust underlying regions of volcanic activity, and presumably must at times solidify without being erupted as lavas at the surface.   Igneous rocks of this nature have

[8] Jaggar, *op. cit.*, pp. 150, 411, 1947.

[9] Cf. H. S. Yoder and H. P. Eugster, Synthetic and natural muscovites, *Geochim. et Cosmochim. Acta*, vol. 8, pp. 266–269, 1955.

[10] N. L. Bowen and O. F. Tuttle, The system $NaAlSi_3O_8$-$KAlSi_3O_8$-H_2O, *Jour. Geol. ogy*, vol. 58, pp. 489–511, 1950; O. F. Tuttle and N. L. Bowen, in Annual Report of the Director of the Geophysical Laboratory, *Carnegie Inst. Washington Year Book*, no. 52, p. 50, 1953; O. F. Tuttle, The origin of granite, *Scientific American*, April, 1955.

[11] E. S. Larsen, The temperature of magmas, *Am. Mineralogist*, vol. 14, pp. 81–94, 1929.

been described from many dissected volcanic provinces, such as the island of Mull in western Scotland or the Pacific island of Tahiti, and, as might be expected, are mineralogically and structurally similar to, though usually not identical with volcanic rocks of proved igneous origin. Starting, therefore, with volcanic rocks, it is possible to draw up mineralogical and structural criteria by means of which many igneous rocks may be recognized as such. Differences in one or more of the many variable factors influencing magmatic evolution in the plutonic (deep-seated) as contrasted with the volcanic (surface) environment can be expected to be reflected in corresponding differences in rock fabric and to a less extent in mineral assemblages. It is, therefore, customary to distinguish between volcanic and plutonic igneous rocks. Some writers recognize an intermediate (hypabyssal) class to include rocks that have crystallized at moderate depth. The need for such a class is doubtful. Many of the "hypabyssal" rocks are identical, or nearly so, with surface lavas, and it would seem more satisfactory to broaden the volcanic class to include surface lavas and near-surface intrusive rocks of similar structure and mineralogy. This is the course adopted in this book. The term *plutonic* is reserved for those igneous rocks that have crystallized at sufficient depth to impart distinctive characters to be discussed in a later section.

CHARACTERISTICS OF VOLCANIC ROCKS

Mode of Occurrence. Typical volcanic rocks occur as flows extruded on the earth's surface. Individual flows range from a few inches to several hundred feet in thickness and seldom attain a length greater than 70 miles. In volcanic complexes resulting from fissure eruptions, areas of several hundred thousand square miles may be covered by flows aggregating thousands of feet in total thickness. Characteristic features commonly displayed by lava flows are a scoriaceous, blocky, or ropy upper surface; a baked oxidized red layer beneath the under surface; columnar jointing with a regular trend normal to cooling surfaces or platy jointing parallel either to cooling surfaces or to the plane of flow, especially near the surface. Near-surface intrusive rocks that fall within the volcanic class may occur in the form of more or less cylindrical vertical necks, representing magma that has solidified in vents of eroded volcanoes or as tabular sheets. The latter are known respectively as dikes or sills, according to whether they trend across or parallel to the dominant structure (*e.g.,* bedding) of the invaded rocks.

Chemical Composition. Oxygen is the most abundant element in volcanic and other igneous rocks, and the chemical composition of an igneous rock, as determined by analysis, can be expressed in terms of a number of oxides. Most important of these is SiO_2, the percentage of which by weight ranges from 35 to 75 per cent in typical volcanic rocks. This wide variation in silica content is the basis for grouping igneous rocks in general into four categories, viz. acid (SiO_2 more than 66 per cent); intermediate (SiO_2 52 to 66 per cent); basic (SiO_2 45 to 52 per cent); ultrabasic (SiO_2 less than 45 per cent).[12]   However convenient these terms may be for expressing qualitatively an important chemical character of igneous rocks, the corresponding, arbitrarily defined divisions cannot be regarded as a satisfactory framework for a general classification.

Al_2O_3, the oxide next in abundance to silica, ranges from 12 to 18 per cent in most volcanic rocks, with a maximum value (rarely exceeding 20 per cent in highly alkaline lavas) in rocks of intermediate silica content. Values as low as 8 to 9 per cent have been recorded in ultrabasic lavas (limburgites and oceanites). Iron oxides, magnesia, and lime, which together total 20 to 30 per cent in typical basic lavas, fall off sympathetically as the silica percentage rises, and total less than 5 per cent in many acid rocks (rhyolites).   The percentage of Na_2O typically lies between 2.5 and 4 per cent, the lower values being typical of basic rocks (basalts).

TABLE 1.   AVERAGE COMPOSITION OF FIVE CLASSES OF VOLCANIC ROCKS*

Constituent	Rhyolite	Trachyte	Andesite	Phonolite	Basalt
SiO_2	72.77	60.68	59.59	57.45	49.06
TiO_2	0.29	0.38	0.77	0.41	1.36
Al_2O_3	13.33	17.74	17.31	20.60	15.70
Fe_2O_3	1.40	2.64	3.33	2.35	5.38
FeO	1.02	2.62	3.13	1.03	6.37
MnO	0.07	0.06	0.18	0.13	0.31
MgO	0.38	1.12	2.75	0.30	6.17
CaO	1.22	3.09	5.80	1.50	8.95
Na_2O	3.34	4.43	3.58	8.84	3.11
K_2O	4.58	5.74	2.04	5.23	1.52
H_2O	1.50	1.26	1.26	2.04	1.62
P_2O_5	0.10	0.24	0.26	0.12	0.45
Total	100.00	100.00	100.00	100.00	100.00

* R. A. Daly, Igneous Rocks and the Depths of the Earth, pp. 9–28, McGraw-Hill, New York, 1933.

Percentages as high as 8 per cent are encountered in highly alkaline rocks of intermediate silica content (phonolites). K_2O rises from between 0.5 and 1 per cent in basic lavas to about 4 per cent in typical acid volcanics

[12] These percentages are quoted from A. Holmes, The Nomenclature of Petrology, Murby, London, 1928.   The upper limit of the basic group is drawn at 55 per cent SiO_2 by F. H. Hatch and A. K. Wells, The Petrology of the Igneous Rocks, p. 131, G. Allen, London, 1939.

(rhyolites); rocks with 5 per cent or more of potash (trachytes, phonolites) can be regarded as highly potassic, while values as high as 10 to 15 per cent are extreme and rare. The range in chemical composition of common lavas is illustrated in Table 1 by R. A. Daly's average compositions of representative classes of lava.

Mineral Composition.   The principal constituents of volcanic rocks are crystalline silica, a relatively small number of silicates and aluminosilicates of Ca, Mg, Fe, Na, and K, oxides of Fe and Ti, and glass (silicate melt undercooled to a condition of extreme viscosity).   Under volcanic as contrasted with plutonic conditions, the magma cools rapidly, and its viscosity tends to increase correspondingly.   Such conditions apparently favor the formation and preservation of metastable phases including not only glass but certain crystalline minerals characteristic of, or peculiar to, volcanic rocks.   Neglecting accessory and rare minerals, the main constituent crystalline minerals of volcanic rocks are as follows:

Crystalline silica, SiO_2:
1. Quartz, common.
2. Tridymite and cristobalite, much less common, metastable at the time of crystallization.

Feldspars:
Feldspars [13] may be described in terms of three end members: potash feldspar, $KAlSi_3O_8$; soda feldspar, $NaAlSi_3O_8$; lime feldspar, $CaAl_2Si_2O_8$. At temperatures of crystallization from magmas there is complete substitution of Na^+ for K^+ in the alkali-feldspar series, and of $Ca^{++}Al^{3+}$ for Na^+Si^{4+} in the plagioclase series.   At lower temperatures isomorphous substitution is restricted, especially in the alkali feldspars, and unmixing may occur, giving intergrowths of two phases.   These intergrowths are termed perthites when distinguishable by ordinary microscopic means. Optically homogeneous feldspars that can be identified as intergrowths of two phases only by the use of X-ray techniques are termed cryptoperthites. High- and low-temperature polymorphs of all three end members of the feldspar series are known.   It is thought that transition from low-temperature to high-temperature forms involves disordering of the lattice with

[13] For the complex phase relations within feldspar series and for optical and X-ray criteria for recognizing varieties of feldspars, see F. Laves, Phase relations of the alkali feldspars, *Jour. Geology*, vol. 60, pp. 436–450, 549–574, 1952; O. F. Tuttle, Optical studies on alkali feldspars, *Am. Jour. Sci.*, Bowen vol., pp. 553–567, 1952; W. S. MacKenzie and J. V. Smith, The alkali feldspars, *Am. Mineralogist*, vol 40, pp. 707–747, 1955, vol. 41, pp. 405–427, 1956; I. D. Muir, Transitional optics of some andesines and labradorites, *Mineralog. Mag.*, vol. 30, pp. 545–568, 1955; J. R. Smith and H. S. Yoder, Variations in X-ray powder diffraction patterns of plagioclase feldspars, *Am. Mineralogist*, vol. 41, pp. 632–647, 1956.

respect to location of Al^{3+} and Si^{4+} ions within the framework of oxygen tetrahedra.

There are two series of common feldspars in volcanic rocks:

1. Sanidine-anorthoclase series: alkali feldspars ranging from monoclinic potassic types (sanidines with small 2V) to triclinic sodic types (anorthoclases with 2V 40° to 50°); intermediate varieties are sanidine cryptoperthites (monoclinic) and anorthoclase cryptoperthites (triclinic); optic axial plane is normal to {010} throughout the series.   High-temperature sanidine, with axial plane parallel to {010}, forms from sanidine or orthoclase on heating above 1000°C.   Comparable natural volcanic sanidines are rare.   The sodic phase in volcanic cryptoperthites is high-temperature albite.

2. Plagioclase series: triclinic soda-lime feldspars ranging from albite to anorthite.   Optical and X-ray data show that the volcanic plagioclases do not form a unique high-temperature series comparable with plagioclases formed by dry synthesis in the laboratory.   Rather they represent various states of inversion from high- to low-temperature forms.[14]   Oligoclase of potassic lavas may contain appreciable potash.

Pyroxenes:

It is customary to describe common pyroxenes in terms of four end members: diopside, $CaMgSi_2O_6$; hedenbergite, $CaFeSi_2O_6$; enstatite, $MgSiO_3$; ferrosilite, $FeSiO_3$.   In most monoclinic pyroxenes Al^{3+}, Fe^{3+}, Ti^{3+}, and Ti^{4+} substitute for $(Mg, Fe)^{++}$ and for Si^{4+}, and there is a sodic series of pyroxenes ranging from aegirine, $NaFeSi_2O_6$, to diopsidic augite.   The pyroxenes of volcanic rocks fall into five series:

1. Augites, aluminous monoclinic pyroxenes, ranging from diopsidic varieties to ferroaugites rich in $CaFeSi_2O_6$ (the usual range of 2V is 45° to 60°, except in titaniferous augites where values between 30° and 40° are common).

2. Hypersthenes, orthorhombic nonaluminous pyroxenes, approximating $(Mg, Fe)SiO_3$.   These are low-temperature forms, rapidly inverting to other polymorphs when heated dry to temperatures above the range 955° (iron-rich hypersthene) to 1140°C. (enstatite).   The upper limits of stability of the hypersthene series, presumably below this range of temperature, are not known, nor are the stability relations of the high-temperature forms protoenstatite and clinoenstatite-clinohypersthene.[15]

---

[14] Muir, *op. cit.*, 1955; Smith and Yoder, *op. cit.*, 1956.

[15] Cf. W. R. Foster, High-temperature X-ray diffraction study of the polymorphism of $MgSiO_3$, *Am. Ceramic Soc. Jour.*, vol. 34, pp. 255–259, 1951; J. F. Schairer, F. R. Boyd, and H. P. Eugster, Phase-equilibrium relations in the quadrilateral $MgSiO_3$-$CaMgSi_2O_6$-$CaFeSi_2O_6$-$FeSiO_3$, *Carnegie Inst. Washington Year Book*, no. 55, 1955–1956, pp. 208–210, 1956.

3. Pigeonites and ferropigeonites, monoclinic pyroxenes approximating clinohypersthene, $(Mg, Fe)SiO_3$, with limited substitution of Ca^{++} for Mg^{++}. (2V 0 to 25°, optic axial plane normal to {010}.)

4. Subcalcic augites, intermediate between augites and pigeonites in composition. (2V 10° to 30°, optic axial plane = {010}.) Subcalcic augites have been interpreted as the metastable equivalent of augite plus pigeonite or augite plus hypersthene, for an immiscibility gap is believed to exist between $Ca(Mg, Fe)Si_2O_6$ and $(Mg, Fe)SiO_3$ throughout the range of temperatures at which most basaltic and andesitic magmas crystallize.[16]   The formation of subcalcic augites is favored by rapid cooling from high temperatures.[17]   They are therefore rather limited in their occurrence and are found mainly in the groundmass of basaltic rocks.

5. Aegirine, $NaFeSi_2O_6$, and aegirine-augites intermediate between aegirine and the augites.

Amphiboles:

1. Common hornblendes: calciferous amphiboles which as a first approximation may be considered as close to pargasite, $NaCa_2(Mg, Fe)_4$ $AlAl_2Si_6O_{22}(OH)_2$. There may be some substitution of Fe^{3+} for Al^{3+} and of F^- for $(OH)^-$. Oxyhornblende, rich in ferric iron, is a variety characteristic of some volcanic rocks.

2. Riebeckite, $Na_2Fe_3^{++}Fe_2^{3+}Si_8O_{22}(OH)_2$.

3. Sodic hornblendes (arfvedsonite, barkevikite and others) intermediate between riebeckite and common hornblendes.

Micas:

1. Muscovite, $KAl_2AlSi_3O_{10}(OH)_2$, cannot exist at temperatures greater than 650°C. except at water pressures exceeding 700 bars [18] (*i.e.*, at depths of 1½ miles or more). Therefore it cannot crystallize from volcanic melts and is absent from lavas except as "sericitic" products of post-magmatic reactions.

2. Biotites can be regarded as phlogopite, $KMg_3AlSi_3O_{10}(OH)_2$, with Fe^{++} substituting for Mg^{++}, and Fe^{3+} and additional Al^{3+} substituting to a less extent for Mg^{++} and Si^{4+}. At low water pressures, even phlogopite is unstable above 1000°C., whereas iron-rich biotites decompose at considerably lower temperatures; therefore biotites have a much more

---

[16] For recent discussion on this relationship see: A. Poldervaart and H. H. Hess, Pyroxenes in the crystallization of basaltic magma, *Jour. Geology,* vol. 59, pp. 472–489, 1951; L. Atlas, The polymorphism of $MgSiO_3$ and solid-state equilibria in the system $MgSiO_3$-$CaMgSi_2O_6$, *Jour. Geology,* vol. 60, pp. 125–147, 1952; H. Kuno, Ion substitution in the diopside-ferropigeonite series of clinopyroxenes, *Am. Mineralogist,* vol. 40, pp. 70–93, 1955.

[17] Kuno, *op. cit.,* pp. 87, 88, 1955.

[18] Yoder and Eugster, *op. cit.,* pp. 225–280 (especially p. 267), 1955.

limited distribution in volcanic than in plutonic rocks.[19]   Decomposition (resorbtion) of biotite phenocrysts is well known in siliceous lavas.   Micas approximating phlogopite are known in certain potassic basic lavas and dike rocks.

Olivine:

(Mg, Fe)$_2$SiO$_4$: Common olivines of the basic and ultrabasic rocks are rich in magnesium.   Fayalite, Fe$_2$SiO$_4$, is known in trachytes and rhyolites.

Feldspathoids:

The term feldspathoid includes a number of alkaline (mainly sodic) aluminosilicates which occur as constituents of igneous rocks and which are poorer in silica than corresponding alkali feldspars.   They are found only in rocks rich in Na$_2$O or K$_2$O as compared with SiO$_2$.

1. Leucite, KAlSi$_2$O$_6$, found only in volcanic rocks.

2. Nepheline, NaAlSiO$_4$, with substitution of K for Na.

3. The sodalite group: sodalite, Na$_8$Al$_6$Si$_6$O$_{24}$Cl$_2$; noselite, Na$_8$Al$_6$Si$_6$O$_{24}$SO$_4$; hauyne, similar to noselite but with some substitution of Ca for Na and a corresponding increase in amount of SO$_4$.

4. Various sodic zeolites, *e.g.*, analcite, NaAlSi$_2$O$_6$H$_2$O; natrolite, Na$_2$Al$_2$Si$_3$O$_{10}$2H$_2$O.

Iron and titanium oxides:

1. Hematite Fe$_2$O$_3$, ilmenite (Fe, Mg)TiO$_3$, and intermediate compounds.

2. Magnetite Fe$_3$O$_4$, ulvospinel Fe$_2$TiO$_4$, and intermediate compounds; commonly nonstoichiometric and with partial substitution of Mg, Al, and Cr for Fe.

3. Rutile TiO$_2$.

To emphasize the influence of the silica content of magmas upon igneous paragenesis, Shand [20] distinguished between minerals that are respectively saturated and unsaturated with silica.   Saturated minerals are those that are compatible with excess silica under magmatic conditions and are therefore commonly associated with quartz.   They include feldspars, pyroxenes, amphiboles, micas, magnetite, ilmenite, and a large number of common accessory minerals such as zircon, sphene, and apatite. Unsaturated minerals, *i.e.*, those incompatible with excess silica under

[19] H. S. Yoder and H. P. Eugster, Phlogopite synthesis and stability range, *Geochim. et Cosmochim. Acta*, vol. 6, pp. 179–182, 1954.

[20] *E.g.*, S. J. Shand, *Eruptive Rocks*, p. 127, Wiley, New York, 1943.

magmatic conditions, include olivine, feldspathoids, and melilites, as well as one or two rare constituents of volcanic rocks such as garnets close to andradite. Association of unsaturated minerals with quartz, e.g., olivine-quartz in the groundmass of some basalts or leucite-quartz in some phonolitic lavas, is due to failure of crystals and melt to reach a state of equilibrium during rapid chilling under volcanic conditions. Such disequilibrium assemblages, like single metastable minerals, are criteria of volcanic as distinct from plutonic conditions of crystallization.

For convenience in general description of igneous rocks, quartz and feldspars and feldspathoids (the light constituents) are sometimes grouped together as *felsic* minerals as distinct from the *mafic* constituents (dark, heavy, rich in Mg and Fe).

Fabric (Structure, Texture). The mutual relationships in space of the various components of a rock (crystals, parts of crystals, multigranular aggregates, or microscopically irresolvable groundmass materials) constitute what has variously been termed the *fabric, structure,* or *texture* of the rock. *Fabric* should be interpreted as a whole; and fine distinction between usage of terms such as *texture* and *structure* seems unwarranted, especially as there is still no uniformity as to such usage. In this book all three terms will be used synonymously.

It has been shown experimentally that crystals separating from a viscous melt, only slightly undercooled with respect to its freezing temperature, tend to grow slowly from comparatively few nuclei, and to attain a larger size than the many crystals which swarm rapidly in a greatly undercooled but still not too viscous melt (cf. pages 47 and 48). Judging from the prevalence of glassy residues in the groundmass of volcanic rocks, it seems that in general the degree of undercooling of a magma freezing under volcanic conditions increases rather than decreases as freezing proceeds. Large crystals in a volcanic rock are, therefore, assumed to have started to form earlier than markedly smaller crystals of the same, or even of other, minerals in the same rock. One of the commonest and most characteristic structures of volcanic rocks is *porphyritic* structure, in which large crystals (phenocrysts) of one or more minerals are set in a finely crystalline or glassy groundmass. In many rocks minerals which occur as phenocrysts are also present in this fine-grained base. The sharp break in grain size between phenocrysts and groundmass is correlated with some corresponding change in conditions prevailing during freezing of the magma. Such a break occurs where slow crystallization of magma deep within the crust has given way to rapid crystallization following uprise of the magma and extrusion at the surface (or injection into cooler rocks of the upper crust), with attendant quick chilling and reduction of pressure. In some otherwise porphyritic dike rocks, however, a glassy or uniformly fine-grained fabric along the margins of the intrusion indicates that the magma was injected in a completely liquid state. In such cases the phenocrysts are thought to represent initial slow crystallization from the slightly undercooled magma immediately after intrusion, while the finer groundmass corresponds to more rapid spontaneous crystallization as the magma became undercooled sufficiently to pass from a metastable to a most unstable condition. That phenocrysts of volcanic rocks do in general belong to the earlier stages of magmatic crystallization is confirmed by prevalence of high-temperature minerals among phenocrysts. Laboratory experiments have shown, for example, that olivine crystallizes early from melts that later yield magnesian pyroxene, and that anorthite is the high-temperature member of the plagioclase series. Olivine and anorthite-rich plagioclase are common phenocryst minerals in rocks containing pyroxene and more sodic plagioclase as the chief constituents of the groundmass. The occurrence of leucite phenocrysts in a nepheline-sanidine groundmass has similar significance. But to assume an early origin for all phenocrysts of igneous rocks still is not justified.

Another very general assumption in interpreting fabric of volcanic rocks is that idiomorphic crystals with sharply defined outlines finished crystallizing earlier than allotriomorphic grains of another mineral with irregular boundaries. This is borne out by the general tendency for phenocrysts to have idiomorphic outlines, and for almost invariable idiomorphism of crystals of any mineral against enclosing glass. But there are certain exceptions to this rule. Olivine phenocrysts in basalt are commonly rounded in outline, while the quartz phenocrysts in acid volcanics of the quartz porphyry and rhyolite groups are in many cases not only rounded but deeply embayed. Here the phenocrysts seem to have crystallized early and then, under changing conditions, to have reacted with or become partially redissolved in the magma, so that an initially idiomorphic form has been partially or completely destroyed by corrosion. In the case of olivine this interpretation is confirmed by the observed behavior of olivine crystallizing from melts in systems such as anorthite-forsterite-silica.

One of the notable exceptions to the principles just discussed is provided by the *ophitic* structure of many diabases (basic volcanics occurring in thick flows and in sills). Here idiomorphic laths of plagioclase and coarser irregularly bounded grains of enclosing augite are generally believed to have crystallized more or less simultaneously. If this is the case we must assume some special conditions, the nature of which is still uncertain, which favored plagioclase in exerting its own crystal outlines against augite. Idiomorphism seems, too, a doubtful criterion of early crystallization in the case of certain consistently idiomorphic accessories, such as apatite, which from their high content of "volatile" components (P_2O_5, OH, Cl) are unlikely to have crystallized before the main crop of anhydrous silicates. Origin by replacement is more likely here. Most petrographers to some extent interpret the criteria of idiomorphism and grain size to suit personal prejudices as to sequence of crystallization.

Glass is a common and highly characteristic constituent of volcanic rocks. In basic rocks, the parent magma of which has a relatively low viscosity, it is typically present in subordinate amount in the groundmass or is altogether lacking. But in acid lavas it is one of the most important constituent materials; and some rocks of the rhyolite and trachyte families (obsidians and pitchstones) are made up almost entirely of glass within which swarms of embryonic crystals (crystallites) may be seen beneath the microscope. Persistence of glass implies extreme undercooling of still-liquid magma to a point where low temperature so retards migration of ions that crystallization virtually ceases. This condition is particularly likely to occur in acid magmas rich in alkali feldspar, for these are notably more viscous than basic lavas rich in iron and lime. On the other hand, even basic magmas may solidify as glass if chilling is unusually rapid, as when flowing lavas encounter sea water, or when magma at the margins of a dike or sill is locally chilled by sudden contact with invaded cooler rocks. Beneath the microscope or under a hand lens many glassy rocks show minute, curved, sometimes partially concentric cracks due to shrinkage of the cooling glass. This is termed *perlitic* structure. It is highly developed in perlites—rhyolitic glasses rich in water absorbed from a wet environment of extrusion. In other glassy rocks, especially in members of the rhyolite family, fibers of feldspar (sometimes accompanied by quartz) have developed radially from scattered nuclei to build up spherical bodies (*spherulites*) ranging from a fraction of a millimeter to several centimeters in diameter. Under the microscope spherulites can in many instances be shown to have grown after flow of the magma had ceased, *i.e.*, after it had reached an essentially glassy condition. Since glass at ordinary temperatures is a metastable material, it is liable to crystallize (devitrify) with passage of time, especially if subjected to raised temperatures resulting from burial. Hence glassy rocks are rare, though not unknown, in Paleozoic as compared with Tertiary formations. In their place are uniformly fine-grained (aphanitic or cryptocrystalline) volcanic rocks, within which persistent perlitic cracks may survive as evidence of a former glassy state.

Many volcanic rocks are riddled with small spheroidal or tubular cavities attributed to escaping bubbles of water vapor and other gases. This is *vesicular* structure. Water is an important constituent of magmas. Goranson [21] has shown that granitic magma at 900°C. and 4,000 bars pressure (corresponding to a depth of 15 km.) can hold 9 per cent of dissolved water. At 500 bars (2 km.) the same magma at the same temperature can hold less than 4 per cent water. While it is still uncertain whether

[21] R. W. Goranson, The solubility of water in granite magmas, *Am. Jour. Sci.*, vol. 22, pp. 481–502, 1931.

the water content of plutonic magmas normally is of this order, it is at least conceivable that release of pressure attendant upon uprise and extrusion of an initially deep-seated water-saturated magma would cause rapid boiling off of excess water, and so could account for vesicular structure of volcanic rocks. General absence of vesicular structures from plutonic rocks is less readily understood. Perhaps plutonic magmas are generally sufficiently fluid to allow complete escape of gas bubbles which, in the more viscous undercooled lavas, become permanently trapped to form vesicles.

Many volcanic rocks, especially those that have crystallized from acid viscous magmas, show a tendency for parallel alignment of various elements in the fabric. This alignment is usually correlated with flow of the partly crystalline magma.[22] Well-known instances are subparallel alignment of feldspar laths in the groundmass of lavas (*trachytic* texture of dacites, trachytes, and phonolites; pilotaxitic texture of basalts); parallel orientation of alternately crystalline and glassy bands in finely banded rhyolites; parallel alignment of tubular vesicles in pumice. The direction of flow may be either parallel or at right angles to the direction of linear parallelism of prismatic or lathy crystals. It can be established with certainty in oriented sections in which rotation of phenocrysts or the presence of swirls of microcorrugation of flow laminae can be detected. In such rocks the flow direction is normal to the axis of rotation or corrugation.

CHARACTERISTICS OF PLUTONIC ROCKS

Occurrence. Plutonic rocks are igneous rocks the distinctive fabric (and in some cases mineral composition) of which indicates that crystallization proceeded under conditions of slow cooling compared with the average rate at which volcanic rocks cool at the earth's surface. Daly[23] cites a number of estimates, ranging from 400 to 6,000 m., for the thickness of cover rocks overlying large bodies of plutonic rocks. No doubt the depth required for plutonic conditions varies. Probably most plutonic rocks exposed today at the earth's surface crystallized at depths ranging from not less than several hundred meters to about 20 km. The continued slow cooling essential for development of the plutonic fabric is controlled only partly by depth. Of equal or greater importance, especially in the shallower plutonic masses, are shape and size of the magma body (particu-

[22] Cf. R. H. Clark, The significance of flow structure in the microporphyritic ophitic basalts of Arthur's Seat, *Edinburgh Geol. Soc. Trans.*, vol. 15, pp. 69–83, 1952; A. Spry, Flow structure and laminar flow in Bostonite dykes at Armidale, New South Wales, *Geol. Mag.*, vol. 90, pp. 248–256, 1953.

[23] R. A. Daly, *Igneous Rocks and the Depths of the Earth*, p. 126, McGraw-Hill, New York, 1933.

larly the ratio of volume to area of cooling surface) and temperature of host rock.

Bodies of plutonic rock vary greatly in form and extent. The smallest are dikes and veins a few inches wide; the larger masses outcrop continuously over areas measured in thousands of square miles. It is beyond the scope of this book to discuss forms of plutonic intrusions, but some of the commoner types, classified conventionally according to form and relation to invaded rocks, are noted briefly.

1. *Sills.* Tabular bodies concordant with the major structure, *e.g.,* bedding or foliation, of the invaded rocks. Large sills hundreds of feet thick and extending laterally for many miles are usually of basic composition.

2. *Laccoliths.* Sheet-like bodies with flat base and domed roof above which the invaded strata have been arched concordantly at the time of intrusion. Most laccoliths are composed of acid rather than basic rocks

3. *Phacoliths.* Curved lensoid masses injected along and concordant with the arches and troughs of folded strata.

4. *Lopoliths.* Roughly sheet-like or lensoid bodies with upper and lower surfaces concave upwards, the general configuration being connected with down-sagging of the floor rocks under the load of the thickening intrusions. Some of the largest known plutonic intrusions are lopoliths composed mainly of basic rocks.

5. *Dikes.* Tabular, often vertical or steeply dipping sheets, cutting across the trend of structure (*e.g.,* bedding) of the invaded rocks.

6. *Ring dikes.* Steeply dipping dikes of arcuate or annular outcrop, formed by uprise of magma along major steeply conical or cylindrical fractures, in some cases several miles in diameter.

7. *Batholiths.* Large intrusive bodies with steeply dipping walls and lacking any visible floor. Batholiths are typically composed of acid rocks (granite, granodiorite, and related rocks). The largest outcrop continuously over many thousands of square miles. Opinion is divided as to whether some batholiths are of igneous, metamorphic, or partly igneous and partly metamorphic origin.

8. *Stocks* (*bosses.*) Similar in form and composition to batholiths but smaller in size (with outcrops less than about 40 square miles in extent).

9. *Plutons.* A term which embraces all intrusive bodies of igneous rock. It is convenient when the intrusion conforms to none of the above definitions or its configuration has not been determined.

Chemical and Mineral Composition. The range of chemical composition is much the same in plutonic as in volcanic rocks. A minor difference between the two classes is a tendency for the ratio of ferric to ferrous iron to be on the whole lower in plutonic rocks—a character that may be correlated with oxidizing reactions connected with volcanic conditions. The prevalence of glass, in some cases relatively rich in water, as a major constituent of some volcanic rocks raises the water content of such rocks considerably above that of any corresponding plutonic type.

Mineralogically there is also general similarity between corresponding plutonic and volcanic rocks. But several distinct points of dissimilarity are noted:

1. A number of almost monomineralic plutonic rocks such as dunite (olivine rock), anorthosite (plagioclase rock), and pyroxenite (pyroxene rock) have no exact counterparts among lavas.

2. Plutonic crystallization occurs at higher pressures (and water pressures) and extends into a lower range of temperature than does crystallization under volcanic conditions. This is reflected in certain mineralogical differences between the two corresponding classes of rocks:

a. Alkali feldspars of plutonic rocks fall into two series, microcline-albite and orthoclase-albite.[24] The sodic member of each is low-temperature albite, which seems to be stable below 700°C. At the potassic end microcline, which is triclinic, is stable at temperatures below a value variously estimated at between 500° and 700°C. Orthoclase (monoclinic) seems to be structurally intermediate between microcline and its high-temperature polymorph sanidine; it is distinguished optically from the latter by its higher 2V (40° to 80°). In both microcline and orthoclase, substitution of Na^+ for K^+ is limited, so that optically recognizable perthites (microperthites) are characteristic of the middle ranges of both alkali feldspar series. The sodic phase of these perthites is low-temperature albite, as contrasted with high albite in volcanic cryptoperthites.

b. Plutonic plagioclases have been generally considered a series stable at low temperatures. More probably they, like volcanic plagioclases, represent various states of inversion between high- and low-temperature forms. But inversion in general appears to be more complete than in volcanic plagioclase, and in consequence the optical and X-ray characteristics of plagioclases from a given type of plutonic environment—such as that of large stratiform intrusions—are relatively constant.[25]

c. Hypersthenes are commoner and pigeonites much more restricted in plutonic than in volcanic rocks.

d. Leucite is virtually unknown in plutonic rocks.

e. The upper limits of temperature at which micas and hornblendes remain stable are raised with increasing water pressure.[26] These minerals are therefore much more common in plutonic than in volcanic rocks.

[24] O. F. Tuttle, Optical studies on alkali feldspars, *Am. Jour. Sci.*, Bowen vol., pp. 553–567, 1952; W. S. MacKenzie, The orthoclase-microcline inversion, *Mineralog. Mag.*, vol. 30, pp. 354–366, 1954; J. R. Goldsmith and F. Laves, The microcline-sanidine stability relations, *Geochim. et Cosmochim. Acta*, vol. 5, pp. 1–19, 1954; MacKenzie and Smith, *op. cit.*, pp. 707–732, 1955.

[25] Smith and Yoder, *op. cit.*, 1956.

[26] Yoder and Eugster, *op. cit.*, 1954; *op. cit.*, 1955.

Muscovite, as a magmatic mineral, is virtually restricted to acid plutonic rocks.

3. Rapid cooling under volcanic conditions permits formation of metastable phases such as glass, tridymite and subcalcic augite and of mineral assemblages such as leucite-quartz or olivine-quartz. These are absent from equivalent plutonic rocks.

Fabric. The most characteristic features of the plutonic fabric are its holocrystalline state (glass being absent) and relatively coarse equigranular texture. Both are attributed to a rate of cooling so slow that the magma has crystallized completely without at any stage undercooling sufficiently to allow free development of nuclei and consequent rapid spontaneous crystallization (cf. p. 48). A mosaic of interlocking mutually interfering grains is thus built up. In basic rocks the structure is typically *gabbroid* (allotriomorphic granular) in that completely allotriomorphic grains dominate the fabric. In most acid and intermediate plutonic rocks there is a distinct tendency on the part of some essential minerals (notably micas, hornblende, feldspars) to develop as grains with subidiomorphic outlines. The resultant structure is *granitoid* (hypidiomorphic granular). It is now realized that some details of the granitoid fabric are due to post-magmatic processes of diffusion and unmixing in a completely solid rock.[27] Albitic rims on crystals of potash feldspar at contacts with plagioclase seem to have originated thus.

A special feature of some acid plutonic rocks is *graphic* or *micrographic* structure. Large grains of feldspar, usually orthoclase or microcline, individually enclose many small imperfectly developed bodies of quartz which maintain constant crystallographic orientation throughout a large part or all of a given host crystal. Such intergrowths possibly represent products of eutectic crystallization of the two minerals concerned, for the structure is closely similar to that of a metallic eutectic. But it is also probable that many intergrowths, especially in pegmatites, are the result of partial replacement of one mineral by the other. In some granitic rocks potash feldspar is replaced marginally by an intergrowth of vermicular quartz in plagioclase. This is called *myrmekite.*

Many of the common minerals of igneous rocks crystallize in partially or completely isomorphous series. In general the degree to which isomorphous substitution of one ion for another is possible decreases with falling temperature, so that a solid solution initially stable with respect to unmixing at high temperatures of crystallization becomes progressively less stable as the temperature falls. Under plutonic conditions, cooling after crystallization may be slow enough to allow establishment of equilibrium by unmixing of the unstable solid solution into two or more stable

[27] O. F. Tuttle, Origin of the contrasting mineralogy of extrusive and plutonic salic rocks, *Jour. Geology,* vol. 60, pp. 107–124, 1952.

solid solutions with more limited isomorphous substitution. Unmixing phenomena in minerals of plutonic rocks are exemplified by alkali feldspars and some pyroxenes. Perthites are potash feldspars (orthoclase or microcline) with microscopically intergrown bodies of albitic plagioclase which commonly are the result of unmixing from original homogeneous potash-soda feldspars. Hypersthene of some norites and diabases shows a lamellar structure resulting from unmixing of diopside laminae during slow cooling or perhaps also during inversion of pigeonite to hypersthene.[28] Rapid cooling under volcanic conditions prohibits unmixing.

Some degree of banded (gneissic) structure and parallel alignment of tabular or prismatic crystals or elongated mineral clots is common in plutonic rocks. The structure may be primary, the result of flow of the partly crystalline magma at the time of intrusion or of the influence of convection currents or spasmodic turbulence affecting the magma after intrusion; or it may result from the sorting of crystals of different sizes and densities as they settled under gravity in a still body of liquid magma. Alternatively it may be a secondary (metamorphic) structure imposed during deformation of the completely crystalline rock. These possibilities will be discussed later, but it may be noted at this point that the origin and significance of gneissic structure in any given instance may be difficult or impossible to determine with certainty. Moreover there are rocks of metasomatic (metamorphic) origin whose mineralogy and structure are identical with those of truly igneous rocks, and whose banded fabric is inherited from a foliated or bedded structure in the parent rock.

Open cavities are much rarer in plutonic than in volcanic rocks. There is no plutonic counterpart of vesicular structure. If at some stage steam boils off from a largely liquid magma under plutonic conditions, the ease with which the gas bubbles seem to have escaped indicates a low viscosity for the melt phase—which is rather surprising in view of the relatively low temperature and high silica content of many such magmas. Perhaps low viscosity is here to be correlated with a low degree of undercooling. Alternately, it may be that the water content of the magma is steadily depleted by crystallization of hydroxyl-bearing minerals (micas, hornblendes) under plutonic conditions, so that boiling is restricted to a very late stage when crystallization of magma is too advanced to permit vesiculation. Some plutonic rocks, especially granites, have sparsely scattered, angular, crystal-lined cavities, which conventionally are attributed to shrinkage incurred during freezing of the magma. The fabric of such rocks is termed *miarolitic*. It is also possible, however, that miarolitic cavities represent late emission of gas in small quantities toward the close of crystallization.

[28] H. Hess, Pyroxenes of common mafic magmas, *Am. Mineralogist*, vol. 26, pp. 515–535, 573–594, 1941.

OUTLINE CLASSIFICATION OF IGNEOUS ROCKS [29]

In this book common types of igneous rock are defined in terms of fabric and essential constituents (*i.e.*, minerals whose presence is essential for classification of a given rock type). Boundaries between the plutonic and volcanic divisions and between many families (*e.g.*, basalt and andesite, granite and syenite) are gradational. Moreover, there are some families that are difficult to place in any classification. Such are the lamprophyres and diabases which, though typically intrusive, have mineralogical and fabric characters analogous to those of volcanic rocks. They are here classed with the latter group.

To avoid repetition, and as an aid to memorizing rock classifications, it is customary to bracket corresponding volcanic and plutonic rocks having similar mineral and chemical composition as being mutually equivalent. If this procedure is adopted it must be remembered that certain of the pairs so bracketed, *e.g.*, gabbro (plutonic) and basalt (volcanic), are not exactly equivalent either in mineral composition or in range of chemical composition. Moreover, there are plutonic rocks such as the peridotites and anorthosites which have no volcanic counterparts. Again, whereas the coarse-grained fabric of acid plutonic rocks permits recognition of a number of families (granites, quartz monzonites, granodiorites, tonalites) on the basis of the nature of feldspar present, corresponding volcanic rocks include so many that are in large part glassy or cryptocrystalline that definition of the same number of equivalent volcanic classes is scarcely justified except on the basis of chemical analyses.

In igneous rocks the most striking mineralogical variations, and therefore the most satisfactory mineralogical criteria for founding a classification, are those connected with the following characters:

1. Relative abundance of mafic and felsic groups of minerals
2. Presence or absence of quartz
3. Presence or absence of unsaturated minerals (mainly feldspathoids or olivines)
4. Nature of the feldspars
5. In rocks rich in Mg and Fe, nature of the mafic minerals

Certain of these variables have a sympathetic relation in that they may be correlated in common with some variable chemical character of the parent magma. For example, as the SiO_2 content of the magma increases there is a distinct tendency, subject to exceptions, for the mafic/felsic

[29] For discussion of this topic, see A. Harker, *The Natural History of Igneous Rocks*, pp. 360–377, Macmillan, London, 1909; N. L. Bowen, *The Evolution of the Igneous Rocks*, pp. 321, 322, Princeton University Press, Princeton, N. J., 1928; A. Knopf, *Geol. Soc. America 50th Anniversary Vol.*, pp. 336–339, 1941; H. Williams, F. J. Turner, and C. M. Gilbert, *Petrography*, pp. 25–36, Freeman, San Francisco, 1955.

ratio to decrease, for the feldspar to become more alkaline, for unsaturated minerals to disappear, and ultimately for free quartz to appear in the corresponding increasingly siliceous series of rocks. It is this tendency for sympathetic variation of mineralogical characteristics that makes it possible to frame two-dimensional tabular classifications of igneous rocks, such as appear in many standard textbooks. These still have obvious defects. For example, it is impossible to fit into any such scheme the well-defined class of lamprophyres.

The following outline classification defines the principal families and classes of igneous rocks sufficiently closely for the purposes of petrogenic discussion.

A. Families of Volcanic Rocks
 1. Basalt family. Basic volcanic rocks with calcic plagioclase, augite, and in many cases olivine as essential constituents.
   a. *Olivine basalt.* Containing noteworthy olivine. *Olivine diabase* is coarse-grained equivalent, typically with ophitic texture. *Picrite-basalt* or *oceanite* is a very basic type rich in olivine phenocrysts; *ankaramite* is very rich in augite.
   b. *Tholeiitic basalt.* Poor in or lacking olivine, and in some cases containing minor quartz. Diabase is coarse-grained equivalent, typically with ophitic texture.
   c. *Trachybasalt.* Olivine basalt with high Na_2O or K_2O content expressed by minor barkevikitic hornblende, biotite, or orthoclase. *Mugearite* is an allied type with oligoclase (instead of labradorite), augite, and olivine.
   d. *Spilite.* Highly sodic olivine-poor basalt, with albite or oligoclase the sole or principal feldspar. *Albite diabase* is coarse-grained equivalent. Evidence of hydrothermal alteration is common in both types of rock.
 2. Nepheline basalt family. Ultrabasic and basic volcanic rocks with a high content of alkali. Basaltic minerals (calcic plagioclase, augite, and olivine in any combination) are accompanied by a feldspathoid or melilite or both.
   a. *Nepheline-bearing types.* *Nephelinite, nepheline basalt* (nepheline, olivine, augite); *nepheline basanite* (nepheline, plagioclase, olivine, augite); *nepheline tephrite* (nepheline, plagioclase, augite).
   b. *Leucite-bearing types.* *Leucitite, leucite basalt* (leucite, olivine, augite); *leucite basanite* (leucite, plagioclase, olivine, augite); *leucite tephrite* (leucite, plagioclase, augite).
   c. *Melilite basalt.* Melilite, olivine, augite, with or without nepheline.

  d. *Limburgite.* Olivine and augite in ultrabasic alkaline glassy base.
3. Lamprophyre family. Ultrabasic and basic dike rocks rich in alkali and in ($FeO + MgO$). Typically porphyritic with femic minerals in two generations: biotite and/or barkevikite, with augite, olivine, and rarely melilite as possible additional minerals. Feldspars are plagioclase or orthoclase, always confined to groundmass. Some lamprophyres are chemically equivalent to lavas of nepheline-basalt family.
4. Trachyte-phonolite family. Volcanic rocks of intermediate silica content, high in alkali and low in lime.
  a. *Trachyte.* Approximately saturated alkaline lavas with dominant sanidine or anorthoclase, accompanied by minor pyroxene, biotite, or hornblende. Small amounts of either quartz or feldspathoid may occur. *Porphyry* is porphyritic variety.
  b. *Keratophyre.* Sodic trachyte with albite as principal constituent.
  c. *Phonolite.* Undersaturated alkaline lavas with abundant feldspathoid (nepheline, sodalite, or more rarely leucite), sanidine or anorthoclase, aegirine or aegirine-augite. *Tinguaite* is porphyritic variety.
5. Andesite-rhyolite family. Intermediate to acid volcanic rocks, not conspicuously alkaline.
  a. *Andesite, porphyrite.* Rocks consisting of andesine or labradorite and some combination of augite, hypersthene, hornblende. Types transitional to basalts may carry olivine.
  b. *Latite, trachyandesite.* Rocks transitional between andesites and trachytes; plagioclase (andesine or labradorite) and sanidine both abundant.
  c. *Dacite.* Acid volcanic rocks with quartz, andesine, biotite and/or hornblende, and in some cases sanidine. Glassy types can be identified only by chemical analysis.
  d. *Rhyolite, quartz porphyry.* Acid volcanic rocks with quartz, sanidine, oligoclase, and minor ferromagnesians (biotite, hypersthene, hornblende, etc.). Glassy types (*obsidians*) can be identified only by chemical analysis. Sodic rhyolites (*e.g., pantellerites*) contain anorthoclase and sodic ferromagnesian such as aegirine and riebeckite.

B. Families of Plutonic Rocks (approximately equivalent volcanic families are noted in parentheses)
  1. Gabbro family (cf. basalt family of volcanics). Basic plutonic rocks consisting mainly of plagioclase (labradorite or bytownite), pyroxene, and in many cases olivine.

a. *Olivine gabbro.* With plagioclase, augite, olivine; *olivine norite*; with hypersthene in addition to the above.

b. *Gabbro.* With plagioclase and augite, rarely hornblende; *norite*, with plagioclase, hypersthene, augite, rarely biotite. Some gabbros and norites are slightly oversaturated and carry minor quartz.

c. *Anorthosite.* Plagioclase with only minor pyroxene and/or olivine.

d. *Pyroxenite.* Enstatite and/or augite.

e. *Troctolite.* Calcic plagioclase and olivine.

2. Peridotite family. Ultrabasic plutonic rocks consisting mainly of olivine. Pyroxene (typically enstatite) and/or calcic plagioclase are common minor minerals. Accessory chromite or picotite.

3. Alkali-gabbro family (cf. nepheline basalt family of volcanics). Basic plutonic rocks, the high alkali content of which leads to appearance of such minerals as biotite, barkevikite, orthoclase, anorthoclase, nepheline, or analcite in association with calcic plagioclase, augite, and olivine.

a. *Teschenite.* Analcite-bearing olivine gabbro, usually with barkevikite and titaniferous augite.

b. *Theralite* and *essexite.* Olivine gabbros containing orthoclase (or anorthoclase), nepheline, brown hornblende and minor biotite.

c. *Picrite* (transitional to peridotites). Olivine, titaniferous augite, minor plagioclase, and analcite.

d. *Shonkinite* (transitional to syenites). Augite, olivine, biotite, potash feldspar.

e. *Kentallenite* (transitional to monzonite). Plagioclase, potash feldspar, augite, olivine, biotite.

4. Diorite family (cf. andesites among volcanics). Intermediate plutonic rocks consisting essentially of andesine and hornblende. *Augite diorite* is transitional to gabbro. *Quartz-diorite* (*tonalite*), with quartz, biotite, hornblende, and andesine, is an oversaturated rock transitional to the granite family.

5. Granite-granodiorite family (cf. rhyolites and dacites among volcanics). Acid oversaturated rocks consisting mainly of quartz, potash feldspar (usually perthitic), and sodic plagioclase, with hornblende and/or biotite as typical mafic constituents.

a. *Granite.* Potash feldspar dominant, oligoclase generally subordinate; biotite alone, or with hornblende or muscovite.

b. *Soda-granite* and *granophyre.* Albite or albite-oligoclase dominant; sodic amphibole or pyroxene may be present.

   c. Quartz-monzonite or *adamellite.* Potash feldspar and oligoclase-andesine subequal.

   d. Granodiorite. Sodic andesine dominant, potash feldspar subordinate.

   e. Granite vein rocks. *Pegmatite,* commonly coarse-grained and mineralogically complex; *aplite,* fine-grained white rock consisting of quartz, albite, potash feldspar, muscovite.

 6. Syenite family (cf. trachyte-phonolite family of volcanics). Saturated or undersaturated plutonic rocks of intermediate silica content, high in alkali.

   a. Syenite. Saturated rock with potash-soda feldspar, subordinate albite-oligoclase and one or more mafic minerals (hornblende and/or biotite; less commonly, augite). Minor amounts of either quartz or feldspathoid mark transitions to granites and nepheline syenites respectively.

   b. Monzonite. Like syenite but with subequal amounts of potash feldspar and oligoclase-andesine.

   c. Nepheline-syenite group. Undersaturated feldspathoidal rocks. A sodic feldspathoid (nepheline, sodalite, analcite, cancrinite) and potash-soda feldspar are typically accompanied by one or more alkaline mafic minerals (biotite, alkali hornblende, aegirine).

CHAPTER 4

Variation in Associated Igneous Rocks

PETROGRAPHIC PROVINCES

Definition. The term *petrographic province* (= comagmatic province) has been applied to any broad region over which presumably related igneous rocks have been injected or poured out during the same general epoch of igneous activity. Such provinces are only vaguely bounded in space and time, and they vary greatly in extent. The Brito-Arctic (North Atlantic or Thulean) Tertiary province embracing northwest Britain, Iceland, and parts of Greenland extends 2,000 miles in a northwest-southeast direction.[1] The Pliocene province of eastern Otago [2] in southern New Zealand occupies only 2,000 square miles and is distinct from the immediately adjacent mid-Tertiary northeastern Otago province covering about 500 square miles.[3] Within such a province there are usually areas of intimately associated roughly contemporary igneous rocks known as *igneous complexes.* Examples are the well-known complexes of Mull, Ardnamurchan, and Arran in western Scotland, a part of the Tertiary Brito-Arctic province.

The rocks of a given province or complex may be relatively uniform in composition, as is the case with the dominantly basaltic rocks of the Columbia River Plateau, northwestern United States. More usually they are varied both in mineralogy and in chemical composition. Thus the rocks of the eastern Otago province (cf. pages 165 to 170) include olivine basalts, alkaline basalts (basanites, theralites and nephelinites), trachybasalts, trachytes, and phonolites, all abundantly represented within an

[1] J. E. Richey, Some features of Tertiary volcanicity in Scotland and Ireland, *Bull. volcanologique*, sér. 2, tome 1, p. 14, 1937.

[2] W. N. Benson, The basic igneous rocks of Eastern Otago, Part I, *Royal Soc. New Zealand Trans.*, vol. 71, pt. 3, p. 208, 1941.

[3] W. N. Benson, The basic igneous rocks of Eastern Otago, Part IV, *Royal Soc. New Zealand Trans.*, vol. 73, pt. 2, pp. 116–138, 1943.

area covering 250 square miles in the immediate vicinity of the city of Dunedin.[4] Yet in spite of this striking variation, it is impossible to escape the conclusion that these eastern Otago rocks have originated from a common source. They are intimately associated in a restricted province surrounded by a much larger area devoid of igneous rocks of corresponding age. They were erupted within a short span of time—a portion of the Pliocene. Within this period three major cycles of volcanism have been recognized, and the products of each cycle include the complete range from olivine-basaltic to phonolitic and trachytic lavas. Moreover the

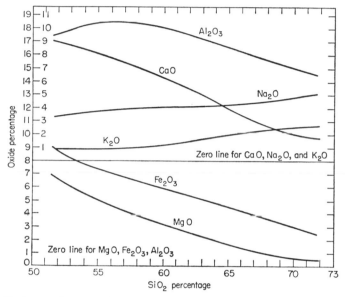

FIG. 1. Variation diagram for volcanic rocks of Crater Lake, Cascade volcanic province, Oregon. (*After H. Williams.*)

lavas of the province have certain chemical characteristics in common, notably a high ratio of total alkalis to lime.

Chemical Characteristics. In most petrographic provinces there is a wide range of chemical variation among associated igneous rocks, but in general this variation tends to be regular rather than of a random nature. The regularity of variation may be illustrated by plotting chemical analyses of representative rocks in the form of what are called variation diagrams.

Harker[5] used a variation diagram (Fig. 1) in which he plotted weight percentages of the principal oxides of each analyzed rock against weight

[4] Benson, *op. cit.*, 1941; W. N. Benson, Cainozoic petrographic provinces in New Zealand, and their residual magmas, *Am. Jour. Sci.*, vol. 239, pp. 537–552, 1941.

[5] A. Harker, *The Natural History of Igneous Rocks*, pp. 88–109, Macmillan, New York, 1909.

percentage of SiO_2—the underlying assumption (at best only an approximation) being that the silica percentage is an index of the stage of magmatic evolution attained. Larsen [6] considered it more satisfactory to plot oxide percentages against ($\frac{1}{3}SiO_2 + K_2O - FeO - MgO - CaO$). Nockolds and Allen [7] have modified this treatment by plotting percentages of component ions (Al^{3+}, Ca^{++}, Na^+, K^+, and so on) against ($\frac{1}{3}Si + K - Ca - Mg$). Curves so constructed (Fig. 2) certainly are smoother than those based on SiO_2 percentage. The reader must decide whether or not this is an advantage in diagrams designed to illustrate the nature of chemical variation in a rock series.

Several kinds of triangular variation diagrams have also been tried. Larsen recommends two that are based on normative composition: quartz-feldspars-femics and orthoclase-albite-anorthite. British petrologists [8] favor plots of alkalis-magnesium-iron and potassium-sodium-calcium, expressed in terms of oxides or of ions (Fig. 3). In constructing any triangular plot the three components are first recalculated to total 100.

Ever since Harker's day the smooth curves of variation diagrams have been interpreted as representing trends of evolution (e.g., differentiation) of magmatic liquids. It was pointed out by Bowen [9] that this interpretation is valid only for diagrams based on analyses of uniformly fine-grained or glassy volcanic rocks. Coarse-grained or porphyritic rocks should be excluded on the ground that their

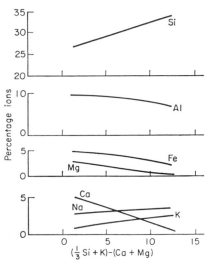

FIG. 2. Variation diagram for volcanic rocks of Crater Lake, Cascade volcanic province, Oregon. (*After S. R. Nockolds and R. Allen.*)

[6] E. S. Larsen, Some new variation diagrams for groups of igneous rocks, *Jour. Geology*, vol. 46, pp. 505–520, 1938.

[7] S. R. Nockolds and R. Allen, The geochemistry of some igneous rock series, *Geochim. et Cosmochim. Acta*, vol. 4, pp. 105–142 (especially p. 116), 1953; vol. 5, pp. 245–285, 1954.

[8] Cf. L. R. Wager and W. A. Deer, The petrology of the Skaergaard intrusion, Kangerdlugssuaq, east Greenland, *Meddelser om Grönland*, vol. 105, no. 4, p. 314, 1939; C. E. Tilley, Some aspects of magmatic evolution, *Geol. Soc. London Quart. Jour.*, vol. 106, pp. 48, 52, 53, 1950; Nockolds and Allen, *op cit.*, 1953, 1954.

[9] N. L. Bowen, *The Evolution of the Igneous Rocks*, pp. 92–124, Princeton University Press, Princeton, N.J., 1928.

compositions are likely to have been determined partly by crystal accumulation.   It is true that analyses of such rocks commonly depart notably from smooth curves based on analyses of fine-grained and glassy rocks of the same province, and it is customary to attribute such departure, regardless of the textures of the rocks in question, to the influence of crystal accumulation.   In Fig. 3 the smooth curves are believed to represent a line of magmatic evolution—a continuous series of related magmatic liquids.   Scattering of points, shown by triangles, near the basic ends of the curves is attributed to the disturbing influence of crystal accumulation, and it is inferred from this that the composition of the parental magma corresponds to the basic ends of the curves.[10]   This seems a reasonable interpretation, but it depends on the assumption that a vari-

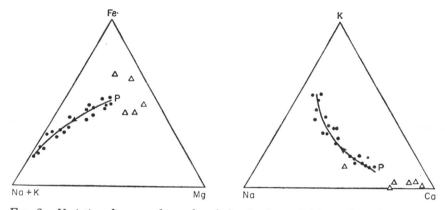

Fig. 3.   Variation diagrams for rocks of the Southern California batholith.   (*After S. R. Nockolds and R. Allen.*)   In each diagram P is the inferred composition of the parent magma.   Triangles represent rocks believed to show influence of crystal accumulation.

ation diagram based on chemical analyses of selected rocks does indeed reflect a line of magmatic evolution starting from a single parent liquid.

Alkali content has long been recognized as one of the more significant chemical characteristics of an igneous rock series in a petrographic province.   That alkali content is significant is obvious in the triangular variation diagrams just discussed.   Tilley [11] has found that a plot of total alkali against silica may even distinguish earlier from later evolutionary series in a single petrographic province—the Hawaiian Islands.   Peacock [12] used an alkali-lime index, given by the weight percentage of SiO_2 for which the weight percentages of CaO and of ($K_2O + Na_2O$) are equal.   This

[10] Nockolds and Allen, *op. cit.*, p. 106, 1953.

[11] Tilley, *op. cit.*, p. 42, 1950.

[12] M. A. Peacock, Classification of igneous rocks, *Jour. Geology*, vol. 39, pp. 54–67, 1931.

index may be read from the point where the respective curves for CaO and ($K_2O + Na_2O$) cross on the variation diagram.   Peacock arbitrarily distinguished four chemical classes of igneous rocks on the basis of alkali-lime index, *viz.*, alkalic (index less than 51); alkali-calcic (51 to 56); calc-alkalic (56 to 61); and calcic (greater than 61).   The basalt trachyte phonolite association of the east Otago province is typically alkalic; associations of basalt with rhyolite and andesite in other provinces are calcic by this criterion.

ROCK ASSOCIATIONS OR KINDREDS

Many petrographic provinces have now been described, some in great detail.   Anyone reading these descriptions cannot fail to be impressed by the tendency of the rocks of petrographic provinces to conform to one or another of a dozen or so standard types of association.   These characteristic associations of rocks have been called rock kindreds.   The kindred typified by association of basalts, trachytes and phonolites has been recorded in widely scattered localities—Tahiti, Hawaii, islands of the mid-Atlantic ridge (St. Helena and Ascension), the eastern Rift Valley of east Africa, and so on.   A totally different kindred, comprised of basalts, andesites, dacites, and rhyolites, appears in Japan, the western margins of the Americas, the Tertiary petrographic provinces of northern New Zealand, and scores of other localities where volcanism and orogeny have been associated in place and time.   The gabbro norite peridotite anorthosite association of the Bushveld complex of South Africa is duplicated in other major stratiform intrusions such as the Stillwater complex of Montana.   The individuality of each of these kindreds is marked by chemical as well as petrographic characteristics.   The variation diagram for each has its own constant and peculiar pattern.

By Kennedy, associations of igneous rocks are classified as either volcanic or plutonic.[13]   A volcanic association may, and frequently does, include intrusive plutonic rocks, but these are genetically related to a cycle of volcanic activity and have been derived from the same parent magmas as the associated strictly volcanic rocks.   Plutonic associations, on the other hand, "comprise the great subjacent stocks and batholiths together with the diverse minor intrusions of such abyssal masses."   Plutonic and volcanic associations are thought by Kennedy to differ fundamentally in their tectonic setting, in the nature of the parent magma, and in the whole process of magmatic evolution.   This thesis will be discussed in a later chapter.   In the meantime it may be noted that the present writers would

[13] W. Q. Kennedy, Crustal layers and the origin of magmas: petrological aspects of the problem, *Bull. volcanologique*, sér. 2, tome 3, pp. 24–41, 1938.

include in the plutonic category not only the dominantly acid batholithic masses, but also many essentially basic complexes such as the gabbro lopoliths and peridotite sills.

A tendency has long been recognized for a particular association of igneous rocks to occur in a given type or types of tectonic environment— the basalt trachyte phonolite kindred on oceanic islands, the andesite rhyolite association along fold arcs, and so on. It is this now well-established relation that underlies Harker's classic attempt to divide the Tertiary and Recent rocks of the world into two major petrographic provinces with the Atlantic and Pacific basins as respective centers. These are regions of generally contrasted tectonics. We still recognize the tendency for undersaturated alkaline rocks such as nepheline basalts and phonolites (Harker's Atlantic rocks) to occur in areas of crustal stability or of regional faulting, and for more calcic oversaturated rocks such as rhyolites and dacites (Harker's Pacific rocks) to be associated with folds such as border the Pacific. But there is such a wide range in tectonic environment (and in associated rock types as well) within both the Atlantic and the Pacific "provinces" as defined by Harker that his broad twofold subdivision and his use of the terms Atlantic and Pacific as applied to rock kindreds have been abandoned.

FACTORS IN MAGMATIC EVOLUTION

General Statement. Taking such facts as these into account, petrologists today are generally agreed that the chemical and mineralogical variation observed in associated igneous rocks is largely the result of some form of magmatic evolution. It is also clear that such evolution tends to follow one or another of several well-defined trends, so that each rock kindred can be regarded as including products of magmatic evolution starting with one type of parent magma and following some particular trend. That either the nature of the parent magma, or the trend of evolution, or both, are in some way connected with the geographic and tectonic setting of the petrographic province is also evident.

Later in this book reasons are stated for believing that basaltic and granitic magmas may both have been generally available in different sectors of the earth's outer crust, as parent materials from which rock kindreds of various types may have been derived. This does not preclude the possibility that other primitive magmas (e.g., hydrated peridotite magma) may have been generated from time to time. Starting with one, two, or a limited number of primitive parent magmas, there are several kinds of evolutionary processes to which the petrologist may appeal to account for the varied ultimate products of igneous activity. These fall into three classes, conventionally referred to respectively as differentiation, assimilation, and mixing of magmas.   It is no longer held that any one of these alone is responsible for magmatic evolution.   Rather the development and modification of magma is envisaged as a complex series of events in which all three may be involved to different degrees in any given instance.

Magmatic Differentiation.   Magmatic differentiation includes all processes by which a broadly homogeneous parent magma breaks up into contrasted fractions which ultimately form rocks of different compositions. The expression "broadly homogeneous" implies that large samples taken at random from the initial magma would be identical as to bulk composition and physical state.   The magma could nevertheless be heterogeneous within a small field, in that it might be partly liquid and partly crystalline before differentiation began.

Behavior of aqueous solutions under laboratory conditions suggests various mechanisms which might conceivably be responsible for differentiation of silicate magmas.   These have been critically discussed by N. L. Bowen [14] from a quantitative standpoint.   His general conclusions are summarized briefly below:

1. Within a still completely liquid magma a composition gradient might develop by sinking of heavy ions or "molecules" under gravity, or by migration of ions wherever a temperature gradient develops within the liquid mass.   This mechanism is dismissed by Bowen as unimportant on account of the extreme slowness of ionic migration in viscous magmas, and also in view of the small magnitude of such gradients even if equilibrium were reached.

2. It is conceivable that a homogeneous liquid magma might, on cooling, split into two immiscible liquid fractions.   However, there is overwhelming evidence, drawn from laboratory experiments, from experience regarding the behavior of slags in metal smelting, and from the fabric of rocks themselves, that at magmatic temperatures silicate melts approximating to igneous rocks of most if not all known kinds are completely miscible in all proportions.   This unmixing mechanism has, therefore, been dismissed by Bowen and others as inapplicable to differentiation.   A possible exception is the development in basic lavas of amygdules rich in iron and silica (green chalcedony, chlorophaeite, carbonates).   S.I. Tomkeieff [15] found it difficult to account for the distribution and great quantity of material of this type in amygdaloidal trachybasalts of the island of Rum in Scotland by appealing to the orthodox hypothesis of infilling of steam vesicles by late gaseous or liquid products of the crystallizing magma.   He suggested, instead, that the amygdaloids are products of unmixing of a liquid fraction

[14] N. L. Bowen, The later stages of the evolution of igneous rocks, *Jour. Geology*, vol. 23, supplement to No. 8, pp. 1–17, 1915; *op. cit.*, pp. 3–24, 293–297, 1928.

[15] S. I. Tomkeieff, The Tertiary lavas of Rum, *Geol. Mag.*, vol. 79, pp. 1–13, 1942.

very rich in silica and iron and totally lacking alkali, from basaltic magma under the influence of high concentrations of water.

3. A somewhat ill-defined mechanism of "gaseous transfer" has been appealed to by a number of petrologists. This hypothesis assumes the presence of a gas phase. This gas (see pages 51 and 64) presumably would consist mainly of volatile material (H_2O, CO_2, etc.) occurring in the form of innumerable bubbles streaming through the liquid magma. These might act as collectors and vehicles of transport for slightly volatile constituents of the magma. Such a process is analogous to the laboratory procedure for estimating CO_2 in a sample of water by bubbling a gas, initially free of CO_2, through the sample, and determining the CO_2 content of the collecting gas as it emerges. Alternatively it is conceivable, at least theoretically, that the magma could reach a critical point (see page 407), in which case solid crystalline phases would be in contact with a single fluid (gaseous) phase, possibly of great mobility. In this case the mobile fluid would have the same composition as had the liquid magma prior to reaching the critical point. It is unlikely that either situation develops under deep-seated conditions, at least until crystallization (and hence differentiation) is far advanced; but near the surface, e.g., under typical volcanic conditions, emission of magmatic gas certainly may occur on a large scale, and may well contribute to the process of differentiation, e.g., in a cupola feeding upward into a volcanic vent.

4. Probably of greater significance than gaseous transfer proper is a mechanism suggested by Kennedy,[16] whereby composition gradients might develop in a liquid magma phase under the influence of dissolved water. "Water will diffuse and distribute itself in a magma so that the chemical potential of water is the same throughout the magma chamber. By this mechanism water tends to be concentrated in the magma chamber in the regions of lowest pressures and temperatures. Alkalis and certain metals will co-ordinate with the water and, similarly, be concentrated in the regions of lowest pressure and temperature."[17] To the authors it seems probable that decreasing pressure and temperature would have opposite effects upon concentration gradients of dissolved water. But, with this modification, Kennedy's suggestion explains in a qualitative way how local concentrations of "volatiles" and alkalis might build up under the influence of temperature and pressure gradients in a cooling magma, without transport of alkali in a discrete gas phase.

5. When once crystallization of the magma begins, various mechanisms of crystal fractionation (i.e., separation of successive crystal fractions from residual liquid fractions) may come into play as possible factors in differ-

[16] C. C. Kennedy, Some aspects of the role of water in rock melts, Geol. Soc. America Special Paper 62, 1955, pp. 489–504.

[17] Ibid., p. 489.

entiation.   That fractional crystallization of some kind is very generally involved in differentiation is strikingly suggested by the nature of the mineral assemblages of common igneous rocks.   From the mass of experimental data on crystallization of such compounds as feldspars, feldspathoids, pyroxenes, olivine, and quartz from silicate melts under controlled conditions, it is clear that minerals which tend to be mutually associated in igneous rocks are those which crystallize over the same range of temperature (e.g., olivine-diopside, olivine-labradorite, orthoclase-oligoclase, fayalite-orthoclase).   There is an equally strong tendency for antipathy between minerals of widely different crystallization range (e.g., oligoclase and olivine, orthoclase and diopside, muscovite and labradorite).   If the rocks which show these mineralogical relationships are indeed products of differentiation, it would seem that fractional crystallization must be the principal factor concerned.   Some of the mechanisms of fractional crystallization that have been suggested by various writers could produce only negligible effects.   Ionic or molecular diffusion through viscous magmas toward an area of localized crystallization (e.g., near the chilled walls) belongs to this category.   On the other hand the processes mentioned below, operating singly or in sequence, not only are to be expected, but together are considered adequate to explain observed phenomena generally attributed to differentiation.

   a. Sinking of crystals of heavy minerals in a less dense liquid (gravitational differentiation) could be effective especially in the earlier stages when the liquid phase is still dominant and has not yet become too viscous to prevent settling of crystals.   Prevalence of layers rich in olivine and augite in differentiated basic sills is generally considered evidence of the effectiveness of crystal settling in this particular instance—especially since it has been shown in the laboratory that in melts of simplified basaltic composition the relative densities of the phases concerned would permit sinking of olivine or of pyroxene crystals.[18]   In the great stratified basic intrusions, such as those of Skaergaard and the Stillwater complex, an over-all bedded structure and textural evidence of sorting analogous to graded bedding of clastic sediments leave little doubt that these intrusions have solidified by accumulation of crystals sinking to the floor through liquid magma.   And there is an obvious concentration of heavy minerals— olivine, pyroxenes and chromite—in the lower layers.   Nevertheless it is still by no means certain that the highly differentiated state of such intrusions is entirely or even largely due to simple gravitational settling of early-formed heavy crystals (cf. page 304).

   b. Gravitational rising of light crystals (e.g., feldspars, leucite) in a denser liquid is also a possibility.   This process has been appealed to, for

[18] N. L. Bowen, Crystallization-differentiation in silicate liquids, Am. Jour. Sci., vol. 39, pp. 175–191, 1915.

example, to explain the occurrence of anorthosite (labradorite rock) in the upper levels of intrusions of gabbro.[19] Its effectiveness is less widely accepted than is the downward concentration of heavy minerals like olivine and augite.

c. If at some stage a gas phase develops, and if gas bubbles then stream upward, it is conceivable that an upward concentration of the lighter crystalline phases may be effected by a process of flotation. This involves the buoying action of rising bubbles that attach themselves to individual crystals. If crystallization is far advanced, as often it must be when boiling occurs, rising gas might blow the residual liquid upward through the crystal mesh. This mechanism has been called "gas streaming" by Shand,[20] who invokes it in explanation of the explosive eruption of trachyte differentiated from olivine-basalt magma.

d. When crystallization of a magma under plutonic conditions is far advanced, the crystals build a continuous mesh in the pores of which is the residual liquid. If the mass is now squeezed by movements in the wall rock, the residual liquid may be pressed out to form a separate body of differentiated magma. Alternatively if the crystal mesh is rifted by tensional forces, the residual liquid tends to fill the voids so formed according to the principle of dilatancy.[21] This process is termed *autointrusion.* One of many examples attributed to it is an irregularly defined body of syenite, 60 ft. in thickness, in a differentiated basic sill in the Shiant Isles, Scotland.[22] Emmons [23] has drawn attention to the possible effects of fracturing, in quartzites and similar brittle wall rocks, in causing outward migration of residual magmas into the voids so formed. He cites, among other examples, the case of differentiated red granophyres in the roof regions of North American gabbro intrusions, where these invade quartzites and similar brittle rocks.

e. Early-formed crystals of heavy minerals (olivine, pyroxene) might well become concentrated locally in a moving body of magma by a mechanism analogous to elutriation. Presumably such a process would be favored by low viscosity of the liquid phase, and its effects should be sought in bodies of basic igneous rock and especially in those with a high content of water. Good evidence has been brought forward by Fuller for gravitational settling of olivine crystals in slowly flowing basaltic

[19] H. von Eckermann, The anorthosites and kenningite of the Nordingra Rödö region, *Geol. fören. Stockholm Förh.,* vol. 60, pp. 243–284, 1938.

[20] S. J. Shand, The lavas of Mauritius, *Geol. Soc. London Quart. Jour.,* vol. 89, p. 11, 1933.

[21] Cf. R. C. Emmons, The contribution of differential pressure to magmatic differentiation, *Am. Jour. Sci.,* vol. 238, pp. 1–21, 1940.

[22] F Walker, The crinanite sills of the Shiant Isles, *Geol. Soc. London Quart. Jour.,* vol. 86, pp. 388–390, 1930.

[23] Emmons, *op. cit.,* 1940.

magma of a late Tertiary volcano in Oregon, "in a manner somewhat analogous to the deposition of sand at the delta of a river"; while Flett and Tyrrell have suggested "elutriation" in flowing teschenitic magma at the time of injection, as a possible mode of origin of the olivine-rich layers of teschenitic sills.[24]

f. The nature of the liquid fraction of a crystallizing magma at a given moment depends on the nature of the parent magma, the prevailing temperature and pressure, the degree to which fractional crystallization has already been effective, and the extent to which equilibrium has been maintained between associated crystals and liquid. We shall see in the next chapter that many igneous minerals stable at high temperature become unstable when in contact with magmatic liquid at lower temperature. Equilibrium is normally reestablished by reaction between liquid and crystals to give some new crystalline phase, a process which is the reverse of incongruent melting. Under some conditions of cooling, this new stable phase may form a rim around the crystal of the unstable phase, which is thus isolated from the liquid. Because diffusion is so much slower in ionic crystals than in liquids, the velocity of reaction at once drops sharply (cf. page 44) and soon becomes infinitesimal as the protective shell of the stable crystalline phase thickens. The familiar occurrence of zoned isomorphous crystals (e.g., plagioclase) and of reaction rims in minerals of igneous rocks shows that this is a common natural phenomenon. This failure to maintain equilibrium cannot by itself cause differentiation except within the minute field of a single zoned crystal, but it may greatly affect the composition of the final residual liquids and corresponding rocks, differentiated by some other means such as "filter pressing" or "gas streaming." It is, therefore, included as one of the mechanisms of differentiation.

The processes enumerated above, taken together, seem adequate to explain most igneous phenomena generally attributed to differentiation. This is not to say that differentiation is in fact proved to be the dominant, nor even necessarily an important, factor in bringing about diversity in associated igneous rocks.

Assimilation. Intrusive magma is seldom in a state of chemical equilibrium with the rocks into which it is injected, although it may be in equilibrium with one or more constituent minerals of the invaded rock. Reaction between magma and wall rock must, therefore, be a normal accompaniment to igneous intrusion. In the course of this reaction the

---

[24] J. S. Flett, The Saline No. 1 teschenite, *Great Britain Geol. Survey Summary of Progress, 1936*, pt. 2, pp. 44–51, 1931; R. E. Fuller, Gravitational accumulation of olivine during advance of basalt flows, *Jour Geology*, vol 47, pp. 303–313, 1939; G. W. Tyrrell, A boring through the Lugar Sill, *Geol. Soc. Glasgow Trans.*, vol. 21, p. 201, 1948.

magma (probably in most cases a silicate melt with suspended crystals of one or more solid phases) becomes modified by incorporation of material originally present in the wall rock. This broad process of modification is termed *assimilation*.[25]

The following general principles, first clearly stated by Bowen,[26] determine the mechanism by which assimilation will proceed in any given instance:

1. A considerable amount of heat, on the average perhaps as much as 100 cal./g., is necessary to melt most common rocks. This heat must be supplied by the magma which causes the melting, and which therefore must cool as assimilation proceeds. If the magma were initially at a temperature barely higher than that at which crystallization begins, melting of solid rock would cause a corresponding amount of crystallization of the magma; furthermore, magma could obviously cause melting of only those minerals that have a lower melting temperature than the magma itself (*e.g.*, basalt at 1200°C. could not melt quartzite, although it could react with it; see below). Wholesale assimilation by a liquid magma thus requires that the magma initially be at a temperature several hundred degrees above its own crystallization range ("superheat"). This is obviously impossible if the magma is itself the product either of partial melting or of differentiation by crystallization (see Chap. 6).

2. Suppose that a magma has started to crystallize, and that the crystals which are forming belong to a reaction series [27]—the usual case with igneous minerals. Then the liquid is effectively supersaturated with any preceding member of the same reaction series (*i.e.*, a mineral which crystallizes at a higher temperature in that series). The liquid is, therefore, incapable of converting such a member into the liquid state. If crystals of this high-temperature member are added to the magma, equilibrium tends to become established by a process of reaction (ionic exchange between liquid and crystals) in the course of which the foreign phase is converted to crystals of that phase with which the liquid is saturated. Consider labradorite crystals coming into contact with a granitic

[25] This usage corresponds with that adopted by S. J. Shand (*Eruptive Rocks*, 2d ed., pp. 90–94, Wiley, New York, 1943). Some petrologists restrict the term to the process of simple solution or melting of solid rock by invading magma (cf. R. A. Daly, *Igneous Rocks and the Depths of the Earth*, p. 288, McGraw-Hill, New York, 1933; F. F. Grout, Formation of igneous-looking rocks by metasomatism, *Geol. Soc. America Bull.*, vol. 52, p. 1533, 1941).

[26] Bowen, *op. cit.*, *Jour. Geology*, pp. 84–91, 1915; *op. cit.*, pp. 174–223, 1928.

[27] In a reaction series (a series of crystalline phases *A,B,C*, . . .) any member (*e.g.*, *A*) separates from a melt of appropriate composition at a given temperature or over a given range of temperatures, but at lower temperatures becomes unstable and reacts with the melt to give the next number (*e.g.*, *B*) of the series. The nature of reaction series and the various reaction series represented among minerals of igneous rocks are discussed in Chap. 6.

melt from which oligoclase is already crystallizing. The plagioclases constitute a reaction series which becomes increasingly sodic toward the low-temperature end. The labradorite crystals, therefore, cannot dissolve or melt. Instead, a complex reaction occurs in which the liquid, the suspended crystals of oligoclase, and the foreign crystals of labradorite all participate. The labradorite is thereby converted to oligoclase, the phase that can exist in equilibrium with the liquid. If the reaction occurs without loss of heat (adiabatically), then the oligoclase crystals originally present become slightly more calcic as reaction proceeds.

3. Suppose now that a magma already containing suspended crystals of a high-temperature member of a reaction series (e.g., olivine) comes into contact with foreign crystals of a low-temperature member of the same reaction series (e.g., hypersthene). Again equilibrium is restored by mutual reaction between the various associated phases. In this case the foreign crystalline phase (hypersthene) is dissolved (melted) into the liquid fraction of the magma, but to supply the necessary latent heat of fusion and to maintain equilibrium in the system, an equivalent amount of the phase with which the liquid was already saturated, namely olivine, crystallizes out.

Assimilation may thus be pictured as a complex process of reciprocal reaction between magma and invaded rock. Certain of the minerals present in the wall rock may become partially or completely melted and in this way incorporated into the liquid fraction of the magma. Others are changed by a process of ionic exchange (reaction) into those crystalline phases with which the liquid was already saturated. Yet other minerals, by chance compatible with the invading magma, persist unchanged as they break out from the transformed partially melted matrix that surrounded them and drift away into the reacting magma. The end product is a contaminated partially crystalline magma. In many cases the proportion of liquid in this contaminated magma diminishes as reaction proceeds. When with continued cooling the mass becomes completely crystalline it has given rise to a contaminated igneous rock which was at no time entirely liquid, and which is made up of material contributed partly by the original magma and partly by the wall rocks. It may then be impossible to recognize any sharp boundary between intrusive and invaded rocks. Toward the intrusive contact the metamorphosed wall rock becomes increasingly affected by metasomatism as a result of chemical exchange with the adjacent magma, until it acquires a composition close to, or identical with, that of the contaminated igneous rock into which it merges.

Mixing of Magmas.[28] As early as 1851 Bunsen suggested that mixing of two contrasted parent magmas—the one basaltic, the other rhyolitic—

[28] Cf. Harker, *op. cit.*, pp. 333–336, 1909.

would account for the range of variation observed in lavas of the basalt andesite rhyolite series in Iceland and elsewhere. This theory was later elaborated by Durocher,[29] who explained all igneous rocks as hybrid derivatives of two primary world-wide magmas. Growing petrographic knowledge soon demonstrated the total inadequacy of this mechanism to explain the wide range of known igneous mineral assemblages. The chemical and mineralogical variation encountered even in a single common association of rocks (such as olivine basalt, nepheline basalt, melilite basalt, mugearite, trachyte, phonolite) is far too complex, when considered in detail, to be compatible with the simple linear variations which would result from mixing of any two end members. Mingling of magmas is no longer considered to be the principal factor in magmatic evolution.

This is not to say that mixing of magmas never occurs. Rocks of unusual composition in which a large number of crystalline phases seem to be associated in a state of disequilibrium may well, in some cases, be products of mixing of two partially crystalline magmas. A possible instance is afforded by the "kaiwekites" of Otago, New Zealand. These are lavas in which phenocrysts of plagioclase, anorthoclase, augite (with aegirine rims), olivine, and brown hornblende are set in a groundmass of oligoclase, anorthoclase, and augite. They are interpreted as basalt-trachyte hybrids.[30] Of much greater significance than petrographic curiosities such as "kaiwekites," is the evidence of magmatic mixing supplied by some of the commonest types of lava, notably andesites and dacites. In the lavas of the San Juan volcanic province of Colorado—perhaps the most extensive series of volcanic rocks yet subjected to modern petrographic investigation—the variation in composition of associated plagioclase phenocrysts is too complex to be accounted for by simple differentiation. It is, however, compatible with a mechanism of mixing of two magmas (perhaps themselves of differentiated origin) containing suspended crystals. Moreover the authors of the San Juan papers conclude that "there must have been a very thorough mixing of very large masses of magma" to account for the uniform distribution of phenocrysts of feldspar in extensive lava flows.[31] From this instance, and from the very general occurrence of plagioclase phenocrysts of widely varying composition in andesites and dacites, it seems reasonable to postulate that the main function of magmatic mixing in the evolution of a magma series is to reblend magmas of common origin which have already attained separate identities as a result of differentiation or assimilation.

[29] Cf. Harker, *op. cit.*, pp. 333–336, 1909.
[30] Benson, *op. cit.*, *Am. Jour. Sci.*, vol. 239, p. 546, 1941.
[31] E. S. Larsen and coauthors, Petrologic results of a study of the minerals from the Tertiary volcanic rocks of the San Juan region, Colorado, *Am. Mineralogist*, vol. 23, pp. 255–257, 429, 1938.

CHAPTER 5

Crystallization of Igneous Minerals from Silicate Melts

INTRODUCTORY STATEMENT

Microscopic study of igneous rocks, with special reference to textural relations of constituent minerals, has yielded much information as to the order in which different mineral phases separate during freezing of magmas. On the basis of petrographic experience Rosenbusch [1] drew up a series of empirical rules which he thought to be generally applicable—admittedly with many exceptions—to crystallizing magmas. According to Rosenbusch, iron oxides and minor constituents of the magma tend to crystallize first as idiomorphic crystals of accessory minerals (iron ores, zircon, apatite, sphene, etc.); the main constituents tend to crystallize in order of increasing silica content, so that the residual liquid is at all stages more siliceous than the essential minerals which have already crystallized; furthermore there is a tendency for early separation of silicates of Mg and Fe, then silicates containing Ca, still later silicates of Na and K, and finally free silica as quartz.

So numerous are the exceptions to these "rules" that some writers consider them worthless.[2] Some of the anomalies, as pointed out by Shand, reflect inadequacy of the microscopic method, based as it is upon apparent relations between crystals of different minerals as revealed in random sections of rocks. Taking this factor into account and remembering, too, the number of crystalline phases represented among igneous minerals and the complexity and diversity of igneous rocks, we should be surprised indeed

[1] H. Rosenbusch, Ueber das Wesen der körnigen und porphyrischen Struktur bei Massengesteinen, Neues Jahrb., 1882, Band 2, pp. 1–17.

[2] For a critical appraisal of Rosenbusch's "rules," see S. J. Shand, Eruptive Rocks, pp. 114–126, Wiley, New York, 1943. See also N. L. Bowen, The order of crystallization in igneous rocks, Jour. Geology, vol. 20, pp. 457–468, 1912.

to find any simple rules that could be applied generally to the order of separation of minerals from magmas.   That some semblance of regularity can be discerned must therefore be significant.   But the implications involved can be discussed adequately only in the light of experimental evidence.

Largely as a result of laboratory experiments by French mineralogists, extensive data on melting and crystallization of igneous minerals and rocks were already available over sixty years ago.[3]   Other data of great value to petrology were collected by J. H. L. Vogt in the course of his classic researches on slags (1884 to 1906).[4]   Vogt's work in particular led to the now accepted conclusion that the laws of solutions are generally applicable to silicate melts, i.e., to magmas.   Thus Harker, when he wrote his classic *Natural History of Igneous Rocks* in 1909, was able to frame a coordinated hypothesis of magmatic crystallization in terms of established chemical principles and experimentally determined data.   Meanwhile the Geophysical Laboratory of the Carnegie Institution of Washington had been founded.   Here during the past five decades, precise studies have been made of the crystallization of igneous minerals from melts analogous to simplified magmas in systems of two, three, or four components.   Much of this work was summarized and applied to igneous petrogenesis by N. L. Bowen, first in his classic paper on The Later Stages in the Evolution of Igneous Rocks,[5] and later in *The Evolution of the Igneous Rocks* (1928). These accounts, extended and modified in the light of later studies, particularly those relating to hydrous systems, must form the basis of any physico-chemical treatment of igneous petrogenesis.[6]

At the outset it is emphasized that caution must be exercised in applying experimental results to the problems of magmatic crystallization.   Magmas are multicomponent systems, far more complex than the melts investigated in the laboratory.   Moreover, in magmas undercooling and failure to maintain equilibrium are common modifying factors.   The presence of water and other "volatile" materials in natural melts undoubtedly affects the course of crystallization profoundly, e.g., by lowering the freezing temperatures (perhaps through several hundred degrees) and by causing the appearance of hydrous mineral phases (biotite, hornblende, etc.) not encountered in the laboratory product.

[3] *E.g.,* F. Fouquet and A. Michel-Lévy, *Synthèse des minéraux et des roches,* Masson et Cie, Paris, 1882.

[4] For a brief summary of Vogt's work, see A. Harker, *The Natural History of Igneous Rocks,* pp. 174–179, Macmillan, New York, 1909.

[5] *Jour. Geology,* vol. 23, supplement to no. 8, pp. 1–91, 1915.

[6] Cf. N. L. Bowen, *The Evolution of the Igneous Rocks,* Princeton University Press, Princeton, N.J., 1928; Recent high-temperature research on silicates and its significance in igneous geology, *Am. Jour. Sci.,* vol. 33, pp. 1–21, 1937; T. F. W. Barth, *Die Entstehung der Gesteine* (Barth, Correns, and Eskola), pp. 16–36, Springer, Berlin, 1939.

CHEMICAL CONDITIONS GOVERNING SOLID-LIQUID EQUILIBRIUM [7]

Systems of One Component. The equilibrium condition for systems of one component in two phases has already been mentioned briefly (page 21). The temperature at which a solid is in equilibrium with a liquid of its own composition is generally known as the melting point T_m of the substance. At this point, the chemical potential of the component is the same in both the solid and the liquid phases. If the temperature is increased beyond T_m, the chemical potential of the liquid decreases faster than that of the solid, the entropy of the liquid being greater than that of the solid. At all temperatures beyond the melting point, the liquid phase is therefore the stable one. Similarly, if the pressure is increased, the chemical potential of each phase increases proportionally to its molar volume, and since the volume of the liquid is generally greater than that of the solid, the chemical potential of the liquid in most cases increases faster and the solid becomes the stable phase. If temperature and pressure are increased simultaneously, equilibrium between the liquid and solid can be maintained only if the effect of temperature is offset exactly by the effect of pressure. This is expressed by Eq. (2–39) which is more commonly written under the equivalent form (2–40) and is known as the Clausius-Clapeyron equation.

As both the heat of melting Δh and the change in volume attendant on melting Δv depend on pressure and temperature, it has been suggested that, as the temperature and the pressure are increased, conceivably a point might be reached at which dT/dP might become zero, and then negative. This would mean that there would be a maximum melting point corresponding to a definite value of pressure beyond which the melting point would decrease with further increasing pressure. It has been suggested also that Δh and Δv might simultaneously become zero at a conceivably "critical" point at which the melting curve would stop and beyond which the distinction between liquid and solid states would vanish. The matter has been carefully investigated by Bridgman,[*] who has followed experimentally a number of melting curves up to very high pressures without finding any trace of such an effect. The melting point apparently increases indefinitely with increasing pressure. The temperatures and heats of melting under normal conditions for a large number of minerals may be found in the *Handbook of Physical Constants*.[†]

---

[7] For a comprehensive account of the phenomena, principles, and graphical representation of heterogeneous fusion equilibria in multicomponent silicate systems, see W. Eitel, *Silicate Melt Equilibria*, Rutgers University Press, New Brunswick, N.J., 1951.

[*] P. W. Bridgman, *The Physics of High Pressure*, pp. 201–207, G. Bell, London, 1931.

[†] *Geol. Soc. America Special Paper 36*, pp. 140–174, 237–238, 1942.

There are many compounds, including a number of common minerals, which, when heated to a certain temperature (incongruent melting point), decompose to give two phases, one of which is then a liquid. That is, the liquid which forms at this temperature does not have the composition of the original substance, and a new solid phase is thus formed at the same time as the liquid. For instance, protoenstatite ($MgSiO_3$) when heated to 1557°C. forms a liquid richer in SiO_2 than protoenstatite and a solid (forsterite, Mg_2SiO_4) correspondingly richer in MgO. Such behavior is known as *incongruent melting*. Strictly speaking, substances which melt incongruently do not form one-component systems. We shall return to them later (see page 96).

Two-component Systems. Study of equilibria in two-phase, two-component systems covers a wide range of phenomena known as "solubility" or "melting." The difference between the two terms is one of usage, not of principle. When we study equilibrium between solid NaCl and an aqueous salt solution we refer to the solubility of salt; when solid albite is in equilibrium at high temperature with a solution of albite in liquid diopside, we refer to the "melting" of albite. Similarly, when we refer to solutions we make no essential distinction between "solvent" and "solute," though the most abundant of the two components is conventionally assumed to be the solvent.

The "solubility" of a solid in a given solvent is, of course, the concentration of this substance in a saturated solution, that is, in a solution which may exist in equilibrium with an excess of the solid. Equilibrium between a solid and a solution is reached when the chemical potential of the substance is the same in both phases. It is important to consider how this could occur. To do so, consider first the equilibrium between the solid and a liquid of its own composition. At the melting point T_m, the two chemical potentials are equal. At any temperature T below the melting point, the chemical potential of the liquid becomes greater than that of the solid by an amount given by Eq. (2–29) integrated from T_m to T

$$\mu_1^l - \mu_1^s = \int_T^{T_m} (s_1^l - s_1^s) \, dT$$

where the indices l and s refer respectively to the pure liquid and solid. In particular, if $s_1^l - s_1^s$ is constant in the temperature interval from T to T_m, we have

$$\mu_1^l - \mu_1^s = \frac{\Delta h}{T_m} (T_m - T) \tag{5-1}$$

an approximate but useful relation, where Δh stands for the heat of melting of component 1.

Now let us suppose that the liquid phase consists of a solution of component 1 in a solvent 2. Then at any temperature T the chemical potential μ_1 of 1 is given by Eq. (2–38) and is

$$\mu_1 = \mu_1^l + RT \ln N_1 f_1$$

where μ_1^l is the chemical potential of pure liquid 1. Hence equilibrium between solid 1 and the solution may exist at temperature T if, and only if,

$$\mu_1 = \mu_1^s$$

that is, if

$$\mu_1^s = \mu_1^l + RT \ln N_1 f_1$$

or, by (5–1), if

$$RT \ln N_1 f_1 = - \frac{\Delta h}{T_m} (T_m - T) \tag{5–2}$$

If the solution of 1 in 2 is ideal, $f_1 = f_2 = 1$ and we have

$$N_1 = \exp\left[- \frac{\Delta h}{RTT_m} (T_m - T) \right] \tag{5–3}$$

where N_1 is the "solubility" of 1 in 2 at the temperature T. It is important to notice that this solubility in an ideal solution is independent of the nature of the solvent. If Δh and T_m are known, Eq. (5–3) may be used to plot a solubility or melting curve showing as a function of temperature the change in the molar fraction of 1 in a saturated solution. Obviously, for $T = T_m$, $N_1 = 1$, so that the curve starts from a point representing the melting point of pure 1.

Bowen [8] has compared the curves drawn from Eq. (5–3) with experimental solubility curves in the systems albite-diopside and anorthite-diopside. He finds excellent agreement between these curves in the neighborhood of the melting point of the pure components, but at lower temperatures some discrepancies appear. Apparently Eq. (5–1) is no longer valid at these comparatively low temperatures, for to the heat of melting must be added a term representing the heat of mutual solution of the two components. In other words, the solution ceases to be ideal. Nonideal behavior is likely to appear whenever the solution contains molecules of contrasted types, that is, whenever the forces acting between unlike molecules (1—2) are different from forces between like molecules (1—1) or (2—2). If the energy of the bond 1—2 is less than that of either bonds 1—1 or 2—2, energy will be released when, on mixing the two liquids 1 and 2, bonds 1—2 are formed at the expense of bonds 1—1 or 2—2. In this case heat is given off, and the heat of solution ΔH is said to be negative. If, on the contrary, bonds between unlike molecules are weaker than between like pairs, energy must be supplied to form the former at the expense of the latter, and the heat of solution must be posi-

[8] N. L. Bowen, op. cit., pp. 176–182, 1928.

tive.   Such a solution will have a tendency to unmix.[9]   Now if heat of mixing is involved, there is a corresponding change in free energy, and the chemical potentials of the two components, which are related to the free energy of the solution by Eq. (2–15),

$$G = n_1\mu_1 + n_2\mu_2$$

are different from what they would be if the solution were ideal.   To take this change into account, we must introduce a correcting term known as the activity coefficient.   But the fundamental result remains valid that the melting curve is still determined essentially by the temperature, the heat of melting of the pure components, and the heat of mixing (this latter term depending on the nature of the two components and on the magnitude of intermolecular forces).   If these were known, or if the solution were nearly ideal, the melting curves or "phase" diagrams could be predicted very simply from thermodynamic considerations.

Phase diagrams of binary systems fall into two large groups which may be described respectively as the eutectic type (e.g., anorthite-diopside) and the solid-solution type (e.g., albite-anorthite) (see pages 99 to 102). Such differences as exist between these two types arise mainly from differences in the behavior of the solid phases concerned.   It must be remembered in this connection that, contrary to views once held by chemists, solids usually do not have a fixed composition that can be represented completely by a formula with stoichiometric coefficients that are small integers.   Most solid phases actually contain an excess of one or several of their components over the simple stoichiometric formula, as well as a certain amount of foreign matter.   This excess amount of material that is soluble in the solid phase may be very small, as in quartz, or may be very large, as in anorthite which may take up in solution such large amounts of albite that it may be made continuously into pure albite.   Such extended mutual solubility arises when corresponding ions of the two components are nearly identical in size and when the two pure components are capable of existing as crystals having similar space lattices.   If these conditions are not met, the mutual solubility becomes small.   All gradations are known between these two extremes; that is, certain solids may be miscible only within a certain range of composition or within a certain range of temperature (e.g., nepheline-anorthite which forms solid solutions containing from about 20 per cent to approximately 45 per cent

[9] Such solution would not form at all, i.e., the two liquids would be immiscible, if it were not for the fact that the process of mixing involves a decrease in orderliness and therefore an increase in entropy.   Hence the free energy $\Delta G = \Delta H - T \Delta S$ may be negative in spite of ΔH being positive, and the process of mixing of the two liquids may occur spontaneously.   We note also that since mixing is essentially an irreversible process ΔS may be positive even if there is no heat effect involved [see Eq. (2–4)].

anorthite by weight in the temperature range 1302° to 1352°C.). A phase diagram for components which are mutually soluble in all proportions has the appearance shown in Fig. 5a (albite-anorthite), page 101. It shows two curves which give respectively the composition of the solid and that of the liquid phase in mutual equilibrium at any given temperature within the melting range. Bowen [10] has compared the experimentally determined curves for this system with the curves theoretically derived on the assumption that both the liquid and solid solutions are ideal (no heat of mixing), and has found excellent agreement. Similarly, the systems forsterite-fayalite and clinoenstatite-ferrosilite form ideal solutions, at least at high temperature. In such ideal systems, the effect of component 1 on the chemical potential of component 2 is the same in both the liquid and solid solutions, since it depends only on the mole fraction [see Eq. (2–38)]. Thus the liquidus and solidus curves are determined entirely by the temperature and by heats of melting of the pure components. If these heats of melting happened to be the same, the two curves would coincide. If either the solid or liquid solutions or both are not ideal, the curves may show a minimum or maximum temperature for some composition between the two end members (e.g., gehlenite-akermanite).

The simple eutectic type of phase diagram (e.g., anorthite-diopside) corresponds to the case where there is no miscibility of the solid phases. It consists of two liquidus curves which start respectively at the melting points of the two pure components. Each of these curves is represented by an equation of the type (5–2), and since these curves for each component slope in opposite directions, they must intersect at some lower temperature. This intersection point is the eutectic point and represents the composition of the liquid which has the lowest melting point. That is, if any mechanical mixture of the two solid components is heated to the eutectic temperature, a liquid will form which has this eutectic composition. This is an interesting example of the reactions which may occur at an interface between two solids; obviously the molecules of component 2 in the solid state exert such a disrupting influence on the lattice of component 1 that this lattice breaks down at a temperature well below the melting point of pure 1, and vice versa. This indicates that the intermolecular bonds 1—2 must be much stronger in the liquid than in the solid-phase state. Actually these forces must be repulsive in the solid state, since the two solids are not mutually soluble.

As already noted, there are other two-component systems which behave in a manner intermediate between the two extremes just described (e.g., anorthite-nepheline).

An interesting feature of systems containing SiO_2 is that in spite of its

[10] N. L. Bowen, Melting phenomena in plagioclase feldspars, Am. Jour. Sci., vol. 35, p. 590, 1913.

relatively high melting point, the proportion of SiO_2 in the eutectic is invariably high. This behavior is related to its relatively low entropy of melting. Indeed, reference to Eq. (5–2) shows that if $\Delta h/T_m$ is small, N will be relatively large even if $T_m - T$ is large, *i.e.*, the proportion of component 1 in the liquid remains large at relatively low temperatures. The low entropy of melting of quartz, in turn, is probably related to the comparatively small difference in orderliness between solid and liquid silica, *i.e.*, liquid silica presumably still possesses much of the orderly arrangement of linked [SiO_4] tetrahedra characteristic of the solid phase.

Incongruent Melting. Incongruent melting, as mentioned earlier, occurs when a solid 1 splits at a definite temperature into a solid 2 and a liquid 3. This solid 2 melts gradually as the temperature is raised above the incongruent melting point, so that there is a certain range of temperature in which a solid and a liquid phase coexist. Conversely, when a liquid of composition 1 is cooled from a high temperature, the first crystals to form consist of 2. These crystals separate gradually over a certain range of temperature in which the composition of the residual liquid must, of course, also change. Finally, a unique temperature is reached, which is the incongruent melting point of 1, and at this the liquid 3 reacts with solid 2 which is made over into solid 1. There are many petrologically important examples of incongruent melting. For instance, orthoclase ($KAlSi_3O_8$) melts incongruently at 1150°C. to form leucite ($KAlSi_2O_6$) and a liquid richer in SiO_2 than orthoclase; complete melting of the leucite so formed occurs only at about 1533°C. Similarly, mullite ($Al_6Si_2O_{13}$) melts incongruently at 1810°C. to form corundum (Al_2O_3) and liquid; monticellite ($CaMgSiO_4$) forms periclase (MgO) and liquid at 1503°C.; hematite (Fe_2O_3) forms from acmite ($NaFeSi_2O_6$) at 990°C., and as mentioned before, protoenstatite ($MgSiO_3$) melts incongruently at 1557°C. to form forsterite (Mg_2SiO_4) and liquid. In this latter case, the temperature interval in which forsterite exists in equilibrium with the liquid is small, as the system becomes wholly liquid below 1600°C.

It may be noted that in all these cases, and others, the liquid which forms at the incongruent melting point is richer in SiO_2 than the original mineral. This again is related to the comparatively low entropy of melting of quartz.

Multicomponent Systems. The only difference between two-component and multicomponent systems is that the latter are much more difficult to study experimentally and to represent graphically; otherwise the principles involved are exactly the same. Any solid phase forms from a multicomponent solution when the chemical potential of this component in the solution becomes larger than that of the corresponding solid at the same temperature; the chemical potential in the liquid depends, as we know, on temperature and composition (neglecting for the time being the effect of pressure, which we may assume to be constant). Broadly speaking, the order of crystallization from an ideal multicomponent solution would be that of decreasing effect of temperature on solubility [see Eq. (2–46)]; any component whose solubility remains constant or increases as the temperature is lowered through a certain range would not precipitate at all in this range.   It should be noted that this has nothing to do with the magnitude of the solubility; component 1 may be far more soluble than component 2 and yet precipitate before 2 when the solution is cooled through a certain range.   If the solution is not ideal, the chemical potential of any component is affected by changes in the molar fractions of the other components, so that the solubility of 1 might either increase or decrease with decreasing temperature simply because some other component crystallized in this range.

Processes of crystallization in magmas and hydrothermal solutions are further complicated by the fact that changes in temperature are usually associated with changes in pressure, as when a magma rises to the surface. The order of crystallization will then be found to depend on the relative changes in pressure and temperature, and a solution or magma of given composition may show different sequences of crystallization under different conditions of cooling and expansion.   The effect of change in pressure is particularly marked when the magma is associated with a gas phase of large molar volume, for relatively small changes in pressure will affect notably the chemical potentials of the components of the gas phase.

Nonequilibrium Conditions.   A characteristic feature of melting or crystallization phenomena in systems of more than one component is that they do not occur entirely at one temperature.   Rather, they extend over a range of temperatures which, in the case of silicate melts, may cover several hundred degrees.   The expression "melting point of a rock" is therefore misleading, and instead we should refer to its "melting range." Melting and crystallization phenomena in multicomponent systems must therefore be thought of as continuous processes involving gradual changes in the composition of the liquid and also, in many cases, changes in the composition of the solid phases which have already crystallized (e.g., incongruent melting, solid solutions).   Now such changes do not occur instantaneously; it takes time to supply the required components by diffusion to the faces of a growing crystal.   A close approach to equilibrium therefore can be maintained only if the rate of cooling is carefully adjusted to the rate at which the various processes involved in crystallization can occur.   But the rate of cooling is determined by physical factors which depend mainly on the environment and have very little to do with the properties of the magma itself.   Hence we may expect that notable departures from equilibrium will occur as crystallization proceeds.   We must remember, furthermore, that the rate of most reactions decreases exponentially with decreasing temperature (see pages 48 to 49), so that if a reaction has not had time to run to completion at high temperature, there is little likelihood that it will be able to do so at a lower temperature. Hence the conditions of disequilibrium remain "frozen" in the rock. A well-known example of this is, of course, the glassy groundmass of many volcanic rocks.

Fractional Crystallization.   It is not only a matter of rate of cooling versus rate of crystallization or of reaction which produces such effects, for a reaction may be prevented from running to completion merely by separating the reactants at an appropriate time.   For instance, if early-formed crystals settle by gravity and accumulate in some portion of the magma, they may become unable to react with the bulk of the remaining liquid.   Or the liquid itself may be squeezed out by mechanical action (filter-press action) and become unable to react with the previously formed crystals.   The term *fractional crystallization* is commonly used in connection with such processes by which the magma splits more or less effectively into fractions which become unable to react mutually as they should if static conditions had been maintained.   Petrographic examples of such arrested reactions are common: olivine cores surrounded and partially replaced by clinopyroxene rims, augites rimmed with hornblende, zoned crystals of plagioclase (see page 102), etc.   The interpretation in cases such as the first mentioned is that the reaction

$$\text{Olivine} + \text{liquid} \rightarrow \text{clinopyroxene}$$

failed to run to completion at the incongruent melting point, either because of rapid cooling or because of lack of a sufficient amount of liquid in contact with the olivine crystals.

A noteworthy feature of such fractional crystallization is that it tends to increase the diversity of products that can be obtained under different conditions from a given magma; it also tends to extend the range of temperature over which crystallization takes place.   Examples of this will be given later when discussing particular systems.   It will be shown also in specific instances how fractional crystallization may lead to an assemblage of phases (*e.g.*, olivine and quartz) which is metastable at ordinary temperature with respect to some reaction (*e.g.*, olivine + quartz → clinopyroxene) which would normally have occurred if static conditions had been maintained.   Igneous rocks are therefore by no means always stable assemblages of minerals.   The final products of crystallization that can be obtained from a given magma will depend to a large extent on the kinetics of the processes involved and not only, as often stated, on the conditions for thermodynamic equilibrium as determined from laboratory experiments.   It is indeed a bold extrapolation "from crucible to magma." The study of processes occurring in crucibles remains nevertheless an essential step in the understanding of processes that are operative in magmas, and it is to this study that we turn next.

CRYSTALLIZATION OF FELDSPARS

The System Nepheline-Silica.[11] The system nepheline-silica, with the intermediate compound albite, constitutes a compound eutectic system (Fig. 4) with two binary eutectics, namely nepheline-albite (E_1) and albite-tridymite (E_2). All melts originally to the left of the albite point (*i.e.*, undersaturated in silica) move toward the nepheline-albite eutectic. The high albite content (76 per cent) of the eutectic mixture is consistent

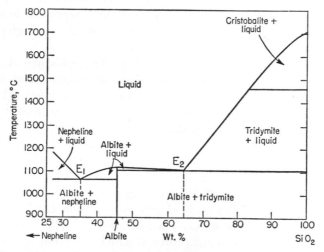

FIG. 4. Phase diagram for part of the system nepheline-silica. (*After T. F. W. Barth.*)

with the observed tendency for nepheline to crystallize early, as phenocrysts, in porphyritic soda-rich nepheline syenites or tinguaites in which albite may be considerably more abundant than nepheline. Melts originally oversaturated in silica (to the right of the albite point) move toward the albite-silica eutectic. Fractional crystallization can bring about variation in relative proportions of albite and either nepheline or tridymite in various crystalline fractions. But it cannot cause nepheline to appear in any fraction separating from a melt originally containing silica in excess of the composition $NaAlSi_3O_8$ (albite). Nor can tridymite crystallize from any melt initially having less silica than $NaAlSi_3O_8$.

Complicating factors of little or no petrogenic significance (and hence

[11] J. W. Greig and T. F. W. Barth, The system nephelite-albite, *Am. Jour. Sci.*, vol. 35A, pp. 94–112, 1938; Barth, *op. cit.*, p. 21, 1939.

omitted from further discussion) are the inversion cristobalite $\rightleftarrows$ tridymite (1470°C.) and the existence of high-temperature polymorphs of $NaAlSiO_4$. Also the compositions of nepheline and albite crystallizing in the subsystem $NaAlSiO_4$-$NaAlSi_3O_8$ depart slightly from the ideal formulas. Nepheline contains a slight excess of SiO_2 while albite is very slightly deficient in SiO_2.

The main effect of adding water to the system is to lower the melting curves by a few hundred degrees.[12] The nepheline-albite eutectic (E_1 in Fig. 4) is $Ab_{76}Ne_{24}$, 1068°C., in the dry system; it is $Ab_{72}Ne_{28}$, 870°C., under 1,000 bars water pressure; and the eutectic temperature at 2,000 bars is 720°C. These temperatures are still too high to allow analcite to appear in contact with melts, except at much higher water pressures. Jadeite, a phase intermediate between albite and nepheline in composition, could crystallize from melts in this system only at very high pressures consistent with depths below the earth's crust.[13]

The System Albite-Anorthite.[14] The plagioclase feldspars, at temperatures of crystallization from artificial melts and probably at magmatic temperatures, too, constitute a solid solution series with complete substitution of NaSi for CaAl. That these solid solutions, inverting from disordered to more ordered states of AlSi distribution, become unstable at ordinary temperatures does not affect the petrogenic significance of the behavior of plagioclase crystallizing from artificial melts.[15] This behavior is illustrated in Fig. 5a. The melting point of anorthite is 1550°C.; that of albite is 1100°C. The upper curve PS is a freezing-point curve (liquidus) giving the temperature at which a melt of any composition can exist in equilibrium with plagioclase crystals. The lower curve WZ is a melting-point curve (solidus) giving the temperature at which crystalline plagioclase of any composition can coexist with a plagioclase melt. Horizontal lines connect melts (on PS) and crystalline phases (on WZ) that can coexist at the same temperature (cf. page 95).

A liquid L ($Ab_{50}An_{50}$ at 1500°C.) cools to 1450°C. (P), at which point crystals W ($Ab_{20}An_{80}$) begin to separate. Continued separation of crystals of this or somewhat more sodic composition would enrich the liquid in albite. But there is only one composition (X) for stable crystals in equilibrium with a melt such as Q. Therefore as the temperature falls

[12] W. S. MacKenzie, in Annual report of the director of the Geophysical Laboratory, *Carnegie Inst. Washington Year Book,* 1953–1954, p. 119, 1954; 1954–1955, p. 123, 1955.

[13] E. C. Robertson, F. Birch, and G. J. F. MacDonald, Experimental determination of jadeite stability relations to 25,000 bars, *Am. Jour. Sci.,* vol. 255, pp. 115–137 (especially Fig. 8), 1957.

[14] Bowen, *op. cit.,* 577–599, 1913; *op. cit.,* pp. 33–35, 1928.

[15] M. J. Buerger, The role of temperature in mineralogy, *Am. Mineralogist,* vol. 33, p. 119, 1948.

and the composition of the liquid becomes progressively more sodic as it moves from P to Q, the early formed crystals (W) react with the liquid and themselves become more sodic, presumably by exchange of appropriate ions $(CaAl \rightleftharpoons NaSi)$ with the surrounding melt. If equilibrium is maintained, by the time the liquid reaches the composition $Ab_{60}An_{40}$ at 1420°C. (point Q) the crystals have attained composition $Ab_{25}An_{75}$ (X). These crystals are homogeneous. They include early-formed originally more calcic crystals, as well as individuals developing about new nuclei. Crystallization, continuously accompanied by reaction between crystals

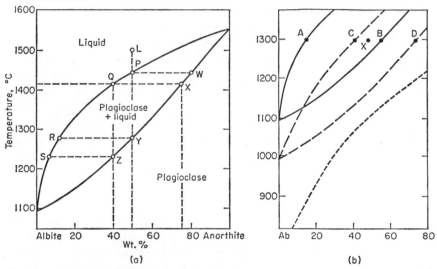

FIG. 5.   Phase diagrams for the system albite-anorthite.   (a) The anhydrous system at atmospheric pressure.   (*After N. L. Bowen.*)   (b) Anhydrous system (full lines) compared with system at water pressure of 5,000 bars (finely dashed line, lower right).   (*After H. S. Yoder, D. B. Stewart, and J. R. Smith.*)   Broken lines represent hypothetically some intermediate water pressure.

and melt, proceeds with falling temperature, so that the composition of the liquid follows the curve PQR while that of the crystals follows WXY. At 1285°C. the mass of homogeneous crystals reaches the composition $Ab_{50}An_{50}$ (Y) just as the last of the liquid, $Ab_{86}An_{14}$ (R), is used up.

The effect of fractional crystallization is illustrated by supposing that at 1420°C. all crystalline material $(Ab_{25}An_{75})$ were removed from the melt Q. On further cooling the melt would change down the curve QRS $(Ab_{60}An_{40}$ to $Ab_{92}An_8)$, and would be completely used up at the point S at the same moment as the homogeneous crystalline mass attained the composition $Ab_{60}An_{40}$ (Z). Effects of fractional crystallization are:

1. To extend the course of crystallization of residual liquids, and thus to increase the effectiveness of later fractional crystallization

2. To enrich the residual liquid, and hence the late crystal fractions, in albite, the low-temperature end member of the series

3. To lower the temperature at which crystallization ceases

If crystals of a given composition, *e.g.*, $Ab_{50}An_{50}$, are heated to their melting point (1285°C.), they melt incongruently, and the first liquid to form is of much more siliceous (cf. page 96) and so, too, more sodic composition, *viz.*, $Ab_{86}An_{14}$ (*R*). Melting, like freezing, is accompanied by continuous reaction between liquid and crystals. Both phases become progressively more calcic in the process. The last small quantity of remaining solid approaches the composition $Ab_{20}An_{80}$, which theoretically is reached just when melting is complete. Note that an albitic liquid can be formed by partial fusion (cf. assimilation) of more calcic crystals, or by fractional crystallization (cf. differentiation) of more calcic liquid.

If equilibrium is maintained between liquid and crystals during either crystallization or fusion, the crystals at any stage should be homogeneous. Plagioclase crystals of igneous rocks are seldom homogeneous. Zonary variation in optical properties is usually conspicuous. This is generally interpreted as expressing zonary variation in chemical composition (*i.e.*, in albite content), correlated with incomplete equilibrium between crystals and magma during crystallization. On this assumption, the anorthite content of any crystal should decrease from within outward. The optical behavior of igneous plagioclase commonly indicates that this is actually so. But almost equally commonly the composition appears to oscillate between alternately more and less albitic zones respectively. Cases of reversed zoning in which crystals appear to become progressively poorer in albite from within outward are not rare. It is therefore generally assumed that during magmatic crystallization crystals of plagioclase (and of other minerals too) very commonly fail to maintain equilibrium with the magma in which they are immersed, and moreover that crystals not uncommonly are subject to oscillating temperatures or pressures in consequence of which oscillatory or even reversed zoning may develop. Fluctuation in water pressure could be particularly effective in causing oscillatory zoning in plagioclase, for the melting temperature of plagioclase is greatly lowered by water pressures of a few thousand bars (Fig. 5*b*). A system *X* consisting of crystals *B* in equilibrium with dry melt *A* at 1300°C. and zero water pressure would no longer be in equilibrium at a water pressure of one or two thousand bars (broken curves, Fig. 5*b*). It would tend to adjust to the new conditions by reaction involving increase in the proportion of melt (which would change toward *C*) and decrease in the quantity of crystals (which would change toward *D*). If adjustment were incomplete, reversed zoning would result. Finally there is a possibility that some zonary structure of an irregular nature might be in part a result of unmixing of anorthite-rich and albite-rich phases during cooling of the crystalline rock.

The rate at which a magma cools, and hence the degree of undercooling, is presumably an important factor controlling zoning in igneous plagioclase. If cooling is very slow, departure from equilibrium is minimal and crystals are homogeneous; with more rapid cooling, adjustment between crystals and melt is incomplete, and zoned crystals are formed. With very rapid cooling (cf. the groundmass of many lavas) the melt becomes strongly undercooled and gives rise directly to unzoned crystals of the same composition as the melt. Thus a melt of composition $Ab_{50}An_{50}$,

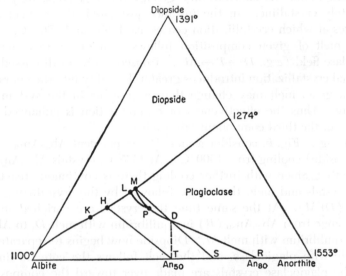

FIG. 6. Phase diagram for the system diopside-albite-anorthite. (*After N. L. Bowen.*)

undercooled to 1285°C., could crystallize as unzoned crystals also of composition $Ab_{50}An_{50}$.

The System Diopside-Albite-Anorthite.[16] Diopside with albite or with anorthite alone forms binary eutectics. The system diopside-albite-anorthite illustrates the crystallization of an isomorphous series (plagioclase) from melts in a multicomponent system. It has an added petrogenic significance in that some of the melts concerned are analogous in composition with greatly simplified anhydrous basaltic magmas. The behavior of plagioclase in the ternary system (Fig. 6) is found to be much

[16] N. L. Bowen, The crystallization of haplobasaltic, haplodioritic and related magmas, *Am. Jour. Sci.*, vol. 40, pp. 161–185, 1915; *op. cit.*, pp. 45–58, 1928.

the same as that observed in the system albite-anorthite, although modified in the following respects:

1. Plagioclase of a given composition crystallizes at notably lower temperatures from melts containing diopside. For example, crystals $Ab_{20}An_{80}$ separate at 1450°C. from a pure melt $Ab_{50}An_{50}$, as compared with 1375°C. in the case of a melt containing 85 per cent $Ab_{50}An_{50}$ plus 15 per cent diopside. By further crystallization and reaction the plagioclase reaches the composition $Ab_{25}An_{75}$ at 1420°C. in the first case and at 1300°C. in the second.

2. Presence of diopside likewise lowers the temperature at which a melt containing albite and anorthite in a given ratio (*e.g.*, $Ab_{50}An_{50}$) becomes completely crystalline. In the instance just cited, the respective temperatures at which crystallization ceases are 1285° and 1200°C.

3. A melt of given composition follows a unique course across the plagioclase field (*e.g.*, $D \to P \to M$). Consequently, as discussed below, fractional crystallization introduces greater flexibility into the course along which a given melt may change than is possible in the system albite-anorthite. Thus the effectiveness of differentiation is enhanced by the presence of the third component (diopside).

Referring to Fig. 6, consider a melt D (85 per cent $Ab_{50}An_{50}$ + 15 per cent diopside) cooling from 1500°C. At 1375°C. crystals $Ab_{20}An_{80}$ begin to separate. Since with further cooling there is continuous reaction between crystals and melt, the course followed by the crystallizing melt is curved (DPM). At the same time the crystals are enriched in albite. They change from $Ab_{20}An_{80}$ (R) in equilibrium with melt D, to $Ab_{33}An_{67}$ (S) in equilibrium with melt M. Diopside now begins to separate simultaneously with plagioclase, and the melt follows the cotectic line MH, while the plagioclase crystals are made over toward the composition T ($Ab_{50}An_{50}$). This composition is reached, theoretically, just when the last remaining drop of liquid reaches H at 1200°C. Crystallization is then complete.

Provided equilibrium is maintained at every stage, the melt D must follow the unique course DPM, and no other melt, not even one such as P situated on the curve itself, can follow the same course. The course pursued by an initial melt P will be along the curve PL and thence to K. This of course is true when the melt P is a residual liquid removed (fractionated) from early-formed crystals at the appropriate stage in crystallization of a parent melt D.

One other point of possible significance in petrogenesis should be noted. Once the cotectic line is reached, whether from the field of plagioclase or from that of pyroxene, residual liquids simultaneously become enriched in albite relative to anorthite, and impoverished in diopside compared with plagioclase, as crystallization proceeds. It is a matter of common observation that in many instances the pyroxene content of igneous rocks tends to decrease as the associated plagioclase becomes increasingly sodic (cf. Fig. 32, page 213).

Recent investigation of albite-rich melts in this system shows that near the albite point the system is not strictly ternary. Highly albitic melts may therefore leave the triangular diagram of Fig. 6, and their behavior thereafter cannot be predicted. Although the range of composition of such liquids is extremely limited, it may well be that in the presence of other components such as TiO_2 or H_2O a wider range of liquids might behave in a similar fashion.

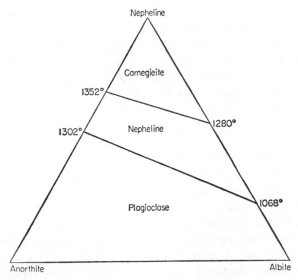

FIG. 7. Phase diagram for the system nepheline-albite-anorthite. (*After J. W. Greig and T. F. W. Barth.*)

The System Nepheline-Albite-Anorthite.[17] Since the binary systems nepheline-albite, nepheline-anorthite, and albite-anorthite have all been investigated, it is possible to predict approximate behavior of melts in the system nepheline-albite-anorthite. The corresponding hypothetical diagram is shown in Fig. 7. Melts in the plagioclase field crystallize in much the same manner as melts in the corresponding field of the system diopside-albite-anorthite, except that plagioclase is ultimately joined by nepheline instead of diopside. Fractional crystallization leads to development of residual liquids progressively enriched in soda. At the same time, however, the proportion of nepheline to albite decreases toward the limiting value (24:76) fixed by the nepheline-albite eutectic.

[17] Greig and Barth, *op. cit.*, pp. 110–112, 1938.

Effects of Water on Crystallization in Plagioclase Systems.[18]   Yoder and associates in the Geophysical Laboratory, Washington, have investigated crystallization in the systems albite-anorthite-water and diopside-albite-anorthite-water at a water pressure of 5,000 bars.   Such pressures are possible in plutonic magmas at depths of 10 miles or more below the earth's surface.   The principal departures from crystallization behavior recorded for dry systems are as follows:

1. Temperatures of crystallization of feldspars are lowered by the order of 300°C. (e.g., temperatures at the apices of Fig. 6 are lowered to 1282°C. for diopside, 1234°C. for anorthite, and 750°C. for albite).

2. The field of plagioclase in Fig. 6 is greatly reduced; the diopside-anorthite eutectic changes from $An_{42}Di_{58}$ in the dry system to $An_{73}Di_{27}$ at 5,000 bars water pressure.   Diopside would invariably be the first phase to crystallize from melts consisting of diopside and plagioclase in proportions found in basaltic rocks, and the tendency for plagioclase to become concentrated in residual liquids resulting from fractionation would be even more general than in the dry system.

3. The water content of the melts in the vicinity of the diopside-anorthite eutectic is 8.8 per cent; that of pure plagioclase melts is 9 per cent to 12 per cent.   These melts are, of course, in equilibrium with an aqueous fluid in the supercritical state.

The System Leucite-Silica.[19]   At atmospheric pressure potash feldspar melts incongruently with formation of a liquid of more siliceous composition and an equivalent amount of crystalline leucite.   Consequently the behavior of potash feldspar must be studied within systems of which leucite and silica are components.   The binary system leucite-silica is illustrated by Fig. 8.   A melt N, undersaturated in silica, cools to O, at which point leucite begins to separate.   With falling temperature and continual crystallization of leucite, the melt follows the curve OPR.   The melt R (orthoclase plus a small definite excess of silica at 1150°C.) is the only liquid phase capable of existing in equilibrium with both leucite and potash feldspar in this system.   Its composition and temperature therefore remain constant while leucite reacts with the melt (i.e., passes into solution) to produce sanidine.   The melt at this stage has a composition, by weight, 74 per cent orthoclase, 26 per cent excess silica, i.e., approximately $3KAlSi_3O_8 + 5SiO_2$.   The reaction may be represented thus:

[18] H. S. Yoder, Diopside-anorthite-water system at 5000 bars, *Geol. Soc. America Bull.*, vol. 66, pp. 1638, 1639, 1955; H. S. Yoder, D. B. Stewart and J. R. Smith, in Annual report of the director of the Geophysical Laboratory, *Carnegie Inst. Washington Year Book*, 1953–1954, pp. 106, 107, 1954; 1955–1956, pp. 190–194, 1956.

[19] G. W. Morey and N. L. Bowen.   The melting of potash feldspar, *Am. Jour. Sci.*, vol. 4, pp. 1–21, 1922; Bowen, *op. cit.*, 1928, p. 241; J. F. Schairer and N. L. Bowen, The system anorthite-leucite-silica, *Soc. géol. Finlande Bull.*, vol. 20, pp. 72–75, 1947.

$$5KAlSi_2O_6 + 3KAlSi_3O_8 + 5SiO_2 \rightarrow 8KAlSi_3O_8$$

Leucite	(Melt)	Potash feldspar
(1,090 gm.)		(2,224 gm.)

Sanidine thus crystallizes about twice as rapidly as leucite goes into solution. Crystallization ceases while some leucite still remains, and the end product is crystalline leucite and sanidine in the ratio $BC:AC$.

A melt of composition $KAlSi_3O_8$ (P in Fig. 8) would behave similarly except that reaction at R would cease with simultaneous elimination of both leucite and melt. Leucite would also be the first crystalline phase to separate from melts with a slight excess of silica, such as O. In this

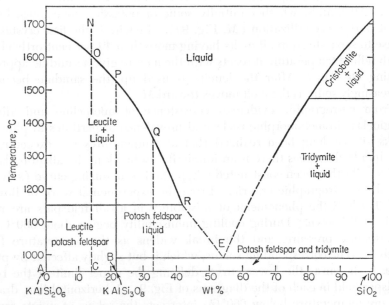

Fig. 8. Phase diagram for the system leucite-silica. (*After J. F. Schairer and N. L. Bowen.*)

case some liquid would remain after all the leucite was used up during reaction at R. The temperature would now fall once more and the melt would change down the curve RE with continued separation of sanidine. At E eutectic crystallization of sanidine and tridymite terminates the process of crystallization.

Since P and Q correspond to residual liquids formed at successive stages during crystallization of an initially silica-deficient liquid O, the possibilities of fractional crystallization are illustrated by the three respective crystalline end products, *viz.*, sanidine-leucite, sanidine, sanidine-silica. Note how fractional crystallization in a system involving an incongruently melting compound may result in the crystallization of an additional mineral phase (*e.g.,* tridymite) and at the same time lower the temperature at which crystallization is complete. A further fact of petrological interest is the appearance of the feldspathoid leucite in the early stages of crystallization of some liquids (between *P* and *R*) with excess silica.

The System Albite-Orthoclase.[20]   Since potash feldspar melts incongruently, it is impossible to treat the behavior of albite-orthoclase melts precisely in terms of a two-component system. Significant data are available from the as yet incompletely investigated system $NaAlSiO_4$-$KAlSiO_4$-SiO_2. At temperatures of crystallization from dry melts (1063° to 1150°C.), potash and soda feldspars form a complete solid-solution series, with a minimum temperature (1063°C.) at about the composition Ab_{65}-Or_{35}. All liquids, whether initially sodic or potassic, move toward this point during crystallization (*M*, Fig. 9*a*). Leucite is the first crystalline phase to separate from all melts having more than 49 per cent orthoclase. With falling temperature it reacts with the melt to give a sanidine approximating $Or_{85}Ab_{15}$. After the leucite is used up, the sanidine becomes increasingly sodic as the melt moves toward *M*.

From petrographic evidence—coexistence of microcline and albitic plagioclase in metamorphic rocks and prevalence of perthites in plutonic rocks—it has long been realized that at temperatures not exceeding a few hundred degrees there is an immiscibility break in the alkali-feldspar series. Barth [21] even constructed a hypothetical unmixing curve (solvus) based on petrographic criteria. From the experimental work of Bowen and Tuttle [22] the phenomena of miscibility in alkali feldspars are now fairly well known. During cooling immiscibility begins about 660°C. at atmospheric pressure, and the break widens as the temperature falls further. The corresponding solvus, which is but slightly affected by pressure, is shown as the lower curve, defining the upper limit of the two-feldspar field in each of the diagrams of Fig. 9. A horizontal line, drawn for any temperature below 660°C., intersects the solvus at points representing the respective compositions of the two feldspars that can coexist in equilibrium at that temperature.

The three diagrams of Fig. 9 illustrate the powerful influence of water pressure in narrowing the field of leucite, in reducing the field of homogeneous K-Na feldspar, in lowering temperatures of crystallization, and in

[20] J. F. Schairer and N. L. Bowen, Preliminary report on equilibrium relations between feldspathoids, alkali feldspar and silica, *Am. Geophys. Union Trans.*, pt. 1, pp. 325–328, 1935 (especially p. 328); Bowen, *op. cit.*, pp. 11, 12, 1937; Barth, *op. cit.*, pp. 24, 25, 1939; N. L. Bowen and O. F. Tuttle, The system $NaAlSi_3O_8$-$KAlSi_3O_8$-H_2O, *Jour. Geology*, vol. 58, pp. 489–511, 1950; Yoder, Stewart, and Smith, *op. cit.*, pp. 190–194, 1956.

[21] Barth, *op. cit.*, p. 24, 1939; The feldspar geologic thermometers *Neues Jahrb. Mineral.*, vol. 82, pp. 143–154, 1951.

[22] Bowen and Tuttle, *op. cit.*, p. 497, 1950.

displacing the composition of the low-temperature point on the liquidus toward the albite end of the series. At a water pressure of 5,000 bars [23] (more than twice the maximum water pressure shown in Fig. 9) the lowest temperature on the liquidus is 695°C., and the corresponding melt has a composition $Or_{29}Ab_{71}$, as contrasted with 1063°C. and $Or_{35}Ab_{65}$ for dry melts (Fig. 9a); the field of leucite, already greatly reduced at 2,000 bars (Fig. 9c), is now eliminated. The one-feldspar field diminishes notably

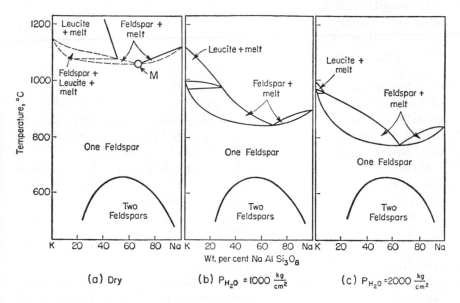

FIG. 9. Phase diagrams for the system $NaAlSi_3O_8$-$KAlSi_3O_8$. (*After N. L. Bowen and O. F. Tuttle.*) (a) In absence of water; broken lines not precisely determined. (b) In presence of water: pressure of $H_2O = 1,000$ kg./cm.². (c) In presence of water: pressure of $H_2O = 2,000$ kg./cm.².

as the liquidus is lowered with increasing water pressure; at 5,000 bars—the maximum water pressure that could develop at depths of a dozen miles—the solvus cuts the solidus between Or_{30} and Or_{60}. This means that two feldspars will crystallize from any melt between Or_{30} and Or_{60} at 5,000 bars water pressure. From other melts, and indeed from melts of any composition at water pressures of 2,000 or 3,000 bars, feldspar crystallizes as a single homogeneous phase. With subsequent slow cooling to points on or below the solvus this homogeneous feldspar may unmix to a perthitic intergrowth of two phases, the one more potassic, the other more sodic.

[23] Yoder, Stewart, and Smith, *op. cit.*, p. 191, 1956; in Annual report of the director of the Geophysical Laboratory, *Carnegie Inst. Washington Year Book*, no. 56, pp. 208, 209, 1957.

The System Albite-Orthoclase-Anorthite.[24] Because of experimental difficulties connected with the extreme viscosity of anhydrous alkali-feldspar melts, the system albite-orthoclase-anorthite is still not completely investigated. Much, however, may be inferred from the known crystallization behavior of melts on the three sides of the triangle; and petrography supplies a good deal of evidence regarding behavior of melts in the interior of the diagram.

Figure 10a is an equilibrium diagram for the anhydrous system albite-orthoclase-anorthite as determined experimentally at atmospheric pressure by Franco and Schairer.[25] There are two fields characterized by early separation of feldspar and leucite respectively. The nature of the feldspars coexisting with liquids of different specific compositions within the feldspar field has not been determined. For this reason, and because of the complicating influence of incongruent melting of potash feldspars, crystallization of ternary feldspar melts cannot be discussed quantitatively in terms of Fig. 10a. Data determined by H. S. Yoder [26] for crystallization of feldspars at 5,000 bars water pressure are shown in Fig. 10b.

Figure 11 illustrates in a qualitative way the behavior of feldspar melts at low pressures (volcanic conditions) as inferred from petrographic data. The field of leucite and the distribution of isothermals are the same as for Fig. 10. The field of feldspars is divided by the curve WYZ, representing the intersection of the solidus (melting-point curve) with the solvus (unmixing curve). If the composition of the melt lies to the left of WYZ, crystallization will be accomplished by separation of a single homogeneous feldspar which will change its composition by reaction with the diminishing liquid as the temperature falls, but at no stage will more than one feldspar coexist with the liquid in a state of equilibrium. The area of the two-feldspar field increases as the solidus surface is depressed to progressively lower temperatures with increase in water pressure. To the right of and above WYZ crystallization of one feldspar changes the composition of the liquid toward a three-phase boundary where two feldspars of changing composition separate simultaneously. The curve AB in Fig. 11 is an imaginary boundary of this kind. Its low-temperature end (A) probably is not far from E, the low-temperature point on the alkali-feldspar liquidus.

The boundary WYZ as shown on Fig. 11 is also hypothetical, but we know from petrography something of its general nature. X and Z are the

[24] Bowen, op. cit., pp. 248–253, 1928; Barth, op. cit., pp. 25, 26, 1939; R. R. Franco and J. F. Schairer, Liquidus temperatures in mixtures of the feldspars of soda, potash and lime, Jour. Geology, vol. 59, pp. 259–267, 1950; Yoder, Stewart, and Smith, op. cit., pp. 190–194, 1956; op. cit., pp. 206–214, 1957.

[25] Franco and Schairer, op. cit., 1950.

[26] Yoder, Stewart, and Smith, op. cit., p. 211, 1957.

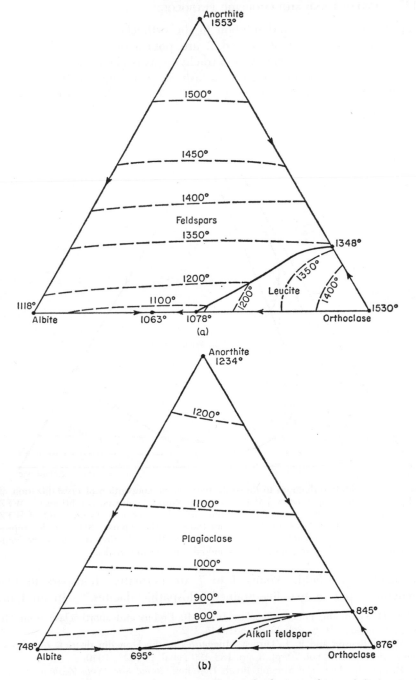

Fig. 10. Phase diagram for the system albite-orthoclase-anorthite. (*a*) At atmospheric pressure. (*After R. R. Franco and J. F. Schairer.*) (*b*) At 5,000 bars water pressure. (*After H. S. Yoder.*)

conjectured limits of solid solution in the orthoclase-anorthite series at high temperatures. Points 1 and 2 are potash oligoclases ("anorthoclases") occurring as phenocrysts in trachytic lavas (kenytes),[27] and point 3 is the normative composition of a potash andesine which is the principal constituent of a New Zealand mugearite.[28]   These points must lie within

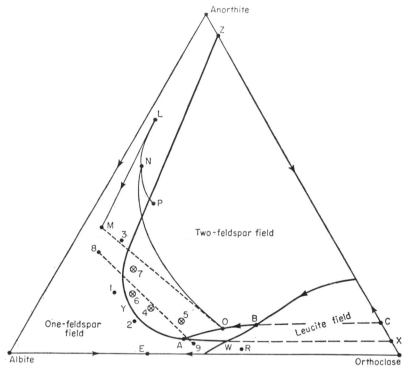

Fig. 11.   Qualitative diagram to illustrate the writers' conception of crystallization of feldspars at high temperatures and low water pressures (volcanic conditions).   WYZ is the hypothetical intersection of the solidus and solvus surfaces; to the right of WYZ temperatures on the solidus are lower than those on the solvus.   Solid circles represent compositions of volcanic feldspars described in the text; circled crosses are compositions of total feldspar (normative or modal) in volcanic rocks.

the one-feldspar field.   Points 4 to 7 are normative feldspars in two trachytes [29] and in two fine-grained feldspathic dacites.[30]   In all four,

[27] E. D. Mountain, Potash-oligoclase from Mt. Erebus and anorthoclase from Mt. Kenya, *Mineralog. Mag.*, vol. 20, pp. 331–345, 1925.

[28] W. N. Benson and F. J. Turner, Mugearites in the Dunedin district, *Royal Soc. New Zealand Trans.*, vol. 70, pt. 3, pp. 188–199 (anal. p. 193), 1940.

[29] R. Speight, Dykes of Summit Road, Lyttelton, *Royal Soc. New Zealand Trans.*, vol. 68, pt. 1, pp. 82–99, 1938.

[30] H. T. Ferrar, The geology of the Dargaville-Rodney subdivision, *New Zealand Geol. Survey Bull.*, no. 34, pp. 57, 58, 62 (anal. 13, 15), 1934.

separation of sodic plagioclase phenocrysts has been followed by simultaneous crystallization of plagioclase and sanidine (or anorthoclase). The trachytes contain relatively few plagioclase phenocrysts; their compositions (4, 5) must therefore lie relatively close to the curve AB. This curve also is straddled by compositions 8 and 9, which represent plagioclase and sanidine simultaneously crystallized in the groundmass of a potash basanite.[31]

The changing composition of residual liquids at successive stages of fractionation of a basalt from the Marquesas Islands has been discussed by Barth [32] and is shown qualitively as NO in Fig. 11. Plagioclase, as preserved in zoned crystals in this rock, changes from L to M as the diminishing liquid moves from N to O. Then follows simultaneous crystallization of andesine M and sanidine R as the liquid moves from O toward A on the curve AB. In this case a melt initially in the one-feldspar field has been diverted by strong fractionation into the field of two feldspars. (A similar history must have preceded crystallization of feldspars 8 and 9 in the New Zealand basanite.) Had equilibrium been maintained throughout, the liquid would have followed a course such as NP, and the crystallizing feldspar a course LN. The end product would have been homogeneous feldspar N.

Presence of excess silica in feldspathic melts reduces or even eliminates the field of leucite. This condition is represented in Fig. 11 by extension of the curve ZYW to X, and of AB to C, the position of which probably approximates the potash-feldspar anorthite eutectic as determined for high water pressures.[33]

Crystallization of alkali feldspars under plutonic conditions is illustrated in Fig. 12. It is assumed that high water pressures or presence of excess silica (as in granites and granodiorites) has eliminated the field of leucite. On the orthoclase-anorthite edge, X and Z are limits of miscibility and C is the eutectic composition as determined at 5,000 bars water pressure. E is the low-temperature point on the alkali-feldspar liquidus at the same pressure.[34] The curves XY, WZ, and CE are taken from determinations made by Yoder at 5,000 bars water pressure (cf. Fig. 10b). At this pressure there is a broad gap WY where the solvus cuts the solidus in the alkali-feldspar series. The one-feldspar field is narrower than in Fig. 11, because of the influence of high water pressures (possible only at depth) in lowering crystallization temperatures to values at which mutual

---

[31] D. S. Coombs, Appendix to D. A. Brown, The geology of Siberia Hill and Mt. Dasher, North Otago, *Royal Soc. New Zealand Trans.*, vol. 83, pt. 2, pp. 369–371, 1955.

[32] T. F. W. Barth, The crystallization process of basalt, *Am. Jour. Sci.*, vol. 36, pp. 324, 325, 1936.

[33] Yoder, Stewart, and Smith, *op. cit.*, p. 207, 1957.

[34] Yoder, Stewart, and Smith, *op. cit.*, pp. 192, 193, 1956.

miscibility of calcic and potassic feldspars is even more limited than at volcanic temperatures. A number of illustrative examples drawn from petrographic literature have been plotted on Fig. 12.

1. Most granites fall in the two-feldspar field, mainly because the

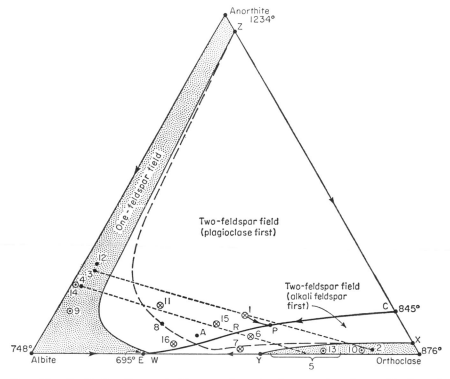

Fig. 12.   Qualitative diagram to illustrate the writers' conception of crystallization of feldspars at moderate temperatures and high water pressures (plutonic conditions). Temperatures are as determined at 5,000 bars water pressure by Yoder, Stewart, and Smith.   XY and WZ are the hypothetical intersection of the solidus and solvus surfaces; above XY and to the right of WZ temperatures on the solidus are lower than those on the solvus.   Solid circles represent compositions of primary plutonic feldspars described in the text; circled points are compositions of feldspars formed by unmixing of primary plutonic feldspars; circled crosses are compositions of total feldspar (normative or modal) in plutonic rocks.   Brooken curve represents hypothetical intersection of solvus and solidus at water pressures consistent with shallow plutonic conditions (less than 1,000 bars).

normative feldspar of typical granites falls in the region immediately above WY.   Granite from Westerly, Rhode Island, contains oligoclase An_{23} and microcline Or_{88} in nearly equal proportions (1 in Fig. 12).[35]   From a melt

[35] F. Chayes, in Annual report of the director of the Geophysical Laboratory, *Carnegie Inst. Washington Year Book*, no. 53, 1953–1954, pp. 130, 131, 136, 1954.

of this composition plagioclase (an andesine) will crystallize first and will soon be joined by potash feldspar when the liquid attains some such composition as P on the curve CE. The liquid now moves along CE toward E, and is used up on attaining some such composition as R. Meanwhile the two feldspars, both of which have become progressively more sodic with advancing crystallization, have reached the final compositions 2 and 3.

2. In a somewhat more alkaline granite from the island of Skye [36] small quantities of high-temperature oligoclase (An_{20}; 4 in Fig. 12) are associated with optically homogeneous cryptoperthite (Or_{60} to Or_{80}; 5 in Fig. 12). This is shown to be sanidine in a state of partial inversion and unmixing to low-temperature potassic and sodic phases. The modal composition of the feldspar (6 in Fig. 12) is consistent with early separation of sanidine followed by simultaneous crystallization of both oligoclase and sanidine. The composition is not far from the boundary of the one-feldspar field; indeed there are local marginal facies of the granite in which only one feldspar—cryptoperthite—is present. This probably has the same significance as the one-feldspar rocks described in the next paragraph.

3. The compositions of some granites and syenites lie in the one-feldspar field. A Norwegian larvikite [37] of this kind consists of initially homogeneous alkali feldspar (8 in Fig. 12) now exsolved to a perthitic intergrowth of oligoclase (9) and orthoclase (10). The granite of Quincy, Massachusetts, described by Tuttle [38] as a typical one-feldspar granite (quartz, 23 per cent; perthite, 66 per cent; riebeckite, 11 per cent) also has a ratio of lime to alkalis that would appear to place it in the twofeldspar field at A in Fig. 12. However, if lime is subtracted to account for normative apatite and calcite and if allowance is made for Na_2O combined in riebeckite, its total feldspar approximates 7 in Fig. 12. Points 7 and 8 are still within the two-feldspar field of Fig. 12. But in a diagram drawn for water pressures of the order of 1,000 bars (i.e., a depth of at least 2 miles or explosive volcanic conditions), the one-feldspar field would be more extensive than in Fig. 12, and its lower boundary would not intersect the alkali-feldspar edge of the diagram. Rocks such as the Norwegian larvikites and the Quincy granite evidently represent magmas that crystallized at relatively shallow levels. Indeed the thickness of cover for the Norwegian intrusions has been estimated, on the basis of stratigraphic evidence, as less than one mile.

[36] O. F. Tuttle and M. L. Keith, The granite problem, *Geol. Mag.*, vol. 91, pp. 61–72, 1954.

[37] I. D. Muir and J. V. Smith, Crystallization of feldspars in larvikites, *Zeitschr. Kristallographie*, vol. 107, pp. 182–195 (specimen 1), 1956.

[38] O. F. Tuttle, Origin of the contrasting mineralogy of extrusive and plutonic salic rocks, *Jour. Geology*, vol. 60, pp. 107–124, 1952.

4. Members of the monzonite family, and many syenites too, fall in the two-feldspar field. A second Norwegian larvikite [39] has normative feldspar 11 in Fig. 12. The modal feldspar is oligoclase An_{26} (12), and anorthoclase—the latter in a state of partial unmixing to orthoclase Or_{75} (13) and oligoclase An_{20} (14).

The picture of feldspar crystallization presented in the preceding pages does not agree in all respects with that given by Tuttle.[40] The disagreement arises partly because available data are still inadequate, so that an element of conjecture necessarily enters into the discussion. The field of one feldspar shown by Tuttle on a CaO-K_2O-Na_2O diagram seems much broader than that of Fig. 12. The discrepancy is largely resolved, however, if allowance is made for the appearance of such minerals as biotite, hornblende, sphene, fluorite, and carbonates in the modes of the rocks concerned. The granites from Pike's Peak, Colorado, are cited by Tuttle [41] as one-feldspar granites containing perthite, but the ratio of lime to alkalis in representative analyses is relatively high and corresponds to points well within the two-feldspar field. Such is point 15 in Fig. 12.[42] This rock, however, contains biotite (10 per cent) and minor fluorite and hornblende. Subtraction of corresponding amounts of K_2O and CaO from the analysis displaces the normative feldspar into the one-feldspar field (16 in Fig. 12) at water pressures of the order of 1,000 bars.

In conclusion the reader is reminded that this section on the system albite-orthoclase-anorthite is a tentative statement consistent with available data. It will have to be modified as new facts emerge from future experiments and from more detailed petrographic observations. In spite of its necessarily imperfect nature some such synthesis is indispensable, for the most important single factor influencing evolution of magmas is the crystallization behavior of feldspars.

CRYSTALLIZATION OF FELDSPATHOIDS

General Statement. In igneous rocks, feldspathoids commonly occur as members of two types of mineral assemblage: (1) associated with alkali feldspars in rocks such as phonolite and nepheline syenite, which by virtue of their high content of ($K_2O + Na_2O$) are deficient in silica with respect to alkali; (2) associated with intermediate and basic plagioclase, and usually olivine and pyroxene, in very basic rocks such as theralites, leucitites, and members of the nepheline-basalt family. Some light is thrown

[39] Muir and Smith, *op. cit.*, pp. 188, 189 (specimen 8), 1956.
[40] *Op. cit.*, 1952.
[41] *Ibid.*, p. 118.
[42] A. Johannsen, *A Descriptive Petrography of the Igneous Rocks*, vol. 2, p. 52 (table 28, no. 1), p. 55 (table 29, no. 1), University of Chicago Press, Chicago, 1932.

on the origin and affinities of these two classes of rock by the behavior of leucite and nepheline crystallizing in systems which include either albite and orthoclase, or anorthite and albite. Some of these systems, *viz.*, nepheline-silica, leucite-silica, nepheline-anorthite-silica, have been discussed in the previous section.

Nepheline and leucite differ in one most significant respect, brought out clearly in the systems referred to above. Nepheline cannot crystallize from melts of soda and potash feldspar with excess silica. Leucite, on the other hand, in consequence of the incongruent melting of potash feldspar, crystallizes from potassic melts containing as much as 26 per cent SiO_2 in excess of the amount corresponding to the formula $KAlSi_3O_8$. From this it follows, as shown in the system leucite-silica, that by appropriate fractional crystallization involving accumulation of leucite crystals, a partially crystalline magma deficient in silica (with respect to total alkali) can develop from an alkaline magma originally oversaturated in silica. No such transition from oversaturated to silica-deficient melts is possible in experimentally investigated systems in which soda is the sole or dominant alkali.

The extent to which the above relation of leucite to orthoclase persists in natural magmas complicated by the presence of pyroxenes, olivines, calcic plagioclase, and water may be predicted to some extent from phenomena of crystallization in the systems discussed in the following paragraphs.

The System Anorthite-Leucite-Silica.[43]   From any melt X, in the triangular area $AQSR$ of Fig. 13, anorthite is the first solid phase to separate. It continues to crystallize with falling temperature while the liquid follows a rectilinear course XT across the anorthite field. At T, anorthite is joined by leucite. These two compounds now crystallize together while the liquid cools down the cotectic boundary TR. At R (the only melt which can coexist with leucite, anorthite, and potash feldspar) reaction between leucite and melt sets in at constant temperature: leucite goes into solution, and potash feldspar and minor anorthite crystallize together. Reaction ceases in one of three ways depending on the position of X in the anorthite field. If X lies in the triangle anorthite-leucite-orthoclase, the liquid phase is eliminated during reaction at R while some leucite still remains. The end product is anorthite plus leucite plus potash feldspar. If X lies on the anorthite-orthoclase join, then reaction at R ceases as leucite and liquid simultaneously are used up. But if X lies to the right of the anorthite-leucite join, some liquid still remains at R when the leucite has completely disappeared; the temperature now falls once more, potash feldspar and minor anorthite separate as the melt follows the curve RE, and finally at E, eutectic crystallization of anorthite, potash feldspar, and

[43] Schairer and Bowen, *op. cit.*, pp. 67–87, 1947.

silica sets in.   By appropriate fractional crystallization, *i.e.*, local accumulation of leucite crystals or concentration of residual liquids, any parent
melt lying in the field $AQSR$ may give rise to any of the three alternative
crystalline end products, namely, anorthite-leucite-sanidine, anorthite-
sanidine, anorthite-sanidine-silica.   From the petrological viewpoint it is
significant that the incongruent melting of orthoclase is not affected by the
presence of large amounts of anorthite in the melt.   In other words, from
melts very rich in anorthite, leucite is still the first potassic phase to crys-

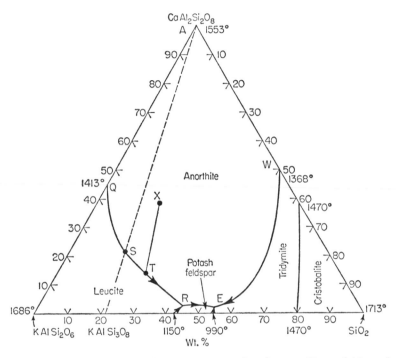

FIG. 13. Phase diagram for the system anorthite-leucite-silica. (*After J. F.
Schairer and N. L. Bowen.*)

tallize, provided the excess of silica in the melt is no greater than would
be necessary to prevent the appearance of leucite in melts containing no
anorthite.   Diopside is precisely similar to anorthite in this respect.   The
phase diagram for the system diopside-leucite-silica [44] resembles Fig. 13
except that an extensive field of diopside occupies roughly the same position as the field of anorthite in Fig. 13.

Some indication of the effects of fractional crystallization upon basic
magmas with appreciable potash but little or no soda is afforded by the

[44] J. F. Schairer and N. L. Bowen, The system leucite-diopside-silica, *Am. Jour.
Sci.*, vol. 35A, pp. 289–309, 1938.

behavior of melts near the anorthite apex of Fig. 13. Prolonged crystallization of anorthite (and/or diopside) greatly enriches the residual liquids in potash. If the liquid is filtered off as the leucite field is approached, it will ultimately crystallize as mixtures of anorthite (and/or diopside) with leucite, potash feldspar, or both. Very late liquids (corresponding to S-R in Fig. 13) are so impoverished in anorthite and diopside as to have a trachytic or even rhyolitic composition. That potash-bearing basic magmas may indeed behave in the same manner as the artificial melts just discussed is suggested, but not proved, by such petrological phenomena as the observed association of leucite with basic plagioclase and diopsidic pyroxene in rocks such as leucite basalt and leucitite; the occurrence of leucite phenocrysts in rocks containing excess silica in the groundmass (interpreted as the quenched equivalent of liquids such as S-R in Fig. 13); and finally the association of potassic trachytes, latites, and leucitophyres with rocks of the leucite-basalt group.

The System Nepheline-Kalsilite-Silica.[45]   The behavior of alkali feldspar melts respectively oversaturated or deficient in silica is illustrated by that portion of the system nepheline-kalsilite-silica which extends horizontally across Fig. 14a on either side of the albite-orthoclase join. Relations between the liquid and crystalline phases are affected by the existence of two solid-solution series. One of these, the alkali feldspar series $(K, Na)AlSi_3O_8$, is further complicated by the incongruent melting of potassic members with separation of leucite. In the other, the "nepheline" series $(Na, K) AlSiO_4$, there is limited substitution of Si for Al (with omission of a corresponding number of Na ions); the composition must be expressed in terms of $NaAlSiO_4$, $KAlSiO_4$, and $NaAlSi_3O_8$. At temperatures of crystallization from dry melts, $NaAlSiO_4$ and $KAlSiO_4$ form a complete solid-solution series, in which at least four structurally distinct phases participate,[46] the two end phases being orthorhombic. At lower temperatures, within the range of magmatic crystallization, the stable phases are nepheline, kalsilite, and intermediate solid solutions. Immiscibility develops below 1000°C.; at 400°C. the gap in the solid solution series is so wide that $Na_3KAl_4Si_4O_{16}$ coexists with nearly pure kalsilite $KAlSiO_4$.

Equilibrium crystallization of all melts above the albite-orthoclase join of Fig. 14a yields a mixture of alkali feldspar and quartz as end product. The course followed by any such melt is strongly influenced by the manner

[45] Schairer and Bowen, op. cit., 1935; Bowen, op. cit., pp. 11, 12, 1927; J. F. Schairer, The alkali feldspar join in the system NaAlSiO₄-KAlSiO₄-SiO₂, Jour. Geology, vol. 58, p. 514, fig. 1, 1950.

[46] The complex phase relations in this series are clearly discussed by O. F. Tuttle and J. V. Smith, The nepheline-kalsilite system, ii, Phase relations, Am. Jour. Sci., vol. 256, pp. 571–589, 1958.

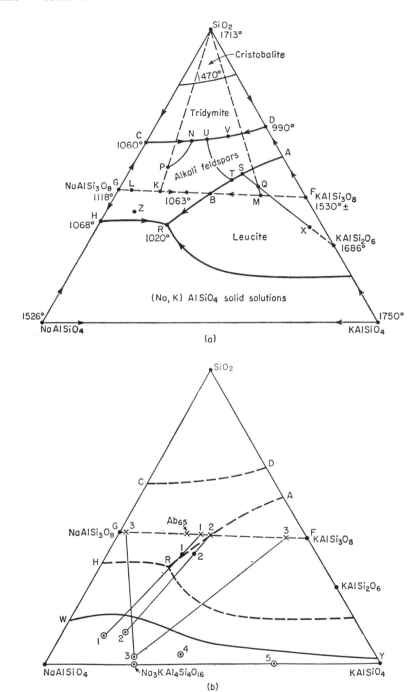

(a)

(b)

in which the composition of the feldspar changes as crystallization proceeds. The latter cannot be read from Fig. 14a, so that the several cases considered below are discussed only qualitatively and illustrate general trends rather than exact courses of crystallization.

From any melt such as P, lying in the feldspar field of the orthoclase-albite-silica triangle, alkali feldspar of some such composition as L separates first. As crystallization continues, the composition of the feldspar moves toward $Ab_{65}Or_{35}$, and the liquid follows a curving course PN toward the silica-feldspar boundary CD. When this is reached, alkali feldspar and silica crystallize together until the melt is used up. The end product is alkali feldspar (K) and tridymite. From liquids below and to the right of AB in Fig. 14a, leucite is the first phase to separate. The liquid Q follows a rectilinear course QS with separation of leucite and then changes down ST while leucite passes into solution and potassic feldspar crystallizes. After complete elimination of leucite by this reaction the melt follows some such course as TU, while the feldspar changes composition toward $Ab_{65} Or_{35}$. At a point U on the feldspar-silica boundary, feldspar is joined by silica. The two separate simultaneously as the melt changes along the curve UV until the whole mass is crystalline. The end product is alkali feldspar M and tridymite.

It is important to note that the presence of albite in moderate quantity restricts the opportunity for leucite to appear at some stage during crystallization of alkali-feldspar melts. For any melt within the triangle AFG the silica/orthoclase ratio is less than the minimum value necessary to prohibit the appearance of leucite in melts free of albite (e.g., in the system leucite-silica). But within a large part of this triangle—the field AGB—alkali feldspar crystallizes from the first, and leucite consequently cannot form at any stage from melts within this field nor from their derivatives. It will be remembered that no such effect attends the presence of anorthite—even in very large amounts—in melts containing potash feldspar.

Alkali feldspar of sodic composition crystallizes first from silica-deficient melts such as Z, within the feldspar field of Fig. 14a. Ultimately when the melt reaches some point on the boundary curve HR, feldspar is joined

---

Fig. 14.   The system nepheline-kalsilite-silica.   (a) Phase diagram for the anhydrous system at atmospheric pressure.   (After J. F. Schairer.)   (b) Mineralogical data relating to natural nepheline parageneses.   Circled points are compositions of nephelines; crosses are compositions of associated alkali feldspars; solid circles are corresponding rock compositions.   1 = hypersolvus phonolite, 2 = hypersolvus nepheline syenite, 3 = subsolvus nepheline syenite.   (After C. E. Tilley.)   4 and 5 = potassic nephelines of east African lavas.   (After T. G. Sahama.)   WY = approximate limit of solid solution of (Na, K)$AlSi_3O_8$ in natural high-temperature nephelines. (After O. F. Tuttle and J. V. Smith.)

by nepheline. Both crystalline phases change in composition as they separate. The melt is at last used up as it moves along the curve HR toward the low-temperature point R. Nephelines in equilibrium with melts in the vicinity of R are found to be sodic types comparable with the typical nephelines of igneous rocks.

Of some petrological interest are silica-deficient melts such as X lying in the field of leucite. As leucite separates, the melt changes from X to S. At S leucite reacts with the melt and potassic feldspar crystallizes, the melt meanwhile following the curve SR. R is an invariant [47] point at which leucite continues to dissolve while potassic feldspar and nepheline simultaneously crystallize. When leucite is eliminated, alkali feldspar and nepheline separate together until the liquid is completely used up. Note that by appropriate concentration of leucite crystals separating from a melt originally containing excess silica, for example, Q, the resulting partly crystalline "magma" may become so enriched in leucite as to be deficient in silica. If such a mass maintained equilibrium during further crystallization its subsequent history would be identical with that of the initially silica-deficient melt X.

The various cases discussed above throw light upon a number of petrographic phenomena:

1. Association of leucite, but never nepheline, with quickly chilled groundmass material which contains free silica. Whereas leucite can exist in equilibrium with some melts containing excess silica (in the field ABF), nepheline cannot crystallize until the associated melt has become deficient in silica.

2. The existence of pseudoleucites, i.e., finely crystalline nepheline and potash feldspar together building trapezohedral pseudomorphs thought to be after leucite. Bowen [48] predicted the existence of the "pseudoleucite reaction" point R from petrographic evidence of this kind and from the known reaction relation between leucite and orthoclase. Experiment subsequently confirmed the prediction. It is now thought that some pseudoleucites are pseudomorphs after potash analcite [49]—a hydrous feldspathoid likely to crystallize from phonolitic magmas at water pressures of 4,000 bars or more, i.e., at depths exceeding ten miles or under explosive volcanic conditions.

3. Great rarity of leucite in alkaline plutonic rocks. Leucite may separate early from a wide range of undersaturated potash-rich melts, but as the temperature falls it tends to be eliminated by reaction with the melt

---

[47] For given constant, pressure. All experimental results reported in this section have been obtained at a constant pressure of one bar.

[48] Bowen, op. cit., p. 245, 1928.

[49] R. T. Fudali, On the origin of pseudoleucite, Am. Geophys. Union Trans., vol. 38, p. 391, 1957.

to give alkali feldspar and nepheline—the characteristic plutonic assemblage. The field of leucite in Fig. 14a is greatly reduced by another condition characteristic of plutonic crystallization—high water pressure.

4. The relatively sodic composition of common igneous nephelines (such as 1, 2, and 3, Fig. 14b), corresponding to phases crystallizing from melts near R in the artificial system.

Experimentally determined phase relations [50] at subsolidus temperatures in the anhydrous system $NaAlSiO_4$-$KAlSiO_4$ confirm mineralogical data relating to natural nepheline parageneses (cf. Fig. 14b):[51]

(a) In alkaline volcanic and some plutonic rocks sodic nephelines containing appreciable $NaAlSi_3O_8$ in solid solution are associated with anorthoclase or other alkali feldspar within the range $Ab_{55}Or_{45}$-$Ab_{65}Or_{35}$ (e.g., tie lines 1, 2, Fig. 14b). This paragenesis represents crystallization at relatively high temperatures above the solvus in both the nepheline and the alkali feldspar series.

(b) Association of albite, microcline, and a nepheline close to Na_3KAl_4-Si_4O_{16} in other plutonic rocks (triangle 3, Fig. 14b) is due to postconsolidational exsolution to phases stable at low temperatures (below the solvus) in the nepheline and alkali feldspar series.

(c) In some highly potassic lavas [52] the nepheline phase, typically associated with leucite and melilite, is a potash nepheline such as 4 or 5 in Fig. 14b, or, rarely, nearly pure kalsilite. Sahama has recorded exsolution of potash-nepheline phenocrysts $Ne_{31}Kl_{69}$ (5 in Fig. 14b) to a perthitic intergrowth of two phases stable at subsolvus temperatures—$Ne_{67}Kl_{33}$ and Ne_3Kl_{97}.

(d) Unmixing of volcanic nephelines close to the line WY in Fig. 14b can be expected to yield such varied subsolvus assemblages as nepheline-kalsilite, nepheline-leucite, nepheline-microcline, nepheline-microcline-albite, nepheline-albite, and others.[53] The nepheline in all these will be close to $Na_3KAl_4Si_4O_{16}$.

Crystallization of Zeolites. Analcite and natrolite are well-known constituents of some alkaline rocks—notably syenites, members of the alkali-gabbro family, and lamprophyres. Further, the texture of some of these

[50] Tuttle and Smith, op. cit., pp. 584–587, 1958.

[51] T. G. Sahama, Parallel intergrowths of nepheline and microperthitic kalsilite from north Kivu, Belgian Congo, Annales Acad. Sci. Fennicae, ser. A, iii, no. 36, 1953; C. E. Tilley, Nepheline alkali feldspar parageneses, Am. Jour. Sci., vol. 252, pp. 65–75, 1954; Problems of alkali rock genesis, Geol. Soc. London Quart. Jour., vol. 113, pp. 325–332, 1958.

[52] T. G. Sahama, Mineralogy and petrology of a lava flow from Mt. Nyiragongo, Belgian Congo, Annales Acad. Sci. Fennicae, ser. A, iii, no. 35, 1953; op. cit., no. 36, 1953.

[53] Tuttle and Smith, op. cit., pp. 586–588, 1958.

rocks indicates a late magmatic origin for these zeolites. They may be classed then as feldspathoids.

In the laboratory analcite has been synthesized at temperatures of 500° to 600°C. at water pressures of 500 to 3,000 bars.[54] These conditions are in the main metamorphic rather than magmatic. But it seems likely that at high water pressures the freezing point of some feldspathoidal melts would be lowered sufficiently to permit crystallization of analcite in place of the pair nepheline-albite. At 600 bars water pressure a glass of nepheline-syenite composition [55] has been found to crystallize to feldspar plus nepheline over the range 625° to 800°C. At 500°C. the same glass gives a potassic feldspar and analcite.

The presence of excess silica greatly lowers the temperature range over which analcite can be formed by hydrothermal synthesis. Thus, even at high water pressures the pair albite-quartz should remain stable down to the lowest magmatic temperatures. Analcite and natrolite are to be expected only in a silica-deficient alkaline environment.

CRYSTALLIZATION OF OLIVINES AND PYROXENES

The System Forsterite-Fayalite.[56] The common olivines of igneous rocks belong to a solid-solution series $(Mg, Fe)_2SiO_4$ with complete miscibility between the end members, forsterite Mg_2SiO_4 and fayalite Fe_2SiO_4. Behavior of melts in the system forsterite-fayalite (Fig. 15) is similar to that of melts in the plagioclase series. Fayalite is the low-temperature member, so that early-formed crystals are rich in magnesium, normally zoned crystals are increasingly rich in iron toward the rim, and fractional crystallization leads to concentration of iron in the residual liquids. The high temperature necessary for complete fusion of magnesian olivines (greater than 1800°C. for $Fo_{100-80}Fa_{0-20}$) is of interest in connection with divergent views as to the existence of liquid peridotite magmas (see pages 313–314).

The System Forsterite-Silica.[57] Between forsterite Mg_2SiO_4 and silica is the incongruently melting pyroxene $MgSiO_3$, which at the high tempera-

---

[54] H. S. Yoder, The jadeite problem, Am. Jour. Sci., vol. 248, pp. 322, 323, 1950; in Annual report of the director of the Geophysical Laboratory, Carnegie Inst. Washington Year Book, no. 53, pp. 121–123, 1954; D. T. Griggs, W. S. Fyfe, and G. C. Kennedy, Jadeite, analcite and nepheline-albite equilibrium, Geol. Soc. America Bull., vol. 66, p. 1569, 1955.

[55] W. S. MacKenzie, in Annual report of the director of the Geophysical Laboratory, Carnegie Inst. Washington Year Book, no. 55, p. 197, 1956.

[56] N. L. Bowen and J. F. Schairer, The system MgO-FeO-SiO_2, Am. Jour. Sci., vol. 29, pp. 161–163, 1935.

[57] N. L. Bowen and O. Andersen, The binary system MgO-SiO_2, Am. Jour. Sci., vol. 37, pp. 487–500, 1914; Bowen, op. cit., pp. 29–31, 1928.

tures of dry melts assumes the form protoenstatite. Crystallization of melts intermediate between $MgSiO_3$ and Mg_2SiO_4 proceeds similarly to crystallization of orthoclase-leucite melts in the system leucite-silica. In this case forsterite separates first, and later, when the liquid has reached R in Fig. 16, reacts with the liquid to give clinoenstatite. Melts originally deficient in silica (*i.e.*, to the right of $MgSiO_3$ in Fig. 16) ultimately give forsterite and pyroxene. Those with excess silica crystallize as a silica-pyroxene eutectic at E. If melts of this latter type originally contain less silica than R, forsterite first separates as a transient phase, which is later eliminated by reaction at R before the remaining liquid completes its course along RE.

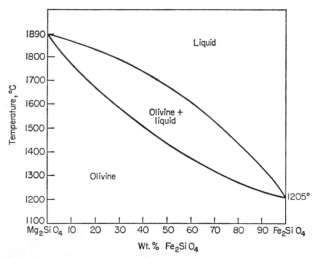

Fig. 15.   Phase diagram for the system forsterite-fayalite.   (*After N. L. Bowen and J. F. Schairer.*)

From the petrogenic standpoint this system is important in that it illustrates a means by which liquids consisting of pyroxene and excess silica may be derived by fractional crystallization from melts initially so deficient in silica as to allow copious separation of olivine. It affords experimental evidence in support of the hypothesis that the observed association of quartz diabase and olivine diabase in basic sills is due to gravitational settling of olivine crystals in the early stages of crystallization.

The System Anorthite-Forsterite-Silica.[58]   Apart from the small range of melts from which spinel crystallizes (the spinel field of Fig. 17), mixtures of anorthite, forsterite, and silica can be treated in terms of a three-component system. The presence of anorthite in no way affects the reac-

[58] O. Andersen, The system anorthite-forsterite-silica, *Am. Jour. Sci.*, vol. 39, pp. 407–454, 1915; Bowen, *op. cit.*, pp. 41–44, 1928.

tion relation between forsterite and pyroxene, except that reaction occurs over a range of temperatures in melts whose compositions vary along the curve QR as reaction proceeds.

From a melt P, forsterite separates as the composition of the melt changes along PS with falling temperature. The melt now follows the boundary SR while forsterite goes into solution and pyroxene separates. At R, an invariant point, temperature and composition of liquid remain constant while forsterite continues to dissolve and anorthite and pyroxene separate. Reaction ceases when the liquid phase is eliminated, and the end product is a crystalline mixture of forsterite, pyroxene, and anorthite. An original liquid T behaves in precisely the same way, except that forsterite and liquid are simultaneously used up at R. In the case of an original liquid U, the course is again similar from U to S and from S to R. Reaction at R is terminated in this case by complete solution of forsterite while some liquid still remains. The latter now changes with falling temperature along RE, while anorthite and pyroxene crystallize. At E, eutectic crystallization of anorthite, pyroxene, and tridymite sets in. An original liquid S follows a different course, since pyroxene is now the first solid phase to appear. The melt changes from S to V with crystallization of pyroxene, and from V to E with separation of pyroxene and anorthite. Again eutectic crystallization at E sets in and continues till the liquid is used up.

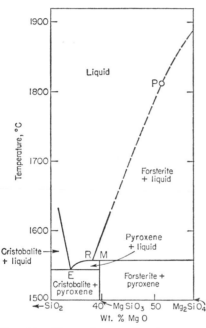

FIG. 16. Phase diagram for the system forsterite-silica. (*After N. L. Bowen and O. Andersen.*)

Now liquids T, U, and S are all derivatives of the melt P first considered. It follows that by appropriate fractional crystallization (*i.e.*, by removal of all forsterite crystals when the liquid attains the composition, T, U, or S, as the case may be) very different end products may develop from the originally ultrabasic melt P. Alternative end products are olivine-pyroxene-anorthite, pyroxene-anorthite, and pyroxene-anorthite-silica. Moreover the relative proportions of olivine in the first, and of silica in the last, of these assemblages vary widely with the degree of fractionation. Once more the effectiveness of fractional crystallization in differentiation is found to be increased where a third component (in this case anorthite) is added to a binary system which includes an incongruently melting compound (pyroxene). In the above instance, in derivatives of the initial melt P a new range of melts (SV) is made available for further fractionation, by removal of olivine crystals when the melt has reached composition S on the olivine-pyroxene boundary.

The System Diopside-Forsterite-Silica.[59] At the temperature of crystallization from dry melts, $CaMgSi_2O_6$ and $MgSiO_3$ form a solid-solution series broken only between En_{40} and En_{55}. Solid solutions with more than

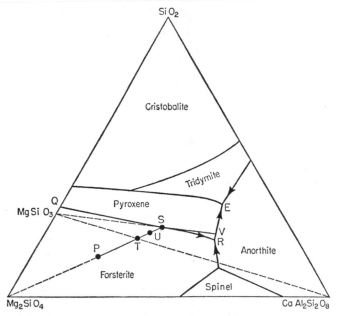

Fig. 17.   Phase diagram for the system anorthite-forsterite-silica.   (After O. Andersen.)

20 per cent $MgSiO_3$ melt incongruently with separation of forsterite. The complicated type of reaction that ensues when forsterite reacts with melts on the olivine-pyroxene boundary is illustrated by following the course of crystallization of an ultrabasic melt P and its residual liquids in Fig. 18.

From the melt P (Fig. 18b) forsterite separates as the first solid phase, and the melt changes along PQ. At Q forsterite begins to react with the liquid and pyroxene of some such composition as S crystallizes. As the melt now follows the boundary QR, the pyroxene and forsterite simultaneously react with the melt to give a progressively less magnesian pyroxene. When the melt reaches R the pyroxene is entirely made over to

[59] Bowen, op. cit., pp. 49–53, 1928.

composition T.   At this point the melt is eliminated, and the end product is pyroxene T and forsterite.   If all previously formed forsterite were removed when the melt reached T, the course followed along TQR would

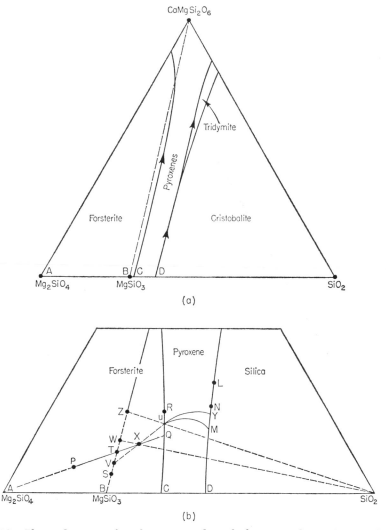

(a)

(b)

FIG. 18.   Phase diagrams for the system diopside-forsterite-silica.   (*After N. L. Bowen.*)   (*a*) True-scale diagram.   (*b*) Diagram distorted to illustrate crystallization of pyroxenes.

be the same as before, but in this case the liquid and forsterite would become exhausted simultaneously just as the pyroxene attained composition T.   Consider now the case of a residual liquid X (from which all olivine has been removed), lying in the forsterite field but containing excess silica. Forsterite separates in small quantity as the liquid changes along XQ. As it further changes from Q to U, the liquid reacts with forsterite, and pyroxene, S to V, separates. When the melt reaches U, the last of the forsterite has been eliminated, and the melt embarks on a curving course UM while pyroxene separates. The pyroxene meanwhile reacts continuously with the melt and becomes progressively richer in diopside. This continues as the melt moves along the boundary MN with simultaneous separation of silica and pyroxene. When the mass consists of homogeneous pyroxene W and a small excess of crystalline silica, the melt (now at N) is exhausted. The course UM is unique. It can be followed only by a melt U initially in equilibrium with a specific quantity of pyroxene V. The pure melt U, if removed from contact with earlier formed pyroxene V, would take some such course as UY (also unique), would then follow the curve YL, and so would ultimately give rise to pyroxene Z together with a somewhat greater quantity of silica than in the previous case.

The main petrological implication is that a reaction relation of the type olivine $\rightarrow$ pyroxene persists unchanged even where one of the phases concerned (in this case pyroxene) is a member of a solid-solution series. The effectiveness of the two mechanisms of reaction in modifying the nature and the behavior of late liquids (*i.e.*, in magmatic differentiation) is cumulative. In spite of the complexity of the melts concerned and the prevalence of solid solutions among the crystallizing phases, we may expect that magnesian olivines that have separated early from basaltic magmas will later react with the magma over some lower range of temperature, with simultaneous crystallization of pigeonite or hypersthene.

The System Diopside-Forsterite-Anorthite.[60] Figure 19 is an equilibrium diagram for the system diopside-forsterite-anorthite. For all melts except those lying within the triangle XDZ, crystallization begins with one of the phases forsterite, diopside, or anorthite; then two of these separate simultaneously as the liquid moves along one of the boundary curves AE, DE, or BE; and finally crystallization of all three phases sets in at E. Since the diopside so formed is slightly aluminous (as are all igneous diopsides), the final liquid lies outside the plane of Fig. 19, and E is not strictly a eutectic point. But to all intents and purposes it may be considered such.

Melts in the triangle XDZ cannot be treated adequately in terms of the three-component system. Their behavior is complicated by crystallization, then ultimate elimination, of spinel. This behavior we cannot afford to neglect, for spinel is a common constituent of ultrabasic and basic

[60] E. F. Osborn and D. B. Tait, The system diopside-forsterite-anorthite, *Am. Jour. Sci.*, Bowen vol., pp. 413–433, 1952.

igneous rocks.   From a liquid P forsterite is the first phase to crystallize, changing the composition of the melt along PQ.   As the temperature drops below that of Q, forsterite and spinel separate simultaneously.   Consequently the liquid leaves the plane of the diagram (Fig. 19) and enters the tetrahedron diopside-anorthite-forsterite-silica.   Its composition moves across a forsterite-spinel divariant surface within this tetrahedron until it reaches a forsterite-anorthite-spinel univariant line.   Down this line the liquid moves with falling temperature, spinel going into solution and forsterite and anorthite crystallizing together.   The spinel is used up just as the liquid once more enters the plane of Fig. 19 at point D.   It now

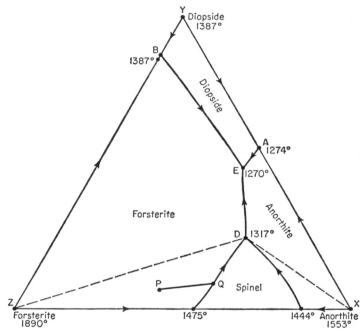

FIG. 19.   Phase diagram for the system diopside-forsterite-anorthite.   (*After E. F. Osborn and D. B. Tait.*)

follows the curve DE, and crystallization ceases with simultaneous separation of diopside, anorthite, and forsterite at E.

Spinel is rare in volcanic rocks except as accessory inclusions in olivine of basalt.   On the other hand pleonaste or picotite is characteristic of ultrabasic plutonic rocks and of nodular olivine-rich inclusions in basalts. Typical assemblages are olivine-enstatite-picotite, olivine-enstatite-diopside-picotite, and anorthite-pleonaste.   It is possible to write an equation

$$CaAl_2Si_2O_8 + 2Mg_2SiO_4 \rightleftarrows CaMgSi_2O_6 + 2MgSiO_3 + MgOAl_2O_3$$

| Anorthite | Olivine | Diopside | Enstatite | Spinel |

As this equation involves a 7 per cent reduction in volume from left to right, there must be a range of temperatures at which increase of pressure will cause the reaction to run left to right. It therefore is tempting to attribute to plutonic conditions (high pressures and temperatures below those of the experimentally investigated system) the pyroxene-spinel assemblages of ultrabasic rocks. This explanation receives additional support from occurrences of spinel-pyroxene coronas at olivine-bytownite contacts in troctolites.[61] But we must also take into account the possible influence of other components in the natural silicate systems. Presence of Na_2O in basaltic rocks would favor the left-hand assemblage—labradorite-olivine; but in ultrabasic plutonic rocks Cr_2O_3 must be reckoned an essential component that might well extend the field of spinel sufficiently to permit crystallization of pyroxene-picotite rather than olivine-anorthite.

Additional Experimental Data on Crystallization of Pyroxenes. During the decade 1930 to 1940 investigations in such systems as CaO-FeO-SiO_2, MgO-FeO-SiO_2, and Na_2SiO_3-Fe_2O_3-SiO_2 yielded valuable data on the crystallization of pyroxenes containing iron.[62] These systems are too complex to be reviewed in any detail in this book, but some of the petrologically significant findings may be listed:

1. In the series $(Mg, Fe)SiO_3$, substitution of Fe for Mg is possible up to about 87 per cent. Pure $FeSiO_3$ is not a stable compound. Its stable equivalent is fayalite, Fe_2SiO_4, and free silica.

2. At low temperatures the stable form of $(Mg, Fe)SiO_3$ is orthorhombic—the enstatite-hypersthene series. There is still some uncertainty as to the relation of orthorhombic pyroxenes to other known polymorphs, namely, experimentally synthesized protoenstatite (orthorhombic) and clinoenstatite (monoclinic), and to the naturally occurring pigeonites (monoclinic). Most probably the stable high-temperature form is protoenstatite, and clinoenstatite is metastable.[63] The temperature of inversion of enstatite has been recorded variously as 990° to 1140°C. Hypersthenes invert at substantially lower temperatures.

3. In the series $(Mg, Fe)\ SiO_3$, incongruent melting with separation of olivine and complementary enrichment of the residual melt in silica is

[61] F. F. Osborne, Coronite, labradorite, anorthosite and dykes of andesine anorthosite, New Glasgow, P.Q., *Royal Soc. Canada Trans.*, ser. 3, sec. 4, vol. 43, pp. 85–112, 1949.

[62] N. L. Bowen, J. F. Schairer, and H. W. V. Willems, The ternary system Na_2SiO_3-Fe_2O_3-SiO_2, *Am. Jour. Sci.*, vol. 20, pp. 405–455, 1930; N. L. Bowen, J. F. Schairer, and E. Posnjak, The system CaO-FeO-SiO_2, *Am. Jour. Sci.*, vol. 26, pp. 193–284, 1933; Bowen and Schairer, *op. cit.*, pp. 151–217, 1935.

[63] W. R. Foster, High-temperature X-ray diffraction study of the polymorphism of $MgSiO_3$, *Am. Ceramic Soc. Jour.*, vol. 34, pp. 255–259, 1951; L. Atlas, The polymorphism of $MgSiO_3$ and solid-state equilibria in the system $MgSiO_3$-$CaMgSi_2O_6$, *Jour. Geology*, vol. 60, pp. 125–147, 1952.

limited to magnesian types, *i.e.*, those with not more than 20 per cent $FeSiO_3$. Olivines which take part in the corresponding reaction are also magnesian, but these have a somewhat wider range of composition (up to roughly 30 per cent Fe_2SiO_4). Now basaltic olivines are almost invariably magnesian types with less than 20 per cent Fe_2SiO_4. Moreover the possible effect of other molecules, such as plagioclase, present in natural magmas still has to be taken into account. It was found, for example, that the presence of anorthite greatly increases the amount of excess olivine that can separate from a magnesian pyroxene melt. The evidence therefore still strongly supports Bowen's contention that separation of olivines from basaltic magma is effective in enriching the residual liquids (late differentiates) in silica.

4. In the system $MgO-FeO-SiO_2$ there is a wide range of melts of moderate to high $FeSiO_3$ content, from which pyroxene separates first and later reacts with the melt to give olivine. Such olivines are always iron-rich types resembling those found in trachytes and phonolites rather than basaltic olivines.

5. At magmatic temperatures there is a miscibility gap [64] in the solid-solution series $Ca(Mg, Fe) Si_2O_6-(Mg, Fe)SiO_3$. Subcalcic augites (often referred to loosely as "pigeonites") which bridge this gap and occur in the groundmasses of some basalts are almost certainly metastable from the moment of crystallization. Their stable equivalent is a mixture of augite and pigeonite, or at lower temperatures typical of plutonic rocks, augite and hypersthene.

6. Crystallization of olivines and pyroxenes normally leads to enrichment of the residual liquids in iron as compared with magnesium. This accords with the observed outward increase in iron content of zoned pyroxenes and olivines of igneous rocks. Crystallization in the diopside-clinoenstatite series enriches the residual melts in calcium relatively to magnesium. In the series $CaFeSi_2O_6-FeSiO_3$ the reverse relation holds good: residual melts are enriched in iron relatively to calcium.

7. The soda pyroxene acmite $NaFeSi_2O_6$ melts incongruently at the low temperature of 990°C., with separation of hematite and development of a melt correspondingly enriched in soda and silica.

CRYSTALLIZATION OF MICAS AND HORNBLENDE

General Statement. The common igneous minerals of whose melting and crystallization behavior least is known are the micas and the amphiboles. There are several reasons for this. The solid phases are among

---

[64] F. R. Boyd and J. F. Schairer (*Geol. Soc. America Bull.*, vol. 68, p. 1703, 1957) have demonstrated the existence of such a gap at 1390°C. where the solvus cuts the solidus at En_{40} and En_{55} in the dry system enstatite-diopside.

the most complex and variable in mineralogy. Isomorphous substitution of the principal ions in their lattices is widespread and far from simple. Polymorphism—especially in micas—is extensive and is difficult to evaluate except by specialized X-ray techniques. Further difficulties arise from the presence of the hydroxyl group in mica and hornblende lattices. Water is one of the essential components of any system within which mica or hornblende is a possible phase, and the stability of any individual phase combination is conditioned by water pressure just as much as by temperature.

As yet it is not possible to depict the crystallization of even the simplest of micas or amphiboles within any completely investigated multicomponent hydrous system. However a start has been made—notably in the Geophysical Laboratory, Washington—on end members that are simple analogues of naturally occurring micas and hornblendes of a more complex nature.[65] Among the phases so far investigated are muscovite, phlogopite, annite, anthophyllite, pargasite, and tremolite. These have been synthesized or destroyed (sometimes with accompanying melting) under controlled conditions of temperature, water pressure, and (for iron-bearing phases) partial pressure of oxygen. Much of this work concerns metamorphic rather than magmatic crystallization. Results that possibly are of significance in igneous petrology are summarized below.

Micas. *Muscovite.* The coexistence of the dioctahedral mica muscovite and the trioctahedral micas of the phlogopite-biotite series in granites and in a wide range of metamorphic rocks shows that isomorphism between the two mica groups, at least below about 600°C., must be limited or nonexistent. Moreover there are no authentic records of any mica having a composition intermediate between muscovite and biotite.

Figure 20 shows the univariant equilibrium curve (AB) for destruction of muscovite according to the equation

$$KAl_2AlSi_3O_{10}(OH)_2 \rightleftarrows KAlSi_3O_8 + Al_2O_3 + H_2O$$

      Muscovite   Sanidine   Corundum   Water

Only at much higher temperatures, of the order of 900°C. at water pressures of a few thousand bars, would a liquid phase appear as a result of melting of sanidine. In the presence of excess silica—a condition typical

[65] F. R. Boyd, H. P. Eugster, H. S. Yoder, and others, in Annual report of the director of the Geophysical Laboratory, *Carnegie Inst. Washington Year Book*, no. 53, pp. 108–111 (amphiboles), 1954; no. 54, pp. 115–119 (amphiboles), 124–129 (micas), 1955; no. 55, pp. 158–161 (micas), 197–198 (micas), 198–200 (amphiboles), 1956; H. S. Yoder and H. P. Eugster, Phlogopite synthesis and stability range, *Geochim. et Cosmochim. Acta*, vol. 6, pp. 157–185, 1954; Synthetic and natural muscovites, *Geochim. et Cosmochim. Acta*, vol. 8, pp. 225–280, 1955; J. V. Smith and H. S. Yoder, Experimental and theoretical studies of the mica polymorphs, *Mineralog. Mag.*, vol. 31, pp. 209–235, 1956.

of the muscovite-bearing igneous rocks—muscovite breaks down at temperatures perhaps 50°C. lower than those of curve *AB* in Fig. 20. The curve of incipient melting of granitic rocks in the presence of water may be constructed, as a first approximation, from the minimal liquidus temperatures in the system albite-orthoclase-silica at various water pressures (cf. Fig. 14). The curve is shown as *CD* in Fig. 20. From Fig. 20, bearing in mind that both curves represent data determined experimentally in simple ideal systems, we may make several inferences, all of which are compatible with the data of petrography:

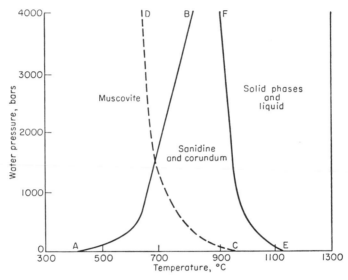

FIG. 20. Curve of univariant equilibrium (*AB*) for reaction muscovite ⇌ sanidine + corundum + water. (*After H. S. Yoder and H. P. Eugster.*) *CD* is curve for incipient melting of "granite." *EF* is an estimated curve for incipient melting of sanidine.

1. Muscovite can crystallize from a melt only at water pressures of the order of 2,000 bars or more. It is to be expected in deep-seated plutonic rocks formed from magmas saturated in water. It cannot crystallize from a melt under volcanic conditions.

2. Primary magmatic muscovite is to be expected only in those granitic rocks whose compositions lie close to the low-temperature region of the liquidus in the system albite-orthoclase-silica. Other liquids in this system, and also all quartzo-feldspathic liquids containing appreciable anorthite, will crystallize completely before muscovite can appear as a magmatic phase. Primary muscovite—so identified from its coarse grain size and interstitial mode of occurrence—is in fact almost restricted to granites high in both potash and soda and low in lime.

3. In the systems K_2O-Al_2O_3-SiO_2, the field of potash feldspar adjoins that of mullite which lies on the Al_2O_3 side of the feldspar-silica join.[66] The liquidus surface rises steeply from the feldspar-mullite boundary into the mullite field.   Much the same thing is seen in the system Na_2O-Al_2O_3-SiO_2.   This means that granitic liquids, no matter what their origin, cannot acquire more than a small excess of normative corundum unless superheated far above the liquidus temperatures of quartz-feldspar mixtures.   Thus it is easy to understand why muscovite (a mineral which appears in the norm as orthoclase plus corundum), though common enough in lime-poor granites, is typically but a minor constituent of these rocks.   Abundance of muscovite in a granite is an indication of possible non-magmatic, *i.e.*, metamorphic origin.

Biotite.   Yoder and Eugster [67] have synthesized and destroyed the biotite end member phlogopite, $KMg_3AlSi_3O_{10}(OH)_2$, at water pressures up to 5,000 bars, and the stability of the corresponding ferrous end member annite has been investigated at water pressures up to 2,000 bars and partial pressures of oxygen ranging from 10^{-26} to 10^{-16} bars.[68]   From this work have emerged several results of significance in igneous petrology (cf. Fig. 21).

The curve of univariant equilibrium for the reaction

$$2KMg_3AlSi_3O_{10}(OH)_2 \rightleftarrows 3Mg_2SiO_4 + KAlSi_2O_6 + KAlSiO_4 + 2H_2O$$

Phlogopite   Forsterite   Leucite

is shown as AB in Fig. 21.   This defines the stability field of pure phlogopite in a silica-deficient environment.   Common igneous biotites are iron-bearing micas that can be regarded, as a first approximation, as intermediate in composition between phlogopite and annite.   Moreover the biotites of igneous rocks typically, though not invariably, are associated with free quartz.   The petrologically significant reaction is

$$K(Mg, Fe)_3AlSi_3O_{10}(OH)_2 + 3SiO_2 \rightleftarrows 3(Mg, Fe)SiO_3 + KAlSi_3O_8 + H_2O$$

Biotite   Pyroxene   Sanidine

Pure magnesian phlogopite breaks down in the presence of excess silica at temperatures some 120°C. below those of AB in Fig. 21.   At 2,000 bars water pressure annite breaks down to sanidine, magnetite, and water at 490° to 827°C. (K to E, Fig. 21; BFC, Fig. 22); in the presence of excess silica it may be converted to sanidine, fayalite, silica, and water at tem-

---

[66] J. F. Schairer and N. L. Bowen, The system K_2O-Al_2O_3-SiO_2, *Am. Jour. Sci.*, vol. 253, pp. 681–746, 1955.

[67] Yoder and Eugster, *op. cit.*, 1954.

[68] H. P. Eugster, Stability of hydrous iron silicates, in Annual report of the director of the Geophysical Laboratory, *Carnegie Inst. Washington Year Book*, no. 55, pp. 158–161, 1956; no. 56, pp. 161–164, 1957.

peratures between 615° and 585°C. (F, Fig. 21; FG in Fig. 22).[69]   These figures are some 300° to 600°C. lower than the breakdown temperature of magnesian phlogopite in a silica-deficient environment at the same water pressure.   We conclude that the stability fields of igneous biotites are bounded by a family of curves occupying a broad area between AB and a more or less parallel curve through K in Fig. 21.

There is yet another complication, for the work of Eugster has shown that the stability of ferrous silicates is affected not only by temperature, total pressure, and water pressure, but also by the partial pressure of

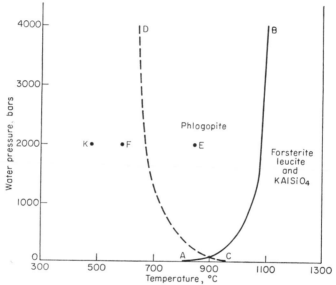

Fig. 21.   Curve of univariant equilibrium (AB) for reaction phlogopite $\rightleftarrows$ forsterite + leucite +$KAlSiO_4$ + water. (*After H. S. Yoder and H. P. Eugster.*) CD is curve for incipient melting of "granite." E, F, and K are points on the curves for breakdown of annite (*e.g.*, F for reaction annite + quartz $\rightleftarrows$ sanidine + fayalite + water).

oxygen P_{O_2}.   This depends on a number of compositional factors, such as the initial ratio Fe_2O_3/FeO in the melt and the concentrations of sulfur and carbon, oxidation of which raises the partial pressures of SO_2 and CO or CO_2 and correspondingly reduces the pressure of oxygen.[70]   Clearly variation of P_{O_2} may notably affect the stability of iron-bearing silicates crystallizing from a melt.   In a closed system at given temperature and water pressure, oxidation or reduction of such iron oxides as may be pres-

[69] The temperature of each reaction at 2,000 bars P_{H_2O} depends on the partial pressure of oxygen (cf. Fig. 22).

[70] E.g., A. J. Ellis, Chemical equilibrium in magmatic gases, *Am. Jour. Sci.*, vol. 255, pp. 416–431, 1957.

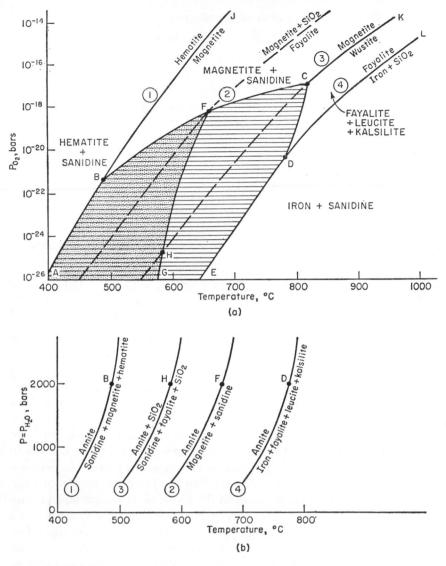

FIG. 22. Influence of partial pressure of oxygen on stability of annite. (*After H. P. Eugster.*) (*a*) Isobaric diagram for total pressure $P = 2,000$ bars. Stability field of annite is ruled; that of annite + quartz is stippled. In place of annite, hematite + sanidine is stable to the left of *ABJ;* magnetite + sanidine in the field *JBCK,* and so on. Curves 1, 2, 3, 4 represent reactions of oxidation (right to left) used experimentally to control P_{O_2}. FHG is the curve for the reaction annite + quartz $\rightleftarrows$ fayalite + sanidine + water. (*b*) Reaction curves for destruction of annite. The nature of each reaction (and the form of the curve) is determined by a range of values of P_{O_2} corresponding to mineral assemblages on the similarly numbered reaction curves of the upper diagram. Thus the left-hand curve (annite $\rightleftarrows$ sanidine + magnetite + hematite + H_2O) is conditioned by P_{O_2} values corresponding to the associated assemblage hematite + magnetite, curve (1) of the upper diagram. B, H, F, and D are identical on the two diagrams.

137

ent regulates P_{O_2} to some constant value. In Fig. 22a reactions of this kind are represented by curves (1) (Hematite $\rightleftarrows$ Magnetite $+ O_2$); (2) (Magnetite $+$ Quartz $\rightleftarrows$ Fayalite $+ O_2$); (3) (Magnetite $\rightleftarrows$ Wüstite $+ O_2$); and (4) (Fayalite $\rightleftarrows$ Quartz $+ \alpha$-Iron $+ O_2$). Annite is stable in the field $ABCDE$. Annite plus silica constitutes a stable assemblage only in the much more limited stippled field $ABFG$. Note how these fields are bounded in part by the curves for oxidation or reduction of associated iron-oxide phases. The curves FG, BFC, and CD are of a different nature. Thus FG represents the reaction

$$2KFe_3AlSi_3O_{10}(OH)_2 + 3SiO_2 \rightleftarrows 3Fe_2SiO_4 + 2KAlSi_3O_8 + 2H_2O$$

Annite   Quartz   Fayalite   Sanidine   Water

Curve BFC, representing the breakdown of annite to sanidine and magnetite, involves oxidation of iron and so is sensitive to P_{O_2}, as indicated by its relatively flat slope

$$KFe_3AlSi_3O_{10}(OH)_2 + \tfrac{1}{2}O_2 \rightleftarrows KAlSi_3O_8 + Fe_3O_4 + H_2O$$

Annite   Sanidine   Magnetite

In Fig. 22b curves are given for a series of reactions—all involving destruction of annite—at partial pressures of oxygen controlled by associated iron oxides and silicates. Figure 22 shows clearly how important is the role of iron oxides in rocks as indicators of the range of P_{O_2} that prevailed during crystallization of some mineral assemblages. Unfortunately among common igneous and metamorphic minerals those that have received least petrographic attention are the "iron-ores."

Most igneous biotites contain a considerable amount of iron, but have compositions nearer to phlogopite than to annite. Also to be taken into account is the content of ferric iron, which in many biotites is between one quarter and one half that of aluminum.[71] Thus the temperatures at which igneous biotites break down in the presence of quartz and magnetite—the usual case for granites and granodiorites—must correspond to a curve well to the left of AB in Fig. 21. These temperatures are still a few hundred degrees higher than those at which muscovite becomes unstable at corresponding water pressures. The fact that biotite is so much more widely distributed than is muscovite in igneous rocks accords with these inferences from experimental data. At the low water pressures characteristic of volcanic conditions, biotite, too, is unstable at magmatic temperatures (area BOC, Fig. 23). The occurrence of resorbed biotite phenocrysts (granular aggregates of iron oxides and pyroxenes) in volcanic rocks is readily explained in terms of what we know from experiment. In Fig. 23, AB is the hypothetical stability curve of some igneous

[71] *E.g.*, E. S. Larsen, F. A. Gonyer, and J. Irving, Petrologic results of a study of the minerals from the Tertiary volcanic rocks of the San Juan region, Colorado, 4, Micas, *Am. Mineralogist*, vol. 22, pp. 898–905, 1937.

biotite in association with quartz and magnetite. Let us assume that crystals of biotite form at 850°C. and 2,000 bars water pressure in a plutonic environment. *EFG* is a curve illustrating possible change of temperature and pressure during slow (plutonic) cooling. Biotite remains stable to the point where crystallization is complete (*F*), and throughout subsequent cooling of the solid mass (along *FG*). But if as a result of upward injection and extrusion of the magma the pressure drops sharply (along *EH*) while the magma is still hot, biotite becomes unstable (below *BO*) and is resorbed to a mixture of iron oxides and anhydrous silicates.

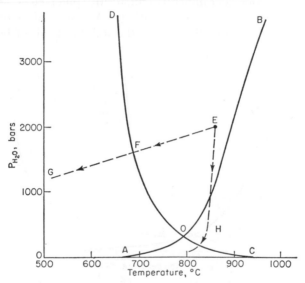

Fig. 23.   Hypothetical curve (*AB*) for breakdown of an igneous biotite (stable to left of *AB*) in presence of quartz and magnetite.   *CD* is the curve of incipient fusion of granite.

Resorption of biotite could also result from a sharp increase in partial pressure of oxygen (*e.g.*, to values above *BFC*, Fig. 22*a*).

Hornblende.[72]   The common igneous and metamorphic hornblendes are calciferous amphiboles which, as a first approximation, may be considered in terms of three end members:

Tremolite $Ca_2(Mg, Fe)_5Si_8O_{22}(OH)_2$

Edenite $NaCa_2(Mg, Fe)_5AlSi_7O_{22}(OH)_2$

Pargasite $NaCa_2(Mg, Fe)_4AlAl_2Si_6O_{22}(OH)_2$

[72] F. R. Boyd, Amphiboles, in Annual report of the director of the Geophysical Laboratory, *Carnegie Inst. Washington Year Book*, no. 53, pp. 108–111, 1954; no. 55, pp. 198–200, 1956.

In the last two, Na^+ occupies a lattice site which is vacant in tremolite; at the same time Al^{3+} substitutes for Si^{4+}. Other substitutions, not indicated above, are F^- for $(OH)^-$ and Fe^{3+} for Al^{3+} in four-coordination.

Figure 24 shows the curve of univariant equilibrium (AB) for the metamorphic reaction.

$$Ca_2Mg_5Si_8O_{22}(OH)_2 \rightleftarrows 3MgSiO_3 + 2CaMgSi_2O_6 + SiO_2 + H_2O$$

| Tremolite | Enstatite | Diopside | Quartz | Water |

DEF represents the breakdown of magnesian pargasite to a mixture of diopside and silica-deficient minerals (forsterite, spinel, etc.); above E the breakdown process is incongruent melting. As with the biotites, doubtless the presence of iron greatly reduces the stability fields of both amphi-

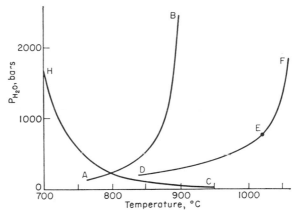

FIG. 24.   Curves for breakdown of pure magnesian tremolite (AB) and pure magnesian pargasite (DEF). Amphiboles are stable to the left of AB and DEF respectively. CH is the curve of incipient fusion of granite. $(After\ F.\ R.\ Boyd.)$

boles.   And pargasite has been shown by Boyd [73] to behave very differently in a silica-saturated environment. Nevertheless it is clear that at water pressures exceeding two or three hundred bars, hornblendes may be expected to crystallize as stable phases from granitic magmas (whose curve of incipient melting is shown as CH in Fig. 24). Much as with biotite, the wider incidence of hornblende in plutonic than in volcanic rocks and the phenomenon of resorption of hornblende phenocrysts in lavas are readily understood in the light of experimentally determined facts. Indeed the hypothetical curve AB of Fig. 23, though drawn to illustrate the breakdown of some igneous biotite, might serve just as well to represent in a qualitative way the behavior of an igneous hornblende.

[73] $Op.\ cit.$, p. 200, 1956.

CRYSTALLIZATION OF MELILITES

Melilites of igneous rocks arc essentially akermanite-gehlenite solid solutions, $Ca_2Mg(Si_2O_7)$-$Ca_2Al(AlSiO_7)$, with limited substitution of (NaAl) for (CaMg) and, to a less extent, of Fe for Mg. They are constituents of a rather rare group of very basic volcanic rocks which shows a strong tendency to occur in association with feldspathoidal basic lavas such as nepheline basalts.

A number of systems in which melilites crystallize have been investigated in the laboratory.[74]   It is scarcely appropriate, in view of the rarity of melilite in igneous rocks, to discuss any of this work in detail; but a few points of possible petrologic significance may be noted.

1. In the akermanite-gehlenite solid-solution series, crystallization proceeds from either end to a minimum freezing point in the vicinity of $Ak_{72}Ge_{28}$. The presence of iron in the akermanite series $Ca_2(Mg, Fe)Si_2O_7$ appreciably lowers the melting temperature, so that residual melts from crystallizing akermanite tend to become enriched in iron. Presence of soda likewise reduces the melting temperature of melilite.

2. Melilites crystallize from melts whose compositions are equivalent to a wide range of diopside-nepheline mixtures.   In such cases the melilites appear to be soda-bearing akermanites analogous in composition to natural melilites.   Crystallization from the artificial melt is typically accompanied or preceded by separation of magnesian olivine.   Since both compounds are extremely basic, this implies considerable enrichment of residual liquids in silica as crystallization proceeds.   As pointed out by Bowen, these observations are significant as showing that magmas need not necessarily attain an unusually high content of lime to allow melilite to crystallize.   Rather it would seem that crystallization of melilite and olivine is favored by concentration of nepheline and diopside "molecules" in a basic magma—a conclusion which finds some support in the well-known mutual association of melilite basalts and rocks of the nepheline basalt family.

3. In the system CaO-FeO-Al_2O_3-SiO_2, those melts that are in any way analogous to magmas move during crystallization toward low-temperature points at which melilites tend to be associated with iron olivines, spinels, and anorthite.

[74] J. B. Ferguson and H. E. Merwin, The ternary system CaO-MgO-SiO_2, *Am. Jour. Sci.*, vol. 48, pp. 81–123, 1919; Bowen, *op. cit.*, pp. 260–268, 1928; E. F. Osborn and J. F. Schairer, The ternary system pseudowollastonite-akermanite-gehlenite, *Am. Jour. Sci.*, vol. 239, pp. 715–763, 1941; J. F. Schairer, The system CaO-FeO-Al_2O_3-SiO_2: I. *Am. Ceramic Soc. Jour.*, vol. 25, pp. 241–274, 1942; Some aspects of the melting and crystallization of rock-forming minerals, *Am. Mineralogist*, vol. 29, pp. 86–90, 1944.

CRYSTALLIZATION OF MELTS ENRICHED IN SODA AND IRON

Fractional crystallization in the plagioclase series leads to progressive enrichment of residual melts in soda. In the olivine and pyroxene series, fractionation leads to concentration of iron. Crystallization of basaltic magma could thus give rise to late differentiates of two contrasted types: sodic liquids resulting from strong fractionation in the feldspar series alone, and iron-rich liquids formed where fractionation has been effective only with respect to the ferromagnesian minerals. But the case where fractional crystallization affects both series of minerals to a comparable degree must also be common. Under such conditions is soda concentrated with respect to iron or does the reverse hold good? Petrographers

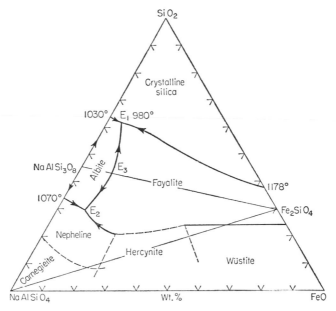

Fig. 25. Phase diagram for the system $NaAlSiO_4$-FeO-SiO_2. (*After N. L. Bowen and J. F. Schairer.*)

have disagreed on this point. However, the investigations of Bowen and Schairer [75] on crystallization of mixtures of nepheline, albite, fayalite, and silica have yielded highly suggestive evidence in support of the first alternative.

Crystallization behavior of appropriate melts in the system nepheline-FeO-SiO_2 is depicted in Fig. 25. We are concerned with melts consisting

[75] N. L. Bowen and J. F. Schairer, Crystallization equilibrium in nepheline-albite-silica mixtures with fayalite, *Jour. Geology*, vol. 46, pp. 397–411, 1938.

of fayalite and albite with varying amounts of nepheline or free silica. Melts so low in silica as to permit separation of spinel or wüstite need not be considered. All other melts proceed to one or other of the two ternary eutectics—albite-fayalite-silica, E_1, and albite-nepheline-fayalite, E_2. Mixtures of fayalite and albite alone will crystallize ultimately at the albite-fayalite eutectic point E_3. It is important to note that E_1, E_2, and E_3 represent mixtures very rich in albite and comparatively poor in fayalite. Within this system at any rate, fractional crystallization leads to sodic, not ferruginous, late differentiates.

CHAPTER 6

Crystallization of Basaltic and Granitic Magmas in the Light of Experimental Data

REACTION SERIES IN IGNEOUS ROCKS [1]

Crystallization of any solid-solution series involves continuous reaction between crystals and the melt from which they separate as the temperature falls. Minerals of this type, *e.g.*, plagioclases, olivines, pyroxenes, and alkali feldspars, constitute what Bowen has termed *continuous reaction series*. A rather similar reaction, but limited to a melt of unique composition and temperature, accompanies crystallization of protoenstatite in the presence of forsterite in the binary system forsterite-silica. The minerals concerned, in this case forsterite → pyroxene, constitute what Bowen has called a *reaction pair*. Instances are known where a crystalline phase *B* melts incongruently with separation of a phase *A*, but also will itself react at some definite lower temperature with a melt of appropriate composition to give a third phase *C*. The series *A* → *B* → *C*, consisting of two reaction pairs with a common member, is a *discontinuous reaction series*. It has been shown that when a third component (*e.g.*, anorthite) is added to a binary system which includes a reaction pair (*e.g.*, forsterite → protoenstatite), reaction is no longer restricted to a melt of unique composition and temperature but takes place continuously with melts of changing composition over an appreciable range of temperature. Reaction between magmas and mineral phases, whether the latter constitute continuous or discontinuous reaction series, must be a continuous process operating over considerable intervals of temperature.

One of the most important generalizations emerging from laboratory investigation of silicate melts is the prevalence of the reaction relation be-

---

[1] N. L. Bowen, *The Evolution of the Igneous Rocks*, pp. 54–62, Princeton University Press, Princeton, N.J., 1928.

tween igneous minerals and the melts from which they crystallize.   That this relation holds good for natural magmatic crystallization cannot be doubted.   The very existence of solid-solution series in so many groups of igneous minerals, and the frequent development of reaction rims (coronas) of one mineral around central cores of another, are evidence of reaction.   So, too, is the tendency sometimes shown for a sequence of minerals to begin and to cease crystallizing in the same order, as deduced from textural relationships.   Crystallization of olivine from basaltic magma may be cited as an instance.   It habitually crystallizes early, and hence develops phenocrysts of large size; but it is usually absent from the ground-mass, and the phenocrysts themselves tend to show rounded outlines attributed to magmatic corrosion.   Presumably olivine ceases to crystallize at an early stage and enters into reaction with the enclosing melt.

It is safe to say that crystallization of magmas is strongly influenced by the reaction principle and that cotectic crystallization of two or more phases in variable amount is also a common phenomenon; but simple eutectic crystallization of two or more phases at fixed temperature from a melt of unique composition must indeed be rare, except perhaps in rocks of the granite family.   It follows that differentiation by fractional crystallization must be far more effective than it could be with melts dominated by eutectic crystallization.

From combined experimental and petrographic evidence, Bowen concluded that crystallization of "ordinary subalkaline" magmas is dominated by two independent but simultaneously active reaction series:

1. The essentially discontinuous series, magnesian olivine → pyroxenes → hornblendes → biotites

2. The continuous plagioclase series, bytownite → labradorite → andesine → oligoclase → albite

This has proved to be too simple a generalization.[2]   The sequence of ferromagnesian silicates crystallizing from a given magma commonly suggests simultaneous operation of two rather than a single reaction series, though the two series may converge in their end products.   In many basalts simultaneous early crystallization of bytownite, olivine, and diopsidic augite ushers in the development of three reaction series; later the lineages heralded by olivine and diopside may converge via intermediate pyroxenes to a single series characterized by ferriferous pyroxenes or by hornblendes. In magmas containing appreciable potash, the plagioclase series and an alkali-feldspar series may crystallize side by side, but ultimately the two series tend to converge, since in each the low-temperature member is rich in soda.

[2] Cf. T. F. W. Barth, *Theoretical Petrology*, p. 117, Wiley, New York, 1952; A. Poldervaart and H. H. Hess, Pyroxenes in the crystallization of basaltic magma, *Jour. Geology*, vol. 59, p. 480, 1951.

The course followed as the minerals of two or more series simultaneously react with a magma is subject to variation depending on such factors as the following: continuous reaction within olivine and pyroxene series, involving progressive concentration of iron in the residual melt; the influence of variable concentrations of water on the appearance of hornblende, micas, and zeolites; the relative concentration of iron compared with alkali, achieved through different degrees of fractionation in different reaction series; the state of oxidation of iron and the corresponding value of the partial pressure of oxygen; the influence of relative concentration of alkali, with respect to silica and to alumina, upon crystallization of alkaline minerals such as feldspathoids, aegirine, and riebeckite; the relative abundance of soda and potash resulting from varying fractionation in the feldspar series.   In spite of the wide latitude thus possible for the course of magmatic evolution, the general data of petrography testify to the dominance of a few reaction series in shaping actual magmatic trends.   Herein lies an explanation of the regular tendencies toward development of standard mineral assemblages and standard textural relations which Rosenbusch discerned when he founded his "laws" of order of crystallization.   The innumerable exceptions to such regularity partly reflect variations in magmatic evolution arising from modifying factors such as those enumerated above.

FRACTIONAL CRYSTALLIZATION OF BASALTIC MAGMA

Basalt is the most widely distributed of volcanic rocks.   For this and other reasons, it is believed by many petrologists that basalt magmas constitute the primary material from which other (though not necessarily all other) magmas are derived.   This question will be discussed further in later chapters.   In the meantime, granting that basaltic magma is widespread and that it is a parent material from which other magmas may evolve, it is appropriate to outline its course of fractional crystallization in the light of experimental data.[3]

Olivine, pyroxene, and calcic plagioclase are the first silicate minerals to separate from basaltic magma.   Olivine always crystallizes early. Common presence of sharply idiomorphic large crystals of both olivine and diopsidic augite in chilled glassy basalts of what we shall later term the alkaline basalt magma-type, suggests that these two minerals normally crystallize side by side and are not a reaction pair.   In other basalts— those of what we shall term the tholeiitic magma-type—olivine is closely followed by enstatite or a pigeonitic augite, and the two constitute a reaction pair.   Opinion of petrographers is divided as to whether plagioclase

[3] Cf. Bowen, op. cit., pp. 63–91, 1928; H. S. Yoder and C. E. Tilley, in Annual report of the director of the Geophysical Laboratory, Carnegie Inst. Washington Year Book, no. 55, pp. 169–171, 1956; no. 56, pp. 156–161, 1957.

is normally preceded by pyroxene or vice versa.   Both alternatives seem possible, and the issue hinges largely on how one interprets the ophitic texture of diabases.   From the melting behavior of typical basalts in the laboratory [4] it would seem that, throughout much of the cooling history of basaltic magma, olivine, pyroxene, and calcic plagioclase are simultaneously present.   A general sequence of fractional crystallization, subject to many minor variations, may be stated as follows:

1. In the early stages, when the total quantity of crystalline material is relatively small, sinking of large heavy crystals under gravity is the most effective mechanism of differentiation.   The minerals most generally affected, as borne out by observations on thick basic sills, are olivine and pyroxene.   The residual liquid is thus enriched in silica which may even become concentrated in excess of the amount corresponding to pyroxene-plagioclase mixtures.   There is also an increase in the FeO/MgO ratio, since early olivines and pyroxenes tend to be more magnesian and less ferruginous than the melts with which they are in equilibrium.   Another possibility is concentration of early-formed basic plagioclase crystals (whether by sinking or floating or by some other means) as may have occurred in the development of the "porphyritic central magma" of Mull in Scotland (see page 225).

2. In the later stages of crystallization, when "filter-pressing" of residual liquid from a largely crystalline mass becomes active, the composition of the melt so expelled depends on the degree of fractionation in both the plagioclase and the pyroxene series.   The FeO/MgO ratio, and probably the $(FeO + MgO)/CaO$ ratio, are still rising under the influence of the pyroxene series.   By separation of early-formed plagioclase, on the other hand, the final liquids are being greatly enriched in soda, potash, and silica, and impoverished in alumina and lime.

3. If water becomes sufficiently concentrated in the later stages to allow separation of hornblende and/or biotite, the content of silica in the residue will be further augmented, since both these minerals are basic.

The sequence of mineral assemblages given below is thus compatible with origin as successive crystalline fractions separated during the freezing of basaltic magma:

1. Dominant magnesian olivine, with magnesian and/or diopsidic pyroxene, and basic plagioclase: peridotite and olivine gabbro families

2. Pyroxene and basic plagioclase with either olivine or minor quartz: olivine basalt, quartz basalt, and related plutonic families

3. Hornblende and intermediate plagioclase: andesite and diorite families

4. Hornblende, biotite, sodic plagioclase, potash feldspar, quartz: rhyolite and related plutonic families

[4] Yoder and Tilley, *op. cit.*, p. 157, 1957.

One of the outstanding conclusions reached by Bowen and his coworkers is that andesitic, rhyolitic, and probably trachytic magmas are all *possible* products of fractional crystallization of basaltic magma.   It is in fact possible to explain the mineralogy of almost all common igneous rocks, including such diverse associations as basalt-trachybasalt-trachyte, basalt-andesite-rhyolite, and gabbro-ferrogabbro, in terms of such a process.   Moreover, as Bowen has aptly pointed out, the marked tendency for certain minerals consistently to be associated (*e.g.*, hornblende-andesine, olivine-labradorite, oligoclase-orthoclase) and for others to be mutually antipathetic (*e.g.*, quartz-bytownite, olivine-muscovite, orthoclase-labradorite) seems highly significant in this connection.   The commonly associated minerals may be interpreted as belonging to the same temperature range, so that they accumulate together in the same crystal fraction; the antipathetic minerals, as belonging to widely separated fractions in the crystallization sequence.

It would nevertheless be a mistake to conclude that fractional crystallization of uncontaminated basaltic magma along diverging lines of descent is the principal mechanism responsible for variation among igneous rocks.   Differential fusion of basaltic rocks could give rise to much the same liquid fractions in reverse sequence.   Moreover, the possible effects of assimilative reaction between magmas and solid rocks remain to be considered.   And it must still be borne in mind that even if it can be shown that granitic liquids may evolve from a basaltic parent magma, there is also good evidence suggesting that granitic rocks may develop by fusion or by metasomatic metamorphism of preexisting nonbasaltic rocks. Indeed, the great preponderance of relatively acid rocks and paucity of basic rocks in many granodiorite-granite and andesite-rhyolite provinces renders basaltic parentage most unlikely.

ASSIMILATIVE REACTION BETWEEN BASALTIC MAGMA AND SOLID ROCKS [5]

In Chap. 4 certain principles believed to control assimilative reaction between magmas and solid rock (as typified by enclosed xenoliths) were outlined.   If we consider basaltic magma from which olivine, calcic plagioclase, and pyroxene have already started to separate, the crucial factor which determines the course of assimilation of solid inclusions is the relation of the xenolith minerals to the reaction series established for igneous rocks.   Two commonly occurring cases may be considered:

Reaction between Basaltic Magma and Acid Igneous Rocks.   Rocks such as granite, granodiorite, rhyolite, and dacite are composed principally

[5] Bowen, *op. cit.*, pp. 199–201, 205–215, 220–223, 1928.

of sodic plagioclase, hornblende, biotite, potash feldspar, and quartz. The first three of these lie toward the low-temperature end of the two main reaction series mentioned earlier in this chapter. The relation of quartz and potash feldspar to the other minerals is not so clearly one of reaction; but they certainly crystallize later from magmas which could be, and in some instances probably are, low-temperature residues derived from basaltic magma. Basaltic magma is therefore capable of melting granitic and allied xenoliths. The amount of solid granite that can be incorporated into the liquid portion of the magma depends upon the amount of heat that can be supplied by crystallization of a thermally equivalent quantity of those phases with which the magma is saturated. So there is a definite limit to the extent of assimilation in any case. When assimilation is well advanced, the quantity of crystalline material (largely plagioclase and pyroxene) that has separated from the magma has been greatly increased. The remaining liquid fraction includes much that has been contributed directly from the granite. In composition this liquid resembles the "granitic" residuum that might in any case be expected to develop, though in much smaller quantity, from uncontaminated basaltic magma. Thus the main effect of assimilation upon the liquid line of descent seems to be an increase in the quantity of potential acid differentiate.

The above statement is based upon the general picture of magmatic evolution dominated by reaction series whose behavior may be predicted with some confidence in the light of laboratory observations. Furthermore it assumes that assimilation, in this case melting of granite, has proceeded to completion. For petrographic confirmation of the course of assimilation we must turn to occurrences of basaltic rocks enclosing incompletely assimilated granitic xenoliths in various stages of reaction with the host rock. A clear instance has been described by A. Knopf [6] from Owens Valley, California, where olivine basalt invades granodiorite and encloses numerous xenoliths of it. Glass of rhyolitic composition has developed by direct fusion throughout many of the xenoliths—especially along quartz-orthoclase contacts. Here and there it encloses newly crystallized microlites of hypersthene. The remaining grains of quartz and feldspar clearly show the effects of partial fusion; biotite apparently has suffered little or no melting but, being unstable at the high temperature and low pressure of its new environment, has been replaced largely by magnetite (cf. page 138). Xenocrysts of quartz and of feldspar show reaction rims of pyroxene against the enclosing basalt, and many of the feldspars are partially fused. Apart from the xenocrysts, the mineralogy of the basalt is normal—olivine and augite phenocrysts in a plagioclase-augite-magnetite groundmass. Here petrography is completely in accord

[6] A. Knopf, Partial fusion of granodiorite by intrusive basalt, Owens Valley, California, *Am. Jour. Sci.*, vol. 36, pp. 373–376, 1938.

with the thesis that basaltic magma may assimilate granitic material by direct fusion.

Finally there is experimental evidence obtained by heating powdered granite in the presence of water under pressure, showing that at depths of several kilometers granitic rocks containing a small percentage of water may become largely molten at temperatures of the order of 700°C.[7] Basaltic magmas, even when not superheated, must maintain temperatures considerably greater than this.

Reaction of Basaltic Magma with Sedimentary Rocks. The principal minerals of common sedimentary rocks are quartz, the clay minerals, calcite, and feldspars (mainly potash and soda-lime varieties). Basaltic magma presumably reacts with quartz and feldspar—especially when these are mutually associated in feldspathic sandstones—in much the same manner as that just discussed. Calcite and the clay minerals, though not members of the igneous reaction series, habitually form at low temperatures, and may be expected to play a similar role to that of low-temperature members of reaction series. On theoretical grounds alone, it would seem likely therefore that basaltic magma should be capable of dissolving (melting) considerable quantities of sedimentary material, the necessary heat being supplied by crystallization of minerals from the magma itself.

Bowen argued that incorporation of sedimentary material into the liquid magma need not necessarily lead to ultimate appearance of unusual mineral phases when the contaminated product crystallizes. Lime, alumina, and silica are all major constituents of, and hence must be readily soluble in, basaltic magma. Presumably considerable additional amounts of any one could be accommodated by redistribution of ions among the ionic groups comprising the liquid magma. This might subsequently be reflected only in differences in composition of pyroxene, olivine, and plagioclase, or in the relative proportions of these minerals, as compared with those in the uncontaminated basalt. Bowen also suggested that where sedimentary xenoliths are partially digested by basaltic magma, a low-melting "granitic" fraction, approximately equivalent to a mixture of quartz and alkali feldspar, is first withdrawn into the liquid phase, leaving a recrystallized residue representing the "excess" of silica, alumina, or lime originally present in the sediment. Petrographic observations, as illustrated below, on the whole confirm Bowen's generalizations.

Reaction with Aluminous Xenoliths.[8]   Modern geological literature con-

[7] Cf. R. W. Goranson, Some notes on the melting of granite, *Am. Jour. Sci.*, vol. 23, pp. 227–236, 1932.

[8] H. H. Thomas, in Tertiary and post-Tertiary geology of Mull, Loch Aline and Oban, *Scotland Geol. Survey Mem.*, pp. 274–278, 1924; Bowen, *op. cit.*, pp. 208, 209, 1928; H. H. Read, On corundum-spinel xenoliths in the gabbro of Haddo House, Aberdeenshire, *Geol. Mag.*, vol. 68, pp. 446–453, 1931; J. C. Brice, Geology of Lower Lake Quadrangle, California, *California Division of Mines Bull.*, 166, pp. 42, 43, 1953.

tains many accounts of basic igneous rocks—often noritic in composition—associated with and enclosing aluminous xenoliths probably of sedimentary origin. The xenoliths are extraordinarily rich in alumina and may even contain over 60 per cent Al_2O_3 by weight. Such associations as the following are common: mullite-glass, corundum-sillimanite, corundum-spinel-anorthite. These may be interpreted as representing the aluminous "excess," which remains after material equivalent to granite in composition has been sweated out of the altered shale in the first stages of differential fusion. Bowen suggests that by further reaction this excess alumina ultimately becomes incorporated into the hybrid magma, and that some such regrouping of magmatic material as the following thereby occurs:

$$Al_2O_3 + SiO_2 + CaMgSi_2O_6 \rightarrow CaAl_2Si_2O_8 + MgSiO_3$$

  Xenolith   Magmatic

Accommodation of the additional alumina in the magma could thus increase the magnesia content of crystallizing pyroxene and the anorthite content of the plagioclase. The resultant contaminated rock could thus be a norite with no mineralogical peculiarities obviously suggesting hybrid origin. In other cases, however, and especially where the original xenoliths had a high magnesia as well as alumina content, cordierite and spinel might crystallize directly from the contaminated magma. Cordierite norites are indeed surprisingly widespread.

Reaction with Siliceous Xenoliths.[9] Petrographic literature contains many references to quartz-rich xenoliths enclosed in basalt or in diabase. Convincing evidence of partial fusion is afforded by the corroded outlines of xenoliths of this type and by the development of glass around the margins and as intimately penetrating veinlets. In some the glass appears from its refractive index to be close to rhyolite in composition.[10] In others it is apparently basaltic glass modified to varying degrees by reaction with the adjacent xenolith. Various observers note the direct crystallization of tridymite, augite, hypersthene, cordierite, or plagioclase from glasses associated with partially assimilated quartz xenoliths. Another highly characteristic feature is the development of reaction rims of augite around quartzose xenoliths, even where the latter have been partially converted to glass.

Petrographic data again agree remarkably well with Bowen's generalizations. Material equivalent to granite in composition is converted into the liquid (glassy) state, especially in sandstone and schist xenoliths where cementing materials, alkali feldspars, and white mica are easily available

⁹ Bowen, *op. cit.*, pp. 214–215, 1928; W. N. Benson, The basic igneous rocks of eastern Otago, Pt. IV, Sec. C, *Royal Soc. New Zealand Trans.*, vol. 75, pt. 2, pp. 288–318, 1945; Brice, *op. cit.*, pp. 41–44, 1953.

¹⁰ *E.g.*, Benson, *op. cit.*, pp. 296, 297, 1945.

to supply the "granitic" fraction. Stored up in the xenolith, at least temporarily, is the excess of silica in the form of quartz. The many described instances of quartz xenoliths in basaltic rocks testify to the refractory nature of pure quartz even in a basaltic environment. Bowen has suggested, however, that even pure quartz is capable of being incorporated into the liquid phase of the magma, by reaction involving such internal grouping of ions or "molecules" as the following:

$$SiO_2 + Mg_2SiO_4 \rightarrow 2MgSiO_3$$
$$SiO_2 + Al_2O_3 + CaMgSi_2O_6 \rightarrow CaAl_2Si_2O_8 + MgSiO_3$$

The common occurrence of hypersthene in the products of reaction between quartz and basalt seems significant in this connection.

Reaction with Limestone Xenoliths.[11]   In 1910 Daly put forward his now classic hypothesis that by assimilation of limestone normal basaltic magma may become desilicated by ensuing crystallization of various lime silicates such as melilite and garnet, and that silica-poor alkaline residual magmas may then develop by differentiation.   That such reactions may be possible under natural conditions is suggested, but not proved, by the association of melilite basalts and alkaline lavas and by the occurrence of lime garnets in various alkaline rocks.   On the other hand Bowen argued that basaltic magma may absorb much lime without causing any special lime-rich minerals to crystallize.   As an alternative result he envisaged increased precipitation of the silica-poor minerals olivine and magnetite and crystallization of pyroxenes notably higher in lime than would otherwise have been the case.   Possible adjustments leading to such results were suggested:

$$CaO + 3MgSiO_3 \rightarrow CaMgSi_2O_6 + Mg_2SiO_4$$
$$CaO + 2FeSiO_3 + Fe_2O_3 \rightarrow CaFeSi_2O_6 + Fe_3O_4$$
$$CaO + 4MgSiO_3 + Al_2O_3 \rightarrow CaAl_2Si_2O_8 + 2Mg_2SiO_4$$

There is a good deal of experimental evidence indicating that precipitation of melilite is favored not so much by a lime-rich as by an alkaline silica-deficient composition of the magma in question.   Melilite reaction rims around augite and olivine crystals immersed in an alkaline base are well known in alnoites and allied rocks, but development of melilite rims around limestone xenoliths in normal basic magmas has yet to be demonstrated.

[11] Bowen, *op. cit.*, pp. 206–207, 1928; H. A. Brouwer, The association of alkali rocks and metamorphic limestone in a block ejected by the volcano Merapi (Java), *K. Akad. Wetensch. Amsterdam Verh.*, vol. 48, pp. 166–189, 1946; C. E. Tilley, The gabbro-limestone contact zone of Camas Mor, Muck, Inverness-shire, *Comm. géol. Finlande Bull. 140,* pp. 97–104, 1948; Some trends of basaltic magma in limestone syntexis, *Am. Jour. Sci.,* Bowen vol., pp. 529–545, 1952.

Tilley has described specially interesting examples of assimilative reaction between basic magma and limestone at Scawt Hill in Ireland and on the Scottish island of Muck. In each occurrence the limestone has been metamorphosed to lime-silicate assemblages in which melilite is prominent. The magma in immediate contact with limestone has been enriched in diopsidic pyroxene to give pyroxenite as the ultimate product. The reaction given by Tilley [12] is

$$3NaAlSi_3O_8 + 2Mg_2SiO_4 + 4CaO \rightarrow 4CaMgSi_2O_6 + 3NaAlSiO_4$$

Albite   Olivine   Diopside   Nepheline

Magmatic

With concentration of nepheline in the residual liquid and further assimilation of lime, a sequence of contamination products develop, including nepheline-diabases and theralites, both commonly containing melilite, and some with wollastonite as well. These at first sight present some analogies with the well-known basic volcanic association of nephelinite, melilite-nephelinite, nepheline-basanite, etc. But the superficial resemblance is outweighed by significant differences such as a general absence of magnesian olivine, a common association of plagioclase with melilite, and a marked enrichment of late liquid fractions in iron, in the products of basalt-limestone syntexis. So Tilley [13] concludes that while nepheline, melilite, and wollastonite may appear in basaltic rocks contaminated by reaction with limestone, their presence in alkaline basic rocks of the nepheline-basalt family is to be attributed to some other completely independent petrogenic process.

A particularly clear instance of reaction between semibasic magma and limestone with production of lime silicates and feldspathoids has been described by Brouwer from the volcano Merapi in Java. Blocks of limestone immersed in pyroxene-andesite magma have been converted to wollastonite, plagioclase, melilite, augite, and grossularite in various combinations. They are streaked on a microscopic scale with veinlets of glassy leucite phonolite and trachyte and with aggregates of leucite-plagioclase-augite, leucite-orthoclase-augite, leucite-augite-wollastonite, etc.

These occurrences show beyond doubt that under some conditions basic magmas by reaction with limestone precipitate lime silicates such as augite, melilite, and wollastonite and thereby become sufficiently desilicated to cause feldspathoids to crystallize. Whether such conditions are commonly realized remains to be proved. And whether reactions of this kind have any bearing on the genesis of nepheline basalts, melilite basalts, and related rocks is doubtful. In view of the great geological interest roused

[12] Tilley, op. cit., p. 538, 1952.
[13] Tilley, op. cit., p. 540, 1952.

by Daly's hypothesis almost half a century ago, and remembering the wide distribution of basalts and limestones on the earth's surface, we may well be surprised that described instances of such reaction between basaltic magma and limestone are so few.

MELTING OF BASIC ROCKS

In current theories of petrogenesis primary magmas of several kinds— granitic, granodioritic and basaltic—have been variously explained as products of partial or complete fusion of initially solid ultrabasic or basic rocks in the deeper parts of the crust or in the earth's mantle itself (cf. Chap. 15). Great interest centers, therefore, round some preliminary experiments by Yoder and Tilley [14] on the melting of common basic rocks in the presence of water. Their observations on a tholeiitic basalt with 9 per

TABLE 2. MELTING BEHAVIOR OF THOLEIITIC BASALT *

Temperature, °C.	$P_{H_2O} = 1$ bar	$P_{H_2O} = 5,000$ bars	Temperature, °C.
750	↑ Plagioclase, pyroxene, olivine, iron-ore ↓	Melting begins: Hornblende, plagioclase, sphene, magnetite, minor glass	750–800
		Hornblende, sphene, magnetite, glass	815
		Hornblende, magnetite, glass	875
		Pyroxene, hornblende, magnetite, glass	925
		Pyroxene, olivine, magnetite, glass	975
1090	Melting begins: Plagioclase, pyroxene, olivine, iron-ore, glass	Olivine, magnetite, glass	1090
		Melting complete	1125
1170	Pyroxene, olivine, iron-ore, glass		
1190	Olivine, iron-ore, glass		
1235	Melting complete		

* H. S. Yoder and C. E. Tilley, in Annual report of the director of the Geophysical Laboratory, Carnegie Inst. Washington Year Book, no. 55, pp. 169–171, 1956.

[14] Yoder and Tilley, op. cit., pp. 169–171, 1956.

cent normative olivine, from Kilauea, Hawaii, are summarized in Table 2. The basalt contains less than 0.1 per cent water. Its melting range at atmospheric pressure, 1090° to 1235°C., agrees with temperatures observed at the erupting volcano. At a water pressure of 5,000 bars, such as might develop at depths of a dozen miles, the initial melting temperature has dropped some 300°C. and is only about 150°C. higher than that at which granite begins to melt at the same pressure. But, as contrasted with granite, the plutonic melting range of basalt is very broad—some 350°C. between the point at which glass is first observed and that at which the last silicate, olivine, disappears.

The melting behavior of basalt under deep plutonic conditions (5,000 bars) brings out a possible origin for an association of rocks well known in regionally developed migmatite complexes—namely, amphibolite and oligoclase granite. Such an association—solid amphibolite and "granitic" melt—would be stable at 5,000 bars water pressure and temperatures around 800°C. Pyroxene gabbros and diabases, on the other hand, represent crystallization of basic magma at low water pressures; this implies either a very low water content in the parent magma, or crystallization at shallow depths. Otherwise the final product should be hornblende gabbro or amphibolite.

CRYSTALLIZATION BEHAVIOR OF GRANITIC MAGMA

Since granites and granodiorites are by far the most abundant and widely distributed plutonic rocks, it is appropriate to review the behavior of crystallizing granitic magma in the light of experimental data. The common minerals of granitic rocks—quartz, sodic plagioclase, potash feldspar, hornblende, and micas—cannot be arranged in simple reaction series based on laboratory observations. The sequence andesine → oligoclase → albite is exceptional in that it has been confirmed by experiment. There is also petrographic evidence that hornblende at times reacts with acid magma to give biotite. Otherwise the sequence of crystallization, and hence the trend of fractionation, depends on such variables as the ratio of potash to soda in the magma (itself likely to be affected by the degree of previous fractionation in the feldspar series), the relative concentration of silica, and especially the water content of the magma, which in turn is conditioned by pressure and hence by depth. This latter factor not only influences the crystallization of hydroxyl-bearing minerals like hornblende and biotite, but assumes great importance in the closing stages of magmatic crystallization when the concentration of water may reach the upper limit possible at a given pressure, so that an aqueous fluid phase then separates from the silicate melt. The various possibilities which are then opened up will be considered later in connection with the genesis and evolution of pegmatites.   Our ideas of the course of fractional crystallization of granitic melts come more from petrography than from experiment and therefore will not be reviewed at this point.

Concerning assimilative reaction between granitic magma and solid rocks, it is possible to generalize in the light of laboratory observations as well as petrography.   Whether the origin of granitic magma is attributed to differentiation (fractional crystallization) or to differential fusion of pre-existing rocks, it is most unlikely that large bodies of granitic melts ever become superheated above the upper limit of crystallization temperatures. The very origin of the magma would generally preclude this possibility. Following Bowen,[15] we may conclude that granitic magma cannot liquefy basic or semibasic igneous rocks, for these are composed of minerals higher in the igneous reaction series than minerals crystallizing from granitic melts.   It is possible that some fraction of water-saturated mixed sediments might be capable of direct solution in granitic magma, but this fraction would itself be essentially granitic in composition.

In general, granitic magma may be expected to react with solid rocks of almost any kind, thereby converting them to appropriate mixtures of those mineral phases with which the magma is already saturated, namely, sodic plagioclase, biotite, hornblende, or potash feldspar.   Any heat necessary for such reactions will be supplied by equivalent crystallization of similar minerals from the melt.   Many such cases have been described, and there is general agreement with Bowen's conception of assimilative reaction. The mechanism by which the ions concerned are exchanged between xenolith and magma is somewhat obscure.   In the absence of fusion it is remarkable how completely large xenoliths enclosed in granite have been converted to an assemblage of granitic minerals.   It has been suggested by Nockolds [16] that the necessary ionic exchange is effected through the medium of a mobile aqueous fluid (perhaps a simple ionic solution of such compounds as sodium and potassium silicates), which is assumed to separate from the magma and penetrate the xenolith freely.   It is difficult to envisage such a mechanism clearly.   But it is worth noting that hydroxyl-bearing minerals such as biotite are among the most conspicuous products of reaction, and that high concentrations of apatite, or less commonly calcite, have been recorded in granitized xenoliths.   Moreover it may be significant that, in spite of the relatively low temperatures involved, effects of assimilative reaction are generally more striking and more extensively developed at the margins of granitic intrusions than at contacts of basic plutonic bodies, and it is in granitic magma that water tends especially to become concentrated as crystallization advances.   A

[15] Bowen, *op. cit.*, pp. 197, 198, 215, 216, 1928.

[16] S. R. Nockolds, Some theoretical aspects of contamination in acid magmas, *Jour. Geology*, vol. 41, pp. 561–589, 1933.

possible example of extensive reaction between rhyolite lava and solid basalt at Yellowstone Park has been described by Fenner,[17] who attributes the observed exchange of material, whereby basalt has become greatly modified, to gaseous emanations coming from the rhyolite lava. Whether actual solution of basalt by the acid magma has occurred, as is claimed by Fenner, is doubtful; for there is an alternative possibility that the observed phenomena are due to mixing of rhyolite and basalt lavas.

Consider now the specific case of reaction between biotite-granite magma and basic rock such as gabbro or amphibolite.[18] Within the basic rock the initially calcic plagioclase becomes converted to oligoclase. Ferromagnesian silicates—olivine, augite, or hornblende—show progressive changes parallel to the normal igneous reaction sequence, namely, olivine → pyroxene → hornblende → biotite. Completely granitized xenoliths are usually rocks of dioritic aspect consisting of oligoclase, biotite, or hornblende, and variable amounts of introduced potash feldspar and quartz. They may be rich in apatite, and in sphene formed at the expense of original ilmenite. Marginally the completely granitized xenoliths tend to disintegrate, and the xenocrysts and microaggregates of biotite and plagioclase so formed then become strewn through the adjacent contaminated granite, which in consequence may be darker in color than the uncontaminated granite. Reciprocal reaction may also affect noticeably the nature and relative proportions of the minerals separating from the liquid phase. For example, diffusion of lime from the xenoliths into the liquid may cause hornblende to crystallize rather than biotite; impoverishment of the liquid in potash, which has been steadily transferred into the xenolith to form biotite, may lead to failure of potash feldspar in the marginal variant of the granite. The ultimate products of assimilative reaction range from dark rocks of dioritic composition to what seem like normal granites. It may be difficult to decide in any given case what minerals have crystallized directly from a liquid phase, and what rocks correspond to magmas that at some stage have been largely molten. Petrographic, chemical, and field evidence very strongly suggests that rocks of dioritic and grano-

[17] C. N. Fenner, Contact relations between rhyolite and basalt of Gardiner River, Yellowstone Park, *Geol. Soc. America Bull.*, vol. 49, pp. 1441–1484, 1938; R. E. Wilcox, The rhyolite-basalt complex on Gardiner River, Yellowstone Park, Wyoming, *Geol. Soc. America Bull.*, vol. 55, pp. 1047–1079, 1944 (and discussion by C. N. Fenner, pp. 1081–1096).

[18] Cf. H. H. Thomas and W. C. Smith, Xenoliths of igneous origin in the Trégastel-Ploumanac'h granite, *Geol. Soc. London Quart. Jour.*, vol. 88, pp. 274–296, 1932; F. J. Turner, The metamorphic and plutonic rocks of Lake Manapouri, *Royal Soc. New Zealand Trans.*, vol. 67, pt. 1, pp. 83–100, 1937; S. R. Nockolds, Contribution to the petrology of Barnavave, Carlingford. I. The junction hybrids, *Geol. Mag.*, vol. 72, pp. 289–315, 1935; M. Härme and M. Laitala, An example of granitization, *Comm. géol. Finlande Bull.*, 168, pp. 95–98, 1955; R. R. Compton, Trondhjemite batholith near Bidwell Bar, California, *Geol. Soc. America Bull.*, vol. 66, pp. 36, 37, 1955.

dioritic composition may form by contamination of granitic magma during reaction with basic rocks. Such a result accords with the general principles deduced by Bowen from the behavior of silicates in contact with melts under controlled conditions. It has even been argued—though this is perhaps an extreme view—that all diorites and granodiorites are contaminated products of reactions of this type.[19]

Reaction between acid magmas and sedimentary rocks, or their metamorphosed equivalents, likewise leads to a parallel development of minerals in the xenolith and in the contaminated magma.[20] Shale or pelitic hornfels becomes converted to an aggregate of micas, alkali feldspars, and quartz, among which unaltered aluminous minerals such as andalusite and cordierite (corundum or spinel in silica-deficient rocks) may persist. The contaminated granites formed by reaction with shale tend to be enriched in biotite and may even contain minor amounts of garnet and cordierite. In part these minerals have been derived from disintegrating xenoliths, but in some rocks they appear to have crystallized also from the liquid phase in the contaminated zone. Experimental data in the system $K_2O\text{-}Al_2O_3\text{-}SiO_2$ show that granitic magmas, even if superheated by 200°C., could not dissolve more than a small excess of Al_2O_3.[21]

There are relatively few records of reaction between acid magmas and siliceous xenoliths. This is readily understood for rocks of strictly granitic composition. These tend to approximate mixtures of quartz and alkali feldspar (around Qu_{40}, Or_{30}, Ab_{30}) that melt at minimum temperatures.[22] Melting temperatures rise very steeply on the silica side of the quartz-feldspar cotectic line in the corresponding diagram; unless liquids of this composition were greatly superheated they could dissolve little additional quartz. Somewhat less siliceous magmas in the granodiorite quartz-monzonite range could, on the other hand, dissolve considerable amounts of silica if the necessary heat were provided by crystallization of some other phase. The Ballachulish granodiorite of Scotland [23] furnishes an interesting example of this kind of reaction. The granodiorite consists of andesine, biotite, hornblende, and subordinate microcline and quartz. The xenoliths show corroded outlines indicating marginal melting, and

[19] S. R. Nockolds, The production of normal rock types by contamination and their bearing on petrogenesis, Geol. Mag., vol. 71, pp. 31–39, 1934.

[20] Cf. A. Brammall and H. F. Harwood, The Dartmoor granites, Geol. Soc. London Quart. Jour., vol. 88, pp. 171–237, 1932; S. R. Nockolds, The contaminated tonalite of Loch Awe, Argyll, Geol. Soc. London Quart. Jour., vol. 90, pp. 302–321, 1934.

[21] J. F. Schairer and N. L. Bowen, The system $K_2O\text{-}Al_2O_3\text{-}SiO_2$, Am. Jour. Sci., vol. 253, p. 741, 1955.

[22] Cf. F. Chayes, in Annual report of the director of the Geophysical Laboratory, Carnegie Inst. Washington Year Book, no. 55, 1955–1956, pp. 212, 213, 1956.

[23] I. D. Muir, Quartzite xenoliths from the Ballachulish granodiorite, Geol. Mag., vol. 90, pp. 409–428, 1953.

some have been partially replaced by microcline and sodic plagioclase. Surrounding any typical xenolith are three zones composed of materials precipitated from the reacting magma: an inner corona consisting of augite and oligoclase, an outer corona of hornblende and more calcic oligoclase, and a marginal zone of contaminated granodiorite enriched in quartz and somewhat depleted in dark constituents. Augite is not one of the ferromagnesian phases that was being precipitated from the granodiorite magma at the time it encountered the xenoliths. Its appearance as an alternative to hornblende perhaps reflects some local disturbance of otherwise prevalent conditions related to water content of the magma.

Reaction between granitic magmas and calcareous rocks may result in replacement of the latter by lime-bearing silicates such as diopsidic pyroxene, hornblende, plagioclase, grossularite, epidote minerals, and even wollastonite. Some of these may crystallize from the adjoining magma as well.[24] Two patterns of reaction and magmatic contamination are well known. The one is characterized by dioritic and the other by alkali-enriched granitic or syenitic contamination products in the zone of reaction.

The first pattern is illustrated by the development of zoned skarns successively replacing one another by inward encroachment upon marble xenoliths enclosed in granodiorite at Ballynacarrick in northwestern Ireland.[25] The marble xenoliths are enclosed in skarn zones characterized by the following mineral assemblages, in sequence from within outward:

1. Grossularite, clinozoisite, quartz, minor diopside
2. Clinozoisite, quartz; minor diopside and grossularite
3. Oligoclase, quartz; minor diopside, microcline, clinosoisite, and sphene
4. Oligoclase, quartz, microcline, hornblende; minor biotite, sphene, diopside, and clinozoisite

Zone 4 has the mineral composition of hornblende granodiorite and, as biotite progressively replaces hornblende, merges into the enclosing biotite granodiorite.

The second pattern of reaction is characterized by migration of silica and alumina into the xenolith, leaving the magma depleted in these components and correspondingly enriched in alkalis.[26] Limestone or dolo-

[24] Cf. Nockolds, op. cit., Geol. Soc. London Quart. Jour., pp. 313, 314, 319, 320, 1934.

[25] A. R. Gindy, Progressive replacement of limestone inclusions in granite at Ballynacarrick, Co. Donegal, Geol. Mag., vol. 90, pp. 152–158, 1953.

[26] C. E. Tilley, An alkali facies of granite at granite-dolomite contacts in Skye, Geol. Mag., vol. 86, pp. 81–93, 1949; op. cit., pp. 542–544, 1952; S. R. Nockolds, On the occurence of neptunite and eudialite in quartz-bearing syenites from Barnavave, Carlingford, Ireland, Mineralog. Mag., vol. 29, pp. 27–33, 1950; I. D. Muir, A local potassic modification of the Ballachulish granodiorite, Geol. Mag., vol. 90, pp. 182–192, 1953.

mite xenoliths are replaced by zoned skarn assemblages of calc-silicates such as grossularite, vesuvianite, diopside, and wollastonite. The contaminated granite consists largely of alkali feldspars, clinopyroxene and quartz in various proportions. In some cases soda is withdrawn from the magma to form plagioclase in the xenoliths; the granite may then become modified to a highly potassic augite syenite in the reaction zone. But where soda as well as potash builds up in the zone of contamination, the final product may be an alkali granite containing aegirine-augite or aegirine-hedenbergite.

It was suggested by Shand and by Daly that reaction between granite magma and limestone leaves the magma so depleted in silica that feldspathoidal differentiates could ultimately form from the contaminated magma. There is no conclusive experimental evidence on this subject. Several cases have been recorded where nepheline syenites appear to have formed as local variants of granite in the vicinity of limestone. But in view of the widespread distribution of both granites and limestones and of the great interest aroused among geologists by the speculations of Daly and Shand, much more numerous instances would be expected if this were a normal petrogenic process. Development of dioritic, granodioritic, or mildly alkaline syenitic or pyroxene-granite hybrids at granite-limestone contacts is a more familiar phenomenon.

DISTRIBUTION OF TRACE ELEMENTS AMONG DIFFERENTIATED FRACTIONS OF MAGMAS

There is a growing volume of information relating to the distribution of trace elements in series of igneous rocks.[27] Attention has been concentrated upon rock series generally believed to be products of fractional crystallization of a common parent magma. Experimental data on the behavior of trace elements during laboratory crystallization of silicate melts comparable to magmas are virtually lacking. Nevertheless a good deal is known regarding the more general problem of how minor elements distribute themselves among associated phases—solid and liquid—in a

[27] L. R. Wager and R. L. Mitchell, The distribution of Cr, V, Ni, Co and Cu during the fractional crystallization of a basic magma, Internat. Geol. Congr., 18th, London, 1948, Rept., Pt. ii, pp. 140–150, 1950; The distribution of trace elements during strong fractionation of basic magma, Geochim. et Cosmochim. Acta, vol. 1, pp. 129–208, 1951; Trace elements in a suite of Hawaiian lavas, Geochim. et Cosmochim. Acta, vol. 3, pp. 217–223, 1953; S. R. Nockolds and R. Allen, The geochemistry of some igneous rock series, Geochim. et Cosmochim. Acta, vol. 4, pp. 105–142, 1953; vol. 5, pp. 245–285, 1954; A. E. Ringwood, The principles governing trace element distribution during magmatic crystallization, Geochim. et Cosmochim. Acta, vol. 7, pp. 189–202, 242–254, 1955.

condition of equilibrium. This information has been discussed from a petrological standpoint.[28]

Suppose that a trace element, say barium, is introduced into a silicate melt from which anorthite is crystallizing. The liquid and solid phases may both be considered, for the sake of simplicity, as solutions of Ba-feldspar (celsian) in Ca-feldspar; as the concentration of Ba is very small, both solutions will further be assumed to be ideal. Equilibrium between liquid and solid requires that the chemical potential of celsian be the same in both phases; then, by (2–38a)

$$\mu_l^0 + RT \ln N_l = \mu_s^0 + RT \ln N_s$$

or

$$\frac{N_s}{N_l} = e^{\mu_l^0 - \mu_s^0 / RT}$$

where N_l and N_s are, respectively, the molar fractions of celsian in the liquid and solid phases, and μ_l^0 and μ_s^0 are the chemical potentials of pure liquid celsian and pure solid celsian at temperature T.   The ratio N_s/N_l is the distribution coefficient, which is independent of concentration (μ_l^0 and μ_s^0 being constants, at given P and T), but which depends markedly on temperature. This relation is exactly the same as that between the compositions along the liquidus and solidus curves in any ideal system, for instance, albite-anorthite, or forsterite-fayalite.

The distribution between two solid phases 1 and 2 (e.g., anorthite 1-pyroxene 2) crystallizing simultaneously would be obtained by the same reasoning:

$$N_1/N_2 = e^{\mu_2^0 - \mu_1^0 / RT}$$

where μ_1^0 and μ_2^0 are, respectively the chemical potentials of pure solid celsian and pure Ba-pyroxene at temperature T. The fact that Ba-pyroxene is unstable will be reflected in its relatively high chemical potential which, in turn, will lead to a high ratio N_1/N_2; we thus expect Ba to be concentrated preferentially in anorthite. The assumption that all solutions present in the system are ideal will, in general, prove incorrect; activities must thus be substituted for molar fractions, i.e., molar fractions must be multiplied by corresponding activity coefficients, which are difficult to determine experimentally or to predict theoretically. A large ac-

[28] V. M. Goldschmidt, The principles of distribution of chemical elements in minerals and rocks, Chem. Soc. London Jour., pp. 655–673, 1937; D. M. Shaw, The camouflage principle and trace-element distribution in magmatic minerals, Jour. Geology, vol. 61, pp. 142–152, 1953; H. Neumann, J. Mead, and C. J. Vitaliano, Trace element variation during fractional crystallization as calculated from the distribution law, Geochim. et Cosmochim. Acta, vol. 6, pp. 90–99, 1954; B. Mason, Principles of Geochemistry, pp. 129–138, Wiley, New York, 1958.

tivity coefficient corresponds to a low solubility; such coefficients are thus determined by the same factors which govern solid solubility and isomorphous replacements (ionic radii, charges, lattice structure of host phase, etc.).

It is noteworthy that even when the trace element forms ideal solutions, crystallization at constant temperature [29] (eutectic, for instance) will cause a change in concentration of the trace element in the crystallizing phase as crystallization proceeds,[30] if the distribution coefficient differs from unity. Consider, for instance, a system in which the trace element is initially present in the melt at a concentration of 10 p.p.m., its distribution coefficient being 5 in favor of the solid phase or phases. The first crystals to form must thus have a concentration of 50 p.p.m. Crystallization thus depletes the melt, in which concentration falls as crystallization proceeds; concentration in the solid phase, which must always be 5 times that in the melt, also decreases. For instance, assuming that equilibrium between crystals and liquid is continuously maintained, when 50 per cent of the liquid has solidified, the concentration y in the melt will be such that

$$0.5y + 0.5\,(5y) = 10 \text{ p.p.m.}$$

or

$$y = 3.33 \text{ p.p.m.}$$

the corresponding concentration in the solid phases being 16.67 p.p.m. The last liquid to disappear would have a concentration of 2 p.p.m., that of the solid phases at that stage being, of course, 10 p.p.m. Fractional crystallization could, therefore, lead to the separation of phases enriched with respect to the trace element. As the distribution coefficients will be different for different phases, and we will vary with temperature and pressure, it would seem unlikely that the distribution of trace elements in differentiated igneous rock series of like parentage should consistently conform to a common pattern. Obviously considerable differences are to be expected between respective distribution patterns of trace elements in plutonic and in volcanic rock series of similar parentage.

CONCLUDING STATEMENT

From the observed behavior of feldspars, feldspathoids, pyroxenes, olivines, free silica, and less important igneous minerals (spinels, hematite, melilites, sphene) as they crystallize from melts of known composition

---

[29] Strictly speaking, this is impossible, for the trace element is an additional component; and as no additional phases form, the variance increases by one unit. Crystallization must now extend over a temperature interval which, however, will be very small if the additional component is present only as traces.

[30] Neumann, Mead, and Vitaliano, *op. cit.*, pp. 90–91, 1954.

under controlled conditions, which is also supplemented by experimental data relating to the stability of micas and amphiboles at magmatic temperatures and pressures, certain generalizations have been drawn up regarding the crystallization of basaltic and granitic magma and the manner in which such magmas may be expected to react with solid rocks. The validity of these generalizations in specific instances has been confirmed by the petrographic, chemical, and field characters of some igneous rocks.   The extent to which the same principles apply to origin and evolution of magmas in general can be estimated only by examining the characteristic features and relationships of the various kindreds or suites of igneous rocks.   This is attempted in the chapters which now follow.

CHAPTER 7

The Alkaline Olivine-Basalt
Volcanic Association

INTRODUCTION: VOLCANIC ASSOCIATIONS IN GENERAL

We follow Kennedy [1] in defining a volcanic association as including all associated igneous rocks, intrusive as well as strictly volcanic, that are genetically related to a cycle of volcanic activity. Contrary to Kennedy's usage we exclude from the volcanic category certain associations of basic rocks which have only subordinate volcanic members and are made up largely of plutonic rocks.

The volcanic associations treated in this book are as follows:

1. The alkaline olivine-basalt association, widely developed in continents and in ocean basins

2. The tholeiitic basalt quartz-diabase association, characteristically found in a nonorogenic continental environment, but also known to occur in oceanic islands

3. The leucite-basalt potash-trachybasalt association of nongeosynclinal continental regions

4. The spilite keratophyre association of geosynclines

5. The basalt andesite rhyolite association characteristic of orogenic belts

Most associations conform to one or another of these types, but some are of a transitional nature. And there are petrographic provinces such as those of Hawaii and Montana in which more than one association, each localized in space and in time, make up a complex geographic and chronological sequence. Some associations arc volcanic and others are plutonic, but there are also provinces in which a volcanic phase precedes or follows plutonic activity.

[1] W. Q. Kennedy, Crustal layers and the origin of magmas: petrological aspects of the problem, *Bull. volcanologique*, sér. 2, tome 3, pp. 24–41, 1938.

The order of treatment here adopted is necessarily arbitrary. It is merely a convenient grouping, for descriptive purposes, of phenomena that cannot be defined precisely nor simply classified. Volcanic associations cannot be arranged in any generally applicable genetic sequence. It has yet to be demonstrated—though it may well be true—that any one association is normally derived from another. But all volcanic associations have this in common: one of the essential members—and usually the predominant member—is basalt. This is one of the main reasons for the prevalent view that in many igneous rock kindreds basaltic magma plays a parent role. It is therefore appropriate to start a survey of volcanic associations by reviewing the field occurrence, petrography, and chemistry of an association of world-wide distribution dominated by basaltic rocks. This is the association of alkaline olivine basalts, nepheline and melilite basalts, trachytes, and phonolites. The islands of the deep ocean basins constitute one of its typical environments. But it is also widely developed on every continent.

THE ALKALINE OLIVINE-BASALT ASSOCIATION ON THE CONTINENTS

The Pliocene Volcanic Association of Eastern Otago, New Zealand.[2] As a type example of the olivine-basalt trachyte phonolite series in a continental environment, the east Otago Pliocene petrographic province has been selected. The region is situated at lat. 46° S. and long. 170½° E. and covers about 2,000 square miles. Throughout this area during late Tertiary (approximately Pliocene) times, hundreds of flows of varied composition, together with locally important pyroclastics, were erupted from numerous centers. Intrusive rocks—sills, small laccoliths, plugs, dikes—are of minor extent compared with the dominant lava flows. Postvolcanic erosion has thoroughly dissected the complex, cutting deep into the prevolcanic basement, which everywhere consists of highly siliceous quartz-albite-epidote-muscovite schist, partly mantled (especially toward the east) by dominantly quartzose sediments ranging in age from Cretaceous to mid-Tertiary.

Three decades of intensive mapping by W. N. Benson, following a preliminary account of P. Marshall,[3] have revealed to an unusually complete

[2] W. N. Benson, Cainozoic petrographic provinces in New Zealand and their residual magmas, *Am. Jour. Sci.*, vol. 239, pp. 537–552, 1941; The basic igneous rocks of eastern Otago and their tectonic environment, *Royal Soc. New Zealand Trans.*, vol. 71, pt. 3, pp. 208–222, 1941; vol. 72, pt. 1, pp. 85–100, pt. 2, pp. 160–185, 1942; D. A. Brown, The geology of Siberia Hill and Mt. Dasher, North Otago, *Royal Soc. New Zealand Trans.*, vol. 83, pt. 2, pp. 347–372, 1955.

[3] P. Marshall, Geology of Dunedin (New Zealand), *Geol. Soc. London Quart. Jour.*, vol. 62, pp. 381–424, 1906.

degree the volcanic and related tectonic history of the province.   Benson recognized two petrographically and tectonically distinct subprovinces, together with a third peripheral region beyond the bounds of the main province and almost devoid of igneous rocks (cf. Fig. 26).   The main characters of all three regions are summarized below.

The Relatively Stable Peripheral Region.   Outside the two main sub-

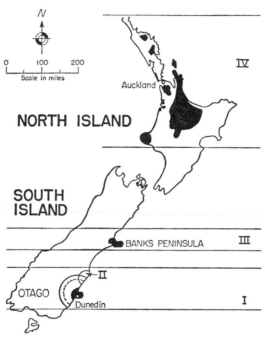

I. East Otago province (central subprovince in solid black).
II. Northeast Otago (mid-Tertiary) province.
III. Banks Peninsula province.
IV. Auckland province.

FIG. 26.   Distribution of late Tertiary and Quarternary volcanic rocks in New Zealand.   (*After W. N. Benson.*)

provinces is a peripheral region of about 1,500 square miles, within which deformation during Pliocene-Pleistocene times was confined to local broad warping with little faulting.   Igneous rocks are represented by insignificant remnants of olivine-basalt flows and dikes.

The Moderately Deformed Subprovince.   This is an arcuate area of 1,900 square miles constituting nine-tenths of the east Otago province proper (cf. Fig. 26).   Here Pliocene and later deformation has been more intense than in the peripheral region and "is characterized by much narrower fault-folds and more closely spaced faults."[4] The total area of volcanic rocks today is 43 square miles, but at one time it may have been several times greater. At any one locality there are never more than four flows in a sequence, and the observed thickness of the volcanic section never exceeds 300 ft. The commonest rocks are olivine basalts with diopsidic augite both as phenocrysts and in the groundmass; some flows contain half-digested inclusions of quartzose schist or xenolithic nodules of picotite-bearing lherzolite.[5] Less abundant, but still widely distributed, are augite-rich basalts (ankaramites), zeolite-bearing basalts, and nepheline-poor basanites, in some of which traces of brown hornblende almost completely replaced by augite and iron oxides have been noted. Nepheline-rich very basic rocks (nephelinites, theralites), limburgites, potash trachybasalts, and andesine basalts (mugearites) are rare. The range of chemical composition may be seen in Table 3.

The parent magma was silica-poor somewhat alkaline olivine basalt (Table 3, analyses 1 and 7) close to Daly's average basalt of Saint Helena (Table 11, analysis 1). Differentiates of the andesine basalt (mugearite) type could well be formed by removal of diopsidic augite (about 20 per cent) and olivine about (10 per cent) from the parent basalt. But liquids at the same time so low in silica and high in alkali as the nepheline-bearing rocks (Table 3, analyses 2, 3, and 4) cannot be accounted for by simple withdrawal of any of the early-formed minerals (labradorite, olivine, augite) in any combination. There is a vague indication that concentration of ionic groupings of the feldspathoid type in basic magmas may be correlated with concentration of "volatile" constituents, for the P_2O_5 content of nephelinic rocks is very much higher than that of the olivine basalts.

A clear instance of the manner in which the basic alkaline magma, once formed, itself undergoes differentiation is afforded by a gravitationally differentiated sill of theralite $\frac{1}{2}$ mile in diameter and 50 ft. in thickness.[6] Early-formed minerals here are bytownite, olivine (starting with the distinctly iron-rich composition Fa_{33}), and diopsidic titanaugite rather rich in iron (molecular ratio FeO/MgO = about 0.4). From the residual liquid which became concentrated especially in the upper part of the sill crystallized iron-rich olivine (Fa_{50} to Fa_{65}), titanaugite enriched further in iron and bordered by aegirine-augite, andesine, anorthoclase and very abundant nepheline. Deuteric alteration of olivine to iddingsite, and of nepheline and anorthoclase to zeolite, testify to the abundance of water in the original magma and to the tendency for iron and alkalis to maintain high concentration even in the aqueous residues of differentiation. The pos-

[4] Benson, *op. cit.*, p. 86, 1942.
[5] Brown, *op. cit.*, pp. 366–368, 1955.
[6] Benson, *op. cit.*, pp. 160–185, 1942.

TABLE 3.   COMPOSITION OF VOLCANIC ROCKS, MODERATELY DEFORMED SUBPROVINCE,
EAST OTAGO, NEW ZEALAND

Constituent	1	2	3	4	5	6	7
SiO_2	47.1	43.03	40.95	43.14	48.34	49.0	46.16
TiO_2	2.3	2.66	1.42	2.04	2.20	1.84	2.45
Al_2O_3	14.7	13.82	14.62	17.77	16.14	18.1	15.32
Fe_2O_3	3.5	3.70	4.87	2.38	3.43	3.9	3.50
FeO	9.0	8.95	8.58	7.90	9.80	4.3	9.67
MnO	0.2	0.21	0.21	0.15	0.23	0.14	0.21
MgO	8.7	8.59	8.51	3.52	3.16	3.0	7.25
CaO	10.2	9.23	11.60	9.51	6.54	6.1	9.58
Na_2O	3.1	3.91	4.10	6.24	5.24	6.2	3.74
K_2O	0.9	2.14	1.55	2.10	2.01	3.3	1.34
H_2O-		0.78	0.66	0.88	0.30	1.55	
H_2O+		2.37	1.34	2.89	0.77	1.75	
P_2O_5	0.3	0.73	1.26	1.22	1.50	0.79	0.65
CO_2		Trace	0.04	Trace	0.17	Trace	
NiO		0.01	0.02	Trace			0.02
BaO		0.06	0.09	0.09	0.06		0.11
SrO		0.08	0.03	0.04	0.06		0.04
Cr_2O_3		0.04	0.03				0.03
V_2O_3		Trace	0.03	0.03	Trace		0.01
Cl		Trace	0.09	Trace	0.11		
F				0.10	0.05		
S		0.04	0.05	0.07	0.04		0.04
Total	100.0	100.35	100.05	100.07	100.15	99.97	100.12

Explanation of Table 3

1. Average composition (water-free) of 8 olivine basalts with SiO_2 range 44.38 to 46.92 per cent. (W. N. Benson, The basic igneous rocks of eastern Otago and their tectonic environment, *Royal Soc. New Zealand Trans.*, vol. 72, p. 113, Nos. 3–10, 1942.)
2. Alkaline basalt (atlantite) with biotite and anorthoclase. (Benson, *op. cit.*, p. 113, No. 11, 1942.)
3. Olivine nephelinite. (Benson, *op. cit.*, p. 113, No. 20, 1942.)
4. Olivine theralite, upper portion of differentiated sill, Waihola. (Benson, *op. cit.*, p. 113, No. 18, 1942.)
5. Mugearite (oligoclase-andesine basalt.) (Benson, *op. cit.*, p. 113, No. 2, 1942.)
6. Zeolitic basanite, Mt. Dasher. (D. A. Brown, The Geology of Siberia Hill and Mt. Dasher, North Otago, *Royal Soc. New Zealand Trans.*, vol. 83, pt. 2, p. 365, No. 8, 1955.)
7. Average composition (water-free) of volcanic rocks of the moderately deformed subprovince, taking into account relative abundance of different analyzed types. (W. N. Benson, *Royal Soc. New Zealand Trans.*, vol. 74, pt. 1, p. 104, No. 14, 1944.)

sibility that trachyte and phonolite liquids could be formed as late residues expelled from crystallizing basic alkaline (theralitic) magma is clear, even though the liquids in question were not segregated as separate differentiates.

The Strongly Deformed Central Subprovince. This is a relatively small centrally situated area of 170 square miles, two-thirds of which is still thickly mantled with volcanic rocks. Folding, closely spaced faulting, and regional down-warping during Pliocene-Pleistocene times reached a greater intensity than in the surrounding subprovinces. This is the region, too, of greatest volcanic activity and of greatest petrographic and chemical variety among the products of volcanism. Lavas and interbedded pyroclastics locally reached thicknesses of the order of 1,500 ft. About one-third of the area of volcanic rocks is occupied by rocks in an advanced stage of magmatic evolution, namely, phonolites. Three cycles of igneous activity, each characterized by eruption of olivine basalt, trachybasalt, phonolite, and trachyte in varied order, have been traced. Evidence of explosive eruption of trachytic breccias and basaltic fire-fountain products is widespread. Crustal folding, which proceeded rather gently and intermittently during the period of volcanism, increased in vigor as volcanic activity waned, but phonolitic magma was still available for injection along a late major fault which formed along the axis of the main anticlinal fold of the region.

The following are the main petrographic types represented:[7]

Olivine basalts:

Olivine averages Fa_{23}, but outer zones of crystals may reach Fa_{45}; pyroxene is mainly diopsidic titanaugite ($2V = 50°$ to $60°$), rarely a type somewhat richer in iron ($2V = 40°$ to $50°$).

Basanites and atlantites:

Resemble the olivine basalts, but somewhat more alkaline; nepheline usually not obvious. (Nepheline-rich basic rocks are rare. A single example of nephelinite pegmatoid, very similar to the Waihola theralite, occurs as a small dike.)

Trachybasalts and trachyandesites:

Olivine more ferrous than that of olivine basalts (average $= Fa_{31}$; sometimes ranging to Fa_{76}); pyroxene is diopsidic titanaugite, sometimes rimmed with aegirine-augite; brown hornblende phenocrysts, common; sanidine in the groundmass of some rocks.

Mugearites:

---

[7] All compositions of olivines and pyroxenes are optically determined. (W. N. Benson and F. J. Turner, Mineralogical notes from the University of Otago, No. 2, *Royal Soc. New Zealand Trans.*, vol. 69, pt. 1, pp. 56–72, 1939. Note that hypersthenes recorded in this paper are actually groundmass olivines with prismatic habit.)

Andesine, olivine (Fa_{32} to Fa_{66}), diopsidic augite, iron ore.

Phonolites and tinguaites:

Sanidine, nepheline (in some cases accompanied by sodalite), aegirine, cossyrite; some rocks contain xenocrysts of basaltic origin—plagioclase, diopsidic titanaugite, magnesian olivine, the two latter rimmed with aegirine; one abnormally potassic rock contains pseudoleucite.

Anorthoclase trachytes:

Sometimes contain augite, aegirine-augite or iron-rich olivine.

Xenoliths and xenocrysts of many types, widely distributed, e.g.,

1. Fragments of schist and quartz of accidental origin, especially in the trachytic breccias.

2. Nepheline-aegirine syenites in trachyte breccias and phonolitic lavas.

3. Basaltic xenoliths and xenocrysts abundantly crowding some phonolitic lavas, with which they have reacted to give a hybrid olivine-bearing phonolite or "trachydolerite."

4. Hornblende-rich xenoliths of essexitic gabbro, nepheline-aegirine syenite, and coarse xenocrysts of hornblende.[8] These are widely distributed in basalts, trachybasalts, phonolites, and trachytes.   The amphibole ranges from brown basaltic hornblende in the basic rocks to more richly titaniferous kaersutite and barkevikite in the trachytes, phonolites, and nepheline syenites.   Partial resorption to a mixture of augite, iron ores, and apatite is usual.

The range of composition of rocks of the central, strongly deformed, subprovince is illustrated in Table 4.   The alkali-lime index for the rock series is variously estimated at 50.1 (for the whole series), 46.5 (for rocks with feldspathoids), and 52.5 (for nonfeldspathoidal rocks).[9]

Petrogenesis.   Correlation between extent of volcanic activity, degree of magmatic variation, and intensity of contemporary crustal deformation has already been noted, but no explanation is attempted.   In contrast with the recorded association of similar lavas with supposedly tensional tectonic environments in other parts of the world, the east Otago volcanic province is located in an environment of rather mild compressional movements.

There is no obvious relation between distribution of phonolitic and trachytic centers and any particular lithologic unit in the carefully mapped underlying sedimentary and metamorphic rocks.   The latter, which everywhere constitute the basement proper, are extremely uniform in composition within and far beyond the limits of the volcanic province.   While the parent magma admittedly may have reacted in the depths with the basement schists, subsequent differentiation must be assumed as the immediate

[8] W. N. Benson, Mineralogical notes from the University of Otago, no. 3, *Royal Soc. New Zealand Trans.*, vol. 69, pt. 3, pp. 283–308, 1939.

[9] Benson, *op. cit., Am. Jour. Sci.*, p. 548, 1941.

TABLE 4. COMPOSITIONS OF VOLCANIC ROCKS, STRONGLY DEFORMED SUBPROVINCE
(DUNEDIN DISTRICT)
PLIOCENE PETROGRAPHIC PROVINCE OF EASTERN OTAGO, NEW ZEALAND

Constituent	1	2	3	4	5
SiO_2	48.18	52.37	57.88	56.40	66.04
TiO_2	2.31	1.27	1.55		
Al_2O_3	16.10	18.49	18.07	15.84	18.38
Fe_2O_3	6.02	2.53	2.22	6.48	1.05
FeO	8.71	5.17	2.54	3.54	
MnO		0.23	0.11		
MgO	3.90	2.34	0.83	0.21	0.69
CaO	9.89	5.25	3.94	1.52	0.96
Na_2O	3.49	5.47	5.77	5.80	7.22
K_2O	1.56	3.20	3.94	5.78	5.09
ZrO_2		0.00	0.06		
H_2O-		1.59	0.77	} 3.96	} 1.50
H_2O+		1.06	1.02		
P_2O_5	0.36	0.46	0.56	0.13	
CO_2		0.03	0.43		
NiO		0.02	Trace		
BaO		0.09	0.10		
SrO		0.02	0.03		
Cl		0.09	Trace		
S		0.05	0.02		
Total	100.52	99.73	99.84	99.66	100.93

Explanation of Table 4

1. Average composition (water free) of 19 basalts. (W. N. Benson, Cainozoic petrographic provinces in New Zealand, *Am. Jour. Sci.*, vol. 239, p. 541, F, 1941.)
2. Hornblende trachybasalt. (W. N. Benson and F. J. Turner, Mineralogical notes from the University of Otago, No. 2, *Royal Soc. New Zealand Trans.*, vol. 69, pt. 1, p. 71, No. 26, 1939.)
3. Augite trachyandesite. (Benson and Turner, *op. cit.*, p. 71, No. 30, 1939.)
4. Nepheline-rich cossyrite phonolite. [P. Marshall, Geology of Dunedin (New Zealand), *Geol. Soc. London Quart. Jour.*, vol. 62, p. 405, A, 1906.]
5. Anorthoclase trachyte. (Marshall, *op. cit.*, p. 399, A, 1906.)

cause of such varied end products as trachyte, phonolite, trachyandesite, and rarely nepheline-theralite pegmatoid. Mixing of partly crystalline differentiates and reaction between late liquids (phonolitic, trachytic) and early-formed crystalline phases (augite, olivine, labradorite) have led in a number of instances to development of extensive hybrid flows such as some of the trachybasalts, trachyandesites, and olivine-bearing trachytes ("kaiwekites").

Following a line of investigation successfully applied by Bowen to the alkaline lavas of east Africa, Benson has shown that the salic members (trachyandesites, phonolites, trachytes) of the Otago association conform remarkably to the chemical requirements of a differentiation series dominated by fractional crystallization of alkali feldspars. Laboratory investi-

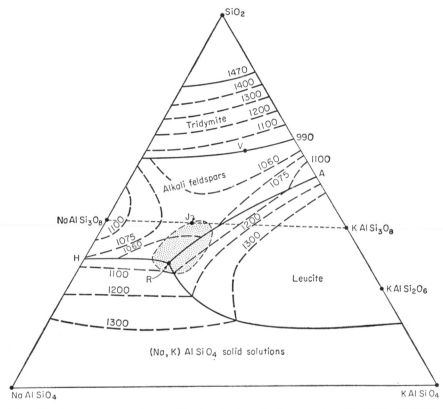

FIG. 27. Temperatures of crystallization in the anhydrous system NaAlSiO₄-KAlSiO₄-SiO₂ at atmospheric pressure. (*After J. F. Schairer.*) Minimal temperatures, marking the axis of a low-temperature trough are at R (1020°C.), J (1063°C.), and near V (less than 990°C.). Stippled area marks the range of composition of salic portions (less anorthite) of trachytes and phonolites of eastern Otago volcanic province, New Zealand. (*After W. N. Benson.*)

gation of the anhydrous system nepheline-kalsilite-silica had revealed the existence of a pronounced low-temperature trough (Fig. 27) toward which all liquids within the system must move as crystallization proceeds. Twenty-two of the analyzed lavas of eastern Otago approximate to mixtures in this three-component system, in that in each case the sum of the salic constituents of the norm, minus the anorthite content, exceeds 80 per cent of the normative composition of the whole rock.   With one exception (an abnormally potassic pseudoleucite phonolite) the plotted compositions of these salic rocks (with nepheline + kalsilite + silica recalculated to 100) fall within the low-temperature trough bounded by the the 1100° isothermal of the experimentally determined system (Fig. 27).   This is very strong evidence indicating that the phonolitic and trachytic rocks of Otago crystallized from liquid residues of fractional crystallization.   More recently it has been found [10] that the configuration of the low-temperature trough above the join $NaAlSi_3O_8$-$KAlSi_3O_8$ is notably different at high water pressures consistent with plutonic conditions of crystallization. This does not invalidate the conclusions just cited, for these refer to volcanic crystallization of silica-deficient magmas below the alkali-feldspar join.

While a differentiated origin seems satisfactorily established for the more salic members of the east Otago association, it is also conceivable that fractional resorption of accumulated amphibole crystals may have played a preliminary part in developing an alkaline but still very basic magma from the parent olivine basalt.   Brown hornblende with 39 per cent SiO_2 and $3\frac{1}{2}$ per cent ($Na_2O + K_2O$) has crystallized freely from the Otago magma at some stage of its evolution.   Benson concluded that this mineral begins to crystallize from olivine-basalt magma at depth, and that crystals so formed later become incorporated into more alkaline differentiates by hybridization.

Broader Environmental Considerations.   So far we have considered as a unit the Pliocene association of a small petrographic province occupying 2,000 square miles in southern New Zealand.   This province proves to be distinctly limited both geographically and in time.   The islands of New Zealand constitute a continental fragment roughly 1,000 miles long, lying well within the "andesite line," with a basement of graywackes (or their metamorphosed equivalents) overlain by basin deposits of Cretaceous and Tertiary mixed marine sediments.   During the later half of Tertiary time there developed over this limited segment of the Pacific borderland a number of distinct petrographic provinces, of which the Pliocene east Otago province is one.   At least four standard rock associations, each with its well-known counterparts in other parts of the world, are represented (cf. Fig. 26):

1. East Otago (Pliocene)[11]: olivine basalt with highly differentiated trachyte, phonolite, and nepheline basalt.   Similar in type to the association characteristic of oceanic islands, but alkaline types are much more plentiful than in the latter.

[10] See, for example, F. Chayes, in Annual report of the director of the Geophysical Laboratory, *Carnegie Inst. Washington Year Book*, no. 55, pp. 210–214, 1956.
[11] Described in preceding pages.

2. Northeast Otago (mid-Tertiary),[12] immediately north of and slightly overlapping province 1. Slightly alkaline olivine-basalt tholeiite association with quartz diabase as typical differentiate, but occasionally giving rise to alkaline pegmatoids with zeolites and anorthoclase. Differs from

TABLE 5.   AVERAGE COMPOSITIONS OF BASALTS FROM TERTIARY VOLCANIC PROVINCES OF NEW ZEALAND [*]

Constituent	1	2	3	4	5	6
SiO_2	48.18	47.26	52.73	49.14	49.33	52.50
TiO_2	2.31	2.48	2.21	3.17	0.90	0.91
Al_2O_3	16.10	14.91	15.14	15.71	15.64	16.01
Fe_2O_3	6.02	2.53	2.40	3.64	5.47	2.78
FeO	8.71	9.59	7.57	8.17	6.47	6.51
MnO		0.18	0.13	0.15	0.16	0.17
MgO	3.90	8.05	6.70	6.03	7.75	7.06
CaO	9.89	9.98	8.17	8.38	10.22	10.99
Na_2O	3.49	3.13	3.34	3.69	3.14	2.38
K_2O	1.56	1.20	1.01	1.39	0.98	0.54
P_2O_5	0.36	0.69	0.39	0.73	0.40	0.14
Total	100.52	100.00	99.79	100.20	100.46	99.99

[*] W. N. Benson, Cainozoic petrographic provinces in New Zealand and their residual magmas, Am. Jour. Sci., p. 541, 1941; The basic igneous rocks of east Otago and their tectonic environment, Part IV, Royal Soc. New Zealand Trans., vol. 74, pt. 1, p. 104, 1944.

Explanation of Table 5

1. East Otago (Pliocene) province; 19 basalts from highly deformed central subprovince.
2. East Otago (Pliocene) province; 5 basalts from moderately deformed subprovince.
3. Northeast Otago (mid-Tertiary); 5 analyses variously weighted.
4. Banks Peninsula (Pliocene); 14 basalts.
5. Auckland province; 14 Pliocene-Pleistocene basalts from northern subprovince.
6. Auckland province; 3 postrhyolite (Pleistocene) basalts from southern subprovince.

the typical plateau-basalt diabase association (tholeiitic magma type of Kennedy) in more basic composition of dominant basalt, in local alkaline pegmatoids, and perhaps in composition of pyroxene (diopsidic augite dominant; subcalcic augite, $2V = 30°$ to $45°$, subordinate).

3. Banks Peninsula [13] (Pleistocene), roughly 200 miles north of the east

[12] W. N. Benson, The basic igneous rocks of eastern Otago and their tectonic environment, Part IV, Royal Soc. New Zealand Trans., vol. 74, pt. 1, pp. 71–123, 1944.

[13] Benson, op. cit., Am. Jour. Sci., 1941.

Otago province.   Somewhat alkaline olivine basalt, olivine andesites, and greatly subordinate phonolitic trachyte and tridymite trachyte.   Alkali-lime index, 51.   Closely similar to associations found on many oceanic islands.

4. Auckland province [13] (Tertiary), occupying northern half of North Island, about 700 miles north of the east Otago province.   Calc-alkaline andesite dacite rhyolite association identical with the typical volcanic series of fold-mountain arcs.   Alkali-lime index, 62.7.   In the northern sector of this province, Pliocene-Pleistocene eruptions gave rise uniformly to somewhat alkaline olivine basalts with which nepheline-theralite pegmatoid is associated in one instance; these perhaps represent initiation of a new province.   In the southern sector Pleistocene and still active volcanoes are andesitic.

The three southern provinces are dominated by olivine basalt, which also was erupted extensively in Pliocene-Pleistocene times in the northern part of the North Island.   As far as can be judged from averages of many excellent analyses, these dominating basalts show distinct chemical difference from province to province (cf. Table 5).   We conclude that, in the later part of Tertiary time, alkaline basaltic magmas have been available for eruption at widely scattered points along the full length of New Zealand.   The mid-Tertiary magma of northeast Otago is notably more siliceous than the others and in this respect resembles continental "plateau" basalts of the tholeiitic kindred.

Differentiated Intrusions in the Alkaline Olivine-Basalt Association. Some of the most striking illustrations of differentiation of basic magma within igneous intrusions are afforded by sills, sheets, and laccoliths of alkaline diabase (teschenite or theralite) belonging to the alkaline olivine-basalt association of continental regions.   Gravitational settling of early-formed crystals of olivine and pyroxene in place has long been accepted as the mechanism of differentiation of this kind.   But as more detailed petrographic, mineralogical, and field data have become available, this simple explanation has proved repeatedly to be not entirely satisfactory. By some writers it has been rejected completely.   Thus Wilkinson [14] finds that fractional crystallization is responsible for vertical variation (Table 6) in a teschenite sill in New South Wales and that the early crystalline fraction is concentrated near the base; but he believes that gravitational settling of crystals played no significant part in the differentiation process. The chemical variation, whatever its cause, is slight.

Several examples of differentiated basic alkaline sills are described below.   Another—the theralite sill of Waihola, east Otago province—has already been referred to briefly (page 167).

---

[14] J. F. G. Wilkinson, The petrology of a differentiated teschenite sill near Gunnedah, New South Wales, *Am. Jour. Sci.*, vol. 256, pp. 1–39, 1958.

The Teschenite Sills of Shiant Isles, Scotland.[15]   The Shiant Isles, minor islands of the Hebrides of Scotland, are made up largely of a horizontal sill of Tertiary age whose maximum thickness exceeds 500 ft. The general composition of the sill is teschenitic,[16] but it varies strongly in a vertical

TABLE 6.   CHEMICAL VARIATION IN TESCHENITE SILL, GUNNEDAH, NEW SOUTH WALES

Constituent	1	2	3	4	5	6
SiO_2	44.78	45.15	44.97	45.66	44.15	44.01
TiO_2	2.49	2.71	2.62	2.79	4.07	3.53
Al_2O_3	14.03	17.29	14.64	16.71	14.13	15.90
Fe_2O_3	4.15	2.59	3.12	2.71	3.83	4.15
FeO	9.15	8.36	9.28	7.93	9.43	9.57
MnO	0.14	0.10	0.12	0.12	0.10	0.08
MgO	9.57	4.76	7.45	4.60	4.80	2.97
CaO	8.12	10.25	8.33	10.01	9.62	8.17
Na_2O	3.30	4.00	3.80	3.94	3.84	4.05
K_2O	1.77	1.76	1.81	1.82	1.83	2.40
H_2O+	2.05	2.86	2.88	2.84	3.26	4.04
H_2O-	0.14	0.13	0.23	0.21	0.20	0.31
P_2O_5	0.62	0.44	0.65	0.58	0.70	1.21
Total	100.31	100.40	99.90	99.92	99.96	100.39

Explanation of Table 6

Note: All analyses taken from J. F. G. Wilkinson, The petrology of a differentiated teschenite sill near Gunnedah, New South Wales, *Am. Jour. Sci.*, vol. 256, pp. 22, 23, 1958.

1. Teschenite 20 ft. from lower contact.
2. Teschenite 120 ft. from lower contact.
3. Teschenite 270 ft. from lower contact.
4. Teschenite 420 ft. from lower contact.
5. Teschenite 500 ft. from lower contact (near roof).
6. Teschenite from roof of intrusion.

sense. A generalized downward vertical section—of which zone (1) occurs on a small island, and zones (2) to (5) constitute a continuous section on the main island—is as follows:

1. Teschenite (originally called "crinanite") with rhythmic layering and igneous lamination characterized by planar orientation of tabular feldspar

[15] F. Walker, The geology of the Shiant Isles, *Geol. Soc. London Quart. Jour.*, vol. 86, pp. 355–398, 1930; H. I. Drever, A note on the field relations of the Shiant Isles' picrite, *Geol. Mag.*, vol. 90, pp. 159, 160, 1953; R. Johnston, The olivines of the Garbh Eilean sill, Shiant Isles, *Geol. Mag.*, vol. 90, pp. 161–171, 1953.

[16] J. F. G. Wilkinson, The terms teschenite and crinanite, *Geol. Mag.*, vol. 92, pp. 282–290, 1955.

and prismatic pyroxene crystals.[17]   Total thickness, 200 ft. Main constituents of the laminated rock are andesine, augite, and analcite.   In laminated portions dark layers rich in augite merge upward into light layers composed largely of andesine and analcite—the genetic counterpart of anorthosite layers in rhythmically banded gabbros of stratified basic intrusions (see page 298).   A feldspathoidal syenite (alkali feldspars, analcite, nepheline, aegirine-augite, aegirine) occurring as small veins is believed to be a late differentiate.

2. Teschenite (ophitic olivine diabase carrying some analcite).   Vertical thickness, 350 to 400 ft.   The top has been removed by erosion, and the extent to which this zone may overlap with (1) is uncertain.   Mineral composition: olivine 11 per cent, augite 25 per cent, plagioclase 58 per cent, iron ores 4 per cent, zeolites 2 per cent.   Mean specific gravity, 2.95. Plagioclase is zoned from An_{80} (cores) to An_{35} (margins); olivine is strongly zoned from Fa_{20-30} (cores) to Fa_{85-90} (margins).

3. Olivine diabase (picrodolerite). Vertical thickness about 100 ft. Average mineral composition: olivine 31 per cent, augite 17 per cent, plagioclase 50 per cent, iron ores 2 per cent, traces of zeolites and barkevikite.   Specific gravity, 3.02.   Olivine, which increases to 44 per cent at the base, is Fa_{20-30} with narrow rims Fa_{30-50}.   Plagioclase is normally zoned An_{80} to An_{30} near the top, An_{80} to An_{55} near the base.   Texture changes from poikilitic near the base to ophitic near the top.   Transition from olivine diabase upward to teschenite is gradual.

4. Picrite.   Vertical thickness at least 10 to 15 ft.   Mineral composition: unzoned olivine (Fa_{15} to Fa_{20}) 59 per cent, augite 10 per cent, plagioclase (An_{80}) 26 per cent, iron ores 2 per cent, zeolites 3 per cent.   Specific gravity, 3.10.   Olivine is poikilitically enclosed in plagioclase or in pyroxene, both of which are thought to have crystallized later.   Upper margin of picrite is sharp, and in places is strongly inclined to the horizontal.

5. Porphyritic chilled basal facies with olivine phenocrysts, Fa_{19}.

The picrite of the Shiant Isles sill has been widely cited as an ultrabasic rock formed by gravitational settling of early-formed magnesian olivine crystals in a magma which, judging from the prevailing nonporphyritic fabric of the overlying rocks, was originally injected in an almost completely liquid condition.   The calcic composition of the plagioclase, the magnesian composition of the olivine, and the unzoned condition of both minerals in the picrite, show that this rock became completely crystalline while the main mass of magma above was still at least partially molten. However, the sharp contact between picrite and olivine diabase, and its locally steep dip are scarcely compatible with simple gravitational differentiation.   It has been suggested as an alternative that upward concen-

---

[17] H. I. Drever, A note on the occurrence of rhythmic layering in the Eilean Mhuire sill, Shiant Isles, *Geol. Mag.*, vol. 94, pp. 277–280, 1957.

tration of zeolitic material drawn from a crystalline picritic fraction may have played some part in the differentiation process.[18]   Whatever the mechanism may have been, the trend of differentiation toward enrichment in iron and soda is indicated by the respective compositions of the outer zones of olivine and plagioclase crystals in the teschenite.   Pegmatoid augite-plagioclase-ilmenite veins poor in olivine and rich in zeolite cut the teschenite and are interpreted as end products of differentiation.

TABLE 7.   CHEMICAL COMPOSITIONS (WATER-FREE) OF ROCKS FROM DIFFERENTIATED TESCHENITE SILLS OF SHIANT ISLES (ANALYSES 1, 3, AND 5) AND CHEMICALLY EQUIVALENT LAVAS OF OCEANIC ISLANDS (ANALYSES 2, 4, AND 6)

Constituent	1	2	3	4	5	6
SiO_2	48.4	49.58	41.9	46.3	60.1	60.9
TiO_2	2.9	3.17	0.8	1.8	0.5	0.4
Al_2O_3	15.5	13.19	9.1	6.1	16.3	17.9
Fe_2O_3	1.2	2.40	0.6	5.9	5.0	2.5
FeO	9.3	9.49	12.9	7.5	2.6	2.9
MnO	0.4	0.12	0.4	0.1	0.3	0.2
MgO	6.7	8.30	26.7	23.8	0.6	0.2
CaO	12.5	10.69	5.8	6.5	2.1	2.0
Na_2O	2.5	2.25	1.4	1.5	7.7	7.6
K_2O	0.4	0.55	0.2	0.4	4.4	5.0
P_2O_5	0.2	0.26	0.2	0.1	0.4	0.4
Total	100.0	100.00	100.0	100.0	100.0	100.0

Explanation of Table 7

1. Teschenite, Shiant Isles, main sill.   (F. Walker, The geology of the Shiant Isles, *Geol. Soc. London Quart. Jour.*, vol. 86, p. 371, No. I, 1930.)
2. Average of 11 analyses of Hawaiian basalts.   (R. A. Daly, *Geol. Soc. America Bull.*, vol. 55, p. 1365, 1944.)
3. Picrite, Shiant Isles, main sill.   (Walker, *op. cit.*, p. 371, No. III, 1930.)
4. Picrite basalt, Mauna Loa, Hawaii.   (H. S. Washington, *Amer. Jour. Sci.*, vol. 6, p. 122, 1923.)
5. Syenitic segregation in laminated augite-andesine-analcite rock, Shiant Isles. (Walker, *op. cit.*, p. 387, No. I, 1930.)
6. Phonolite, average of 5 bodies, Saint Helena.   (R. A. Daly, *Am. Acad. Arts Sci. Proc.*, vol. 62, p. 73, 1927.)

Composite Intrusions of Morotu, Sakhalin.[19]   In the Morotu district of the island of Sakhalin, there are numerous composite sheets, laccoliths and dikes of alkaline dolerite (teschenite) and syenite or monzonite.

[18] H. I. Drever, The origin of some ultramafic rocks, *Dansk geol. Fören. Medd.*, vol. 12, pp. 227–229 (especially 229), 1952.

[19] K. Yagi, Petrochemical studies on the alkali rocks of the Morotu district, Sakhalin, *Geol. Soc. America Bull.*, vol. 64, pp. 769–810, 1953.

Dolerite is the dominant rock of any intrusion, and it typically grades into a centrally situated mass of monzonite, syenite, or both rocks. The series dolerite-monzonite-syenite is believed to represent a line of differentiation from a parent magma having the chemical composition of alkaline olivine basalt. Field relations, however, preclude the possibility that differentia-

TABLE 8.   CHEMICAL COMPOSITIONS OF ROCKS FROM MOROTU DISTRICT, SAKHALIN, AND CHEMICALLY COMPARABLE ROCKS FROM ELSEWHERE

Constituent	1	2	3	4	5	6	7
SiO_2	46.62	53.20	58.86	46.50	49.14	54.14	56.07
TiO_2	2.59	1.82	0.53	2.25	3.17	1.81	1.24
Al_2O_3	16.28	17.81	20.19	16.42	15.71	17.82	18.16
Fe_2O_3	6.29	3.08	3.63	4.05	3.64	3.90	3.60
FeO	5.40	4.96	1.40	6.73	8.17	5.34	3.37
MnO	0.04	0.10	0.13	0.12	0.15	0.08	0.13
MgO	4.51	1.62	0.33	6.35	6.03	1.88	1.20
CaO	9.12	4.52	1.64	8.86	8.38	4.94	2.64
Na_2O	4.25	6.56	7.42	3.10	3.69	6.24	5.38
K_2O	1.79	2.19	4.39	1.54	1.39	2.72	3.75
H_2O-	0.10	0.47	0.07	0.43		0.03	2.22
H_2O+	3.62	2.84	3.29	3.40		0.24	1.79
P_2O_5	0.30	0.43	0.07	0.25	0.73	0.61	0.38
Total	100.09	99.60	99.95	100.00	100.20	99.75	99.93

Explanation of Table 8

1. Teschenite (dolerite), Morotu, Sakhalin. (K. Yagi, Petrochemical studies on the alkali rocks of the Morotu district, Sakhalin, *Geol. Soc. America Bull.*, vol. 64, p. 790, No. 1507, 1953.)
2. Monzonite, Morotu, Sakhalin. (Yagi, *op. cit.*, p. 791, No. 902, 1953.)
3. Syenite, Morotu, Sakhalin. (Yagi, *op. cit.*, p. 793, No. 1501, 1953.)
4. Average of 4 teschenites (dolerites), Morotu, Sakhalin. (Yagi, *op. cit.*, p. 803, 1953).
5. Average of 14 basalts (calculated water-free), Banks Peninsula, New Zealand. (Table 6, No. 4.)
6. Mugearite (oligoclase andesite), Haleakala, Hawaii. (G. A. Macdonald and H. A. Powers, Contribution to the petrology of Haleakala volcano, Hawaii, *Geol. Soc. America Bull.*, vol. 57, p. 119, No. 6, 1946.)
7. Trachyte, Banks Peninsula, New Zealand. (R. Speight, The dykes of the summit road, Lyttelton, *Royal Soc. New Zealand Trans.*, vol. 68, pt. 1, p. 92, No. 2, 1938.)

tion is due to gravitational settling of crystals in place. Some typical rocks from a single intrusion (the Takara Bridge laccolith) are as follows (for chemical analyses see Table 8).

1. Coarse-grained ophitic teschenite (dolerite): plagioclase An_{40}, anor-

thoclase, diopsidic titan-augite, titan-biotite, olivine, iron ore, analcite; minor barkevikite, sphene, apatite (Table 8, analysis 1).

2. Monzonite, with hypidiomorphic granular texture and abundant miarolitic cavities: plagioclase An_{30}, microperthite, soda augite, aegirine-augite, titan-biotite, analcite; minor iron ore, chlorite, arfvedsonte (Table 8, analysis 2).

3. Syenite with fine-grained hypidiomorphic granular texture: microperthite, anorthoclase, albite, aegirine, aegirine-augite, analcite; minor biotite, iron ore (Table 8, analysis 3).

Comparison of the three principal rock types and petrographic evidence of reaction trends in each permit the course of differentiation to be reconstructed. Crystallization of feldspar begins with labradorite An_{55}. The plagioclase becomes increasingly sodic and is followed by a sodic anorthoclase. Thus in one monzonite, cores of plagioclase An_{30} are mantled with alkali feldspar $Or_{20}Ab_{75}An_5$; in a syenite a few crystals of oligoclase An_{16} are enclosed in alkali feldspar $Or_{37}Ab_{61}An_2$. The trend of crystallization of pyroxenes is diopsidic augite $\rightarrow$ titan-augite $\rightarrow$ sodic augites$\rightarrow$ aegirine augite $\rightarrow$ aegirine. Crystallization of amphiboles, biotites, and analcite point to concentration of water in the later stages; in fact at the shallow depths at which the Morotu intrusions solidified, analcite must have appeared only at the hydrothermal stage. The trend of differentiation from teschenite through monzonite to syenite closely parallels the trend from alkaline olivine basalt to mugearite and trachyte in other provinces (in Table 8 compare analyses 1 and 4 with 5; 2 with 6; 3 with 7). It involves increase in silica and alkalis and decrease in lime, magnesia, and iron. Note that while the ratio FeO/MgO increases, there is no absolute enrichment in iron relative to alkalis.

There are a number of other recorded instances of analcite syenites occurring within sills and sheets of somewhat alkaline diabase.[20] By some writers they are attributed to multiple intrusion of basic magma and derivative syenitic magma, drawn at successive stages of differentiation from a subjacent reservoir. Others believe that intrusion of originally homogeneous basic magma was followed by differentiation in place, involving rifting of the crystal mesh in the later stages and consequent segregation of the final syenitic residuum sucked into the voids so developed. The latter mechanism is rendered all the more probable where, as is commonly the case, the syenite bodies occur *within* the basic sills rather than as separate intrusive bodies. There are major provinces, however, where to account for the wide range of variation among con-

[20] Cf. J. Gilluly, Analcite diabase and related alkaline syenite from Utah, *Am. Jour. Sci.*, vol. 14, pp. 198–211, 1927; G. W. Tyrrell, On some dolerite sills containing analcite syenite in central Ayrshire, *Geol. Soc. London Quart. Jour.*, vol. 84, pp. 540–569, 1928.

temporaneous intrusions, preintrusive differentiation must also be invoked. This is so in the Terlingual-Solitario region of southwestern Texas,[21] where many dikes, sills, and laccoliths of analcite-bearing rocks, ranging from teschenitic to monzonitic and syenitic types, and commonly of composite lithology, outcrop over an area of several hundred square miles.

The Teschenite-Picrite Sill of Lugar, Scotland.   In the Carboniferous olivine-basalt province of Scotland, surface eruption of olivine basalts, mugearites, trachyandesites, phonolites, and rhyolites was followed by an intrusive phase of activity during which were injected basic alkaline sills showing striking evidence of differentiation.   Some are teschenite-syenite bodies similar in many respects to those described in the preceding section.   A more alkaline type is exemplified by the teschenite-picrite sill of Lugar in Ayrshire, made classic through the accounts of G. W. Tyrrell.[22] A complete vertical section (168 ft. thick) described in the latest available account shows the following downward sequence, as revealed in a continuous bore:

1. Chilled fine-grained marginal zone of teschenite: plagioclase, titan-augite, comparatively abundant biotite, olivine and analcite, minor nepheline.   Thickness about 2 ft.

2. Teschenite: plagioclase (andesine-labradorite) and augite make up about one-third of the composition, the remainder being analcite, alkali feldspar, nepheline, olivine, biotite, and iron ore.   Within this layer there are local coarse nepheline teschenite, segregations of alkali syenite, and a band of olivine theralite a few inches thick.   Thickness about 10 ft.

3. Olivine theralite (normal theralite); composition (Tyrrell, 1917), andesine 20 per cent, nepheline 15 per cent, augite 35 per cent, olivine 15 per cent, biotite 5 per cent, barkevikite 5 per cent, minor iron ore and apatite, and in some cases analcite.   The theralite is interrupted by bands a few inches thick of coarse nepheline-rich pegmatoid theralite (= lugarite).   Thickness 28 ft.

4. Melanocratic olivine theralite richer in barkevikite and biotite and poorer in olivine than the normal theralite; analcite increases at expense of nepheline.   Locally, especially in the lower levels, the rock contains little or no feldspar.   Thickness 20 ft.

5. Olivine theralite of normal type with very abundant olivine; passing downward into picrite-theralite.   Thickness 12 ft.

6. Hornblende picrite; olivine, titan-augite, barkevikite; minor labradorite and analcite.   Thickness 33 ft.

---

[21] J. T. Lonsdale, Igneous rocks of the Terlingua-Solitario region, Texas, *Geol. Soc. America Bull.*, vol. 51, pp. 1539–1626, 1940.

[22] G. W. Tyrrell, The picrite-teschenite sill of Lugar (Ayrshire), *Geol. Soc. London Quart Jour.*, vol. 72, pp. 84–131, 1917; A boring through the Lugar sill, *Geol. Soc. Glasgow Trans.*, vol. 21, pp. 157–202, 1948.

7. Hornblende-augite peridotite, with dominant olivine; either barkevikite or augite may be next in abundance to olivine.   Thickness 31 ft.

8. Picroteschenite: very abundant augite, labradorite, olivine, zeolites, nepheline, minor biotite and barkevikite; in places this passes into teschenitic pyroxenite.   Thickness 21 ft.

9. Teschenite passing into fine-grained bleached carbonated type at the base.   Thickness 11 ft.

Notable concentration of olivine and less conspicuous accumulation of pyroxene in the lower half of the sill (picrite and peridotite layers), and an obvious downward increase in specific gravity in the series teschenite (2.8), theralite (2.95), picrite (3.0), peridotite (3.1) constitute strong evidence of gravitational sinking of olivine and pyroxene crystals, with compensating upward displacement of aqueous alkaline residues from which nepheline, analcite, and alkali feldspar were later to crystallize. Yet the full picture is too complex to allow of such simple explanation. This applies especially to the complicated detail now demonstrated within the upper half of the sill.   Nor is the great quantity of olivine present in the massive zones of picrite and peridotite compatible with a mechanism involving sinking of olivine in a magma whose composition is represented by the chilled teschenite borders.   Moreover the theralite layers are mineralogically and chemically similar to rocks which have formed by gravitational differentiation of what Tyrrell considers to be another type of magma ("crinanite" or kylite) in other sills of the Carboniferous province of Scotland.

The hypothesis favored by Tyrrell combines gravitational differentiation with multiple injection.   It is suggested that a picroteschenite magma was emplaced as a sill in a position other than that which it now occupies, and that, by sinking of olivine, downward gradation from analcite-rich teschenite to teschenite to picrite and peridotite was established between chilled teschenitic border phases.   While sufficient liquid remained for the bulk of the mass still to retain some degree of mobility, it was slowly injected into its present position under the propelling force accompanying a fresh surge of magma, this time of kylitic ("crinanitic") composition, which locally forced its way above the picrite layers, thereby thickening the sill to its present dimensions, and causing development of the olivine-theralite zones the lower parts of which are thought to represent mixtures of the kylite and picroteschenite magmas.   The data so far available are compatible with this hypothesis.

Whatever may have been the relative roles of crystal sinking and multiple injection in the evolution of the Lugar sill, the series of rocks there represented affords a graphic illustration of differentiation of a parent magma, equivalent in composition to rather alkaline olivine basalt or trachybasalt, into ultrabasic accumulations of crystals (picrite, peridotite) on the one hand and a basic but more strongly alkaline fraction more or less equivalent to nepheline basanite on the other. Again close comparisons may be made with strictly volcanic lavas of the olivine-basalt phonolite association elsewhere (Table 9). Note however that the picrites of

TABLE 9. COMPOSITIONS (WATER-FREE) OF ROCKS FROM THE LUGAR SILL (ANALYSES 1, 3, AND 5) AND CHEMICALLY SIMILAR LAVAS (ANALYSES 2, 4, 6, AND 7) FROM ALKALINE OLIVINE-BASALT PROVINCES ELSEWHERE

Constituent	1	2	3	4	5	6	7
SiO_2	46.5	44.6	42.7	41.9	44.6	44.3	40.7
TiO_2	2.5	3.2	2.6	1.6	2.7	2.8	2.7
Al_2O_3	15.9	14.8	14.7	14.9	7.6	9.5	8.8
Fe_2O_3	2.9	3.0	3.4	4.9	6.3	3.2	6.0
FeO	7.2	9.1	8.6	8.8	9.6	9.7	8.3
MnO	0.3	0.2		0.2	0.5	0.2	0.2
MgO	5.2	8.6	11.2	8.7	12.0	14.7	16.4
CaO	12.6	10.4	8.6	11.8	10.3	12.5	11.8
Na_2O	4.1	3.7	5.9	4.3	4.3	1.8	3.1
K_2O	2.5	1.7	1.2	1.6	1.5	0.5	1.2
P_2O_5	0.3	0.7	1.1	1.3	0.6	0.8	0.8
Total	100.0	100.0	100.0	100.0	100.0	100.0	100.0

Explanation of Table 9

1. Chilled teschenite near upper contact, Lugar sill. (G. W. Tyrrell, The picrite-teschenite sill of Lugar [Ayrshire], *Geol. Soc. London Quart. Jour.*, vol. 72, p. 104, No. 1, 1917.)
2. "Zeolitized atlantite" (alkaline basalt), Tokomariro Beach, east Otago, New Zealand. (W. N. Benson, *Royal Soc. New Zealand Trans.*, vol. 72, pt. 1, p. 113, No. 14, 1942.)
3. Average of theralites, Lugar sill. (Tyrrell, *op. cit.*, p. 118, No. 2, 1917.)
4. Olivine nephelinite, eastern Otago, New Zealand. (Benson, *op. cit.*, vol. 72, pt. 1, p. 113, No. 20, 1942.)
5. Picrite, Lugar sill. (Tyrrell, *op. cit.*, p. 114, No. 1, 1917.)
6. Picritic basalt, Upolu, Samoa. (G. A. Macdonald, *Geol. Soc. America Bull.*, vol. 56, p. 1357, No. 1, 1944.)
7. Picrite-basalt, Kauai, Hawaiian Islands. (R. A. Daly, *Geol. Soc. America Bull.*, vol. 55, p. 1369, No. 5, 1944.)

the Lugar sill, though broadly to be regarded as counterparts of the volcanic ankaramites and oceanites, are distinctly richer in alkali. This is probably due to separation of barkevikite under the particular conditions, namely, high water pressure in a closed magma chamber, afforded by the sill.

The Role of Water. The most striking examples of differentiated basic sills are those—such as the teschenite bodies just described—in which the magma was at the same time rich in water and distinctly high in total alkali compared with average olivine basalt. That water is retained to the end and eventually gives rise to aqueous solutions of alkaline salts is obvious from the widespread occurrence of interstitial and in some cases deuteric zeolites, from the free development of pegmatoid segregations (lugarite and allied rocks), and from the high concentration of zeolites normally present in the late differentiates of most of these sills. There is also an amazingly close detailed resemblance, as to mineralogy, chemistry, and fabric, between massive theralites of sills and the small-scale pegmatoid segregations (seldom more than a few inches thick) which occur sporadically in basanites and olivine basalts of many olivine-basalt provinces.[23] A correlation may thus be made between high total alkali, high water content, and high potentiality to differentiate in bodies of basic magma. This is compatible with, but does not prove, the widely prevalent hypothesis that concentration of water and other "volatiles" in some way is concerned in the differentiation of alkaline magmas from a parent olivine-basalt magma. There is little evidence to show whether the function of the water is merely to reduce viscosity and so to facilitate separation of crystals from liquid at various stages, or whether water in some way causes ions to group into feldspathoid rather than into feldspar patterns within the liquid phase, or again whether water promotes the transfer of alkali from one part of the magma to another.[24] It is even possible that the most effective role of water is to lower crystallization temperatures to ranges within which some "alkaline" mineral assemblage tends to crystallize in place of a "normal" assemblage that would crystallize at higher temperatures from a less hydrous melt. Alternative development of the pairs nepheline-pyroxene and plagioclase-olivine might be controlled by some such condition. Speculation as to the relative merits of these various possibilities is unwarranted until more precise physical data are available.

[23] For example, see Benson, *op. cit.*, vol. 72, pp. 173–179, 1942.

[24] The process of "alkali-volatile diffusion differentiation" invoked by S. I. Tomkeieff (Petrochemistry of the Scottish Carboniferous-Permian igneous rocks, *Bull. volcanologique*, sér. 2, tome 1, pp. 59–87, 1937) to explain derivation of trachytic and phonolitic end products from olivine-basalt magma in the Scottish Carboniferous province would come within this category. So too does the thesis of G. C. Kennedy (Some aspects of the role of water in rock melts, *Geol. Soc. America Special Paper*, no. 62, pp. 489–504, 1955) that alkalis tend to coordinate with water dissolved in the melt phase.

THE ALKALINE OLIVINE-BASALT ASSOCIATION
IN THE OCEAN BASINS

Two Pacific Island Provinces—Tahiti and Samoa. Tahiti [25] is the largest island (350 square miles) of the Society group, which occupies an arc 300 miles in length extending in a northwesterly direction between lat. 16° and 18° S., and long. 148° and 152° W. It consists of two dissected volcanoes, the highest peaks of which exceed 7,000 ft. in elevation.

The dominant rocks (95 per cent of the lavas, according to Williams) are "basanitoids"—sodic basic lavas chemically equivalent to basanites but lacking feldspathoids. While phenocrysts of magnesian olivine and of augite are abundant, porphyritic feldspars are generally lacking and some of the glassy ("limburgitic") rocks are almost devoid of feldspar of any kind. Ankaramites and oceanites, very rich in augite and olivine, respectively, are common. Interesting but minor components of the Tahitian cones are phonolite, phonolitic trachyte, feldspathoidal trachyandesite ("tahitite"), and mica-bearing trachybasalt. In these rocks the feldspars include anorthoclase and sodic plagioclase; feldspathoids, when present, are sodic (nepheline, hauyne, analcite); aegirine-augite is common; hornblende and biotite are rare; and neither leucite nor melilite has ever been recorded.

Erosion has exposed a central core or plug of alkaline plutonic rocks in the center of each volcano. The larger core is approximately a mile in diameter and must have solidified at a depth of 1 to 1½ miles below the summit of the fully developed volcano. The component rocks belong to the alkaline basic and intermediate families and include theralites, essexites, and nepheline syenites and monzonites, individually equivalent to various types of Tahitian lavas. In basic types various combinations of labradorite, barkevikite, titan-augite, and olivine (any of which may predominate) are usually associated with minor nepheline, orthoclase, or analcite. In the nepheline monzonites and syenites, orthoclase, nepheline, plagioclase (sometimes labradorite), and augite occur together.

The chemical composition of typical Tahitian lavas is illustrated in Table 10 and also by the variation diagrams shown in Fig. 28. The alkali-lime index, read from the variation diagram Fig. 28a has a value of 46. However, it should be noted that in the average compositions given by Lacroix (Table 10) the ($K_2O + Na_2O$) content of average "tahitite" ($SiO_2 = 48.91$ per cent) is nearly twice the lime content, whereas the total ($K_2O + Na_2O$) of a glassy andesitic basalt ($SiO_2 = 45.88$ per cent) is less than one-half the amount of CaO in the same rock. The primary magma

[25] H. Williams, Geology of Tahiti, Moorea, and Maiao, B. P. Bishop Mus. Bull. 105, 1933.

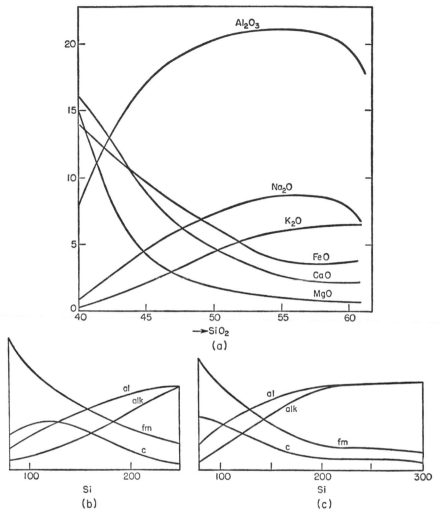

FIG. 28.   Variation diagrams for Tahitian and allied lavas.   (*After H. Williams.*)
(*a*) Diagram for 66 analyses of Tahitian lavas: oxide percentages (by weight) plotted
against SiO_2 percentages; (*b*) Niggli variation diagram for lavas of Tahitian type
(*C. Burri*); (*c*) Niggli variation diagram for lavas of Hawaiian type (*C. Burri*).   In
Niggli diagrams $al = Al_2O_3 + Cr_2O_3$; $fm = FeO + Fe_2O_3 + MnO + MgO$; $c = CaO +
BaO + SrO$; $alk = Na_2O + K_2O + Li_2O$; $al + fm + c + alk = 100$.

would appear to be alkaline basalt (cf. average basanitoid, Table 10,
analysis 1).   It closely resembles certain nepheline basanites of the
Hawaiian Islands.   The trachytic end products of differentiation in the
two regions are almost identical chemically (compare analysis 5, Table 10,
with analysis 8, Table 18).

Samoa [26] is situated at approximately lat. 14°S., long. 172° to 173°W., 1,200 miles west of Tahiti. Dominant olivine basalts are accompanied by picrite-basalts (oceanites), andesites, oligoclase andesites, and late local quartz trachytes and limburgite-basalt. The trachytes are more siliceous than typical Hawaiian or Tahitian trachytes and may contain over 70 per cent SiO_2. The alkali-lime index of the series is 51 (cf. Fig. 29).

TABLE 10. CHEMICAL COMPOSITION OF TYPICAL TAHITIAN LAVAS *

Constituent	1	2	3†	4	5
SiO_2	44.26	43.25	48.91	55.47	61.73
TiO_2	3.46	2.27	2.43	1.34	0.87
Al_2O_3	14.30	8.33	19.00	19.00	18.76
Fe_2O_3	4.61	4.90	3.40	3.22	2.00
FeO	7.79	7.58	4.08	2.22	1.54
MnO	0.21	0.13	0.34	0.24	0.09
MgO	8.34	19.02	2.04	1.68	0.94
CaO	11.26	12.36	5.78	3.71	1.61
Na_2O	3.48	1.36	7.07	7.80	6.98
K_2O	1.59	0.55	3.92	4.87	5.40
P_2O_5	0.70	0.25	0.47	0.45	0.08
Total	100.00	100.00	97.44†	100.00	100.00

* A. Lacroix, *Acad. Sci. Paris Mém.*, vol. 59, pp. 16–21, 1928.
† To this composition should be added $H_2O = 2.0$, $Cl = 0.16$, $SO_3 = 0.40$ per cent.

Explanation of Table 10

1. Average basanitoid. (A. Lacroix, *Acad. Sci. Paris Mém.*, vol. 59, p. 20, 1928.)
2. Average ankaramite. (Lacroix, *op. cit.*, p. 21, 1928.)
3. Average tahitite = feldspathoidal trachyandesite. (Lacroix, *op. cit.*, p. 18, 1928.)
4. Average phonolite. (Lacroix, *op. cit.*, p. 16, 1928.)
5. Average phonolitic trachyte. (Lacroix, *op. cit.*, p. 16, 1928.)

Volcanic Rocks of the Atlantic and Indian Oceans. Olivine basalts, with minor phonolites and trachytes as constant associates, are just as characteristic of the volcanic islands of the Atlantic and Indian oceans as of the intra-Pacific region.

The island of Saint Helena,[27] situated at lat. 16°S. and long. 6°W., 1,200 miles from the African coast and nearly 2,000 miles from South America, lies some 500 miles east of the crest of the mid-Atlantic swell.

[26] G. A. Macdonald, Petrography of the Samoan Islands, *Geol. Soc. America Bull.*, vol. 56, pp. 1333–1362, 1944.
[27] R. A. Daly, The geology of Saint Helena Island, *Am. Acad. Arts Sci. Proc.*, vol. 62, pp. 31–92, 1927.

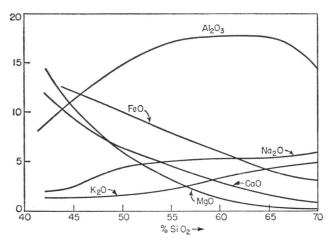

FIG. 29.   Variation diagram for the lavas of Samoa.   (*After* G. A. *Macdonald.*)

TABLE 11.   CHEMICAL COMPOSITION OF LAVAS FROM SAINT HELENA
AND CORRESPONDING TYPES FROM PACIFIC ISLANDS

Constituent	1	2	3	4	5
SiO_2	47.62	48.44	50.02	60.23	61.73
TiO_2	3.57	4.29	1.82	0.42	0.87
Al_2O_3	16.27	13.27	18.37	17.74	18.76
Fe_2O_3	3.87	4.06	4.25	2.49	2.00
FeO	8.17	8.27	6.78	2.87	1.54
MnO	0.11	0.15	0.05	0.16	0.09
MgO	6.33	8.21	3.26	0.17	0.94
CaO	8.65	7.72	6.75	1.92	1.61
Na_2O	3.82	3.47	4.81	7.50	6.98
K_2O	1.14	1.55	2.00	4.93	5.40
P_2O_5	0.45	0.57	0.34	0.40	0.08
H_2O-			0.23	0.35	
H_2O+			0.94	0.82	
Total	100.00	100.00	99.62	100.00	100.00

Explanation of Table 11

1. Basalt, average of 4 flows, Saint Helena. (R. A. Daly, The geology of Saint Helena Island, *Am. Acad. Arts Sci. Proc.*, vol. 62, p. 71, 1927.)
2. Basalt, average of 5 flows, Tutuila, Samoa. (Daly, *op. cit.*, p. 71, 1927.)
3. "Trachydoleritic basalt," Saint Helena. (Daly, *op. cit.*, p. 64, 1927; compare with average oligoclase andesite, Hawaii, Table II, No. 5.)
4. Phonolite, average of 5 bodies, Saint Helena. (Daly, *op. cit.*, p. 73, 1927.)
5. Average phonolitic trachyte, Tahiti. (A. Lacroix, *Acad. Sci. Paris, Mém.*, vol. 59, p. 16, 1928.)

It is a highly eroded volcanic doublet. Basalt flows, some rich, some poor in olivine, make up 99 per cent of its bulk. Rocks of trachybasaltic composition are rare, but there are a number of dikes, necks, and thick flows of nepheline-poor phonolite and of phonolitic trachyte. Chemical analyses of typical rocks match corresponding analyzed rocks of the intra-Pacific region almost exactly, as shown in Table 11. If the single analyzed trachy-basaltic rock (Table 11, analysis 3) is representative of the differentiation series, the alkali-lime index must be in the vicinity of 50.

Ascension Island,[28] about 800 miles northwest of Saint Helena, is situated on the crest of the mid-Atlantic swell. The association of lavas here closely parallels those of Saint Helena and Samoa. As in Samoa, the trachytic members grade into rhyolites and obsidians with as much as 71 to 72 per cent SiO_2, in which soda predominates over potash. Xenoliths

TABLE 12.   PETROGRAPHIC EVOLUTION OF KERGUELEN LAVAS

(After A. B. Edwards)

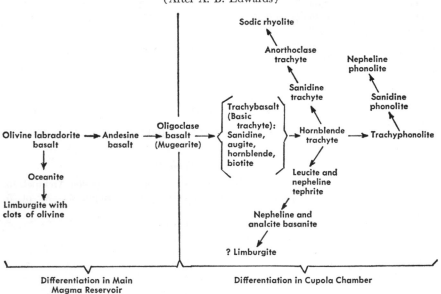

of plutonic rocks are widely distributed and were interpreted by Daly as indicating the existence of a continental basement along the site of the mid-Atlantic swell. But Tilley [29] has shown that these are alkali granites chemically equivalent to the soda-rhyolite lavas.

[28] R. A. Daly, The geology of Ascension Island, *Am. Acad. Arts Sci. Proc.*, vol. 60, pp. 3–80, 1925.
[29] C. E. Tilley, Some aspects of magmatic evolution, *Geol. Soc. London Quart. Jour.*, vol. 106, p. 43, 1950.

The Kerguelen archipelago [30] is the largest of several volcanic island groups situated in the southern Indian Ocean.   Its location is lat. 49°S., long. 70°E.   Edwards describes these islands as mainly basaltic, with minor quantities of limburgite and basanite and numerous but subordinate flows of trachyte and phonolite.   At Kerguelen, the petrographic and

TABLE 13.   CHEMICAL COMPOSITION OF KERGUELEN LAVAS

Constituent	1	2	3	4	5	6
SiO_2	46.00	43.74	50.41	62.74	54.10	46.48
TiO_2	1.62	1.61	1.63	0.42	0.40	1.74
Al_2O_2	12.42	13.57	16.26	16.46	21.66	18.33
Fe_2O_3	0.53	4.14	3.95	1.83	1.87	3.51
FeO	12.72	8.98	7.08	3.70	1.72	5.98
MnO	0.27	0.45	1.10	0.19	0.35	0.27
MgO	11.08	10.85	3.22	0.07	0.13	2.55
CaO	10.80	9.09	6.67	1.92	2.09	8.50
Na_2O	2.64	2.35	3.63	5.29	7.68	3.61
K_2O	0.42	2.47	2.28	5.68	6.55	4.46
H_2O-	0.63	0.77	1.41	0.78	0.38	2.00
H_2O+	0.60	1.22	0.98	1.01	2.38	1.05
P_2O_5	0.19	0.50	1.01	0.05	0.24	1.88
CO_2	Trace	0.05	Trace		Trace	Nil
Cl	Nil	Nil	Trace		0.14	Nil
S	Nil	0.01	Nil		Nil	Nil
BaO	Nil	0.02	0.02		0.02	0.03
Total	99.92	99.82	99.65	100.14	99.71	100.39

Explanation of Table 13

1. Olivine basalt.   (A. B. Edwards, Tertiary lavas from the Kerguelen Archipelago, *B.A.N.Z. Antarctic Expedition* [*D. Mawson*] *1929–1931, Rept.*, ser. A., pt. 5, p. 86, No. 2, 1938.)
2. Limburgite.   (Edwards, *op. cit.*, p. 86, No. 1, 1938.)
3. Oligoclase basalt.   (Edwards, *op. cit.*, p. 83, No. 1, 1938.)
4. Trachyte.   (Edwards, *op. cit.*, p. 77, No. 4, 1938.)
5. Sanidine phonolite.   (Edwards, *op. cit.*, p. 76, No. 1, 1938.)
6. Leucite tephrite.   (Edwards, *op. cit.*, p. 80, No. 1, 1938.)

chemical variety is somewhat greater than is usual in oceanic islands. The rock types represented and their genetic relationships as envisaged by Edwards are shown in Table 12.   The range of chemical composition is illustrated in Table 13.   Mineralogical features typical of the Kerguelen lavas are as follows: Olivines range from Fa_{15} (in olivine basalts and olivine clots) to types approaching fayalite (in the trachytes).   Pyroxenes are diopsidic and titaniferous augites, grading into aegirine-augite

[30] A. B. Edwards, Tertiary lavas from the Kerguelen Archipelago, *B.A.N.Z. Antarctic Expedition* (*D. Mawson*) *1929–1931, Rept.*, ser. A, pt. 5, pp. 72–100, No. 2, 1938.

and aegirine in the more sodic rocks; corroded aegirine is also present in the basanites and limburgites. Brown hornblende occurs not only in the phonolites and trachytes, but in basanites and oligoclase basalts as well; it is usually in process of replacement by pyroxene and apatite as a result of reaction with the enclosing magma at low water pressures during extrusion. While nepheline is the feldspathoid of phonolites and basanites, analcite is present in a number of alkaline basalts, and leucite was recorded in a single rock classed as leucite tephrite. Xenoliths are locally abundant—dunite, pyroxene peridotite, and gabbro in the basic lavas, hornblende peridotite in the basanites and trachybasalts. A unique feature of the southwestern part of Kerguelen is the presence of minor plutonic rocks (syenites, monzonites, diorites, gabbros) believed by Lacroix to be intrusive into the lavas.

GENERAL CHARACTERISTICS OF THE ALKALINE OLIVINE-BASALT ASSOCIATION

The volcanic rocks of the eastern Otago province of New Zealand illustrate one of many recorded instances where olivine basalts and derivative lavas occur in a continental setting. Other examples discussed in recent literature include the volcanic provinces of the Midland Valley of Scotland (Carboniferous-Permian), the Oslo district of Norway (Permian lavas and essexite necks), the Bohemian Mittelgebirge (Tertiary), Mozambique (Tertiary), the eastern Rift zone of east Africa (Tertiary and Pleistocene), south-eastern Australia (Tertiary), and Timor (Permian).[31] The olivine basalts, which on account of their general pre-

[31] For illustrative petrographic and chemical detail the reader is referred to the following: A. Holmes, The Tertiary volcanic rocks of the district of Mozambique, Geol. Soc. London Quart. Jour., vol. 72, pp. 222–279, 1917; W. C. Smith, A classification of some rhyolites, trachytes and phonolites from part of Kenya Colony, Geol. Soc. London Quart. Jour., vol. 87, pp. 212–258, 1931; H. Knorr, Differentiations und Eruptionsfolge in Böhmischen Mittelgebirge, Min. Pet. Mitt., vol. 42, pp. 318–370, 1932; M. MacGregor and A. G. MacGregor, British Regional Geology, The Midland Valley of Scotland, Great Britain Geol. Survey, pp. 49–68, 1936; F. H. Hatch and A. K. Wells, The Petrology of the Igneous Rocks, 9th ed., pp. 323–330, G. Allen, London, 1937; S. I. Tomkeieff, op. cit., 1937; A. B. Edwards, The Tertiary volcanic rocks of central Victoria, Geol. Soc. London Quart. Jour., vol. 94, pp. 243–320, 1938; Petrology of the Tertiary older volcanic rocks of Victoria, Royal Soc. Victoria Proc., vol. 51, pp. 73–98, 1939; The petrology of the Cainozoic basaltic rocks of Tasmania, Royal Soc. Victoria Proc., vol. 62, pp. 97–120, 1950; Benson, op. cit., Am. Jour. Sci., 1941; W. P. de Roever, Olivine basalts and their alkaline differentiates in the Permian of Timor, Geol. Expedition Lesser Sunda Islands (H. A. Brouwer), vol. 4, pp. 209–289, 1942; T. F. W. Barth, Studies on the igneous rock complex of the Oslo region, Part II, Norske vidensk.-akad. Oslo Skr., I. Math.-naturv. Kl., no. 9, 1945; H. Kuno, Plateau basalt lavas of eastern Manchuria, Seventh Pacific Sci. Cong. Proc., vol. 2, pp. 375–383, 1953; R. H. Clark, A petrological study of the Arthur's Seat volcano, Royal Soc. Edinburgh Trans., vol. 58, pt. 1, pp. 37–70, 1956.

dominance are assumed to represent the parent magma in every case, are closely similar, both chemically and mineralogically, to the olivine basalts of oceanic islands, and exhibit much the same range of variation (cf.

TABLE 14.   COMPARISON OF "PARENT" OLIVINE BASALTS IN OCEANIC (ANALYSES 7 AND 8) AND IN CONTINENTAL (ANALYSES 1 TO 6) PROVINCES

Constituent	1	2	3	4	5	6	7	8
SiO_2	48.18	47.26	50.6	47.1	49.1	42.7	48.44	44.26
TiO_2	2.31	2.48	1.5	2.9	2.9	2.1	4.29	3.46
Al_2O_3	16.10	14.91	15.4	15.8	16.6	17.2	13.27	14.30
Fe_2O_3	6.02	2.53	4.4	4.0	4.3	8.1	4.06	4.61
FeO	8.71	9.59	6.9	8.3	7.3	5.9	8.27	7.79
MnO		0.18	0.2	0.2	0.2		0.15	0.21
MgO	3.90	8.05	7.5	7.4	5.7	6.1	8.21	8.34
CaO	9.89	9.98	8.7	9.4	8.3	11.2	7.72	11.26
Na_2O	3.49	3.13	3.0	3.1	3.5	4.2	3.47	3.48
K_2O	1.56	1.20	1.4	1.3	1.7	2.5	1.55	1.59
P_2O_5	0.36	0.69	0.4	0.5	0.4		0.57	0.70
Total	100.52	100.00	100.0	100.0	100.0	100.0	100.00	100.00

Explanation of Table 14

Note: All compositions are calculated water-free.
  1. East Otago Pliocene, highly deformed subprovince; 19 olivine basalts. (Benson, Cainozoic petrographic provinces in New Zealand and their residual magmas, *Am. Jour. Sci.*, vol. 239, p. 451, 1941.)
  2. East Otago Pliocene, moderately deformed subprovince; 5 olivine basalts. (Benson, *op. cit.*, *Am. Jour. Sci.*, p. 451, 1941.)
  3. Newer volcanic series, Victoria, Australia; 27 olivine basalts. (A. B. Edwards, The Tertiary volcanic rocks of central Victoria, *Geol. Soc. London Quart. Jour.*, vol. 94, p. 309, Nos. 11, 12, 1938.)
  4. Carboniferous of Scotland; average olivine basalt (27 analyses). (S. I. Tomkeieff, Petrochemistry of the Scottish Carboniferous-Permian igneous rocks, *Bull. volcanologique*, sér. 2, tome 1, Table 1, No. 13, 1937.)
  5. Carboniferous of Scotland; average lava. (Tomkeieff, *op. cit.*, Table 5, 1937.)
  6. Bohemian Mittelgebirge; feldspar basalt, Scharfenstein type. (H. Knorr, Differentiations und Eruptionsfolge in Böhmischen Mittelgebirge, *Min. Pet. Mitt.*, vol. 42, p. 353, No. 29, 1932.)
  7. Tutuila, Samoa; average of 5 olivine basalts. (R. A. Daly, *Am. Acad. Arts Sci. Proc.*, vol. 62, p. 71, 1927.)
  8. Tahiti; average basanitoid. (A. Lacroix, *Acad. Sci. Paris Mém.*, vol. 59, p. 20, 1928.)

Table 14). Olivine basalt with alkaline affinities has been erupted in quantity, again and again, in provinces widely scattered over the continents and the ocean basins alike. Individual provinces, *e.g.*, that of eastern Otago, may be somewhat restricted in area, but regional flooding of large areas with alkaline basalt is also known, *e.g.*, in Palestine and Syria (cf. page 204).

In many respects the olivine-basalt association exhibits identical characters in both environments. There is the same general series of associated olivine- and augite-rich basalts (oceanites, ankaramites) and intrusive equivalents (picrites), trachybasalts, oligoclase basalts (mugearites), trachytes, and phonolites. Mineralogical detail of corresponding rocks is much the same: olivine ranges from highly magnesian types (commonly zoned with outward enrichment in iron) in the basalts to varieties rich in iron in phonolites and trachytes; pyroxenes are diopsidic titan-augites in the basalts, aegirine-augite and aegirine in trachytes and phonolites; brown hornblende commonly appears in the trachybasalts; the typical feldspathoid of the alkaline differentiates is nepheline or analcite. But there are also differences between the continental and the oceanic associations: the relative quantity of highly differentiated lavas (phonolite, trachyte, alkali rhyolite) in some continental provinces is notably greater than in oceanic provinces; rhyolitic end products figure more prominently in the continental assemblage as a whole than in the lavas of oceanic regions; leucite, a mineral of the greatest rarity in derivatives of oceanic basalt, is prominent in the basic alkaline rocks (tephrites and basanites) of some continental series, *e.g.*, in the Bohemian province. Here is a transitional assemblage showing affinities with the highly leucitic volcanic association to be considered later. Finally, in the more ancient continental provinces deep erosion into the subvolcanic basement in many instances has revealed necks, sills, and similar intrusive masses of co-magmatic plutonic rocks in much greater variety than is usual in oceanic volcanoes, only the summits of which have been exposed to erosion. The older essexitic necks of the Oslo district of Norway have thus been interpreted by Barth as plutonic equivalents of contemporary (Permian) olivine basalts and trachytes.

It has yet to be demonstrated that the alkaline olivine-basalt association habitually occurs in any particular tectonic setting. The tendency, noted by earlier writers, for alkaline lavas (including the phonolites and nepheline basalts of the present series) to occur in regions of crustal stability or of faulting, rather than in zones of strong orogeny, is on the whole borne out by recent observations. But the picture is by no means as simple as was once thought. The Rift Valleys of east Africa, especially the western zone, are no longer generally attributed to simple tensional faulting.[32] Again, the alkaline volcanic series of the east Otago Pliocene province was erupted during a period of rather mild local folding and compressional faulting of the uppermost crust, which reached its

[32] Cf. N. L. Bowen, Lavas of the African Rift Valleys and their tectonic setting, *Am. Jour. Sci.*, vol. 35A, pp. 19–33, 1938.

greatest intensity in the area where alkaline differentiates are most abundant and varied.

It is characteristic of this volcanic association, as developed in continental regions, that igneous associations of quite different types may precede or follow it within a given province or may appear simultaneously in closely adjacent provinces. The relation of the Pliocene olivine-basalt association of eastern Otago to other Tertiary volcanic series in New Zealand illustrates this phenomenon clearly. In the Midland Valley of Scotland olivine basalts, trachytes, phonolites, and rhyolites, accompanied and followed by sills and dikes of basic alkaline rocks (teschenites, essexites, theralites, monchiquites) afford a good example of long-continued development of the olivine-basalt association throughout much of Carboniferous and Permian times; but in the same area are Devonian lavas belonging to the basalt andesite rhyolite association; and the lengthy period of igneous activity dominated by the alkaline olivine-basalt association was either interrupted toward its close, or else closely followed, by intrusion of swarms of quartz-diabase sills belonging to yet a third kindred.[33] Even in some mid-oceanic provinces—notably in Hawaii—alkaline olivine-basalts, nepheline basalts, and melilite basalts appear as a late alkaline association preceded by more extensive outpourings of tholeiitic magma (see pages 218, 219).

PETROGENESIS

Digression on Hypotheses of Origin of Alkaline Rocks. Since Harker [34] divided the Tertiary and Recent igneous rocks of the world into his Pacific (calc-alkaline) and Atlantic (alkaline) "suites," English-speaking petrologists have tended to distinguish between calc-alkaline or calcic rocks on the one hand and alkaline rocks on the other. Nowadays rocks are usually classed as alkaline when the content of ($K_2O + Na_2O$) is sufficiently high, as compared with SiO_2 or Al_2O_3, for specially alkaline (usually sodic) minerals such as feldspathoids and aegirine to appear. Feldspathoids are correlated with a high proportion of alkali with respect to silica; aegirine, aegirine-augite, and nonaluminous sodic amphiboles (riebeckite, arfvedsonite) express a high ratio of alkali to alumina.

The most widely distributed and most abundant igneous rocks, namely, granites, granodiorites, basalts, and andesites, are devoid of alkaline min-

[33] For different views on the sequence of igneous rocks in this region see Hatch and Wells, *op. cit.*, p. 329, 1937; S. I. Tomkeieff, *op. cit.*, pp. 78–83, 1937; A. G. MacGregor, Problems of Carboniferous-Permian volcanicity in Scotland, *Geol. Soc. London Quart. Jour.*, vol. 104, pp. 133–153, 1948.

[34] A. Harker, *The Natural History of Igneous Rocks*, pp. 90–93, Macmillan, New York, 1909. *See also* J. Iddings, The origin of igneous rocks, *Washington Philos. Soc. Bull.*, vol. 12, pp. 183, 184, 1892.

erals of either type. Partly on this account and partly by reason of the attraction offered by the varied alkaline mineral assemblages to descriptive petrographers, it became customary to regard alkaline rocks as "abnormal" types. Abnormality demanded explanation, and rival hypotheses of genesis of alkaline rocks accordingly developed during the decades 1910 to 1930. Petrographic research so stimulated has proved most fruitful, but some misconceptions have arisen from arbitrary statement and oversimplification of the problem. It is not necessarily the case, nor indeed is it likely, that all rocks arbitrarily defined as alkaline must have related origins. Nor are all alkaline rocks to be regarded as abnormal merely because certain rare and striking alkaline assemblages can be shown to have originated in some instances under an "unusual" combination of circumstances.

The reader interested in the origin of alkaline rocks in general is referred to excellent summaries of the problem by Shand, Smyth, and Bowen.[35] To view the origin and evolution of lavas of the alkaline olivine-basalt association in proper perspective, it is appropriate to enumerate some of the suggested modes of origin of alkaline rocks:

1. Magmatic differentiation connected with, and in some way influenced by, epeirogenic tectonic movements—large-scale fracturing not accompanied by folding (Harker).

2. Desilication of basaltic or granitic magmas by assimilation of limestone involving crystallization of lime silicates like melilite and garnet; subsequent differentiation of the resulting contaminated magma assisted by escaping CO_2 (Daly; Shand).

3. Separation of leucite from trachytic magma, and concentration of the crystals to give a partly crystalline magma of phonolitic (nepheline syenite) composition (Bowen and Morey).

4. Internal reconstitution of granitic liquid magmas in the presence of high concentrations of water, and compounds of Cl, S, and C, which are assumed to promote a regrouping of ions into feldspathoid and olivine "molecules" rather than feldspar and metasilicate "molecules" in the differentiated liquids (Smyth; Bowen). If this process actually is effective it probably involves increased substitution of Al for Si in the aggregated lattices of [SiO_4] tetrahedra that build up in the liquid magma under these conditions.

5. Accumulation of hornblende and biotite crystals, which become par-

---

[35] S. J. Shand, The problem of the alkaline rocks, *Geol. Soc. South Africa Trans.*, vol. 25, pp. xix–xxxii, 1923; The present status of Daly's hypothesis of the alkaline rocks, *Am. Jour. Sci.*, vol. 234A, pp. 495–507, 1945; C. H. Smyth, The genesis of alkaline rocks, *Am. Philos. Soc. Proc.*, vol. 66, pp. 535–580, 1927; N. L. Bowen, *The Evolution of the Igneous Rocks*, pp. 227–273, Princeton University Press, Princeton, N. J., 1928.

tially or completely fused by reaction with hot basaltic magma into which they are supposed to sink, thereby giving basic alkaline magmas of the lamprophyre and nepheline-basalt families (Bowen).

6. Development of primary highly potassic magmas (of the leucite basalt and leucitophyre types) by differential fusion of the peridotite substratum, followed in some cases by mixing with olivine-basalt magma (Holmes).

7. Development of leucite-bearing rocks by metasomatism and partial fusion of preexisting mixed rocks under the influences of potash- and aluminum-bearing fluids rising from the deeper part of the lithosphere (Holmes).

8. Fractional crystallization of primary olivine-basalt magma to give trachytic, phonolitic, and basanitic differentiates (Daly, Barth, Kennedy, Edwards).

Evolution of the Alkaline Olivine-Basalt Association. Over much of the earth's surface, at least in Tertiary and Recent times, vast supplies of alkaline olivine-basalt magma have been available for eruption. Because of its wide geographic distribution, this magma cannot have been drawn from a single source of homogeneous liquid. So it is not surprising to find some degree of variation among the olivine-basalt provinces of the world, and in some instances a chemical gradation toward the more siliceous, less alkaline basalts that are generally distinguished as tholeiitic. From the great preponderance of olivine basalts among the rocks of any province of the kind now under consideration, it is reasonable to assume that these represent the parent magma from which the less plentiful associated rocks of other compositions have been derived. Such, in fact, is current orthodox petrologic opinion.

Olivine- and augite-rich rocks—picrite-basalts, oceanites, ankaramites, picrites, and so on—evidently represent accumulations of early-formed minerals crystallizing from the parent magma. The mechanism of differentiation commonly invoked is gravitational settling in a largely liquid medium. This no doubt plays an important role, but it fails to account completely for the mutual relations of the differentiated layers in most composite intrusions.

The trachytes and phonolites of olivine-basalt provinces typically are products of late, often explosive, eruptions from scattered independent centers. These rocks conform mineralogically, chemically, and in their small total bulk, to what may be expected in the final residue of fractional crystallization of undersaturated basaltic magma. Bowen [36] showed that in the phonolitic, trachytic, and alkaline rhyolitic lavas of the east African Rift Valleys the relative proportions of K_2O, Na_2O, and SiO_2 are

[36] N. L. Bowen, Recent high temperature research on silicates and its significance in igneous petrology, *Am. Jour. Sci.*, vol. 33, pp. 1–21, 1937.

precisely those of the low-temperature residual liquids of the experimentally investigated system $NaAlSiO_4$-$KAlSiO_4$-SiO_2. The same has been found to be true for the trachytes and phonolites of other provinces.[37] And finally the minor monzonitic and syenitic components of composite basic intrusions are chemically identical with the mugearites and trachytes of strictly volcanic provinces. The case for derivation of mugearite, trachyte, and phonolite by fractional crystallization of alkaline olivine-basalt magma is satisfactorily established.

It is clear, too, that differentiation in the later stages is commonly so efficient that the small fraction of trachytic melt is by some means effectively segregated to give individual bodies that tend to be expelled with explosive force. The phenomenon is essentially volcanic, and the site where the trachytic residue normally collects is located high in the volcanic column. Some mechanism involving boiling of an increasingly aqueous residual liquid under reduced pressure is strongly suggested.[38] It may be that bubbles of gas streaming up toward the vent drive the liquid residue in the same direction (as pictured by Shand) or merely stir the partly crystalline mass sufficiently to aid the normal upward migration of light trachytic liquid through the growing crystal mesh.

Some aspects of chemical variation [39] in lavas believed to represent successive fractions in the differentiation series alkaline olivine basalt → mugearite → trachyte (or phonolite) are illustrated in Fig. 30. The behavior of trace elements in the same series [40] may be summarized as follows: at the basic end V, Cr, Ni, and Co are all high, Ni > Co; Li and Rb are low. Ni and Cr diminish very rapidly, Co and V less so, so that in the middle of the series Co > Ni. Sr and Ba rise to a pronounced maximum and drop off at the acid end. Li and Rb increase steadily. The acid end of the series is marked by high Li, Rb, and Zr, appreciable Mo, and negligible V, Cr, Ni and Co.

The broad picture of differentiation of trachytic and phonolitic liquids from a parent alkaline olivine-basalt magma is clear enough. There is still no general agreement, however, as to the relative influence of such factors as initial composition of magma, sequence of mineral crystallization, diffusion of alkali coordinated with water, and gaseous transfer of alkali on the recognized variants of the course of magmatic evolution. By numerous petrologists the prevalence of undersaturated (phonolitic)

---

[37] Benson, op. cit., Am. Jour. Sci., 1941.

[38] Cf. S. J. Shand, The lavas of Mauritius, Geol. Soc. London Quart. Jour., vol. 89, p. 11, 1933; Edwards, op. cit., p. 96, 1938; R. A. Daly, Volcanism and petrogenesis as illustrated in the Hawaiian Islands, Geol. Soc. America Bull., vol. 55, p. 1384, 1944.

[39] S. R. Nockolds and R. Allen, The geochemistry of some igneous rock series: Part II, Geochim. et Cosmochim. Acta, vol. 5, pp. 246, 248, 1954.

[40] Nockolds and Allen, op. cit., pp. 253–276, 1954.

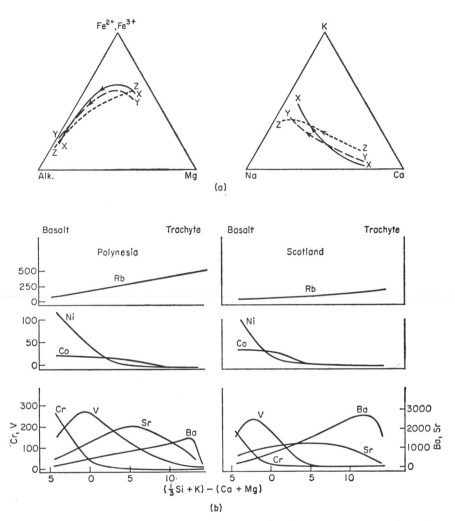

Fig. 30.   (a) Variation in the alkaline olivine basalt → mugearite → trachyte (or phonolite) series.   XX = Scottish Tertiary province; YY = Hawaiian alkaline series; ZZ = Polynesian alkaline series.   (After S. R. Nockolds and R. Allen.)   (b) Variation in trace-element content of two alkaline olivine basalt → trachyte series: Polynesian Islands (Marquesas, Society, Gambier, and Pitcairn); and Scottish Tertiary province.   The content of each element, in parts per million, is plotted against an index of differentiation (⅓ Si + K) − (Ca + Mg).   (After S. R. Nockolds and R. Allen.)

rocks among trachytic late differentiates is attributed to undersaturation of the parent basaltic magma in silica.[41] This undersaturation normally persists into the final differentiates. Both Lehmann and Kennedy attach great importance to the nature of pyroxene as a controlling factor. According to Kennedy, initially low silica content of the magma causes separation of abundant olivine, which reduces the Mg/Ca ratio in the liquid to such an extent that when pyroxene begins to form it is essentially diopsidic. This in turn leads to storing of alkali plus alumina in the residual liquid, so that the final fractions are highly alkaline and undersaturated in silica. (By contrast, Kennedy considers that, in continental tholeiitic basalt magmas, the silica content is high enough to limit separation of olivine, so that the pyroxenes which separate are magnesian pigeonites, and the end liquids—still oversaturated—are calc-alkaline, *i.e.*, rhyolitic.) It seems likely, as clearly stated by Barth, that variation in the nature of the final liquid residuum depends mainly on slight original variation in the parent basaltic magma. Notable departure in either direction from the saturated condition in the parent magma will normally persist in the end differentiates. There is also the possibility suggested by Edwards, that trachytic differentiates become converted into phonolitic liquids by addition of soda transported upward by streaming gas boiling from the magma after the roof of the magma chamber is fractured. Simultaneous oxidation of iron in pyroxene and addition of soda under these conditions could also account for the observed conversion of augite to aegirine and iron oxides in phonolites and trachytes.

In the alkaline olivine-basalt association, there is complete chemical gradation from olivine basalt to oligoclase basalt (mugearite), but transitional types between the latter and trachytes are rare or altogether absent. For example, Washington's analyses of Hawaiian lavas show an unfilled gap between 53 and 58 per cent SiO_2.[42] It would seem that while differentiation is generally efficient in the early stages (by gravitational settling of crystals) and the final stage of crystallization (by gas streaming or some similar process), its effects are counteracted during the intermediate stages by some other mechanism. Convection is a likely process which could have just such an effect on magma occupying a vertically elongated cupola beneath the vent of a growing volcano.[43] That such

[41] E. Lehmann, Beziehungen zwischen Kristallisation und Differentiation in basaltischen Magmen, *Min. Pet. Mitt.*, vol. 41, pp. 8–57, 1931; W. Q. Kennedy, Trends of differentiation in basaltic magmas, *Am. Jour. Sci.*, vol. 25, pp. 239–256, 1933; T. F. W. Barth, The crystallization process of basalt, *Am. Jour. Sci.*, vol. 31, pp. 321–351, 1936.

[42] T. F. W. Barth, *Die Entstehung der Gesteine* (Barth, Correns, Eskola), pp. 65, 66, Springer, Berlin, 1939.

[43] Cf. A. Holmes, The problem of the association of acid and basic rocks in central complexes, *Geol. Mag.*, vol. 68, pp. 241–255, 1931; Edwards, *op. cit.*, pp. 94, 95, 1938; Daly, *op. cit.*, pp. 1381–1383, 1944.

convection is at times effective, even when the residual liquids have reached the trachytic stage, is shown by the persistent occurrence of trachytic minerals such as aegirine, brown hornblende, and anorthoclase (in various stages of resorption) in the basic alkaline lavas of the Kerguelen archipelago, the east Otago province of New Zealand, and elsewhere.

Somewhere in this complex series of events, differentiation may lead to local development of sodic basic or ultrabasic lavas of the nepheline-basalt and melilite-basalt groups. The mechanism is not clear. At Kerguelen, the basic alkaline lavas are considered by Edwards to be differentiates (complementary to phonolite and trachyte) that separated from trachybasaltic, i.e., basic trachytic, magma in local cupolas. High concentration of "volatiles" is invoked, and convection is also believed to have been active. The picture envisaged by Edwards accords well with a mechanism mentioned by Bowen, namely, that crystals of hornblende, brought into contact with hot magma by convection, could by differential resorption (fusion) give rise to basic liquids notably enriched in the nepheline and pyroxene molecules; but this is not a likely mode of origin for nepheline basalts containing no trace of hornblende phenocrysts. Bowen has also shown that liquids rich in nepheline and pyroxene, in contact with olivine and pyroxene crystals, provide a medium form which melilite is liable to crystallize. The nepheline basalts, melilite basalts, and tephrites are too low in silica to be intermediate between basalt and phonolite, although this relation has been considered possible by some writers.[44] Edwards' scheme of complementary differentiation of trachy-basalt into trachyte and phonolite on the one hand and nepheline and melilite basalts on the other seems compatible with petrographic data at present available at Kerguelen, but it is not necessarily applicable to occurrences of nepheline basalt elsewhere. No entirely satisfactory hypothesis for the origin of nepheline-basalt magma has yet been put forward.

Assimilation probably plays only a minor role in the evolution of oceanic magmas. Hot largely liquid magma rising into the upper levels of the volcanic column possibly reacts with, and causes local differential fusion of, the rock walls of the cupola and so becomes enriched in material equivalent to trachyte. Assimilation of calcareous reef deposits or foraminiferal oozes is too remote a possibility to explain the repeated occurrence of feldspathoidal rocks according to the classic Daly hypothesis. In the continental environment there is more opportunity for assimilative reaction between magma and crustal rocks. Nevertheless the alkaline olivine-basalt association maintains its character uniformly without regard to lithological variation in the basement. There is a

[44] Cf. Barth, op. cit., p. 68, Fig. 44, 1939.

strong impression that the basaltic magma and its course of evolution have been but little affected by the immediately underlying basement.

Significance of Peridotite Nodules in Alkaline Basalts. In most alkaline olivine-basalt provinces, abundant nodules of peridotite occur locally in individual flows of olivine basalt, nepheline basalt, or melilite basalt.[45] Their composition is that of dunite or harzburgite: dominant magnesian olivine, abundant enstatite, minor but ubiquitous chrome-diopside, and picotite. In their coarse allotriomorphic-granular texture and the strong preferred orientation commonly shown by the olivine, they also resemble plutonic peridotites. By some writers the peridotite nodules are regarded as segregations, crystallized at an early stage from magmas corresponding to the rocks in which they are now enclosed.[46] This mode of origin is highly unlikely in view of the oriented character of the olivine fabrics and the presence in the nodules of minerals such as enstatite and picotite otherwise absent from the enclosing rocks; nor do the nodules resemble the known products of segregation (picrites) from alkaline basaltic magma. To meet these objections early very deep-seated ("intratelluric") crystallization of the nodules has been invoked,[47] and this is supported independently by numerous instances where xenoliths of eclogite and other metamorphic rocks are associated with olivine nodules. There is yet another possibility—that the nodules are fragments of deep-seated peridotite masses, perhaps even of a continuous peridotite "shell" beneath the earth's crust.[48] It has become increasingly clear that volcanic eruptions commonly tap magma that has risen rapidly from great depths. It is therefore not unreasonable to suppose that olivine nodules of alkaline basalts, in spite of their tendency to be concentrated in individual flows

[45] T. Ernst, Der Melilith-Basalt des Westberges, *Chemie der Erde*, vol. 10, pp. 631–666, 1936; C. Andreatta, Basalti della valle dell'alpone e loro inclusi peridotitici, *Soc. geol. italiana Boll.*, vol. 57, pp. 239–264, 1938; F. J. Turner, Preferred orientation of olivine crystals in peridotites, *Royal Soc. New Zealand Trans.*, vol. 72, pp. 280–300, 1942; G. A. Macdonald, Hawaiian petrographic province, *Geol. Soc. America Bull.*, vol. 60, pp. 1533, 1575, 1949; A. Harumoto, Melilite-nepheline basalt, its olivine-nodules, and other inclusions from Nagahama, Japan, *College Sci. Univ. Kyoto Mem.*, ser. B, vol. 20, pp. 69–88, 1952; C. S. Ross, M. D. Foster, and A. T. Myers, Origin of dunites and of olivine-rich inclusions in basaltic rocks, *Am. Mineralogist*, vol. 39, pp. 693–737, 1954; Brown, *op. cit.*, pp. 366–368, 1955; H. Kuno, K. Yamasaki, G. Iida, and K. Nagashima, Differentiation of Hawaiian magmas, *Japanese Jour. Geology and Geography*, vol. 28, no. 4, pp. 214, 215, 1957; J. F. Lovering, The nature of the Mohorovicic discontinuity, *Am. Geophys. Union Trans.*, vol. 39, pp. 947–955, 1958 (and discussion by H. Kuno and J. F. Lovering, *Am. Geophys. Union Trans.*, vol. 64, pp. 1071–1073, 1959).

[46] Macdonald, *op. cit.*, 1949.

[47] Cf. Andreatta, *op. cit.*, 1938; F. J. Turner and J. Verhoogen, *Igneous and Metamorphic Petrology*, 1st ed., p. 140, McGraw-Hill, New York, 1951; Harumoto, *op. cit.*, pp. 81, 82, 1952.

[48] Ernst, *op. cit.*, 1936; Ross, *et al.*, *op. cit.*, 1954; Kuno, *op. cit.*, 1959.

of a volcanic complex, may well have been transported from a very deep source—perhaps the outer part of the earth's mantle itself. This possibility, which the writers now are inclined to favor, is compatible with that section of current petrological opinion which places the source of primary olivine-basalt magma in the outer part of the mantle (cf. pages 233).

CHAPTER 8

Tholeiitic Flood Basalts and Intrusive Quartz Diabases

FLOOD BASALTS AND THEIR PARENT MAGMA TYPES

The term *plateau basalt*, long familiar in geological literature, and Tyrrell's more appropriate expression *flood basalt* are applied synonymously to basaltic lavas occurring as vast composite accumulations of subhorizontal flows which, erupted in rapid succession over great areas, have at times flooded sectors of the earth's surface on a regional scale.[1] It is generally believed that they are for the most part products of fissure eruptions. In some regions of flood basalts, dissection into the prevolcanic basement has revealed the existence of immense swarms of dikes and sills of diabase, chemically similar to the surface lavas with which they are clearly connected. Their great number, regular alignment over large areas, individual continuity over distances of many miles, and generally undeformed condition convey a vivid picture of basic magma welling up from the depths in a continuous flood along tension fractures, some of which must have extended deep into the crust to tap so copious a supply of basic liquid magma. While this broad picture of fissuring, injection, and surface eruption is further strengthened by the consistently parallel disposition of the flows and prevailing lack of quaquaversal patterns such as might indicate eruptive centers, absence of authentic instances of transition between individual flows and individual dikes remains an unexplained anomaly.

Many writers have commented upon the remarkable similarity of flood basalts from provinces of different age and situation. But as more chemi-

[1] For general characters of flood basalts and related dike and sill swarms see G. W. Tyrrell, Flood basalts and fissure eruptions, *Bull. volcanologique*, sér. 2, tome 1, pp. 89–111, 1937.

cal and petrographic data have become available it has now become clear that flood basalts tend to conform to either one of two standard compositions rather than to a single universal type. These two types of parent magma have been distinguished by W. Q. Kennedy as the tholeiitic and the olivine-basalt magma-types respectively.[2] For the second of these Tilley[3] considers more appropriate the term "alkali olivine-basalt" or "alkali series." This usage in a general way will be adhered to in this book. The alkaline olivine-basalt magma-type is the assumed parent of the alkaline olivine-basalt trachyte association, whether this is represented principally by flood basalts (e.g., in Palestine and Syria) or by products of more localized central eruptions (as in Tahiti or Samoa). The tholeiitic magma-type is the assumed parent magma of a distinctly different association of flood basalts and related diabases which form the main subject of this chapter. This particular type of flood basalt will henceforth be referred to as tholeiitic basalt or tholeiite, and corresponding rocks of associated dike and sill swarms will be grouped collectively as the quartz-diabase family.

While the alkaline olivine-basalt and tholeiitic kindreds develop in most parts of the world as distinct and clear-cut associations, there are some important provinces (e.g., Hawaii and the Brito-Arctic Tertiary region) in which lavas having affinities with both magma-types are mutually associated. This case will be discussed in the later part of this chapter.

THE THOLEIITE QUARTZ-DIABASE ASSOCIATION

Tholeiitic Flood Basalts and Associated Lavas.[4] *Field Occurrence.* The Deccan plateau (Cretaceous-Eocene) of western India and the Columbia-Snake River plains of Washington, Oregon, and Idaho (Mio-

---

[2] W. Q. Kennedy, The parent magma of the British Tertiary province, *Great Britain Geol. Survey Summary of Progress, 1930*, pt. 2, pp. 61–73, 1931; Trends of differentiation in basaltic magmas, *Am. Jour. Sci.*, vol. 25, pp. 239–256, 1933.

[3] C. E. Tilley, Some aspects of magmatic evolution, *Geol. Soc. London Quart. Jour.*, vol. 106, pp. 40–46, 1950.

[4] H. S. Washington, Deccan traps and other plateau basalts, *Jour. Geology*, vol. 33, pp. 765–804, 1922; T. M. Broderick, Differentiation in lavas of the Michigan Keweenawan, *Geol. Soc. America Bull.*, vol. 46, pp. 503–558, 1935; Kennedy, *op. cit.*, 1933; Tyrrell, *op. cit.*, 1937; A. B. Edwards, The Tertiary tholeiitic magma in Western Australia, *Royal Soc. Western Australia Jour.*, vol. 24, pp. 1–12, 1938; Some basalts from the North Kimberley, Western Australia, *Royal Soc. Western Australia Jour.*, vol. 27, pp. 79–89, 1942; C. D. Campbell, Petrology of the Columbia River basalts, *Northwest Science*, vol. 24, pp. 74–83, 1950; Washington geology and resources, *Washington State College Research Studies*, vol. 21, pp. 114–153 (especially 142–144), 1953; R. N. Sukheswala and A. Poldervaart, Deccan basalts of the Bombay area, India, *Geol. Soc. America Bull.*, vol. 69, pp. 1475–1494, 1958; W. D. West, The petrography and petrogenesis of forty-eight flows of Deccan trap, *Nat. Inst. Sci. India Trans.*, vol. 4, pp. 1–56, 1958.

cene) are classic instances of tholeiitic flood-basalt provinces. Other examples are the Keweenawan lavas of Lake Superior, the Stormberg lavas (Jurassic) of South Africa, the Triassic "traps" of New Jersey and the Paraná basalts (Jurassic) of South America, all of which have been dissected sufficiently to expose contemporaneously injected swarms of diabase dikes and sills. The Tertiary tholeiitic lavas of Western Australia belong to the same general kindred and may at one time have covered areas of comparable extent.

Perhaps the most striking character of provinces of this type is the enormous volume and extent of uniformly basaltic (tholeiitic) lava erupted. The Keweenawan lavas of Lake Superior are more than 15,000 ft. thick. The Deccan basalts, now covering 200,000 square miles, must at one time have occupied more than twice that area. Their maximum thickness, in the vicinity of Bombay, is given by Washington as nearly 10,000 ft., and the average as 2,000 ft. or more. In the Columbia-Snake River province the total area of basaltic rocks is about 200,000 square miles and the average thickness exceeds 3,000 ft. with a maximum of at least 6,000 ft. Volumes of the order of 50,000 to 200,000 cubic miles must have been erupted in these colossal outpourings. The extent of the South African, South American, and Australian basalt floods was apparently of a comparable order. In every case the environment is continental. Intercalated sediments are mainly of fresh-water or aeolian origin. The lavas themselves take the form of thin flows (averaging 15 ft. and not exceeding 50 ft. in the Deccan province, commonly 100 to 200 ft. in the Columbia River area), while tuffs and other explosion products are rare and local.

Petrographic Character of Tholeiites Contrasted with Alkaline Basalts. Considering the great extent and theoretical significance of tholeiitic flood basalts, petrographic data, especially those relating to composition of pyroxenes, feldspars, and olivines, are meager indeed. Some caution must be exercised therefore in evaluating the generalizations put forward by Kennedy, Barth, and others as to mineralogical distinctions between tholeiites and alkaline olivine basalts in general. The common tholeiitic flood basalt consists largely of labradorite and monoclinic pyroxenes, with minor iron ores and little or no olivine. In many rocks there are small quantities of residual glass which may be densely charged with dusty iron ore or may be altered to ferruginous chlorophaeite (sideromelane). Abundance of zeolites, chlorophaeite, various hydrated silicates containing iron (chlorites, celadonite, etc.), carbonates, and chalcedony testifies to the activity of late magmatic solutions charged with silica, carbon dioxide, iron, soda, and lime. According to Kennedy[5] two distinctive and persistent criteria of the tholeiitic basalts, as contrasted with alkaline

[5] Kennedy, *op. cit.*, p. 241, 1933.

olivine basalts, are prevalence of pigeonitic pyroxenes and the common presence of an acid residuum "which may be glassy but is dominantly quartzo-feldspathic."

More recent work [6] has confirmed and expanded this distinction as follows: In tholeiitic basalts and their coarse-grained diabasic equivalents, olivine, if present, is magnesian and unzoned; two pyroxenes are commonly associated, *viz.*, augite with pigeonite or with hypersthene or even with subcalcic augite (in a chilled groundmass); pyroxenes of late differentiates are strongly enriched in iron. By contrast, the olivine of alkaline olivine basalts is strongly zoned, with pronounced late enrichment in iron; but the pyroxene is relatively uniform in composition—a titan-augite closely approximating the diopside-salite series and seldom richer in ferrous ions than $Ca_{45}Mg_{25}Fe_{30}$. While it would seem likely, from evidence afforded by pegmatoids and by differentiated sills, that the glassy residuum of tholeiites should be more siliceous and less alkaline than that of olivine basalts, there is no obvious petrographic indication in most tholeiitic flood basalts that such is actually the case. In both classes of basalt the residuum is commonly partly or completely replaced by the ferruginous mineraloid chlorophaeite or is charged with opaque black particles. In some alkaline olivine basalts, however, the groundmass contains recognizable crystals of alkali feldspar.

Rhyolites and, more rarely, andesites or trachytes have been reported as minor members of the tholeiitic flood-basalt association,[7] but their occurrence commonly is purely local and their quantity insignificant compared with that of the tholeiites.

Chemical Composition of Tholeiites Contrasted with Alkaline Basalts. The composition of average tholeiitic flood basalt from each of several provinces is given in Table 15, analyses 1 to 7. Uniformity of composition throughout the whole group is obvious. Moreover comparison of average analyses 1, 2, and 3 (all representing Deccan basalts) shows that the way in which analyses have been selected and grouped is partly responsible for such variation as is shown in Table 15. The rhyolites which tend to be associated in minor amount with tholeiites are so different from the more alkaline frequently undersaturated trachytes and phonolites of olivine-basalt provinces that existence of some fundamental chemical dif-

[6] *E.g.*, H. H. Hess, Pyroxenes of common mafic magmas, *Am. Mineralogist,* vol. 26, pp. 515–535, 573–594, 1941; Edwards, *op. cit.,* 1938; *op. cit.,* 1942; A. Poldervaart and H. H. Hess, Pyroxenes in the crystallization of basaltic magma, *Jour. Geology,* vol. 59, pp. 472–489, 1951; J. F. G. Wilkinson, Clinopyroxenes of alkali olivine-basalt magma, *Am. Mineralogist,* vol. 41, pp. 724–743, 1956.

[7] *E.g.*, L. L. Fermor, On the lavas of Pávágad Hill, *India Geol. Survey Rec.,* vol. 34, pp. 148–166, 1906; F. Walker and A. Poldervaart, Karroo dolerites of the Union of South Africa, *Geol. Soc. America Bull.,* vol. 60, pp. 591–706, 1949 (especially pp. 693, 694); Sukheswala and Poldervaart, *op. cit.,* pp. 1491, 1492, 1958; West, *op. cit.,* p. 36, 1958.

ference between tholeiitic and alkaline olivine-basalt magmas seems likely. Kennedy, impressed also by the constant development of undersaturated alkaline pegmatoids (theralitic, teschenitic) in olivine basalts and of saturated quartz-dioritic pegmatoids in tholeiites, drew up what he considered to be the average compositions of fundamentally distinct tholeiitic and olivine-basalt magma-types. These averages are reproduced in columns 8 and 9 of Table 15. The tholeiitic type is said to be distinguished by notably higher SiO_2, higher alkali (especially K_2O), lower Al_2O_3, and much lower values for the MgO/CaO and $MgOFeO$ ratios. Comparison of average tholeiite magmas (e.g., Table 15, analyses 1 to 7) with average olivine-basalt magmas of both oceanic and continental provinces (e.g., Table 6, analyses 1 and 2; Table 11, analyses 1 and 2; Table 14, analyses 1 to 8) fails to sustain all these supposed differences. Contrary to Kennedy's generalization, most lavas of the alkaline olivine-basalt family are richer in both Na_2O and K_2O than typical tholeiitic flood basalts. The MgO/CaO ratio is clearly lower in tholeiitic basalts than in most alkaline olivine basalts; yet in the analysis of Scottish teschenite cited in Table 7, analysis 1, this ratio is just as low as in any tholeiitic basalt, in spite of the obvious relation of teschenites and their associated alkaline differentiates to the alkaline olivine-basalt magmatype. The percentage of SiO_2 is certainly higher in most tholeiitic magmas than in most alkaline olivine-basalt magmas; yet in the typical olivine-basalt trachyte phonolite province of Victoria, Australia (newer Tertiary volcanic series), the parent olivine-basalt magma (Table 15, analysis 11) computed from more than usually numerous data, has a silica content of 50 per cent and, except for the MgO/CaO ratio, is almost identical with Kennedy's tholeiitic magma-type. So R. A. Daly, in direct opposition to the thesis put forward by Kennedy, repeatedly expressed the view that a world-wide basaltic substratum of the lithosphere is the source of tholeiitic and olivine-basalt magmas alike and that there is no essential consistent chemical difference between the two types.[8]

To the present authors it seems established that there are real chemical differences between the tholeiitic and alkaline olivine-basalt magmatypes, although these are not as rigid nor always of the same nature as pictured by Kennedy. A tendency for SiO_2 to be higher and for Na_2O, K_2O, and MgO to be lower in tholeiites than in alkaline olivine basalts can be discerned in any set of representative analyses from petrographic provinces of both types. This tendency may be obscured, but is not invalidated, by identity of individual tholeiites with individual olivine basalts as regards one or more of the chemical characters just mentioned

---

[8] R. A. Daly, The geology of Ascension Island, Am. Acad. Arts Sci. Proc., vol. 10, No. 1, pp. 72–74, 1925; Volcanism and petrogenesis as illustrated in the Hawaiian Islands, Geol. Soc. America Bull., vol. 55, pp. 1391–1396, 1944.

TABLE 15.   COMPOSITIONS OF THOLEIITIC FLOOD BASALTS (ANALYSES 1 TO 8)
COMPARED WITH THOSE OF OLIVINE BASALTS (ANALYSES 9 TO 11)

Constit-uent	1	2	3	4	5	6	7	8	9	10	11
SiO₂	50.61	49.51	51.69	49.98	50.66	50.75	50.39	50	45	48.44	50
TiO₂	1.91	2.34	0.63	2.87	1.30	1.15	0.96			4.29	
Al₂O₃	13.58	13.05	14.72	13.74	14.28	13.80	14.80	13	15	13.27	15
Fe₂O₃	3.19	3.96	2.83	2.37	3.41	4.65	3.38	} 13	} 13	4.06	} 11.5
FeO	9.92	10.39	10.87	11.60	8.58	6.20	8.31			8.27	
MnO	0.16	0.22	0.11	0.24	0.12	0.10	0.20			0.15	
MgO	5.46	5.71	4.18	4.73	6.92	7.10	6.03	5	8	8.21	8.5
CaO	9.45	10.18	8.20	8.21	8.60	8.90	10.93	10	9	7.72	8.5
Na₂O	2.60	2.25	3.25	2.92	2.92	2.85	2.93	2.8	2.5	3.47	3
K₂O	0.72	0.51	0.93	1.29	0.72	0.85	0.57	1.2	0.5	1.55	1.2
H₂O+	} 2.13	1.99	2.01	} 1.22	} 2.28	2.35	0.28				
H₂O−		0.32	0.58			1.85	1.39				
P₂O₅	0.39	0.37	0.42	0.78	0.17	0.20	0.04			0.57	
Total	100.12	100.80	100.42	99.95	99.96	100.75	100.21	95	93	100.00	97.7

Explanation of Table 15

1. Average of 11 Deccan basalts. (H. S. Washington, Deccan traps and other plateau basalts, *Jour. Geology*, vol. 33, p. 797, No. 1, 1922.)
2. Average of 4 Deccan basalts, low in the volcanic formation. (L. L. Fermor, On the chemical composition of the Deccan trap flows of Linga, *India Geol. Survey Rec.*, vol. 68, p. 355, 1934.)
3. Average of 3 Deccan basalts, high in the volcanic formation. (Fermor, *op. cit.*, p. 355, 1934.)
4. Average of 6 Oregon basalts. (Washington, *op. cit.*, p. 797, No. 2, 1922.)
5. Average of 8 New Jersey basalts. (Washington, *op. cit.*, p. 797, No. 5, 1922.)
6. Basalt, Zuurberg, Cape, South Africa. (A. L. Du Toit, *The Geology of South Africa*, Oliver & Boyd, Edinburgh, p. 446, No. 24, 1926.)
7. Average of three Tertiary tholeiites, Western Australia. (A. B. Edwards, The Tertiary tholeiitic magma in Western Australia, *Royal Soc. Western Australia Jour.*, vol. 24, p. 7, No. 1, 1938.)
8. Tholeiitic magma-type. (W. Q. Kennedy, Trends of differentiation in basaltic magmas, *Am. Jour. Sci.*, vol. 25, p. 241, 1933.)
9. Olivine-basalt magma-type. (Kennedy, *op. cit.*, p. 241, 1933.)
10. Average (water-free) of 5 basalts, Tutuila, Samoa. (R. A. Daly, The geology of Saint Helena Island, *Am. Acad. Arts Sci. Proc.*, vol. 62, p. 71, 1927.)
11. Parent olivine-basalt magma (23 analyses) of olivine-basalt trachyte phonolite province of Victoria, Australia. (A. B. Edwards, *Geol. Soc. London Quart. Jour.*, vol. 94, p. 313, 1938.)

(*e.g.*, in SiO_2 or in Na_2O content).   Such instances suggest a transitional relation rather than a sharp cleft between the two magma-types.   Although the chemical distinction may be too subtle to be read directly from chemical analyses, it is perhaps more obviously and consistently visible in the norm (or alternatively in the Niggli values) calculated from the analyses.   A high value for normative hypersthene is characteristic of tholeiites.   High SiO_2 and low $(Na_2O + K_2O)$ and MgO, all favor appearance of normative quartz and exclusion of olivine and nepheline from the norm (the corresponding criterion in the Niggli system is a positive value for the quartz number).   The norms of the great majority of analyzed tholeiites show appreciable amounts of quartz; a few contain small amounts of olivine, while even traces of normative nepheline are very rare indeed.   Typical alkaline olivine basalts, on the other hand, are distinctly undersaturated so that olivine and in many cases nepheline appear conspicuously in the norms.   Sporadic occurrence of quartz trachyte and rhyolite with alkaline basalts and phonolites on oceanic islands does not invalidate the generalization that the less basic lavas of this association are normally saturated (trachytic) or undersaturated in silica (phonolitic).   Nor does the rare appearance of trachyte and oceanite in the Stormberg lavas of South Africa prevent recognition of the essentially tholeiitic nature of the Stormberg series as a whole.

Diabases of Continental Dike and Sill Swarms.[9]   Occurrence over vast areas is just as characteristic of intrusive diabase sills and dikes as it is of their strictly effusive equivalents, the tholeiitic flood basalts.   In South Africa horizontal or gently inclined sediments of the Karroo system (especially those of the Permo-Carboniferous Ecca series) were invaded in Jurassic times by a multitude of diabase intrusions which today outcrop intermittently over an area of at least 1,500,000 square miles.   Other spectacular occurrences of diabase sills and dikes cutting late Paleozoic sediments are well known in the Southern Hemisphere, notably in Tasmania and in Antarctica.   Of the total area (26,000 square miles) of the island of Tasmania, almost one-quarter is still covered, and perhaps one-

[9] A. Holmes and H. F. Harwood, The age and composition of the Whin sill and the related dikes of the north of England, *Mineralog. Mag.*, vol. 21, pp. 493–542, 1928; The tholeiite dikes of the north of England, *Mineralog. Mag.*, vol. 22, pp. 1–52, 1929; S. I. Tomkeieff, A contribution to the petrology of the Whin sill, *Mineralog. Mag.*, vol. 22, pp. 100–119, 1929; R. A. Daly and T. F. W. Barth, Dolerites associated with the Karroo system, South Africa, *Geol. Mag.*, vol. 67, pp. 97–110, 1930; G. W. Tyrrell and K. S. Sandford, Geology and petrology of the dolerites of Spitzbergen, *Edinburgh Geol. Soc. Proc.*, vol. 53, pt. 3, pp. 284–321, 1933; T. C. Phemister, A review of the problems of the Sudbury irruptive, *Jour. Geology*, vol. 45, pp. 2–8, 1937; F. Walker, Differentiation of the Palisade diabase, New Jersey, *Geol. Soc. America Bull.*, vol. 51, pp. 1059–1106, 1940; F. Walker and A. Poldervaart, The Hangnest sill, South Africa, *Geol. Mag.*, vol. 78, pp. 429–450, 1941; A. B. Edwards, Differentiation of the dolerites of Tasmania, *Jour. Geology*, vol. 50, pp. 451–480, 579–610, 1942; Walker and Poldervaart, *op. cit.*, 1949; F. Walker, H. C. C. Vincent, and R. L. Mitchell, The chemistry and mineralogy of the Kinkell tholeiite, Stirlingshire, *Mineralog. Mag.*, vol. 29, pp. 895–908, 1952; F. Walker, The dolerites of the Cape Peninsula, *Geol. Soc. South Africa Trans.*, vol. 59, pp. 1–16, 1957; W. Carey (editor), *Dolerite, a Symposium*, University of Tasmania, 1958.

half was formerly covered, by diabase sills distributed through a vertical thickness of 8,000 ft. of subhorizontal Permo-Carboniferous and Triassic-Jurassic sediments. In each of these provinces and in the Triassic province of New Jersey thick sills of great extent, injected parallel to the bedding of gently inclined or horizontal strata, are characteristic. The Mount Wellington sill of Tasmania exceeds 1,000 ft. in thickness and at one time had a total volume of the order of 3,000 cubic miles. The classic Palisade sill of New Jersey, in places over 1,000 ft. thick, outcrops continuously for a distance of 50 miles along the west bank of the Hudson River facing the City of New York. In the South African Karroo [10] true sills are less common than inclined sheets, whose curving outcrops suggest a basin-like form in many instances. Vertical dikes are extremely numerous in the South African province, where they characteristically range from 5 to 30 ft. in width and outcrop individually for distances of 5 to 40 miles. Long solitary dikes are much the commonest form of intrusive body in the Permo-Carboniferous tholeiite province of northern England and the Midland Valley of Scotland. Typically these are 60 to 100 ft. wide and many miles (in one 80 miles) in length. Their regular east-west trend is remarkably uniform. But sills are also represented in this region, the classic example being the Whin sill which has been traced, in nearly horizontal strata, for nearly 80 miles in northern England. Its thickness ranges from a few feet to over 250 ft.

Both chemically and petrographically the diabases of this association have been investigated in much greater detail than the tholeiitic basalts. Their general uniformity, regardless of age or geographic situation, is amazing, and their resemblance to the tholeiitic flood basalts is very close. Consistent mineralogical characters, as recorded in recent accounts of the Whin sill, the Hangnest sill and other Karroo diabases, the Palisade sill, and the sills and dikes of Tasmania, are as follows:

1. Olivine is completely lacking in many of the largest intrusions (*e.g.*, Whin sill; Mount Wellington sill of Tasmania). When present it is in small amounts except where, as in the lower levels of the Palisade sill, it appears to have accumulated by gravitational settling of crystals separating sparsely from a large body of magma. It is distinctly richer in iron than that of olivine basalts: Fa_{27} to Fa_{40} in many Karroo diabases; Fa_{30} to Fa_{35} in the Palisade sill. However, the iron content seldom reaches the high values typical of olivines in the upper portions of differentiated sills of teschenite.[11]

[10] These and other field data relating to the South African diabases are cited from A. L. DuToit, *Geology of South Africa*, pp. 283–291, Oliver and Boyd, Edinburgh, 1926.

[11] Cf. J. F. G. Wilkinson, The olivines of a differentiated teschenite sill near Gunnedah, New South Wales, *Geol. Mag.*, vol. 108, pp. 441–455, 1956.

2. Plagioclase consistently makes up 50 to 60 per cent of the composition wherever this has been micrometrically estimated for undifferentiated diabase. The average composition is An_{50} to An_{60}, with a range from An_{70-75} at the cores to An_{20-30} at the rims in zoned crystals. Rarely, for example, in some of the dikes of northern England, much more basic crystals (e.g., An_{90}), showing strong effects of reaction, have been recorded.

3. The content of pyroxene—30 to 45 per cent—is also remarkably consistent. In most diabases two varieties of pyroxene are present. Common alternative pairs are: (1) diopsidic augite ($2V = 45°$ to $55°$) associated with earlier or simultaneously formed bronzite, in chilled contact facies or throughout the lower levels of differentiated sills; (2) at higher levels, and representing a somewhat later period of crystallization, a rather more ferruginous augite associated with pigeonite in such a manner as to suggest broadly synchronous crystallization. Late pyroxenes (from the upper levels) may be strongly enriched in iron.[12]

4. A micrographic intergrowth of alkali feldspar and quartz is typically present in small quantities—up to 7 per cent in the Palisade sill, ranging from 3 per cent to 11 per cent in the Hangnest sill, and reaching 25 per cent in pegmatoid patches of the Whin sill.

5. Minor constituents include iron ores, biotite, hornblende, and apatite. Concentration of iron ore to values as high as 10 per cent in the late differentiates (e.g., in pegmatoids and in diabases from the upper levels of differentiated sills) has been noted by several writers and is attributed to high concentration of water and other volatile materials in residual liquids, developed during the final stages of crystallization.

Differentiation in place, following intrusion of an initially homogeneous essentially liquid diabase magma, has been demonstrated in various of the larger sills, notably the Palisade sill of New Jersey[13] and the Mount Wellington sill of Tasmania,[14] each of which somewhat exceeds 1,000 ft. in thickness. In both intrusions, within a few tens of feet of the upper and lower contacts a fine-grained basaltic phase of the diabase preserves the chemical characters of the original magma and shows that this was almost completely liquid at the time of intrusion. Apart from these contact facies, the sills exhibit a generally similar, though not identical, vertical variation in chemical and mineralogical composition that can readily be explained on the assumption that crystals of early-formed

[12] Wilkinson, op. cit., Am. Mineralogist, p. 728, 1956.

[13] L. V. Lewis, Petrography of the Newark igneous rocks of New Jersey, New Jersey Geol. Survey Ann. Rept. for 1907, pp. 97–167, 1908; The Palisade diabase of New Jersey, Am. Jour. Sci., vol. 26, pp. 155–162, 1908; N. L. Bowen, The Evolution of the Igneous Rocks, pp. 71–73, Princeton University Press, Princeton, N. J., 1928; Walker op. cit., 1940; Hess, op. cit., pp. 525–526, 529, 1941.

[14] Edwards, op. cit., Jour. Geology, 1942.

pyroxene (and in the Palisade sill, olivine) sank under the influence of gravity until the process was arrested by increasing viscosity and diminution in quantity of the liquid phase of the magma. This hypothesis has been generally accepted.[15] Details of the observed vertical variation are reproduced in Figs. 31 to 33. Attention is drawn particularly to the following results of the differentiation process, since they afford a clue to the general course which differentiation of large bodies of saturated basaltic magma might be expected to take at deeper levels within the crust of the continents:

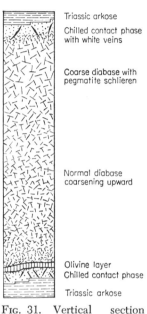

Triassic arkose

Chilled contact phase with white veins

Coarse diabase with pegmatite schlieren

Normal diabase coarsening upward

Olivine layer
Chilled contact phase

Triassic arkose

Fig. 31. Vertical section through the Palisade sill, New Jersey. (*After F. Walker.*)

1. In the Palisade sill olivine accumulated in sufficient quantity to form an olivine-diabase layer, 15 ft. thick, 30 to 60 ft. from the base of the sill, *i.e.*, immediately above the lower chilled border layer. The total composition of the magma (crystals plus liquid) in this zone was thereby sufficiently changed to cause a shower of more ferruginous olvine (Fa_{30} to Fa_{35}) to develop locally. In places the olivine-diabase layer contains 25 per cent of olivine.

2. Pyroxenes are notably concentrated in the lower third of the sills, where diopsidic augite is associated with bronzite-hypersthene. In the Tasmanian sill the latter shows a gradual increase in iron ($FeSiO_3 : 15 \rightarrow 30$ per cent) at successively higher levels, while the augite maintains a constant composition approximately to $Ca_{35}Mg_{60}Fe_5(SiO_3)_{100}$. The most ferruginous composition reached by hypersthene in other sills corresponds to between 30 and 35 per cent $FeSiO_3$.

3. Magnesian pyroxenes and any olivine that may have formed in the upper two-thirds of the sills have not survived; either they have sunk into the lower levels or they have been eliminated higher up by reaction with the liquid phase. Two monoclinic pyroxenes, both showing enrichment in iron at successively higher levels, are mutually associated in the upper portions of the intrusions. Augite changes progressively (upward) in composition from $Ca_{35}Mg_{60}Fe_5$

---

[15] For discussion of the feasibility of extensive crystal settling in a sill of diabase magma see J. C. Jaeger and G. Joplin, Rock magnetism and the differentiation of dolerite sills, *Geol. Soc. Australia Jour.*, vol. 2, pp. 1–19, 1955; F. Walker, The magnetic properties and differentiation of dolerite sills, a critical discussion, *Am. Jour. Sci.*, vol. 254, pp. 433–451, 1956.

$(SiO_3)_{100}$ to $Ca_{25}Mg_{50}Fe_{25}(SiO_3)_{100}$, while associated pigeonite changes from $Ca_5Mg_{65}Fe_{30}(SiO_3)_{100}$ to $Ca_{10}Mg_{20}Fe_{70}(SiO_3)_{100}$.

4. Correlated with upward increase in proportion of plagioclase to pyroxene is a gradual increase in the albite content of the former mineral.

5. The grain-size of the rock coarsens notably from below upward—a fact which probably reflects upward concentration of water and other

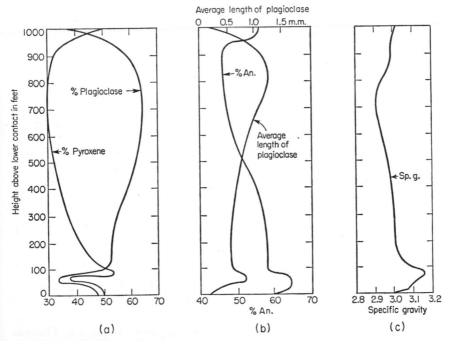

Fig. 32. Petrographic variation in a vertical section through the Palisade sill, New Jersey. (*a*) Percentages (by weight) of plagioclase and of pyroxene; (*b*) mean length of plagioclase crystals (in millimeters), and composition (percentage An) of plagioclase; (*c*) specific gravity of diabase. (*After F. Walker.*)

volatiles as the liquid phase of the magma was gradually displaced upward by sinking crystals.

6. The specific gravity of the whole rock increases downward (cf. Fig. 32*c*). In the last stages of crystallization of quartz-diabase sills, pegmatoid streaks are formed at various levels but are specially numerous in the upper parts of the intrusions. These have crystallized from aqueous residues but differ chemically from the normal diabase only in being slightly richer in alkali and iron and correspondingly poorer in magnesia. They consist of andesine-labradorite, augite, intergrown alkali feldspar and quartz, iron ore, biotite, and hornblende. Fine-grained veins of

quartz-albite-pyroxene rock represent the final highly sodic residues expelled from the almost completely consolidated mass.

Although gravitational differentiation has been demonstrated in quartz-diabase bodies such as the Palisade and Mount Wellington sills, the range of chemical and mineralogical variation so achieved within a vertical distance of a thousand feet is not spectacular (Figs. 32 and 33). The extreme range of silica content in the Palisade sill is from 48.3 per cent (olivine diabase) to 52.2 per cent (upper quartz diabase); in the Mount

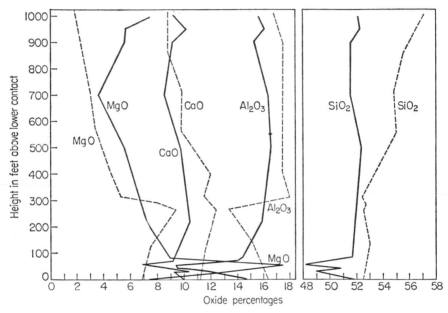

FIG. 33.   Oxide profiles (vertical) for diabase sills of Mount Wellington, Tasmania (broken lines), and Palisade, New Jersey (full lines). (*After A. B. Edwards, with additional data from F. Walker, Geol. Soc. America Bull., vol. 51, Table 3, p. 1080, 1940.*)

Wellington sill it is of much the same order (52.5 per cent near the base to 57 per cent near the top). The micropegmatitic mesostasis which is so characteristic of this family of rocks indicates clearly the granophyric composition toward which the final residues from fractional crystallization were trending. But the quantities of these acid residues are very small, they are sodic rather than potassic, and there is a general lack of dioritic products bridging the gap between quartz diabase and granophyre. A diabase sill near Dillsburg, Pennsylvania,[16] contains a thick zone of sodic hedenbergite granophyre (Table 17, analysis 3) formed by

[16] P. E. Hotz, Petrology of granophyre in diabase near Dillsburg, Pa., *Geol. Soc. America Bull.*, vol. 64, pp. 675–704, 1953.

TABLE 16. CHEMICAL COMPOSITION OF REPRESENTATIVE QUARTZ
DIABASES AND THEIR DIFFERENTIATION PRODUCTS

Constituent	1	2	3	4	5	6	7	8	9	10	11
SiO_2	48.28	52.32	52.50	57.05	52.65	51.91	50.52	52.25	51.9	50.61	50.1
TiO_2	0.82	0.97	0.60	0.75	0.58	1.25	2.39	1.10	1.3	1.91	1.0
Al_2O_3	9.36	16.54	13.39	16.95	16.23	15.31	13.76	14.60	15.1	13.58	15.1
Fe_2O_3	2.14	1.58	0.90	2.60	0.51	0.98	3.87	0.84	1.3	3.19	3.1
FeO	11.54	8.66	8.72	8.64	8.21	9.31	8.50	9.89	9.0	9.92	7.4
MnO	0.12	0.12	0.21	0.10	0.15	0.08	0.16	0.45	0.2	0.16	0.2
MgO	17.48	5.43	9.16	1.77	6.64	7.52	5.42	6.95	6.6	5.46	7.2
CaO	7.00	9.68	12.25	8.60	11.34	9.71	9.09	9.71	10.0	9.45	10.0
Na_2O	1.59	2.32	0.80	2.09	1.58	2.30	2.42	2.21	2.1	2.60	2.0
K_2O	0.41	1.03	0.70	1.56	0.90	0.79	0.96	0.96	0.9	0.72	0.8
H_2O+	0.99	0.84	0.21	0.12	0.48	0.93	1.51	0.71	0.9	} 2.13	1.3
H_2O-	0.06	0.40	0.48	0.24	0.85	0.15	0.76	0.32	0.5		1.6
P_2O_5	0.11	0.06	Trace	0.10	0.01	0.18	0.26	0.22	0.2	0.39	0.2
CO_2		0.16					0.58				Trace
Total	99.90	100.11	99.92	100.57	100.13	100.42	100.20	100.21	100.0	100.12	100.0

Explanation of Table 16

1. Palisade sill, New Jersey. Olivine diabase from olivine-rich layer. (F. Walker, Differentiation of the Palisade diabase, New Jersey, *Geol. Soc. America Bull.*, vol. 51, Table 3, 1940.)
2. Palisade sill, New Jersey. Quartz diabase, 500 ft. above lower contact. (Walker, *op. cit.*, Table 3, 1940.)
3. Mount Wellington sill, Tasmania. Diabase enriched in pyroxene (magnesian), 275 ft. above lower contact. (A. B. Edwards, Differentiation of the dolerites of Tasmania, *Jour. Geology*, vol. 50, p. 467, No. 3, 1942.)
4. Mount Wellington sill, Tasmania. Diabase with dioritic affinities, 1,025 ft. above lower contact. (Edwards, *op. cit.*, p. 467, No. 10, 1942.)
5. Average undifferentiated diabase, Tasmania; 6 analyses. (Edwards, *op. cit.*, p. 465, No. 1, 1942.)
6. Average undifferentiated diabase, Palisade sill; taken from Walker, *op. cit.*, 1940. (Edwards, *op. cit.*, p. 465, No. 5, 1942.)
7. Average Whin sill magma; 6 analyses. (A. Holmes and H. F. Harwood, The age and composition of the Whin sill and the related dikes of the north of England, *Mineral Mag.*, vol. 21, p. 539, A, 1928.)
8. Average Karroo diabase, South Africa; 5 analyses. (R. A. Daly and T. F. W. Barth, Dolerites associated with the Karroo system, South Africa, *Geol. Mag.*, vol. 67, p. 101, No. 6, 1930.)
9. Average of columns 5 to 8.
10. Average Deccan basalt. (H. S. Washington, Deccan traps and other plateau basalts, *Jour. Geology*, vol. 33, p. 797, No. 1, 1922.)
11. Average Drakensberg basalt, Stormberg series, South Africa; 4 analyses. (F. Walker and A. Poldervaart, Karroo dolerites of the Union of South Africa, *Geol. Soc. America Bull.*, vol. 60, p. 694, Table 27a–d, 1949.)

differentiation in place. Granophyric rocks are conspicuous too in the marginal portions of various diabase sills invading the Karroo sediments in South Africa, but in some at least (e.g., the Hangnest sill) gravitational differentiation has been ineffective, and the granophyres prove to be products of reaction between the diabase magma and the invaded silt-stones, the chemical composition of which happens to be appropriate for ready conversion into potassic granophyre.

TABLE 17. LATE DIFFERENTIATES OF QUARTZ DIABASE

Constituent	1	2	3	4
SiO$_2$	66.80	72.4	61.69	66.04
TiO$_2$	0.18	0.2	1.46	1.03
Al$_2$O$_3$	12.10	13.1	12.61	12.72
Fe$_2$O$_3$	0.97	1.1	2.98	2.48
FeO	1.50	1.6	8.32	6.55
MnO	n.d.	n.d.	0.15	0.11
MgO	0.50	0.5	0.77	0.54
CaO	2.62	2.8	4.04	2.65
Na$_2$O	2.40	2.6	5.71	4.62
K$_2$O	4.20	4.6	0.57	2.26
H$_2$O+	5.75		0.83	0.84
H$_2$O−	3.00		0.15	0.19
P$_2$O$_5$	0.39	0.4	0.50	0.22
BaO	0.29	0.3		
F	0.01			
S	0.4	0.4		
Total	100.71	100.0	99.78	100.25

Explanation of Table 17

1. Residual glass in tholeiitic diabase, Kinkell, Scotland. (F. Walker, H. C. G. Vincent, and L. R. Mitchell, The chemistry and mineralogy of the Kinkell tholeiite, Sterlingshire, *Mineralog. Mag.*, vol. 29, p. 900, No. 3, 1952.)
2. The same as 1 recalculated water-free.
3, 4. Granophyres in diabase, Dillsburg, Pennsylvania. (P. E. Hotz, Petrology of granophyre in diabase near Dillsburg, Pa., *Geol. Soc. America Bull.*, vol. 64, p. 690, 1953.)

The first four columns of Table 16 show the range of chemical composition (exclusive of pegmatoids and aplitic veins) resulting from differentiation within the Palisade and Mount Wellington sills. The average compositions of the diabase magma for four different provinces are given in columns 5 to 8, followed by an average of all four in column 9. This latter probably represents the mean composition of tholeiitic magma more accurately than average values computed from analyses of "plateau basalts"—especially as some of these are members of the alkaline olivine-basalt association. Apart from complementary differences in Fe_2O_3 and Al_2O_3, this average diabase of the tholeiitic association agrees closely with Washington's average Deccan basalt (Table 16, analysis 10), and even more with average Drakensberg basalt (analysis 11). The figures cited in Table 16 consistently confirm the conclusion expressed on page 207 that the tholeiitic magma-type tends to be distinctly higher in SiO_2 and lower in MgO and in $(Na_2O + K_2O)$ than the alkaline olivine-basalt magma-type.

Compositions of some late differentiates of quartz diabase are given in Table 17. Analyses 1 and 2 are of a residual glass separated from a Scottish diabase. They are essentially dacitic, and like most dacites have K_2O in excess of Na_2O. By contrast granophyres associated with diabase sills (analyses 3 and 4) typically are sodic rocks.

ASSOCIATION OF THOLEIITIC AND ALKALINE OLIVINE-BASALT MAGMAS

Introduction. There are some volcanic provinces, e.g., the Hebridean province of Britain, and the Hawaiian Islands, where tholeiitic lavas are broadly associated with alkaline olivine basalts, nepheline basalts, and trachytes. The relation of the two suites of rocks in any such province is not a random one. Rather the two tend to alternate with or to succeed each other in localized subprovinces; hence for any one restricted area over a limited period of time only one of the two main magmas is erupted.

The Hawaiian Volcanic Province.[17] The eight principal islands of Hawaii lie on an arc about 300 miles in length stretching from long. $155\frac{1}{2}°W.$, lat. $19\frac{1}{2}°N.$, to long. $160°W.$, lat. $22°N.$ The islands are the projecting summits of immense, largely submerged shield volcanoes, with a total exposed area of over 6,000 square miles. The largest island,

[17] H. S. Washington, Petrology of the Hawaiian Islands, *Am. Jour. Sci.*, vol. 5, pp. 465–502, 1923; vol. 6, pp. 100–126, 338–367, 1923; vol. 15, pp. 199–220, 1928; T. F. W. Barth, Mineralogical petrography of the Pacific lavas, *Am. Jour. Sci.*, vol. 21, pp. 377–405, 491–530, 1931; G. A. Macdonald, petrographic sections in *Div. Hydr. Terr. Hawaii Bull.* 5, pp. 63–91, 1940 (Oahu); *Bull.* 6, pp. 149–173, 1940 (Kahoolawe); *Bull.* 9, pp. 187–208, 1946 (Hawaii); *Bull. 11*, pp. 89–110, 1948 (Molokai); *Bull. 12*, pp. 41–51, 1947 (Niihau); Hawaiian petrographic province, *Geol. Soc. America Bull.*, vol. 60, pp. 1541–1596, 1949; H. Winchell, Honolulu series, Oahu, Hawaii, *Geol. Soc. America Bull.*, vol. 58, pp. 1–48, 1947; Tilley, *op. cit.*, vol. 106, pp. 39–41, 1950; H. A. Powers, Composition and origin of basaltic magma of the Hawaiian Islands, *Geochim. et Cosmochim. Acta*, vol. 7, pp. 77–107, 1955; I. D. Muir and C. E. Tilley, Contribution to the petrology of Hawaiian basalts, 1, *Am. Jour. Sci.*, vol. 255, pp. 241–253, 1957; H. Kuno, Discussion of paper by J. F. Lovering, The nature of the Mohorovicic discontinuity, *Jour. Geophys. Res.*, vol. 64, p. 1071, 1959.

Hawaii itself, has been built up by five volcanoes, the two highest of which exceed 13,600 ft. in elevation.   Two are still active.

The dominant rocks of Hawaii are olivine basalts; locally associated with these in the later stages of volcanism are mugearites ("oligoclase andesites"), nepheline and melilite basalts, and trachytes.   Until recently the whole association was considered to belong to the alkaline olivine-basalt type that is so characteristic of other mid-oceanic islands.   But it has been clearly demonstrated [18] that the main volcanic edifice is built of tholeiitic lavas and their olivine-enriched picrite-basalt differentiates. The following stages of volcanic evolution are exemplified by different islands:

1.   Copious outpouring of highly fluid tholeiitic basalt—the "primitive" basalts of Hawaiian geologists—building a chain of great shield volcanoes with summits emerging above sea level.   The most active volcanoes, Kilauea and Mauna Loa, still exemplify this stage of activity.   So also do the older rocks (Koolau series) of Oahu.   The main rock types,[19] in order of abundance, are olivine basalts, basalts with little or no olivine, picrite basalts, and hypersthene basalts.   The olivine is magnesian; monoclinic pyroxene phenocrysts are diopsidic augite and are never abundant; groundmass pyroxene is pigeonite or subcalcic augite; plagioclase is labradorite.   Chemically all rocks are characterized by relatively high silica and lime and low alkalies.   For representative analyses see Table 18, analyses 1 to 4.   Most rocks, in spite of the presence of abundant modal olivine, contain normative quartz, and those that do have normative olivine are only slightly undersaturated in silica.   Especially characteristic, even in picrite basalts, is a high content of normative hypersthene.   There is a general absence of peridotite xenoliths such as occur plentifully in lavas of the third stage; but gabbroic inclusions have been recorded.

2.   A declining stage of activity, marked by caldera collapse and building of cinder cones.   This is exemplified today by Mauna Kea and Haleakala.[20]   Rocks belong to the alkaline olivine-basalt association: [21] olivine basalts with phenocrysts of augite and plagioclase, picrite basalts rich in both olivine and augite, mugearites ("oligoclase andesites"), and some trachytes.   Relatively high alkalies and low silica are characteristic. In the norms of most rocks olivine and some nepheline are present, hyper-

[18] Tilley, op. cit., 1950; Powers, op. cit., 1955.

[19] C. K. Wentworth and H. Winchell, Koolau basalt series, Oahu, Hawaii, Geol. Soc. America Bull., vol. 58, pp. 49–78, 1947; G. A. Macdonald, Petrography of the island of Hawaii, U. S. Geol. Survey Prof. Paper 214-D, 1949; Powers, op. cit., pp. 78, 79, 1955.

[20] G. A. Macdonald and H. A. Powers, Contribution to the petrography of Haleakala volcano, Hawaii, Geol. Soc. America Bull., vol. 57, pp. 115–124, 1946.

[21] Cf. Powers, op. cit., pp. 79, 93, 1955.

sthene absent or present only in small amount. Representative analyses
are shown in Table 18, analyses 5 to 8. In their high lime content, reflecting abundant augite, the picrite basalts of this stage of activity resemble
the differentiated picrite layers of teschenitic sills such as that of Lugar
(p. 183, Table 9, analysis 5). Xenoliths of peridotite and of gabbro occur
in some basalts of this series.

3. A late stage of renewed volcanic activity, following prolonged erosion
in several islands. Lavas of this stage are the Honolulu series of Oahu,[22]
and the Koloa series of Kaui.[23] They include picrite basalts, olivine
basalts, nepheline basanites, and nepheline-melilite basalts. For analyses
of typical lavas, see Table 18, analyses 9 and 10. Inclusions of ultrabasic
and plutonic rocks, of which dunite and harzburgite are by far the most
plentiful, are abundant in many flows; eclogite xenoliths occur locally.[24]

To those who find it necessary to postulate a single parent magma for
the Hawaiian province, the tholeiitic magma typified by the lavas of
Kilauea and Mauna Loa is the obvious choice.[25] It is by far the dominant magma-type in the islands, it is remarkably uniform in composition,
and its eruption precedes the alkaline olivine-basalt eruptions in time.
It is clear, then, that a primary tholeiitic magma has been available in
great quantity throughout the main shield-building phase of every
Hawaiian volcano. And simple fractionation of olivine, crystallizing
early from such a magma, can account for the formation of associated
picrite basalts on the one hand and olivine-free and hypersthene basalts
on the other. It is improbable, however, that all the tholeiitic lavas of
the Hawaiian Islands have been drawn from a single continuous reservoir.
Powers[26] has argued convincingly that each major eruptive cycle of a few
hundred years' duration represents the generation and eruption of a
unique batch of magma, which in its primitive state was tholeiitic. The
depth of origin, as inferred from the foci of earthquakes preceding eruption, may be between 30 and 35 miles—well below the base of the
crust.

The mode of origin and the evolution of the alkaline olivine-basalt
magmas in the later stages of eruption are debatable. Macdonald and
Tilley would derive such magmas from a primitive tholeiitic magma by
fractional crystallization.[27] But Powers raises objections to so simple a
scheme of petrogenesis:[28]

[22] Winchell, op. cit., 1947.
[23] Macdonald, op. cit., Geol. Soc. America Bull., pp. 1555–1558, 1949.
[24] Cf. Powers, op. cit., p. 93, 1955; Kuno, op. cit., 1959.
[25] Macdonald, op. cit., Geol. Soc. America Bull., pp. 1567–1570, 1949.
[26] Powers, op. cit., pp. 91–98, 1955.
[27] Macdonald, op. cit., Geol. Soc. America Bull., pp. 1586–1587, 1949; Tilley,
op. cit., p. 45, 1950.
[28] Powers, op. cit., p. 95, 1955.

TABLE 18.   CHEMICAL COMPOSITION OF HAWAIIAN LAVAS

Constituent	Tholeiitic types				Alkaline basalts and related lavas					
	1	2	3	4	5	6	7	8	9	10
SiO$_2$	50.45	50.97	50.42	45.97	42.30	47.98	51.99	62.19	36.75	42.86
TiO$_2$	2.33	2.14	2.97	1.75	2.41	3.53	3.02	0.37	2.41	2.94
Al$_2$O$_3$	14.94	13.72	11.62	5.98	10.52	15.32	16.30	17.43	11.98	11.46
Fe$_2$O$_3$	3.38	2.39	2.71	5.86	4.22	2.49	2.75	1.65	6.05	3.34
FeO	7.55	7.61	9.07	7.39	9.70	8.86	7.44	2.64	7.45	9.03
MnO	0.08	0.11	0.10	0.11	0.06	0.12	0.11	0.32	0.08	0.13
MgO	7.67	10.18	10.11	23.55	14.90	6.16	3.19	0.40	12.08	13.61
CaO	9.17	8.51	9.74	6.47	12.08	10.28	6.67	0.86	13.81	11.24
Na$_2$O	2.84	2.56	2.09	1.50	1.56	3.56	5.64	8.28	4.75	3.02
K$_2$O	0.35	0.61	0.39	0.42	0.42	1.08	2.13	5.03	0.91	0.93
H$_2$O$-$	0.23	0.08		0.04	0.45	0.25	0.07	0.14	0.36	0.12
H$_2$O$+$	0.73	0.61		0.64	0.87	0.62	0.29	0.39	1.61	0.44
CO$_2$								0.02		
P$_2$O$_5$	0.27	0.28	0.27	0.21	0.33	0.22	1.25	0.14	1.41	0.52
Cr$_2$O$_3$	0.05	0.07			0.11				0.03	0.04
BaO						0.06		0.03	0.13	0.04
Total	100.04	99.84	99.49	99.89	99.93	100.53	100.85	99.89	99.81	99.72

Explanation of Table 18

1. Average of 10 basalts, Koolau Series, Oahu. (C. K. Wentworth and H. Winchell, Koolau basalt series, Oahu, Hawaii, *Geol. Soc. America Bull.*, vol. 58, p. 71, Table 4, A, 1947.)
2. Hypersthene basalt with phenocrysts of olivine hypersthene and labradorite, Koolau Series, Oahu. (Wentworth and Winchell, *op. cit.*, p. 71, Table 4, No. 9980, 1947.)
3. Average of 24 analyses of lavas from Mauna Loa, Hawaii. (G. A. Macdonald, Petrography of the island of Hawaii, *U. S. Geol. Survey Prof. Paper* 214-D, p. 74, No. 2, 1949.)
4. Picrite basalt, Mauna Loa, Hawaii. (Macdonald, *op. cit.*, p. 63, No. 3, 1949.)
5. Picrite basalt, Haleakala, Hawaii. (G. A. Macdonald and H. A. Powers, Contribution to the petrography of Haleakala volcano, Hawaii, *Geol. Soc. America Bull.*, vol. 57, p. 119, No. 1, 1946.)
6. Olivine basalt, Polulu Series, Kohala, Hawaii. (Macdonald, *op. cit.*, p. 87, No. 5, 1949.)
7. Mugearite ("oligoclase andesite"), Hawi Series, Kohala, Hawaii. (Macdonald, *op. cit.*, No. 11, 1949.)
8. Trachyte, Hualalai, Hawaii. (Macdonald, *op. cit.*, p. 78, No. 9, 1949.)
9. Nepheline-melilite basalt, Honolulu Series, Oahu. (H. Winchell, Honolulu series, Oahu, Hawaii, *Geol. Soc. America Bull.*, vol. 58, p. 30, No. 17, 1947.)
10. Nepheline basanite, Honolulu Series, Oahu. (Winchell, *op. cit.*, p. 30, No. 13, 1947.)

The difference in chemistry between a given decadent suite [akaline] and a given primitive suite [tholeiitic] . . . cannot be expressed in quantities of phenocryst olivine, hypersthene, augite and plagioclase, either added to or subtracted from either suite; part of the difference is always an excess or lack of silica. No satisfactory explanation has yet been proposed for this desilication based on fractional crystallization alone. . . . The only explanation for the compositions of these rocks that offers much promise seems to call for fractional crystallization superimposed on actual removal of free silica, or addition of soda and lime, or a combination of both.

There is still another possibility which the present writers tentatively prefer: It may be that in the course of a single Hawaiian eruptive cycle the nature of the primary magma changes at its source, so that generation and eruption of tholeiitic magma is followed in each cycle by generation and eruption of alkaline olivine-basalt magma. This is the view of Kuno,[29] who derives the two magmas from two different parent materials melting under different conditions at different levels beneath the crust.

The Volcanic Association of the Brito-Arctic Province. The petrographic province variously referred to as the Brito-Arctic, Thulean, or North Atlantic province [30] covers a large area extending from northwest Britain 2,000 miles northwestward to the eastern coast of Greenland. Within it also lie the Faeroe Islands, Iceland, Jan Mayen,

[29] H. Kuno, K. Yamasaki, C. Iida, and K. Nagashima, Differentiation of Hawaiian magmas, *Japanese Jour. Geology and Geography*, vol. 28, no. 4, pp. 179–218, 1957; H. Kuno, Origin of Genozoic petrographic provinces of Japan and surrounding areas, *Bull. volcanol.*, sér. 2, tome 20, pp. 59–60, 1959.

[30] Selected references:

1. *Scotland:* A. Harker, The Tertiary igneous rocks of Skye, *United Kingdom Geol. Survey Mem.*, 1904; E. B. Bailey, H. H. Thomas, and others, The Tertiary and post-Tertiary geology of Mull, Loch Aline and Oban, *Scotland Geol. Survey Mem.*, 1924; G. W. Tyrrell, The geology of Arran, *Scotland Geol. Survey Mem.*, 1928; J. Richey, H. H. Thomas, and others, The geology of Ardnamurchan, northwest Mull and Coll, *Scotland Geol. Survey Mem.*, 1930; Kennedy, *op. cit.*, 1931; J. E. Richey, British regional geology: Scotland, The Tertiary volcanic districts, *Scotland Geol. Survey Mem.*, 1935; J. E. Richey, The dykes of Scotland, *Edinburgh Geol. Soc. Trans.*, vol. 13, pp. 419–432, 1939; G. P. Black, The tertiary volcanic succession of the isle of Rhum, Inverness-shire, *Edinburgh Geol. Soc. Trans.*, vol. 15, pp. 39–51, 1952.

2. *Northern England:* Holmes and Harwood, *op. cit.*, pp. 1–52, 1929.

3. *Arctic and Subarctic regions:* G. W. Tyrrell, The petrography of Jan Mayen, *Royal Soc. Edinburgh Trans.*, vol. 54, pt. 3, pp. 747–765, 1926; Tyrrell and Sandford, *op. cit.*, vol. 53, pp. 284–321, 1933; F. Walker and C. F. Davidson, A contribution to the geology of the Faeroes, *Royal Soc. Edinburgh Trans.*, vol. 58, pp. 869–897, 1936; A. Noe-Nygaard, Sub-glacial volcanic activity in ancient and recent times, *Fol. Geog. Danica*, Tm. 1, No. 2, 1940; G. P. L. Walker, Geology of the Reydarfjördur area, eastern Iceland, *Geol. Soc. London Quart. Jour.*, vol. 114, pp. 367–394, 1959.

4. *Ireland:* E. M. Patterson and D. J. Swaine, A petrochemical study of Tertiary tholeiitic basalts: the middle lavas of the Antrim plateau, *Geochim. et Cosmochim. Acta*, vol. 8, pp. 173–181, 1955.

Spitzbergen, and Franz Josef Land. Together these constitute the broken edges and scattered remnants of a basalt-flooded plateau which is believed largely to have foundered in the late Tertiary. The marginal remnant, which comprises the Hebrides and adjoining western coast of Scotland, is classic ground in igneous petrology, for it has been the subject of exhaustive field, petrographic, and chemical investigations, the results of which are embodied in a number of memoirs of the Geological Survey of Scotland. It is impossible in a short space to do justice to the wealth of critical petrological data provided by these works. But it is appropriate to outline briefly the chemistry, possible relationships, and evolutionary development of the basaltic magmas which conspicuously dominate the whole province.

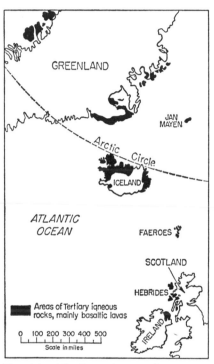

Some parts of the Brito-Arctic province, notably the Arctic island of Jan Mayen, are built up entirely of alkaline olivine basalts (trachybasalts), with associated oceanites, trachyandesites, and trachytes, just as in many island provinces of the ocean basins in general. Tyrrell describes the Jan Mayen rocks as forming a "mildly alkaline series," with an alkali-lime index of 52, and distinctly richer in potash than is usual in oceanic lavas.

FIG. 34. The Brito-Arctic volcanic province. (*After G. W. Tyrrell and M. A. Peacock.*)

There are other areas where the lavas or dike rocks are essentially tholeiitic. Such are the swarm of Tertiary tholeiite dikes that trend west-northwest to northwest across the north of England and merge into the contemporaneous swarm of Mull and the adjoining Scottish coast. The early Cretaceous diabases and basalts of Spitzbergen, regarded by Tyrrell and others as local forerunners of igneous activity which in Eocene times spread over the whole Brito-Arctic province, are also uniformly tholeiitic.

More typical of the Brito-Arctic region in general is close association and alternation of great thicknesses of alkaline olivine basalts and of tholeiitic lavas. Thus in the Faeroes, Walker and Davidson have recorded a generalized sequence of 15,000 ft. of basaltic lavas; of this the upper

2,000 ft. is built of alkaline olivine basalts and subordinate oceanites, the middle portion is predominantly tholeiitic, and the lower 6,000 ft. is composed of both types in alternation. In eastern Iceland both types of basaltic lava are present in association with subordinate rhyolite (8 per cent of the total) and andesite (3 per cent).

In the Hebridean-Irish region, too, alkaline olivine-basalt and tholeiitic lavas were poured out in profusion to build a basalt plateau, remnants of which still occupy an area of over 2,000 square miles. Locally, for example in the island of Mull, their total thickness may reach 6,000 ft. While it is possible that the plateau lavas were erupted from a network of fissures, it seems more probable that they represent dissected remnants of a series of shield volcanoes of the Hawaiian type. Certainly such a volcano was responsible for the plateau lavas of Mull. Although the rocks of the plateau group are dominantly basaltic (alkaline olivine basalts and tholeiites), they include local flows of mugearite, trachyte, andesite, and rhyolite; in Skye there is a series of these acid and intermediate lavas and tuffs 2,000 ft. in total thickness. Following eruption of the plateau lavas was a prolonged period of alternating explosive and intrusive activity, localized at a number of independent centers (e.g., Skye, Mull, Ardnamurchan, and the Irish center of Slieve Gullion) located along a north-south line over 200 miles in length (cf. Fig. 35). Detailed mapping of these centers in each case has revealed a complex series of events making up this part of the igneous cycle: subsidence of earth blocks a few miles in diameter along annular fractures, which are then injected with magma in various stages of evolution to give massive ring dikes; related surface manifestations such as development of calderas of subsidence and eruption of lava floods, development of concentric conical fractures fanning out upward from magma cupolas situated at depths of two or three miles, and injection of magma to form swarms of closely spaced cone-sheets; drilling of explosion vents, especially in connection with eruption of acid magmas; shifting of individual centers and repetition of one or more of the above phases of activity. Rocks of this whole period vary from gabbros and local ultrabasic intrusives to abundantly developed granophyres, granites, rhyolites, and trachytes. Here, as also in Iceland, the relative abundance of thoroughly acid granitic and rhyolitic rocks and rarity of intermediate (e.g., andesitic) types are striking. The final stages of activity in the Hebridean-Irish region were marked by widespread injection of dikes and sills of alkaline olivine basalt (diabase and teschenite) and tholeiite, in swarms which, while specially concentrated at individual centers of earlier intrusion, spread on all sides far beyond the limits of the area previously covered by the plateau lavas. The tholeiitic dikes of northern England are related in this way to the dike swarm emanating from the intrusive center of Mull (cf. Fig. 35).

Petrographically and chemically, the Tertiary igneous rocks of the Hebridean region fall into the following principal series (special terminology of the Scottish Geological Survey memoirs is noted in italics):

1. Alkaline olivine-basalt flows and equivalent intrusive rocks (olivine diabases, gabbros, teschenites or crinanites) together constituting the *plateau magma-type*. Mugearites, trachytes, and local segregations of

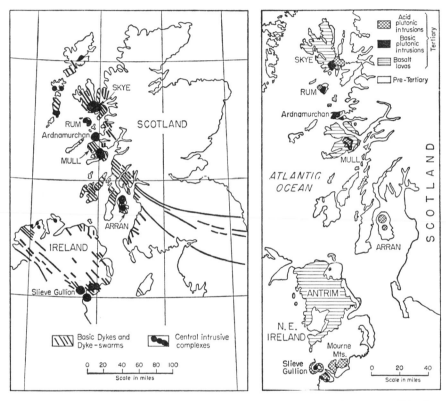

FIG. 35.   (a) Tertiary dike swarms and plutonic centers of north Britain; (b) Tertiary lavas and plutonic rocks of western Scotland and northern Ireland.   (*After J. E. Richey.*)

alkali syenite (*alkaline magma-series*) are regarded as differentiates of the alkaline olivine-basalt magma, and complete the analogy with the normal alkaline olivine-basalt association of other provinces.

2. Tholeiitic basalts and diabases occurring as flows, dikes, cone sheets, and sills (*nonporphyritic central magma-type*).

3. Andesitic and rhyolitic lavas and breccias, and intrusive quartz porphyries, granophyres, and granites of the various central complexes (*acid magma-series*).   Acid members greatly predominate.

4. Porphyritic basalts, with abundant phenocrysts of basic plagioclase, and equivalent feldspathic gabbros (*porphyritic central magma-type*).

5. Ultrabasic olivine gabbros and peridotites (*eucrite-allivalite magma-series*).

Opinions differ as to the relationships of these various "magma-types" and "magma-series." Most writers think that the acid magma-series is the result of progressive differentiation of the tholeiitic (nonporphyritic central) magma by fractional crystallization and that the alkaline magma-series is similarly derived from the alkaline olivine-basalt (plateau) magma. Bowen suggested—and his view was accepted by the authors of the Ardnamurchan memoir—that the porphyritic central magma was derived from the olivine-basalt type by separation and accumulation of crystals of calcic plagioclase. Accumulation of olivine is likewise believed to be responsible for the eucrite-allivalite series, and from work in Ardnamurchan it would seem that the porphyritic central magma is the direct parent in this case. This general scheme of magmatic evolution, as outlined in the Ardnamurchan memoir, may be summarized thus:

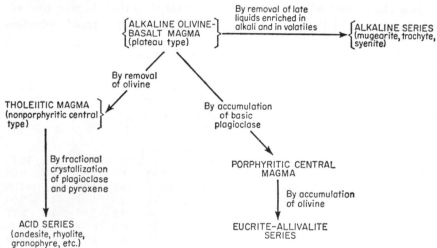

An alternative scheme favored by Kennedy is given for comparison:

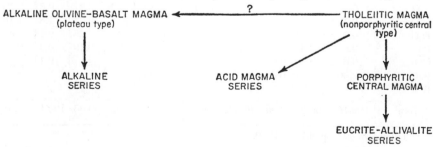

Tilley,[31] arguing from analogy between the basaltic rocks of the Hebridean province and those of Hawaii, derives the plateau magma-type from a tholeiitic parent type (nonporphyritic central).

The above should be regarded as alternative hypotheses of petrogenesis, subject to confirmation or disproof in the light of data which may later be collected in other provinces where similar rocks occur. In the meantime, we note the occurrence, in close association, of tholeiitic and alkaline olivine-basalt magmas, each accompanied by what we have learned elsewhere to recognize as its characteristic suite of related rock types; and we bear in mind the possibility that the two magmas may be of independent origin.

In Table 19 compositions of some typical Brito-Arctic basaltic rocks are given for comparison with rocks of purely alkaline olivine-basalt or purely tholeiitic associations in other parts of the world.

PETROGENESIS OF THE THOLEIITE QUARTZ-DIABASE ASSOCIATION

It is clear, from what has already been stated in this chapter, that extensive sectors of the continents widely varying in geographic situation

TABLE 19.   COMPOSITIONS OF BASIC IGNEOUS ROCKS OF BRITO-ARCTIC TERTIARY PROVINCE

Constituent	A				B				C	
	1	2	3	4	5	6	7	8	9	10
SiO_2	45.9	45.5	44.68	49.24	49.2	50.41	50.5	49.24	47.9	46.40
TiO_2	3.4	3.1	2.51	1.84	2.9	1.30	1.8	2.46	1.3	3.05
Al_2O_3	16.0	15.0	16.37	15.84	14.4	15.14	13.7	14.12	19.8	16.30
Fe_2O_3	3.3	3.5	4.31	6.09	3.4	2.71	3.8	1.51	4.5	3.60
FeO	7.8	10.4	8.11	7.18	10.1	7.95	9.2	12.68	4.3	7.17
MnO	0.3	0.2	0.32	0.29	0.4	0.17	0.3	0.08	0.2	0.23
MgO	5.9	8.1	6.59	3.02	5.4	6.57	5.2	5.79	5.0	6.00
CaO	10.3	8.7	8.70	5.26	9.4	11.30	9.7	10.62	12.3	11.04
Na_2O	3.2	2.6	3.28	5.21	2.0	2.29	2.8	2.17	2.2	2.14
K_2O	2.2	0.4	0.21	2.10	1.0	0.82	1.1	0.37	0.3	0.29
H_2O+	0.4	1.7	1.69	1.61	} 1.6	1.01	} 2.2	0.32	1.8	1.10
H_2O-	0.2	1.0	2.99	1.08		0.72		0.28	0.1	2.40
P_2O_5	0.6	0.3	0.15	1.47	0.2	0.15	0.3	0.19	0.3	0.23
CO_2	0.6	0.1	0.06			0.07			0.1	
BaO	0.06		0.02	0.09		0.03				
Rest						0.11				
Total	100.16	100.6	99.99	100.32	100.00	100.75	100.6	99.83	100.1	99.95

[31] Tilley, op. cit., pp. 43–45, 1950.

have at different times been copiously injected and flooded with a completely liquid basaltic magma of uniform, slightly oversaturated composition. This is the tholeiitic magma. So uniform are its characters that any hypothesis of origin reconstructed for an individual province must apply also to all other tholeiitic provinces in order to be accepted as satisfactory. Current opinions as to its origin and subsequent evolution include the following: [32]

Hypothesis of Bowen.[33] According to Bowen, basaltic magmas originate in quantity by differential fusion of an ultrabasic (probably peridotite) layer beneath the earth's crust. The parent magma welling up from this source presumably could vary somewhat in composition.

[32] The literature dealing with the origin and evolution of primary basaltic magma is so extensive that it is impossible to treat this topic historically and at the same time concisely. Thus the views summarized under the heading Hypothesis of Kennedy include some previously put forward by E. Lehmann and much that has subsequently been confirmed or modified by Tomita, Kuno, Edwards, and others.

[33] Bowen, op. cit., pp. 74–78, 315–320, 1928; Magmas, Geol. Soc. America Bull., vol. 58, pp. 263–280, 1947 (especially pp. 268–269, 271–274).

Explanation of Table 19

A. Olivine basalts and related rocks (cf. plateau magma-type and alkaline magma-series):
  1. Average trachybasalt, Jan Mayen; 4 analyses. (G. W. Tyrrell, The petrography of Jan Mayen, Royal Soc. Edinburgh Trans., vol. 54, p. 706, Nos. 3–6, 1926.)
  2. Average basalt of plateau type, Mull; 3 analyses. (Mull Memoir, p. 15, Nos. i–iii, 1924.)
  3. Analcite-olivine diabase (crinanite), Arran, Hebridean province. (G. W. Tyrrell, Arran Memoir, p. 121, No. 1, 1928.)
  4. Mugearite, Skye. (A. Harker, Skye Memoir, p. 263, 1904.)
B. Tholeiitic basalts and related rocks (cf. nonporphyritic central magma-type):
  5. Average diabase, Spitzbergen; 4 analyses. (G. W. Tyrrell and K. S. Sandford, Geology and petrology of the dolerites of Spitzbergen, Edinburgh Geol. Soc. Proc., vol. 53, p. 312, 1933.)
  6. Olivine tholeiite, Kielderhead, Northumberland. (A. Holmes and H. F. Harwood, The tholeiite dikes of the north of England, Mineral. Mag., vol. 22, p. 16, 1929.)
  7. Average of tholeiitic (nonporphyritic central) rocks of Hebridean province; 8 analyses. (Mull Memoir, p. 17, Nos. i–viii, 1924; computed by Tyrrell and Sandford, op. cit., p. 312, 1933.)
  8. Pleistocene subaerial basalt flow, Laxárnes, Iceland. (A. Noe-Nygaard, Subglacial volcanic activity in ancient and recent times, Fol. Geog. Danica, Tm. 1, No. 2, p. 42, 1940.)
C. Rocks of the porphyritic central type:
  9. Average basaltic lava of porphyritic central type, Mull; 3 analyses. (Mull Memoir, p. 24, Nos. iii–v, 1924.)
  10. Basalt with porphyritic feldspars, Höjvig, Faeroes. (F. Walker and C. F. Davidson, A contribution to the geology of the Faeroes, Royal Soc. Edinburgh Trans., vol. 58, p. 890, B, 1936.)

Bowen originally favored an undersaturated alkaline olivine-basalt magma (plateau magma-type) as the parent material of the Mull province, and from it he derived the tholeiitic magma (nonporphyritic central type) by removal of early-formed crystals of olivine, calcic plagioclase, and to a less extent pyroxene. Efficient removal of olivine, separating in excess during the early stages, could leave the remaining liquid enriched in silica, and so give rise to a tholeiitic magma, from which granophyric and rhyolitic and granophyric differentiates could ultimately develop. In a later paper Bowen [34] stated clearly his contrary belief that "the uniform character of the abundant regional diabase dikes in all terranes, evidently coming from great depths," stamps the corresponding tholeiite magma as "representing primary basaltic magma if anything does."

Hypothesis of Daly.[35] Daly advocated a common origin for oceanic and continental basalts of both types. All were regarded as originating in the vitreous basaltic substratum (abyssolith), originally pictured as a continuous shell encircling the whole globe, but later thought to be to some extent discontinuous across parts of the oceanic sectors of the earth. Daly minimized chemical differences between continental and oceanic basalts; he noted, however, that beneath the oceans the substratum is probably somewhat more femic than beneath the continents and he tentatively suggested that observed variations in composition of primary basaltic magmas might be due in part to differences in the levels at which a gravitationally stratified substratum was tapped. Rhyolitic differentiates of certain continental basalt provinces were correlated with preliminary contamination of the basaltic magma by assimilative reaction with the overlying "granitic" shell prior to extrusion. To the authors, Daly's interpretation of the chemical data seems not completely adequate, in that he compared an average continental plateau basalt (based on analyses of alkaline olivine-basaltic as well as tholeiitic types) with average Hawaiian basalt also based on analyses of both types. Comparison of average Deccan basalt, or any of the average diabases of Table 16, with average olivine basalt from Tahiti (Table 10, analysis 1) or Saint Helena (Table 11, analysis 1) or the east Otago province of New Zealand (Table 4, analysis 1) brings out a distinctly different picture. In other words, the chemical data, like most statistics, are capable of several interpretations. Daly's comparison minimized what other writers still consider to be a fundamental difference between alkaline olivine-basalt and tholeiitic magmas. It showed, however, that the differences between oceanic and

[34] N. L. Bowen, Magmas, *Geol. Soc. America Bull.*, vol. 58, p. 269, 1947.

[35] R. A. Daly, *Igneous Rocks and the Depths of the Earth*, pp. 200–202, McGraw-Hill, New York, 1933; *op. cit.*, pp. 1391–1396, 1944; Nature of the asthenosphere, *Geol. Soc. America Bull.*, vol. 57, pp. 723, 724, 1946.

continental basalts were not so sharp nor so universally prevalent as was thought by many geologists.

Hypothesis of Kennedy.[36] According to Kennedy, a world-wide olivine-basalt shell is succeeded upward, in continental regions only, by a tholeiitic layer, and this in turn by a granitic shell.[37] Periodic local fusion of one or other of these basaltic layers, under assumed appropriate tectonic and physical conditions, is supposed to account for local independent development of alkaline olivine-basalt and of tholeiitic magmas, the latter being restricted always to the continental environment. Fractional crystallization in the alkaline olivine-basalt magma series leads to normal development of the succession olivine basalt → trachyandesite → trachyte → phonolite. In the tholeiitic series the differentiation sequence is tholeiitic basalt → andesite → rhyolite. An essential part of Kennedy's hypothesis is that the original composition of the parent magma is the prime factor controlling differentiation. From alkaline olivine-basalt magma, as a result of its low silica content, olivine separates early, thereby impoverishing the residual liquid in magnesia. Consequently, when pyroxene begins to separate it is essentially diopsidic rather than pigeonitic, and the final liquid fraction that ultimately develops is enriched in alkali rather than lime. The end product is thus alkaline trachyte or phonolite, with a tendency to undersaturation in silica, inherited from the parent olivine-basalt magma. On the other hand, the tholeiitic magma, by virtue of its initially oversaturated condition, tends to give quartz-bearing differentiates, and the calc-alkaline composition of these is attributed to earlier separation of lime-poor pigeonitic pyroxenes which in turn is correlated with nonappearance of olivine in the earliest stages. There are thus three independent parts to Kennedy's hypothesis, any of which may be accepted or rejected without prejudice to the other two:

1. Alkaline olivine-basalt and tholeiitic magmas originate independently as primary magmas.

2. Differentiation in the one case follows the sequence olivine basalt → trachyandesite → trachyte → phonolite; in the other it follows the sequence tholeiite → andesite → rhyolite.

3. Differentiation in both series is a process of fractional crystallization in which the nature of the pyroxene that crystallizes (diopside in alkaline olivine basalts, pigeonite in tholeiites) is the main factor determining the ratio of alkali to lime in the late liquid fractions.[38]

[36] Kennedy, op. cit., 1933; Crustal layers and the origin of magmas, Bull. volcanologique, sér. 2, tome 3, pp. 24–41, 1938.

[37] The pros and cons regarding this hypothetical structure of the crust are considered in Chap. 15.

[38] Pyroxene control of differentiation in basaltic magmas had already been advocated by E. Lehmann (Beziehungen zwischen Kristallization und Differentiation in basaltischen Magmen, Min. pet. Mitt., vol. 41, pp. 8–57, 1931).

Hypothesis of Tomkeieff.[39]   From an analysis of the chemical and petrographic characters of the Carboniferous and Permo-Carboniferous igneous rocks of the Midland Valley of Scotland, Tomkeieff concluded that these fall into three related differentiation series, within each of which fractional crystallization is responsible for most of the observed variation. But an entirely different process termed "alkali-volatile diffusion" is appealed to, to account for derivation of the respective parent magmas of two of the series from the third.   The primitive magma is assumed to be alkaline olivine basalt (from which, in one differentiation series, mugearites, trachytes, etc., are derived by normal fractional crystallization). By upward diffusion of alkalis, assisted and accompanied by volatiles, this olivine basalt becomes converted to teschenitic magma (the parent for the second series) in the upper levels of the differentiation chamber. The residue in the lower levels assumes the composition of a less alkaline, more siliceous quartz diabase (the parent of the third differentiation series).   The process admittedly is vaguely conceived.   But in view of the many variable and largely unpredictable factors that might influence ionic diffusion in a complex multicomponent silicate liquid rich in water, it can scarcely be denied that such a mechanism might operate in the direction postulated by Tomkeieff.   At least it offers an explanation of phenomena which cannot be correlated with simple fractional crystallization of phases so far investigated in the laboratory.   But there are grave objections to it as a broad hypothesis of petrogenesis.   Among the many investigated instances of differentiated bodies of basic igneous rock there should surely be some in which the downward sequence is teschenite, olivine diabase, quartz diabase.   The writers are unaware of any such case having been recorded.   Nor can Tomkeieff's hypothesis be extended to account for the origin and upsurge of vast floods of tholeiitic magma (in most cases unaccompanied by magma of other kinds).

Critical Review.   Final discussion of possible conditions of origin and nature of primary magmas, and hence of basaltic magmas in particular, must be postponed until yet other igneous rock associations have been reviewed.   Even then it will be found that there is room for wide divergence of opinion on this subject.   But at this juncture it is already clear that the floods of basaltic magma which from time to time have been injected into the upper crust and poured out on its surface, in the majority of cases, closely conform to one or the other of two chemically distinct types, the one alkaline olivine basalt, the other tholeiitic basalt.   This is an important generalization on which Kennedy in particular has focused attention.   Its significance is not to be disregarded merely because alkaline

[39] S. I. Tomkeieff, Petrochemistry of the Scottish Carboniferous-Permian igneous rocks, *Bull. volcanologique*, sér. 2, tome 1, pp. 59–87, 1937.

olivine basalts in some provinces approach the tholeiitic type, and *vice versa*; nor because in other cases (*e.g.*, in the Brito-Arctic province) both types, accompanied by their respective trachytic and rhyolitic derivatives, are contemporaneously associated. The existence of the two magma-types is here accepted as established, although it is recognized that magmas of intermediate composition may also occur.

Now arises another question on which current opinion also is divided: are the two magma-types of independent origin, or is one a derivative of the other? That some common factor of origin exists may be suspected from the relative narrowness of the compositional gap between typical alkaline olivine-basalt and typical tholeiitic magmas. There are several possibilities as to the nature and degree of this relationship:

1. Kennedy [40] postulated that in Mull the tholeiitic magma was the primitive material from which the alkaline olivine-basalt magma was derived. Tilley [41] argues the same relationship for the two principal magmas of the Hawaiian Islands, and in support of his view he cites the great preponderance of tholeiitic over alkaline basaltic lavas and the early appearance and continued eruption of tholeiitic lavas through the most active stages of the volcanic cycle (see p. 218). There are two weighty objections to this relationship. First, while it is now known that tholeiitic lavas may occur in oceanic as well as in continental regions, their geographic distribution is more restricted than that of alkaline olivine basalts. The latter occur in profusion on all the continents and are also recognized as constituting "the characteristic and most widespread assemblage of the oceanic islands." [42] In the second place, neither the more basic types of tholeiitic basalt (Table 15, analysis 2) nor the basic olivine- and pyroxene-rich differentiates of quartz-diabase sills (Table 16, analyses 1 and 3), nor yet the primitive picrite basalts of the tholeiitic series (Table 18, analysis 4) closely approximates in composition typical alkaline olivine basalt.

2. The case for alkaline olivine basalt as the primitive magma and tholeiite as the derivative magma rests particularly on the world-wide occurrence of alkaline basalts in great quantities and receives some support from the occurrence of peridotite nodules, possibly derived from beneath the earth's crust, in alkaline basaltic lavas. It has been weakened by recent recognition that the ocean basins are not exclusively the domain of the alkaline olivine-basalt magma. Various writers, including Bowen, Holmes, and Kuno, have supported the hypothesis that the alkaline basaltic magma gives rise directly to tholeiitic magma by a differentiation

---

[40] Kennedy, *op. cit.*, 1931.
[41] Tilley, *op. cit.*, p. 37, 1950.
[42] Tilley, *op. cit.*, p. 42, 1950.

process involving removal of early-formed minerals, especially olivine.[43] However, differentiation of basaltic magma would enrich the derivative liquid fraction in alkalis as well as in silica; and tholeiite, though more siliceous, is less alkaline than the supposed parent alkaline olivine-basalt magma. The difficulty is not resolved if, instead of differentiation, we suppose that a parent alkaline olivine-basalt magma has become somewhat acidified by reaction with the overlying granitic material of the sial. In any case this latter possibility must be discarded if we accept the hypothesis of genesis of potassic basic rocks of the leucite-basalt and lamprophyre families by granite contamination of alkaline olivine-basalt magma (cf. pages 249 and 250).

3. A third alternative, which could satisfactorily account for the mutual association of, and in some cases the apparently transitional relation between, the two principal types of basaltic magma, is differentiation of a parent magma of intermediate composition into alkaline olivine-basalt and tholeiitic fractions. But association and mutual transition should on this hypothesis be the rule, not the exception, and the dominantly continental distribution of the tholeiitic association would have to be explained on other grounds. It has been suggested tentatively by Barth [44] that the typical world-wide basalt of a crustal substratum has a composition corresponding to R. A. Daly's "average plateau basalt," [45] and that this has locally undergone differentiation by "alkali volatile diffusion" (as advocated by Tomkeieff) to give an alkali-enriched alkaline olivine-basalt fraction underlain by an alkali-impoverished tholeiitic fraction. The eruptible component of the assumed basaltic substratum along the mid-Atlantic swell is supposed to be an alkaline olivine basalt originating in this way. The only evidence cited in favor of this hypothesis is that the mid-Atlantic lavas of Gough Island and Saint Helena (typical oceanic alkaline olivine basalts) are more alkaline than "average plateau basalt." Furthermore, we note that the ultimate products of "alkali volatile diffusion" in the mid-Atlantic (e.g., average basalt of Saint Helena) are almost identical with the *parent* basaltic magma from which the teschenitic and tholeiitic magmas are supposed by Tomkeieff to have originated in a similar manner in Scotland.

Clearly there are good reasons for rejecting each of the above alternative direct relationships between alkaline olivine-basalt and tholeiitic

[43] Bowen, *op. cit.*, pp. 74–78, 1928; Holmes and Harwood, *op. cit.*, pp. 537, 541, 1928; H. Kuno, Fractional crystallization of basaltic magmas, *Japanese Jour. Geology and Geography*, vol. 14, pp. 189–208, 1937.

[44] T. F. W. Barth, Lavas of Gough Island, *Sci. Results Norwegian Antarctic Expedition 1927–1928 (Lars Christensen)*, No. 20, p. 17, 1942.

[45] The reader is reminded that Daly's "average plateau basalt" is computed from analyses of typical tholeiitic basalts together with some olivine basalts of the flood type (notably those from the Brito-Arctic province).

magmas. There remains the possibility of a more general and perhaps indirect relationship, and this, for want of a more satisfactory explanation, we tentatively accept.[46] The two magmas could well represent products of either of two processes: (1) large-scale fusion at different levels or at different situations within a variable basaltic substratum; (2) partial fusion of a peridotite layer at somewhat different temperatures. If the existence of a basaltic substratum is assumed, then without necessarily accepting the detailed scheme of petrogenesis postulated by Kennedy (to be discussed in Chap. 15), we need only assume further that the basalt substratum of the earth's crust varies, or has varied, vertically, laterally, and possibly in time, between tholeiitic and alkaline olivine-basalt extremes of composition, and that from time to time conditions favoring generation of magma have developed and have been sustained locally within the substratum. On this assumption it would follow that the tholeiitic component of the substratum is either absent from some large oceanic sectors of the lithosphere or is present only at levels where conditions requisite for large-scale fusion are seldom realized. Beneath the "granitic" shell of the continents, however, either alkaline olivine-basalt or tholeiitic magma could develop, according to local conditions. If, on the other hand, the existence of a basaltic layer in the crust is denied, alternative development of tholeiitic and alkaline olivine-basalt magmas beneath the continents could be attributed (though of necessity somewhat vaguely) to differential fusion of peridotite over a wider range of temperature than would normally be the case beneath the simpler crustal structure of ocean basins.

Once tholeiitic magma is generated in quantity and ascends toward the surface, its subsequent evolution is relatively clear. An early crystalline fraction enriched in magnesian olivine and in pyroxenes is differentiated from a liquid fraction somewhat higher in silica and alkalis, and with a notably higher FeO/MgO ratio, than the parent magma. The oversaturated nature of the tholeiitic magma is preserved and even enhanced in its derivative residual liquids; and it is this, according to some writers,[47] that determines the mineralogical peculiarities—especially the character of the pyroxenes—in the tholeiitic differentiation series. That extreme differentiation may lead at last to liquids of dacitic or granophyric composition is proved by analyses of the glassy mesostasis isolated from each of two tholeiites.[48] These contain over 6 per cent ($Na_2O + K_2O$), 67 per cent

[46] Kuno, Yamasaki, Iida and Nagashima (op. cit., pp. 212–216, 1957) also advocate independent origins for the two contrasted magma-types of Hawaii; and Kuno (op. cit., 1959) explains similarly the contrasted basic magmas of Japan.

[47] Cf. Wilkinson, op. cit., Am. Mineralogist, pp. 724, 735–740, 1956.

[48] F. Walker, The late Paleozoic quartz dolerites and tholeiites of Scotland, Mineralog. Mag., vol. 24, p. 150, No. 4, 1935; F. Walker, H. C. C. Vincent, and R. L. Mitchell, op. cit., p. 900, Nos. 4, 5, 1952.

SiO_2, and 31 to 36 per cent normative quartz. Late liquids in the tholeiitic line of descent are also represented by bodies of granophyre in quartz-diabase sills.[49] Some chemical trends attributed to fractional crystallization of tholeiitic magma are shown graphically in Fig. 36.

Differentiation in quartz-diabase sills seldom leads to separation of large bodies of acid (granophyric) residual liquid. The restricted nature of differentiation within such bodies is noteworthy. Edwards[50] attributes it to the insufficient vertical distance afforded by floored magma

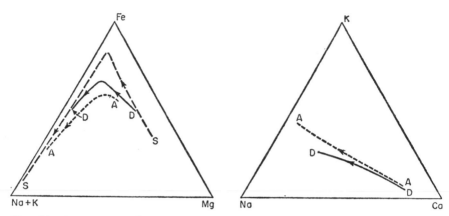

Fig. 36.   Variation in a diabase → granophyre series, Dillsburg, Pennsylvania. (*After P. E. Hotz.*) DD = diabase → granophyre.   For comparison are given SS (Skaergaard ferrogabbro series, cf. p. 294) and AA (alkaline basalt → trachyte series, Hawaii).

bodies for effective separation of sinking olivine and pyroxene from crystallizing feldspar.   He suggests that the process is more efficient in subjacent chambers of batholitic proportions and that small quantities of andesitic and rhyolitic magma could thus be generated and concentrated there.   It is also possible that some kind of reaction between tholeiitic magma and the rocks of the sial, followed by appropriate differentiation, may be responsible for the subordinate rhyolites and andesites reported in some tholeiite provinces, or even that these are products of independent fusion of the lower part of the sial.

[49] *E.g.*, Hotz, *op. cit.*, 1953.
[50] Edwards, *op. cit.*, *Jour Geology*, pp. 603–609, 1942.

CHAPTER 9

Potash-rich Basic Volcanic Rocks
and the Lamprophyres

THE LEUCITE-BASALT POTASH-TRACHYBASALT ASSOCIATION

General Statement. A number of interesting and perplexing questions
arise in connection with the highly potassic continental volcanic association dominated by basic and ultrabasic lavas containing leucite. Its occurrence is very restricted as compared with the olivine-basalt trachyte
kindred, but it is represented at widely scattered points on all the continents and its chemical and petrographic individuality is both remarkable
and constant. Well-known instances include the Roman province of
western Italy (as exemplified by the lavas of Vesuvius), the Laacher See
district of Germany, the Birunga and Toro-Ankole fields of Uganda, the
Leucite Hills area of Wyoming, and the West Kimberley Province of
Western Australia.[1]

[1] W. Cross, The igneous rocks of the Leucite Hills and Pilot Butte, Wyoming, Am.
Jour. Sci., vol. 4, pp. 115–141, 1897; R. Brauns, Die phonolitischen Gesteine des
Laacher Seegebietes, Neues Jahrb., Beilage-Band 46, Abt. A, pp. 1–116, 1922; Die
chemische Zusammensetzung der Basaltlaven des Laacher Seegebietes, Neues Jahrb.,
Beilage-Band 56, Abt. A, pp. 468–498, 1928; A. Holmes and H. F. Harwood, Petrol-
ogy of the volcanic fields east and southeast of Ruwenzori, Uganda, Geol. Soc. London Quart. Jour., vol. 88, pp. 370–442, 1932; A. Rittmann, Die geologische bedingte
Evolution und Differentiation des Somma-Vesuvius Magma, Zeitschr. Vulkanologie,
vol. 15, pp. 8–94, 1933; A. Holmes and H. F. Harwood, The petrology of the volcanic
area of Bufumbira, Uganda Geol. Survey Mem. 3, pt. 2, 1937; A. Wade and R. T.
Prider, The leucite-bearing rocks of the West Kimberley area, Western Australia,
Geol. Soc. London Quart. Jour., vol. 96, pp. 39–98, 1940; A. Holmes, Petrogenesis of
katungite and its associates, Am. Mineralogist, vol. 35, pp. 772–792, 1950; R. A.
Higazy, Trace elements of volcanic ultrabasic potassic rocks of southwestern Uganda
and adjoining part of the Belgian Congo, Geol. Soc. America Bull., vol. 65, pp. 39–70,
1954.

The Bufumbira Area of Uganda.[2] The Bufumbira area of equatorial east Africa, which forms the subject of the memoir by Holmes and Harwood, is a portion of the Birunga field which covers some 1,500 square miles in the extreme southwest of Uganda and the adjoining Belgian Congo. The field is underlain by a thick series of late Precambrian arenaceous, argillaceous, and locally ferrugineous sedimentary rocks extensively invaded by granites. Volcanic activity dates from Pliocene to Recent times. The Western Rift, on which the Birunga field is located, is a product of faulting which began in mid-Tertiary times and is still in progress. No clear relation has been traced between possible alignments of the many volcanic cones and the trend of the Rift faults, although the trend of the volcanic range as a whole is across the rifted valley. Petrographically the lavas of the Bufumbira volcanoes fall into three main series all characterized by dominance of potash over soda:

1. Leucite basalts (= leucitite, mikenite) consisting of olivine, augite, and leucite. Rocks of the same series (ugandites) with a potassic glassy base but lacking leucite microscopically resemble limburgites.

2. Leucite basanites (= kivites) consisting of olivine, augite, calcic plagioclase, and leucite. Again there are partially glassy types (murambites) superficially resembling limburgite.

3. Potash trachybasalts (= absarokite, shoshonite) consisting of olivine, augite, labradorite, and potash feldspar. Some types are relatively rich in feldspar and grade into hypersthene trachyandesite (latite). Potash trachytes are rare associates.

Xenoliths of peridotite and pyroxenite are fairly common in many of the basic lavas (leucite basalts, leucite basanites). The most widely distributed type is augite peridotite grading on the one hand into pyroxenite and into dunite on the other. Biotite pyroxenites (augite-biotite-garnet, augite-biotite-ilmenite-sphene) are somewhat less abundant. Ultrabasic xenoliths are almost or completely lacking in the more siliceous lavas some of which, however, enclose feldspathic xenoliths; e.g., monzonite (augite, plagioclase, alkali feldspar, biotite) in trachyandesite and trachyte; kentallenite (augite, olivine, plagioclase, leucite, biotite) in some of the leucite basanites and trachybasalts; plagioclase clots in basanites, trachybasalts, trachyandesites, and trachytes. There are also many instances of completely foreign inclusions of quartz and sedimentary rocks in various stages of reaction with the enclosing lavas.

Southwest of the Bufumbira area but still within the Birunga field are volcanoes that are built up largely of melilite-nepheline basalts, limburgites, and trachybasalts in which soda predominates over potash. These illustrate the remarkable manner in which closely adjacent volcanoes may maintain independent chemical and petrographic characters

[2] Holmes and Harwood, op. cit., 1937.

over long periods of time.[3]   The two great active volcanoes of Nyamlagira and Niragongo and the small cinder cone Nahimbi lie at the apices of a triangle whose sides are roughly ten miles long.   Nyamlagira is built up entirely of potassic lavas of the leucite-basanite group.   Niragongo, on the other hand, is composed of dominantly sodic melilite-nepheline basalts and potassic nepheline-leucite basalts the nepheline [4] of which contains up to 40 per cent $KAlSiO_4$.   Nahimbi has erupted nothing but sodic limburgite.

The range of chemical composition in the Bufumbira lavas is illustrated by selected analyses in Table 20.   The series as a whole is high in total alkali, though perhaps no more so than some recorded instances of the alkaline olivine-basalt trachyte phonolite kindred.   Very striking, however, is the predominance of potash over soda throughout most of the Bufumbira field.   With this may be correlated the prevalence of potassic trachybasalts rather than normal olivine basalts, the absence of which is strongly emphasized by Holmes.   Undersaturation in silica, expressed mineralogically by appearance of olivine and feldspathoids, is typical of the whole series of lavas with the exception of the hypersthene trachyandesites (latites).   The trace-element content of the Birunga lavas [5] is unusual. Even in very basic and ultrabasic types, Ba, Sr, Rb, and Zr—elements normally lacking in ultrabasic rocks—are very high: Ba commonly 0.2 to 0.3, occasionally > 1 per cent; Sr 0.4 to 1 per cent; Rb 150–300; Zr 600–1,100 parts per million.   There is an appreciable content of La and Y, with $Y > La$.   Cr and Ni are rather low, except in magnesian ultrabasic lavas.   As is usual in alkaline rocks, TiO_2 and P_2O_5 are very high.

Any theory of petrogenesis of the Birunga lavas must take into account the extent to which they resemble and differ from contemporary volcanic rocks from other parts of the Rift region of east Africa (cf. Fig. 37):

1. The Toro-Ankole volcanic province [6] is situated in the Western Rift region about 80 to 150 miles north-northeast of the Bufumbira field. Chemical and petrographic peculiarities like those of the Bufumbira rocks are here emphasized to an even more striking degree.   The main rock types are highly alkaline, dominantly potassic, relatively rich in BaO, and strongly undersaturated in silica.   Many are ultrabasic with silica percentages ranging from 35 to 40 per cent; values as high as 45 per cent are

---

[3] J. Verhoogen, Les Volcans Virunga et l'éruption du Nyamlagira de 1938, Soc. géol. Belgique Annales, vol. 62, pp. 326–353, 1939.

[4] T. G. Sahama, Mineralogy and petrology of a lava flow from Mt. Nyiragongo, Belgian Congo, Acad. Scient. Fennicae Annales, Ser. A, iii, no. 35, 1953.

[5] Higazy, op. cit., pp. 58–63, 1954.

[6] Holmes and Harwood, op. cit., 1932; A. Holmes, The petrology of katungite, Geol. Mag., vol. 74, pp. 200–219, 1937; A. D. Combe and A. Holmes, The kalsilite-bearing lavas of Kabirenge and Lyakauli, southwest Uganda, Royal Soc. Edinburgh Trans., vol. 62, pt. 2, pp. 359–379, 1945; Holmes, op. cit., 1950.

rare.   The rocks belong for the most part to the leucite-basalt and melilite-basalt families.   Olivine and augite are accompanied by leucite (or the rare feldspathoid kalsilite, $KAlSiO_4$), melilite, or nepheline.   In many

TABLE 20.   CHEMICAL COMPOSITIONS OF ROCKS FROM BUFUMBIRA FIELD OF UGANDA

Constituent	1	2	3	4	5	6	7	8	9	10	11
SiO_2	44.50	44.21	47.68	49.34	44.41	46.78	48.82	58.89	61.09	46.71	39.88
TiO_2	2.55	2.31	2.54	1.08	3.75	2.47	3.17	0.58	1.75	1.74	2.38
Al_2O_3	11.67	15.03	17.08	18.29	15.63	13.24	15.81	17.61	15.36	12.46	17.07
Fe_2O_3	2.05	3.79	3.45	4.61	3.10	2.64	2.21	2.44	1.34	3.00	5.75
FeO	8.90	7.53	5.26	3.12	9.00	7.85	8.14	3.47	4.60	9.03	5.94
MnO	0.16	0.18	0.14	0.08	0.16	0.18	0.17		0.10		0.28
MgO	13.25	6.61	2.90	2.66	5.39	11.64	5.56	0.65	2.47	9.56	3.98
CaO	10.18	10.87	7.17	9.74	9.73	8.59	8.45	2.11	4.45	11.61	10.54
Na_2O	2.53	3.75	5.33	3.46	2.95	2.20	2.47	4.57	3.00	3.10	5.72
K_2O	2.91	4.07	6.45	7.16	4.24	2.72	4.09	7.86	4.97	1.06	7.03
H_2O+	0.44	0.36	0.37	} 0.58	0.54	0.57	0.41	} 1.80	0.29	} 1.11	0.39
H_2O-	0.12		0.24		0.11	0.26	0.20		0.05		0.16
CO_2	0.05	0.02	0.02			0.03					0.36
P_2O_5	0.62	0.90	0.77	0.44	0.74	0.55	0.60	0.11	0.40	0.33	1.12
Cl	0.05	0.11	0.14		0.10	0.01	Trace		0.01		
F	0.05	0.11	0.08		0.09	0.07	0.04		0.06		
BaO	0.13	0.17	0.22		0.16	0.12	0.08		0.11		
SrO	0.07	0.09	0.10		0.03	0.09	0.05		0.04		
Cr_2O_3	0.12	0.03			Trace	0.11	0.01		Trace		
V_2O_3	0.04	0.05	0.03		0.05	0.04	0.04		0.02		
Rest	0.04	0.05	0.01			0.07	0.02			0.05	
Total	100.43	100.24	99.98	100.56	100.18	100.23	100.34	100.09	100.09	99.76	99.90

Explanation of Table 20

Note: All analyses except no. 11 are taken from A. Holmes and H. F. Harwood, The petrology of the volcanic area of Bufumbira, *Uganda Geol. Survey Mem. 3*, pt. 2, 1937, to which reference is made in parentheses giving page number and analysis letter in each case.

   1. Leucite basalt (ugandite) with glassy base; Muzanza.   (p. 65, E.)
   2. Leucite basalt (olivine leucite); Lutale.   (p. 75, G.)
   3. Leucite basalt (mikenite); Mikeno.   (p. 91, H.)
   4. Leucite tephrite (vesuvite); Kitale.   (p. 99, 60.)
   5. Leucite basanite (kivite); Busamba.   (p. 106, J.)
   6. Leucite basanite (murambite) with glassy base; Murambe.   (p. 138, L.)
   7. Trachybasalt (shoshonitic absarokite); Mgahinga.   (p. 159, O.)
   8. Trachyte; Karisimbi.   (p. 185, 104.)
   9. Trachyandesite (hypersthene latite); Sabinyo.   (p. 194, U.)
   10. Limburgite; Nahimbi (Adolf Friedrich Volcano).   (p. 204, 124.)
   11. Partly glassy melilite-leucite nephelinite (= niligongite); Niragongo.   (T. G. Sahama, *Acad. Scient. Fennicae Annales*, Ser. A, iii, no. 35, p. 15, 1953.)

rocks (katungites) olivine and pyroxene or melilite are the only crystalline constituents and are associated with abundant glass containing material chemically equivalent to feldspathoids.   Feldspars are completely lacking.   In addition to ultrabasic xenoliths (biotite pyroxenite, biotite-augite peridotite, mica rock), there are many fragments of partially fused granite, the glassy portions of which in some cases are chemically equivalent to almost pure leucite.   The content of trace elements is similar to that of the Birunga lavas, except that Cr and Ni are high and La > Y.

2.   On the Western Rift some seventy miles south of the Bufumbira area is the South Kivu volcanic field extending from Lake Kivu southward to

OUTLINE MAP
OF PART OF
EASTERN AFRICA

N

0 100 200 300 400 500
Scale in miles

1 - Birunga field
2 - Toro- Ankole field
3 - South Kivu field

RED
SEA

ABYSSINIA

Lake
Rudolf

KENYA

BELGIAN
CONGO

Western Rift

Lake
Victoria

Eastern Rift

Lake Kivu

Lake
Tanganyika

TANGANYIKA

INDIAN
OCEAN

Fig. 37.   Location of the Bufumbira volcanic province in relation to other volcanic fields, and to the east African Rift Valleys (stippled).

Lake Tanganyika.[7]   The lavas here have for the most part been erupted from two great volcanoes and are thought to be among the oldest products of volcanic activity along the Western Rift.   They are predominantly olivine basalts with alkaline affinities, accompanied by tholeiitic basalts and subordinate soda-potash trachytes.   These are much more sodic than

[7] Holmes and Harwood, *op. cit.*, pp. 273–276, 1937; N. L. Bowen, Lavas of the African Rift Valleys and their tectonic setting, *Am. Jour. Sci.*, vol. 35A, pp. 28–32, 1938; A. Holmes, Basaltic lavas of South Kivu, Belgian Congo, *Geol. Mag.*, vol. 77, pp. 89–101, 1940.

the potash trachytes of Bufumbira. There are also local late potash-rich rhyolites with unusually high contents of silica and in some instances magnetite or graphite—apparently introduced hydrothermally.

3. Some 500 miles east of the Bufumbira region are the extensive volcanic fields situated along the Eastern Rift.[8] Here are alkaline olivine basalts and trachybasalts with associated nepheline and melilite basalts, trachytes, phonolites, and aegirine- and riebeckite-bearing rhyolites. This is the typical alkaline olivine-basalt trachyte association. In contrast with the potassic lavas of the Bufumbira and Toro-Ankole fields, the lavas of the Eastern Rift are, with minor exceptions, dominantly sodic.

The Birunga field is thus seen to be one of several volcanic provinces where highly potassic lavas have been erupted, in the vicinity of the Western Rift. But sodipotassic lavas such as are usually associated with alkaline olivine basalts in other parts of the world have also been erupted consistently from some of the Birunga volcanoes and are the dominant rocks of adjacent volcanic fields on the Western Rift. The much more extensive lavas of the Eastern Rift, as typified by those of Kilimanjaro and Kenya, are in the main sodic or sodipotassic rocks typical of the alkaline olivine-basalt trachyte phonolite association; but here, too, are recorded instances of dominantly potassic rocks of the rhyolite and trachyte families. In view of the doubt that still exists as to the relative roles of tension and compression in the evolution of the east African Rift Valleys, any attempt to correlate the development of sodic magmas on the one hand and potassic on the other with particular tectonic environments must be unsatisfactory. We are justified merely in noting the occurrence of the alkaline olivine-basalt association and of the leucite-basalt association in a sector of the African continent which was contemporaneously affected by faulting on a regional scale.

The West Kimberley Area of Western Australia.[9] Highly potassic leucite-rich rocks have been found at nineteen localities within an area of 100 by 80 miles in the West Kimberley region, Western Australia (lat. $17\frac{1}{2}°$ to 19°S., long. 124° to 126° E.). The rocks occur as plugs, dikes, and restricted flows penetrating or erupted upon a great thickness of middle and late Paleozoic sediments (sandstones, limestones, shales).

Chemical and petrographic characters of the whole suite are striking and extreme. In every rock type abundant leucite (30 to 60 per cent of the modal composition) is associated with a potash-magnesia amphibole, phlogopite, diopside, or some combination of these, and an indeterminate groundmass very rich in silica and magnesia. Outstanding chemical

---

[8] W. C. Smith, A classification of some rhyolites, trachytes, and phonolites from part of Kenya Colony, *Geol. Soc. London Quart. Jour.*, vol. 72, pp. 222–279, 1931; Bowen, *op. cit.*, pp. 24–26, 1938.

[9] Wade and Prider, *op. cit.*, 1940.

peculiarities (cf. Table 21) are excess of K_2O over Al_2O_3, very low Na_2O, high TiO_2 and P_2O_5, and unusually high BaO (averaging 0.9 per cent). In spite of abundance of leucite and complete lack of feldspar, analyses show sufficient silica for quartz to appear in the norm.

These chemical peculiarities are also shown, though to a somewhat less degree in the leucitic lavas of Leucite Hills, Wyoming. There are less obvious analogies with nonfeldspathic lavas described by Combe and Holmes [10] from a group of volcanoes in the Toro-Ankole province of Uganda, but these latter consistently are richer in Al_2O_3, poorer in K_2O, and strongly undersaturated in silica. There is also some resemblance to kimberlites, biotite pyroxenites, and certain mica lamprophyres.

The Potassic Provinces of Montana, Wyoming, and Arizona. Highly potassic lavas and associated near-surface intrusive rocks, in some instances carrying leucite or pseudoleucite, are widely distributed along the eastern fringe of the Rockies from the Canadian to the Mexican border.[11] These igneous rocks are of various ages within the Tertiary era. Over much of the region they are basic rocks moderately high to high in total alkali, and throughout a number of the constituent provinces (e.g., Yellowstone province) are sodic rather than potassic. Only a limited number of localized provinces or subprovinces are conspicuously potassic. Thus H. Williams found that the highly potassic rocks of the Navajo area of northeastern Arizona are in this respect sharply distinct from the nearby highly sodic basic volcanics of the Hopi subprovince. Both constitute remnants of a volcanic field that must formerly have covered thousands of square miles, situated between the sodic provinces of Utah-Colorado to the north and San Francisco mountains to the south. Over forty years ago Pirsson likewise recognized that the potassic rocks of the Highwood province, Montana, pass laterally into contemporary rocks of a much more sodic character.

The Pliocene rocks of the Navajo province described by Williams (cf.

[10] *Op. cit.*, 1945.

[11] For the general distribution of alkaline igneous rocks in this region the reader is referred to J. P. Iddings, *Igneous Rocks*, vol. II, pp. 400–414, 463–481, Wiley, New York, 1913. For details of petrology and chemistry of individual provinces, see H. Williams, Pliocene volcanoes of the Navajo-Hopi country, *Geol. Soc. America Bull.*, vol. 47, pp. 111–172, 1936; E. S. Larsen and B. F. Buie, Potash analcime and pseudoleucite from the Highwood Mountains of Montana, *Am. Mineralogist*, vol. 23, pp. 837–849, 1938; C. S. Hurlbut and D. Griggs, Igneous rocks of the Highwood Mountains, Part I, *Geol. Soc. America Bull.*, vol. 50, pp. 1043–1112, 1939; E. S. Larsen, C. S. Hurlbut, C. H. Burgess, and B. F. Buie, Igneous rocks of the Highwood Mountains, Parts II–VII, *Geol. Soc. America Bull.*, vol. 52, pp. 1733–1868, 1941; E. S. Larsen, Petrographic province of Central Montana, *Geol. Soc. America Bull.*, vol. 51, pp. 887–948, 1940. For structural details, see C. R. Appledorn and H. E. Wright, Volcanic structures in the Chuska Mountains, Navajo Reservation, Arizona-New Mexico, *Geol. Soc. America Bull.*, vol. 68, pp. 445–468, 1957.

Constituent	1	2	3	4	5	6	7	8	9
SiO_2	45.82	51.19	51.07	50.23	51.80	47.50	44.74	37.80	35.51
TiO_2	7.34	4.89	2.13	2.27	1.80	1.85	2.10	3.05	4.88
Al_2O_3	6.86	8.53	9.93	11.22	11.10	9.62	11.82	9.47	6.83
Fe_2O_3	6.07	6.12	2.72	3.34	3.55	3.37	3.89	5.64	9.68
FeO	1.98	1.38	1.19	1.84	3.42	4.74	7.06	5.21	2.70
MnO	0.10	0.06	0.04	0.05	Trace	Trace	0.19	0.20	0.22
MgO	10.90	7.15	10.31	7.09	8.15	13.00	14.28	11.30	11.67
CaO	4.70	5.82	4.87	5.99	7.95	9.00	9.61	15.20	16.00
Na_2O	0.84	0.58	0.82	1.37	2.25	1.96	2.16	1.70	1.56
K_2O	8.82	9.02	9.92	9.81	5.97	3.28	2.51	2.23	3.30
H_2O+	0.75	1.99	2.19	1.72	1.90	3.90	0.40	2.20	3.11
H_2O-	2.40	1.26	2.04	0.93	1.00	0.90	0.10	2.45	1.31
CO_2	0.08						0.21	1.70	1.47
P_2O_5	1.83	0.79	1.53	1.89	0.95	1.05	0.66	1.68	1.18
Cl	Trace			0.03					
F	Trace			0.50					0.27
BaO	1.27	0.60	0.57	1.23			0.07		0.27
SrO				0.24			0.05		0.24
Cr_2O_3	0.07			0.10					0.02
V_2O_3	0.03								0.04
Rest	0.07	0.33	0.33	0.77			0.16		0.15
Total	99.93	99.71	99.66	100.62	99.84	100.17	100.01	99.83	100.41

Explanation of Table 21

1. Wolgidite (leucite, K-Na-Mg amphibole, phlogopite, diopside, serpentinous groundmass), Mount North, West Kimberley. (A. Wade and R. T. Prider, The leucite-bearing rocks of the West Kimberley area, Western Australia, *Geol. Soc. London Quart. Jour.*, vol. 96, p. 75, No. 2, 1940.)
2. Cedricite (leucite, diopside, phlogopite, K-Na-Mg amphibole, groundmass), Mount Gytha, West Kimberley. (Wade and Prider, *op. cit.*, p. 75, No. 4, 1940.)
3. Orendite (leucite, sanidine, phlogopite, amphibole), Leucite Hills, Wyoming. (W. Cross, 1897, cited by Wade and Prider, *op. cit.*, p. 80, No. 2, 1940.)
4. Wyomingite (phlogopite, leucite, diopside), Leucite Hills, Wyoming. (W. Cross, 1897, cited by Wade and Prider, *op. cit.*, p. 80, No. 1, 1940.)
5. Mica lamprophyre (orthoclase, diopside, biotite), Shiprock, Navajo province, Arizona. (H. Williams, *Geol. Soc. America Bull.*, vol. 47, p. 166, No. 11, 1936.)
6. Leucite basalt (leucite, augite, olivine, minor phlogopite and orthoclase), Todilto Park, Navajo province, Arizona. (Williams, *op. cit.*, p. 166, No. 6, 1936.)
7. Murambite (olivine, augite, leucite, glass) Bufumbira area, Uganda. (A. Holmes and H. F. Harwood, The petrology of the volcanic area of Bufumbira, *Uganda Geol. Survey Mem. 3*, pt. 2, p. 141, M, 1937.)
8. Nepheline monchiquite (augite, nepheline, orthoclase, iron ore, minor olivine and biotite), Wildcat Peak, Navajo province, Arizona. (Williams, *op. cit.*, p. 166, No. 1, 1936.)
9. Katungite (melilite, olivine, leucite, biotite, glass), Katunga, Uganda. (A. Holmes, The petrology of katungite, *Geol. Mag.*, vol. 74, p. 207, E, 1937.)

Table 21, analyses 5 and 6) are sanidine-rich trachybasalts, leucite basalts, and chemically equivalent dike rocks (lamprophyres rich in biotite, diopside, and orthoclase). Ultrabasic melilite-bearing lamprophyres similar to alnoite or monchiquite are rare (Table 21, analysis 8). It is interesting to note the chemical identity of analyzed lamprophyres—typical minettes—from this region, with associated flows of trachybasalt, and the chemical resemblance between several of the Navajo rocks and lavas from the Bufumbira area of Uganda (cf. Table 21, analyses 5 to 9). Granitic xenoliths are extremely plentiful in the Navajo volcanics but are almost absent in the sodic rocks of the adjacent Hopi area. Also present are xenoliths of pyroxene-peridotites, such as are commonly observed in basic volcanic rocks elsewhere, and of various metamorphic rocks including magnesian types.

In the broad Tertiary volcanic region of central Montana, the most potassic rocks are those of the Highwood province. Here E. S. Larsen recognizes four subprovinces, each with its characteristic volcanic rock association:

1. A series of quartz latites rather higher in K_2O and MgO and lower in Al_2O_3 than most calc-alkaline lavas. They contain plentiful quartz, oligoclase, and hornblende—minerals rare or lacking in rocks of the other subprovinces—and are the oldest rocks of the Highwood region.

2. An intrusive stock of monzonite and syenite.

3. A small intrusion of ultrabasic rocks (monticellite peridotite and alnoite), very rich in lime and in some parts containing as much as 1.26 per cent BaO.

4. Basic potassic rocks and their derivatives.

It is the rocks of this last subprovince that fall within the scope of the present discussion. Most abundant are flows and dikes of mafic phonolite, and flat laccoliths and dikes of chemically equivalent shonkinite. These rocks are composed principally of augite, olivine, leucite (or pseudoleucite), and analcite (or other zeolites); sanidine, in some cases with a high content of barium, and biotite are important in some rocks, nepheline is rare, and plagioclase typically absent. Certain of the laccoliths, notably that of Shonkin Sag (made classic through the account of Weed and Pirsson [12]), are composite bodies reminiscent of the differentiated teschenite sills described earlier in this chapter. The Shonkin Sag intrusion is composed principally of shonkinite (augite, olivine, biotite, apatite, magnetite in a matrix of alkali feldspar, zeolites, and carbonate). Pseudoleucite appears conspicuously in a chilled marginal phase. Toward the top of the intrusion is a lensoid sheet of alkali syenite containing the same minerals as are present in the shonkinite but in greatly different

[12] W. H. Weed and L. V. Pirsson, Geology of the Shonkin Sag and Palisade Butte laccoliths, Am. Jour. Sci., vol. 12, pp. 1–17, 1901.

proportions: alkali feldspar and zeolites make up 70 per cent of the total composition, and augite is conspicuously rimmed with aegirine, while olivine is present only in accessory amount. The syenite undoubtedly is a differentiate derived from a parent shonkinite magma. There is still some divergence of opinion as to whether differentiation occurred in place or whether the laccolith is a product of multiple injection of magma drawn off at successive stages from a differentiating body in the vicinity. The increasing number of recorded instances where lensoid bodies of alkali syenite occur within sills or laccoliths of basic alkaline rocks renders the hypothesis of multiple injection correspondingly less plausible.

Table 22 brings out the correspondence, apart from reversal in relative proportions of potash and soda, between differentiated rocks of the Shonkin Sag laccolith and equivalent rocks in basic alkaline sills of the Scottish Carboniferous province.

Origin of Leucite-rich Volcanic Rocks.   In the light of the petrographic and chemical data outlined above, it is appropriate now to review severally the hypotheses of origin of leucitic lavas that appear in current literature.

1. Daly's limestone-assimilation hypothesis, originally proposed to account for the origin of undersaturated alkaline magmas in general, has been invoked, in modified form, by Rittmann [13] to explain the evolution and observed sequence of lavas erupted from the Somma-Vesuvius volcano. The parent magma is assumed to have been trachytic. It is suggested that, by assimilative reaction with the thick dolomitic limestones that underlie the volcano, the magma became impoverished in silica. Sinking of ferromagnesian silicates, upward concentration of potash by gaseous transfer, and differential removal of soda by gases escaping into the wall rock are additional factors in a complex process of magmatic evolution deduced from the observed sequence of lava types in time: trachytes and feldspathic tephrites (older Somma), plagioclase-rich leucite basalts (younger Somma), basic plagioclase-poor leucitites (Vesuvius). The hypothesis finds further support in the great abundance of xenoliths and fragments of metamorphosed limestone in the ejecta of Monte Somma, and in the occurrence of numerous xenoliths of syenites and nepheline syenites carrying melanite, vesuvianite, and bytownite. Rittmann's hypothesis ingeniously explains most of the numerous data available for the Somma-Vesuvius volcano, but it is clearly inapplicable to other leucitic provinces where trachytes are subordinate or lacking, where limestone is absent from the basement rocks, and where (as in the Birunga field) leucitic lavas devoid of calcareous xenoliths well up quietly and are erupted with no suggestion of explosive emission of gas. In the case of Somma-Vesuvius, limestone assimilation has been clearly demonstrated, but

[13] Rittmann, op. cit., 1933.

whether it has brought about extensive desilication of the magma in general has not been proved. Moreover the nature of the parent magma remains uncertain, for, by analogy with the sequence of eruption observed in other provinces, it is by no means unlikely that long-continued differ-

TABLE 22.   COMPOSITIONS OF ROCKS OF SHONKIN SAG LACCOLITH AND OF ROCKS FROM DIFFERENTIATED SILLS OF SCOTLAND (CARBONIFEROUS)

Constituent	1	2	3	4
SiO_2	45.77	44.6	50.72	56.44
TiO_2	0.76	2.7	0.79	1.16
Al_2O_3	8.94	7.6	18.60	15.54
Fe_2O_3	3.63	6.3	3.52	3.27
FeO	7.13	9.6	3.24	3.67
MnO	0.13	0.5	0.09	
MgO	12.96	12.0	2.14	1.73
CaO	11.56	10.3	4.26	4.16
Na_2O	1.40	4.3	4.33	5.81
K_2O	4.60	1.5	7.87	4.27
H_2O+	0.95		3.43	2.06
H_2O-	0.18			0.44
CO_2				0.97
P_2O_5	1.52	0.6	0.15	0.83
BaO	0.26		0.70	
SrO	Trace		0.10	
S			0.04	
Total	99.79	100.00	99.98	100.35

Explanation of Table 22

1. Lower shonkinite, Shonkin Sag, Montana. (C. S. Hurlbut and D. Griggs, Igneous rocks of the Highwood Mountains, Part I, *Geol. Soc. America Bull.*, vol. 25, p. 1071, No. 3, 1939.)
2. Picrite, Lugar Sill, Scotland (calculated water-free). (G. W. Tyrrell, The picrite-teschenite sill of Lugar (Ayrshire), *Geol. Soc. London Quart. Jour.*, vol. 72, p. 114, No. 1, 1917.)
3. Aegirine syenite, Shonkin Sag, Montana. (Hurlbut and Griggs, *op. cit.*, p. 1071, No. 8, 1939.)
4. Analcite syenite, Hawford Bridge sill, Scotland. (G. W. Tyrrell, On some dolerite sills containing analcite syenite in central Ayrshire, *Geol. Soc. London Quart. Jour.*, vol. 84, p. 559, No. 1, 1928.)

entiation within the magma reservoir may have preceded the cycle of surface eruption. In that event the earliest trachytic lavas would represent products of extreme differentiation, and the leucite basalts and leucitites of later eruptions could correspond more nearly to the parent magma. Assimilation of limestone accounts for one chemical characteristic of basic potassic lavas—the combination high alkali, low silica—, but it fails completely to explain the observed high content of Sr, Ba, and Rb and the presence of La and Y, in all of which elements limestones are strikingly deficient.[14]

2. In his earlier papers on the leucite-bearing lavas of east Africa, Holmes derived the various nonfeldspathic members from a primary peridotite magma by a process of crystal differentiation. High pressure, favoring early separation of eclogite and dunite, was suggested as an essential factor leading to the development of a residual alkaline ultrabasic liquid having the composition of kimberlite or its approximate chemical equivalent in the leucite-basalt series. Reaction between such a magma and basaltic or possibly granitic rocks was assumed to account for further evolution of feldspathic lavas in this series. Subsequently, in the Bufumbira memoir, Holmes rejected this hypothesis on the grounds that the eclogite fragments in African kimberlites prove to be much older than, and therefore genetically unrelated to, the enclosing kimberlite. The general mechanism of early separation and removal of olivine crystals from a primary mica-peridotite magma is nevertheless still advocated by Wade and Prider in the case of the West Australian leucite rocks. However, the possibility that liquid peridotite magmas can exist within the outer crust of the lithosphere is seriously challenged by experimental evidence on the melting behavior of olivines. This topic will be dealt with in a later chapter.

3. In more recent papers Holmes[15] proposed a complex series of "transfusion" processes to account for generation of potash-rich ultrabasic magmas. A stream of "emanations," highly "energized" and rich in alkalis and various other materials, was assumed to rise from the depths and to react with materials of the crust encountered in the ascent. The products of reaction were said to range from essentially solid metasomatically altered rocks to partially or completely liquid magmas. For example pyroxene peridotite, by introduction of K_2O, Al_2O_3, and water, would be converted into biotite pyroxenite; by further introduction of CaO, Fe_2O_3, TiO_2, P_2O_5, Na_2O, K_2O, CO_2, and H_2O and increase in energy, the biotite pyroxenite might be changed into the ultrabasic potassic magma corresponding to katungite. In the Bufumbira area, partial fusion of the more or less granitic basement rocks under the action of incoming emanations rich in alkali and alumina was postulated to account for the assumed parent magma of the province—a very basic sodipotassic

[14] Higazy, op. cit., pp. 63, 64, 1954.

[15] Holmes, op. cit., 1937; Holmes and Harwood, op. cit., pp. 261–272, 1937; Combe and Holmes, op. cit., p. 378, 1945; A. Holmes, Leucitized granite xenoliths from the potash-rich lavas of Bunyaruguru, Southwest Uganda, Am. Jour. Sci., vol. 243A, pp. 313–332, 1945.

nepheline-melilite basalt; the other basic and much more potassic magmas of the region are then brought into being as the parent magma reacts with peridotite, biotite pyroxenite, and various sedimentary rocks of the basement. In deducing these and similar lines of magmatic and metasomatic evolution, Holmes closely studied the xenoliths present in the various lavas. They were interpreted as remnants, in different stages of metasomatism, of solid materials which, by "transfusion" in the depths, contributed to the generation of the lavas subsequently erupted at the surface. Though the highly variable "flux of emanations" postulated by Holmes is adequate to explain in detail the chemical variety in the rocks of the east African volcanic provinces, this is no proof that such a flux ever ascended from the depths as the prime agent in generation of magmas. It is equally possible that, in the great variety of rocks developed in such a field as Bufumbira, we see an expression of the bewilderingly complex and largely unpredictable variations in physical conditions (and in mutual relations of solid, liquid, and gaseous phases) that must prevail in a magma chamber walled by heterogeneous rocks of variable composition and structure (cf. page 52). If the "flux of emanations" has played an important role in the origin and evolution of the potassic lavas of east Africa, it must have played a comparable part in the development of the sodic olivine-basalt magma which is so widely represented over the African Rift region.

4. Holmes's latest hypothesis [16] attempts to explain the origin of potash-rich ultrabasic magmas such at katungite, and is a development from hypothesis 3. The primary magma, pictured as rising from the substratum, has the composition of carbonatite. This accords with the widespread occurrence of Tertiary carbonatites in eastern Uganda and Kenya, and with certain chemical characteristics shared by carbonatites and magmas of the katungite family: deficiency in silica; high Sr, La, and Y associated with Ca; Cr and Ni associated with Mg; V and Ti associated with Fe. The carbonatite magma, coming into contact with the granitic rocks of the basement, is supposed to react strongly to give two complementary products: biotite pyroxenite, and katungite magma. To the granitic parent are attributed the Si, K, Rb, Ba, and Ra of katungite and its associates. While Holmes's hypothesis explains much of the peculiar geochemistry of the potash-rich basic volcanic suite, it leaves two major problems unresolved. Why is the total volume of carbonatite so small as contrasted with that of alkaline basic lavas in the east African volcanic fields? And what is the origin of that most perplexing of magmas, the carbonatite magma itself?

5. According to Bowen, differential fusion of biotite (an ultrabasic potassic mineral), brought into contact with hot basaltic magma by gravi-

[16] Holmes, *op. cit.*, 1950; Higazy, *op. cit.*, 1954.

tational sinking, could modify the compositions of the magma in the direction of mica lamprophyre or leucite basalt. However, biotite is not a mineral which separates in great quantity from normal magmas, nor does its tabular crystal habit favor sinking in a viscous liquid. It is unlikely that it could accumulate on such a scale as to convert olivine-basalt magma to leucite-basalt magma over a large area. A contributory role in causing local concentration of K_2O (and BaO) in an already potassic basic magma in process of crystallization is possible.

6. E. S. Larsen treats the origin of the highly potassic rocks of the Highwood province as part of the broad problem of petrogenesis for the whole central Montana region.[17] All petrographic and chemical variation within this region is explained in terms of magmatic differentiation modified by assimilative reaction between magma and granitic rocks of the basement. The primary magma of the Montana region is believed to have been olivine basalt. By separation and removal of calcic plagioclase and hypersthene in the depths, this gave rise to still basic but increasingly alkaline magmas which were drawn off, in various stages of development, to constitute the respective primary magmas of the several central Montana provinces. That of the Highwood province was equivalent to shonkinite (mafic phonolite). Its highly alkaline condition compared with that of other provinces in the same broad region is attributed to "a long period of quiet differentiation in depth," in an environment long free from noteworthy deformation. Development of alkali-syenitic end products is correlated with crystal settling (mainly diopsidic augite and olivine) under gravity. The general absence of granitic rocks in this province, as contrasted with adjacent areas, is interpreted as due to lack of reaction with the essentially granitic basement rocks. The mechanism suggested by Larsen accounts adequately for almost all the many chemical and petrographic data which have been collected for this region. But no satisfactory explanation has been found for the very high content of BaO and SrO in the rocks of the Highwood province, and in view of the consistent appearance of high BaO and SrO in potassic basic lavas of other provinces, this remains a serious anomaly in Larsen's scheme of petrogenesis.

7. Waters [18] pictures concentrations of hornblende and biotite developing as solid residues left behind while complementary anatectic granitic liquids are progressively expelled upward from the crustal basement with passage of time. Local melting of such residues could then give rise to magmas at the same time rich in potash, iron, and magnesia but low in silica.

[17] Larsen, *op. cit.*, 1940; Larsen and coauthors, *op. cit.*, pp. 1862–1867, 1941.

[18] A. C. Waters, Volcanic rocks and the tectonic cycle, *Geol. Soc. America Special Paper*, no. 62, pp. 703, 717–720, 1955.

8. It has been suggested by Williams that in the Navajo-Hopi province of Arizona an originally sodic ultrabasic magma having the composition of monchiquite (= nepheline basalt or nephelinite) reacted with the potash feldspar of granites encountered in the underlying basement, and so attained the potassic composition which prevails throughout the Navajo subprovince. The field and petrographic evidence in this case is most convincing. In the Hopi subprovince granitic xenoliths are rare or absent and the rocks are essentially sodic—monchiquites, limburgites, analcite basalts. In the Navajo subprovince granitic xenoliths are exceedingly numerous, and the volcanic rocks are potassic—mica lamprophyres and sanidine-rich trachybasalts. A generally similar mechanism of reaction between basaltic magma and granite accounts well for the composition of potash trachybasalts of Tertiary age in the Sierra Nevada of California.[19] These rocks are richer in SiO_2 (53.9 per cent) and contain less K_2O (2.6 per cent) than most basic potassic lavas, but they show consistently high Ba and Sr and appreciable Zr and Y.

9. If it be assumed that a considerable section of the ultramafic subcrustal rock of the earth's mantle is close to the melting temperature, then a body of magma formed locally at some depth could move upward under the influence of gravity by a process of simultaneous melting of the roof and crystallization at the bottom of the magma chamber. No considerable change in volume or in gross composition of the magma need be involved. Such a process is pictured by P. G. Harris.[20] He suggests that the rising magma would become progressively enriched in all minor elements (including K, Ba, Rb, Zr, and others) unable to substitute in the mineral phases of the mantle (probably olivine, pyroxene, and spinel). For these the distribution coefficient $K \left(\dfrac{\text{concentration in solid phase}}{\text{concentration in liquid phase}} \right)$ would be low. This is an ingenious explanation of the origin of basic and ultrabasic magmas rich in K and in certain minor elements otherwise concentrated in granitic rocks. However, if we accept differential fusion of the mantle as the normal mode of origin of basaltic magma, then, according to the above argument, potassic basic magmas should be the rule rather than the exception.

The authors favor the general hypothesis that the highly potassic magmas of the leucite-basalt association are products of assimilative reaction between alkaline olivine-basalt magma or nepheline-basalt magma and the "granitic" rocks of the continental basements. Several

[19] W. B. Hamilton and G. J. Neuerburg, Olivine-sanidine trachybasalt from the Sierra Nevada, California, Am. Mineralogist, vol. 41, pp. 851–873, 1956.

[20] P. G. Harris, Zone refining and the origin of potassic basalts, Geochim. et Cosmochim. Acta, vol. 12, pp. 195–208, 1957.

well-established facts, difficult to account for by other hypotheses, are cited in support of this view:

1. General abundance of granitic xenoliths in rocks of the leucite-basalt and mica-lamprophyre families.

2. Close association of mica lamprophyres, similar in composition to leucite basalt, with areas of earlier granitic rocks.

3. Demonstration by Holmes [21] that granite xenoliths immersed in leucite-basalt (ugandite) and melilite-basalt (katungite) magmas are converted to aggregates of leucite and augite, to pure leucite, or to leucite glass.

4. Association of leucite basalts with broad areas of alkaline olivine basalt and derivative sodic lavas.

5. Availability, in the granites and metasediments of the "granitic" basement, of such elements as Ba, Sr, Rb, Zr, and Ra—abundant in basic potassic magmas but otherwise lacking in ultrabasic rocks.

6. Restriction of the leucite-basalt association to the continents, *i.e.*, to regions underlain in depth by "granitic" rocks.

7. The occurrence of leucite basanites and leucite tephrites in some provinces, such as the Tertiary province of the Bohemian Mittelgebirge, otherwise characterized by sodic lavas of the alkaline olivine-basalt trachyte association. Such occurrences are interpreted as transitional between the alkaline olivine-basalt and the leucite-basalt associations.

The nature of the reaction product in any province—whether ultrabasic katungite, or the more familiar leucite basalt, or again a nonfeldspathoidal potash trachybasalt—would depend on the nature of the "granitic" basement and on the physical conditions of assimilation. Reactions involving crystallization or destruction of biotite and hornblende have proved to be very sensitive to variation in temperature, partial pressure of water and partial pressure of oxygen. A pre-Cambrian basement which in addition to widespread "granites" includes such varied ingredients as amphibolites, biotite-rich metasediments, and ferruginous metacherts, permits a wide latitude for variations in the assimilation process, and provides an adequate source for the array of trace elements that characteristically appear in all basic potassic magmas.

LAMPROPHYRES [22]

Characteristics of Lamprophyres. The lamprophyre class and its subdivisions have been variously defined according to mineral and chemical

[21] Holmes, *op. cit.*, 1945; Combe and Holmes, *op. cit.*, 1945.

[22] H. H. Thomas, in Tertiary and post-Tertiary geology of Mull, etc., *Scotland Geol. Survey Mem.*, pp. 377–382, 1924; N. L. Bowen, *Evolution of the Igneous Rocks,* Princeton University Press, Princeton, N. J., pp. 258–273, 1928; A. Knopf, Igneous

composition, texture, association with other igneous rocks, and supposed mode of origin. The development of the lamprophyre concept since its formulation by Rosenbusch over sixty years ago has been graphically summarized by Knopf,[23] who then defines lamprophyres as mesocratic and melanocratic rocks (*i.e.*, those in which dark minerals make up more than about one-third of the composition by volume), with strongly porphyritic texture due solely to recurrence of ferromagnesian minerals in two generations of euhedral crystals. Feldspars are confined to the groundmass, which is fine-grained to aphanitic. Nomenclature of the group is confused and needlessly complicated. The following are some common mineral assemblages, with corresponding rock names:

Hornblende-augite-plagioclase (camptonite)
Biotite-hornblende-plagioclase (kersantite)
Biotite-orthoclase (minette)

Olivine may be present, especially in the augite-bearing lamprophyres, some of which (monchiquites) contain analcite instead of feldspar. Rarer types may contain melilite (alnoites) or even nepheline. Chemically the lamprophyres are basic to ultrabasic rocks with a high content of both $(FeO + MgO)$ and $(Na_2O + K_2O)$, expressed mineralogically by abundance of biotite and barkevikitic hornblende. Lamprophyre magmas contain unusually high concentrations of water and other volatiles judging from the typically high H_2O, P_2O_5, CO_2, S, etc., in chemical analyses, and such evidence of deuteric activity as bleached crystals of biotite, carbonatized feldspars, talc-carbonate pseudomorphs after olivine, and disseminated pyrite. A persistent chemical characteristic is the presence of notably high barium, strontium, or both.

Lamprophyres typically occur as narrow dikes. These may be members of dike swarms (which may also include aplites, porphyrites, diabases, etc.) associated with intrusions of granite or granodiorite. There are cases where the tendency for radiating lamprophyric dikes to converge on individual centers of granite intrusion is clear beyond doubt.[24]

geology of the Spanish Peaks region, Colorado, *Geol. Soc. America Bull.*, vol. 47, pp. 1727–1784, 1936; H. G. Smith, New lamprophyres and monchiquites from Jersey, *Geol. Soc. London Quart. Jour.*, vol. 92, pp. 365–383, 1936; Williams, *op. cit.*, pp. 111–172, 1936; F. H. Hatch and A. K. Wells, *The Petrology of the Igneous Rocks*, pp. 249–255, G. Allen, London, 9th ed., 1937; E. Bederke, Zum Problem der Lamprophyre, *Akad. Wiss. Göttingen Nachricht.*, Math.-phys. Kl., pp. 53–57, 1947; I. Campbell and E. T. Schenk, Camptonite dikes near Boulder Dam, Arizona, *Am. Mineralogist*, vol. 35, pp. 671–672, 1950; P. Eskola, Ein Lamprophyrgang in Helsinki und die Lamprophyrprobleme, *Tschermaks min. u. petr. Mitt.*, Bd. 4, pp. 329–337, 1954; C. Oftedahl, Studies on the igneous rock complex of the Oslo region, XV, *Norske Vidensk.-Akad. Oslo Skr.*, Mat.-Naturv. Kl., no. 2, 1957.

[23] Knopf, *op. cit.*, pp. 1745–1749, 1783, 1784, 1936.

[24] Knopf, *op. cit.*, 1936; S. Kaitaro, On central complexes with radial lamprophyre dikes, *Soc. Géol. Finlande Compt. rend.*, no. 29, pp. 55–65, 1956.

Thus at Spanish Peaks, Colorado, some 500 dikes, the majority of which are augite porphyries and lamprophyres, radiate in all directions from two large stocks of porphyritic granitic rocks. The mid-Paleozoic lamprophyre dikes of northern England are similarly related to the Shap granite intrusion; and the same is true of the Åva granite-monzonite stock in Finland.

The relation between lamprophyres and associated granites in other provinces may be of a much broader type. In New Zealand the two groups of rocks are both confined to the western part of the South Island —a belt some 300 miles in length. At the northern end of this province both groups of rock are well represented, but at the southern end very extensive masses of granite are almost unaccompanied by lamprophyres, while in the middle sector granites are virtually lacking but lamprophyres are numerous. In New Zealand, camptonites are the dominant type where granites are absent, but where lamprophyres and granites are associated, the former include both camptonitic and biotite-bearing types. The granites are not younger than middle Cretaceous; some of the associated lamprophyres are not older than Eocene.

There are other granitic provinces where lamprophyres are almost lacking. Such is the Southern California batholith, where associated basic dike swarms are of dioritic composition.[25] Though lamprophyres are known in the Sierra Nevada, they are not numerous and are virtually restricted to camptonitic types.

Lamprophyres are associated in some regions with plutonic rocks other than those of the granite granodiorite class. Indeed, Rosenbusch defined various groups of lamprophyre according to their supposed respective associations with certain classes of plutonic rock. Thus, among the hornblendic lamprophyres, camptonites were defined as associates of syenites and nepheline syenites, while similar rocks associated with granites were termed spessartites. The validity of such generalizations is doubtful. Finally it should be noted that lamprophyres also occur widely as minor members of the alkaline olivine-basalt volcanic association. For example, camptonites have been described from the Brito-Arctic Tertiary province, from the Tertiary volcanic associations of Montana, Wyoming, and Arizona, and even from the oceanic alkaline olivine-basalt province of Tahiti. It was noted earlier (page 250) that there is a marked tendency for potassic lamprophyres of such volcanic associations to be restricted to regions underlain by granitic rocks.

Though lamprophyres typically occur as narrow dikes, there are coarse-grained plutonic rocks (shonkinites, teschenites, theralites) and also strictly volcanic lavas (nepheline basalts, leucite basalts, limburgites) of

[25] E. S. Larsen, Batholith of Southern California, *Geol. Soc. America Mem.* 29, pp. 103, 104, 1948.

almost identical chemical composition. This similarity is sometimes overlooked on account of textural and mineralogical difference between the rocks in question and equivalent lamprophyres. It is the reason for treating the lamprophyres and the leucite-basalt volcanic association in the same chapter.

Compositions of some typical lamprophyres (not including extreme types) are given in Table 23.

TABLE 23.   COMPOSITIONS OF TYPICAL LAMPROPHRYES

Constituent	1	2	3	4	5	6
SiO_2	39.00	40.70	41.94	47.40	46.37	51.50
TiO_2	3.60	5.81	4.15	3.31	1.80	1.85
Al_2O_3	11.72	8.99	15.36	17.40	11.98	11.55
Fe_2O_3	4.11	4.51	3.27	4.65	5.05	2.38
FeO	8.19	8.37	9.89	5.87	5.16	4.72
MnO	0.16	0.17	0.25	0.14	0.10	0.10
MgO	12.24	9.69	5.01	4.45	8.38	7.90
CaO	11.80	9.94	9.47	8.30	9.33	9.10
Na_2O	3.04	1.58	5.15	5.41	2.84	2.55
K_2O	0.43	1.97	0.19	2.58	4.34	5.65
H_2O+	3.30	4.16	3.29		2.17	1.10
H_2O-	1.30	1.42			0.71	0.45
P_2O_5	1.03	1.65		0.49	1.34	0.96
CO_2		0.64	2.47		0.24	
S		0.09				
BaO					0.31	
SrO		0.36				
F		0.15				
Cl		0.02				
Rest		0.09*				
Total	99.92	100.31	100.44	100.00	100.12	99.81

* $NiO + Cr_2O_3 + V_2O_3$.

Explanation of Table 23

1. Monchiquite, Hopi Buttes, northeast Arizona.   (H. Williams, Geol. Soc. America Bull., vol. 47, p. 166, No. 2, 1936.)
2. Limburgite, dike from Kawarau gorge, Otago, southern New Zealand.   (C. O. Hutton, Royal Soc. New Zealand Trans., vol. 73, pt. 1, p. 63, No. 1, 1943.)
3. Camptonite, Campton Falls, New Hampshire.   (H. Rosenbusch and A. Osann, Elemente der Gesteinslehre, p. 338, No. 9, Erwin Nägele, Stuttgart, 1923.)
4. Average camptonite, Tahiti.   A. Lacroix's analyses, recalculated to 100 per cent free of H_2O and CO_2.   (H. Williams, B. P. Bishop Mus. Bull. 105, p. 43, No. 11, 1933.)
5. Biotite-augite lamprophyre, Spanish Peaks, Colorado.   (A. Knopf, Geol. Soc. America Bull., vol. 47, p. 1781, No. 11, 1936.)
6. Minette, Shiprock, northeast Arizona.   (Williams, op. cit., p. 166, No. 9, 1936.)

Petrogenesis of Lamprophyres.   Several hypotheses—none entirely satisfactory nor universally applicable—have been advanced to account for the origin and evolution of lamprophyric magmas.

1. From the common close association of lamprophyres and aplites in regions of granite intrusion grew the old belief that the two contrasted rocks are complementary products of differentiation from a common parent granitic magma.[26]   This concept depends purely on a common but by no means universal type of field association, which can equally well be explained in another way.   Moreover, it fails to account for the appearance of lamprophyres and of basic alkaline lavas of identical chemical composition in volcanic associations dominated by alkaline olivine basalts.   Nor is it easy to reconstruct differentiation of lamprophyres from granitic magma, in terms of the experimentally determined data of silicate crystallization.   Today any direct comagmatic relation between granite and associated lamprophyres is generally denied.[27]   Instead the lamprophyre magmas are thought to have risen from an independent source along fractures formed in response to previous emplacement of granitic bodies.   In some instances a considerable time interval can be shown to separate the two episodes of intrusion.

2. Beger [28] suggested that lamprophyre magmas form by remelting of crystals of ferromagnesian minerals accumulating at the base of a body of differentiating magma.   Bowen [29] proposed a hypothesis that is essentially a modification of this idea.   If hornblende or biotite—common igneous minerals rich in alkali and water and very low in silica—were brought into contact with liquid alkaline olivine-basalt magma, the ensuing reaction would result in enrichment of the liquid phase in alkali and water and in simultaneous crystallization of augite and olivine.   Starting with the partly crystalline sodic and potassic magmas so produced, Bowen showed how cooling under various conditions with varying degrees of fractional crystallization could lead to such diverse rock types as camptonites, monchiquites, and nepheline basalts; melilite basalts, melilite lamprophyres, and phonolites; biotite lamprophyres and members of the leucite-basalt family.   The main difficulty raised by Bowen's hypothesis concerns the general improbability of the assumed situation, in which hot olivine-basalt magma has ready access to hornblende or biotite (the latter especially) concentrated by crystal differentiation.   This difficulty is only

[26] E.g., G. W. Tyrrell, *The Principles of Petrology*, p. 121, Dutton, New York, 1926.

[27] Cf. Smith, *op. cit.*, pp. 379, 380, 1936; Bederke, *op. cit.*, p. 56, 1947; E. S. Hills, The Wood's Paint dike swarm, Victoria, *Sir Douglas Mawson Anniversary Volume*, p. 97, University of Adelaide, 1952; Eskola, *op. cit.*, p. 336, 1954; Kaitaro, *op. cit.*, p. 64, 1956.

[28] P. J. Beger, Der Chemismus der Lamprophyre, in P. Niggli, *Gesteins- und Mineralprovinzen*, Band 1, pp. 571–574, Berlin, 1923.

[29] Bowen, *op. cit.*, pp. 258–273, 1928.

partly removed by frequent records of coarse crystals of barkevikitic hornblende in trachybasalts of the alkaline olivine-basalt association (cf. pages 170 and 191).

3. Differentiation of quite another kind is advocated by Eskola.[30] A basic magma becomes enriched—by some process not completely understood—in CO_2 and water. This delays crystallization of Mg- and Fe-bearing silicates, so that both elements become concentrated, along with alkalis, in late-magmatic lamprophyric liquids. The main difficulty raised by Eskola's hypothesis concerns the chemistry of the differentiation process. Petrographic and experimental evidence has shown conclusively that in the normal course of crystallization of relatively anhydrous basaltic magma, olivine and pyroxenes separate early. Introduction of water into the system favors appearance of hornblende and, if potash is available, biotite, rather than olivine and pyroxene. But such data as have emerged from experiments on fusion of basalt at high water pressures (cf. page 154) have not demonstrated any tendency for retention of Mg and Fe in the low-melting fractions. However, the effect of water on crystallization of basaltic melts has as yet been imperfectly explored, and that of CO_2 is almost wholly conjectural.

4. Hypotheses of another kind invoke some kind of assimilative reaction between some more normal magma and an appropriate wall rock to give lamprophyric magmas. The parent magma could be basic [31] or acid.[32] The contamination process could be reaction of the kind pictured by Bowen, enhanced by the heated condition of recently injected granite and adjacent country rocks; [33] or it could involve partial or complete melting of the invaded rock.[34] Thus Oftedahl explains the potassic lamprophyres as local products of complete fusion of a shaly roof permeated by hot gases emanating from immediately underlying granitic magma. These and other hypotheses of assimilation account for the peculiar chemical characters of lamprophyres (including high content of BaO, SrO, H_2O, CO_2, and P_2O_5) and the high incidence of quartzose and granitic xenoliths in these rocks. It is compatible, too, with the wide range of lithological variation commonly displayed in lamprophyre dike swarms and even within individual dikes.

It must be admitted that no completely satisfactory explanation of the origin of lamprophyres has yet been proposed. The group is so diversified that more than one mode of origin is probable. The authors, like Bowen, would derive lamprophyres ultimately from alkaline olivine-ba-

[30] Eskola, op. cit., pp. 334–336, 1954.
[31] Bederke, op. cit., 1947; Oftedahl, op. cit , pp. 13, 14, 1957.
[32] Oftedahl, op. cit., pp. 3, 11–12, 1957.
[33] Bederke, op. cit., p. 56, 1947; Kaitaro, op. cit., p. 64, 1956.
[34] Smith, op. cit., p. 380, 1936; Oftedahl, op. cit., 1957.

salt magma.   The camptonites and monchiquites are so close in composition to volcanic rocks of the nepheline-basalt family that the same origin must be presumed for both families of rocks.   The nepheline basalts are known as constant associates of alkaline olivine basalts in both oceanic and continental environments.   It has been generally assumed that the dominant alkaline olivine basalts must play a parent role in the evolution of nepheline basalts; but this is not certain, nor has a satisfactory course of evolution been clearly demonstrated.   High concentrations of CO_2 and of water in the silicate-melt phase of the magma are possibly essential conditions.   Whether the observed appearance of hornblende in basic lavas of this series is a prerequisite for, or the result of, evolution of basic alkaline magma is not clear.

The tentative view that selective fusion of potassic minerals of granite and mica schist during reaction with basic magma under special conditions leads to development of leucite-basalt magmas has already been discussed (pages 249, 250).   It applies equally to the origin of mica lamprophyres.   It would explain, too, the almost ubiquitous presence of quartz xenoliths in lamprophyres.   One of the special conditions may well be previous concentration of CO_2, and perhaps water as well, in the basic magma, but from what we know of the behavior of volatile-enriched silicate melts (pages 412 to 416), it is obviously impossible to reconstruct the course of reaction in such a complex system.   The hypothesis, too, is admittedly a mere speculation, but it is on the whole compatible with the data of petrography and geochemistry and with such experimental observations as are available.

The typical porphyritic panidiomorphic texture of lamprophyres presumably is determined by crystallization in the presence of a silicate-melt phase only.   Carbonatization and other deuteric effects commonly shown by lamprophyres are correlated with late appearance of gas or solutions rich in CO_2 and in water.   It would appear from the work of Bowen and Tuttle [35] that where olivine of lamprophyres has been directly replaced by talc without intermediate serpentinization, a rather high temperature of replacement (above about 500°C.) is indicated.   A necessary condition is that the vapor phase (consisting largely of water and perhaps CO_2) should be capable of adding silica or removing magnesia.

[35] N. L. Bowen and O. F. Tuttle, The system $MgO-SiO_2-H_2O$, *Geol. Soc. America Bull.*, vol. 60, p. 452, 1949.

Volcanic Associations of Orogenic Regions

GENERAL STATEMENT

In fold-mountain regions, such as the western mountain zones of the United States or the Alpine areas of Europe, there is everywhere evidence of igneous activity broadly contemporaneous with orogeny. When the tectonic and igneous histories of such regions are worked out in detail, it becomes clear that during a single major orogenic cycle the type of igneous activity and the nature of the corresponding rock associations have varied considerably. Moreover, this variation tends, in many instances, to conform to a single broad pattern:

1. Eruption of dominantly basic (including spilitic) lavas, during the geosynclinal phase of the tectonic cycle.

2. Injection of ultrabasic and basic plutonic intrusions during the early stages of folding; in some cases this overlaps phase 1 above.

3. Development of granodioritic and granitic batholiths during and following the main period of folding.

4. Surface eruption of basalts, andesites, and rhyolites during and following elevation of the folded mass. This phase is typically separated by a lengthy period of time from the main phase of folding and plutonic activity.

In the present chapter we shall concern ourselves only with the two volcanic associations (1 and 4 above), which mark the early and the final stages, respectively, of orogeny. But in framing any general hypothesis of petrogenesis we must also view these in relation to the other two igneous (plutonic) associations mentioned. Also to be taken into account are any other igneous associations that may have developed contemporaneously in nonorogenic or lightly deformed areas adjoining the orogenic belt, as well as products of preorogenic igneous activity where such may be deciphered.

THE SPILITE KERATOPHYRE ASSOCIATION

Igneous Activity Synchronous with Geosynclinal Sedimentation.   Most eroded geosynclines show evidence of igneous activity approximately synchronous with at least the later part of the filling and sinking of the trough.   Prominent among the products of such activity, and almost confined to the geosynclinal environment, are submarine lavas, tuffs, and equivalent intrusives of sodic composition.   These constitute the spilite keratophyre association.   This association may also include basalts of normal composition.   Moreover, in many other geosynclines the volcanic component of the filling consists of basalts, andesites, or rhyolites unrelated to the spilitic kindred.[1]

It has long been recognized that ultrabasic intrusives (serpentinites, peridotites) tend to be associated geographically with spilitic rocks, but the period of their intrusion typically is distinctly later than that of spilitic eruption.   The ultrabasic rocks may therefore be treated either as part of the spilitic association or as part of a distinct plutonic kindred.   The latter alternative is adopted here.   But the reader is reminded that in some geosynclinal terranes, e.g., in the Franciscan-Knoxville of California, intrusion of ultrabasic bodies was partly synchronous with submarine eruption of spilitic and normal basaltic lavas during sinking and filling of the trough.

The description of the spilite keratophyre association that now follows refers to those numerous cases where geosynclinal volcanic and equivalent near-surface intrusive rocks are for the most part of distinctly sodic composition.

General Characteristics of the Spilite Keratophyre Association.   The most characteristic rocks of the spilite keratophyre association are spilites—basic lavas consisting principally of highly sodic plagioclase (albite or oligoclase) and augite or its altered equivalent (actinolite, chlorite-epidote, chlorite-hematite, etc.).   Olivine typically is absent or is represented sparingly by serpentine pseudomorphs.   Evidence of hydrothermal activity (e.g., alteration of pyroxene, infilling of vesicles with epidote, calcite, and so on) is usually conspicuous, while persistence of relict patches of labradorite or andesine within crystals of albite shows conclusively that in some cases the present condition of the feldspars is a result of albitization of initially more calcic plagioclase.   Many spilites are pillow lavas.   This is not essential, however; nor, by any means, are all pillow lavas spilitic.   The larger pillows commonly show concentric

---

[1] G. W. Tyrrell, Flood basalts and fissure eruptions, *Bull. volcanologique*, sér. 2, tome 1, pp. 90–92, 1937; A. Knopf, The geosynclinal theory, *Geol. Soc. America Bull.*, vol. 57, pp. 649–670, 1948.

zoning due to radial variation in composition and fabric (cf. Table 24).[2] The central zone is coarse grained (diabasic) and richer in lime and somewhat lower in soda than the border. The latter is a variolite, with spherules of acicular albite in a "chloritic" base. Between the pillows is a "chloritic" ultrabasic matrix very high in ferrous iron and magnesia and low in alkali; it sometimes contains scattered phenocrysts and microlites of albite. The "chloritic" base of the variolite zone and the material of the matrix probably represent glass in various stages of reaction with sea

TABLE 24. CHEMICAL COMPOSITIONS OF ZONES IN SPILITE PILLOWS

Constituent	1	2	3	4
SiO_2	48.38	50.40	30.24	29.46
TiO_2	1.21	1.55	0.58	3.01
Al_2O_3	12.73	14.16	16.83	16.95
Fe_2O_3	3.17	1.63	3.95	5.23
FeO	6.52	8.54	18.72	15.53
MnO	0.17	0.16	0.28	0.10
MgO	7.96	8.58	16.73	16.08
CaO	9.48	5.90	1.92	2.97
Na_2O	3.92	4.28	0.27	0.84
K_2O	0.08	0.16	0.02	0.51
H_2O+	6.14	4.49	10.18	9.64
H_2O-	0.16	0.13	0.27	0.11
P_2O_5	0.15		0.11	Trace
CO_2			0.08	
Total	100.07	99.98	100.18	100.43

Explanation of Table 24

Note: All analyses from M. Vuagnat, Variolites et spilites, *Archiv. Sci.*, vol. 2, fasc. 2, p. 235, 1949.
  1. Center of pillow, Newborough, Anglesey, Wales.
  2. Border (variolitic) of same pillow as 1.
  3. Matrix surrounding same pillow as 1 and 2.
  4. Matrix surrounding a spilite pillow, Alp Champatsch, Switzerland.

water. In its low silica and high water content it resembles both chlorophaeite (an alteration product of magnesian minerals in diabase and basalt) and palagonite (hydrated basaltic glass).[3] Like chlorophaeite it has a high content of $(Fe + Mg)$; like palagonite it is rich in Al.

[2] M. Vuagnat, Variolites et spilites, *Archiv. Sci.*, vol. 2, fasc. 2, pp. 223–236, 1949; Sur les pillow lavas dalradiennes de la péninsule de Tayvallich (Argyllshire), *Schweizer. min. pet. Mitt.*, vol. 29, pt. 2, pp. 524–536, 1949.

[3] M. A. Peacock, The distinction between chlorophaeite and palagonite, *Geol. Mag.*, vol. 67, pp. 170–178, 1930.

Common associates of spilitic lavas are intrusive chemically equivalent albite diabases, as well as flows, minor intrusions, and tuffs of highly sodic keratophyre and quartz keratophyre. In many of these rocks, too, albite seems to be of secondary origin; in one group, the magnetite keratophyres, concentration of iron appears to have been achieved through the activity of late-magmatic fluids. The weilburgites of Germany are composed entirely of alkali feldspar set in a "chloritic" base. The feldspar is mostly albite; but potash weilburgites with potash feldspar are also known. There is no indication that the albite is secondary; and Lehmann attributes both albite and "chlorite" to primary crystallization at low temperatures from a special water-rich magma.[4]

Chemical features characteristic of the whole rock association are a high content of Na_2O, low K_2O, and rather low Al_2O_3. But not uncommonly a typically spilitic suite of rocks is intimately associated with, and indeed includes, basalts and diabases of normal composition with labradorite as the constituent feldspar, as well as basaltic rocks which clearly have been affected by lime metasomatism involving replacement of plagioclase and infilling of vesicles by hydrous lime-aluminum silicates, notably epidote, prehnite, and various zeolites.

While it is agreed by most writers that late-magmatic metasomatism plays an important role in the evolution of spilitic rocks in general, several points bearing on the significance of such metasomatism may appropriately be noted:

1. In certain spilites and albite diabases there is nothing to show that the albite of individual spilitic rocks is other than a product of direct crystallization from a magma.[5]

2. Spilitic rocks, by reason of their early appearance in geosynclinal belts, are particularly liable to be affected by regional metamorphism, and many of the greenschists of orogenic zones are doubtless of such origin. Since low-grade metamorphism corresponding to the greenschist facies (cf. pages 534 to 537) leads to the same assemblage of minerals (albite, epidote, chlorite, actinolite, calcite) as may also result from low-temperature metasomatism of basic rocks, the general presence and local concentration of albite in partially metamorphosed "greenstones" is no criterion of spilitic parentage. Writers on the spilite problem, whose experience has been drawn largely from lavas showing incipient effects of regional metamorphism, have not always appreciated the character of

---

[4] For various opinions on the genesis of weilburgites, see: E. Lehmann, Beitrag zur Beurteilung der paläozoischen Eruptivgesteine Westdeutschlands, *Deutsche geol. Gesell. Zeitschr.*, vol. 104, I, pp. 219–237, 1952; The significance of the hydrothermal stage in the formation of igneous rocks, *Geol. Mag.*, vol. 89, pp. 61–69, 1952; H. Hentschel, Zur Petrographie des Diabas-Magmatismus im Lahn-Dill Gebiet, *Deutsche geol. Gesell. Zeitschr.*, vol. 104, pp. 237–246, 1952.

[5] Lehmann, *op. cit., Geol. Mag.*, p. 62, 1952.

unmetamorphosed spilites such as those of New South Wales and New Zealand.

3. There are many recorded instances where shales invaded by albite diabases of the spilitic kindred have been converted to adinoles by contact metamorphism involving introduction of soda. However, similar soda metasomatism commonly occurs in connection with intrusion of normal diabases. Furthermore, from the widespread occurrence of jaspers and manganiferous sediments in close association with spilites, it would seem that silica, iron, manganese, and perhaps magnesium are the main constituents of late-magmatic solutions emanating from spilitic rocks. A number of writers believe that silica in particular is commonly expelled in such vast amounts as to allow its precipitation, either by chemical or by organic agencies, to give those thick extensive beds of chert with which spilitic lavas so frequently are broadly associated.

4. A very low content of potash is one of the characteristic chemical features of spilites. Yet there are rare instances where highly potassic counterparts of spilites have formed in a similar geosynclinal environment, apparently by "adularization" of plagioclase in initially basaltic lavas. Such are the "poeneites" (potash spilites) associated with normally sodic spilites, olivine basalts, and subordinate keratophyres and soda rhyolites in the Permian of Timor, and the potash weilburgites of Germany. The poeneites of Timor are products of early igneous activity in the first stages of geosynclinal evolution and were followed by normal spilites and keratophyres in Mesozoic times as the geosyncline reached its full development.[6]

From the characteristically close association of spilitic rocks with marine sediments, and from the prevalence of pillow structure among typical spilites, it is clear that these are mostly submarine lavas poured out on the sea floor or injected into unconsolidated sediments. It is equally obvious from the great thicknesses of associated arkosic sandstones and graywackes consistently present, that volcanic activity of this type occurs during the development and slow sinking of geosynclines or unstable basins. So it is doubtful whether the bedded radiolarian cherts that are so conspicuous in spilitic provinces represent abyssal deposits as was believed to be the case by Steinmann and others. Of somewhat later origin, but so consistently present in such terranes as to suggest some genetic connection with the spilitic lavas, are ultrabasic and basic intrusive rocks represented for the most part by serpentinites and gabbros. These will be considered later as constituting a distinct plutonic association.

[6] W. P. De Roever, Olivine basalts and their alkaline differentiates in the Permian of Timor, Geol. Expedition Lesser Sunda Islands (H. A. Brouwer), vol. IV, pp. 209–289, 1942.

The chemical character of spilites and associated rocks is illustrated in Table 25. All are highly sodic, and most are conspicuously low in potash. Other features of spilites as contrasted with other basaltic rocks are rela-

TABLE 25. CHEMICAL COMPOSITIONS OF SPILITES AND ASSOCIATED ROCKS

Constituent	1	2	3	4	5	6	7	8
SiO_2	51.22	48.22	52.94	53.30	53.15	75.10	75.04	56.95
TiO_2	3.32	2.68	2.54	2.41	1.50	0.22	0.10	0.89
Al_2O_3	13.66	14.82	12.81	15.16	14.39	12.84	13.39	17.87
Fe_2O_3	2.84	0.56	3.76	2.54	1.28	0.70	1.61	4.49
FeO	9.20	9.25	9.29	8.71	9.33	1.36	0.37	6.00
MnO	0.25	0.23	0.21	0.28	0.14	0.04	0.05	0.08
MgO	4.55	5.58	3.65	4.14	4.74	0.30	0.18	0.93
CaO	6.89	8.81	6.22	2.97	7.04	0.32	0.40	2.30
Na_2O	4.93	4.95	5.25	5.55	4.58	5.12	6.36	8.80
K_2O	0.75	0.44	0.18	0.32	1.01	2.39	0.83	0.38
H_2O+	} 1.88	2.54	2.33	3.14	2.02	0.95	1.07	0.71
H_2O-		0.15	0.21	0.18	0.19	0.27	0.24	0.38
P_2O_5	0.29	0.23	0.36	0.51	0.19	0.04	0.08	
CO_2	0.94	1.40	Nil	Nil	0.10	0.03	0.10	0.91
BaO		Nil	Trace	Nil		0.05	Nil	
Rest		0.40	0.16	0.40		0.09		0.04
Total	100.72	100.26	99.91	99.61	99.66	99.82	99.82	100.73

Explanation of Table 25

1. Average spilite, 19 analyses: 3 from England, 7 from New South Wales, 5 from Kiruna, Sweden, 4 from Karelia, Finland. (N. Sundius, *Geol. Mag.*, vol. 67, p. 9, 1930.)

2. Spilite, Nundle, New South Wales. (W. N. Benson, *Linnean Soc. New South Wales Proc.*, vol. 40, p. 139, No. 117, 1915.)

3. Spilite, Great King Island, New Zealand. (J. A. Bartrum, *Geol. Mag.*, vol. 73, p. 417, No. 3, 1936.)

4. Albite diabase, Poorman Mine, Oregon. (J. Gilluly, *Am. Jour. Sci.*, vol. 29, p. 235, No. 6, 1935.)

5. Spilite, somewhat high in potash, eastern Oregon. (Gilluly, *op. cit.*, p. 235, No. 8, 1935.)

6. Quartz keratophyre, Great King Island, New Zealand. (Bartrum, *op. cit.*, p. 417, No. 1, 1936.)

7. Quartz keratophyre, eastern Oregon. (Gilluly, *op. cit.*, p. 235, No. 2, 1935.)

8. Magnetite keratophyre, Nundle, New South Wales. (Benson, *op. cit.*, p. 139, No. 1086, 1915.)

tively high TiO_2 and somewhat low Al_2O_3, CaO, and MgO. By some writers the chemical data have been interpreted as indicating a sharp distinction between spilites and "normal" tholeiitic and alkaline olivine basalts.[7]   Others record instances of rocks intermediate between spilites and normal basalts and hold that there is complete gradation between the two groups of rocks.[8]   This may be so.   As chemical data become more numerous, they can be expected to reveal transitions between even the most clearly defined rock associations and rock families.   The spilitic association, whatever its relation to the other basaltic kindreds may be, is one of striking individuality, maintained in widely scattered provinces of all ages and recognized wherever the rocks of geosynclinal terranes have been petrographically investigated.   The interesting petrogenic problem which it presents will be discussed after we have reviewed briefly some typical occurrences.

Typical Spilitic Provinces.   *The Devonian Spilites of Southwest England and Germany.*[9]   The spilites and associated rocks of Devon and Cornwall are those upon which Dewey and Flett based their original definition of the spilitic "suite" (rock association).   Here the dominant igneous rocks are spilitic pillow lavas and tuffs interbedded with mudstones, limestones, and radiolarian sediments deposited in a geosyncline continuous with the Hercynian trough of Germany and central Europe. Flows of quartz keratophyre and minor intrusions of albite diabase, hornblende-albite diorite, camptonite, and picrite complete the igneous suite. In Carboniferous times the geosynclinal filling was folded, partially metamorphosed, and invaded extensively by granitic intrusions.

In Germany, too, the igneous rocks associated with Devonian geosynclinal sediments include spilites, albite diabases, keratophyres, weilburgites, and picrites, but albitization is not nearly so widespread as in contemporary British rocks, and many rocks of this province are normal basalts and diabases without obvious spilitic affinities.   Here, too, Carboniferous folding (Variscan orogeny) was accompanied by metamorphism of many of the earlier volcanics and by plutonic intrusion of peridotite, gabbro, and finally granites.

The Devonian and Carboniferous Spilites of New South Wales.[10]   In northeastern New South Wales a geosyncline several hundred miles in

---

[7] Cf. H. Dewey and J. S. Flett, Some British pillow lavas and the rocks associated with them, *Geol. Mag.*, vol. 8, pp. 202–209, 241–248, 1911; N. Sundius, On the spilitic rocks, *Geol. Mag.*, vol. 67, pp. 1–17, 1930.

[8] Cf. R. A. Daly, *Igneous Rocks and the Depths of the Earth*, pp. 419–420, McGraw-Hill, New York, 1933; J. Gilluly, Keratophyres of eastern Oregon and the spilite problem, *Am. Jour. Sci.*, vol. 29, pp. 225–252, 336–352, 1935 (especially pp. 248–251).

[9] Dewey and Flett, *op. cit.*, 1911; W. N. Benson, The tectonic conditions accompanying the intrusion of basic and ultrabasic rocks, *Nat. Acad. Sci. Mem.*, vol. 19, Mem. 1, pp. 12, 19–20, 1924.

[10] W. N. Benson, The geology and petrology of the Great Serpentine Belt of New South Wales, Parts III, IV, V, *Linnean Soc. New South Wales Proc.*, vol. 38, pp. 662–724, 1914; vol. 40, pp. 121–173, 540–624, 1915; *op. cit.*, pp. 37, 38, 1926.

length became filled, during Devonian times, with many thousands of feet of marine sediments, prominent among which were radiolarian claystones. All are believed to have been laid down in fairly shallow water. Contemporary submarine volcanism produced vast quantities of spilitic pillow lavas and tuffs some of which were certainly intrusive into the soft muds then covering the sea floor. Locally important are rocks of the keratophyre family, including magnetite keratophyres as well as highly siliceous quartz-bearing types. Inclined sills of albite diabase dating from this period are uniformly sodic across thicknesses as great as 1,500 ft. and like the majority of the spilitic rocks, show clear albite associated with, and ophitically enclosed in, almost unaltered augite. Metasomatic introduction of iron into cherts adjacent to keratophyre and spilite has caused local development of jasper.

During the early Carboniferous, deposition of claystones (no longer radiolarian) continued near the western margin of the geosyncline, and eruption of great quantities of keratophyre tuff took place. This was followed in the middle Carboniferous by extensive injection and extrusion of normal diabases, andesine basalts, andesites, soda rhyolites, quartz keratophyres, and quartz latites. These eruptions accompanied faulting and minor folding on the relatively stable block margining the Devonian geosyncline on the west. The geosynclinal filling itself (Devonian radiolarian claystones and spilitic rocks) was at the same time intensely folded and invaded by a series of plutonic intrusions commencing with steep sills of harzburgite and minor gabbros (the great serpentine belt) and concluding with a long series of granites, the latest of which were of Permian or Triassic age.

The Eocene Spilites of Olympic Peninsula, Washington.[11] In the Olympic Peninsula of northwestern Washington, argillites, graywackes, volcanic rocks, and limestones of early-middle Eocene age make up a total thickness of at least 30,000 ft. The volcanic rocks, which in places form accumulations several thousand feet thick, are mainly spilitic pillow lavas, but normal basalts and diabases are also present. Chemical gradation has been demonstrated from slightly sodic basalt and diabase (Na_2O = 3.02 to 3.45 per cent) through typical albitic spilites to extremely altered rocks (Na_2O = 8 per cent) consisting largely of albite, zeolites, chlorite, calcite, and quartz. Pyroxene of the spilites in many cases shows little sign of alteration. Fossils such as *Globigerina* tests and sharks' teeth in the tuffaceous layers show that the volcanic rocks are of submarine origin and may have been extruded under a considerable depth of water.

Intimately associated with the volcanics of Olympic Peninsula are red

[11] C. F. Park, The spilite and manganese problem of the Olympic Peninsula, Washington, *Am. Jour. Sci.*, vol. 244, pp. 305–323, 1946.

calcareous sediments formed by admixed oxidized volcanic material and calcium carbonate chemically precipitated from the sea water; jasper occurring as veinlets between pillows and replacing both lavas and the red sediments mentioned above; and irregular bodies of manganese minerals (hausmannite, rhodonite, rhodochrosite, and various complex sili-

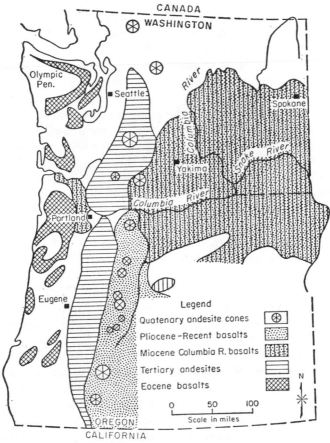

Fɪɢ. 38.   Post-Mesozoic volcanic rocks of Washington and Oregon.   (*Simplified, after A. C. Waters.*)

cates) veining and replacing the red rocks and even the tops of spilitic flows.

It has been shown by Waters [12] that the non-spilitic basalts of the Olympic Peninsula are tholeiitic.   Moreover the spilitic province is but part of a much larger coastal province, extending over 200 miles south into Oregon (cf. Fig. 38) and characterized throughout by an enormous

[12] A. C. Waters, Volcanic rocks and the tectonic cycle, *Geol. Soc. America Special Paper*, no. 62, pp. 704–707, 1955.

development of undifferentiated tholeiitic submarine lavas and tuffs. Indeed the Olympic spilites can properly be regarded as local variants of much more extensive tholeiites. Their development in a geosynclinal environment is to be contrasted with the complete absence of spilites in the almost equally extensive terrestrial tholeiitic basalts of the Miocene Columbia River province a hundred miles east. Contemporaneous with the tholeiitic basalts and spilites of the coastal province are the earliest flows and tuffs of the Cascade andesitic province (Upper Eocene to Lower Pliocene) immediately to the east (Fig. 38).

Spilites and keratophyres of northern New Zealand.[13]   At several localities in northernmost New Zealand spilitic pillow lavas and flows of keratophyre are interbedded with marine sediments, mainly graywackes, probably of mid-Mesozoic age.   The principal constituents of the spilites are low-temperature albite, diopsidic augite, an iron-rich chlorite, and iron ore.   The prismatic habit of albite and pyroxene, absence of relict plagioclase of a more calcic nature, and the completely fresh condition of the two minerals attest to a primary magmatic origin for both albite and augite.   It is even thought that the chlorite has crystallized directly from an iron-rich residual magmatic liquid.   There are close chemical analogies between the New Zealand spilites and tholeiitic basalts.   Except for a deficiency in K in the more basic spilites, there is much the same spread in the ratios Fe:Mg:(K + Na) and Ca:Na:K in the two series.

The keratophyres are composed of quartz and alkali feldspars with small amounts of chlorite, biotite, leucoxene, and accessory zircon.   Only the feldspars are phenocrystic.   In the typical sodic keratophyres the feldspar is mainly low-temperature albite; but there are also potassic rocks, indistinguishable chemically from normal rhyolites, in which albite shows all stages of replacement by potash feldspar—orthoclase cryptoperthite. Even within a single flow, variation in the ratio K:Na is remarkable.   It is concluded that the rocks were initially normal rhyolites with about equal soda and potash and that the present distribution of Na and K is the result of recrystallization and autometasomatism induced, at moderately elevated temperatures, by solutions which promoted internal migration of alkalis and silica through the buried volcanic pile.

Petrogenesis of the Spilite Keratophyre Association.[14]   *The Spilite Problem.*   No attempt will be made here to recapitulate the history of

[13] J. A. Bartrum, Spilitic rocks in New Zealand, *Geol. Mag.*, vol. 73, pp. 414–423, 1936; M. H. Battey, Alkali metasomatism and the petrology of some keratophyres, *Geol. Mag.*, vol. 92, pp. 104–126, 1955; The petrogenesis of a spilitic rock series from New Zealand, *Geol. Mag.*, vol. 93, pp. 89–110, 1956.

[14] Dewey and Flett, *op. cit.*, 1911; A. K. Wells, The nomenclature of the spilitic rocks, *Geol. Mag.*, vol. 59, pp. 346–354, 1922; vol. 60, pp. 62–74, 1923; Sundius, *op. cit.*, 1930; Gilluly, *op. cit.*, 1935; Park, *op. cit.*, 1946; F. J. Turner, Mineralogical and

growth of ideas concerning the genesis of the spilitic kindred, for this has already been summarized at considerable length by Gilluly.[15] Opinion today is still divided upon a number of questions raised in Gilluly's paper: (1) the possible existence of a distinct spilitic parent magma; (2) magmatic versus metasomatic origin of the albite; and (3) the relative roles of residual igneous fluids and of introduced fluids in metasomatism of spilitic rocks.   These questions may now be considered in turn.

The Parent Magma.   Keratophyres are so consistently associated with spilites, and both groups of rocks typically are so conspicuously sodic, that common parentage is highly probable in the majority of cases.   Thus arises the problem whether or not there exists a primary magma of spilitic composition which could differentiate to a keratophyre magma. A high CaO/Na_2O ratio (typically between 3 and 5, using weights per cent) is normal for common basalts of both tholeiitic and alkaline types. If it be assumed that magma of spilitic composition ($CaO/Na_2O = 1.5$) is usually generated independently of these "normal" basaltic magmas, then it must further be assumed that in orogenic belts, and there alone, very special local conditions prevail which may lead to the development of this special magma.   For example, it is barely conceivable that in these particular zones either there exists, at levels conducive to periodic fusion, a localized crustal zone of the composition necessary to generate spilitic magma, or that special physical and chemical conditions from time to time cause a spilitic magma to be generated from rocks which elsewhere remain unfused or give rise to "normal" magmas.   Both these possibilities are highly speculative and neither commends itself to the authors.   It would seem much more likely that spilites are in some way derived from magma of alkaline olivine-basalt or tholeiitic composition, either by differentiation or contamination of a "normal" basaltic magma, or else by metasomatic introduction of soda into rocks which first crystallized as "normal" basalts.   The probability of some such mode of origin is strengthened by the prevalence of "normal" basalts in geosynclinal terranes.   To weigh the relative merits of these two alternatives it is first necessary to consider the question of the origin of albite in spilites and the evidence for soda metasomatism.

Albitization in Spilites.   There is most convincing petrographic evidence that in many spilites and keratophyres albite has replaced a preexisting feldspar of more calcic composition, relict inclusions of which

structural evolution of metamorphic rocks, *Geol. Soc. America Mem. 30*, pp. 120–124, 1948; M. Vuagnat, Le rôle des roches basiques dans les Alpes, *Schweizer min. pet. Mitt.*, vol. 31, pt. 1, pp. 309–322, 1951; M. H. Battey, *op. cit.*, 1955, 1956; J. J. Reed, Petrology of the lower Mesozoic rocks of the Wellington district, *New Zealand Geol. Survey Bull.*, no. 57, pp. 34–39, 1957.

[15] Gilluly, *op. cit.*, pp. 234–244, 1935.

may still persist within the albite crystals. Confirmatory indications of soda metasomatism are albite-filled vesicles, veinlets of albite, and in some cases widespread development of zeolites.[16]   It has been shown experimentally that, at temperatures below about 330°C. and confining pressures in the vicinity of 220 bars, anorthite readily reacts with sodium carbonate solution and silica in a closed system to give the assemblage albite-calcite.[17]   It should be noted that if labradorite of basalt or diabase is replaced under like conditions by albite, then, since the reaction proceeds without appreciable volume change, it involves introduction of both Na and Si, and complementary removal (presumably in solution) of Ca and Al according to some such equation as

$$NaCaAl_3Si_5O_{16} + Na^+ + Si^{4+} \rightarrow 2NaAlSi_3O_8 + Ca^{++} + Al^{3+}$$
Labradorite   Albite

It is not surprising, in view of resulting variability in relative concentration of Na^+, Ca^{++}, Al^{3+}, and Si^{4+} ions in the solutions taking part in this process, that local introduction of hydrous silicates of Ca and Al (epidote, prehnite, and certain zeolites), and even epidotization of feldspars, have taken place in many lavas closely associated with albitized rocks of the spilite association.   It is probable, too, that variation in temperature and pressure in submarine lavas accumulating to variable thickness and at different depths also determines whether initially calcic plagioclase becomes replaced by albite, zeolites, or epidote, or remains unaltered.   Albitization of plagioclase is commonly accompanied by development of chlorite, calcite, epidote, and actinolitic amphibole at the expense of augite in spilitic rocks.

Two alternative mechanisms, currently invoked to explain albitization of spilites in their characteristic geosynclinal environment, merit attention:

1. The marine environment of spilites may be directly responsible.   Sea water is a possible source of Na^+ ions and of water necessary for soda metasomatism.   It is conceivable, even probable, that sea water entrapped, vaporized, and streaming upward through hot but largely solidified submarine basic lavas, could bring about the type of alteration observed in spilites and could also contribute Ca, Fe, Mn, Mg, and Si to the surrounding sea in sufficient quantity to account for the chemically precipitated cherts, jaspers, manganese ores, and possibly limestones associated with many spilites.   Much the same effect might be achieved by sea water absorbed from invaded wet sediments into intrusive basic magma.   Com-

---

[16] Park, op. cit., pp. 316, 323, 1946.

[17] P. Eskola, U. Vuoristo, and K. Rankama, An experimental illustration of the spilite reaction, Comm. géol. Finlande Bull. 119, pp. 61–68, 1937.   See also Turner, op. cit., p. 122, 1948.

paction and incipient metamorphism of the basal sediments of a geosyncline during sinking and early compression must involve reduction in
porosity,[18] and hence expulsion (presumably upward) of vast quantities
of aqueous saline solutions originally held in the pores. Solutions of such
origin encountering magma or still heated igneous rocks in their upward
passage could contribute notably to their metasomatism. These possibilities are in no way invalidated by certain rare keratophyres, and possibly spilites too, that have been erupted subaerially. Metasomatism of
a type very generally brought about by the action of sea water presumably
could also be reproduced under favorable conditions by other agencies in
another environment, e.g., by saline ground water in sediments of desert
basins.

2. It is also possible to appeal to autometasomatism—chemical alteration of an igneous rock by residual aqueous fluids derived from its own
parent magma—under special, but admittedly obscure, conditions in some
way controlled by geosynclinal sinking synchronous with volcanism. It
might be supposed that a specially sodic (spilitic) magma tends to develop under such conditions and to give rise, as crystallization proceeds,
to sodic residual solutions necessary for albitization and related processes.
Or alternatively the basic magma might be of "normal" basaltic composition, the critical factor for albitization being one connected with the
marine environment, e.g., quick surface chilling, and consequent retention of volatile materials which otherwise would have escaped. A variant
of the autometasomatism hypothesis is Battey's suggestion that keratophyres may develop from normal rhyolites by redistribution and differential concentration of the two alkalis through the agency of aqueous
solutions diffusing at moderately elevated temperatures through the consolidated volcanic pile.[19]

Primary Albite in Spilites. Although the secondary nature of albite
and associated low-temperature minerals of some spilites has been demonstrated beyond reasonable doubt, there is a strong and persistent opinion
in current literature to the effect that in other spilites the albite and
even the chlorite are primary magmatic minerals.[20] This is strongly suggested by textural relations between albite and augite or chlorite in spilites lacking any trace of metamorphism. Indeed if this textural evidence
is rejected, it follows that volcanic textures in general are of dubious
genetic significance. Finally the data that are beginning to accrue from

---

[18] The porosity of Californian Tertiary sandstones with argillaceous cement com-
pacted at depths of the order of 8,000 to 10,000 ft. is in many cases still as high as 15
to 20 per cent.

[19] Battey, *op. cit.*, 1955.

[20] Vuagnat, *op. cit.*, *Archiv. Sci.*, p. 234, 1949; Lehmann, *op. cit.*, *Geol. Mag.*, 1952;
Battey, *op. cit.*, pp. 90–93, 1956; Reed, *op. cit.*, p. 38, 1957.

laboratory experiments on hydrous systems are not inconsistent with the possibility that albite and chlorite could crystallize from hydrous melts at temperatures of the order of 650°C.[21] Whether of primary or secondary origin, the albite of spilites is the low-temperature polymorph.

Evolution of Spilites. The authors tentatively favor a rather complex scheme of petrogenesis starting with normal basaltic magmas and leading by converging lines of descent to spilites. Differentiation of the parent magmas, assimilative reaction with rocks situated in the basal levels of the geosyncline, concentration of soda in late-magmatic aqueous extracts, and chemical activity induced by entrapped sea water and rising connate waters squeezed up from deeply buried sediments, are all factors of possible significance in the evolution of keratophyres and spilites.

Judging from the common association of mugearites with alkaline olivine basalts, fractional crystallization of the basalt magma may very generally lead to development of a sodic liquid differing from average spilite mainly in its higher content of alumina and in the presence of significant amounts of potash (Table 26). Such a liquid could be formed merely by removal of early-formed olivine, pyroxene, and possibly basic plagioclase in appropriate quantities. This could well be the first step in the evolution of a spilitic from a basaltic magma. Some spilites, indeed, contain sufficient potash and alumina to approach rather closely the composition of mugearites (Table 26, no. 3). Tholeiitic magmas must also be considered as possible parents of spilites. Battey [22] has emphasized certain chemical similarities between tholeiitic and spilitic series, and suggests that under the influence of high concentrations of water the trend of differentiation of a tholeiitic magma may lead into a spilite-keratophyre line.

If simple fractional crystallization alone were responsible for evolution of spilitic rocks from a basaltic parent, we should expect to find rocks of strictly spilitic composition in differentiated basic sills, especially where (as in teschenite and theralite intrusions) these sills are rich in soda and water. Such cases have not been recorded; we must turn to some other influence, peculiar to the geosynclinal environment. In this connection it seems unlikely that the occurrence of a sodic igneous association in an environment where soda-rich sediments and sodic waters both abound can be pure coincidence. Geosynclines are sites for accumulation of vast thickness of feldspathic sandstones (graywackes), the major constituents of which are quartz and oligoclase. Analyses of such rocks, as contrasted with those of other sandstones and shales, persistently show great predominance of soda over potash (Table 26, analyses 4 to 6). In view of the relatively low temperature and high albite content of the quartz-

---

[21] Battey, *op. cit.*, p. 102, 1956.
[22] Battey, *op. cit.*, pp. 101, 103–105, 1956.

albite eutectic, it is reasonable to suppose that basaltic magma, held for a long period in contact with graywackes or their metamorphic equivalents the albite-quartz-epidote schists, could incorporate into the liquid state material chemically equivalent to an albite-quartz mixture. Water

TABLE 26. COMPOSITION OF SPILITES, MUGEARITE, AND GRAYWACKES

Constituent	1	2	3	4	5	6
SiO_2	51.22	51.35	48.6	71.72	66.82	65.95
TiO_2	3.32	2.74	1.94	0.35		0.55
Al_2O_3	13.66	16.34	16.1	13.23	18.83	13.52
Fe_2O_3	2.84	4.64	7.6	0.30	2.58	1.94
FeO	9.20	6.19	4.0	3.58	2.87	4.18
MnO	0.25	0.20	0.34			
MgO	4.55	3.73	3.6	1.81	0.67	1.88
CaO	6.89	6.61	6.2	1.80	2.67	3.10
Na_2O	4.93	5.01	4.5	2.72	2.11	3.73
K_2O	0.75	1.94	1.76	1.29	1.34	2.01
H_2O+	} 1.88		2.9	2.53	} 2.13	3.13
H_2O-			0.22	0.15		
P_2O_5	0.29	1.00	0.34	0.09		
CO_2	0.94		1.45	0.32		
Rest				0.04		
Total	100.72	99.75	99.6	99.93	100.02	99.99

Explanation of Table 26

1. Average spilite. (N. Sundius, *Geol. Mag.*, vol. 67, p. 9, 1930.)
2. Average oligoclase andesite (= mugearite) of Hawaiian Islands. (G. A. Macdonald, in Geology and ground water resources of the island of Molokai, Hawaii, *Div. Hydr. Territ. Hawaii Bull.* 11, p. 102, 1947.)
3. Somewhat potassic spilite, Wellington, New Zealand. (J. J. Reed, *New Zealand Geol. Survey Bull.* 57, p. 37, No. 2, 1957.)
4. Fresh Franciscan sandstone (graywacke) Carbona quadrangle, California. (N. L. Taliaferro, Franciscan Knoxville problem, *Am. Assoc. Petroleum Geologists Bull.*, vol. 27, p. 136, No. 111, 1943.)
5. Average of 5 graywackes, Balclutha, New Zealand. (P. Marshall, *New Zealand Geol. Survey Bull. 19*, 1918.)
6. Average of 2 analyses of schistose graywacke, Marlborough, New Zealand. (J. Henderson, *New Zealand Geol. Survey Ann. Rept. 1934–1935*, p. 15, Nos. 4, 5.)

presumably would facilitate such a contamination process. Prevalence of albite-quartz veins and laminae in low-grade schists of graywacke parentage testifies to the ease with which the components of albite diffuse and segregate in the presence of water at temperatures far below those that prevail in reservoirs of molten basalt.

Superposed on the effects of differentiation and contamination, and operating in the same general direction, there could be the possible effect of connate and sea waters.   In the deeper levels these perhaps could diffuse into the still liquid magma and might there be effective in causing local upward concentration of soda by some process of "alkali-volatile diffusion" akin to that which many writers believe to play an important part in the evolution of alkaline magmas in general.   There is ample evidence, too, that near the sea floor metasomatism, including albitization, of solid but still heated rock may also be effected through the agency of externally derived waters.

Finally there is the possibility that some spilites and keratophyres have formed by a kind of low-grade regional metamorphism and metasomatism of a submarine volcanic series.   Thus Waters [23] suggests that the Eocene spilites of Washington were formed from tholeiites by the action of siliceous and sodic waters stewed from an underlying mass of wet sediments in the first stages of metamorphism.   And Battey [24] attributes the local development of keratophyres from rhyolites to redistribution of alkalis in a buried mass of rhyolitic lavas.   These lines of petrogenesis, coupled with a primary magmatic origin for other spilites, exemplify convergence of igneous and metamorphic processes toward a single broad mineral facies.

The scheme outlined above appears adequate to account for the range in chemical composition of associated basalts, spilites, and keratophyres, for the varied nature of metasomatism exhibited by such rocks, for the apparently primary nature of the albite in some spilites, and for localization of the sodic spilitic association in an environment of sodic sediments saturated with saline water.

THE BASALT ANDESITE RHYOLITE ASSOCIATION

General Statement.   Eruption of andesitic lavas and tuffs, with which olivine basalts are usually associated at some stage of the igneous cycle, is a highly characteristic accompaniment of orogeny, especially in its later stages.   This volcanic association is not limited to areas of strong folding. But it is confined to continental sectors of the earth's surface, and its most typical development is in connection with moderate to strong orogenic movements such as have continually affected the Pacific margins through the later part of geologic time.   Rocks of the andesite rhyolite kindred have been erupted everywhere along the chains of active and recently active volcanoes that mark the fold arcs of the Pacific borders, so that abundant data are available regarding mode and sequence of eruption of such rocks as well as their chemical and petrographic character.

[23] Waters, *op. cit.*, pp. 706, 707, 713, 1955.
[24] Battey, *op. cit.*, 1955.

The San Juan Volcanic Province of Colorado.[25]   Along the Rocky Mountain belt of the United States, sedimentation proceeded during Paleozoic and Mesozoic times under geosynclinal conditions without noteworthy contemporary volcanism. The Laramide orogeny which terminated this state of affairs in late Cretaceous times was accompanied by outbursts of volcanic and related intrusive activity along the length of the rising Rocky Mountain folds. This activity continued intermittently throughout Tertiary and locally into Pleistocene time, and its products are to be seen today along the various ranges of the Rocky Mountain system and in the borderlands immediately to the east. We have already considered in Chap. 9 two provinces of this period, namely, the potassic provinces of the Highwood region, Montana, and the Navajo Butte region, Arizona. We now turn to a third volcanic province, this one andesitic, namely, that of the San Juan Mountains and surrounding country of southwestern Colorado, situated within the Rocky Mountain zone approximately 100 miles northeast of the Navajo province referred to previously (Fig. 39).

FIG. 39.   Relative positions of (1) Highwood volcanic province of Montana, (2) Navajo volcanic province of Arizona, (3) San Juan volcanic province of Colorado, (4) Cascade province of California, Oregon, and Washington.

In the San Juan region a granitic-metamorphic pre-Cambrian basement is overlain by varied Paleozoic and Mesozoic sediments with a total thickness of the order of 10,000 ft. of which between 3,000 and 5,000 ft. accumulated in upper Cretaceous times under geosynclinal conditions. The volcanic rocks are of Tertiary age and occupy an area 100 miles in diameter, throughout much of which they exceed 5,000 ft. in total thickness. Volcanism was intermittent and was punctuated by numerous long periods of erosion. In the following

[25] W. Cross and E. S. Larsen, A brief review of the geology of the San Juan region of Southwestern Colorado, *U. S. Geol. Survey Bull.* 843, 1935; E. S. Larsen and co-authors, Petrologic results of a study of the minerals from the Tertiary volcanic rocks of the San Juan region, Colorado, *Am. Mineralogist,* vol. 21, pp. 679–701, 1936; vol. 22, pp. 889–905, 1937; vol. 23, pp. 227–257, 417–429, 1938; E. S. Larsen, Some new variation diagrams for groups of igneous rocks, *Jour. Geology,* vol. 46, pp. 505–520, 1938.

downward stratigraphic sequence, given by Larsen, it should be noted that the Miocene volcanics of the Potosi series are by far the most extensive and aggregate between 5,000 and 6,000 cubic miles in total bulk.

Quaternary andesite: one small body.

Erosion to mountain topography.

Pliocene (?) andesite, andesite-basalt, and rhyolite.

Erosion to peneplain.

Miocene latite-andesite.

Erosion to mountain topography.

Miocene (Potosi series) andesites, quartz latites, rhyolites, and sub- ordinate andesitic basalts; several internal erosion intervals sepa- rating conformable sequences of lavas in which dominantly quartz-latite lavas and tuffs are succeeded upward by dominant andesites.

Erosion to mountain topography.

Upper Cretaceous to Eocene andesite (dominant), latite, and rhyolite; all occur locally and several internal erosion intervals can be recog- nized.

The great bulk of the San Juan volcanic accumulation is made of andes- itic and rhyolitic (including quartz latite) rocks in approximately equal amounts. Basalts transitional to andesites are subordinate. The miner- alogy of these rocks, as revealed by unusually full optical data and chemi- cal analyses both of minerals and rocks, is too complex to be correlated with any simple scheme of magmatic evolution along a single line of descent. Some of the most conspicuous characters apparently do reflect serial derivation from basic magma by fractional crystallization (cf. Chap. 6). Thus a plot of oxide percentages of analyzed rocks upon a variation diagram yields points that conform well to smooth curves similar in configuration to those representing associated lavas of other andesite rhyolite provinces (Fig. 40). Moreover, the mineral assem- blages of typical San Juan rocks, especially if certain phenocryst minerals are omitted from consideration, in most respects correspond well with the successive crystal fractions that would develop from basaltic magma. They illustrate the continuous reaction series bytownite → oligoclase and (over the limited range Fa_{20} to Fa_{30}) forsterite → fayalite. The general tendency for members of the sequence olivine, pyroxene, hornblende, biotite to appear and disappear in the same order and for sanidine and quartz to appear characteristically with low-temperature members of proved reaction series, corresponds well with the two main reaction series which were believed by Bowen to dominate crystallization of calcalkaline and calcic magmas in general. Characters that seem to reflect this "orthodox" element in the genesis of San Juan magmas are summarized below:

1. Basalts [26] ($SiO_2 = 52$ to 54 per cent): olivine (Fa_{21} to Fa_{23}), calcic plagioclase (An_{70}), pigeonitic pyroxene.

2. Andesites ($SiO_2 = 55$ to 60 per cent): medium labradorite (average An_{60}), diopsidic augite; in basic members, olivine (Fa_{25} to Fa_{30}) may be present.

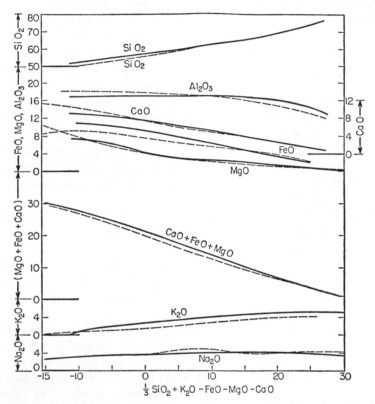

Fig. 40. Variation diagrams for San Juan volcanic series (full lines) and for Modoc lavas of California (broken lines). (*After E. S. Larsen.*)

3. Andesite-latites ($SiO_2 = 58$ to 61 per cent): medium labradorite (An_{63} in groundmass), diopsidic augite, hypersthene, hornblende, biotite.

4. Rhyolitic quartz latites ($SiO_2 = 62$ to 67 per cent): andesine (average An_{40} to An_{50}), quartz, biotite; any of hornblende, diopsidic augite, and sanidine may be present.

[26] The values given for SiO_2 in this paragraph are weight percentages, recalculated according to Larsen's method for constructing the variation diagram. The rock composition, free of H_2O, TiO_2 and P_2O_5, and with total iron calculated as FeO, is recalculated so that the sum $[SiO_2 + Al_2O_3 + (FeO + MnO) + MgO + (CaO + BaO + SrO) + Na_2O + K_2O] = 100$. In a typical basalt $SiO_2 = 49.3$ per cent in the analysis, 51.2 per cent in the recalculated composition.

5. Rhyolites ($SiO_2 = 68$ to 77 per cent): sanidine, oligoclase-andesine (average An_{27} to An_{35}; orthoclase content 8 to 12 per cent), quartz, biotite.

Contrasting with the above are equally persistent mineralogical peculiarities which cannot be correlated with simple linear differentiation and which seem to reflect a crosscurrent in the course of magmatic evolution:

1. In many lavas two markedly different varieties of plagioclase phenocrysts are mutually associated. Thus in one andesite nearly uniform crystals An_{62} occur side by side with zoned crystals in which a core An_{28} is enclosed by a broad rim An_{41}. Other examples are the association of unzoned crystals respectively An_{46} and An_{29} in quartz latite; dominant phenocrysts An_{11} with a few phenocrysts An_{30} in rhyolite obsidian. The authors have found similar variability of plagioclase phenocrysts to be a widespread feature of andesite rocks from other provinces.

2. Phenocrysts of quartz and sanidine, both strongly corroded or mantled respectively with pyroxene or plagioclase, are sparingly but persistently present in many of the basalts and andesites—even those carrying olivine and calcic plagioclase.

3. Although there is a clear tendency for *average* plagioclase to be increasingly sodic along the general series basalt → andesite → quartz latite → rhyolite, the composition of plagioclase *phenocrysts* shows little relation to the chemical composition of the rocks in which they occur. The range for phenocrysts in andesite-basalts is An_{20} to An_{80}. More than half this range is shared by that of phenocrysts in the rhyolitic quartz latites, *viz.*, An_{12} to An_{55}.

4. The normal tendency for the FeO/MgO ratio to increase in residual liquids during fractional crystallization in the olivine and pyroxene series is reflected by observed outward increase in iron in zoned crystals of pyroxene, and in the marked rise in FeO/MgO from $1\frac{1}{2}$ in average San Juan basalt to 7 in siliceous members of the rhyolite group. Yet pyroxene phenocrysts of andesites consistently are richer in iron than those of quartz latites and rhyolites.

These anomalous features seem to have a common genetic significance; they express a tendency for mutual association of minerals belonging to different stages of magmatic evolution in many San Juan lavas. Such an association is readily understood if, following Larsen, we assume thorough mixing of partially crystalline magmas of common parentage, but at different stages of differentiation and possibly modified to different degrees by assimilative reaction with adjoining rocks. "The uniform distribution of the phenocrysts in widespread lava flows, and groups of flows, shows that there must have been a thorough mixing of very large masses of magma." [27]

[27] Larsen, *op. cit.*, p. 429, 1938.

Several less important elements in the genesis of San Juan lavas, which probably have wide application to andesites and rhyolites of other provinces, may be noted briefly. Tridymite and cristobalite have formed widely as metastable substitutes for quartz, especially where crystallization of the groundmass has been rapid; also gaseous transfer of silica has been responsible for deposition of cristobalite in vesicles after complete solidification of the rock. Appearance of hornblende and/or biotite, rather than pyroxene, seems due not so much to the ratio of such metallic ions as Ca, Mg, and Fe in the magma as to abundance of water. Resorption of hornblende and biotite to give pseudomorphous aggregates of iron ore, pyroxene, and plagioclase—a process involving oxidation of iron to the ferric condition—is favored by near-surface conditions, including relatively low water pressure. Much of this resorption occurred after eruption of the enclosing magma as lava.

The Volcanic Province of the Cascade Range, Northwestern United States.[28]   The extent to which a volcanic association dominated by andesitic rocks may vary in a lateral sense as well as in time is well illustrated in the Tertiary-Quaternary province which comprises the Cascade Range of Washington, Oregon, and northern California. This region extends for over 600 miles in a north-south direction, from the northern tip of the Sierra Nevada to the Canadian border (Fig. 41). Within it lies the chain of lofty Pleistocene volcanoes which constitute the high peaks of the Cascade Range. The pre-Tertiary basement is concealed throughout most of the province by the vast accumulation of Tertiary and Quaternary flows and tuffs. But at the southern end the volcanic pile of Mount Lassen and adjacent volcanoes overlaps onto the ancient (Silurian to Jurassic) complex of folded sediments and varied plutonic rocks, which emerges both to the northwest and to the southeast, in the Klamath Mountains and the northern end of the Sierra respectively. Elsewhere there are indications that Cretaceous and Eocene sediments lie immediately beneath the Cascade volcanics.

[28] H. Williams, Geology of the Lassen Volcanic National Park, California, *California Univ., Dept. Geol. Sci., Bull.*, vol. 21, no. 8, pp. 195–385, 1932; Newberry volcano of central Oregon, *Geol. Soc. America Bull.*, vol. 46, pp. 253–304, 1935; The geology of Crater Lake National Park, Oregon, *Carnegie Inst. Washington Pub. 540*, 1942; T. P. Thayer, Petrology of later Tertiary and Quaternary rocks of the north-central Cascade Mountains in Oregon, *Geol. Soc. America Bull.*, vol. 48, pp. 1611–1652, 1937; J. Verhoogen, Mount St. Helens, a Recent Cascade volcano, *California Univ., Dept. Geol. Sci., Bull.*, vol. 24, no. 9, pp. 263–302, 1937; C. A. Anderson, Volcanoes of the Medicine Lake highland, California, *California Univ., Dept. Geol. Sci., Bull.*, vol. 25, no. 7, pp. 347–422, 1941; S. R. Nockolds and R. Allen, The geochemistry of some igneous rock series, *Geochim. et Cosmochim. Acta*, vol. 4, pp. 105–142, 1953; Waters, *op. cit.*, pp. 709–713, 1955.

The following generalized sequence of events may be traced in the broad volcanic history of the Cascade province (cf. Fig. 38):

1. Closely following Eocene movements of uplift which caused a general westward retreat of the sea, the Western Cascade series of volcanic rocks was erupted along a north-south belt flanking the present crest of the Cascade Range on the west. These locally reached a total thickness of the order of 10,000 ft. by the close of the Miocene. The compositional range is from olivine basalt to rhyolite. Basaltic andesites and pyroxene andesites constitute 75 per cent of the total; tholeiitic and hypersthene basalts are well represented. [More or less contemporary volcanic rocks in adjacent provinces include (1) the Eocene spilite keratophyre association of Olympic Peninsula, west of the northern Cascades (see pages 264 to 266) and contemporary tholeiitic basalts of western Oregon; (2) Eocene and Oligocene andesites, basalts, and rhyolites, followed by the Miocene flood basalts of the Columbia River plateau, east of the Cascade province.]

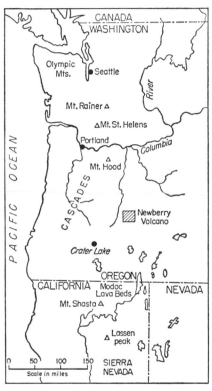

FIG. 41.   The Cascade volcanic province of northwestern United States.

2. At the close of the Miocene, an important compressional movement tilted and broadly folded the earlier volcanic rocks of the Cascade province and ushered in the uplift which culminated in the Pleistocene elevation of ranges throughout the whole Cordilleran region. The late Miocene movements were accompanied by intrusion of dioritic stocks into the Western Cascade volcanics and closely followed by extrusion of andesitic lavas and tuffs (Keechelus group) and still further intrusive activity.

3. Pliocene uplift along north-south fractures and accompanying extrusion of great volumes of olivine basalt and basaltic andesite from shield volcanoes along the crest of the Cascade Range combined to build an elongated high plateau by late Pliocene times. These rocks bulk more largely than any other group in the Cascade volcanic mass.

4. Volcanic activity continued unabated through the Pleistocene and at some points even into Recent times. This was the period of growth of the chain of imposing volcanoes which today constitute the peaks of the High Cascades. These belong to several petrologically distinct types:

a. Essentially andesitic cones, including most of the great volcanoes of the northern Cascades. The dominant rocks are hypersthene-augite andesites with plentiful porphyritic plagioclase. Olivine basalts and basaltic andesites may also be plentiful (e.g., Mount St. Helens); dacites are insignificant or absent.

b. Composite cones of andesite, dacite, and rhyolite. These are most strikingly developed in the southern section of the Cascades and are exemplified by Mount Mazama (Crater Lake) and the Lassen group of volcanoes. Eruption of hypersthene-augite andesites characterizes the earlier stages of evolution of volcanoes of this type. But, although the more acid rocks—dacites and rhyolites—tend to be more conspicuous among the later products of volcanism, little correlation is possible between chemical composition and order of eruption of the various rock types. Thus during the later history of the Lassen group acid (dacite, rhyolite) and basic (andesite, basalt) eruptions alternated irregularly. At Mount Mazama, following mature development of a hypersthene-andesite cone of great size, olivine basalts and dacites (flows, domes, and tuffs [29]) were emitted from parasitic cones on the flanks of the volcano; then came violent ejection of dacite tuff followed by basic hornblende-rich scoria, development of a collapse caldera, and finally extrusion of hypersthene andesite to build a small cone on the caldera floor.

c. Basaltic shield volcanoes interspersed among the more massive andesitic cones of types a and b.

d. Composite cones of basalt and rhyolite, situated on the western margin of the basalt-floored "interior platform" where this abuts against the Cascade Range proper. Such are the volcanoes of Medicine Lake, California, and Newberry, Oregon. The latter is a cone 20 miles in diameter at the base, rising 4,000 ft. above the general level of the interior platform. It is composed mainly of weakly oversaturated augite basalts and only slightly less abundant rhyolites. Andesites are almost absent. Late activity continuing into sub-Recent times involves simultaneous extrusion of rhyolite and olivine basalt both within the summit caldera and from fissures and parasitic cones on the flanks.

[29] One of the most characteristic features of andesite rhyolite provinces in general is the presence of extensive massive sheets of welded rhyolitic or dacitic tuff, in many cases composed largely of glassy material (cf. P. Marshall, Acid rocks of the Taupo-Rotorua volcanic district, *Royal Soc. New Zealand Trans.*, vol. 64, pp. 1–44, 1935; C. M. Gilbert, Welded tuff in eastern California, *Geol. Soc. America Bull.*, vol. 49, pp. 1829–1862, 1938).

The Huzi (Fuji) Province of Japan.[30]   Pleistocene and Recent volcanoes are thickly clustered along great arcs extending the full length of Japan, and far beyond to Kamchatka in the northeast and to Formosa in the southwest.   Throughout this whole region of contemporary volcanism and orogeny the dominant volcanic rocks are andesites.   But to the west, in the stable continental regions of Manchuria and the Sea of Japan, Pleistocene and Recent lavas include feldspathoidal types and belong to the alkaline olivine-basalt series.

The Huzi province is located some 70 miles southwest and south of Tokyo.   It occupies a Tertiary rift (the Fossa Magna of Honsyu) which cuts north-northwest obliquely across the dominant northeasterly trend of the islands and the main volcanic arcs of Japan; it is prolonged south-southwest for several hundred miles along the Izu Islands chain.   Most of the surface volcanic rocks and all the great volcanoes of this region are of Quaternary age.   Underlying them are Tertiary lavas and pyroclastics, dominantly andesitic and basaltic in composition.

According to Kuno,[31] the Quaternary volcanic rocks fall into two series:

1. *A pigeonitic series* (Table 27, analyses 1–4): "tholeiitic" olivine basalts; pyroxene andesites in which the groundmass pyroxene is all monoclinic (pigeonitic augite or pigeonite), though hypersthene, augite and olivine are commonly associated as phenocrysts.   Chemically[32] these rocks have a low ratio $MgO:(FeO + Fe_2O_3)$, low alkalis and high normative quartz.   The "tholeiitic" basalts resemble the tholeiites discussed in Chap. 8 in their high content of normative hypersthene and in being oversaturated in silica; but they are consistently much higher in alumina.   In this and other respects they resemble the porphyritic central magma-type of Mull.

2. *A hypersthenic series* (Table 27, analyses 5, 6): pyroxene andesites in which hypersthene is present, with or without augite, in the groundmass, while phenocrysts typically include olivine and both pyroxenes; hornblende-pyroxene andesites; hornblende-pyroxene dacites; hornblende and biotite rhyolites.   Chemically this series is said to be characterized

---

[30] S. Tsuboi, Volcano Ōshima, Idzu, *Tokyo Imp. Univ. Jour. Coll. Sci.*, vol. 43, pt. 6, pp. 1–46, 1920; H. Tsuya, On the volcanism of the Huzi volcanic zone, *Earthquake Research Inst. Bull.*, vol. 15, pt. 1, pp. 215–357, 1937; H. Kuno, Petrology of Hakone volcano and the adjacent areas, Japan, *Geol. Soc. America Bull.*, vol. 61, pp. 957–1020, 1950; Cenozoic volcanic activity in Japan and surrounding areas, *New York Acad. Sci. Trans.*, ser. 2, vol. 14, pp. 225–231, 1952; Geology and petrology of Ōmuro-yama volcano group, north Izu, *Tokyo Univ. Fac. Sci. Jour.*, sect. 2, vol. 9, pt. 2, pp. 241–265, 1954.

[31] Kuno, *op. cit.*, pp. 992, 993, 1950; *op. cit.*, pp. 228–230, 1952; *op. cit.*, pp. 256–261, 1954.

[32] Kuno, *op. cit.*, pp. 241, 264, 1954.

by higher MgO:($FeO + Fe_2O_3$), higher Al_2O_3, and higher alkalis than rocks of the same silica content in the pigeonitic series.

At any given center of eruption one or other series tends to predominate. Thus only the pigeonitic series is represented in some volcanoes of the Izu Islands; in others about 10 per cent are of the hypersthenic lineage. At Hakone and adjacent centers of eruption the pigeonite series is domi-

TABLE 27.   CHEMICAL COMPOSITION OF VOLCANIC ROCKS OF THE HUZI (FUJI) PROVINCE, JAPAN

Constituent	1	2	3	4	5	6
SiO_2	48.73	49.62	56.70	67.37	57.74	71.16
TiO_2	0.63	0.87	1.24	0.72	0.28	0.36
Al_2O_3	16.53	20.37	15.90	15.28	17.70	14.72
Fe_2O_3	3.37	2.61	2.94	1.13	1.95	1.12
FeO	8.44	6.71	7.74	3.86	5.32	2.15
MnO	0.29	0.17	0.18	0.15	0.09	0.08
MgO	8.24	4.05	3.34	1.20	5.00	1.30
CaO	12.25	11.97	7.77	4.46	9.14	3.50
Na_2O	1.21	1.89	3.13	4.67	1.59	3.61
K_2O	0.23	0.31	0.74	0.92	0.72	2.06
H_2O+		0.39	0.36	0.14	0.09	0.27
H_2O-		0.71	0.09	0.09	0.16	0.10
P_2O_5	0.10	0.07	0.16	0.18	0.04	0.08
Total	100.02	99.74	100.29	100.17	99.82	100.51

Explanation of Table 27

1. "Parental magma" of pigeonitic series, Izu and Hakone province (average of 2 analyses, aphyric olivine basalts, slightly over-saturated). (H. Kuno, *Tokyo Univ. Fac. Sci. Jour.*, sect. 2, vol. 9, pt. 2, p. 262, 1954.)
2. Olivine basalt, pigeonitic series (close to assumed parent magma) Hakone. (H. Kuno, *Geol. Soc. America Bull.*, vol. 61, p. 1000, No. 1, 1950.)
3. Aphyric andesite, pigeonitic series, Hakone. (Kuno, *op. cit.*, pp. 1004–1006, No. 22, 1950.)
4. Augite-hypersthene andesite, pigeonitic series, Hakone. (Kuno, *op. cit.*, pp. 1004–1006, No. 23, 1950.)
5. Olivine andesite, hypersthenic series, Ōmuro-yama. (Kuno, *op. cit.*, p. 259, No. 10, 1954.)
6. Olivine-bearing pyroxene-hornblende dacite, hypersthenic series, Amagi volcano. (H. Tsuya, *Earthquake Research Inst. Bull.*, vol. 15, pt. 1, p. 257, No. 40, 1937.)

nant but is accompanied by two-pyroxene andesites of the hypersthene series (20 per cent of the total). Ōmuro-yama, 25 miles south of Hakone, is built up largely of andesites of the hypersthenic series, but basalts and andesites of the pigeonitic series account for 10 per cent of the total bulk. A broader tendency for respective concentration of the two series in different parts of the Huzi province has been established and two corresponding subzones recognized. There is also a correlation between the distribution of the hypersthenic series (especially those rocks carrying hornblende and biotite) and the incidence of negative gravity anomalies—presumably indicating local thickening of the "granitic" crust.

A rather basic, alumina-rich "tholeiitic" basalt (Table 27, analyses 1, 2)—in spite of its relatively small bulk in the Huzi province—has been assumed by Kuno [33] to be the parent magma of both series. The pigeonitic series is considered to represent a line of evolution determined by fractional crystallization in the olivine-pyroxene series. To account for the hypersthenic series Kuno invokes reaction with the "granitic" crust, accompanied and followed by fractional crystallization of the contaminated magma. In support of this view he cites the tendency for the hypersthenic series to occur in areas of negative gravity anomaly, and the widespread occurrence of granitic xenoliths in andesites of the hypersthenic series.[34]

Petrogenesis of the Basalt Andesite Rhyolite Association. Viewed broadly, this association, in whatever part of the world it may occur, comprises basalts, andesites, dacites, and rhyolites in various combinations. It is clear, however, that the range and trend of chemical variation is not precisely the same in all provinces, and even varies notably both in place and time from one center of eruption to another within a given province. Within the two western American provinces just considered we have recognized such variants as:

Olivine basalt, andesite, quartz latite, rhyolite
Olivine basalt, andesite
Olivine basalt, andesite, dacite, rhyolite
Olivine basalt, basaltic andesite, rhyolite

Tables 27 and 28 illustrate the range in composition of common rocks in the basalt andesite association as represented in the Huzi province of Japan and in the Cascade province of northwestern United States. Other chemical data are shown in variation diagrams, Figs. 40, 42, and 43.

While tholeiitic basalts are represented in some provinces, the basic members of the andesite association more commonly are basalts (e.g., Table 28, nos. 1–3) that differ markedly from both of Kennedy's principal basaltic magma types. Chemically they are strongly reminiscent of the porphyritic central type of Mull. Chemical features that tend to recur in basic members of the andesite association are high Al_2O_3, low TiO_2 and ($FeO + Fe_2O_3$), and very low K_2O. For the association as a whole the alkali-lime index is high; in various units of the Cascade province it

[33] Kuno, *op. cit.*, p. 993, 1950.
[34] Kuno, *op. cit.*, pp. 229, 230, 1952; *op. cit.*, pp. 242, 261–264, 1954.

ranges from 58 to 63.7,[35] values corresponding to Peacock's calc-alkaline and calcic classes.

Bowen explained all the chemical, petrographic, and associational characteristics of the basalt andesite rhyolite association by a simple hypothesis of fractional crystallization of basaltic magma;[36] and for the next two decades Bowen's views were widely though not universally ac-

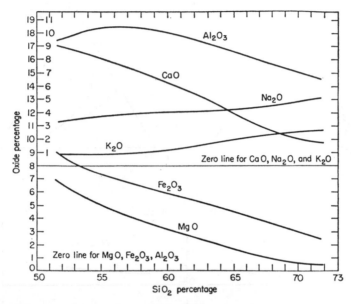

Fig. 42.   Variation diagram for volcanic rocks of Crater Lake, Cascade volcanic province, California.   (*After H. Williams.*)

cepted.[37] Even the immense mass of accurate petrographic, mineralogical, and chemical data relating to this association in the San Juan and Central Montana provinces was correlated by Larsen[38] with diverging lines of fractional crystallization, influenced by local tectonic environments, and variously modified by intermittent mixing of the products or by minor assimilative reaction with the granitic rocks of the underlying

[35] Williams, *op. cit.*, p. 297, 1935; Anderson, *op. cit.*, p. 401, 1941.

[36] N. L. Bowen, *The Evolution of the Igneous Rocks*, pp. 92–132, Princeton University Press, Princeton, 1928.

[37] For a radically different scheme of petrogenesis, involving melting of crustal rocks to give a primary granitic magma which then mixes with basaltic magma and so becomes basified, see A. Holmes, The origin of igneous rocks, *Geol. Mag.*, vol. 69, pp. 543–558, 1932.

[38] Larsen and coauthors, *op. cit.*, 1937, 1938; Petrographic province of Montana, *Geol. Soc. America Bull.*, vol. 51, pp. 887–948, 1940.

basement. During the past decade or so opinion against hypotheses of pure fractional crystallization has mounted steadily. The intimate field relations of basalt and andesite and the existence of complete petrographic transition between the two rock types show that in the evolution of the andesite-dacite line some essential role has been played by basaltic

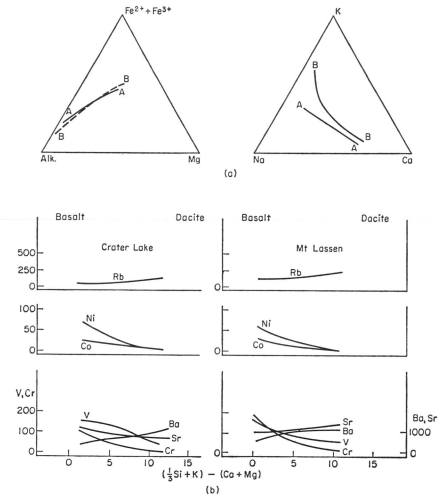

(a)

(b)

Fig. 43. (a) Variation in the basalt-andesite-dacite series of the Cascade province, Oregon and California. *AA* = Crater Lake and Mt. Shasta; *BB* = Mt. Lassen. (*After S. R. Nockolds and R. Allen.*) (b) Variation in trace-element content of the basalt-andesite-dacite series in rocks of two volcanoes in the Cascade province (Crater Lake, Mt. Lassen). The content of each element, in parts per million, is plotted against an index of evolution (⅓ Si + K) − (Ca + Mg). (*After S. R. Nockolds and R. Allen.*)

TABLE 28.   CHEMICAL COMPOSITION OF ROCKS OF THE BASALT ANDESITE RHYOLITE
ASSOCIATION OF THE CASCADE PROVINCE, NORTHWESTERN UNITED STATES

Constituent	1	2	3	4	5	6	7	8	9	10
SiO_2	47.10	50.70	51.51	55.83	60.09	63.10	67.70	68.64	72.35	73.59
TiO_2	0.90	1.30	1.12	0.84	0.54	0.54	0.30	0.49	0.25	0.31
Al_2O_3	18.52	18.05	17.52	18.01	17.85	18.22	16.32	15.62	13.98	14.03
Fe_2O_3	Trace	1.62	1.51	2.63	2.03	1.36	0.27	1.28	0.60	0.42
FeO	7.91	6.96	6.69	4.07	3.45	3.33	3.20	2.20	1.78	1.43
MnO	Trace	0.40	0.15	0.08	Trace	Trace	Trace	Trace		0.02
MgO	10.89	7.60	6.74	5.12	3.50	2.30	1.25	1.14	0.30	0.36
CaO	11.98	9.70	8.78	7.40	6.28	5.24	3.35	3.05	1.30	1.38
Na_2O	2.33	2.70	3.34	3.64	4.17	4.06	3.89	4.52	5.04	4.04
K_2O	Trace	0.68	0.77	1.22	1.31	1.16	3.22	2.23	3.92	4.34
H_2O+	0.10		0.49	0.72	0.26	0.10	0.22	0.54	0.05	0.12
H_2O-	0.18	0.10	0.15	0.26	0.12	0.40	0.05	0.26	0.45	0.06
P_2O_5	0.09	0.23	0.31	0.11	0.23	0.14	0.06	0.13	Trace	0.09
BaO					0.05	0.05				
SrO					0.05	0.04				
Rest					0.05		0.20			0.05
Total	100.00	100.04	99.08	99.93	99.98	100.10	100.03	100.10	100.02	100.24

Explanation of Table 28

1. Olivine basalt (? Pliocene, Warner basalt) with 24 per cent normative olivine;
   Medicine Lake highland. (C. A. Anderson, Volcanoes of the Medicine Lake
   highland, California, *California Univ., Dept. Geol. Sci., Bull.*, vol. 25, p. 387,
   No. 1, 1941.)
2. Olivine basalt (late Pleistocene); Lava Top Butte, Newberry volcano. (H. Williams, Newberry volcano of central Oregon, *Geol. Soc. America Bull.*, vol. 48,
   p. 295, No. 3, 1935.)
3. Olivine basalt (? late Pliocene) with 0.12 per cent normative quartz; near summit of Outerson Mountain, Oregon. (T. P. Thayer, Petrology of later Tertiary
   and Quaternary rocks of the north-central Cascade Mountains in Oregon, *Geol.
   Soc. America Bull.*, vol. 48, p. 1622, 1937.)
4. Basaltic andesite (Pleistocene) with 6.3 per cent normative quartz; near summit
   of Crater Peak, southern Oregon. (H. Williams, The Geology of Crater Lake
   National Park, Oregon, *Carnegie Inst. Washington Pub. 540*, p. 149, No. 10,
   1942.)
5. Hypersthene andesite (Pleistocene) with 11.7 per cent normative quartz; Crater
   Lake, southern Oregon. (Williams, *op. cit.*, p. 150, No. 15, 1942.)
6. Pyroxene andesite (Pleistocene) with 18.7 per cent normative quartz; Mount
   St. Helens, Washington. (J. Verhoogen, Mount St. Helens, a Recent Cascade
   volcano, *California Univ., Dept. Geol. Sci., Bull.*, vol. 24, p. 293, No. 4, 1937.)
7. Porphyritic dacite (sub-Recent) with 21.8 per cent normative quartz; Medicine
   Lake highland. (Anderson, *op. cit.*, p. 396, No. 3, 1941.)
8. Dacite (Pleistocene) with 24.9 per cent normative quartz; Crater Lake. (Williams, *op. cit.*, p. 151, No. 20, 1942.)
9. Rhyolite obsidian (Pleistocene); Newberry volcano. (Williams, *op. cit.*, p. 295,
   No. 10, 1935.)
10. Rhyolitic obsidian (sub-Recent) with 29.7 per cent normative quartz; Medicine
    Lake highland. (Anderson, *op. cit.*, p. 399, No. 7, 1941.)

285

magma. But the nature of this role is a topic of current debate among petrologists.[39] Salient features of the problem are as follows:

1. Preponderance of some rock type in a volcanic association is a generally accepted criterion for recognizing a corresponding primary parent magma. Now basalts are ubiquitous and usually plentiful in andesitic provinces. But they are less abundant than pyroxene andesites. Moreover in some provinces, and at some stage in the history of most provinces, great volumes of andesite, dacite, or rhyolite have been erupted over large areas with little or no accompanying basalt. On quantitative grounds alone, it would seem unlikely that basalt is the parent and andesite the derivative magma.

2. A purely chemical criterion for identifying the composition of a parent magma in a differentiation series is supplied by the variation diagram. A curve is assumed to represent a liquid line of descent when the plotted points approximate it closely. The composition of the parent magma is indicated by the composition (toward the basic end of the series) at which the plotted points become scattered and no longer approximate a smooth curve. In diagrams for basalt-andesite-rhyolite provinces, the parent magmas so identified do not correspond to the most basic basalts in the series.[40] Generally they are oversaturated basaltic andesites with 53 to 54 per cent SiO_2.

3. Strong fractional crystallization of a tholeiitic magma leads, in its earlier stages, to a marked increase both in total iron and in the ratio FeO:MgO in the liquid differentiates. This increase imparts a characteristic form to curves on the Alk-Fe-Mg variation diagram (cf. Fig. 36, page 234). A similar though less obvious trend marks the initial stages in differentiation of olivine-basalt magma. The corresponding curve for the basalt-andesite-dacite series (Fig. 43) is strikingly different from the curves for both basaltic series. It would seem, therefore, that andesites and dacites either are not differentiates of basaltic magma or represent a line of evolution radically different from that responsible for differentiation observed in tholeiitic intrusions.

4. A further obstacle to accepting the differentiation hypothesis concerns the alternate, and in some cases simultaneous, eruption of fine-grained olivine basalt and glassy rhyolite in comparable amounts from the same volcano, without appearance of lavas of intermediate (andesitic) composition. This condition is illustrated by the Newberry volcano. A

---

[39] L. R. Wager and W. A. Deer, The petrology of the Skaergaard intrusion, Kangerdlugssuaq, east Greenland, *Meddelelser om Grönland*, vol. 105, no. 4, pp. 313–324, 1939; H. H. Hess, An essay review of "The petrology of the Skaergaard intrusion," *Am. Jour. Sci.*, vol. 238, pp. 376–378, 1940; C. E. Tilley, Some aspects of magmatic evolution, *Geol. Soc. London Quart. Jour.*, vol. 104, pp. 50–54, 1950; Kuno, *op. cit.*, pp. 229, 230, 1952; Waters, *op. cit.*, pp. 712–715, 1955.

[40] Nockolds and Allen, *op. cit.*, p. 139, 1953.

distinct but much narrower compositional break between the more basic and more siliceous members of volcanic series is by no means uncommon, and could be explained by assuming that some special mechanism of fractional crystallization, such as gas streaming, is effective in separating the last liquid fractions from the mass of early-formed crystals. Such are the breaks between andesite and dacite in the present association, and between mugearite (or trachyandesite) and trachyte in the lavas of oceanic islands.   To account for the basalt rhyolite combination of such volcanoes as Newberry, it seems necessary to assume either some drastic mechanism of differentiation, *e.g.*, unmixing of magma into immiscible rhyolitic and basaltic liquid fractions, or independent origin and uprise of the two kinds of magma.

With such anomalies in mind, the authors suggest that some of the features of andesites and rhyolites ascribed by orthodox opinion to fractional crystallization of basaltic magma in the depths are really due to differential fusion of crustal rocks.[41]   Students of tectonic geology have brought forward convincing geophysical evidence (relating to gravity and seismology) that downward thickening of the crust to form a mountain "root" accompanies or precedes orogenic folding. Thickening of the crust with its relatively high radioactive content could conceivably raise locally the temperature at the base of the root to the point of partial or complete fusion (see page 668).   Such melting would affect rocks generally believed to be more basic than average granite, and so could produce magmas of rhyolitic, andesitic, or even basaltic composition. Where folding is active, there is ample opportunity for filter-pressing and segregation of the magmas so formed and for mixing and blending of magmas en route to the surface.   These processes, modified by differentiation wherever magma temporarily is held in a closed chamber, are surely complex enough to account for the wide variation observed in the products of eruption at the surface.   Nor is it difficult to imagine why volcanic series in which andesites and rhyolites are so conspicuous are confined to the continents and attain their most spectacular development along the Pacific margin, where for long ages the rocks of the sial and adjacent underlying basic material have almost continuously been kneaded together.

In conclusion, brief mention may be made of local alkaline deviations from the normal andesite rhyolite association.   On the hypothesis outlined above, rather rare early concentration of potash (perhaps preceded by assimilation of granite), leading to development of a latite-rhyolite line of descent, is perhaps to be expected.   So, too, is the much rarer branching of this line to give occasional trachytic end members.   Rare appearance of soda rhyolites containing riebeckite, arfvedsonite, or aegirine seems to

[41] Cf. Holmes, *op. cit.*, pp. 545–550, 1932; Waters, *op. cit.*, pp. 712–714, 1955.

be correlated with local abundance of Fe_2O_3 and poverty in Al_2O_3 rather than with unusually high content of soda.   There are, however, recorded cases of andesitic associations characterized by unusually high alkali, and especially soda, in rocks of intermediate silica content—andesine basalts and oligoclase andesites.   The lavas of Deception Island at the southern end of the South Shetlands arc belong to this category.[42]

[42] G. W. Tyrrell, Report on rocks from west Antarctica and the Scotia arc, *Discovery Repts.*, vol. 23, pp. 61–64, 1945.

CHAPTER 11

Basic and Ultrabasic Plutonic Associations

INTRODUCTORY STATEMENT

The term *plutonic association* is here applied to igneous rock series most of whose members are of plutonic origin. Clearly there are rock associations the alternative classification of which as plutonic or volcanic is a matter of personal choice. This applies particularly to some of the basic and ultrabasic associations here placed in the plutonic category. Thus Kennedy groups the great differentiated lopoliths of the Bushveld type as members of a volcanic association partly because they occur in regions not affected by contemporaneous orogeny, and partly because they in some instances occur within essentially volcanic provinces and seem to be equivalent chemically to strictly volcanic series dominated by basaltic lavas.[1] To the authors it appears desirable, in the face of a tendency on the part of some writers to regard some "plutonic" rocks as products of metamorphism, to admit, where the evidence is strong enough to warrant it, that chemically equivalent volcanic and plutonic associations do exist.

A number of volcanic associations in which basic lavas greatly predominate have already been considered. We now turn to several plutonic associations in which the principal rock types are of basic or ultrabasic composition:

1. Gabbros, peridotites, and associated rocks of layered funnel intrusions and lopoliths
2. Peridotites and serpentinites
3. Anorthosites and associated rocks

[1] W. Q. Kennedy, Crustal layers and the origin of magmas, *Bull. volcanologique*, sér. 2, tome 3, pp. 25, 26, 1938.

GABBROS, PERIDOTITES, AND ASSOCIATED ROCKS
OF LAYERED INTRUSIONS

The Skaergaard Intrusion of Greenland. The account by Wager and Deer [2] of the gabbro intrusion of Skaergaard Peninsula in eastern Greenland is one of the most detailed contributions to igneous petrology. For this reason the Skaergaard intrusion has been selected as a type illustration of a plutonic rock association represented elsewhere by layered intrusions of much greater size.

The Eocene basic intrusive and volcanic rocks of eastern Greenland belong to the northwestern fringe of the Brito-Arctic Tertiary province (see pages 221 to 223). The rocks of the Skaergaard mass may therefore be regarded as constituting a minor but complete plutonic association within a very much broader petrographic province, the great bulk of whose rocks are strictly volcanic. The sequence of events reconstructed by Wager and Deer at Skaergaard is as follows:

1. Deposition of a thin mantle of Cretaceous and early Eocene sediments on a peneplained metamorphic complex.

2. Outpouring of basalts of early Eocene age (olivine basalts and tholeiitic types are both present).

3. Deposition of middle Eocene sediments, with contemporaneous (or immediately antecedent) intrusion of basic magma to give the Skaergaard intrusion and gabbro sills.

4. Crustal flexuring and accompanying injection of a swarm of dikes of diabase and augite and hornblende lamprophyres.

5. Intrusion of syenite magma—post-mid-Eocene.

The original form of the Skaergaard intrusion, omitting distortion subsequently imposed by the flexural movements noted under 4 above, was a steep-sided funnel or truncated cone, tapering downward, with its axis plunging southward at 45° (Fig. 44). The surface outcrop covers approximately 60 km.[2] There is a thin envelope of rocks collectively referred to as the "border group," particularly well exposed along the steeply dipping margins. Within 30 m. of the walls these are chilled fine-grained gabbros; but through the greater part of their extent the rocks of the marginal border group are coarse gabbros with pronounced fluxional structures parallel to the contact indicating upward or downward

---

[2] L. R. Wager and W. A. Deer, The petrology of the Skaergaard intrusion, Kangerdlugssuaq, east Greenland, *Meddelelser om Grönland*, vol. 105, no. 4, 1939 (see also review of the above by H. H. Hess, *Amer Jour. Sci.*, vol. 238, pp. 372–378, 1940); L. R. Wager, Layered intrusions, *Dansk. Geol. Foren, Medd.*, vol. 12, pp. 335–349, 1953; G. M. Brown, Pyroxenes from the early and middle stages of fractionation of the Skaergaard intrusion, East Greenland, *Mineralog. Mag.*, vol. 31, pp. 511–543, 1957.

magmatic flow.[3] Within the envelope formed by the border group the main mass of the intrusion shows a remarkable layered structure into which enter three independent elements:

1. Most conspicuous is small-scale rhythmic layering developed throughout the lower two-thirds (about 6,000 ft.) of the intrusion. This feature, commonly observed in basic and ultrabasic intrusions elsewhere, is due to variation in relative proportions of the main constituents (plagioclase, clinopyroxene, orthopyroxene, and olivine). Individual layers commonly show gravity stratification: feldspar—the lightest constituent— is concentrated toward the top and dark heavy minerals toward the base, which tends to be sharply defined. The layers maintain a gentle inward dip sharply discordant with the steeply dipping walls and the foliation of

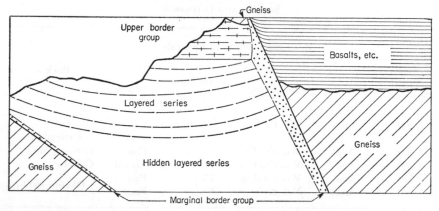

Fig. 44. Section through Skaergaard intrusion of Greenland. (*After L. Wager and W. A. Deer.*)

the border group. The over-all structure resembles that of a pile of saucers, concave upward.

2. Less conspicuous but petrogenically of great significance is "cryptic layering" of the whole mass.[4] This is characterized by a steady change in the chemical composition of each of the four main mineral phases in a vertical sense (Table 29). Passing upward through the mass, one encounters phases progressively enriched in the low-temperature end members of solid-solution series: pyroxenes and olivine become richer in iron, plagioclase in soda. Cryptic layering is further emphasized by sudden disappearance of some phases (magnesian olivine, then orthopyroxene) and by incoming of others (iron ore, iron olivines, apatite, and quartz) along an upward traverse.

[3] L. R. Wager and G. M. Brown, Funnel-shaped layered intrusions, *Geol. Soc. America Bull.*, vol. 68, p. 1072, 1957.

[4] Wager, *op. cit.*, pp. 335, 336, 1953.

3. Except toward the top of the section, many of the rocks display a foliated or laminated condition resulting from preferred orientation of tabular crystals. This is termed "igneous lamination."

The hypothesis put forward by Wager and Deer, and backed by strong detailed evidence, is that, following injection of a magma of more or less tholeiitic composition, crystallization began near the roof, but the crystals were steadily transported downward by convection currents, and accumulated as a sedimentary crystal mush on the floor of the magma chamber. Gravitational settling of heavy dark minerals within the layer of mush carpeting the floor over a given period is thought to be responsible for the rhythmic layering so widely prevalent in the lower levels of the

TABLE 29.   VARIATION OF MINERAL COMPOSITION WITH DEPTH IN THE
SKAERGAARD INTRUSION

Height above lowest exposed level, m.	Mean composition of minerals				Rock
	Plagioclase	Clinopyroxene	Ortho-pyroxene	Olivine	
2,500*	An_{30}	$Wo_{30}En_2Fs_{68}$	Absent	Fa_{96}	Quartz ferrogabbro
1,800	An_{40}	$Wo_{27}En_{23}Fs_{50}$	Absent	Fa_{60}	Ferrogabbro
1,200	An_{45}	$Wo_{32}En_{32}Fs_{36}$	Fs_{57}†	Absent	Gabbro
500	An_{56}	$Wo_{42}En_{40}Fs_{18}$	Fs_{45}	Fa_{37}	$\left.\right\}$ Olivine gabbros
0	An_{61}	Not given	Fs_{41}	Fa_{34}	

* Roof.
† A minor constituent.

layered series. The intrusion as a whole is thus pictured as slowly solidifying from the base upward. The border group, with its fluxional banding parallel to the contacts, represents magma convecting upward or downward along the margins of the intrusion and slowly freezing through loss of heat outward through the walls.

Table 29 summarizes the main features of vertical variation in lithology and mineralogy. The sequence upward from olivine gabbro to hortonolite- or fayalite-bearing ferrogabbro represents very completely a trend of differentiation resulting from strong fractional crystallization of a tholeiitic basaltic magma. It is only a partial sequence, however, for it is estimated that 60 per cent of the total thickness of the Skaergaard intrusion still remains unexposed below the lowest level at present accessible. Relatively small volumes of hedenbergite granophyre and sodic acid granophyre, locally invading the upper levels of the gabbro mass, are be-

TABLE 30.   CHEMICAL COMPOSITION OF REPRESENTATIVE ROCKS OF THE
SKAERGAARD INTRUSION

Constituent	1	2	3	4	5	6	7
SiO_2	47.92	46.37	48.15	44.81	48.27	58.81	75.03
TiO_2	1.40	0.79	2.64	2.55	2.20	1.26	0.31
Al_2O_3	18.87	16.82	18.02	13.96	8.58	12.02	13.17
Fe_2O_3	1.18	1.52	2.52	3.75	4.06	5.77	1.56
FeO	8.65	10.44	9.50	16.66	22.89	9.38	0.58
MnO	0.11	0.09	0.12	0.17	0.26	0.21	0.01
MgO	7.82	9.61	5.25	5.54	1.21	0.72	0.15
CaO	10.46	11.29	10.17	8.53	7.42	5.03	0.69
Na_2O	2.44	2.45	3.46	3.35	2.65	3.91	4.24
K_2O	0.19	0.20	0.14	0.33	0.34	2.39	3.85
H_2O+	0.41	0.29	0.20	0.34	1.13	0.21	0.28
H_2O-	0.10	0.09	0.02	0.19	0.37	0.19	0.13
P_2O_5	0.07	0.06	0.05	0.08	0.65	0.71	0.02
CO_2	0.06		0.03				
SrO	0.20		0.07				
BaO	0.02		0.01				
S	0.27		0.14				
Total	100.17	100.02	100.49	100.26	100.03	100.61	100.02

Explanation of Table 30

Note: All analyses are quoted from L. R. Wager and W. A. Deer, The petrology of the Skaergaard intrusion, Kangerdlugssuaq, east Greenland, *Meddelelser om Grönland*, vol. 105, no. 4, 1939, references to which are given in parentheses.

1. Chilled marginal olivine gabbro; average of two analyses (p. 147, No. XIIIa). Plagioclase, An_{60}, 62 per cent, diopsidic augite 22 per cent, hypersthene 1 per cent, olivine 14 per cent, ore 1 per cent.
2. Noritic olivine gabbro 500 m. above lowest exposed horizon (p. 93, No. II). Plagioclase, An_{56}, 55 per cent; augite, $Wo_{42}En_{40}Fs_{18}$, 21 per cent; hypersthene, Fs_{45}, 5 per cent; olivine, Fa_{37}, 17.5 per cent; ore 1.5 per cent.
3. Gabbro, 1,200 m. above lowest exposed horizon (p. 96, No. III). Plagioclase, An_{45}, 60 per cent; augite, $Wo_{32}En_{32}Fs_{36}$, 33 per cent; hypersthene, Fs_{57}, 2 per cent; ore 5 per cent.
4. Hortonolite ferrogabbro, 1,800 m. above lowest exposed horizon (pp. 102, 103, No. V). Plagioclase, An_{40}, 56 per cent; ferroaugite, $Wo_{27}En_{23}Fs_{50}$, 20 per cent; olivine (hortonolite), Fa_{60}, 16 per cent; ore 8 per cent.
5. Fayalite ferrogabbro, 2,500 m. above lowest exposed horizon (p. 106, No. VIII). Plagioclase, An_{30}, 29 per cent; ferroaugite, $Wo_{30}En_2Fs_{68}$, 34 per cent; fayalite, Fa_{96}, 17 per cent; quartz and perthite, 11 per cent; ore 8 per cent; apatite 1 per cent.
6. Hedenbergite granophyre (p. 211, No. XXII), from lensoid mass 200 m. thick in upper part of the intrusion.
7. Acid granophyre (p. 208, No. XXIII) from sill near roof of the intrusion.

lieved to represent successive residual liquids squeezed away when the intrusion as a whole had almost completely solidified. One of the most interesting facts which emerges from chemical investigation of selected minerals from the Skaergaard gabbros is the regular progressive enrichment in iron shown by pyroxenes and olivines with advancing crystallization. These trends are shown in Fig. 45.

The range of composition of typical rocks from the Skaergaard intrusion is illustrated in Table 30. The initial magma,[5] now represented by chilled marginal olivine gabbros (Table 30, analysis 1), is chemically very close to some of the alumina-rich basalts of andesitic provinces (Table 27, analysis 1; Table 28, analysis 1). It has affinities, too, with the porphyritic central magma type of Mull (Table 19, analysis 9) and with tholeiitic magmas in general. In the Skaergaard parent magma

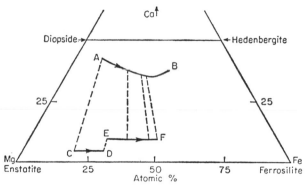

FIG. 45. Trend of crystallization of pyroxenes in Skaergaard intrusion. A–B, calcium-rich clinopyroxenes; C–D, orthopyroxenes; E–F, pigeonites. Ties are drawn between coexisting analyzed pyroxenes. (After G. M. Brown.)

Al_2O_3 is high, the iron is in a highly reduced state, and potash very low. Differentiation, illustrated by upward passage of gabbros into ferrogabbros, involves marked progressive enrichment of the liquid fractions in iron (X–Y in Fig. 46a). In the final stages granophyric liquids enriched in soda develop (U–V in Fig. 46a). Though some of these granophyres are iron-rich, those believed by Wager and Deer to represent the ultimate product of differentiation are in no way abnormal rocks.

The distribution of trace elements in the Skaergaard rocks has been investigated in great detail by Wager and Mitchell.[6] Some of their data

[5] Cf. also L. R. Wager and R. L. Mitchell, The distribution of trace elements during strong fractionation of basic magma—a further study of the Skaergaard intrusion, east Greenland, Geochim. et Cosmochim. Acta, vol. 1, pp. 129–208 (especially pp. 138–139), 1951.

[6] Wager and Mitchell, op. cit., 1951. This is a classic comprehensive study of the progressive partition of trace elements between the crystalline and liquid fractions of a differentiating magma.

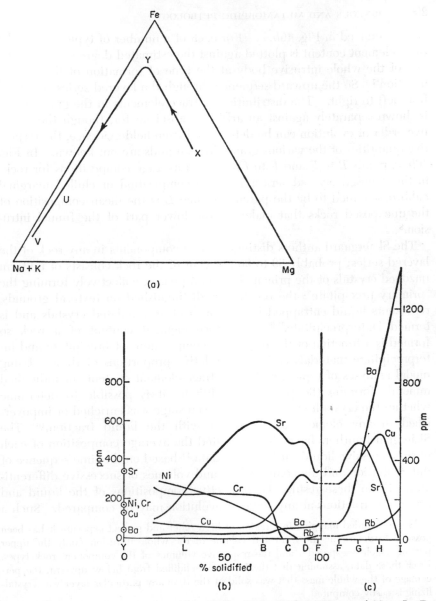

FIG. 46. Chemical variation in rocks of the Skaergaard intrusion. (*After L. R. Wager and R. L. Mitchell.*) (*a*) Fe:Mg:(Na + K) diagram: X–Y, gabbros; U–V, granophyres. (*b*) Trace-element content of main rock types in parts per million, plotted against percentage of the whole intrusion solidified. Compositions of rocks formed at successive stages of solidification are: Y = chilled marginal gabbro; L = mean composition of rocks not exposed (60 per cent of total mass); B = hypersthene olivine gabbro (lower part of exposed layered series); C = middle gabbro; D = hortonalite ferrogabbro; F = fayalite ferrogabbro. (*c*) Continuation of Fig. 46*b* beyond point F, horizontal scale arbitrary. G = basic hedenbergite granophyre; H = hedenbergite granophyre; I = acid granophyre.

are reproduced in Fig. 46b, c. For each of a number of typical rocks the trace-element content is plotted against the estimated degree of solidification of the whole intrusive body at the time of formation of the rock in question.[7] So the upward sequence through the layered series is plotted from left to right. The distribution of trace elements in the granophyres is shown separately against an arbitrary abscissa; for, though their relative order of evolution can be determined from field evidence, the respective quantities of the various granophyric liquids are not known. In Fig. 46b, c, points B to F and F to I represent average compositions for rocks in the exposed layered series; Y is the composition of chilled marginal gabbro, assumed to be the parent magma; L is the mean composition of the unexposed rocks that make up the lower part of the funnel intrusion.[8]

The Skaergaard authors distinguish two components in any rock of the layered series: probably 80 to 90 per cent of the rock consists of uniform unzoned crystals of the principal mineral phases, collectively forming the "primary precipitate"; the remainder, distinguished on textural grounds, represents liquid entrapped in the mush of accumulated crystals and is termed "interprecipitate."[9] The trace element content of a rock so formed is a function of the respective compositions of precipitate and interprecipitate materials and of the relative proportions of these. Using modal analyses of typical rocks and trace-element contents of individual minerals forming the primary precipitate, it is possible to determine whether the crystalline fraction at a given stage was enriched or impoverished in any element as compared with the liquid fraction.[10] The Skaergaard authors have also computed the average composition of each of six successive liquids, using other data[11] based on the time sequence of differentiation and the compositions and volumes of successive differentiates. From these results the respective compositions of the liquid and crystalline fractions at any stage of evolution may be compared. Such a

[7] From detailed mapping in a region of high relief and perfect exposure it has been possible to determine the form and volume of the whole intrusion (only the upper part of which is exposed), and the respective volumes of the component rock types. From these data, assuming that the intrusion solidified from below upward, the percentage of the whole mass that was solid at the time any particular layer was crystallizing is easily computed.

[8] The mean content of any element in the unexposed basal part of the intrusion can be computed from these data: total volume of the intrusion; composition of initial magma (= chilled marginal gabbro); total quantity of the element in the exposed rocks of the intrusion (40 per cent of the whole) (cf. Wager and Mitchell, op. cit., pp. 174–177, 1951).

[9] Wager, op. cit., pp. 342–345, 1953.

[10] Wager and Mitchell, op. cit., p. 143, 1951.

[11] Wager and Deer, op. cit., pp. 217 224, 1939; Wager and Mitchell, op. cit., p. 174, 1951.

comparison is shown graphically for vanadium in Fig. 47.[12]   Vanadium
enters into the lattices of pyroxenes and iron ores, but not into olivines.
The early olivine gabbros (to the left of B, Fig. 47) contain less vanadium
than the liquids from which they crystallized.   But with increasing
crystallization of pyroxene and the incoming of iron ores (especially magnetite) this condition is reversed and the vanadium content of the crystalline fraction rises to a peak.   The liquid now becomes so impoverished
that the content of vanadium in the last crystalline fractions (granophyres) also drops to near zero.

   The Bushveld Complex of South Africa.[13]   The Bushveld complex of
Transvaal, South Africa, has been described as "the world's most spectacular igneous assemblage."[14]   Its most important component is a vast
lopolith of basic igneous rocks covering an area of the order of 20,000

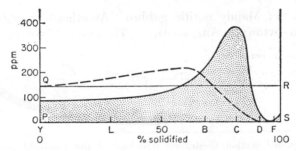

FIG. 47.   Vanadium content of Skaergaard rocks and corresponding liquids.   Full
curve = rocks Y to F (same as for Fig. 46b).   Broken curve = liquids at successive
stages of solidification of the whole mass.   Note that the total stippled area equals
the area of the rectangle PQRS.

square miles and in places as much as 5 miles in vertical thickness.   The
plutonic association developed within the lopolith is the topic now under
discussion, but the nature of other components of the broad complex
should be noted.   The floor of the lopolith is composed of late pre-
Cambrian sediments with which some andesites are interstratified.   These
are injected with basic sills of quartz-bearing noritic gabbro and diabase
probably drawn from the same source as the magma of the lopolith.   The
lopolith is roofed with a thick layer of granophyre and granite, which in

---

[12] Cf. Wager and Mitchell, *op. cit.*, pp. 185–186, 1951.
   [13] R. A. Daly, The Bushveld igneous complex of the Transvaal, *Geol. Soc. America
Bull.*, vol. 39, pp. 703–768, 1928; A. L. Hall, the Bushveld igneous complex of the
Central Transvaal, *South Africa Geol. Survey Mem.* 28, 1932; B. V. Lombaard, On the
differentiation and relationships of the rocks of the Bushveld complex, *Geol. Soc. South
Africa Trans.*, vol. 37, pp. 5–52, 1935; C. F. J. Van der Walt, Chrome ores of the
western Bushveld complex, *Geol. Soc. South Africa Trans.*, vol. 44, pp. 79–112, 1942;
Wager, *op. cit.*, pp. 340–341, 1953.
   [14] A. Knopf, in *Geol. Soc. America 50th Anniversary Vol.*, p. 347, 1941.

turn is roofed by acid volcanic rocks ("felsites"), remnants of which still exceed 8,000 ft. in thickness.   Whether the granites were emplaced before or after injection of the lopolith magma is not certain, but all the rocks mentioned above, including the felsites, seem to belong to the same broad cycle of igneous activity and are of late pre-Cambrian age.

Throughout the Bushveld lopolith an amazingly regular stratified arrangement of the component rocks, resembling that of sedimentary series, is conspicuous.   Moreover the influence of gravity on this arrangement is clear, for there is an over-all tendency for rocks of high specific gravity to predominate in the lower levels.   The generalized downward sequence given by Lombaard is as follows:

4. Upper zone: Granodiorites, quartz diorites, diorites and gabbros. Anorthosite, when present, consists of andesine An_{45} to An_{50}.   Thickness, 9,000 ft.

3. Main zone: Mainly noritic gabbro.   Anorthosite members consist of labradorite-bytownite An_{65} to An_{75}.   Thickness, 17,000 ft.   Thin con-

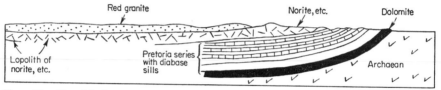

Fig. 48.   Simplified section illustrating structure of the Bushveld complex, South Africa; scale, 1:3,000,000.   (After R. A. Daly.)

tinuous bands of titaniferous iron ore occur near the top of the main zone.

2. "Critical" zone: Norites, hypersthene and bronzite pyroxenites, harzburgite, anorthosite; chromite-rich bands ($Cr_2O_3 = 35$ to 45 per cent), from a few inches to 4 ft. thick, persist with extraordinary regularity for distances as great as 40 miles.   Anorthosites consist of bytownite An_{75} to An_{85}.   Thickness 1,000 to 5,000 ft.

1. Basal zone: Norite with bronzitite and anorthosite bands.   Thickness uncertain.

Rhythmic layering is prominent and horizontally persistent, especially in the critical zone.   Widespread cryptic layering has also been established, for variation in composition of the principal minerals may be correlated to some extent with height above the floor.   Thus rhombic pyroxene shows a progressive increase in iron content from Fs_{15} at the base of the critical zone to Fs_{50} in the main zone.   At the same time it diminishes in quantity and diallage becomes more conspicuous (95 per cent of the pyroxene in the lower part of the critical zone is rhombic; but near the top of the main zone this percentage has decreased to between 20 and 40 per cent).   Plagioclase ranges from bytownite (critical zone) to andesine-labradorite (upper zone).   In the oxide bands of the critical and main zones, chromium and aluminum are concentrated in the lower, vanadium and iron in the middle, and titanium in the upper bands.[15]

Detailed examination of any extensive vertical section reveals frequent local reversals of the mineralogical trends enumerated above. Moreover, while heavy pyroxene-rich rocks predominate in the lower levels and more

TABLE 31. CHEMICAL COMPOSITIONS OF MEMBERS OF THE BUSHVELD LOPOLITH, TRANSVAAL

Constituent	1	2	3	4	5
SiO_2	51.45	52.05	55.40	41.20	60.80
TiO_2	0.34	0.21	0.10	0.25	1.06
Al_2O_3	18.67	17.24	2.35	1.20	13.14
Fe_2O_3	0.28	0.65	Nil	4.00	3.23
FeO	9.04	6.65	9.65	8.20	7.78
MnO	0.47	0.13	0.15	0.10	0.22
MgO	6.84	8.98	30.85	34.40	0.40
CaO	10.95	11.37	0.65	1.35	4.45
Na_2O	1.58	1.83		0.20	3.69
K_2O	0.14	0.40		0.30	3.18
H_2O+	0.34	0.52	Nil	5.20	1.42
H_2O-	0.03	0.11	0.10	0.25	0.51
P_2O_5	0.09	0.12	0.30		0.23
Cr_2O_3			0.60	3.20	
NiO			0.15	0.35	
Total	100.22	100.26	100.30	100.20	100.11

Explanation of Table 31

Note: Page references to A. L. Hall, The Bushveld igneous complex of the Central Transvaal, South Africa Geol. Survey Mem. 28, 1932, are given in parentheses.
  1. Hypersthene gabbro (norite), chilled floor phase of lopolith (p. 310, No. I).
  2. Typical norites (noritic gabbros) mainly from main zone; average of 5 analyses (p. 304, No. VI).
  3. Pyroxenite from lower zone (p. 315, No. I).
  4. Harzburgite, critical zone (p. 327, No. IV).
  5. Quartz syenites (granodiorites, quartz diorites) of upper zone; average of 5 analyses with SiO_2 ranging from 54.8 to 65.4 per cent (p. 307, No. VI).

feldspathic rocks are conspicuous toward the top of the intrusion, bands of anorthosite and of pyroxenite may occur at any horizon. Again, certain rock types, the mineral composition of which conforms to origin by crystal accumulation, nevertheless habitually cut across the more normal stratified igneous rocks. Such are the pipes of dunite, 300 ft. in diameter,

[15] H. D. B. Wilson, Geology and geochemistry of base metal deposits, Econ. Geol., vol. 48, p. 382, 1953.

which cut vertically across "bedded" pyroxenite in the critical zone, and which in some cases enclose central vertical zones of platinum-bearing hortonolite rock.

Thus, while in broad outline the density relations and mineralogy of the main zones of the Bushveld lopolith are compatible with a mechanism of gravitational differentiation of a crystallizing sheet of basic magma, there are details that are rather difficult to reconcile with such a process. It is for this reason that Lombaard [16] suggests "emplacement of the lopolith by intermittent supplies of magma which was continually changing its composition, and which was generated by refusion of the crystalline products in the same order as they were formed [by gravitational accumulation] in the intra- or sub-crustal reservoir," followed by further gravitational differentiation of each draft of magma after its arrival within the lopolith. To the present authors this hypothesis of multiple injection seems even more difficult of acceptance. It is a little disconcerting to find that even for the largest of igneous sheets, it may be necessary to invoke preintrusive differentiation within an elusive but presumably larger and still more deep-seated magma chamber to account for the observed data! A more plausible explanation is that of Wager,[17] who invokes the simultaneous operation of a series of convection cells in a single sheet of magma. The layered series developed within different cells would be similar but not identical, and lateral migration of one cell could interrupt the rhythmic and cryptic layering that had been developing in an adjacent cell.

Analyses of typical members of the Bushveld lopolith are given in Table 31. The composition of the chilled floor phase of the Bushveld intrusion (Table 31, analysis 1) closely resembles that of the chilled marginal gabbro of the Skaergaard mass (Table 30, analysis 1). The latter is more basic and somewhat richer in Na_2O and TiO_2. Ferrogabbros of the Skaergaard type are lacking in the Bushveld assemblage, but the rocks of the upper zone of the Bushveld intrusion (Table 31, analysis 5) show notable enrichment in iron compared with MgO and CaO, and chemically resemble the hedenbergite granophyres of Skaergaard (Table 30, analysis 6).

The Stillwater Complex of Montana.[18]   The Stillwater complex of Montana is a steeply dipping basic sheet, intrusive into pre-Cambrian

[16] Lombaard, op. cit., p. 52, 1935.

[17] Wager, op. cit., pp. 341, 348, 1953.

[18] A. L. Howland, J. W. Peoples, and E. Sampson, The Stillwater igneous complex, Montana Bur. Mines and Geol. Misc. Contr., 7, 1936; H. H. Hess, Primary banding in norite and gabbro, Am. Geophys. Union Trans., 19th Ann. Meet., pt. 2, pp. 264–268, 1938; Extreme fractional crystallization of a basaltic magma, Am. Geophys. Union Trans., 20th Ann. Meet., pt. 3, pp. 430–432, 1939; op. cit., 1940; J. W. Peoples and A. L. Howland, Chromite deposits of the eastern part of the Stillwater complex, Stillwater, Montana, U.S. Geol. Survey Bull. 922–N, pp. 377–382, 1940.

schists and gneisses of the Beartooth plateau on the eastern fringe of the Rockies. It is believed originally to have been a lopolith or thick sub-horizontal sheet, the roof and upper part of which were eroded, and the lower part of which was locally invaded by granite, before the opening of the Cambrian. Burial beneath Paleozoic and Mesozoic sediments, tilting of the whole mass, and erosion of the cover followed in turn. The present outcrop, probably faulted at each end, is 30 miles long. The floor is exposed, and above it is a thickness (originally vertical) of some 17,000 ft. of banded rocks, the nature and sequence of which are amazingly similar to those of the Bushveld assemblage.

A local chilled zone of diabasic norite (300 ft.) at the base is followed upward by an ultrabasic zone (2,500 ft.) consisting mainly of bronzite pyroxenite and harzburgite with minor dunite, all of which may show a banded structure. Bands of chromite are present in the harzburgite. Then come norites, gabbros, and anorthosites in irregularly alternating bands, with the more feldspathic members particularly conspicuous in the upper three or four thousand feet. The principal minerals in all these rocks are diopsidic augite, bronzite-hypersthene, and labradorite-by-townite. Rhythmic banding like that of the Skaergaard intrusion is developed throughout the lower two-thirds of the norite-gabbro zone. Internal gravitational differentiation of individual bands into upper feldspathic and lower pyroxenic phases is characteristic. Individual bands vary from a fraction of an inch to several feet in thickness.

From the petrography of an extensive vertical cryptically layered series of representative rocks, Hess [19] has reconstructed a general course of crystallization of the Stillwater magma, on the assumption that there was a single period of injection followed by gradual building of the floor upward by accumulation of sinking crystals:

1. First, as shown by rocks just above the chilled basal zone, magnesian olivine (Fa_{14}) appears. It is soon joined by bronzite (Fs_{13}), by minor diopsidic augite ($Wo_{37}En_{56}Fs_7$), and then by bytownite (An_{86}).

2. As crystallization proceeds, pyroxenes and such olivine as remains both show enrichment in iron. At a stage corresponding to a height of 6,000 to 7,000 ft. above the floor, diopsidic augite separates abundantly and is associated with hypersthene Fs_{20}, medium bytownite, and some olivine (Fa_{20}).

3. Still later (about 15,000 ft. above the floor) the association is sodic bytownite (An_{75}), augite ($Wo_{41}En_{47}Es_{12}$), hypersthene (Fs_{26}). The latter mineral, from this level to the highest exposed, has a lamellar structure, interpreted as the result of exsolution of diopside attendant on inversion of pigeonite to hypersthene.

[19] H. H. Hess, Pyroxenes of common mafic magmas, Part I, *Am. Mineralogist*, vol. 26, pp. 515–535, 1941 (especially pp. 529, 530).

4. At the latest recorded stage (equivalent to a measured height of 17,000 ft. above the floor and an estimated distance of 5,000 to 15,000 ft. below the original roof), the plagioclase is labradorite (An_{61}) and is associated with augite and ferruginous hypersthene (Fs_{30} to Fs_{40}).

Petrogenesis. *General Characteristics of Layered Basic Intrusions.* The layered gabbro-norite-peridotite pluton has become one of the most clearly recognized types of major igneous intrusion. Many of these bodies are of pre-Cambrian or early Paleozoic age, *e.g.*: Duluth (Minnesota and Wisconsin),[20] pre-Cambrian; Caribou Lake (Ontario), pre-Cambrian;[21] Stillwater (Montana), pre-Cambrian; Bushveld (South Africa), pre-Cambrian; Bay of Islands (Newfoundland),[22] Ordovician; Aberdeenshire (Scotland),[23] pre-Devonian. But there are smaller though still very thick Tertiary bodies of this class, such as the Skaergaard intrusion of Greenland, and the layered ultrabasic complex of the isle of Rum in western Scotland.[24]

It has long been thought that the typical form of the layered basic intrusion is a lopolith—"a large, lenticular, centrally sunken, generally concordant, intrusive mass, with its thickness approximately one-tenth to one-twentieth of its width or diameter."[25] But the Skaergaard mass proves to be funnel-shaped, and it now seems likely that a triangular or funnel-shaped cross section is characteristic of lopoliths and even of highly elongate layered bodies like the Great Dyke of Southern Rhodesia, which is 300 miles long, 4 to 7 miles wide.[26] In general the layering dips gently toward the center and is discordant with respect to the walls and floor, which dip inward more steeply; thus it is impossible to deduce the configuration of the sides and floor of the intrusion from attitudes of layering within the mass.

Where a chilled border phase is present its composition is basic, typically that of a gabbro just saturated or slightly undersaturated in silica.

[20] F. F. Grout, Internal structures of igneous rocks, with special reference to the Duluth gabbro, *Jour. Geology,* vol. 26, pp. 439–458, 1926; R. B. Taylor, The Duluth gabbro complex, Duluth, Minnesota, *Geol. Soc. America Guide Book for Field Trip,* pp. 42–66, 1956.

[21] G. M. Friedman, Structure and petrology of the Caribou Lake intrusive body, Ontario, Canada, *Geol. Soc. America Bull.,* vol. 68, pp. 1531–1564, 1957.

[22] J. R. Cooper, Geology of the southern half of the Bay of Islands igneous complex, *Newfoundland Dept. Nat. Resources, Geol. Sect., Bull. 4,* 1936.

[23] F. H. Stewart, The gabbroic complex of Belhelvie in Aberdeenshire, *Geol. Soc. London Quart. Jour.,* vol. 102, pp. 465–498, 1947.

[24] G. M. Brown, The layered ultrabasic rocks of Rum, Inner Hebrides, *Royal Soc. London Philos. Trans.,* ser. B, no. 668, vol. 240, pp. 1–53, 1956.

[25] F. F. Grout, The lopolith: an igneous form exemplified by the Duluth gabbro, *Am. Jour. Sci.,* vol. 46, p. 518, 1918.

[26] H. D. B. Wilson, Structure of lopoliths, *Geol. Soc. America Bull.,* vol. 67, pp. 289–300, 1956; Wager and Brown, *op. cit.,* pp. 1071–1074, 1957.

It has chemical affinities with the tholeiitic rather than the alkaline olivine basalts. The principal rock types consist of any combination—even monomineralic and bimineralic extremes—of the minerals olivine, orthopyroxene, monoclinic pyroxene, plagioclase; local bands are very rich in one of the oxides chromite, magnetite, or ilmenite.

The most obvious and widespread element in the layered condition is rhythmic banding. This reflects the influence of gravity in a general tendency for rock density to increase downward in the layered series, and in textural features reminiscent of those of sedimentary rocks—notably "graded bedding" within individual bands. Independent of but usually associated with rhythmic banding is cryptic layering, by virtue of which minerals stable at progressively lower temperatures occur at successively higher levels in the mass. Fluxional banding reflecting local trends of magmatic flow may also be present.

Two components enter into the texture of any typical rock of the layered series. There is a primary precipitate of one or several mineral phases each of which is uniform in composition and in grain; and there is interprecipitate material believed to have crystallized from liquid entrapped in the interstices of the primary aggregate. The interprecipitate material may appear as zones of different composition margining the crystals of the primary precipitate that have acted as nuclei. Where no such nuclei are present, e.g., where an augite interprecipitate develops in a monomineralic plagioclase precipitate, the interprecipitate crystals of new phases may grow to a very large size. This is the origin of the ophitic and poikilitic textures commonly displayed by gabbros and peridotites of layered intrusions. Idiomorphic primary plagioclase crystals, with marginal interprecipitate zones of more sodic composition, may be enclosed in coarse interprecipitate augite or olivine or iron ore in different parts of the same rock. In the peridotites of Rum a single coarse interprecipitate crystal of pyroxene or of plagioclase several centimeters in diameter may enclose as many as 10,000 primarily precipitated olivine grains.

Mechanism of Differentiation. Since not even the most enthusiastic devotee of the hypothesis of plutonic emplacement by solid diffusion could doubt the strictly igneous (magmatic) origin of gabbro-peridotite intrusions, the petrogenesis of such rocks has important broad implications. In them is seen a picture of the kind and degree of lithologic variation that tends to develop within large masses of basic magma under plutonic conditions in a relatively undisturbed tectonic environment.

From the nature and regularity of the rhythmic and cryptic layering, it is reasonable to conclude that sinking of heavy high-temperature minerals (e.g., magnesian olivines, bronzite) played an important part in magmatic evolution, and that crystallization to completion proceeded from the lower levels toward the top of the chamber. Wherever long sections transversed to the major stratification have been measured in detail, the picture that emerges nevertheless is not consistent with simple gravitational settling or floating of crystals in a stagnant body of magma. Reversals of the normal sequence, and oscillations in density and mineralogy of rock types, are common, and in a given intrusion are conspicuous and persistent at particular horizons.

The over-all simplicity of the rock sequence and the complicated variation which modifies and to some degree obscures it are reminiscent of the variation encountered in a sedimentary series—e.g., where, in a series of interbedded shales, sandstones, and conglomerates, a general downward coarsening of the formation as a whole can be detected. In the rock succession of a stratified lopolith, as in a clastic sedimentary series, we see the result of sinking and accumulation of mineral grains of varying size and density in a fluid medium subject to fluctuating turbulent flow. In spite of the relative ease with which problems of aqueous sedimentation may be studied in the field and in the laboratory, the process is so complex that reconstruction of the conditions of deposition of a given sedimentary formation on the basis of its petrographic and field characters is still attended by much uncertainty. Igneous "sedimentation" in a body of magma is subject to just as many variable factors, but experimental data are much less adequate than in the case of aqueous sedimentation. It is not surprising, therefore, that opinions differ widely as to the relative roles of gravitational settling or floating of crystals and of various possible flow movements in the magma body in bringing about the heterogenous condition of basic lopoliths. Some of these opinions may now be noted:

1. According to Hall (Bushveld complex) and to Hess (Stillwater complex), the bulk of the magma was injected in a single act of intrusion and the present stratified condition is due mainly to gravitational differentiation of the mass *in situ*.[27] Some of the thin ore bands of great lateral extent are attributed to separate injection of appropriate sulfide- or chromite-rich magmas. Hess explains the small-scale rhythmic banding as due to "short epochs of mild but irregular turbulence," strong enough temporarily to hold plagioclase in suspension without interrupting the downward movement of the heavier pyroxene crystals. It has been suggested by Coats that a banded condition would develop, without any such disturbance, by differential sinking of plagioclase and pyroxene in a magma less dense than either mineral.[28]

2. In the case of the Skaergaard intrusion, Wager and Deer [29] attribute the banded condition to settling of crystals in the bottom current of a

[27] Hall, *op. cit.*, p. 353, 1932; Hess, *op. cit.*, pp. 266, 267, 1938; p. 431, 1939.

[28] R. R. Coats, Primary banding in basic plutonic rocks, *Jour. Geology*, vol. 44, pp. 407–419, 1936.

[29] Wager and Deer, *op. cit.*, p. 332, 1939; Wager, *op. cit.*, pp. 345–348, 1953.

magmatic convection cell.   Crystals growing in the convecting magma as it moves across the top and down the sides of the funnel come to rest as they sweep across the floor.

3.   Lombaard and Cooper support the view that only multiple injection of magmas drawn at intervals from a concealed reservoir of differentiating magma can account for some of the large-scale oscillations observed in layered lopoliths.[30]

4.   Yoder [31] has shown experimentally that the composition of the diopside-anorthite eutectic is sensitive to changes in water pressure.   He suggests that periodic fluctuations in water pressure—such as might occur if the roof of an intrusion periodically fractured—could account for alternate precipitation of plagioclase and pyroxene from a melt near the cotectic range of composition.

To the present writers the remarkable similarity of the vertical sequences in different intrusions strongly suggests that some single universally effective factor is responsible for the main trend of differentiation. Such a factor is the force of gravity.   It would control the convection pattern in a thick body of magma, and under its influence crystals would become sorted in sinking to rest within the convecting liquid.   The mechanism of convection and gravitational differentiation proposed by Wager and his associates is adequate to explain the regular characteristics of layered basic intrusions.   Local intermittent turbulence and lateral shifting of convection cells in a multiple-cell system of a large intrusion could explain observed departures from regularity and reversals or oscillations in the layered sequence.

Trends in Differentiation.   The over-all composition of well-preserved layered intrusions and the composition of chilled marginal facies show that the parent magma of basic layered intrusions is basaltic, with tholeiitic rather than alkaline affinities.

By sinking and accumulation of early-formed crystals of magnesian olivine or of pyroxene (mainly bronzite) great thicknesses of ultramafic rocks may develop during the early stages of differentiation.   This does not mean that great volumes of mobile, partly crystalline, ultramafic magma necessarily become available for injection as separate bodies; for it has not been shown that a mushy condition persists for a vertical distance of more than a few feet at any stage during the upward growth of the ultramafic layer.   Two alternative trends are exhibited during this phase of differentiation.   In some cases (*e.g.*, Bushveld, Stillwater) the ultramafic differentiate is dominantly bronzite-pyroxenite, and chromite-

---

[30] Lombaard, *op. cit.*, p. 52, 1935; Cooper, *op. cit.*, pp. 44–46, 1936.

[31] H. S. Yoder, The system diopside-anorthite-water, in Annual report of the director of the Geophysical Laboratory, *Carnegie Inst. Washington Year Book*, pp. 106, 107, 1954.

rich bands are conspicuous. In others (*e.g.*, Duluth, Aberdeenshire, Rhum) it is largely dunite and troctolite, with only minor pyroxenite and no chromite rock.[32] Partial serpentinization of peridotites is common and appears to date from a period shortly following complete solidification, and in some cases deformation, of the rocks in question.

Although crystal fractionation has clearly been strong, the great bulk of the residual magma from which olivine and pyroxene have been removed in the early stages still had a composition within the range of gabbro or basalt. In the Bushveld mass considerable volumes of more siliceous rock (diorite, granodiorite) were developed during the final stages of cooling, but it is not certain to what extent this trend in magmatic evolution was determined by differentiation on the one hand or assimilative reaction with overlying sediments and felsites on the other. From the widespread tendency for veins, lenses, and outlying fringes of granophyre and granite to appear at the top of stratified lopoliths, it seems likely that the ultimate product of differentiation, extracted only by filter-pressing of the almost completely solidified mass, is a granitic liquid.[33]

The course of magmatic evolution outlined above conforms in a general way to Bowen's experimentally deduced conception of the normal trend of fractional crystallization of initially basaltic magma. But in the Skaergaard intrusion progressive enrichment in iron so outweighed all other effects of differentiation that the liquids remaining when the great bulk (90 per cent) of the mass had solidified had the composition of ferro-gabbro extremely rich in iron. This accords with the trend of magmatic evolution long ago advocated by Fenner[34] as the normal course of fractional crystallization of basic magma. Wager and Deer uphold Fenner's conclusion. They also note, however, that the small amount of residual liquid filter-pressed out in the very last stages is granophyric.

If, as is clearly demonstrated in the Skaergaard mass, fractional crystallization of basaltic magma leads to development of highly ferruginous magmas, why is it that corresponding rocks—ferrogabbros and ferro-basalts—are so very rare? Wager and Deer[35] reply that simple strong fractionation of basaltic magma is a process which rarely operates in petrogenesis, and that the common calc-alkaline igneous rocks (andesites, rhyolites, and their plutonic equivalents) are products of some other process. They appeal to mixing of basic and acid magmas or to assimila-

---

[32] Cooper, *op. cit.*, p. 40, 1936.

[33] E.g., see Wager and Deer, *op. cit.*, pp. 308–309, 1939; R. C. Emmons, The contribution of differential pressures to magmatic differentiation, *Am. Jour. Sci.*, vol. 238, pp. 1–21, 1940 (especially pp. 9–12).

[34] C. N. Fenner, The crystallization of basalts, *Am. Jour. Sci.*, vol. 18, pp. 225–253, 1929; The residual liquids of crystallizing magmas, *Mineralog. Mag.*, vol. 22, pp. 539–560, 1931.

[35] Wager and Deer, *op. cit.*, pp. 334, 335, 1939.

tive reaction between basic magma and granite, with subsequent mixing and differentiation. The present authors consider it likely that some such blending of acid and basic magmas may play an important part in the evolution of the andesite rhyolite volcanic kindred. But it is also clear, as pointed out recently by Bowen,[36] that fractional crystallization of basaltic magma is a flexible process. One trend, possibly determined by poverty of the magma in water [37] or by the influence of a floor on free movement of crystals,[38] or even perhaps by low partial pressures of oxygen connected with high sulfur content in the magma, leads to absolute enrichment of residual liquids in iron. But much more commonly increase in the content of alkali and silica outpaces and overshadows the simultaneous rise in the FeO/MgO ratio, and the trend of differentiation is toward granitic liquids of more normal composition.

THE PERIDOTITE SERPENTINITE ASSOCIATION OF OROGENIC ZONES

Ultramafic Rocks as Members of Plutonic Associations. Ultramafic rocks figure prominently in several types of plutonic association, of which the first two listed below have by far the widest distribution and development:

1. Massive, rhythmically banded layers in the lower levels of stratified basic intrusions (Bushveld, Stillwater, Rhum, etc.).

2. Peridotites and serpentinites of the "alpine type" [39] occurring in folded geosynclinal sediments of orogenic belts.

3. Peridotites and serpentinites occurring as minor associates—commonly interpreted as differentiates—of granite-granodiorite-diorite intrusive complexes.[40]

4. Zoned stock-like masses of peridotite, pyroxenite, hornblendite, and gabbro.[41]

[36] N. L. Bowen, Magmas, *Geol. Soc. America Bull.*, vol. 58, pp. 263–380, 1947.

[37] *Ibid.*, p. 273.

[38] A. B. Edwards, Differentiation of the dolerites of Tasmania, *Jour. Geology*, vol. 50, pp. 603–606, 1942.

[39] W. N. Benson, The tectonic conditions accompanying the intrusion of basic and ultrabasic plutonic rocks, *Nat. Acad. Sci. Mem.*, vol. 19, mem. 1, p. 6, 1926; H. H. Hess, Serpentines, orogeny and epeirogeny, *Geol. Soc. America Special Paper*, no. 62, pp. 391–408, 1955.

[40] *E.g.*, S. R. Nockolds, The Garabal Hill-Glen Fyne igneous complex, *Geol. Soc. London Quart. Jour.*, vol. 96, pp. 451–511, 1941; E. S. Larsen, Batholith of Southern California, *Geol. Soc. America Mem. 29*, pp. 40–41, 1948.

[41] M. S. Walton, The Blashke Island ultrabasic complex, Alaska, *New York Acad. Sci. Trans.*, vol. 13, pp. 320–323, 1951; A. Aho, Geology and petrogenesis of ultrabasic nickel-copper-pyrrhotite deposits at the Pacific Nickel property, Southwest British Columbia, *Econ. Geology*, vol. 51, pp. 444–481, 1956; J. C. Ruckmick and J. A. Noble, Origin of the ultramafic complex at Union Bay, southeastern Alaska, *Geol. Soc. America Bull.*, vol. 70, pp. 981–1018, 1959.

5. Picrites, hornblende peridotites, and mica peridotites genetically connected with intrusions of alkaline basic magma. The African kimberlites,[42] occurring as breccia-filled pipes, perhaps belong to this category, for they have affinities with lamprophyres, alnoites, and carbonatites.

Here we shall concern ourselves solely with ultramafic intrusions of the "alpine type." These are so constantly present throughout dissected fold mountains the whole world over, and exhibit such consistent characters as regards petrography, form of intrusion, and tectonic relationships, that it is possible to generalize rather widely on these matters. And it is in connection with this association that the problem of the nature of ultramafic magma becomes most intriguing.

Occurrence. Alpine peridotites and serpentinites most typically appear in the form of steeply inclined sheets or lenses concordant with the structure of the enclosing (usually strongly folded) rocks. Individual intrusions may in certain instances extend without noteworthy interruption for many miles along the strike. Thus along the full length of the southwest Pacific island of New Caledonia a dozen early Tertiary ultramafic masses, the largest of which measures 70 by 25 miles, cover a total area estimated as 2,600 square miles;[43] in southern New Zealand[44] vertical sheets of peridotite and serpentinite, believed to be of late Paleozoic age, include at least three individual masses whose continuous outcrops, dissected to a relief of 5,000 to 7,000 ft., each exceed 40 miles in length and locally attain a width (thickness) of 3 to 5 miles.

More usually the ultramafic masses are of much smaller size—narrow lenses ranging from 100 yards to a few miles in length—and occur in swarms which may consist of hundreds of subparallel masses, arranged en échelon or in linear series along the length of arcuate or sinuous "ultramafic belts" which follow the axes of broadly contemporary fold-mountain chains. Such is the late Paleozoic "serpentine belt" of New South Wales,[45] 250 miles long, itself but part of a discontinuous linear series of serpentinite intrusions extending for over a thousand miles through eastern Australia. A well-known American example is the swarm of small ultramafic bodies

---

[42] J. Verhoogen, Les pipes de kimberlite du Katanga, *Ann. Ser. Mines Katanga*, tome 9, pp. 1–45 (especially pp. 29–32), 1939.

[43] Benson, *op. cit.*, p. 41, 1926.

[44] Benson, *op. cit.*, pp. 42, 43, 1926; C. O. Hutton, Preliminary note on the occurrence of an ultrabasic intrusion in the Livingstone Range, western Otago, *Royal Soc. New Zealand Trans.*, vol. 68, pp. 349–350, 1937; W. N. Benson and J. T. Holloway, Notes on the geography of the ranges between the Pyke and Matukituki rivers, northwest Otago, *Royal Soc. New Zealand Trans.*, vol. 70, pt. 1, pp. 1–24, 1940.

[45] Benson, *op. cit.*, pp. 37–39, 1926; G. D. Osborne, The structural evolution of the Hunter-Manning-Myall province, New South Wales, *Royal Soc. New South Wales Monograph 1*, pp. 63–70, 1950; J. F. G. Wilkinson, Some aspects of the alpine-type serpentinites of Queensland, *Geol. Mag.*, vol. 90, pp. 305–321, 1953.

that extends for 1,600 miles along the Appalachian axis of eastern United States from the St. Lawrence estuary to North Carolina (Fig. 49). In the latter state alone, several hundred individual masses, rarely exceeding 100 yards in width, have been mapped.[46] No less impressive is the multitude of sills, sheets, and plugs of serpentinized peridotite that invade the Franciscan sediments of the Coast Range of California and Oregon.[47] Some of these are several thousand feet in width (thickness),

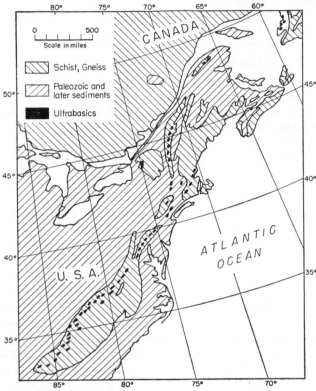

Fig. 49.   The Appalachian peridotite belt.   (*After J. M. Pratt and L. V. Lewis.*)

but most are much smaller. According to Hess,[48] ultramafic belts tend to be compound: each consists of two parallel linear series about 120 miles apart, located along opposite flanks of a central zone of intense orogenic deformation. He cites a similar disposition of serpentinite belts in island arcs such as those of the West Indies and the eastern Pacific, which are

[46] J. H. Pratt and J. V. Lewis, Corundum and the peridotites of North Carolina, *North Carolina Geol. Survey Repts.*, vol. 1, 1905.

[47] Cf. N. L. Taliaferro, Franciscan Knoxville problem, *Am. Assoc. Petroleum Geologists Bull.*, vol. 27, pp. 109–219, 1943 (especially pp. 153–158, 202–212).

[48] Hess, *op. cit.*, pp. 391, 394–395, 1955.

believed to represent the early stage of development of fold-mountain chains.

Broad concordance between the direction of elongation of tabular and lensoid ultramafic intrusions and the trend of bedding or foliation of the enclosing rocks is highly characteristic. The larger masses commonly are sills located along the junctions between stratigraphically distinguishable formations. Yet small-scale discordance between intrusive contacts and bedding is commonly prevalent. Thus Taliaferro,[49] writing of the larger sill-like bodies of serpentinite that invade the Franciscan sediments of California, states: "Although essentially concordant, the sill-like bodies, both large and small, commonly transgress the bedding and thin and thicken in short distances. It is not uncommon to find a single thick sill on one side of a syncline represented by a number of thin sills separated by included leaves of sediments on the other side."

Attention has already been drawn to the almost constant association of alpine peridotites and serpentinites with geosynclinal sediments (graywackes and cherts) and volcanics (basalts, spilites, keratophyres). Most writers agree that emplacement of the ultramafic bodies normally occurs during the earliest stage of folding which terminates sedimentation and ushers in orogeny.[50] Not infrequently the ultramafic rocks become completely converted to magnesian schists as the climax of deformation develops. It is not surprising, therefore, to find that major intrusions of peridotite and serpentinite tend to be located along zones of strong dislocation or at least to be bounded by faults of great magnitude. There are several possibilities here. The ultramafic "magma"[51] may have been squeezed up along major dislocations developed in the geosynclinal basement during the onset of folding as pictured by Suess, Benson, Hess, and others.[52] But when once a large ultramafic body many miles in length has been so emplaced, and especially when it has been largely converted to serpentinite, its steeply dipping boundary surfaces may become the locus of postorogenic faulting. Such is probably the case with some of the larger peridotite-serpentinite bodies of southwestern New Zealand, and it has been clearly demonstrated for the great serpentinite block of Unst in the Shetland Islands.[53] Under comparable conditions large

---

[49] Taliaferro, *op. cit.*, p. 153, 1943.

[50] Hess, *op. cit.*, pp. 391, 395, 1955.

[51] The term *magma* is here used for the time being to denote ultramafic igneous material injected at high temperature without prejudice as to whether it was largely liquid or largely crystalline at the time of intrusion.

[52] Cf. Benson, *op. cit.*, pp. 2, 75–76, 1926; Taliaferro, *op. cit.*, p. 153, 1943; H. H. Hess, Major structural features of the western North Pacific, *Geol. Soc. America Bull.*, vol. 59, pp. 432–433, 1948.

[53] H. H. Read, The metamorphic geology of Unst in the Shetland Islands, *Geol. Soc. London Quart. Jour.*, vol. 40, pp. 662–669, 1934.

bodies of serpentinite, traversed internally by innumerable surfaces of rupture and differential slip, may even migrate bodily upward along fault zones as "cold intrusions," carrying with them enclosed blocks of the adjacent country rocks. This type of serpentinite intrusion has long been recognized by Taliaferro and others in the Coast Ranges of California.

Mineralogy. In the ultramafic alpine plutonic association, peridotites and serpentinites greatly predominate over all other rock types. Small masses of gabbro and troctolite are not uncommon. Pyroxenites, though widely distributed, typically take the form of veins and narrow dikes cutting the dominant peridotite and serpentinite. Also commonly present, though always of insignificant total bulk, are dikes of diabase, diorite, albitite, and a variety of calc-silicate rocks (consisting of such minerals as grossularite, hydrogrossular, idocrase, clinozoisite, and zoisite). These latter are certainly of metamorphic (hydrothermal) origin.

The mineralogy of the unaltered peridotites is simple.[54] They consist principally of magnesian olivine, usually accompanied by enstatite and to a less extent by diopsidic augite; chromite or picotite is a constant accessory. Plagioclase, hornblende, and biotite typically are absent. Probably the commonest rock type is olivine-rich harzburgite, grading into dunite as the content of enstatite falls off. Traces of incipient shearing—undulose extinction in olivine, and lamellae in both olivine and enstatite—are almost invariably obvious, while strongly deformed schistose peridotites and even dunite-mylonites are by no means rare.[55] The banded fabric of some peridotites and serpentinites, rendered conspicuous by parallel streaks rich in chromite, may also reflect deformation of the nearly or completely crystalline rock. In other rocks—notably those of layered intrusions—it is rhythmic banding of primary origin.[56]

Careful microscopic scrutiny of most peridotites reveals at least incipient replacement of olivine by one of the serpentine minerals, and complete transition to pure serpentine rocks (serpentinites) is very common. In the absence of shearing the latter typically consist of a mesh of chrysotile veinlets enclosing cores of indefinitely crystalline serpentine (serpophite) or relics of undestroyed olivine. That replacement has proceeded according to the equal-volume principle, i.e., that no substantial change in vol-

---

[54] C. S. Ross, M. D. Foster, and A. T. Myers, Origin of dunites and of olivine-rich inclusions in basaltic rocks, *Am. Mineralogist*, vol. 39, pp. 704–719, 1954; Hess, *op. cit.*, p. 394, 1955.

[55] F. J. Turner, Preferred orientation of olivine crystals in peridotites, *Royal Soc. New Zealand Trans.*, vol. 72, pt. 3, pp. 280–300, 1942; C. E. Tilley, The dunite-mylonites of St. Paul's Rocks (Atlantic), *Am. Jour. Sci.*, vol. 245, pp. 488–491, 1947; Ross et al., *op. cit.*, p. 724, 1954.

[56] E.g., A. T. V. Rothstein, The Dawros peridotite, Connemara, Eire, *Geol. Soc. London Quart. Jour.*, vol. 108, pp. 9, 10, pl. 1, 2, 1957.

ume has been involved, is clear in many instances where outlines of individual grains of pyroxene and olivine, and other details of original igneous texture, still persists in serpentinite "pseudomorphs" after peridotite. In many alpine serpentinites there is evidence of strong deformation synchronous with or postdating serpentinization of the parent peridotite. The serpentine mineral in most such rocks is antigorite. Associated with these antigorite serpentinites there may be a considerable variety of antigorite-chlorite rocks, talc-tremolite-carbonate schists, nephrites, etc., some of which represent original dikes of pyroxenite. Others, especially talc and talc carbonate schists, are clearly products of local hydrothermal alteration of the main serpentinite body itself.

Metamorphism of Adjacent Rocks. In marked contrast with the striking effects of high-temperature contact metamorphism habitually shown in the vicinity of gabbro-norite intrusions, the metamorphic influence of even large bodies of peridotite of the alpine type usually is astonishingly slight. This is readily understood in the case of "cold intrusions" of serpentinite. But even such extensive masses of unaltered peridotites as are known in southern New Zealand have failed to convert the adjoining sediments and schists to high-temperature hornfelses such as are constantly found in the aureoles of granitic and gabbro intrusions. Instead there is but a meager development of such rocks as epidote hornfels corresponding to relatively low temperatures within the metamorphic range. This suggests a relatively low temperature of intrusion of peridotite magmas; but alternatively it might reflect some other condition, such as inward flow of water from the country rock into the hot intrusion.

In a few peridotite serpentinite provinces, of which the Franciscan of the California Coast Ranges is perhaps the most spectacular known example, sodic mineral assemblages of the glaucophane-schist facies occur locally as derivatives of the geosynclinal sediments and lavas within which the ultramafic bodies are found. These have very generally been attributed to hydrothermal metamorphism genetically connected with intrusion of the ultramafic magma. The puzzling metamorphic problems connected with this association will be referred to in Chap. 20. In the meantime certain points, possibly significant with regard to petrogenesis of ultramafic magmas, are noted.[57]

1. In general, glaucophane schists tend to occur in regions of folded geosynclinal rocks wherein ultrabasic and basic intrusives are also widely distributed. The Franciscan glaucophane schists of California afford a clear example of this association.

2. There are many geosynclinal terranes where glaucophane rocks are completely absent though intrusions of peridotite and serpentinite are

[57] Cf. Taliaferro, *op. cit.*, pp. 159–182, 1943.

conspicuous; *e.g.*, the peridotite belts of New Zealand and New South Wales.

3. In California, and in other provinces too, it has been established that some bodies of glaucophane schist have formed by local hydrothermal metamorphism of the invaded rocks (graywackes, cherts, basalts) near contacts with intrusive serpentinite. Elsewhere within the same provinces outcrops of glaucophane schist and outcrops of serpentinite are not obviously related. It seems as if neither type of rock need necessarily be accompanied by the other.

4. In California, glaucophane and quartz-jadeite schists, whether associated with serpentinite or not, commonly possess a foliated and at times lineated fabric indicative of deformation at the time of metamorphism. Elsewhere within the same broad region the generally unmetamorphosed Franciscan graywackes are traversed by shear zones showing distinctly foliated structure but no development of glaucophane.

The above generalizations are compatible with the widely accepted hypothesis that chemically active solutions, given off from intrusive ultramafic "magma" or from bodies of peridotite in process of serpentinization, are responsible for genesis of glaucophane schists whether at the intrusive contacts themselves or at more remote points situated on zones of dislocation. However, they are also compatible with the alternative possibility —which we tentatively accept—that glaucophane metamorphism and intrusion of ultramafic rocks are independent events in the early stages of orogeny in geosynclines.

Nature and Origin of the Ultramafic Magma. From the mineralogical and field evidence outlined in the preceding section, there can be little doubt as to the following points in connection with the petrogenesis of peridotites and serpentinites of the alpine type:

1. With the exception of "cold intrusions" of serpentinite, emplacement of the ultramafic bodies involved intrusion of highly magnesian ultrabasic "magma" along stratigraphically or structurally controlled surfaces of weakness in the invaded rocks.

2. The end product of consolidation of the intrusive "magma" in some cases (including some of the largest known ultramafic bodies) was dunite or dunite-harzburgite. It is highly likely that all serpentinite bodies of the alpine association at one stage in their history consisted largely of crystalline olivine, with pyroxenes (especially enstatite) as additional, but often minor, constituents.

3. Contact temperatures, even adjacent to large bodies now scarcely affected by serpentinization, correspond to the lower grades of metamorphism—probably not more and possibly less than 500°C.

This last conclusion, based on evidence afforded by the metamorphic condition of the invaded rocks, seems incompatible with any hypothesis

involving intrusion of a liquid or largely liquid magma; for under laboratory conditions magnesian olivines of the type that occurs in dunites begin
to melt at about 1600°C. and become completely molten only at 1800°C.[58]
Even allowing for possible lowering of temperature by several hundred degrees in the presence of water and additional silica, we are forced to conclude that peridotite melts can exist only at very high temperatures. Yet
rocks adjoining alpine peridotite intrusions, in spite of vigorous searching
by many observers, fail to reveal any indication that such temperatures
were ever approached. On these grounds we reject the classic hypothesis
of Vogt that peridotite liquids develop by remelting of olivine crystals
accumulating under gravity during the early stages of crystallization of
basaltic magma. Additional support to the conclusion that peridotite
liquids are never generated in the outer crust of the earth is afforded by
complete lack of lavas of corresponding composition.

In spite of weighty evidence pointing to nonexistence of peridotite
liquids, there are certain field relationships that at first sight are difficult
to reconcile with this doctrine. In
particular, Hess has drawn attention
to the occurrence, in rare instances,
of what may be interpreted as fine-
grained chilled border facies of
dunite intrusions and to the existence of narrow branching dikelets
of peridotite composed of fresh unstrained interlocking crystals of
olivine.[59] Other writers mention the occurrence of narrow pipes of

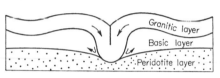

FIG. 50. Vertical transverse section
through a fold arc to illustrate H. H.
Hess' view of origin of ultrabasic magmas.
(*After H. H. Hess.*)

dunite cutting, and apparently intrusive, into pyroxenite, while many
recorded instances of enstatite pyroxenite veining peridotite suggest
at first that pure enstatite magmas may be capable of existence as
mobile liquids. Hess attempted to resolve the difficulties so raised by
assuming that the parent magma of peridotites and serpentinites is a highly
aqueous ultramafic liquid approximating to serpentine in composition.
He suggested, too, that such magmas are generated by differential fusion
of an assumed peridotite substratum under local impact of a very greatly
thickened segment of the overlying granitic crust where this is folded
downward at the onset of orogeny (cf. Fig. 50). This hypothesis would
explain a number of observed facts, *viz.*: absence of high-temperature
metamorphism at peridotite contacts; absence of peridotite lavas (on the

[58] N. L. Bowen and J. F. Schairer, The problem of the intrusion of dunite in the
light of the olivine diagram, *16th. Internat. Geol. Cong. Repts.* (1933), pp. 391–396,
1936.

[59] H. H. Hess, A primary peridotite magma, *Am. Jour. Sci.*, vol. 35, pp. 321–344,
1938.

assumption that peridotite magmas could retain their high water content only at high pressures); association of peridotite belts with arcs of negative gravity anomaly (? thickened granite) in active orogenic zones such as the island arcs of the East and West Indies.[60] However, the hypothesis stands on the assumption that aqueous ultramafic melts can be generated and sustained on a large scale at temperatures too low to permit noteworthy fusion of the impinging down-buckled granitic mass. Stimulated by the hypothesis of Hess, Bowen and Tuttle [61] have investigated the system $MgO\text{-}SiO_2\text{-}H_2O$ in the laboratory at temperatures ranging up to 900°C. and pressures as great as 30,000 lb./in.^{2} (corresponding to a depth of about 7 km.). No liquid phase was observed within this range of conditions, or even at 1000°C. and pressures up to 15,000 lb./in.^{2} In consequence Bowen and Tuttle conclude that "there is . . . no likelihood that any magma can exist that can be called a serpentine magma, and certainly no possibility of its existence below 1000°C." The serpentine magma hypothesis of Hess must be rejected as incompatible with experimental facts.

Over twenty years ago, Bowen [62] proposed a mechanism of peridotite intrusion which has been rather widely accepted. According to Bowen, peridotite "magmas" at the time of intrusion consist largely of olivine crystals. Gravitational settling of olivine, separating from basaltic magma, is a well-established mechanism capable of producing "magmas" of this type. But would accumulations of olivine crystals be sufficiently mobile to be capable of intrusion under plutonic conditions? Bowen appeals to the lubricating effect of small quantities of intergranular magmatic liquid, or even of water vapor, to impart the necessary degree of mobility. He cites, in support of this view, the occurrences of narrow borders of porphyritic olivine basalt margining peridotite dikes in Skye. Such cases admittedly are rare, and the Skye dikes do not exemplify peridotites of the alpine type. But they demonstrate that largely crystalline peridotite magmas are capable of injection in the presence of a small quantity of basaltic liquid. Moreover dunite and other peridotites commonly display fabric characters usually interpreted as due to deformation and flow of an essentially crystalline mass: undulose extinction in olivine, and in many rocks a banded or even typically mylonitic fabric.[63] These features suggest that olivine is a mineral susceptible to "plastic" deformation under plutonic conditions and that peridotites of the alpine type have

[60] H. H. Hess, Geological interpretation of data collected on cruise of USS *Barracuda* in the West Indies, *Am. Geophys. Union Trans.*, 18th Ann. Meeting, pp. 69–77, 1937.

[61] N. L. Bowen and O. F. Tuttle, The system $MgO\text{-}SiO_2\text{-}H_2O$, *Geol. Soc. America Bull.*, vol. 60, pp. 439–460, 1949.

[62] N. L. Bowen, *The Evolution of the Igneous Rocks,* p. 167, Princeton University Press, Princeton, N.J., 1928. Cf. also Bowen and Tuttle, *op. cit.*, pp. 454–457, 1949.

[63] F. C. Phillips, Mineral orientation in some olivine-rich rocks from Rum and Skye, *Geol. Mag.*, vol. 75, pp. 130–135, 1938; Turner, *op. cit.*, 1942.

commonly been affected by plastic deformation after consolidation. Laboratory experiments at temperatures and pressures corresponding to depths of about 10 miles tend to confirm this behavior. If it be further supposed that the slowly advancing crystalline peridotite mass absorbs water, especially peripherally, from invaded wet sediments and so becomes partially serpentinized, increased mobility of the intrusive body presumably would result.

Against Bowen's hypothesis it has been argued that if alpine peridotites represent a crystalline fraction differentiated from basaltic magma, they should be expected to be accompanied by other and more siliceous rocks representing the complementary liquid differentiate. Actually such complementary rocks typically are either absent or present in insignificant quantity. To take an individual instance, in the peridotite belt of northwestern Otago in Southern New Zealand, the exposed peridotites and serpentinites total some 300 cubic miles. At least 1,500 to 2,000 cubic miles of complementary differentiate of approximately the same age should be present somewhere in the vicinity. Only trivial amounts of other intrusive rocks (gabbros and diorites) actually are present. But here, as in similar provinces the world over, intrusion of the ultramafic bodies was preceded by outpouring of spilitic and other basic magma in vast quantity. It is at least conceivable, though in our opinion not likely, that the spilitic lavas, notable for consistent poverty in olivine, represent the complementary differentiate for which we are searching. Finally it is possible that some alpine peridotites represent mobile masses of crystalline olivine that have been separated from a peridotite substratum beneath the earth's crust and have been squeezed upward as intrusive bodies.[64]

An ingenious solution of the hitherto perplexing problem of origin of enstatite-pyroxenite veins cutting dunite, and of narrow veins of dunite in pyroxenite, has also been suggested by Bowen and Tuttle's recent work. Water vapor saturated with SiO_2 and streaming through a crack in dunite at temperatures above 650°C. could convert the wall rock to enstatite pyroxenite. The branching form and small width of such veins and the coarse grain of the component enstatite crystals are consistent with this mode of origin. Conversely pyroxenite could be converted locally to dunite by water undersaturated in SiO_2 at similar temperatures.

Serpentinization of Peridotite.[65] Many ultramafic intrusions of the alpine type are serpentinites. There can be no doubt that the component

---

[64] Bowen and Tuttle, *op. cit.*, p. 455, 1949; W. P. De Roever, Sind die Alpinotypen Peridotitmassen vielleicht tektonisch verfrachtete Bruchstücke der Peridotitschale?, *Geol. Rundschau*, vol. 46, pp. 137–146, 1957.

[65] W. N. Benson, The origin of serpentine, *Am. Jour. Sci.*, vol. 46, pp. 693–731, 1918; H. H. Hess, The problem of serpentinization and the origin of certain chrysotile asbestos talc and soapstone deposits, *Econ. Geology*, vol. 28, pp. 634–657, 1933; P.

serpentine minerals have been formed from olivine and pyroxene (enstatite), for innumerable instances of transition from peridotite to serpentinite have been recorded, and in many serpentinites there are relict grains of unaltered olivine and pyroxene or recognizable pseudomorphs of serpentine after one or other of these minerals. The numerous field, chemical, and petrographic data that must be explained by any satisfactory theory of serpentinization have been discussed in some detail by Benson and by Hess in the works cited. These cannot be reviewed fully here. Rather attention is drawn to certain of the more important generalizations as to localization, temperature, and time relations of serpentinization, and possible attendant volume changes, with a view to examining critically various current hypotheses.

Many ultramafic intrusions consist partly of peridotite and partly of serpentinite. In no case is the distribution of serpentinite in such bodies clearly related to the land surface or the water table. This condition is well exemplified in some of the great peridotite belts of southern New Zealand, where in some places fresh dunites outcrop along the mountain crests at elevations of 5,000 to 7,000 ft., while elsewhere deep postglacial canyons have been cut through nothing but serpentinites to a depth of several thousand feet. It is thus clear at the outset that serpentinization of peridotite is a process unrelated to weathering and allied surface phenomena. As regards distribution of serpentinite in relation to configuration of the intrusive mass, opinion is divided. Hess states that "serpentinization in most cases is either evenly distributed throughout the ultrabasic, or it has a haphazard distribution unrelated to the borders of the ultrabasic; but in a few cases serpentinization increases outward from the core of the ultrabasic." [66] Other writers consider peripheral serpentinization of peridotite to be a more general phenomenon than is admitted by Hess; certainly a number of clear instances have been described.[67] On the whole the relations of serpentinite to peridotite in space seem equally compatible with alternative mechanisms of serpentinization involving internally (i.e., magmatically) or externally derived waters respectively.

Serpentinization of olivine, at least to an incipient degree, is very commonly exhibited in a great variety of volcanic and plutonic rocks, including basalts, gabbros, picrites, and the peridotites of stratified lopoliths. In such cases the process is generally believed to have been accomplished

---

Haapala, On serpentine rocks in northern Karelia, *Comm. géol. Finlande Bull.*, no. 114, 1936; F. J. Turner, Mineralogical and structural evolution of metamorphic rocks, *Geol. Soc. America Mem. 30*, pp. 130–132, 1948; Bowen and Tuttle, *op. cit.*, 1949; Wilkinson, *op. cit.*, 1953; Hess, *op. cit.*, pp. 403–406, 1955.

[66] Hess, *op. cit.*, p. 649, 1933.

[67] Cf. Benson, *op. cit.*, pp. 702–705, 1918; Taliaferro, *op. cit.*, pp. 155–156, 1943.

by late-magmatic aqueous solutions acting on the still heated rock. Serpentinization of magnesian olivines of metamorphic rocks must certainly have taken place at temperatures no higher than a few hundred degrees. By analogy, serpentinization of peridotite bodies of the alpine type might well be attributed to activity of aqueous solutions upon moderately heated crystalline peridotite bodies during, or subsequent to, intrusion.

The experimental work of Bowen and Tuttle supports this general thesis. They have shown that a water-bearing magnesian olivine melt, cooled to 1000°C., would yield a mass of olivine crystals with water vapor occupying the intergranular spaces. This mass would cool without chemical change to about 400°C., when serpentine and brucite would begin to replace olivine so long as water remained available. The temperature at which serpentinization can begin is appreciably lower if the olivine contains iron and, in the case of iron-rich hortonolite, is probably so low that serpentinization of this mineral under plutonic conditions is perhaps impossible. Serpentine can form at temperatures as high as 500°C., either by action of pure water on olivine-enstatite mixtures, or from olivine alone if the aqueous solution is rich in CO_2 and so is capable of removing magnesia from the system. Above 500°C. olivine cannot be converted to serpentine by any means. In the presence of aqueous solutions capable of adding SiO_2 or removing MgO, olivines are liable to other types of alteration at higher temperatures:

1. Between 500° and 625°C., olivine → talc
2. Between 625° and 800°C., olivine → enstatite → talc
3. Above 800°C., olivine → enstatite

Before reviewing the various hypotheses of serpentinization in the light of these data, it is appropriate to note the implied volume relations. Serpentinization of olivine by simple addition of water and silica or carbon dioxide, without removal of magnesia, would involve great increase in volume, as illustrated by the classic equations given below:

$$2Mg_2SiO_4 + 2H_2O + CO_2 \rightarrow H_4Mg_3Si_2O_9 + MgCO_3$$

Olivine   Introduced   Serpentine   Magnesite
(280 gm.; 88 cc.)   (276 gm.; 110 cc.)   (84 gm.; 28 cc.)

and

$$3Mg_2SiO_4 + 4H_2O + SiO_2 \rightarrow 2H_4Mg_3Si_2O_9$$

Olivine   Introduced   Serpentine
(420 gm.; 131 cc.)   (552 gm.; 220 cc.)

The microscopic fabric and field relations of undeformed serpentinites show clearly, however, that serpentinization is commonly accompanied by little or no increase in volume. Equations such as those just cited cannot therefore represent truly the course of serpentinization of dunite. A more satisfactory type of reaction would be one in which olivine is replaced by the same volume of serpentine, the excess MgO and SiO$_2$ being removed in solution. This condition is approximated in the equation

$$5Mg_2SiO_4 \;+\; 4H_2O \rightarrow 2H_4Mg_3Si_2O_9 + 4MgO + SiO_2$$

Olivine	Introduced	Serpentine	(160 gm.) (60 gm.)
(700 gm.; 219 cc.)	(72 gm.)	(552 gm.; 220 cc.)	Removed in solution

For such a reaction to proceed, the combined concentration of MgO and SiO$_2$ in the aqueous solution leaving the system must not exceed some limiting value as shown by Eqs. (2–21) and (2–38). Great quantities of water would therefore have to be available. Thus if 700 gm. of olivine were converted to serpentine through the chemical activity of an equal weight of water, 72 gm. of water would remain as serpentine, and the remaining 628 gm. would have to remove 160 gm. MgO and 60 gm. SiO$_2$ from the system. Moreover if solutions so rich in magnesia and silica were continuously expelled from the ultramafic mass at a temperature of 200° or 300°C., noteworthy magnesia metasomatism of the adjoining rocks would be expected. Such effects are seldom conspicuous, though a number of instances of regional silicification in serpentinite belts have been recorded.[68] It is impossible to escape the conclusion that serpentinization of peridotite by equal-volume replacement demands great quantities of available water. The "serpentine magma" hypothesis of Hess avoided this difficulty, by assuming preliminary crystallization of olivine and subsequent reaction between olivine and an almost comparable volume of residual aqueous silicic acid solution to give serpentine:

$$3Mg_2SiO_4 + H_4SiO_4 + 2H_2O \rightarrow 2H_4Mg_3Si_2O_9$$

Olivine	(61 ± cc.)	(36 ± cc.)	Serpentine
(131 cc.)			(220 cc.)

But, as previously stated, Hess' hypothesis must be rejected as incompatible with recently established experimental data.

Bearing in mind the data of serpentinization and also the various possibilities regarding nature and origin of peridotite magma, petrologists today must choose between two alternatives:

1. Peridotite magmas are hydrous magnesian melts, possibly approaching serpentine in composition. Serpentinization is either a late-magmatic or a deuteric (autometasomatic) process—a reaction between still heated olivine and aqueous melts or solutions derived from the crystallizing magma. This is the view once upheld by Lodochnikow, Hess, and others.[69] It seems clearly untenable in the light of the experimental data recorded by Bowen and Tuttle.

[68] Osborne, *op. cit.*, pp. 64–67, 1950.

[69] W. Lodochnikow, Serpentines, serpentinites and the petrological problems connected with them, *Problems Soviet Geol.* 5, pp. 119–150, 1933; Hess, *op. cit.*, 1933; *op. cit.*, Am. Jour. Sci., 1938.

2. Peridotite "magmas" are composed largely of olivine and pyroxene crystals lubricated by interstitial magmatic liquid or water vapor. Serpentinization approximates to an equal-volume replacement and occurs at temperatures of perhaps 200° to 400°C. The necessary water for this reaction, together with dissolved silica and carbon dioxide, may be derived from various sources:

a. In cases of very incomplete serpentinization the relatively small amount of water concerned may be of magmatic origin, and serpentinization may be an autometasomatic process. This is the mechanism stressed especially by Benson [70] and widely accepted by many subsequent writers as an explanation for serpentinization on a large scale. Bowen and Tuttle have shown, however, that autometasomatism of peridotites should involve a rather complex sequence of replacements including conversion of enstatite to talc at high temperature and alteration of olivine to serpentine and brucite at temperatures below 400°C. The fact that olivine and enstatite very generally are both replaced by serpentine (enstatite being the more resistant of the two) suggests that autometasomatism is much less important than has hitherto been realized.[71] Where talc is pseudomorphous after enstatite, autometasomatism is more probable.

b. Serpentinization in some cases may be due to the action of extraneously introduced magmatic water, derived, for example, from nearby intrusive granites.[72] There seem, however, to be many instances (*e.g.*, the Franciscan serpentinites of California and the great peridotite serpentinite bodies of southern New Zealand) in which granites younger than the ultramafic intrusions are not available as a source of magmatic water.

c. Probably the greater part of the immense volume of water (and dissolved CO_2SiO_2, etc.) necessary for complete serpentinization of large ultramafic bodies is contributed by enclosing water-charged geosynclinal sediments, or by gases and solutions driven off and upward from similar rocks undergoing compaction, cementation, and metamorphism in the depths beneath, or even by an upward stream of juvenile water not connected with any magmatic source.[73] Ultramafic intrusions slowly injected along zones of major dislocation in geosynclinal terranes would be particularly liable to permeation by such solutions streaming upward along the same path of minimum resistance. We have already seen that solutions of similar origin may well have played a large part in converting ba-

[70] Benson, *op. cit.*, 1918.

[71] Bowen and Tuttle, *op. cit.*, p. 457, 1949.

[72] Cf. F. E. Keep, The geology of the Shabani mineral belt, *South Rhodesia Geol. Survey Bull.*, no. 12, 1929; T. DuRietz, Peridotites serpentines and soapstones of northern Sweden, *Geol. fören. Stockholm Förh.*, vol. 57, p. 255, 1935. Bowen and Tuttle, *op. cit.*, p. 457, 1949.

[73] Cf. Hess, *op. cit.*, pp. 656–657, 1933; Taliaferro, *op. cit.*, p. 154, 1943; Bowen, *op. cit.*, p. 271, 1947; Bowen and Tuttle, *op. cit.*, p. 455, 1949.

saltic rocks to spilites by soda metasomatism, and in development of glaucophane schists. Possibly significant in this connection is the relatively high content of chlorine and boron in some serpentinites,[74] and a rather marked tendency for tourmaline, axinite, and other boron-bearing minerals to be locally abundant in serpentine rocks. Chlorine and boron are present in insignificant amounts in unserpentinized peridotites but attain relatively high concentrations in sea water.

Growth of scientific theory by testing of rival hypotheses from all angles is well illustrated by two decades of investigation and discussion of the peridotite serpentinite problem since the appearance of Bowen's *Evolution of the Igneous Rocks*. The writings of Bowen and of Hess in this connection are particularly instructive.[75] Field and laboratory work have gradually strengthened the second of the two hypotheses outlined above, at the expense of the first. In 1938, Bowen, while favoring the hypothesis of intrusion of peridotites as crystal mush, cautiously summed up the position thus:[76]

The older concept of the existence in nature of highly magnesian liquid magmas is thus extraordinarily tenacious of life. Its vitality may be the vitality of truth itself. On the other hand it may be the result of a natural reluctance on the part of the observer to abandon opinions once held. In any case the problem is handed back to the student of phase equilibrium who is asked to determine equilibrium relations of such highly aqueous magnesian magmas. . . .

How well the student of phase equilibrium has succeeded in clarifying the problem is made obvious in the recent contribution by Bowen and Tuttle. We are now able to accept as a satisfactory working hypothesis the dual concept of intrusion of peridotite "magma" in a largely crystalline condition, with simultaneous or subsequent serpentinization of its constituent minerals (olivine and enstatite) through the activity of aqueous solutions or vapors derived for the most parts from surrounding geosynclinal sediments or from intrusive bodies of granitic magma. But this, like any other hypothesis, is subject to future modification or rejection should it prove incompatible with facts yet to be discovered.

THE PRE-CAMBRIAN ANORTHOSITE ASSOCIATION

General Characteristics. There are two types of anorthosite, each belonging to a distinct association of plutonic rocks:

[74] T. G. Sahama, Spurenelemente der Gesteine im südlichen Finnisch-Lappland, *Comm. géol. Finlande Bull.*, no. 135, pp. 32, 55, 1945; J. W. Earley, On chlorine in serpentinized dunite, *Am. Mineralogist*, vol. 43, pp. 148–155, 1958.

[75] Hess, *op. cit.*, 1933; *op. cit.*, 1937; *op. cit.*, *Am. Jour. Sci.*, 1938; *op. cit.* pp. 401–406, 1955; Bowen, *op. cit.*, 1928; *op. cit*, 1947; Mente et malleo atque catino, *Am. Mineralogist*, vol. 23, pp. 128–130, 1938; Bowen and Tuttle, *op. cit.*, 1949.

[76] Bowen, *op. cit.*, p. 129, 1938.

1. Bytownite anorthosites occurring as layers within stratified basic sheets and lopoliths, as for example in the Bushveld and Stillwater complexes.

2. Andesine or labradorite anorthosites occurring as large independent intrusions in pre-Cambrian terranes. The best known examples are those of Scandinavia and of eastern North America (the Adirondacks and Quebec).

The discussion which follows refers only to anorthosites of this second class. These consistently display a number of distinctive characteristics: [77]

1. The anorthosite bodies are limited to pre-Cambrian terranes.

2. They take the form of large intrusions with domed roofs and may reach batholithic proportions. The areal extent of the anorthosite massif of the Adirondacks is 1,200 square miles, while in southern Norway anorthosite covers about 1,000 km.^{2} within a continuous anorthosite granite complex of more than twice that area.[78]

3. The principal constituent mineral is plagioclase approximating to andesine-labradorite but ranging from An_{35} to An_{60}. Hypersthene and augite, less commonly accompanied by olivine, make up less than 10 per cent of the composition in anorthosite proper, 10 to $22\frac{1}{2}$ per cent in gabbroic anorthosite, and $22\frac{1}{2}$ to 35 per cent in anorthositic gabbro.

4. Many anorthosites are very coarse-grained. Cataclastic and locally mylonitic fabrics are common. Uniform shattering of individual component grains is widely prevalent.

5. Complete transition from anorthosite to noritic gabbro has been traced in most large bodies of anorthosite. There can be no doubt that these rocks have a common origin. Pyroxene-bearing granites (charnockites), syenites, and monzonites (mangerites) are also associated with anorthosites; but whether these have been derived from the same magmatic source as the anorthosites or whether their presence is due rather to independent intrusion of anorthositic and syenite-granite magmas in a particular environment is debatable.

6. Lavas of anorthositic composition are unknown. But there are rare instances of dike rocks exactly equivalent in composition to gabbroic anorthosite and possessing a porphyritic texture and fine-grained locally vesicular base which show that magmas of corresponding composition can

---

[77] Cf. R. A. Daly, *Igneous Rocks and the Depths of the Earth*, p. 411, McGraw-Hill, New York, 1933; A. F. Buddington, Adirondack igneous rocks and their metamorphism, *Geol. Soc. America Mem.* 7, pp. 208–209, 1939; D. W. Higgs, Anorthosite and related rocks of the San Gabriel Mountains, Southern California, *Univ. California Publ. Geol. Sci.*, vol. 30, no. 3, pp. 171–222, 1954.

[78] T. F. W. Barth, The large pre-Cambrian intrusive bodies in the southern part of Norway, *16th Internat. Geol. Cong. Rept.* (1933), pp. 297–309, 1936.

exist in a partially liquid highly mobile condition within the accessible portion of the earth's crust.[79]

The Anorthosite Massif of the Adirondacks. The great anorthosite massif of the Adirondacks, northern New York State, has been the subject of much detailed investigation which has culminated in the last two decades in the structural and petrographic studies of Balk and of Buddington.[80] It illustrates well the difficult problems of petrogenic interpretation that necessarily arise in connection with pre-Cambrian terranes where the sequence of geological events has been blurred by repeated igneous intrusion and metamorphism. However the Adirondack rocks display so well the characteristics of the plutonic association now under discussion that any satisfactory hypothesis of their origin and evolution must apply equally well to pre-Cambrian anorthosite massifs in general.

The main Adirondack intrusion invades highly metamorphosed sediments of the Grenville series and covers an area of 1,200 square miles. The anorthositic series, of which it is composed, is the earliest series of intrusive rocks known to be developed on a large scale in this region. According to Buddington,[81]

Both the main mass and the outlying intrusive rocks show a similar development of finer-grained border facies more mafic than the core, of segregations of more mafic rock containing included blocks of anorthosite, and locally a series of intrusions with similar sequence of composition in the order: anorthosite, gabbroic anorthosite, gabbro, and very mafic gabbro rich in apatite and ilmenite-magnetite, as conformable segregations, sheets, and dikes. All the rocks of the anorthositic series belong to the "saturated" class. The foliation of the main anorthosite body indicates several domical surfaces in the roof like that of a laccolithic group, but no positive direct evidence for a base has been found. Several anorthosite masses, one of considerable size, have the form of approximately conformable sheets or sills in the Grenville formations. The main anorthosite body created considerable disturbance of structure in its immediate vicinity and developed [within itself] a primary banding and flowage structure most obvious in the border facies.

At contacts with the Grenville formations, predominantly gabbroic anorthosites enclose numerous sheets and lenses of highly metamorphosed invaded rock. Extensive skarns are found at contacts with limestones of the Grenville series. In fact contact phenomena at the borders of the anorthosite body are strongly reminiscent of those familiarly developed in the marginal zones of granitic batholiths.

[79] H. von Eckermann, The anorthosite and kenningite of the Nordingrå-Rödö region, Geol. fören. Stockholm Förh., vol. 60, pp. 243–284, 1938.

[80] R. Balk, Structural survey of the Adirondack anorthosite, Jour. Geology, vol. 38, pp. 289–302, 1930; Buddington, op. cit., pp. 19–52, 201–230, 1939; Geology of the Saranac Quadrangle, New York, New York State Mus. Bull. 346, pp. 74–77, 1953.

[81] Buddington, op. cit., p. 1, 1939.

The complex series of events which followed emplacement of the anorthosites is reconstructed by Buddington as follows: intrusion of unsaturated olivine gabbros; intrusion of diorites (pyroxene diorite and biotite-quartz diorite); intrusion of pyroxene-quartz syenite magma which by differentiation yielded pyroxene-rich syenite, quartz syenite, and biotite granite; orogeny and folding of the Grenville series; intrusion of younger granites and far-reaching deformational metamorphism with which the cataclastic structure of the anorthosites, as well as other rocks, is correlated.

In Table 32, chemical analyses of Adirondack anorthosites are compared with those of similar rocks from other regions.

Petrogenesis, with Special Reference to Adirondack Anorthosites.   One of the principal difficulties attending petrogenic interpretation of anorthosites concerns the relation of these rocks to associated pyroxene-bearing plutonics, namely charnockites, pyroxene syenites, and noritic gabbros. Some petrologists, impressed by the common association of pre-Cambrian anorthosites with such rocks, recognize an anorthosite charnockite kindred or magmatic stem.[82]   In conformity with this concept, Kolderup and Barth consider the extensive anorthosites and closely associated bronzite granites of western and southern Norway to be comagmatic,[83] and in the case of the Adirondack massif, Bowen and Balk derived anorthosite and pyroxene-quartz syenite from the same parent magma.   Buddington, on the other hand, while recognizing the close affinity of Adirondack olivine gabbros and syenites with the charnockitic rocks of India and elsewhere, maintains that these have originated independently from the adjacent anorthosites, which are placed in an earlier cycle of igneous activity. From this divergence in interpretation of field data a corresponding divergence in petrogenic hypothesis arises.   The conflict cannot be resolved satisfactorily on the basis of evidence now available.   But it is perhaps well to remember at this stage that common mutual association of rocks is no infallible criterion of origin from a common source.   No one, for example, would consider graywacke and peridotite to be derivatives of a common magma merely because these two kinds of rock so frequently are found side by side in folded geosynclines.

Today opinion is still divided on the question of whether pre-Cambrian anorthosites are accumulations of plagioclase crystals separated from parent magmas of radically different (gabbroic) composition or whether

[82] *E.g.*, V. M. Goldschmidt, Stammestypen der Eruptivgesteine, *Norske Vidensk.-akad. Skr., I. Math.-Naturv. Kl., 10,* p. 8, 1922; G. W. Tyrrell, *The Principles of Petrology,* pp. 139, 140, Methuen, London, 1926.
[83] C. F. Kolderup, The anorthosites of western Norway, *16th Internat. Geol. Cong. Rept. (1933),* 1935; Barth, *op. cit.,* 1936.

they were injected as mobile, largely liquid feldspathic magma. Bowen [84] has argued against the possible existence of bodies of liquid anorthosite within the outer accessible crust of the earth. Absence of anorthositic lavas and the high temperature (about $1400°C$.) necessary to maintain

TABLE 32. CHEMICAL COMPOSITIONS OF ANORTHOSITES AND RELATED ROCKS

Constituent	1	2	3	4	5	6	7
SiO_2	54.54	53.34	53.22	45.25	53.40	52.13	54.8
TiO_2	0.67	0.72	0.69	6.88	0.77	0.50	1.3
Al_2O_3	25.61	22.50	20.03	11.84	23.96	24.15	25.2
Fe_2O_3	1.00	1.26	0.70	1.59	0.91	0.90	1.1
FeO	1.26	4.14	3.98	14.12	3.02	3.43	2.1
MnO		0.07	0.09	0.23			
MgO	1.03	2.21	4.08	6.42	1.88	2.42	1.0
CaO	9.92	10.12	12.33	10.23	9.85	10.36	8.7
Na_2O	4.58	3.79	3.57	2.14	4.17	4.31	5.4
K_2O	1.01	1.19	0.55	0.47	0.80	1.25	0.7
H_2O+	} 0.55		0.73	0.14	} 0.69	} 0.60*	
H_2O-			0.06	0.03			
P_2O_5		0.13	0.02	0.16	0.18	0.03	0.1
CO_2		0.41	0.38	0.26			
Rest					0.43	0.05	
Total	100.17	99.88	100.43	99.76	100.06	100.13	100.4

* Water content before recalculation = 2.87.

Explanation of Table 32

1. Average of 4 analyses of anorthosite (Marcy type) from core of Adirondack massif. (A. F. Buddington, Adirondack igneous rocks and their metamorphism, Geol. Soc. America Mem. 7, p. 30, A, 1939.)
2. Average of 7 analyses of gabbroic anorthosite (Whiteface type) from border zones of Adirondack massif. (Buddington, op. cit., p. 30, B, 1939.)
3. Anorthositic gabbro from sheet in Adirondack massif. (Buddington, op. cit., p. 36, No. 25, 1939.)
4. Pyroxene gabbro facies of coarse anorthositic gabbro, Adirondack massif. (Buddington, op. cit., p. 36, No. 27, 1939.)
5. Gabbroic anorthosite (composite sample of Whiteface facies), Adirondack massif; postulated as similar to primary anorthosite magma of Adirondacks. (Buddington, op. cit., p. 235, d, 1939.)
6. Kenningite (porphyritic dike rock of anorthositic composition), Sweden; average of 2 analyses recalculated to same water content as (5). (Buddington, op. cit., p. 235, e, 1939.)
7. Anorthosite, average of 2 analyses, southern Norway. [T. F. W. Barth, The large pre-Cambrian intrusive bodies in the southern part of Norway, 16th Internat. Geol. Cong. Rept. (1933), p. 301, No. 2, 1936.]

[84] Bowen, op. cit., 1928.

andesine-labradorite in a completely liquid state have been cited in support of this view.   It is generally agreed that "sedimentary" accumulation of plagioclase crystals in basic magmas plays a significant role in the origin of anorthosite layers in stratified lopoliths.   At first sight, therefore, a similar mechanism might seem possibly to have been effective in the evolution of large pre-Cambrian masses of anorthosite.   Moreover if gravitational accumulation of plagioclase were followed by filter-pressing of the resultant largely crystalline differentiate, possibly while in the course of intrusive flow, it would not be difficult to account for the almost monomineralic composition and persistently cataclastic fabric of anorthosites so formed, and granitic and syenitic associates could be regarded as formed from complementary residual liquids expelled during this operation.   This general mode of origin has been invoked by a number of recent writers, notably Bowen, Balk, and Barth.   The composition of the parent magma is variously considered to have been gabbroic (Bowen), dioritic (Balk), or granodioritic (Barth).

An entirely different hypothesis, put forward by Buddington,[85] is based in particular on observations on the Adirondack anorthosites but is also applicable to similar rocks of other pre-Cambrian terranes.   Of fundamental importance in this connection is Buddington's view that olivine gabbros, syenites, and granites are all of independent origin and postdate the period of anorthosite intrusion.   If this is the case, the only comagmatic associates of the Adirondack anorthosites are gabbroic anorthosites and saturated gabbros and norites.   All these rocks are regarded by Buddington as derivatives of a parent gabbroic anorthosite magma, the composition of which (cf. Table 32, analysis 5) would approximate to 80 per cent plagioclase ($Ab_{50}An_{50}$) plus 20 per cent hypersthene.   This could exist in the completely molten state at about 1350°C.   But presumably it would be mobile enough for intrusion at considerably lower temperatures if allowance were made for the presence of a small content of water and for partial crystallization.   This hypothesis affords a ready explanation of the great size of pre-Cambrian anorthosite bodies, of the injection phenomena observed at their contacts, and of the gabbroic anorthosite composition so consistently recorded in the border facies of the Adirondack mass.   It is significant, too, that the rare dike rock kenningite—the only anorthositic rock whose fabric clearly shows that it was injected in a largely liquid state—has a composition almost identical with that postulated by Buddington for his parent magma (cf. Table 32, analysis 6).   Complementary differentiates from gabbroic anorthosite magma would be on the one hand anorthosites with low pyroxene content, and on the other a series of saturated rocks ranging from gabbro and norite to highly mafic gabbro (Table 32, analyses 3 and 4).

[85] Buddington, op. cit., pp. 208–221, 1939.

During the past two decades new geological and physicochemical evidence has tended to support Buddington's interpretation of anorthosites as differentiates of liquid gabbroic anorthosite magmas. Higgs's studies of the San Gabriel anorthosite-norite mass of California [86] indicate an average bulk composition of 71 per cent andesine (An_{43}), 29 per cent dark minerals—anorthositic gabbro somewhat less feldspathic than the primary gabbroic anorthosite postulated by Buddington. The range of lithologic variation is almost identical with that in the Adirondack anorthosite massif, and there is no reason to postulate any parent magma other than one in the gabbro-anorthosite range. A parent corresponding to the average composition of the whole complex is the most likely. Both Buddington and Higgs invoked the influence of water and other "volatiles" to lower the temperature of gabbroic anorthosite magma to levels consistent with observed effects at anorthosite contacts. Subsequent experiments by Yoder [87] (see page 106) have shown that this effect is potent indeed. At 5,000 bars water pressure, such as could conceivably develop in plutonic magmas at depths greater than 10 miles, the melting range of pure plagioclase An_{50} is about 820° to 1100°C.; the water content of such a melt is about 10 per cent; the composition of the diopside-anorthite eutectic is at 73 per cent anorthite (as compared with 43 per cent in the anhydrous system), and its melting temperature is 1095°C. Allowing for substitution of hypersthene for diopside, an anorthositic gabbro melt approximating the andesine-hypersthene cotectic might be expected to remain completely liquid to temperatures below 1000°C.

To account for the origin of gabbroic anorthosite magma, Buddington suggests that the basic substratum of the earth's crust has a stratified structure analogous with that of stratified lopoliths. He pictures it as being composed, at least locally, of a gravitationally differentiated sequence of layers, viz.: quartz gabbro and gabbro; bytownite anorthosite; olivine gabbro and norite; pyroxenite peridotite and dunite. The gabbroic anorthosite magma is pictured as forming by differential fusion of the bytownite anorthosite layer and upward migration of the liquid fraction so developed. The fact that large anorthosite bodies are limited to the pre-Cambrian is attributed to the steeper geothermal gradient generally assumed for the pre-Cambrian lithosphere as compared with that of later times. In other words, it is thought that, since the Cambrian, temperatures in the bytownite anorthosite layer of the crust have never been high enough for extensive differential fusion. Buddington's hypothesis of

[86] Higgs, op. cit., pp. 197, 198, 1954.

[87] H. S. Yoder, Diopside-anorthite-water system at 5,000 bars, Geol. Soc. America Bull., vol. 68, pp. 1638, 1639, 1955; H. S. Yoder, D. B. Stewart, and J. R. Smith, in Annual report of the director of the Geophysical Laboratory, Carnegie Inst. Washington Year Book, 1955–56, pp. 190–194, 1956.

origin of anorthositic magma must be regarded as speculation unsupported by direct evidence, for there is no additional indication that the crust actually has a structure of the type which he pictures (cf. Chap. 15).

An alternative possibility is that anorthositic magmas originate at great depth, perhaps of the order of 200 km. It is conceivable that at correspondingly great pressures plagioclase of intermediate composition might melt at temperatures lower than either end member; so that melts rich in intermediate rather than in sodic plagioclase would be the first products of differential fusion of feldspathic peridotite.[88]   This, too, is speculation. But however obscure the origins of anorthositic magma may be, the existence of such magmas is compatible with a mass of data relating to occurrence, mineralogy, petrography and associations of pre-Cambrian anorthosites, and with experimental observations on hydrous plagioclase systems.[89]

[88] J. Verhoogen, Petrological evidence on temperature distribution in the mantle of the earth, *Am. Geophys. Union Trans.,* vol. 35, p. 88, 1954.

[89] Metasomatic origin has been suggested for anorthosite bodies in Idaho (A. Hietanen, Anorthosite in Boehls Butte Quadrangle, Idaho, *Geol. Soc. America Bull.,* vol. 67, p. 1770, 1956). The recorded evidence seems to the authors to be consistent with metamorphism of intrusive magmatic anorthosite.

CHAPTER 12

The Granite Granodiorite Plutonic Association

INTRODUCTORY STATEMENT

Granites and granodiorites are by far the most extensively developed rocks of igneous aspect in the plutonic-metamorphic complexes of pre-Cambrian terranes and eroded fold mountains. Chemically these rocks are approximately equivalent to rhyolites and dacites. Their mineralogy and coarse holocrystalline fabric suggest origin by slow crystallization of siliceous magma, and in many instances an intrusive relation to the surrounding rocks has been demonstrated. In orthodox German, British, and American classifications, granites and granodiorites accordingly have long been grouped with igneous rocks, and in this way a belief that all granites are products of crystallization of molten magma has become widely prevalent. Nevertheless, since at depths of a few kilometers, igneous and metamorphic processes merge into one another, and apparently may yield identical end products within a given range of temperature and pressure, it is not unreasonable to suppose that some granitic bodies may be partly or even entirely of metamorphic origin. This possibility was generally recognized a century ago and has always been favored by petrologists of the French school. Outside France, however, the concept of metamorphic granites was generally rejected following Rosenbusch's demonstration in 1877 of the intrusive magmatic origin of certain European granites, and it remained virtually forgotten for half a century. The past two decades have witnessed a significant revision of opinion on this subject. Impressive contributions by distinguished petrologists in all parts of the world and presidential addresses and symposia on "granitization" delivered before British and American geological societies bear witness to a resurgence of geological thought in favor of metamorphic (metasomatic) origin of at least some granitic rocks.[1]   Although British and American writers have made valuable contributions in this connection, it is amply clear from the reviews of Read and Holmes cited above that the modern revived concept of granitization derives mainly from the work of the French and Scandinavian schools of petrology.[2]

Throughout this chapter the term granite will be used in a broad sense to include holocrystalline coarse-grained rocks of plutonic aspect (but not necessarily of igneous origin), composed essentially of quartz, potash feldspar and/or sodic plagioclase, and subordinate biotite, hornblende, or pyroxene, and having a hypidomorphic-granular texture.   The authors believe that such rocks may originate in more than one way.   Some granitic bodies are clearly intrusive into the rocks with which they are associated.   They may be classed as igneous in the broad sense, for the intrusive relation implies the former existence of a liquid phase (silicate melt) permeating the mass in sufficient quantity to impart general mobility necessary for active intrusion.   Such a mass, whether the liquid phase is subordinate or predominant, conforms to our earlier definition of magma.   Where a granitic body can be shown to be of igneous origin, we must ask ourselves what the nature of the mobile magma was at the time of intrusion, how it came into being, and how the space now occupied by granite was provided.

Other granitic bodies are not obviously intrusive into the adjoining rocks.   Contacts are blurred and gradational.   These rocks may possibly be products of metamorphisim (metasomatism) of preexisting solid rocks *in situ.*   In these cases we must examine critically the evidence for metamorphic versus igneous origin, and if we accept the former alternative, we must ask ourselves what chemical changes have occurred, by what

---

[1] The history of development of ideas on origin of granites has been ably summed up by H. H. Read (Meditations on granite, *Geologists' Assoc. Proc.,* vol. 54, pp. 64–85, 1943; vol. 55, pp. 45–93, 1944; *The Granite Controversy,* esp. pp. xi–xix, Murby, London, 1957) and by A. Holmes (Natural history of granite, *Nature,* vol. 155, pp. 412–415, 1945).   Other papers (in English) dealing with the problem include the following: P. Eskola, On the differential anatexis of rocks, *Comm. géol. Finlande Bull. 103,* pp. 12–25, 1933; M. MacGregor and G. Wilson, On granitization and related processes, *Geol. Mag.,* vol. 76, pp. 193–215, 1939; H. H. Read, This subject of granite, *Sci. Progress,* vol. 34, pp. 659–669, 1946; D. L. Reynolds, The sequence of geochemical changes leading to granitization, *Geol. Soc. London Quart. Jour.,* vol. 102, pp. 389–446, 1946; The association of basic "fronts" with granitization, *Sci. Progress,* vol. 35, pp. 205–219, 1947; H. H. Read, A. F. Buddington, F. F. Grout, G. E. Goodspeed, and N. L. Bowen, Origin of granite, *Geol. Soc. Amer. Mem. 28,* 1947; M. Walton, The emplacement of granite, *Am. Jour. Sci.,* vol. 253, pp. 1–18, 1955.

[2] A summary of contributions by Scandinavian, French, and German authorities has been given by T. F. W. Barth (Recent contributions to the granite problem, *Jour. Geology,* vol. 56, pp. 235–240, 1948).

means (a moving aqueous fluid, a diffusing cloud of ions?) these changes were effected, whence came the wave of matter introduced into the granitic body, and what has become of the complementary material so replaced and expelled from the granitized mass. It will be noted that here, too, there is a space problem. In this case, however, space must have been provided not for the whole volume of granite, but for such material as was removed from the mass during granitization.

Rastall [3] has grouped granitic bodies into three categories:

1. The enormous, usually gneissic masses of the pre-Cambrian continental shields
2. Core-batholiths in fold-mountain ranges
3. Intrusive granitic masses of relatively small size, e.g., dikes, ring dikes, laccoliths, sills, etc.

The first he attributed to regional metasomatism, the second to fusion at the bases of geosynclines, and the third to injection of granitic magma in a largely liquid condition. This seems too simple a conception of an exceedingly complex set of phenomena. The three classes of plutons represent successive stages in the Granite Series of Read.[4] Evidence of both magmatic intrusion and metasomatic replacement may be found in the contact zones of many granitic bodies; but whereas the former is clearly dominant in the minor intrusions of Rastall's third group, the imprint of metasomatism (granitization) tends to be more conspicuous among the great bodies of granite and granodiorite exposed in eroded fold mountains and especially in those of the pre-Cambrian basement. However, before the petrogenesis of granites can profitably be explored, we must review the essential features of each class of granite.

GRANITES OF "MINOR INTRUSIONS"

Preliminary Statement. In this section the term "minor intrusions" is used to cover the third of Rastall's three groups. Some masses belonging to this category are of considerable extent, but they are not comparable in size to the great batholiths of fold-mountain cores and pre-Cambrian terranes, and they appear to have consolidated at relatively shallow depth —in some cases less than a mile below the surface. It is in the minor intrusions of granite and granodiorite that evidence of magmatic origin

[3] R. H. Rastall, The granite problem, *Geol. Mag.*, vol. 82, p. 28, 1945. In an important review which appeared recently, A. F. Buddington (Granite emplacement, *Geol. Soc. America Bull.* vol. 70, pp. 671–748, 1959) treats granitic plutons in three categories corresponding more or less with those of Rastall. These he correlates, respectively, with the kata, meso, and epi depth zones of the crust.

[4] H. H. Read, A contemplation of time in plutonism, *Geol. Soc. London Quart. Jour.*, vol. 105, pp. 324–337, 1949.

is most clearly and consistently displayed. The nature of this evidence is illustrated below by referring to a limited number of specific instances.

Granitic Rocks in the Tertiary Volcanic Province of Western Scotland and Northern Ireland. The Tertiary volcanic rocks of western Scotland and northern Ireland (a part of the extensive Brito-Arctic province) have already been discussed on pages 223 to 226. It will be recalled that the predominant igneous rocks of the region are basaltic lavas, but that acid magma was also widely available, especially in the later stages of igneous activity, and today is represented by numerous dikes, sheets, flows, and explosion breccias. General prevalence of glassy and fine-grained types among these rhyolitic rocks bears witness to the largely liquid condition of the acid magma at the time of intrusion or extrusion.

Chemically identical rocks of plutonic habit (granites) are by no means rare, especially as components of ring-dike complexes. Unless we reject completely the apparently well-established concept of the intrusive igneous nature of ring dikes in general, we must admit that the granitic rings and bosses of such complexes were formed by intrusion of mobile, though not necessarily completely liquid, granitic magma. Some of the characteristic features of this type of intrusion are illustrated below by reference to two examples in the Scottish island of Arran: [5]

1. The northern granite of Arran occurs as a boss 7 to 8 miles in diameter, circular in plan. Boundaries against the enclosing schists are extremely regular and dip very steeply. In the schists themselves, the regional northeast trend of the foliation has been strikingly diverted near the granite boundary, to which it now lies parallel around almost the whole circumference. Dips in the schist are uniformly steep and away from the granite contact. The picture of a cylindrical mass of intrusive granitic magma mechanically punching its way upward through the schist cover is abundantly clear (cf. Fig. 51). Within distances as great as 100 yards from the contact, the schists have been indurated or converted to hornfelses containing andalusite, cordierite, or biotite. The Arran boss is a composite intrusion consisting of an outer ring of coarse biotite granite, enclosing an inner and later core of fine-grained granite of similar composition. Contacts between the two types may be remarkably sharp. At its outer boundary against the schists, the coarse granite passes locally into a porphyritic or fine-grained marginal phase, presumably resulting from chilling of a largely liquid magma. The whole composite boss is interpreted as the root region of a ring-dike complex.

2. An independent composite mass of granitic and basic plutonic rocks is the central ring complex of Arran. This is between 2 and 3 miles in

[5] G. W. Tyrrell, The geology of Arran, *Scotland Geol. Survey Mem.*, pp. 149–164, 188–194, 1928; W. R. Flett, The contacts between the granites of northern Arran, *Geol. Soc. Glasgow Trans.*, vol. 20, pp. 180–204, 1942.

diameter, roughly circular in outline, and represents a section through the upper levels of a ring-dike complex. The most conspicuous elements in the complex are a broad outer ring of biotite and biotite-hornblende granite, a central core of similar composition, and an intervening com-

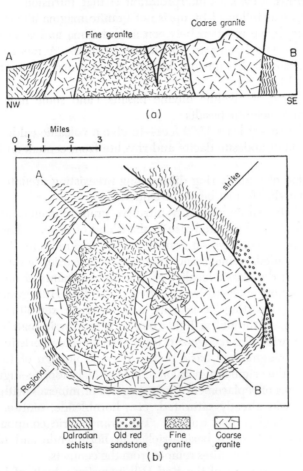

FIG. 51.   (a) Section across northern granite plug of Arran. Note regional southeast dip (on left) diverted to northwest dip against the margin of the plug; (b) simplified map of northern granite plug of Arran.

plete ring of explosion breccia. The composition of the biotite-hornblende granite is compared with that of an Arran pitchstone in Table 33 (analyses 1 and 2). If allowance is made for the high water content of the pitchstone, the two analyses are nearly identical. The outer granite ring locally passes into a fine-grained granophyric or felsitic border facies where it abuts against the neighboring invaded sandstones. Locally bordering and partly enclosed within the outer granite ring are masses of gabbro, freely veined with granite and passing into highly variable horn-blendic rocks (diorites and quartz diorites) within a few hundred yards of the granite contact. These have every appearance of being granite-gabbro hybrids. Tyrrell's interpretation is that intrusion of incomplete gabbro rings was followed by uprise of granite magma along arcuate fissures. Assimilative reaction between acid magma and solid gabbro has converted these in large part to dioritic hybrids. A more detailed sequence of events in the igneous history of the central complex has been reconstructed by King: [6]

1. Outpouring of alkaline olivine basalts (and some mugearites and trachytes) and tholeiitic basalts

2. Subsidence—at least 3,000 feet—to give a volcanic caldera

3. Eruption of andesite dacite and rhyolite from centers on the caldera floor

4. Intrusion of granitic ring dikes along preexisting arcuate faults connected with earlier formation of the caldera

Reaction between undoubtedly intrusive granitic magma and invaded rocks is beautifully illustrated also in the island of Skye.[7] The composite Red Hills intrusion of Skye—a granitic body 6 to 8 miles in diameter—has intricately invaded, veined, and brecciated an earlier series of gabbro intrusions. At the classic locality of Marsco,[8] remnants of one such dike of gabbro are now enclosed by a portion of the Red Hills granite. Intrusion breccias consisting of sharply angular dark xenoliths of altered gabbro, dispersed through a light-colored granophyric matrix, strikingly illustrate the principles recognized by Bowen as characteristic of reaction between acid magma and basic rock. There is no sign of fusion of the xenoliths, but the original mineral assemblage (labradorite-augite-olivine) shows all stages of replacement by aggregates of minerals with which the acid magma was already saturated, viz., hornblende, biotite, oligoclase, and minor orthoclase and quartz. The granophyric component of the intrusion breccia is more basic—richer in hornblende and oligoclase—than uncontaminated granites remote from the contacts.

A mile or so southeast of the Red Hill intrusion a body of hornblende-biotite granite 2 square miles in extent invades and is entirely surrounded by Cambrian dolomites.[9] Contacts, though intricate in outine, are

[6] B. C. King, The Ard Bheinn area of the central ring complex of Arran, Geol. Soc. London Quart. Jour., vol. 110, pp. 323–355, 1955.

[7] A. Harker, The Tertiary igneous rocks of Skye, Geol. Survey Scotland Mem., 1904 (especially pp. 126–144, 169–196).

[8] Ibid., pp. 175–187.

[9] Harker, op. cit., p. 137, 1904; C. E. Tilley, An alkali facies of granite at granite-dolomite contacts in Skye, Geol. Mag., vol. 86, pp. 7–13, 1949.

mainly vertical and are sharply defined. Brecciation is not evident and xenoliths almost totally lacking. Over wide areas in the vicinity of the granite, the dolomites have been converted to brucite and forsterite marbles by what appears to have been normal contact metamorphism. Locally there are narrow zones, an inch or two in width, of garnet- and magnetite-bearing skarns, defining the granite-marble contacts, and in such cases the immediately adjacent granite has been modified to an alkaline variant marked by great abundance of potash feldspar and by the presence of a somewhat sodic diopsidic augite and minor aegirine or arfvedsonite. The relatively low Al_2O_3 and Na_2O and high K_2O of the modified granite are attributed to differential transfer of Al_2O_3 from magma to dolomite, and accompanying copious crystallization of orthoclase within the magma itself.

In contrast with the dolomite contacts, junctions between the Red Hills granite and invaded Torridonian sandstone [10] are described as irregular on a small scale. The granite body frays out into tapering sheets and veins which "dovetail into the strata." "Not infrequently it is impossible in a hand-specimen to draw any sharp line between the two rocks. . . . A slice shows that the intruded magma has insinuated itself for a short distance into the interstices between the sand grains, and the clastic grains of quartz are seen embedded in a delicate micropegmatite." It would seem that at the borders of an intrusive body of granitic magma, the mineral composition and fabric of certain rock types (in this case sandstones) may render these susceptible to *local* granitization. Complete transition between granite and adjacent country rock is thus no certain criterion of metasomatic origin for the granite body as a whole.

The Tertiary Granites of the Central Montana Province. The Tertiary petrographic province of central Montana as defined by Larsen [11] is an area 200 by 400 miles in extent, some 15 per cent of which is occupied by outcrops of Tertiary igneous rocks. It is essentially a volcanic province, although rather shallow-seated intrusive bodies—laccoliths, sills, stocks, and dikes—are numerous. A score of subprovinces have been recognized. The chemical character of these ranges from typically calc-alkaline (*e.g.*, the basalt rhyolite association of Yellowstone and the diorite granodiorite stock of Crazy Mountains) to highly alkaline (*e.g.*, the potassic shonkinite nepheline-syenite association of Highwood Mountains discussed on pages 243 and 244). In spite of marked variation at the basic ends of the various igneous series, the respectve acid end members, whether rhyoltic flows or intrusive granites of plutonic habit, are more uniform in composition.

[10] Harker, *op. cit.*, p. 137, 1904.
[11] E. S. Larsen, Petrographic province of central Montana, *Geol. Soc. America Bull.*, vol. 51, pp. 887–948, 1940.

The occurrence of granitic rocks in central Montana is illustrated by referring to the Castle Mountain subprovince.[12]   The predominant rock is granite, with which are associated minor bodies of diorite, syenite, and lamprophyres, and remnants of rhyolitic and basaltic flows.   The principal igneous mass is a granitic stock (8 by 4½ miles in section) consisting of miarolitic biotite-hornblende granite with a decided tendency to porphyritic development of potash feldspar.   Near the margins this rock is distinctly finer in grain (granite porphyry) and passes gradually into quartz porphyries which locally form sheets injected parallel to the bedding of adjacent sediments.   The latter are strongly uptilted against the granite boundary and have been converted to normal hornfelses for distances which may be as much as a quarter of a mile from the contact.   As regards chemical composition, the granite is almost identical with quartz porphyries and rhyolites of the same province (Table 33, analyses 3 and 4).

Granite Rocks of the White Mountain Magma Series of New Hampshire and Vermont.[13]   The White Mountain magma series is a group of plutonic and volcanic rocks, probably Mississippian in age, developed throughout central and northern New Hampshire and adjacent parts of Vermont and Maine.   In this sector of the Appalachian highlands, Silurian and Devonian sediments were folded and metamorphosed in later Devonian times, with simultaneous emplacement of considerable bodies of granite and granodiorite.   After an interval of erosion, the uprise, intrusion, and extrusion of the White Mountain magma took place.

Rocks of the White Mountain magma series include remnants of very thick accumulations of rhyolitic, andesitic, and basaltic flows and pyroclastics, and numerous intrusive stocks and ring dikes of varied but dominantly acid plutonic rocks, some of which have decidedly alkaline affinities.   Biotite granites and somewhat alkaline granites with riebeckite or hastingsite are by far the most abundant (about 78 per cent) of the exposed plutonic rocks.   Syenite and quartz syenite make up a further 20 per cent, and the insignificant remainder (2 per cent) comprises small bodies of alkali syenite, monzonite, quartz diorite, diorite, norite, and

[12] W. H. Weed and L. V. Pirsson, Geology of the Castle Mountain mining district, Montana, *U.S. Geol. Survey Bull. 139*, pp. 58–65, 82–95, 1896; Larsen, *op. cit.*, pp. 901, 902, 944, 1940.

[13] L. Kingsley, Cauldron subsidence of the Ossipee Mountains, *Am. Jour. Sci.*, vol. 22, pp. 139–168, 1931; R. W. Chapman and C. R. Williams, Evolution of the White Mountain magma series, *Am. Mineralogist*, vol. 20, pp. 502–530, 1935; R. W. Chapman and C. A. Chapman, Cauldron subsidence at Ascutney Mountain, Vermont, *Geol. Soc. America Bull.*, vol. 51, pp. 191–212, 1940; M. P. Billings, Geology of the central area of the Ossipee Mountains earthquake, *Seismol. Soc. America Bull.*, vol. 32, pp. 83–92, 1942; *The Geology of New Hampshire, Pt. 2*, New Hampshire State Planning Commission, Concord, N. H., pp. 129–135, 1956; Buddington, *op. cit.*, p. 683, 1959.

gabbro. The dominant biotite granite and the quartz syenite are chemically almost identical with corresponding volcanic members of the series, namely, quartz porphyries and trachyte respectively (Table 33, analyses 5 to 8). This latter feature, the relatively sharp nature of exposed contacts, and a marked prevalence of ring complexes, together constitute strong evidence that the plutonic bodies were emplaced in the form of

TABLE 33. COMPOSITIONS OF ACID PLUTONIC AND VOLCANIC ROCKS FROM WESTERN ISLES OF SCOTLAND, CENTRAL MONTANA, AND NEW HAMPSHIRE

Constituent	1	2	3	4	5	6	7	8
SiO_2	75.65	73.20	72.48	74.90	72.5	73.7	65.2	65.05
TiO_2	0.28	0.16	0.32	0.15	0.4	N.d.	0.7	0.25
Al_2O_3	11.89	10.75	13.14	13.64	13.7	12.7	16.8	16.80
Fe_2O_3	1.19	0.95	1.66	0.66	0.7	1.3	1.3	4.97
FeO	1.02	1.02	1.02	0.50	1.3	1.4	2.7	1.12
MnO	0.26	0.37	Trace	Trace	0.1	0.4	0.3	Trace
MgO	0.15	0.15	0.15	Trace	0.2	Nil	0.4	0.20
CaO	0.91	0.76	1.04	0.61	1.1	0.7	1.7	1.68
Na_2O	3.44	3.78	4.22	4.22	4.0	3.8	4.7	3.94
K_2O	4.26	4.20	4.88	4.64	5.0	5.5	5.3	5.22
H_2O+	0.40	4.52 ⎱	0.42 ⎱	0.33 ⎱	0.6	0.4	0.7	0.30
H_2O-	0.41	0.18 ⎰	⎰	⎰	0.3	0.1	0.1	0.45
P_2O_5	0.16	0.19			0.1	N.d.	0.1	Trace
CO_2	0.09							
BaO	0.03	0.05						
Total	100.14	100.28	99.33	99.65	100.0	100.0	100.0	99.98

Explanation of Table 33

1. Granite of Central Ring-complex, Arran, Scotland. (G. W. Tyrrell, The geology of Arran, *Scotland Geol. Survey Mem.*, p. 192, No. 9, 1928.)
2. Pitchstone, sill, near Corrygills, Arran, Scotland. (Tyrrell, *op. cit.*, p. 234, No. 14, 1928.)
3. Granite, Elk Peak, Castle Mountain subprovince, Montana. (W. H. Weed and L. V. Pirsson, Geology of the Castle Mountain mining district, Montana, *U.S. Geol. Survey Bull. 139*, p. 84, 1896.)
4. Rhyolite, Fourmile Creek, Castle Mountain subprovince, Montana. (Weed and Pirsson, *op. cit.*, p. 120, 1896.)
5. Biotite granite of White Mountain magma series, New Hampshire; average of 4 analyses. (R. W. Chapman and C. R. Williams, Evolution of the White Mountain magma series, *Am. Mineralogist*, vol. 20, p. 502, Table 1, 1935.)
6. Quartz porphyry of White Mountain magma series, New Hampshire; average of 3 analyses. (Chapman and Williams, *op. cit.*, p. 505, Nos. 4, 5, 6, 1935.)
7. Quartz syenite (Albany type) of White Mountain magma series, New Hampshire; average of 2 analyses. (Chapman and Williams, *op. cit.*, p. 502, Table 1, 1935.)
8. Trachyte of White Mountain magma series, New Hampshire. (Chapman and Williams, *op. cit.*, p. 505, No. 3, 1935.)

highly mobile, largely or completely liquid magmas. In every instance the order of intrusion is from basic to acid, with the prevailing biotite granite last in the sequence.

It was upon observations at the complex syenite-granite stock of Ascutney Mountain, Vermont, that Daly based his original theory of intrusion of granite magma by stoping. Recent reexamination of the area revealed no evidence seriously opposing this hypothesis, but sinking of large dome-like blocks (cauldron subsidence) was shown to be the principal mechanism involved in intrusion. On the other hand, the hypothesis of forceful injection of magma was discarded as incompatible with field evidence.[14]

Summary. There are many recorded instances where "minor intrusions" (*e.g.*, ring dikes, stocks, and laccoliths) of acid plutonic rocks can have originated only by injection of mobile largely liquid magma at relatively shallow depths (in some cases probably less than a mile) below the earth's surface. Some individual intrusions must have occupied volumes at least as great as 10 or 15 cubic miles (*e.g.*, the Red Hills granites of Skye). The strictly igneous origin of rocks belonging to this group is confirmed by such features as chemical identity with acid lavas of the same province, prevalence of porphyritic and fine-grained marginal variants sharply bounded against the invaded rocks, and gradual transition from coarse granites to fine-grained rocks forming clearly intrusive sheets and dikes, injected into the rocks adjoining the main intrusion.

Commonly the dominant rocks are strictly granitic in composition (cf. Table 33). In some provinces, *e.g.*, the White Mountain series of New Hampshire, there may be considerable though subordinate quantities of other acid rocks such as quartz syenite, syenite, or quartz monzonite. Where basic rocks occur in dominantly granitic ring-dike complexes, *e.g.*, in the western isles of Scotland, intermediate rocks other than hybrids characteristically are lacking.

In some terranes circular bosses and ring complexes of relatively undeformed "younger granite" pierce a basement of migmatitic "older granite" or granodiorite. This is perfectly exemplified by the granites of Donegal in northwestern Eire.[15] Here the younger are not the direct

---

[14] Chapman and Chapman, *op. cit.*, 1940.

[15] W. S. Pitcher, The migmatitic older granodiorite of Thorr district, Co. Donegal, *Geol. Soc. London Quart Jour.*, vol. 108, pp. 413–446, 1953; The Russes granitic ring-complex, County Donegal, Eire, *Geologists' Assoc. Proc.*, vol. 64, pt. 3, pp. 153–182, 1953; S. V. P. Iyengar, W. S. Pitcher, and H. H. Read, The plutonic history of the Maas area, Co. Donegal, *Geol. Soc. London Quart. Jour.*, vol. 100, pp. 203–230, 1954; W. S. Pitcher and R. L. Cheesman, Summer field meeting in north-west Ireland, *Geologists' Assoc. Proc.*, vol. 65, pt. 4, pp. 348–350, 1954; E. H. T. Whitten, The Gola granite (Co. Donegal) and its regional setting, *Royal Irish Acad. Proc.*, vol. 58, sect. B, pp. 245–292, 1957.

derivatives of the older granites with which they now happen to be associated at the present level of erosion. Rather they exemplify an earlier and a later stage in the development of Read's Granite Series. If the earth's surface at the time of emplacement is taken as a datum level, clearly the environment in which the older granites evolved was much more deep-seated than the level of intrusion and crystallization of the associated younger granites.

Effects of thermal metamorphism without appreciable change in bulk composition are characteristically shown by the invaded rocks within a distance of a few feet to a few hundred yards of the intrusive contact. At the contact itself, more striking effects of reciprocal reaction between magma and invaded rocks may be obvious: skarns formed by introduction of silica alumina and iron into calcareous rocks; "granitized" modifications of sandstone and shale affected by introduced alkali and alumina; dioritic and granodioritic hybrids formed at the expense of basic rocks. The granite itself may be visibly contaminated in the contact zone as a complementary result of such reactions, e.g., in the case of alkaline granites adjacent to dolomites, or in that of monzonitic and other "basified" variants developed by reaction with basic igneous rocks or argillaceous sediments.

The problem of origin of large volumes of acid magma was raised in earlier discussion of the basalt-andesite-rhyolite association (pages 286 to 287). The same problem now arises in connection with intrusive granitic magmas. Differentiation of a basalt magma could ultimately give rise, through a series of intermediate steps, to a granitic differentiate comparable in composition to the analyzed granites and rhyolites of Table 33. But the quantities of granite observed in the various intrusive complexes described above are much too great, compared with the volume of associated basic rocks, to have originated in this manner, unless immense volumes of complementary basic and ultrabasic differentiates have always remained concealed in the depths beneath. Moreover, there is a striking lack of rocks of intermediate (e.g., dioritic) composition leading up to the development of a granitic differentiate. Once again—this time in the plutonic environment—we encounter evidence strongly suggestive of independent origins for acid and basic magmas.

GRANITES OF BATHOLITHS

General Statement. In the pre-Cambrian basement complexes and in orogenic zones, great masses of granite, often of batholithic proportions, have been emplaced at depths (perhaps ranging up to 20 km.) considerably greater than the typical depth of intrusion of the minor granitic bodies discussed in the preceding section. The surrounding rocks commonly show effects of regional metamorphism, local dislocation and deformation, and, in the vicinity of the granite itself, contact metamorphism which may involve obvious metasomatism. Deformation of the granitic body during and after emplacement is a common phenomenon, particularly in the border zones. Field relations between granite and adjacent rocks therefore tend to be complex and ambiguous. It is upon this situation that much of our present confusion regarding the nature of granite hinges. Different observers, variously prejudiced, are likely to interpret the same set of phenomena as products of totally different processes—magmatic injection, metasomatism, differential fusion, and so on.[16] It is thus exceedingly difficult, from objective perusal of the literature alone, to judge the relative merits of "magmatist" and "transformationist" views on the granite problem.

Although our attention is here focused primarily on granitic rocks of batholiths, the reader is reminded that there is no sharp line of demarcation between batholithic granites and either the granites of stocks and ring complexes or the regionally developed granites of Archaean shields. In general batholithic granites represent the middle range of Read's Granite Series. Some are relatively deep-seated migmatitic syntectonic bodies; others are more massive post-tectonic intrusions penetrating, at the time of emplacement, to within a few kilometers of the earth's surface. Many are composite masses consisting of many small bodies of different composition emplaced within a relatively small span of time.

The aim of the account which follows is to illustrate some of the data which must be correlated in any satisfactory hypothesis of origins of granite. Illustrations are drawn from a few selected papers covering a range of granitic types and presenting the data in such detail that ambiguity of interpretation is reduced to a minimum. The data in question relate to form and size of the granitic bodies, the chemical and mineralogical composition and fabric of the component rocks, field relations between granites and adjacent rocks, and such changes in mineralogy, composition, and fabric of the adjoining rocks as can be correlated with proximity to granite masses.

Form and Size of Major Granite Bodies. Any intrusion whose surface outcrop exceeds 40 square miles may be termed a batholith; but many of the granitic bodies now under consideration are of immensely greater size. For example, erosion of the western fold ranges of North America has revealed a series of vast batholiths of late Mesozoic age, of which the most extensive are the Coast Range batholith of Alaska and British Columbia (1,100 by 80 to 120 miles), the Sierra Nevada batholith of

---

[16] For example, see the discussion following B. C. King's paper, "The textural features of the granites and invaded rocks of the Singo batholith of Uganda," *Geol. Soc. London Quart. Jour.,* vol. 103, pp. 37–64, 1947 (especially pp. 57–64).

California (400 by 40 to 70 miles), and the partially uncovered batholith of southern and lower California (probably 1,000 by 70 miles). Where, as in the fiords of British Columbia and in the High Sierra region of California, the topographic relief is extreme, the acid plutonic rocks which make up these enormous masses extend vertically through distances of the order of 5,000 to 8,000 ft. without any sign of downward narrowing or of the possible existence of a floor. Contacts between plutonic and adjoining rocks are vertical or steeply dipping through distances measured in thousands of feet. These western American granites and granodiorites appear to represent the upper portions of batholithic masses, for large remnants of the roof rocks, as well as smaller pendants enclosed completely by the granitic rocks, are numerous.

In regions of low to moderate relief, especially in the deeply eroded pre-Cambrian shields, the three-dimensional picture of the granitic bodies is more difficult to reconstruct. Extensive areas of granite are interspersed with expanses of folded and metamorphosed country rock. The form of any granitic body, especially as regards its configuration in depth, must be inferred from the detailed map of its outcrop and from observations of its internal fabric (foliation, lineation, jointing) in relation to structure of the surrounding rocks. Clearly there is room for divergence of opinion in such cases. Thus, H. Cloos assumes that the internal fabric of the granite gives a kinematic picture of movements taking place within the plutonic body during its emplacement, and so deduces that at least some granitic "batholiths," e.g., the Variscan (Hercynian) massifs of Germany, are composite irregular sheets and wedges floored at no great depth below the surface. This interpretation, based as it is upon hypothesis, is not universally accepted. But it must be admitted that Cloos and his co-workers have collected a mass of evidence suggesting, though not proving, that many granitic bodies of fold-mountain ranges are more or less stratiform and do not extend indefinitely downward. Similar conclusions have been reached by other workers by mapping in detail the contacts between granite and folded country rock. Thus in the Adirondack pre-Cambrian complex of New York State, Buddington deduces lensoid (phacolithic) and sheetlike forms for various bodies of granite and quartz syenite that occupy 1,800 out of the 2,150 square miles which he terms the "highland igneous complex." [17]

It seems likely, from direct observations in regions of high relief, that the upper surfaces of batholiths are typically irregular. Steep-sided protuberances (bosses) may project upward thousands of feet beyond the mean level of the roof, while swarms of apophyses may fray out from the main granitic body into the adjacent rocks. It is customary to interpret

[17] A. F. Buddington, Origin of the granitic rocks of the Northwest Adirondacks, Geol. Soc. America Mem. 28, pp. 21–43, 1948.

small granitic bodies in regions of low relief in the light of these facts. Groups of independent bosses such as the granitic bodies of southwest England, and migmatite zones wherein metamorphosed country rock is intimately veined and streaked with granite, are both commonly interpreted as upward extensions fringing a continuous unexposed larger intrusion in the depths beneath.[18]   But, as we shall see later, other interpretations of these phenomena are possible.

Composition of Major Bodies of Granitic Rocks. *Preliminary Statement.* Most "granitic" batholiths that have been mapped in detail prove to be heterogeneous. While typical granites may predominate in some of these bodies, others are composed principally of granodiorite and tonalite. Less commonly the plutonic suite has a more alkaline character and includes syenites and minor nepheline syenites as well as granites. The range of composition in some typical granitic associations of large plutonic bodies is illustrated below by reference to particular cases.

Granodiorite Tonalite Type. Granodiorites and tonalites—rocks in which potash feldspar is subordinate to calcic oligoclase or andesine or is completely lacking—are very conspicuous in the batholiths of western North America. Larsen [19] gives the following respective percentage areas occupied by the main petrographic types which make up the Southern California batholith in a sample area of 860 square miles:

1. Gabbros, 14 per cent. Mainly hornblende-rich gabbros, including some quartz-biotite-bearing types; less abundant but still noteworthy hornblende-poor gabbros and norites with plentiful pyroxenes or olivine.

2. Tonalites, 50 per cent.[20] The composition of the most extensive variety (Bonsall tonalite) is andesine 55 to 60 per cent, quartz 20 to 25 per cent, hornblende 10 per cent, biotite 10 per cent, minor orthoclase.

3. Granodiorite, 34 per cent. The predominant type (Woodson Mountain type) has a mean composition: basic oligoclase 41 per cent, quartz 33 per cent, microperthite 20 per cent, biotite 5 per cent, hornblende 1 per cent.

4. Granite, 2½ per cent. The most widely developed type (Roblar granite) has a composition: quartz 39 per cent, perthite 35 per cent, medium oligoclase 22 per cent, biotite 2 per cent.

---

[18] This kind of relation has been well demonstrated for the "younger granites" of Donegal. Here one of several cylindrical bosses grades continuously in a lateral sense into an elongate intrusion 26 miles long (Pitcher and Cheesman, *op. cit.*, pp. 347–349, 1954).

[19] E. S. Larsen, Batholith of Southern California, *Geol. Soc. America Mem. 29,* 1948 (especially p. 138).

[20] Statistical analysis of variation in modal composition throughout the batholith strongly suggests that the tonalite is a hybrid formed by reaction between granite magma and solidified gabbro (F. Chayes, in Annual report of the director of the Geophysical Laboratory, *Carnegie Inst. Washington Year Book,* no. 55, pp. 215–216, 1956).

Chemical compositions of typical rocks of this batholith are given in Table 34, analyses 1 to 3.

The Sierra Nevada batholith of California, like many other large bodies of granite, is composite in character. In a small sample area of 160 square miles Hamilton [21] has mapped ten distinct major plutons. These range from alaskite and granite to quartz diorite; but four-fifths of the exposed rocks are quartz monzonite or granodiorite.

Trondhjemite Mica-Diorite Type.[22] Closely allied to the above associations is one in which trondhjemites are associated with biotite diorites and even biotite norites. This suite is known particularly through Goldschmidt's accounts of its occurrence in connection with the Caledonian folding of southern Norway. The most characteristic rock, trondhjemite, is allied to granodiorite but is more siliceous and notably more sodic (Table 34, analysis 4). It consists essentially of oligoclase and quartz with minor biotite and potash feldspar.

Granite Type. True granites are the dominant rocks of many acid plutonic complexes. With them may be associated quartz syenite or quartz monzonite in subordinate quantity. Not infrequently the marginal zones of a granitic body may be of more basic composition—tonalite or granodiorite.

About 60 per cent of the rocks constituting the Highlands igneous complex (2,150 square miles) of the Adirondacks are synorogenic granites of this class.[23] The rest is a much older, preorogenic, charnockitic quartz syenite and remnants of metamorphic host rocks. The main types of granite are as follows:

1. Greatly predominant (about 80 per cent of the total granite) is a hornblende granite (Table 34, analysis 5) consisting of quartz (24 to 33 per cent), microperthite (61 to 66 per cent), hornblende (6 per cent), a little plagioclase and associated potash feldspar, and accessory biotite, iron ore, apatite, and zircon.

2. Mainly occurring as border facies to the hornblende granite—and believed to represent a felsitic differentiate derived therefrom—are greatly subordinate alaskites. A mean composition for 26 specimens [24] is quartz (34 per cent), microperthite (64 per cent), and accessory biotite, fluorite, and zircon. Modal analyses of 55 samples [25] show that most contain

---

[21] W. B. Hamilton, Variations in plutons of granitic rocks of the Huntington Lake area of the Sierra Nevada, California, *Geol. Soc. America Bull.*, vol. 67, pp. 1585–1598, 1956.

[22] V. M. Goldschmidt, Stammestypen der Eruptivgesteine, *Norske Vidensk.-akad. Skr., Math.-Naturv. Kl. 10*, p. 6, 1922.

[23] Buddington, *op. cit.*, pp. 23, 24, 30–43, 1948; Interrelated pre-Cambrian granitic rocks, northwestern Adirondacks, New York, *Geol. Soc. America Bull.*, vol. 68, pp. 291–306, 1957.

[24] Buddington, *op. cit.*, p. 31, 1948.

[25] Buddington, *op. cit.*, p. 295, 1957.

TABLE 34. CHEMICAL COMPOSITIONS OF GRANITIC AND ASSOCIATED ROCKS OF MAJOR PLUTONIC BODIES

Constituent	1	2	3	4	5	6	7	8	9	10
SiO_2	62.2	73.4	76.54	69.30	70.72	75.98	63	64.76	57.80	62.94
TiO_2	0.7	0.2	0.19	0.23	0.41	0.26	0.8	0.70	1.15	0.95
Al_2O_3	16.6	14.1	11.86	16.81	13.11	11.92	16	17.13	18.82	16.58
Fe_2O_3	1.4	0.7	0.59	0.28	1.85	1.24	2.7	1.87	1.60	2.45
FeO	4.5	1.7	1.22	1.26	1.97	1.26	3.1	1.25	3.50	1.93
MnO	0.06	0.02	0.03	Trace		0.04	0.1	0.19	0.14	0.14
MgO	2.7	0.4	0.30	1.08	0.50	0.29	1.0	0.33	1.48	0.92
CaO	5.7	2.1	1.10	3.34	1.36	0.18	3.3	1.48	3.72	2.50
Na_2O	3.4	3.4	3.06	6.00	3.35	3.67	4.5	5.80	6.48	5.58
K_2O	1.6	3.5	4.29	1.39	5.60	4.69	5.0	5.70	3.97	4.66
H_2O+		0.3	0.22	0.50	0.37	0.14		0.20	0.64	} 0.57
H_2O-	0.6				0.06	0.04		0.21	0.02	
P_2O_5	0.09		0.02	0.03	0.23	0.02	0.4	0.11	0.55	
CO_2				0.15		0.06			0.10	0.31
BaO	0.05	0.05				0.03		0.09	0.30°	0.16
Rest	0.09	0.14	0.06					0.19†	0.12	0.31‡
Total	99.69	100.01	99.48	100.37	99.53	99.82	99.9	100.01	100.39	100.00

° Including 0.13 SrO.   † Including 0.11 ZrO_2.   ‡ Including 0.08 F, 0.04 Cl.

Explanation of Table 34

1. Tonalite; average composition of Bonsall type, Southern California batholith. (E. S. Larsen, Batholith of Southern California, Geol. Soc. America Bull. 29, p. 66, No. 6, 1948.)
2. Granodiorite; average composition of Woodson Mountain type, Southern California batholith. (Larsen, op. cit., p. 80, No. 8, 1948.)
3. Granite; Roblar leucogranite, Southern California batholith. (Larsen, op. cit., p. 98, 1948.)
4. Trondhjemite, Trondhjem, Norway. (V. M. Goldschmidt, Kristiania Vidensk.-selsk. Skr., Math.-Naturv. Kl. 2, p. 75, 1916.)
5. Normal hornblende granite, main batholith of northwest Adirondacks. (A. F. Buddington, Geol. Soc. America Mem. 7, p. 138, No. 133, 1939.)
6. Hornblende alaskite, northwest Adirondacks. (A. F. Buddington, Genl. Soc. America Bull., vol. 68, p. 296, no. 4, 1957.)
7. Quartz syenite, Diana complex, northwest Adirondacks; average composition of mass (275 square miles) based on weighted average of several chemical analyses. (Buddington, op. cit., p. 103, Table 28, C, 1939.)
8. Nordmarkite (quartz-bearing soda syenite), Nordmarka, Oslo district, Norway. (W. C. Brögger, Norske Vidensk-akad. Oslo Skr., Math.-Naturv. Kl. 1, p. 87, 1933.)
9. Larvikite (augite-bearing soda syenite), Oslo district, Norway. (Brögger, op. cit., p. 59, 1933.)
10. Average plutonic rock of Oslo district, Norway, based on weighted average of chemical analyses. (T. F. W. Barth, The igneous complex of the Oslo region, Norske Vidensk.-akad. Oslo Skr., Math.-Naturv. Kl. 1944, 9, p. 18, Table 5, 1945.)

344

between 30 and 40 per cent quartz, while a very few contain as little as 22 or as much as 42 per cent. An analysis is given in Table 34 (analysis 6).

3. Microcline granites make up somewhat less than 20 per cent of the total granites. They consist typically of microcline (about 55 per cent), albitic plagioclase (10 to 15 per cent), quartz (25 to 35 per cent), minor biotite or hornblende, and a little magnetite.

Field evidence demonstrates an intrusive (presumably magmatic) origin for both hornblende granites and alaskites; and it is in just these rocks that the alkali feldspar crystallized as a single phase (now microperthite).

The microcline granites with two feldspars on the other hand seem to be products of granitization of preexisting metasediments and amphibolites, relict lenses and schlieren of which still persist, imparting to the rock a variable and heterogeneous composition.

Granite Alkali-Syenite Type. Quartz syenite not uncommonly is a local minor variant of granite. More rarely syenitic rocks are dominant in batholiths that also include granite and nepheline syenite as less plentiful contrasted associates.

The Oslo district of Norway, largely as a result of the long-continued researches of W. C. Brögger, is a classic region for occurrence of alkali syenites. According to a recent account by Barth [26] this region was flooded in Permian times by a series of trachytic (rhomb porphyry) lavas, basalts, and minor rhyolites, the remnants of which now cover 1,800 km.^{2} Of this area 80 per cent is occupied by the trachytic members. What are believed to have been feeders to the lava floods are a number of plugs of basic alkaline plutonic rocks (essexite series) of varied composition. Slightly younger plutonic rocks thought to belong to the same broad igneous epoch cover a much larger area—5,000 km.^{2} The principal types are the following:

	Per Cent
Augite syenites (larvikites)	33
Quartz-bearing soda syenites (nordmarkites)	28.5
Biotite granites	17
Soda granites	16
Mica syenites	4
Nepheline syenites	1.5

Predominance of syenites (65 per cent) among the plutonic rocks and of trachytic lavas of equivalent composition (80 per cent) in the corresponding volcanic series strongly suggests derivation of volcanic and plutonic

[26] T. F. W. Barth, The igneous rock complex of the Oslo region, II, *Norske Vidensk.-akad. Oslo Skr., Math.-Naturv. Kl. 1944, 9,* 1945.

rocks from a single source.   Magmatic origin for the syenite granite series of the Oslo district is thus indicated.   It is also of interest to note a close similarity between the average chemical composition computed for the Oslo plutonic complex and that computed for one of the charnockitic syenite complexes of the Adirondacks (Table 34, analyses 10 and 7). The syenites and granites of the Oslo district evidently consolidated at no great depth below the surface.   The cover was thin enough to collapse locally and so to permit development of ring-dike complexes like those of New Hampshire and western Scotland.[27]

Charnockite Type.   In every continent there are expanses of Archaean gneiss composed of mineral assemblages dominated by feldspars and pyroxenes and lacking micas and hornblende.   These rocks have been variously termed charnockites and granulites.   Some are undoubtedly metamorphic; these, whether of sedimentary or igneous parentage, may be collectively termed granulites (see page 553).   Other rocks—to which we shall refer as the charnockite series—seem to be direct products of magmatic crystallization under plutonic conditions.[28]   One chemical characteristic they have in common: lack of combined water.   This is generally attributed to crystallization at relatively high temperatures from essentially anhydrous magmas.   Charnockitic series of different regions differ chemically from one another in much the same way as do the various granitic magma types described above.   Two examples are cited below.

In the Musgrave Range of central Australia [29] intrusions of charnockitic rocks invade a metamorphic basement of metasediments (gneisses with garnet, sillimanite, cordierite, and spinel), acid feldspathic granulites and basic pyroxene granulites.   Four groups of intrusive charnockite are recognized:

1. Norites, pyroxenites, and anorthosites, constituting the oldest group

[27] C. Oftedahl, The igneous rock complex of the Oslo region, XIII, *Norske Vidensk.-akad. Oslo. Skr., Math.-Naturv. Kl. 1953, 3,* 1953; Buddington, *op. cit.,* pp. 683–684, 1959.

[28] There is a good deal of confusion as to the nomenclature and petrogenesis of granulites and charnockites.   Some writers think that even the intrusive rocks of plutonic aspect that are here termed the charnockite series have undergone high-temperature metamorphism or metasomatism after intrusion [*e.g.,* P. G. Cooray, The nature and occurrence of some charnockite rocks from Ceylon, *Pan-Indian Ocean Science Cong., Perth, 1954, Proc.,* Sect. C, pp. 52–58; R. A. Howie, The geochemistry of the charnockite series of Madras, India, *Royal Soc. Edinburgh Trans.,* vol. 62, pt, 3, pp. 725–768, 1955 (especially pp. 762–766)].

[29] A. F. Wilson, The charnockitic and associated rocks of north-western South Australia, Pt. I, *Royal Soc. South Australia Trans.,* vol. 71, pt. 2, pp. 195–211, 1947; The charnockite problem in Australia, *Sir Douglas Mawson Anniversary Volume,* pp. 203–224, University of Adelaide, 1952 (especially pp. 207–213); Charnockitic rocks in Australia—a review, *Pan-Indian Ocean Science Cong., Perth, 1954, Proc.,* Sect. C, pp. 10–12.

of the four.   The principal minerals are perthitic andesine, hypersthene, and less plentiful diopsidic augite.

2. Charnockitic adamellite (= quartz monzonite).

3. Ophitic norite: antiperthitic andesine-labradorite, hypersthene, diopsidic augite, and minor microperthite.

4. Hypersthene adamellite (= quartz monzonite) and granodiorite: andesine (50 per cent), microperthite (20 per cent), quartz (15 per cent), hypersthene, diopside.   In variants occurring in cupolas hornblende is plentiful.   There is clear field evidence that rocks of this group (the youngest of the four) were emplaced by magmatic intrusion.

Charnockitic rocks of quite different chemical affinities occur in a series of preorogenic intrusive complexes in the Adirondacks.[30]   The dominant rocks are quartz-pyroxene syenites (Table 34, analysis 7) chemically similar to the computed average magma (Table 34, analysis 10) of the Oslo syenite-granite association.   Details of two quartz-syenite complexes, of which only one (the Tupper Saranac) is strictly charnockitic are given below:

The average modal composition of the Diana complex [31] is feldspar (plagioclase mantled with microperthite) 76.5, quartz 10, augite 4.5, hornblende 3.5, iron ores 3.2, and accessories.   The main rock type is quartz-augite syenite, but there is an upper facies (interpreted as a differentiate) of hornblende-quartz syenite and hornblende granite.   While lack of hypersthene in the syenites and prevalence of hornblende in the granitic facies differentiate the rocks of the Diana complex from typical charnockites, there are clear indications nevertheless that the syenite magma was relatively anhydrous and crystallized at high temperature: [32] prevalence of pyroxene rather than amphibole; presence of accessory pyrrhotite instead of pyrite; high content of TiO_2 in magnetite; high-temperature metamorphic mineral assemblages at contacts.

The Tupper-Saranac complex of the same magma series is strictly charnockitic in its mineralogy.   The average modal composition is computed as feldspar (plagioclase mantled with microperthite) 70.5, quartz 9.2, augite 6.2, hypersthene 4, hornblende 3.8, iron ores 3.4, garnet 1.6, apatite 1, zircon 0.3.   Hornblende is confined to an upper facies of quartz syenite.   Chemically the rocks differ from those of the Diana complex in a higher content of FeO (6.89 compared with 3.1).   It is this character, attributed by Buddington [33] to contamination of the magma by reaction with amphibolite, which causes abundant development of pyroxene, and

[30] Buddington, op. cit., pp. 24–30, 1948; op. cit., pp. 303–305, 1957.

[31] Buddington, op. cit., p. 29, 1948.

[32] Buddington, op. cit., pp. 304, 305, 1957.

[33] Ibid., p. 30.

so imparts a distinctive charnockitic mineralogy to the Tupper-Saranac syenites.

Compositional Evidence of Magmatic Origin of Batholithic Granitic Rocks. Valuable evidence bearing on the problem of magmatic versus metasomatic origin for granitic rocks is afforded by the range of chemical composition encountered within large bodies of such rocks as well as by the average compositions of the dominant rock types. The nature of this evidence is as follows:

The general range of composition in the tonalite granodiorite granite series of major plutonic bodies is matched precisely by that of the volcanic andesite dacite rhyolite series. The plutonic syenites and nepheline syenites likewise match the volcanic trachytes and phonolites. Such chemical similarities are clearly brought out by comparing the variation diagrams for corresponding plutonic and volcanic associations (Fig. 52). They are difficult to reconcile with the thesis that the plutonic rocks are products of metasomatism of a mixed series of parent rocks while the volcanic rocks have crystallized from magmas. Within a given range of physical conditions the same mineral assemblage (*e.g.*, quartz, orthoclase, oligoclase, biotite) could develop by crystallization from a melt, by metasomatism, or by any other appropriate means. But it would be surprising indeed if such radically different mechanisms of petrogenesis were consistently to yield the same set of minerals *in identical proportions*, so that their respective end products (*e.g.*, dacite and granodiorite) should have the same bulk composition, even down to the minor constituents [34] such as the elements Mn, Cr, V, Ba, Li, Zr, and P. Since the andesite dacite rhyolite series is undoubtedly of magmatic origin, it is at least highly probable that similar magmas have played a major role in the genesis of their plutonic counterparts, the tonalites, granodiorites, and granites.

Acid plutonic rocks, like volcanic rhyolites and dacites, consistently display certain chemical features that have also been recognized as characteristic of those liquids which are stable at minimal temperatures in simple silicate systems that have been investigated in the laboratory. High values for the ratios FeO/MgO, Na_2O/CaO, and K_2O/CaO are characteristics of this nature. Even more significant is the ratio of quartz to alkali feldspars in granites [35] as compared with that in low-melting mixtures in the system $NaAlSi_3O_8$-$KAlSi_3O_8$-SiO_2. At high water pressures consistent with crystallization under plutonic conditions the liquidus surface in this system (Fig. 53) shows a low-temperature trough more or less parallel to the alkali feldspar join, and having a very steep wall on the

[34] Cf. S. R. Nockolds and R. Allen, The geochemistry of some igneous rock series, *Geochim. et Cosmochim. Acta,* vol. 4, pp. 105–142, 1953.

[35] F. Chayes, in Annual report of the director of the Geophysical Laboratory, *Carnegie Inst. Washington Year Book,* no. 55, pp. 210–214, 1956.

SiO₂ side.   Compositions along this trough have 25 to 40 per cent excess silica.   Chayes finds that the modal compositions of granites, quartz monzonites, and granodiorites of batholiths consistently fall in the low-temperature trough of Fig. 53.   Moreover the proportion orthoclase/albite in Daly's average syenite is that (S in Fig. 53) for which the melting

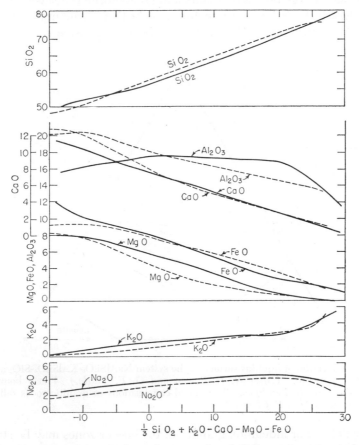

Fig. 52.   Variation diagrams for volcanic rocks of Yellowstone volcanic province (full lines) and plutonic rocks of the Lower California batholith (broken lines). (*After E. S. Larsen.*)

point has been found to be a minimum in the alkali-feldspar series.   In the light of such evidence, it can scarcely be denied that equilibrium between a liquid silicate melt (magmatic liquid) and crystalline phases such as alkali feldspars and quartz has played a dominant role in the evolution of acid plutonic rocks.   This does not imply that granitic rocks have necessarily crystallized from liquid residues of fractional crystallization.   Precisely similar chemical characters may be expected in acid melts formed by partial fusion of common sediments or igneous rocks. What is important regarding the present problem is that acid plutonic rocks have complex chemical characteristics of just the type to be expected in rocks crystallized from siliceous feldspathic magmas.

Structural Relation of Batholiths to Adjacent Rocks. *Introductory Statement.* Large bodies of granite and granodiorite differ widely among themselves in their internal structure and in their structural relations to adjacent country rocks. Contacts may be transgressive or concordant,

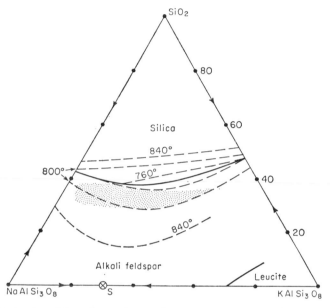

FIG. 53.   Projection of liquidus surface in the system $NaAlSi_3O_8$-$KAlSi_3O_8$-SiO_2 at 1,000 bars water pressure. (*After N. L. Bowen and O. F. Tuttle.*)   Phase boundaries, solid; isothermals, broken lines.   The modal compositions of granitic rocks fall in the stippled area.   (*After F. Chayes.*)

sharply defined or indefinite; granites of the border zones may be strongly foliated or virtually structureless; and so on.   On the whole, there is a tendency for particular structural features to be mutually associated, and on this basis various writers have divided granitic batholiths into two contrasted classes.[36]   Some of the features said to be characteristic of the two classes are as follows:

Class 1: Granitic rocks are highly variable in composition and structure. Outcrops are elongated parallel to the axis of contemporary folding;

[36] *E.g.,* W. R. Browne, Notes on batholiths and some of their implications, *Royal Soc. New South Wales Jour. and Proc.,* vol. 65, pp. 112–144, 1932; A. M. Macgregor, Batholiths of Southern Rhodesia, *Geol. Mag.,* vol. 69, pp. 18–29, 1932.

boundaries are broadly concordant with the structure of the country rock, though discordance in minor detail is not unusual. Near the borders the granite tends to be strongly foliated and may be crowded with aligned xenoliths of country rock. Roof pendants are numerous. Contacts commonly are gradational, apparently as a result of mutual exchange of material between granite and country rock. At the boundaries, granite may pass into a zone of mixed rock (migmatite) consisting of metamorphosed country rock veined and streaked with granite or pegmatite. Typically the country rocks bear the imprint of regional metamorphism.

Class 2: Granites are more uniform in character and are either massive or weakly foliated. Boundaries of batholiths cut across the structure of the enclosing rocks. Contacts tend to be sharp, and migmatite zones are inconspicuous or lacking. There is little direct evidence of exchange of material across the contacts. The country rocks typically have been converted to hornfelses by contact metamorphism, the degree of which is clearly related to proximity to granite boundaries.

This and similar twofold classifications of granitic batholiths reflect a long-sustained controversy regarding the mechanics of emplacement of these bodies.[37] The classic hypotheses of E. Suess, R. Daly, and H. Cloos all assume emplacement by intrusion of acid magma. Cloos and his followers have attributed the distinctive features of batholiths of class 1 to forceful injection of magma obliquely upward along inclined surfaces of structural weakness, the bordering rocks being gradually forced apart as intrusion continues. Such a mechanism would demand regional deformation (orogeny) contemporaneous with intrusion. On this assumption batholiths of class 1 have been termed *synchronous* (with respect to orogeny). They are early members of the Granite Series of Read. Totally different hypotheses of steady uprise of granitic magma without the urge of accompanying deformation have been developed by Suess and Daly, with special reference to batholiths of class 2, which may be regarded as later members of the Granite Series. Daly emphasized the mechanism of piecemeal stoping, whereby the magma is displaced upward, as blocks of roof rock break away and sink slowly into the depths. Suess, on the other hand, envisaged a process of fusion of the roof rocks whereby the ever-increasing mass of magma would bore its way upward through the roof rocks, a view which nowadays is not widely accepted in view of physicochemical considerations discussed on pages 156 to 160.

It is obvious that many variable factors may possibly influence the structural character of a granitic batholith, *e.g.*, mechanism of emplace-

---

[37] Cf. H. Cloos, Das Batholithenproblem, *Fortschr. Geologie u. Palaeontologie*, Ht. 1, 1923; Browne, *op. cit.*, 1932; R. A. Daly, *Igneous Rocks and the Depths of the Earth*, McGraw-Hill, New York, pp. 113–134, 267–286, 1933; R. Balk, Structural behavior of igneous rocks, *Geol. Soc. America Mem.* 5, pp. 121–129, 1937.

ment, time of emplacement in relation to orogeny, physical condition of the mobile materials (liquid magma, partially crystalline magma, "gaseous emanations," or "ionic clouds"), and depth at which the granite assumed its present condition. It is fairly generally agreed, for example, that shallow depth favors development of transgressive nonfoliated granites of class 2. In the face of this complexity of influences, it is scarcely likely that a simple twofold division of granitic batholiths should prove adequate. Some batholiths do indeed conform fairly well to all of the criteria listed as distinctive for one or other of the above two classes. Many more, however, combine some characteristics of both classes. The only satisfactory course is to consider a number of structural properties of batholiths individually and to treat each batholith as a special problem. Some of these properties will now be discussed briefly.

Concordance and Discordance of Boundary Surfaces. Most large batholiths are elongated in horizontal cross section. Where erosion has removed large areas of roof rocks, outcrops of large batholiths show a marked tendency for general elongation parallel to the regional tectonic axis of related folding. Some degree of concordance between the regional trend of granite boundaries and the general strike of adjacent rocks is therefore usual. For example, on a small-scale geological map of North America, there is obvious parallelism between the trend of the great batholiths of California and British Columbia and the north-northwest to northwest trend of adjacent rocks folded and metamorphosed in the preceding Nevadan orogeny. This very broad regional concordance of batholiths is modified, and in some cases outweighed, by an almost equally prevalent discordance in detail as revealed by geological mapping on a normal scale.[38] So general is this cross-cutting relationship that it is considered by Daly and others to be an essential character of batholiths in general. Presumably discordance is likely to be most marked in batholiths emplaced late in the orogenic cycle under relatively static tectonic conditions, i.e., in "subsequent" batholiths. Blackwelder and Baddley,[39] from scrutiny of a hundred reports on batholiths, concluded that in two-thirds of the cases they examined there was a transgressive relation between the boundary of the batholith and the foliation of enclosing metamorphic rocks; in only 10 per cent of the cases examined was the boundary substantially concordant with the foliation of the country rock.

Local folding, deflection, or other disturbance of bedded or schistose country rocks are commonly seen in the immediate vicinity of contacts

[38] E.g., W. H. Taubeneck, Geology of the Elkhorn Mountains, northeast Oregon: Bald Mountain batholith, *Geol. Soc. America Bull.*, vol. 68, pp. 181–238, 1957 (especially p. 189); Buddington, *op. cit.*, pp. 699–700, 1959.

[39] E. Blackwelder and E. Baddley, Relations between batholiths and schistosity, *Geol. Soc. America Bull.*, vol. 36, pp. 208, 209, 1925.

with granitic batholiths or their offshoot stocks.[40]   This strongly suggests forceful intrusion of mobile magma and cannot be reconciled with metasomatic emplacement of the granite mass.

Foliation and Lineation in Granitic Batholiths.   In most bodies of granite, and especially in their marginal portions, traces of parallel structure within the rock fabric may be discerned.   Parallel surfaces determined by alignment (preferred orientation) of tabular or prismatic crystals (mica, hornblende, feldspar) and elongate inclusions, or by segregation of particular minerals in alternating streaks and bands, will be referred to here as foliation surfaces or *s*-surfaces—descriptive terms which avoid the ambiguity that necessarily accompanies a genetic terminology. In some cases a tendency for linear parallelism of elongate crystals or aggregates in one particular direction gives rise to lineation within the plane of foliation.

Interpretation of parallel fabrics in rocks depends upon the well-justified assumption that such fabrics reflect a picture of internal movements associated with rock genesis.   Application of this principle to the parallel structure of bedded sediments, *e.g.*, reconstruction of the changing conditions of aqueous flow that accompanied deposition of a water-laid crossbedded sandstone, is familiar to every geologist.   Earlier in this book (pages 304 to 305) were noted comparable deductions as to the kinematic significance of rhythmic banding in certain gabbros.   In the realm of metamorphic geology (cf. pages 650 to 654) tectonic interpretation of foliation in deformed schistose rocks has been revolutionized by Sander's detailed application of this same principle.   In the case of foliated (gneissic) granites, certain ambiguities arise at the outset from varied possibilities as to whether bodies of granite have risen into place by upward flow of partially crystalline magma, or whether they represent products of metasomatism of solid rocks in place.   We must also bear in mind the possibility that the parallel fabrics of some granites may be due in part to deformation of solid granite after emplacement by one of the mechanisms just mentioned.

It is to the detailed field studies of Hans Cloos [41] and those who have

---

[40] *E.g.*, Taubeneck, *op. cit.*, pp. 187, 230, 1957; A. Knopf, The Boulder bathylith of Montana, *Am. Jour. Sci.*, vol. 255, pp. 81–103, 1957 (especially p. 88).

[41] For details of Cloos' methods and views the English-speaking reader is referred to the following works, in which many original references are cited: E. Cloos, The application of recent structural methods in the interpretation of the crystalline rocks of Maryland, *Maryland Geol. Survey*, vol. 13, pp. 36–49, 1937; Balk, *op. cit.*, 1937; M. P. Billings, *Structural Geology*, pp. 298–312, Prentice-Hall, New York, 1942; F. J. Turner, Mineralogical and structural evolution of the metamorphic rocks, *Geol. Soc. America Mem. 30*, pp. 311–315, 1948; N. R. Martin, The structure of the granite massif of Flamanville, Manche, north-west France, *Geol. Soc. London Quart. Jour.*, vol. 108, pp. 311–341, 1953; Buddington, *op. cit.*, pp. 697–699, 1959.

used his technique that we owe most of our knowledge of the internal structure of granitic bodies in relation to the structure of the adjacent country rocks. Cloos interprets the primary parallel fabric of granites (as distinct from any secondary foliation imprinted later by deformation of the solid rock) as due to flow of granitic magma during emplacement (intrusion) of the body in which it occurs. The foliation is assumed to be parallel to the surfaces of flow in the moving magma. Lineation in the marginal zones of the intrusion is interpreted as lying parallel to the direction of flow, but there is also the possibility, especially in the case of weaker lineation developed far from contacts, that a lineation may be perpendicular to the direction of flow. Mapping of joint planes commonly reveals the presence of persistent sets whose orientations are simply related to foliation and lineation. Particularly useful are open tension joints (Q joints) approximately perpendicular to both foliation and lineation. These have been interpreted as the result of tensions imposed during the last stages of consolidation of flowing magma, and finally relieved by rupture of the completely solidified rock. Of special significance with regard to mode of emplacement is the relation of foliation and lineation in granite to the structure of the country rock. Contacts are said to be conformable when respective structures of granite and wall rock are mutually parallel.

Application of the Cloos method to the study of many large bodies of granitic rocks has demonstrated wide prevalence of a limited number of internal structural patterns [42]—a fact suggesting considerable general regularity in the kinematics of granite emplacement. In most typical steep-sided batholiths, foliation and lineation are most pronounced in the marginal zones and both show a strong tendency to dip steeply parallel to the walls of the batholith. Transverse sections across the regional trend of many such bodies reveal a simple domed or arched structure with the foliation and lineation flattening out in the central portion of the mass. Most large batholiths (e.g., the Sierra Nevada mass [43]) are composite in structure, as well as in lithologic composition, in that more than one domed or arched structure can be seen in a complete transverse vertical section. The simplicity of the over-all picture of internal structure of many smaller batholiths, as contrasted with the complex structure often displayed by the adjoining country rocks, supports the view of Cloos that the internal fabric of the granite developed during upward surge of granitic magma. On the whole, it is incompatible with the hypothesis that granites are products of metasomatism in place.

[42] Cf. Balk, op. cit., pp. 54–56, 113–115, 1937.

[43] Cf. E. Cloos, Der Sierra Nevada Pluton in California, Neues Jahrb. Beilage-Band 76, Abt. B, pp. 355–450, 1936; E. B. Mayo, Sierra Nevada pluton and crustal movement, Jour. Geology, vol. 45, pp. 169–192, 1937.

In general, uprise of mobile partially crystallized magma seems to be the most satisfactory explanation of the internal structural patterns displayed by many granitic batholiths. In particular, prevalence of disconformable contacts is difficult to reconcile with metasomatism in place, but it is readily explained, especially where foliation of granite is parallel to irregular discordant contacts, on the hypothesis of emplacement of batholiths by magmatic flow. A fine illustration is afforded by the protoclastic border of the Colville batholith of Washington,[44] the intricately swirled foliation and locally cataclastic and even mylonitic character of which afford convincing evidence of upward flow of a granodioritic magma already in a very advanced stage of crystallization. The classic Hauzenberg massif of Bavaria [45] (Fig. 54) outcrops over an oval area of 40 square miles and cuts discordantly across the regular trend of foliation in surrounding metamorphic gneisses. The massif is composed of unfoliated but universally lineated granite. Lineation within the granite mass conforms to a simple arched structure with a central north-south axis, and pitches at angles of 25° to 32° at the east and west contacts. From detailed mapping of contacts and internal structure of the Hauzenberg mass, Cloos developed a hypothesis of intrusion of granite magma as a gently dipping sheet, expanding upward and outward as it flowed in a direction transverse to the trend of the lineation. Whether or not this brilliant deduction is generally accepted, it can scarcely be doubted that the internal structure of the granite massif reflects some simple pattern of magmatic flow—not metasomatism in place.

Discordance and disconformity of contacts are readily demonstrated in regions where small bodies of granitic rocks are enclosed in rocks which maintain a regular structural pattern over large areas.[46] The small bodies of quartz diorite and granite which belong to the same series of plutonic rocks as the Hauzenberg massif and outcrop through the Passau Forest to the west of it (Fig. 54) illustrate this condition well. Balk's summary [47] of the structural picture in this area is cited below in detail, for it seems incompatible with any mechanism of granite emplacement other than by intrusion of magma:

The discordant Paleozoic intrusions which penetrate an older, gneissic complex in the Passau Forest, Bavaria, consist of (1) an old series of quartz diorites,

[44] A. C. Waters and K. Krauskopf, Protoclastic border of the Colville Batholith, *Geol. Soc. America Bull.*, vol. 52, pp. 1355–1418, 1941.

[45] For a summary of the work by H. Cloos in this region, and for appropriate references, see Balk, *op. cit.*, pp. 76, 78, 1937.

[46] *E.g.*, E. Cloos, Fabric at granodiorite-schist contact, Bear Island, Maryland, *Tschermaks min. pet. Mitt.*, vol. 4, 1–4, pp. 81–89, 1954.

[47] For a summary of work by H. Cloos, E. Cloos, R. Balk, and H. Scholtz, 1927, see Balk, *op. cit.*, pp. 83, 84, 1937.

(2) an intermediate group of syenite-granites, and (3) a young series of granites. . . . The quartz diorites form discordant lenses and sheets, which dip to the northeast at angles varying from 20 to 60 degrees. The dip is in-

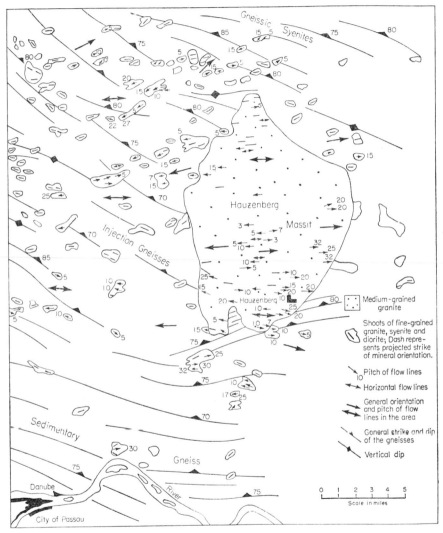

Fig. 54. Structural map of the Hauzenberg massif and adjoining Passau Forest. (*After R. Balk.*)

variably gentler than that of the surrounding gneisses, the folia of which they crosscut. The rocks are all foliated; the flow-planes are parallel to the contact planes. . . . The flow structures are of primary origin for they swing with the curvature of the contacts of the lenses. The younger syenite-granites appear in irregular masses, or dike-like bodies. . . . These rocks are also intensely foliated parallel to the local contacts, disregarding the mineral parallelism of the local gneisses. . . . Scattered through the same region are hundreds of discordant granite dikes, small transgressive sheets, and irregularly formed masses. . . . Through the granite bodies a well-developed linear parallelism trends east-southeast west-northwest, disregarding the orientation of local contacts, and the foliation of older rocks. Foliation in the granites is only locally developed, and remains feebler than the linear element.

From this general structural picture, taking into account the conformable relation of lineation in the Passau granites to that in the granites of the Hauzenberg and Saldenburg massifs which flank the region on either side, Cloos deduced a mechanism of upward surge of granite magma unaffected by local boundary planes, but guided by a regional expansion and sundering of the entire gneiss terrane in an east-west direction. This conclusion, although it has the merit of compatibility with a multitude of structural data, remains a hypothesis.

Conformability of contacts where the foliated steeply dipping borders of batholiths adjoin isoclinally folded or steeply foliated country rocks is compatible with either metasomatic or magmatic origin for the granites in question. The latter is much the more probable, on the other hand, where the internal structure of the granite is domed and the generally flat attitude of foliation in the center of the dome is disconformable to the still steeply dipping structure of roof pendants. The late Paleozoic granites of Devon and Cornwall [48] illustrate this condition. Bodies of nonfoliated granite with abrupt contacts, such as the rapakivi granites of Finland, might equally well be interpreted as of magmatic or of metasomatic origin if structural evidence alone were taken into account.

In conclusion, it would seem that if granitization of solid rocks in place has commonly led to development of granite bodies of batholithic size, we should expect to find numerous instances of conformable discordant contacts. However discordant the boundary between granite and country rock might be, we should expect to see the foliation of the granite passing continuously into the regional structure of the country rock. Large bodies of granite bounded mainly by this type of contact seem to be relatively rare, perhaps because only very detailed mapping can establish such a relationship.

Evaluation of a large mass of evidence such as that relating to structure of granitic batholiths is largely a matter of opinion. It is the opinion of the authors that many granitic batholiths, and plutons of smaller size, have persistent internal structural characters which strongly favor origin

[48] Cf. Balk, *op. cit.*, pp. 60–63, 1937 (and papers cited therein).

by intrusion of partially crystalline magma.[49]   In other cases the structural evidence is inconclusive.   In a few instances the evidence favors metasomatism; but even then there are indications that metasomatism may have been a restricted process operating around the periphery of a mass of intrusive granitic magma.

Mineralogical and Lithological Relations at Borders of Granitic Batholiths.   *Sharp and Transitional Contacts.*   The boundary of a large body of granite is a zone of lithological discontinuity.   On one side lies the assemblage of minerals characteristic of granite; on the other lie the varied assemblages constituting the wall rock.   Much of the discussion on the problem of granitization hinges on the mineralogical and lithological relationships between granite and wall rock in this zone of discontinuity.

Apart from faulted contacts, it has been clearly established that the boundary between granite and country rock is remarkably sharp over long distances in many batholiths.   Sharp contacts appear on the whole to be favored by shallow depth of emplacement.   They also tend to be emphasized where wall rocks have been converted to structureless hornfelses in contact aureoles.   The great batholiths of the Sierra Nevada and Southern California may be cited as examples of very large bodies of granitic rocks consistently bordered by sharp contacts.   Pabst's description of the boundary of the Sierra Nevada mass may be cited in this connection: [50]

Nearly all the contacts of granitic rocks in the Sierra Nevada visited by the writer were found to be exceedingly sharp, there being but slight evidence of contact action and very rarely any large amount of *Mischgesteine*.   In the glaciated regions it is not at all unusual to find bare contacts that can be traced with the pencil point for many rods or even several miles.   In the granitic rocks along these contacts there sometimes occur contact breccias involving the country rock. . . .   More commonly, however, only a few fragments of the encasing rock are found in the granitic rocks. . . .   At a number of contacts the granites show scarcely a trace of xenolithic material.

In the Southern California batholith, Larsen [51] notes that not only the external contacts between granite and wall rock but also internal contacts between the many individual bodies that make up the composite batholith are characteristically sharp.   This probably applies to many composite batholiths.[52]

---

[49] Several instances of American granites whose structures bear out this conclusion are cited by R. H. Jahns in the section on "Discussion," *Origin of Granites, Geol. Soc. America Mem. 28*, pp. 93–95, 1948.   Many examples are discussed by Buddington, *op. cit.*, 1959.

[50] A. Pabst, Observations on inclusions in the granitic rocks of the Sierra Nevada, *California Univ., Dept. Geol. Sci., Bull.*, vol. 17, no. 10, pp. 329, 330, 1928.

[51] Larsen, *op. cit.*, p. 139, 1948.

[52] *E.g.*, Hamilton, *op. cit.*, p. 1588, 1956; Taubeneck, *op. cit.*, p. 186, 1957.

Prevalence of sharp contacts is evidence supporting emplacement of granite by magmatic flow (intrusion). It is difficult to visualize development of a sharp contact scores of miles in length by abrupt halting of a wave of metasomatism. On the other hand, contacts of this nature are readily explained on the assumption that flow of a crystallizing magma ceased when the latter finally froze to somewhat cooler wall rocks margining a magma chamber.

The literature on granitic batholiths also abounds in well-substantiated accounts of granitic rocks merging imperceptibly into adjacent country rocks, especially where the latter are well-foliated schists whose fabric facilitated inward diffusion of such mobile materials as aqueous fluids. Particularly common is passage from pelitic mica schists (often carrying garnet or kyanite), through augen schists streaked with lenses and large individual crystals (porphyroblasts) of potash feldspar, into gneissic granite in which potash feldspar is even more plentiful.[53] It is clearly established that rocks of the transition zones in such cases are products of metasomatism of country rock by reaction with material diffusing from the direction of the granite mass. But this does not necessarily imply that the granite itself is a more extreme product of the same metasomatism. On the contrary, it is possible, and in many occurrences probable, that granitization in the transition zone was the result of introduction of alkalis and other appropriate materials from a granitic magma into chemically and physically receptive wall rocks.

At many granite contacts a zone of mixed rock (migmatite) separates granite from country rock.[54] Migmatites consist of two lithological elements intimately mixed: one is country rock variously altered by metamorphism and metasomatism; the other is granitic. The mutual relations of the two elements are very varied. Subparallel layers and veins of granite may fray out from the granite margin along the foliation of the wall rock to give banded migmatite. Where individual layers in the migmatite are continuous with and petrographically similar to the adjacent granite, injection of magma is strongly suggested. Migmatites of this type have been termed arterites or injection gneisses. Contact breccias in which angular fragments of altered wall rock are profusely strewn

[53] Cf. V. M. Goldschmidt, Die Injektionsmetamorphose im Stavanger-Gebiete, Norske Vidensk.-akad. Oslo Skr., I. Math.-Naturv. Kl. 10, 1921.

[54] The literature on border-zone migmatites is very extensive. A few selected references are as follows: J. T. Stark, Migmatites of the Sawatch Range, Colorado, Jour. Geology, vol. 43, pp. 1–26, 1935; T. F. W. Barth, Structural and petrologic studies in Dutchess County, New York, Part II, Geol. Soc. America Bull., vol. 47, pp. 803–806, 810–813, 825–832, 1936; F. J. Turner, The metamorphic and plutonic rocks of Lake Manapouri, Fiordland, New Zealand, Royal Soc. New Zealand Trans., vol. 67, pt. 1, pp. 89–99, 1937; A. C. Waters, Petrology of the contact breccias of the Chelan Batholith, Geol. Soc. America Bull., vol. 49, pp. 763–764, 1938.

through a granitic matrix, also are generally interpreted as evidence of emplacement of granite by magmatic intrusion.   There are many cases, however, where the granitic element of migmatites forms discontinuous lenses and streaks which may be interpreted alternatively as products of metasomatism or as local streaks of magma sweated out of the wall rock by incipient fusion.   The latter is likely in deep-seated regionally developed migmatites of pre-Cambrian basements which we shall discuss later.

Evidence of Reaction at Granite Contacts.   In the immediate vicinity of granite contacts, chemical and mineralogical composition of both granite and country rock may be conspicuously modified—apparently by reaction involving transfer of materials across the contacts.   These reaction zones have been variously interpreted, according to the alternative "magmatist" and "transformationist" points of view adopted by different observers.

Where granites abut against older basic rocks such as gabbros and amphibolites, the latter are commonly replaced, throughout zones varying from an inch to a foot or more in thickness, by dark rocks consisting of hornblende, biotite, sodic plagioclase, and in some subordinate quartz and orthoclase.   Xenoliths of basic rock may be similarly affected.   The modified initially basic rocks range in bulk composition from gabbro or amphibolite to diorite to granodiorite.   This progressive granitization of basic country rocks is exhibited at contacts of large batholiths and of small granitic bodies alike.[55]   The granite itself is likely to show reciprocal effects of contamination (basification) in contact zones of this type.   It is commonly richer in hornblende biotite and oligoclase, and poorer in potash feldspar and quartz, than is the uncontaminated granite remote from the contact, and in many occurrences it is strewn with hornblende- or biotite-rich xenoliths of modified country rock, or with microscopic aggregates of hornblende biotite and sphene formed by mechanical disintegration of such xenoliths.   The bulk composition of the contaminated granite tends to approach that of granodiorite or tonalite.   Thick zones of granodiorite developed along the margins of otherwise granitic bodies have very generally been interpreted as contaminated rocks formed by just such a process.   All these phenomena are attributed by petrologists of the "magmatist" school to reaction between granitic magma and

[55] Cf. A. Brammall and H. F. Harwood, The Dartmoor granites, *Geol. Soc. London Quart. Jour.*, vol. 88, pp. 171–237, 1932 (especially pp. 202–216).   H. H. Thomas and W. C. Smith, Xenoliths of igneous origin in the Trégastel-Ploumanac'h granite, *Geol. Soc. London Quart. Jour.*, vol. 88, pp. 274–296, 1932; S. R. Nockolds, Contributions to the petrology of Barnavave, Carlingford, I.F.S., *Geol. Mag.*, vol. 72, pp. 289–315, 1935; Turner, *op. cit.*, pp. 83–100, 1937; Waters, *op. cit.*, 1938; A. F. Buddington, Adirondack igneous rocks and their metamorphism, *Geol. Soc. America Mem.*, vol. 7, pp. 108, 185–189, 1939.

country rock, whereby the minerals of the latter have been converted to those crystalline phases (hornblende, biotite, oligoclase, sphene) with which the liquid phase of the magma was in equilibrium at the time of intrusion. The observed mineralogical changes are compatible with the theory of assimilative reaction developed by Bowen as a result of laboratory experiment (cf. pages 156, 157). Moreover the textural and structural relationships of the various rocks here interpreted as modified country rock, xenoliths, and contaminated granite, resemble closely those observed at borders of basic (gabbro) intrusions and universally attributed to reaction between magma—in this case basic—and wall rocks. Some process of ionic diffusion, either through an intergranular film of disordered material or through the crystal lattices themselves, must be invoked to account for marginal granitization of solid basic rock by reaction with granitic magma.[56]   But the existence of a magma can scarcely be denied. Petrographic and textural evidence alike seem to the writers to be incompatible with any hypothesis which attributes the main bulk of the granite to metasomatism of solid rock in place.

At contacts between granite and limestone, the latter rock may be converted to amphibolite (hornblende, diopside, biotite, plagioclase) while the granite shows complementary marginal modification to hornblende granite, granodiorite, or tonalite.[57]   Replacement of limestone by skarns consisting of iron-rich pyroxenes, andradite, hornblende, wollastonite, hematite, etc., clearly involves transfer of silica, alumina, iron, magnesia, and other bases from granite to limestone.[58]   Much of this may have been effected through the agency of volatile residual fluids set free in the last stages of freezing of granitic magma.   The possibility that acid magma may be converted to undersaturated alkaline magma (nepheline syenite) by desilication attendant upon reaction with limestone has already been noted (page 160).   Many undoubted instances of broad mutual association of granite, nepheline syenite, and limestone have been recorded; [59] and in some of these, such as the nepheline shonkinites associated with the Boulder batholith of Montana,[60] there can be no doubt that the feld-

[56] Cf. S. R. Nockolds, Some theoretical aspects of contamination in acid magmas, *Jour. Geology*, vol. 41, pp. 561–589, 1933.

[57] A. Lacroix, Le granit des Pyrénées et ses phénomènes de contact, *Service Carte géol. France Bull. 64*, tome 10, 1898; *Bull. 71*, tome 11, 1900; F. D. Adams and A. E. Barlow, Geology of the Haliburton and Bancroft areas, Ontario, *Canada Geol. Survey Mem. 6*, 1910; H. von Eckermann, The rocks and contact minerals of Tennberg, *Geol. fören. Stockholm Förh.*, vol. 45, pp. 465–537, 1923; Buddington, *op. cit.*, pp. 168–171, 1939.

[58] Turner, *op. cit.*, p. 125, 1948.

[59] Adams and Barlow, *op. cit.*, pp. 227, 228, 332, 1910; S. J. Shand, The present status of Daly's hypothesis of the alkaline rocks, *Am. Jour. Sci.*, vol. 243A, pp. 495–507, 1945.

[60] Knopf, *op. cit.*, pp. 98, 99, 1957.

spathoidal rocks represent magma contaminated by reaction with limestone. But nepheline syenites seldom if ever take the form of well-defined reaction zones comparable with the marginal granodioritic and dioritic hybrids just discussed. Instead, alkaline marginal variants of granites at limestone contacts are commonly nonfeldspathoidal rocks— pyroxene syenites or aegirine-augite granites.[61]

Reaction zones on the whole are not so sharply defined at contacts between granitic rocks and sedimentary rocks of the sandstone-shale family. The main constituents of sandstone—quartz, sodic plagioclase, potash feldspar, and mica—would already be in equilibrium with most granite magmas. Assimilative reaction consequently would not be expected to occur at sandstone-granite contacts. The clay minerals of shales on the other hand are sensitive to temperature changes, so that shales are normally converted by thermal metamorphism to pelitic hornfels (quartz, feldspars, biotite, andalusite, cordierite) for some distance from contacts with large bodies of granite. Here too, the main constituents of the rock (hornfels) abutting against granite would be stable when in contact with granite magma. However, judging from the instability of andalusite and cordierite in the presence of excess potash at all but the highest metamorphic temperatures (cf. Fig. 74, page 519), these minerals would be expected to react with the residual melts of partially crystalline granite magmas to give micas. Effects generally observed at contacts between granite and shale or sandstone are as follows:

1. Thermal metamorphism of sediments to hornfelses.

2. Development of augen (porphyroblasts) of alkali feldspars by metasomatic replacement of minerals within the country rock. Sandstones, perhaps on account of their high permeability, seem particularly susceptible to this type of alkali metasomatism. At some contacts wide zones of metamorphosed sandstone have thus been converted to gneissic granite—a metamorphic rock.[62]

3. Disintegration of metamorphosed sediments, fragments of which become dispersed through the granite and at the same time are progressively broken down to microscopic aggregates and xenocrysts of biotite, andalusite, and cordierite. The border phase of the granite may be considerably enriched in these minerals. Brammall and Harwood's account of the petrography of the Dartmoor granites [63] admirably illustrates this

[61] S. R. Nockolds, On the occurrence of neptunite and eudialite in quartz-bearing syenites from Barnavave, Carlingford, Ireland, *Mineralog. Mag.*, vol. 29, pp. 27–33, 1950; C. E. Tilley, Some trends of basaltic magma in limestone syntexis, *Am. Jour. Sci.*, Bowen vol., pp. 542–544, 1952; I. D. Muir, A local potassic modification of the Ballachulish granodiorite, *Geol. Mag.*, vol. 90, pp. 182–192, 1953.

[62] A. L. Anderson, Contact phenomena associated with the Cassia batholith, Idaho, *Jour. Geology*, vol. 52, pp. 376–392, 1934.

[63] Brammall and Harwood, *op. cit.*, 1932.

condition, which clearly implies the existence of mobile granite magma. They show that contamination (basification) of acid magma by reaction with shale and diabase played an essential role in the evolution of the dominant porphyritic biotite granite of this area.

Significance of "Porphyritic" Feldspars. At some granite contacts the granite locally is profusely strewn with large "phenocrysts" of potash feldspar. These may be simple idiomorphic crystals of perthite; or they may take the form of large composite ovoids, in each of which a single large rounded crystal of potash feldspar is enclosed in a polycrystalline rim of small oligoclase grains (rapakivi structure). Not uncommonly similar large feldspars appear in xenoliths, or in the host rock immediately adjacent to the porphyritic granite.[64] Such occurrences have been explained in orthodox terms of magmatic crystallization: early phenocrysts of potash feldspar have reacted later with contaminated magma in the contact zone, and so have become corroded and then rimmed with oligoclase; the large feldspars of the host rock have crystallized from magma infiltrating inward from the intrusive body. On the other hand petrologists of the "transformationist" school have argued that where feldspars of granite and host rock are found to be identical, metasomatic origin of both is proved.[65]

More information is needed as to whether the fabric of the granite groundmass is continuous with that of mineral grains enclosed in the feldspar "phenocrysts"—as would be expected if the latter were metasomatic. There is a need, too, for precise information—readily obtainable by modern techniques—on the composition, crystallography, and structure of large feldspars in granite and in adjacent country rock or xenoliths. Some interesting data of this kind have been reported by Stewart [66] with regard to rapakivi granite of a contact zone in Maine: the composition of potash feldspar (microcline) in cores of ovoids, in unmantled phenocrysts, in the granite groundmass, and in porphyroblasts of xenoliths is the same, averaging Or_{69}; low-temperature plagioclase of rapakivi mantles averages An_{22} (one analysis gives $Ab_{73}An_{24.2}Or_{2.8}$); rapakivi ovoids are absent from the country rock, but "stages of reaction leading to their development" are preserved in xenoliths. Stewart finds his observations compatible with magmatic crystallization as deduced

[64] *E.g.*, Brammall and Harwood, *op. cit.*, p. 219, 1932; Thomas and Smith, *op. cit.*, pp. 289–292, 1932; E. Spencer, The potash-soda feldspars, Pt. II, *Mineralog. Mag.*, vol. 25, pp. 111–114, 1938; C. E. Wegmann, Geological investigations in southern Greenland, Pt. I, *Meddelelser om Grönland*, vol. 113, no. 2, pp. 98–121, 1938; Read, *op. cit.*, pp. 74–87, 1944.

[65] *E.g.*, Wegmann, *op. cit.*, pp. 117, 118, 1938; Read, *op. cit.*, p. 80, 1944.

[66] D. B. Stewart, in Annual report of the director of the Geophysical Laboratory, *Carnegie Inst. Washington Year Book*, no. 55, pp. 194, 195, 1956.

from phase diagrams for the experimentally investigated system ortho-clase-albite-anorthite-water.

It has been argued by advocates of metasomatic granitization that if the large potash feldspars of granite are indeed of metasomatic origin, then the whole granite body in which they occur must also be a product of replacement of preexisting rock, without participation of a silicate-melt phase. This is not necessarily so. It is conceivable that granite of mag-matic origin, formed by early crystallization of magma adjacent to a cooler country rock, could provide an environment for later metasomatic growth of potash feldspar porphyroblasts. Today Read, who at one time con-tended that the supposed metasomatic status of "porphyritic" feldspars in granite supplied crucial evidence proving the metasomatic nature of many granites,[67] is inclined to minimize the significance of this aspect of the granite problem. Speaking of his observations in South Africa in 1951 he writes: [68]

The *dents de cheval* [large feldspars] at Sea Point, Cape Town, were now taken as local phenomena, not to be applied by themselves to the interpretation of the pluton as a whole. Whether these feldspars could be formed in two different environments or only in one no longer seemed to me to be funda-mental in the granite controversy.

Chemical Gradients at Granite Contacts.   Near contacts with granitic rocks the country rocks usually show obvious effects of metamorphism—mineralogical and structural reconstitution of the rock in an essentially solid state. In many cases, and especially at distances greater than a few feet from the contact, these changes have been effected with little or no change in bulk composition except with respect to minor volatile constitu-ents such as water, carbon dioxide, boron compounds, etc. But there are also instances where metamorphism has been accompanied by notable changes in composition, which in rocks of high permeability may extend long distances from the contact. Development of porphyroblasts of albite or potash feldspar in pelitic and especially in psammitic country rocks—an instance of alkali metasomatism—has already been noted.[69] Other well-known examples involve introduction of iron, magnesia, and silica into the country rock, *e.g.*, the development of skarns at limestone con-tacts, and conversion of psammitic or other quartzofeldspathic rocks (*e.g.*, the leptites of Sweden and Finland) to anthophyllite-cordierite

---

[67] Read, *op. cit.*, pp. 80–87, 1944 (especially pp. 80, 86); The granite problem, *Geol. Soc. America Mem.* 28, pp. 12–14, 1948.

[68] H. H. Read, *The Granite Controversy*, Murby, London, p. xvii, 1957.

[69] Cf. V. M. Goldschmidt, Ueber einen Fall von Natronzufuhr bei Kontaktmetamor-phose, *Neues Jahrb. Beilage-Band 39*, pp. 193–214, 1914; *op. cit.*, 1921.

rocks.[70] Replacement of basic igneous rocks by dark hornblendic or biotite-rich reaction products has been discussed in the previous section. The nature of chemical gradients from granite through metasomatized rocks of the contact zone to unaltered country rocks has been summarized by D. L. Reynolds.[71] She has deduced a sequence of chemical changes by comparing analyses of cores (first stage) with rims (second stage) of individual xenoliths immersed in granite. Comparison of analyses of altered country rocks at decreasing distance from some contacts is likewise interpreted as reflecting a sequence of chemical changes. The essence of Reynolds' observations is that modification of country rocks immediately adjacent to granite contacts in all cases leads ultimately to granitization, in that rocks at an advanced stage of alteration (e.g., xenolith rims) are always closer to granite in composition than are rocks at earlier stages of alteration (e.g., xenolith cores). However, the first stage of alteration, especially in the case of pelitic and semipelitic sediments and basic igneous rocks, is usually one of *desilication*, in that the normative quartz content (and usually the percentage of SiO_2 by weight) decreases during the initial steps of modification. Desilication may involve feldspathization (syenitization) of quartz-rich rocks, or basification (addition of Ca, Mg, and Fe) of many different types of country rock. With these first stages of alteration is usually associated concentration ("geochemical culmination") of some bases (K_2O, Na_2O, MgO, FeO, etc.) and especially such minor constituents as TiO_2, P_2O_5, and MnO, up to proportions exceeding those for either the granite or the unaltered country rock.

There can be no doubt as to the reality and nature of at least some of the chemical gradients described by Reynolds, but their petrogenic significance is uncertain. Some of the more striking changes cited refer to gradients across distances of a few inches or feet. Others, such as the zones of iron-magnesia metasomatism surrounding the Orijärvi granite batholith, are developed across distances of several hundred yards from the granite boundary. In many cases the validity of the assumption that a sequence of metasomatic effects observed in space actually reflects a sequence of stages of alteration in time may be questioned. Phenomena of this type have been interpreted by Wegmann, Backlund, Reynolds, and others from the "transformationist" standpoint. Their hypothesis is that the granites are end products of metasomatism of the country rocks. In

[70] P. Eskola, On the petrology of the Orijärvi region in South-western Finland, *Comm. géol. Finlande Bull. 40*, 1914; N. H. Magnusson, The evolution of the lower Archaean rocks in central Sweden and their iron manganese and sulphide ores, *Geol. Soc. London Quart. Jour.*, vol. 92, pp. 332–359, 1936; Turner, *op. cit.*, pp. 125–126, 1948.

[71] Reynolds, *op. cit.*, pp. 389–446, 1946.

advance of the encroaching granitic "front" go waves of iron, magnesia, and other "unwanted" constituents expelled from the country rock itself as its composition is made over to that of granite. The materials of this wave of emanations are pictured as becoming temporarily fixed to give basic magnesian or iron-rich "fronts" margining the granite.[72] It is even suggested that rocks chemically, mineralogically, and texturally identical with gabbros, occurring on the margins of some granitic bodies, could be of metasomatic origin—products of fixation of a basic front.

The orthodox "magmatist" hypothesis supposes that granites are products of crystallization of a granitic magma. It is highly probable that during the later stages of crystallization of such magma under plutonic conditions, a complex aqueous gas phase may evolve and pass out through the pores of the wall rock. Much of the modification of country rocks adjoining granites, including development of basic or alkaline "fronts," is attributed by "magmatists" to more or less contemporaneous reactions between the country rock and the wave of gas expelled from the granite center. The composition of this gas presumably varies rapidly with distance from its source, as some of its constituents (e.g., boron, phosphorus, iron) are withdrawn and fixed in the modified country rock. The classic theory of contact metamorphism attributes most mineralogical changes observed within contact aureoles to chemical reaction controlled by a combination of factors—high temperature, water pressure, and so on—which changes with passage of time. At the contact itself these changes are accompanied or preceded by striking modification of country rock and magma by mutual reaction. According to this complicated concept of contact phenomena, it would be most unsafe to assume that the first and second stages of Reynolds's geochemical sequences necessarily reflect a simple sequence in time.

The relative merits of the "magmatist" and "transformationist" hypotheses of granite origins must be tested on other grounds than those afforded by chemical gradients at granite contacts. The gradients can be interpreted on the whole adequately in terms of either hypothesis. Reynolds [73] has justly stated with regard to the concept of introduction of iron and magnesium from residual magmatic solutions or gases that "the field observations that can be adduced as evidence of these processes are the very phenomena that have to be explained." But this remark seems equally applicable to introduction of iron and magnesium throughout a basic front spreading in advance of a center of granite metasomatism. Two facts, in the present authors' opinion, support the view that most granites have crystallized from at least partially liquid magmas, whether these them-

[72] Cf. Reynolds, op. cit., pp. 432–436, 1946; Read, op. cit., pp. 9–12, 1948; E. Raguin, Géologie du Granite, 2d ed., Masson et Cie, Paris, pp. 79–93, 1957.

[73] Reynolds, op. cit., p. 433, 1946.

selves are products of extreme metasomatism leading to fusion or of some other process. In the first place phenomena of the same order, though differing in detail, have been observed in the border zones of such large basic intrusions as the Duluth gabbro sheet.[74] Unless the very existence of plutonic igneous rocks of any type be denied, these must be classed as phenomena of magmatic reaction and contact metamorphism. In the second place, if the basic front really represents a zone of concentration of material expelled during granitization of country rock in place, the great extent of many granitic bodies contrasted with the small bulk of associated "basic front" rocks presents a very striking anomaly.

In conclusion we must remember that metasomatism is by no means universal in contact aureoles; nor have its nature and scope—especially in connection with possible redistribution of the main components of the rocks—been fully explored. It is only with the advent of rapid methods of chemical analysis that sampling has become possible on a scale sufficiently detailed to distinguish chemical variation initially present from that imposed by metasomatism and metamorphism. Pitcher and Sinha,[75] from 50 analyses of 220 carefully collected samples, have made what is probably the only adequate study of chemical variation in pelitic rocks of a contact aureole adjoining intrusive granite. They find that most of the observed chemical variation is inherited from initial differences in composition of the parent sediment. Metamorphism was largely isochemical. Throughout the aureole, however, the water content has been significantly reduced, and there is possibly some slight introduction of soda and potash near the contact.

REGIONAL PRE-CAMBRIAN GRANITES

Field Occurrence. The Archaean basements of pre-Cambrian shields are made up principally of granitic rocks, and these also occur extensively in some later pre-Cambrian terranes. They may occur as individual batholiths and smaller intrusions; but more typically they are developed as vast formless seas of gneissic granite in which swim islands and rafts of metamorphic rocks in various stages of conversion to migmatite. Any adequate theory of the origin of granites must take into account this mode of occurrence of granites on a regional scale. It must explain, too, the fact that granites of this class have the same general range of chemical and mineralogical composition and fabric, and show the same relations to

[74] Cf. F. F. Grout, Contact metamorphism of the slates of Minnesota by granite and by gabbro magmas, *Geol. Soc. America Bull.*, vol. 44, pp. 989–1040, 1943; Turner, *op. cit.*, pp. 40–42, 1948.

[75] W. S. Pitcher and R. C. Sinha, The petrochemistry of the Ardara aureole, *Geol. Soc. London Quart. Jour.*, vol. 113, pp. 393–408, 1958.

associated metamorphic rocks, as granitic rocks of batholithic intrusions in general.

We shall now consider three problems relating mainly, though not exclusively, to pre-Cambrian granites: the origin of rapakivi granites; the regional development of migmatites; and the occurrence of granite at what seem to be stratigraphic horizons in metamorphosed sediments.

Rapakivi Granites.   Granites with rapakivi structure have already been noted as marginal variants of normal intrusive granites of all ages.   By far their greatest known development, however, is in the later pre-Cambrian formations of Finland and Sweden, where immense bodies of granite, some exceeding 15,000 km.^{2} in extent, show rapakivi structure throughout the greater part of their bulk.[76]

Common and characteristic features of pre-Cambrian rapakivi granites are as follows: [77] The potash feldspar occurs in large rounded or ovoid crystals which are commonly, though not invariably, enclosed in shells of small crystals of oligoclase.   Quartz may be present in two generations—bipyramidal phenocrysts and anhedral grains in the matrix.   The latter consists of quartz, potash feldspar, plagioclase, biotite, and in some rocks hornblende.   Fluorite and zircon are common accessories.   The fabric of rapakivi granite lacks foliation or lineation; there is no recognizable inhomogeneity that might be identified as relict structure inherited from some pre-exsting state.   Miarolitic cavities are common.   Pegmatites and aplites are scarce.   Contact between granite and country rock may be sharp; eruptive breccias consisting of randomly oriented schist fragments in a granite matrix have been recorded at some contacts.

Among Fennoscandian geologists, Backlund[78] alone has advocated origin of rapakivi granites by metasomatic replacement of the Jotnian sandstones with which they are associated in the field.   The horizontal sheetlike form of the granite masses was thought to have been inherited from the horizontal disposition of the assumed parent sandstone.   Backlund's views were given prominence by Read[79] in support of his own conclusion "that rapakivi-like granitic rocks may arise by a variety of feldspathization."   However, the consensus among Finnish and Swedish

[76] J. J. Sederholm, On migmatites and associated pre-Cambrian rocks of Southwestern Finland, *Comm. géol. Finlande Bull. 58*, pp. 75–95, 1923; On the geology of Fennoscandia, with special reference to the pre-Cambrian, *Comm. géol. Finlande Bull. 98*, pp. 24–26, 1932; H. von Eckermann, The genesis of rapakivi granites, *Geol. fören. Stockholm Förh.*, vol. 59, pp. 503–524, 1937; Read, *op. cit.*, pp. 74–80, 1944; A. Savolahti, The Ahvenisto massif in Finland, *Comm. géol. Finlande Bull. 174*, pp. 83–93, 1956.

[77] Savolahti, *op. cit.*, pp. 83, 84, 1956; cf. also Read, *op. cit.*, p. 74, 1944.

[78] H. G. Backlund, The problems of the rapakivi granites, *Jour. Geology*, vol. 46, pp. 339–396, 1938.

[79] Read, *op. cit.*, p. 79, 1944.

geologists who are most familiar with rapakivi granites is that they are products of crystallization from granitic magmas; [80] and that the composition of the magma and physical conditions were such that potash feldspar and quartz crystallized early and were followed by plagioclase and the dark constituents. The course of crystallization so postulated seems compatible with experimental data relating to crystallization in the ternary feldspar system under plutonic conditions, e.g., $T = 700°$ to $800°C.$; $P_{H_2O} = 5,000$ bars.[81] With moderate changes in melt composition, temperature, or water pressure, the nature of the phases crystallizing from a rather potassic anorthite-poor melt might be expected to oscillate between any combination of three alternatives: potash feldspar alone, oligoclase alone, and both feldspars together. The compositions of feldspars associated in rapakivi ovoids compare favorably with the values (Or_{78} and An_{23}) determined by Yoder for feldspars simultaneously crystallizing at $720°C.$ and 5,000 bars from a melt equally rich in Or and Ab and poor in An.

The problem of origin of rapakivi granite has been approached by Chayes from a statistical standpoint.[82] His conclusion is as follows: In rocks formed by replacement the amount of residual and introduced materials ought to vary inversely. This implies "negative correlation" between the two types of material. If rapakivi granites of Fennoscandia are indeed replacement products of Jotnian or other arkosic sandstones, there should be an inverse relationship ("negative correlation") between normative quartz (an index of residual material) and normative alkali feldspar (an index of replaced material) in any large number of chemical analyses of the granites in question. Forty analyses of Swedish and Finnish rapakivi granites were tested, with this criterion in view. The Swedish rocks were found to accord with the hypothesis of replacement, but analyses of the Finnish rocks were found to depart "by a very large margin" from the requirements of significant correlation. Chayes therefore concludes that "the alkali-emanation hypothesis as advanced by Backlund does not afford a satisfactory explanation of variations in the bulk chemical composition of the rapakivi granites."

Much of what has been stated above regarding pre-Cambrian rapakivi granites applies also to coarsely porphyritic granites of other types and of all ages. As one instance we cite the coarse porphyritic granitic rocks which extend continuously for scores of miles within the Sierra Nevada

[80] Savolahti, op. cit., p. 92, 1956.

[81] H. S. Yoder, D. B. Stewart, and J. R. Smith, in Report of the director of the Geophysical Laboratory, Carnegie Inst. Washington Year Book, no. 56, pp. 206–214, 1957.

[82] F. Chayes, A petrographic criterion for the possible replacement origin of rocks, Am. Jour. Sci., vol. 246, pp. 413–425, 1948.

batholith of California.   The main body of these rocks trends parallel to the axis of the batholith and lies well within it.   Certainly it is not a marginal variant either of the batholith as a whole or of one of its component plutons.   The phenocrysts of potash feldspar are commonly idiomorphic, and they enclose concentric zones of biotite crystals suggesting magmatic crystallization.[83]   Rapakivi structure is rather restricted in occurrence. A tendency to local crowding of porphyroblasts into discontinuous irregular streaks somewhat reminiscent of replacement pegmatites might be construed as evidence of local replacement origin.   But there is no evidence that the granite as a whole is the product of metasomatism of some preexisting rocks in place.

The writers tentatively accept a magmatic origin for extensive bodies of uniformly porphyritic granite in general—the classic instance being the pre-Cambrian rapakivi of Finland and Sweden.   This does not exclude the possibility that some large feldspars of some granites are porphyroblasts of postmagmatic origin.[84]

Regional Development of Migmatites.   The term migmatite has been variously defined.[85]   It is generally agreed that the most typical rocks of this class are those in which a granitic component (granite, aplite, pegmatite, granodiorite, or the like) and a metamorphic host rock are intimately admixed on a scale sufficiently coarse for the mixed condition of the rock to be megascopically recognizable.   We shall here confine our discussion to rocks of this type.   It may be noted at the same time, as emphasized by Read, that migmatites so defined in a restricted sense are commonly associated with and grade into "permeation gneisses," "imbibition rocks," and so on, the mixed nature of which is usually ascribed to metasomatic introduction of K_2O, Na_2O, SiO_2, etc., into preexisting solid rocks.   The granitic component and the host material of such rocks are blended on such a fine scale that it is impossible or difficult to resolve them in a hand specimen.   If rocks of this kind are termed *migmatites*, then how are migmatites to be distinguished from a great variety of other metamorphic rocks which have been affected to some degree by metasomatism, or which have developed a laminated structure by segregation (differentiation) of their various component minerals during recrystallization?   Moreover it is not appropriate to insist on participation of magma as an essential

---

[83] Cf. L. J. G. Schermerhorn, The granites of Trancoso (Portugal), *Am. Jour. Sci.*, vol. 254, pp. 329–348, 1956 (especially pp. 332–335).

[84] Whitten has presented a good case, based on analysis of modal variation and fabric, for late metasomatic origin of large potash feldspar crystals and coarse aggregates of quartz in the Gola granite of the Donegal granite complex (Whitten, *op. cit.*, 1957).

[85] J. J. Sederholm's original definition of migmatites (*Comm. géol. Finlande Bull. 23*, p. 110, 1907) and his subsequently modified views on these rocks have been stated clearly by Read, *op. cit.*, pp. 61–65, 1944.

process in evolution of migmatites, for there are some geologists today who believe that the most typical members of the migmatite family have originated without intervention of any silicate-melt phase. For these reasons, we here restrict the term *migmatite* to cover only those mixed rocks in which a granitic component (in the broad sense of the phrase) is clearly recognizable. This descriptive definition avoids ambiguities which necessarily attend any definition framed in genetic terms ("igneous," "permeation," "injection," etc.). Where, as is usually the case, the granitic component of a migmatite occurs as continuous veins or laminae, Holmquist's term *venite* (= veined gneiss)—also descriptive and non-genetic—has sometimes been used.

We have already noted the occurrence of relatively narrow zones of migmatite bordering some granitic bodies. Elsewhere, and especially in pre-Cambrian terranes, migmatites appear on a regional scale. A glance at the maps which accompany Sederholm's classic accounts of the migmatites of Finland [86] shows that in this region vast areas of granite alternate with great expanses of migmatite and others of metamorphosed sediments and basic rocks. Sederholm estimated that migmatites make up over 20 per cent of the exposed pre-Cambrian rocks of Finland. Again Read [87] has described in Sutherland, Scotland, a migmatite complex occupying 700 square miles. This illustrates a condition commonly displayed in regionally developed migmatites, namely, gradation from areas of predominant host rock sparsely veined with granite, through zones in which the granitic component is conspicuous or dominant, to local occurrences of almost pure granite. In the Sutherland complex, the granitic component is most prominent within a central area, about which is developed a peripheral zone more sparsely veined with granitic rocks. In yet other occurrences, e.g., in the complex of the New Jersey highlands described by Fenner,[88] extensive masses of granite only locally enclose dark hornblendic or biotite-rich bands in sufficiently high proportion to warrant use of the term *migmatite*.

Veins and laminae of granitic rock in migmatites show great variety in form. Two features nevertheless are sufficiently persistent to merit special mention. First is the tendency for granitic laminae and streaks to develop parallel to schistosity surfaces of foliated host rocks. Second is the frequent occurrence, especially where the host rock lacks schistosity,

[86] J. J. Sederholm, On migmatites and associated pre-Cambrian rocks of Southwest Finland, *Comm. géol. Finlande Bull.* 58, 1923; *Bull.* 77, 1926; *Bull.* 107, 1934.

[87] H. H. Read, The geology of central Sutherland, *Geol. Survey Scotland Mem.*, pp. 88–164, 1931. See also Y. Cheng, The migmatite zone around Bettyhill, Sutherland, *Geol. Soc. London Quart. Jour.*, vol. 99, pp. 107–154, 1944.

[88] C. N. Fenner, The mode of formation of certain gneisses in the highlands of New Jersey, *Jour. Geology*, vol. 22, pp. 594–612, 694–701, 1914.

of veins showing what has been termed ptygmatic folding (Fig. 55).[89]
Veins so folded are distinguished by extreme contortion, by irregularity
(even within a hand specimen) of trend of the axes of associated folds,
by lack of obvious relation between the folds of a vein and the internal
structure of the enclosing host rock, and in many cases by absence of any
competency relation between size of folds and thickness of the folded
veins.   There is a widely prevalent idea (cf., for example, the writings of
Sederholm) that the veins originate as planar layers of magma in a host

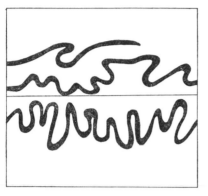

FIG. 55.   Ptygmatic folding of granitic veins in migmatites. (*After H. H. Read.*)

rock, which, rendered mobile by
partial fusion, deforms irregularly by
fluxional movements and so imposes
a highly contorted form upon the
enclosed fluid veins.   There is also
the possibility suggested by Read that
magma has been injected along tor-
tuous fissures that have opened up in
a nonschistose, mechanically homogeneous rock (*e.g.* hornfels) under
disruptive forces that have accompanied igneous intrusion. Injection of
magma along the schistosity of foliated rocks during active deformation
has also been appealed to in certain
instances, *e.g.*, by Buddington for
some migmatites in the Grenville series of the Adirondacks.   Wilson [90]
has shown that ptygmatic folds could develop by intrusion of viscous
granitic magma into even less competent host rock.   Finally there is the
possibility that a veined rock becomes deformed in the plastic state, *i.e.*,
metamorphically, with subsequent recrystallization of the vein filling.   It
is interesting to note that only the last of these interpretations can be
reconciled with nonmagmatic metasomatic origin for the vein materials.
The writers would not rule out the possibility that certain ptygmatically
folded veins have developed by metasomatism.   Highly convoluted veins
of quartz, albite, and such minerals are well known among deformed
rocks whose strictly metamorphic origin cannot be doubted.   But these

[89] Many examples of ptygmatic folding are illustrated in Sederholm's papers (*op. cit.*, 1923, 1926, 1934).   For discussion on significance of ptygmatic folding see Read, *op. cit.*, pp. 110, 111, 1931; P. H. Kuenen, Observations and experiments on ptygmatic folding, *Comm. géol. Finlande Bull. 123*, pp. 11-27, 1938; Buddington *op. cit.*, pp. 164, 165, 1939; Turner, *op. cit.*, pp. 315, 316, 1948; J. D. Godfrey, The origin of ptygmatic structures, *Jour. Geology*, vol. 62, pp. 375-387, 1954; G. Wilson, Ptygmatic structures and their formation, *Geol. Mag.*, vol. 89, pp. 1-21, 1952; Raguin, *op. cit.*, pp. 170 172, 1957.

[90] G. Wilson, *op. cit.*, 1952.

seldom if ever show all the distinctive features of ptygmatic folds as consistently developed in many migmatite terranes. Such migmatites convey a very strong impression that mobile granitic magma has played an essential role in their evolution.

The origin of regional migmatites has been and remains a debatable topic, which reflects the current divergence of opinion between petrologists of the "magmatist" and "transformationist" schools. Migmatites can be interpreted almost equally well from either standpoint. Certain details, e.g., some of the more complex examples of ptygmatic folding, favor the "magmatist"; others, e.g., development of isolated feldspathic lenses and augen, are more compatible with the views of the "transformationist." We therefore note below, without critical comment, some of the mechanisms that have been considered significant in evolution of the migmatites.[91]

1. Injection of magma along surface of weakness (schistosity, joints, irregular surfaces of rupture, etc.) in the host rock. If such an origin can be demonstrated, but not otherwise, the resulting migmatite may be termed an arterite or injection gneiss. Sederholm, especially in his earlier writings, expressed the belief that many of the migmatites of southwest Finland were arterites. Read also appealed to active injection of magma in explaining the origin of the migmatite complex of Sutherland. Viewed from this standpoint, the extensive migmatite terranes of the pre-Cambrian are injection zones on the fringes of subjacent granitic intrusions of immense size.

2. Metasomatism (granitization) of the host rock by ionic exchange of material between the host rock and fluids which penetrate it along paths of minimum resistance.[92] The active fluid, variously referred to as "magma," "magmatic emanation," "granitic juice," "ichor," "solution," and so on, need not necessarily be of magmatic origin. This mechanism is favored particularly by "transformationists" of the French and Fennoscandian schools.

3. Differential fusion of host rocks of varied composition, yielding a low-melting granitic or pegmatitic liquid (magma) distributed through the rock as discontinuous streaks and veins. Subsequent freezing of the only partially fused mass, which may or may not have flowed or been deformed under tectonically induced stresses, would give migmatite. This mode of origin for deep-seated pre-Cambrian migmatites has been fa-

---

[91] Cf. Read, *op. cit.*, pp. 146–150, 1931; pp. 61–74, 1944; Eskola, *op. cit.*, pp. 12–25, 1933; MacGregor and Wilson, *op. cit.*, vol. 76, pp. 193–215, 1939; Cheng, *op. cit.*, 1944; Turner, *op. cit.*, pp. 306–311, 1948.

[92] Cf. C. E. Wegmann, Zur Deutung der Migmatite, *Geol. Rundschau*, vol. 26, pp. 305–350, 1935; Cheng, *op. cit.*, 1944; Holmes, *op. cit.*, p. 414, 1945; Turner, *op. cit.*, pp. 308, 309, 1948.

vored by Holmquist, Eskola, and others.[93]   In literature on granitization the term "mobilization" repeatedly crops up with reference to the evolution of granite-streaked migmatites.   To the present writers, and perhaps to many advocates of granitization as well, "mobilization" implies the development of a mobile silicate-melt phase—usually by incipient melting in place.   In other words "mobilization" is synonymous with what we would describe as transition from the metamorphic to the magmatic condition.

All three of the above mechanisms, and probably others as well, are likely to have been involved in the genesis of migmatites.   In our final discussion of the origin of granites we shall see how some writers have developed general hypotheses assigning a specific role to each.

Granitic Bodies at Stratigraphic Horizons.   What seem to be regular stratigraphic relations between granitic bodies and metasedimentary units have been demonstrated by careful mapping of some pre-Cambrian metamorphic-plutonic complexes.   These relations may be of two kinds:

1. A metasedimentary series, traced along the strike, passes imperceptibly into granitic gneiss.[94]

2. Within a dominantly metasedimentary terrane certain individual formations have the composition and fabric of granite, and show no sign of an intrusive relation to the adjacent metasedimentary units.[95]

Some records of this phenomenon may be discounted on various grounds: the "granite" may be what most petrographers would term a paragneiss, complete with index minerals of high-grade metamorphism (garnet, sillimanite, etc.); or the granite-metamorphic contact may prove to be an unconformity.   But there still remain instances where convincing field data point to conversion of a sedimentary formation into what any petrologist would call granite or granodiorite.

It has been further argued by "transformationists" that the situation just described is proof that granitic bodies may arise by selective metasomatism of sedimentary units whose texture and composition predispose them to the influence of granitizing fluids.   This is indeed a possible explanation; but, as usual, the evidence is open to another interpretation from the standpoint of the magmatist.   There can be no doubt that the temperatures and pressure conditions of regional metamorphism overlap

---

[93] E.g., Eskola, op. cit., 1933.

[94] E.g., T. T. Quirke and W. H. Collins, The disappearance of the Huronian, Canada Geol. Survey Mem., 160, 1930.

[95] E.g., J. H. Rattigan and C. E. Wegener, Granites of the Palmer area and associated granitized sediments, Royal Soc. South Australia Trans., vol. 74, pt. 2, pp. 149–164, 1951; F. D. Eckelmann and J. L. Kulp, The sedimentary origin of the so-called Cranberry and Henderson granites in western North Carolina, Am. Jour. Sci., vol. 254, pp. 288–315, 1956; J. L. Kulp and A. Poldervaart, The metamorphic history of the Spruce Pine district, Am. Jour. Sci., vol. 254, pp. 393–403, 1956.

the melting range of hydrous granitic magmas. It is possible—indeed likely—that in regions where metamorphic temperatures are uniformly high some sedimentary rocks will be subject to conditions within the melting range. Rocks of the graywacke-arkose family—rich in feldspar and clay constituents and approximating granite or granodiorite in composition—are likely subjects for partial or complete melting under conditions that bring about metamorphism of associated quartzites and limestones in the solid state. This is what seems to have occurred in the Archaean rocks of the Toodyay district of Western Australia.[96] Gneissic granite occurs at a definite horizon interbedded with high-grade quartzites, sillimanite schists, and amphibolites. Contacts revealed by detailed mapping are essentially stratigraphic. But evidence of several kinds shows that at one time the granitic formation was at least partially molten (magmatic): locally the granite sends intrusive apophyses into the adjacent quartzites; metamorphic and partially granitized inclusions in the granite are varied in composition and are satisfactorily interpreted as xenoliths derived from adjacent formations; quartz grains in granite are randomly oriented, while those of the adjacent quartzite have a high degree of preferred orientation.

We conclude that the stratigraphic relation of some pre-Cambrian bodies of granite to associated metasediments strongly suggest a sedimentary origin for such granites. This does not necessarily imply metasomatic granitization; it is equally possible, and in some cases is clearly indicated by field and petrographic evidence, that the sediments concerned were converted to granite through a magmatic (at least partially molten) stage.

ORIGIN OF GRANITIC ROCKS

Existence of Large Bodies of Granitic Magma. *Statement of the Problem.* The reader is reminded that the essential characteristics of magmas are mobility (capacity for intrusion or extrusion) and presence in noteworthy proportions of a silicate-melt phase. Much of what some other writers have distinguished on genetic grounds as "migma"—mobile mixtures of crystalline rock material and magmatic liquid injected into or fused out of it—falls within the category of "magma" as the term is employed in this book.[97]

[96] R. T. Prider, The petrology of part of the Toodyay district, *Royal Soc. Western Australia Proc.*, vol. 28, pp. 83–133, 1944 (especially pp. 108–124). One of the writers (Turner) had the privilege of examining the terrane under Professor Prider's guidance in 1956.

[97] The term *migma*, coined by M. Reinhard in 1934, has been used freely in the above sense by a number of modern writers on the granite problem, *e.g.*, T. F. W. Barth (*Die Entstehung der Gesteine*, Barth, Correns and Eskola; Springer, Berlin,

So strong has been the influence of the "magmatist" school led by Rosenbusch that it would not have been thought necessary two decades ago to discuss the possibility that large bodies of granitic magma have been generated from time to time within the outer crust of the lithosphere. The exisence of great masses of granite at the earth's surface was very generally accepted, without further thought, as evidence of previous existence of a corresponding body of magma, and the magma itself was tacitly assumed by many to have been at one time entirely liquid.   Nevertheless French and Scandinavian petrologists have long stressed the role of metasomatism in the genesis of granitic rocks, and in recent years they have progressively minimized the part played by magmas—especially completely liquid magmas.   The extreme view which has developed as a result of this trend of thought is that which appears in the recent writings of Perrin and Roubault, Ramberg, Bugge, and others.[98]   It assigns only an insignificant role to fluids of any kind (silicate melts, aqueous solutions, or gases) in plutonic and metamorphic processes, and it explains the manifold mineralogical and structural characters of granitic and metamorphic rocks as products of ionic diffusion through crystal lattices and along intergranular boundaries in solid rocks.   Faced with this stimulating proposition, we must first critically examine the suggested mechanism of "dry granitization" and then review the evidence which has hitherto led geologists to attribute large bodies of granite to intrusion of magma.

Granitization by Solid Diffusion, Critically Reviewed.   It has been shown in Chap. 2 that the chemical potential of a substance depends essentially on temperature, pressure, and concentration (or on "activity"; see Eqs. 2–30 and 2–38).   Whenever there are gradients in physical conditions there are also gradients in chemical potential, and hence there is a tendency for substances to move from a state of high to a state of low potential (*e.g.,* from high to low pressure, from high to low "activity," and so on).   The essence of the mechanism of granitization postulated by Ramberg and by Bugge is that, in response to these gradients, different types of ions are assumed to migrate selectively, *i.e.,* at different rates and in different directions.   The great granitic batholiths of the continents are interpreted as stable end products of metasomatism, brought

1939, pp. 103, 114) and Read (*op. cit.,* p. 667, 1946).   It has also been used to cover immobile granitized material, as distinguished from magma capable of intrusion (*e.g.,* P. Niggli, 1946, as reported by T. F. W. Barth, *Jour. Geology,* vol. 56, p. 238, 1948).

[98] R. Perrin and M. Roubault, Le granit et les réactions à l'état solide, *Service carte géol. Algérie Bull.,* sér. 5, no. 4, 1939; On the granite problem, *Jour. Geology,* vol. 57, pp. 357–379, 1949; H. Ramberg, The thermodynamics of the earth's crust. I, *Norsk geol. Tidsskr.,* vol. 24, pp. 98–111, 1944; J. A. W. Bugge, The geological importance of diffusion in the solid state, *Norske Vidensk.-akad. Oslo Skr., I. Math.-Naturv. Kl. 13,* pp. 5–59, 1945.

about by predominantly upward flow of light "chemically active" ions (K^+, Na^+, Al^{3+}, Si^{4+}, etc.), presumably compensated by complementary downward flow of heavier less "active" ions (Fe^{++}, Fe^{3+}, Mg^{++}, Ca^{++}, etc.).

The simplicity of this and allied mechanisms renders them attractive to some students of petrology. But this simplicity is, of course, only apparent. For instance, it would be almost impossible, for lack of relevant data, to determine the activity gradients that may exist in a given sector of the earth's crust, for it must be remembered that ions involved in metasomatism are not initially free to migrate. On the contrary, they are presumably tied up in preexisting mineralogical associations. To obtain, for example, the supply of K^+ ions required to granitize a sandstone, requires previous breaking down of the compounds in which potassium was initially bound. Now, as will be discussed in greater detail in Chap. 17, the activity of K^+ in a given mineral is not determined only by the richness of that mineral in potassium, but also by the strength of the bonds by which the K^+ ions are held in that particular lattice. Hence little can be said at present about activity gradients in the earth's crust, and it is equally difficult to prove or to disprove that conditions (gradients) favoring granitization existed at any given time or place.

If, neglecting this prevailing uncertainty regarding fundamental principles of diffusion in rocks, the uncritical student of petrology still accepts granitization by solid diffusion as an established mode of origin for large bodies of granite, he finds it a simple matter to write equations showing what must be added to, and what must be removed from, some assumed parent rock, to convert it into some particular type of granite. Moreover, as pointed out especially by Perrin and Roubault, many petrographic and field phenomena relating to granites (textural relations of the component minerals of granite, phenomena at granite contacts, occurrence of granitic dikes, etc.) are compatible with the concept of granitization by solid diffusion. On the other hand, in the opinion of many petrologists, these same phenomena can be correlated equally well, or even better, with a mechanism of intrusion and solidification of magma. Some of the difficulties raised by the hypothesis of magmatic crystallization (e.g., oscillatory or reversed zoning in crystals of plagioclase) still persist or even are intensified if we turn instead to granitization by solid diffusion.[99]

Any satisfactory theory of the origin of granites must explain the observed chemical, petrographic, and field data of granites, in terms of processes which both are theoretically possible and also have been shown in the laboratory to be effective over a range of temperatures and pressures such as prevail at depths of a few miles below the earth's surface.

[99] Cf. Bowen, The granite problem and the method of multiple prejudices, Geol. Soc. America Mem. 28, pp. 80–85, 1948.

Ionic diffusion in single crystals and in multicrystalline aggregates under the influence of chemical or thermodynamic potentials is a theoretically possible process. It has been shown experimentally to be effective, especially at high temperatures, in bringing about chemical reorganization of crystalline materials, including silicates, within restricted fields (measured in millimeters or centimeters). But there is as yet little information as to absolute or relative velocities of migration of different types of geologically important ions (K^+, Na^+, Al^{3+}, etc.) in different silicate media under controlled temperature, pressure, or electrical gradients, or as to the influence of radius, charge, structure, and density of ions upon their velocities of migration. All experimental evidence to date [100] appears conclusive in showing that solid diffusion is too slow, even at temperatures within the magmatic range, to account for large-scale migration of elements over distances characteristic of batholithic granites. We have no sound basis for supposing that a dike of granite several yards in thickness, and even less a batholith 50 miles wide, can grow at the expense of solid rock of some other composition by diffusion of ions in a solid medium. On the contrary, there is abundant evidence of a petrographic or lithologic nature, and some experimental evidence too (cf. Chap. 17), which suggests that, under a wide range of plutonic and metamorphic conditions, "solid diffusion" is effective only through small distances. The heterogeneous nature of many gneissic or xenolith-laden granites and the persistence of zoned structures in crystals of plagioclase and hornblende are examples of this kind of evidence. If solid diffusion under geological conditions were effective in large-scale granitization, surely it would at the same time render initially zoned crystals homogeneous. Moreover, even if ionic diffusion in solids is later found to be much more effective than is at present admitted by "magmatists," it would still have to be shown that the relative velocities of diffusion of different ions are such as would commonly lead to granitization of rocks of varied composition. This, of course, has not yet been proved, nor even indicated, by laboratory experiment. The authors regard the hypothesis of granitization by solid diffusion as an interesting speculation which cannot profitably be applied to the problem of origin of granite batholiths until many more experimental data become available.

Tested by comparable criteria, the hypothesis of magmatic intrusion, as reviewed below, appears today to rest on much firmer foundations.

Evidence for Emplacement of Granite as Magma. The view that large bodies of granitic magma can develop within or penetrate into the outer part of the earth's crust is supported by a strong body of evidence, some of which is noted briefly below:

[100] See Chap. III in W. S. Fyfe, F. J. Turner, and J. Verhoogen, Metamorphic reactions and metamorphic facies, *Geol. Soc. America Mem. 73,* 1958.

1. Contact phenomena. Fine-grained border variants, reasonably interpreted (by analogy with similar variants at the borders of basic intrusions) as chilled marginal facies of granitic magma, have been recorded in many granitic bodies—especially in those of relatively small size, e.g., ring dikes, laccoliths, and dikes. Mobility of the granitic material at the time of emplacement is indicated by such frequently observed phenomena as marginal dikes and other apophyses penetrating open fractures in the wall rock, contact breccias consisting of sharp xenoliths in a continuous granitic matrix, offsetting and rotation of blocks of wall rock, laminar structure and lineation parallel to contacts, ptygmatic folding in migmatites, and so on. Persistence of sharp contacts for scores of miles along the margins of batholiths such as the Sierra Nevada suggests magmatic intrusion rather than granitization by diffusion. The above positive evidence of magmatic intrusion is in no way invalidated by the convincing evidence, in other cases, of diffusion of material across gradational contacts, for such diffusion could be involved in reaction between mobile magma and permeable wall rock.

2. Chemical composition of granitic rocks. On the whole the composition of granitic rocks accords well with that of low-melting silicate liquids which would be expected to develop by differential fusion of a great variety of rocks (basalt, andesite, shale, sandstone) in the presence of water, or by fractional crystallization of basaltic magma. Experimental evidence, accumulated through half a century of research, shows that the relative proportions of albite, potash feldspar, and excess silica (quartz) in such low-melting fractions are exactly those which are typical of granitic rocks. Other characteristic features of granite, e.g., the high ratios of iron to magnesia, of soda to iron, and of soda to lime, have all been found to be characteristic of low-melting liquids in experimentally investigated silicate systems. The evidence is overwhelming that magmatic liquids formed by partial fusion of rocks of various types, or by fractional crystallization of basic or semibasic magma, have played an essential role in the origin of granitic rocks in general. Contrast this situation with the lack of experimental confirmation of evolution of granitic rocks by solid diffusion.

3. Mineral composition of granitic rocks. Statistical studies by Chayes [101] on the relative proportions of the component minerals of many bodies of granitic rocks have revealed persistent compositional characteristics that are wholly compatible with crystallization from largely liquid magma and cannot be reconciled with origin by metasomatism of a

[101] F. Chayes, Composition of the granites of Westerly and Bradford, Rhode Island, Am. Jour. Sci., vol. 248, pp. 378–407, 1950; Composition of some New England granites, New York Acad. Sci. Trans., ser. 2, vol. 12, no. 5, pp. 144–151, 1950; op. cit., 1956.

heterogeneous rock mass: (a) Many plutons, such as those of New England, are amazingly homogeneous in composition; this may be demonstrated even with the most careless and random sampling.   (b) The range of quartz content is 20 to 40 per cent; how could metasomatism of heterogeneous sediments remove all quartz in excess of 40 per cent but leave at least 20 per cent undestroyed in every rock?   (c) Modal proportions of quartz and alkali feldspars (Fig. 53) invariably approximate closely to compositions in a low-temperature trough of the liquidus surface as determined experimentally under high water pressures (plutonic conditions). The general modal composition of granites is perhaps the strongest evidence that most granites have crystallized from magma.

4. Similarity to acid lavas.   The range of composition of the granite granodiorite series is strikingly parallel with that of rhyolites, dacites, and andesites observed to have been erupted as largely liquid magmas. Some of those ubiquitous mineralogical features of granites which have been cited by Perrin and Roubault as difficult or impossible to reconcile with crystallization from a magma are nevertheless found to be equally persistent in rhyolites and dacites of undoubted magmatic origin.   Crystals of plagioclase with complex oscillatory or reversed zoning, associated with others showing simple normal zoning, and habitual appearance of accessory minerals (apatite, magnetite, sphene, zircon) in idiomorphic crystals suggesting early crystallization, may be quoted in this connection. Although it may be necessary to postulate complex conditions of magmatic crystallization to explain these and allied phenomena, it must be admitted that they persistently appear in, and are characteristic of, lavas of unquestionably magmatic origin.   They constitute strong evidence that crystallization of magmatic liquids has also played an essential role in the origin of the vast bodies of granite in which they also appear with equal persistence.   By contrast it should be noted that in metamorphic rocks—products of diffusion and reaction in an essentially solid medium—zoned structure in plagioclase and the tendency toward idiomorphism such as is normally displayed by feldspars of granitic rocks are comparatively rare; and metamorphic plagioclase, moreover, differs markedly from plagioclase of most granites as regards the type of twinning commonly developed.

Contrasted Mineralogy and Texture of Granites and Rhyolites.   While granites and acid lavas have many mineralogical and compositional characters in common, there are also striking and persistent differences between the two groups of rocks:[102] the temperature of high-low inversion of quartz in granites is appreciably lower than that in rhyolites; the sodic plagioclase of granites is invariably the low form, whereas that of rhyolite phenocrysts may be the high form; the alkali feldspar of granite is per-

[102] O. F. Tuttle, Origin of the contrasting mineralogy of extrusive and plutonic salic rocks, *Jour. Geology*, vol. 60, pp. 107–124, 1952.

thitic orthoclase or microcline, while that of rhyolitic lavas is sanidine, anorthoclase, or a corresponding cryptoperthite. These differences can readily be explained by assuming that granites, like rhyolites, have crystallized from magmas at high temperature, but that during slow cooling following plutonic solidification the minerals of granites have undergone reorganization and partial recrystallization consistent with the changing stability of the solid phases with falling temperature. Obvious effects of such reorganization in many granites are unmixing of perthites and growth and development of low-albite borders around coarse grains of alkali feldspar. Viewed thus, the hypidiomorphic granular texture so characteristic of granitic rocks combines some features inherited from magmatic crystallization with others impressed by at least minor postmagmatic recrystallization.[103]

Conclusion. Repeatedly, throughout the earth's history, magmas of granitic or granodioritic composition have burst as lavas from beneath the earth's surface. There is a large body of evidence suggesting that many if not all great masses of granite formed as a result of intrusion, or generation in place, of granitic magmas on a very large scale, followed by reaction between the freezing magma and the invaded or adjoining rocks. Such an origin is in close harmony not only with field observations but also with a multitude of data observed during laboratory investigation of crystallization phenomena in silicate melts belonging to a large number of selected systems. In the absence of similar experimental support, the hypothesis of origin of granitic batholiths by diffusion of ions through solid rocks cannot be regarded as a serious alternative to the magmatic theory. The "magmatist" must admit, from evidence afforded by transitional contact zones, that under favorable circumstances exchange of material between magma and wall rock may be an important petrogenic process, perhaps leading to granitization of the wall rock for considerable distances from the original contact. In this process, ionic diffusion through, or along the boundaries of, crystal lattices possibly plays a significant though as yet undetermined role. Probably much more effective, however, is a one-way introduction of certain ions (K^+, Na^+, Al^{3+}, Mg^{++}, etc.) into the wall rock, with complementary removal of replaced ions in the same direction, through the medium of aqueous and other mobile fluids continuously expelled from the crystallizing granitic magma. This could account for the phenomena of peripheral metasomatism, mineralization, and metamorphism (including development of "basic fronts") as being normal accompaniments of large-scale intrusion of granitic magma.

[103] For criticism of this interpretation and a reply to criticism see: R. Perrin and M. Roubault, *Jour. Geology,* vol. 61, pp. 275–278, 1953; O. F. Tuttle, *Jour. Geology,* vol. 61, pp. 278–280, 1953.

Origin and Emplacement of Granitic Magma. *Preliminary Statement.* The authors' tentative acceptance of the hypothesis that most large bodies of granitic rocks have formed from granitic magma has been stated above. Several major objections that have been raised against this view must be mentioned briefly before the hypothesis is elaborated. Satisfactory answers must be sought to such questions as these: Can immense batholiths ever have been in a mobile, essentially liquid condition? Were most granites formed from magmas once entirely liquid? What explanation can be offered for the great total bulk of granitic as contrasted with basic plutonic rocks? How has room been made for the injection of immense bodies of granitic magma into the outer part of the earth's crust? Geologists of the "transformationist" school claim that negative or unsatisfactory answers given by "magmatists" to these questions invalidate the hypothesis of magmatic origin of granites. This criticism is largely dispelled, however, if it is admitted that many, if not most, bodies of granitic magma were partially crystalline at the time of intrusion; *i.e.,* that they correspond to the mobile "migma" of "transformationists."

A vast mass of granodiorite and granite such as the Sierra Nevada batholith cannot seriously be attributed to a single upsurge of liquid magma. Batholiths of this magnitude are invariably heterogeneous—both petrographically and structurally—on a large scale. Tens or perhaps even hundreds of separate major injections must have been involved in their development. Even within a single component mappable unit of a composite batholith, some degree of flow structure (banding, lineation, and alignment of tabular crystals, crystalline aggregates, xenoliths, etc.) shows that the magma flowed into place already in a partially crystalline condition. In some intrusions, *e.g.,* the Colville batholith of Washington (page 355), crystallization must have been far advanced at the time of intrusion. Moreover, sharp contacts between component members of composite batholiths often testify to complete solidification of one draft of magma prior to intrusion of the next.

Granitic intrusions vary very greatly in size and structure. There is every gradation from great composite batholiths which seem to represent repeated intrusion of partially crystalline, variously contaminated magma, to the smaller bodies (ring dikes, laccoliths, stocks, and dikes) of more homogeneous granite which could well have been formed from essentially liquid magmas penetrating to within short distances of the earth's surface. At the opposite extreme are the highly heterogeneous migmatite bodies of the deep basement. Some of these have a bulk composition falling within the granite granodiorite range and were evidently at one time mobile enough to invade the overlying rocks bodily as intrusive masses. These rose into place then as magmas, but there is no reason to believe that these particular magmas were ever predominantly liquid.

Granitic magmas formed by differentiation of a parent basic magma would be entirely liquid at the moment when they came into being. They would form as liquid residues strained away from a mass of early-formed crystals. This conclusion applies equally well, whether or not we assume that prior to differentiation, the basic magma became acidified by assimilative reaction with solid granitic rocks of the continental crust. Granitic magma could also form by differential fusion of various rocks (sandstones, shales, granites, or even gabbros). Magmas of this kind would range from completely liquid granite, approximating in composition to a eutectic of quartz and alkali feldspar, to more basic granites and granodiorites (mobilized migmatites) containing a noteworthy amount of unfused crystalline material. These two modes of origin (by differentiation and by partial fusion) are in complete harmony with the experimentally derived data of silicate chemistry. Their relative roles in the genesis of granites may be judged, however, from a purely geological criterion, namely, the great total bulk of granitic rocks in comparison with that of associated basic igneous rocks. A given bulk of basic magma, even if previously contaminated by reaction with acid rocks, can yield only a limited quantity of granitic differentiate. The observed great predominance of granitic rocks in the continental basements accords with the possibility that, at some early stage in the earth's history, a thick shell of basic magma differentiated to give a dominantly granitic, relatively thin, outer crust, underlain by a much greater bulk of basic and ultrabasic rocks. But we must turn to a mechanism of differential fusion to account for the high proportion of granite rocks in the plutonic intrusions that have repeatedly invaded the continental crust throughout recorded geological time. This recurrent invasion of the outer crust by granitic magma on a very large scale, and the continuous growth of a primitive granitic continental crust, could both be explained by assuming that the rocks of the outer part of the earth's crust have been subject, from time to time and at diffierent levels, to temperatures high enough to cause separation of a granitic liquid phase.

There still remains the perplexing "room problem" to which Read and other writers of the "transformationist" school have repeatedly drawn attention. Most writers agree that emplacement of relatively small bodies of granitic magma—as in near-surface ring dikes and laccoliths—is compensated by structural adjustments in the invaded rocks: doming of the roof, subsidence of crustal segments on annular fractures, and so on. Vast stratiform masses of basic magma, e.g., the Bushveld lopolith, have been similarly accommodated. But it is difficult to visualize the emplacement of a batholithic mass of granitic magma, especially if it is assumed (as it is by many geologists) that batholiths widen indefinitely downward. The formidable difficulty raised by this "room problem" is greatly reduced, however, if we suppose that much of the matter making up any large batholith has not been introduced into its present surroundings. This could be so if the granite in question were formed purely by metasomatism or if it crystallized from magmas resulting from partial fusion of rocks *in situ*. From this standpoint batholithic intrusion is merely a phase of regional deformation—the upward squeezing of the more mobile partially liquid portion of a mechanically heterogeneous rock mass. The "room problem" is also rendered less acute if it be accepted that many batholiths in reality are sheet-like in form (cf. page 341).

Statement of the Theory. If the extreme hypothesis of granitization by solid diffusion as propounded by Perrin and Roubault, Bugge, and Ramberg is excluded, we find surprising unanimity of opinion in recent writings on the origin and general course of evolution of granitic magmas. In this connection the reader will find it instructive to compare views expressed within the last two decades by leading authorities of many nationalities. E. Raguin's book on the geology of granite [104] is the logical conclusion to half a century of discussion by the brilliant French school which earlier included Lacroix, Michel-Lévy, and Termier. The conclusions of Eskola, Backlund, and Wegmann [105] are the culmination of a lengthy series of researches upon the pre-Cambrian complexes of Scandinavia and Finland, classical contributions to which were made earlier by Sederholm and Holmquist. The trend of modern Swiss opinion is reflected in papers by Reinhard and by Niggli; [106] that in Britain and the United States by valuable documented summaries by MacGregor and Wilson, Read, Holmes, Bowen, and others.[107] Particularly interesting is Read's statement summarizing the evolution of his own ideas on the granite problem.[108] By all these writers genesis of granitic magma (in the broad sense of the word as here employed) is attributed to differential fusion of mixed rocks in the continental basements.

The hypothesis of Eskola (1933) is summarized below as illustrating the kind of relation generally envisaged between different types of granite and depth. Eskola recognizes three indefinitely bounded depth zones, subject to local upward displacement in orogenic belts or to local inver-

---

[104] Raguin, *op. cit.*, 1957.

[105] Eskola, *op. cit.*, pp. 12–25, 1933; Read, *op. cit.*, pp. 68–73, 1944; Holmes, *op. cit.*, 1945; Barth, *op. cit.*, p. 235, 1948.

[106] M. Reinhard, *Uber die Entstehung des Granits*, Hebling and Lichtenhahn, Basel, 1943; P. Niggli, Das Problem der Granitbildung, *Schweizer min. pet. Mitt.*, vol. 22, pp. 1–84, 1942; Barth, *op. cit.*, pp. 237, 238, 1948.

[107] MacGregor and Wilson, *op. cit.*, 1939; Read, *op. cit.*, 1944; Holmes, *op. cit.*, 1945; N. L. Bowen, Magmas, *Geol. Soc. America Bull.*, vol. 58, pp. 275–277, 1947. See also symposium on origin of granite, *Geol. Soc. America Mem. 28*, 1948.

[108] Read, *op. cit.*, pp. xi–xix, 1957.

sion in regions where there has been extensive lateral migration of magma:

1. The zone of differential anatexis. In most rocks in this deep-seated zone (very basic igneous rocks excepted), the intergranular spaces are filled with a water-rich silicate melt (magma) of aplitic or granitic composition, resulting partly from differential melting of the rock itself and partly from an upward influx of magma, probably itself of palingenic origin, from still greater depths. The residual material in the host rock constitutes the metamorphic component of the resulting migmatite, and its bulk composition (and hence mineral composition) is of course altered by the selective nature of the incipient fusion. The newly formed liquid phase tends to be squeezed upward by crustal movements aided by gravity, and so collects into larger and more continuous masses as it migrates along surfaces of minimal resistance. Much of the magma remains within the rock, however, either as an intergranular liquid or else segregated as veins and streaks in venites. In some cases the partially fused rock itself becomes sufficiently mobile to be injected bodily as a mixture of crystals and liquid, i.e., as a partially liquid magma, so that the granitized migmatite comes to have sharply defined intrusive contacts with other migmatites that have never attained a mobile condition.

2. The zone of injection and potash metasomatism. The upward-migrating, largely liquid granitic magma has now coalesced into intrusive bodies of phacolithic or batholithic proportions, which diminish and fray out upward into a fringe of pegmatite and aplite veins, toward the boundary with the overlying zone. Many of the major granitic intrusions of this zone may perhaps be regarded more properly, however, as mobilized migmatites in an advanced stage of granitization, injected as mixtures of liquid and crystals, in a mushy condition. Close to the intrusions, the host rocks commonly are converted into migmatites by combined injection and impregnation with the invading granitic magma and its derivative, residual, water-rich silicate melts (pegmatite magmas). The resulting veined migmatites are mainly arterites, such as Sederholm has described from Finland. Further from the intrusive contacts the zone of migmatites (arterites and permeation gneisses) may merge into zones of metasomatism representing the advance of potash, soda, or magnesia-iron "fronts," and marking the transition between migmatites and metamorphic rocks proper. These in turn give way to more remote zones of pure contact metamorphism without noteworthy metasomatism.

3. In the uppermost zone, such metamorphism as occurs is mainly regional and may be accompanied by purely hydrothermal metasomatism. Granitic intrusions and pegmatite veins are rare, but quartzose veins of hydrothermal origin are abundant and represent the only material contribution expelled from the magmas congealing in the depths beneath.

Eskola emphasized evolution of granites as a function of depth. His was an easily grasped, idealized, simplified statement of what in fact must be a highly varied phenomenon conditioned by other variables than depth. Other European geologists—especially the French school— have discussed the evolution, deployment, and emplacement of granites as a function of time,[109] especially in relation to orogeny. Wegmann has presented a picture which is more complex and perhaps truer than that given by Eskola in that he relates the development of granites to both place and time. He stresses the variability introduced into the environment of evolving granitic magmas by the interplay of such factors as fluctuating temperature gradients, composition of pore fluids at a given place and time, relative velocities of advancing waves of metasomatism ("fronts") and of waves of rising or falling temperature, deformation preceding or synchronous with evolution of the magma, and so on. Read's summing up of some of Wegmann's ideas (cf. Figs. 116 and 117, pages 670 and 671) may be cited:[110]

A distinction must be drawn, he [Wegmann] maintains, between happenings in the non-migmatitic superstructure and the migmatitic infrastructure lying below. The two zones fold in disharmonic fashion, and the migmatite zone comes to fill the arches in the superstructure, as is seen in Greenland. Between the two zones is a transitional zone which is the site of the so-called regional metamorphism. . . . The folding of the superstructure is of two kinds, by contraction of the segment or by collection of the infrastructure into arches and its heaving up. In these compressed arches of the infrastructure, recrystallization is especially intensive; in many cases obvious granite stocks break through the folded gneisses; when these stocks are investigated in detail, however, many of them prove to be recrystallized developments of strongly folded migmatites. By strong folding of the superstructure, tongues of the infrastructural type are formed, and extreme kneading of such tongues and their recrystallization convert them into granite intrusions in some of which a migmatitic origin is no longer discernible. These tongues of infrastructural nature can form at different phases; when they are overhauled by the migmatite front they are the Early Granites; when they are developed at the maximum of the migmatitization and invade the transitional zone they are the Main Granites, when they are formed during the retreat stage of the fronts [i.e., when the waves of metasomatism and high temperature are receding], they intrude the already solidified part of the migmatite zone and are the Later Granites.

The migmatite front is considered by Wegmann to encroach from below on to the base of the superstructure; heat and material flow in and create an infrastructural condition, whereby geosynclinal sediments become the host-rocks for fresh migmatitic cycles. . . . The source [of the invading material] is

---

[109] For a summary of the views of French geologists (notably A. Demay, E. Raguin, J. Jung, and M. Roques) see Read, op. cit., pp. 308–318, 324–328, 1957.
[110] Read, op. cit., pp. 69, 70, 1944.

usually supposed to be from associated intrusives. Many of these intrusives
. . . are of mixed origin. It must be confessed that in many regions the source
of the invading materials is not yet known. They must come from great depths
at which they must occur in quantity. If they arise from deep melts, two possibilities may be entertained: they may be concentrated by crystallization-differentiation or they may be squeezed out from the depths [presumably products
of fusion].

The ideas of Eskola and of Wegmann, as set forth above, constitute
alternative models of granite petrogenesis developed by two students of
granite and orogeny. Other models of the same phenomenon, differing
mainly in the degree of stress laid upon particular aspects of granite
evolution, may commend themselves more or less to the tastes and prejudices of individual readers. One such is Read's concept of the Granite
Series [111]—a synthesis developed during prolonged meditation and debate
upon the granite problem, and influenced strongly by the French school
of petrology. The Granite Series presents "a unified sequence of plutonic
events that [can] be followed through time and place in an orogenic
belt": [112]

1. The series begins in the depths with the development of migmatites
by granitization involving a "state of extreme chemical mobility" (*i.e.,
incipient fusion*).[113] The product is an irregularly and indefinitely
bounded anatectic batholith or migmatite complex.

2. Sectors of the migmatite in which granitization is advancing "begin
to move," mainly by virtue of "chemical mobility" (*i.e., the melt fraction
locally increases in quantity to a point where it begins to impart mobility
to the sector concerned*). The subautochthonous batholiths so formed
have "risen slightly from their anatectic roots" and, at least locally, cut the
metamorphic country rocks.

3. The granitized core "parts company with its envelope" of regionally
metamorphosed rocks and "true mechanical mobilization has begun"
(*i.e., large volumes of partly liquid, partly crystalline material become
squeezed upward and away from largely solid metamorphic residual portions of the complex*). The granitic product takes the form of intrusive
batholiths with obviously cross-cutting contacts.

4. Finally the granitic material "may free itself completely from its
plutonic associates and move high into the crust, even into nonplutonic
regions" (*i.e., the largely or completely liquid magma squeezed from the*

[111] Read, *op. cit.*, p. 17, 1948; *op. cit.*, pp. 143–151, 1949; *op. cit.*, pp. 324–327,
1957; Buddington, *op. cit.*, pp. 675–676, 1959.

[112] Read, *op. cit.*, p. xvi, 1957.

[113] Expressions in quotation marks are taken from Read, *op. cit.*, pp. 324–327,
1957. Italicized expressions in parentheses convey the present authors' interpretation of the processes involved.

deep zone of anatexis invades the lightly metamorphosed or nonmeta-morphic environment of the upper crust). The products are cross-cutting stocks, ring-dikes, laccoliths—the massifs circonscrits of French petrologists, plutons of the epizone in Buddington's terminology.

As set forth above, the Granite Series comprises a sequence of evolutionary products of crustal fusion developed continuously in time. Early members of the series tend to be deep-seated, syntectonic, and granodioritic in composition; late members tend to invade the upper levels, to be late syntectonic or post-tectonic and to have a more potassic composition (quartz-monzonite or granite proper). Only large regions are adequate to display in full the Granite Series as developed in one orogeny. Such regions provide sections in both place and time.

The general scheme of evolution of granitic magmas via migmatites from the deep-seated rocks of the earth's crust as pictured by Eskola, by Wegmann, and by Read, may be accepted tentatively as a working hypothesis to account for migmatite areas and granite batholiths. The present authors think it probable that largely liquid granitic and granodioritic magmas, formed by almost complete fusion of deep-seated rocks or squeezed upward from deep zones of partial fusion, have repeatedly invaded the upper crust on a large scale. Such intrusive bodies of mobile magma are considered to be the main source of chemically equivalent andesite-dacite-rhyolite lavas erupted in great quantity from volcanoes of orogenic belts. Magmas of this type are so far removed from the deep zone of fusion that no obvious trace of the migmatite stage of their development remains. In addition to these extensive primary granite-granodiorite magmas, we must recognize the existence of small shallow bodies of acid magma which may well be products of differentiation of basaltic magma. The small granitic intrusions which at times are associated with greatly predominant basic volcanic rocks (as in the Hebridean region of the Brito-Arctic province) may have originated thus. Certainly this is the mode of origin of granophyre lenses in the upper levels of some tholeiitic diabase sills. But the great bulk of granitic and granodioritic magmas seem to have no genetic connection with basaltic magma.

CHAPTER 13

Nepheline Syenites, Ijolites, and Carbonatites

INTRODUCTION

Rocks of the nepheline syenite family [1] are widely distributed in sectors of the continents characterized tectonically by crustal stability or simple fracturing. Chemically (cf. Table 35) they are characterized by high alkalis, low silica, low combined CaO, FeO, and MgO, high ratio FeO/MgO, and exceptional richness in "volatile" and rare elements (P, F, Cl; Zr, Ti, Nb, Ta, and rare-earth metals). The principal constituent minerals are feldspathoids (nepheline being by far the commonest), alkali feldspars, and ferromagnesian silicates (aegirine, biotite, hornblende); but minerals which otherwise are rare (e.g., eudialyte, astrophyllite, tantalite, columbite) or of merely accessory status (zircon, sphene, apatite, garnet) may rank as essential minerals in some of the many varieties of nepheline syenite.

The ijolites and their allies (urtites, jacupirangites, etc.) are basic and ultrabasic nonfeldspathic alkaline rocks consisting mainly of nepheline and pyroxenes; apatite, sphene, perovskite, melanite, and other accessory minerals are relatively plentiful.

Intrusions composed entirely of nepheline syenites, ijolites, and allied feldspathoidal rocks are mostly circular or elliptical in plan and of small size, though a very few are known which outcrop continuously over areas of several hundred square miles. Large masses even remotely comparable in size with the great granite granodiorite batholiths or with stratified basic sheets and lopoliths of the Bushveld type are unknown. We are safe in concluding that magmas of nepheline-syenite composition, though not uncommon, typically develop in volumes very much smaller than basaltic and granitic magmas.

[1] Many aspects of petrogenesis of nepheline syenites have been summarized by C. E. Tilley (Problems of alkali rock genesis, *Geol. Soc. London Quart. Jour.*, vol. 113, pp. 323–360, 1958). This paper has a good bibliography.

Associated with feldspathoidal rocks (ijolites and shonkinites more commonly than nepheline syenites) of some provinces are the interesting and perplexing rocks known as carbonatites. These consist principally of carbonates. They are generally thought to be of igneous origin on account of their mineralogy (presence of nepheline, aegirine, etc.), field occurrence (as ring dikes, cone sheets, and other apparently intrusive bodies), and association with alkaline plutonic rocks. Like the latter they occur in stable or fractured continental regions and are rich in rare elements.

MAGMATIC NEPHELINE SYENITES AND IJOLITES

Nepheline Syenites in Alkaline Olivine-Basalt Provinces. The typical differentiated associates of alkaline olivine basalts are trachytes and phonolites. Chemically identical with these are syenites and nepheline syenites which appear in insignificant volume in some alkaline olivine-basalt provinces. Thus in the Dunedin district of East Otago, New Zealand, (see pages 169, 170) there is one small intrusion of nepheline syenite, and other rocks of the same family are represented by cognate xenoliths in some alkaline lavas and by blocks in trachyte breccia. Another common but inextensive mode of occurrence of syenite and nepheline syenite is in lenses and streaks in differentiated basic alkaline sills (teschenite, shonkinite, etc.; see pages 180, 243). It is clear, then, that small quantities of nepheline-syenite magma may develop by differentiation of alkaline olivine-basalt magma; and under plutonic conditions this will crystallize as true coarse-grained nepheline syenite.

The Post-Karroo Plutonic Complexes of Southwest Africa.[2] In Damaraland, southwest Africa, a dozen ring complexes occur in a belt 250 miles long and 100 miles wide. Some of these are principally granitic; in others nepheline syenites and syenites are associated with gabbros and granites (Messum) or with gabbros alone (Okonjeje). In the province as a whole there seem to be at least two independent parent magmas, one granitic and the other basic.

In the Okonjeje complex (Table 35, analyses 1–4) there are two lines of differentiation descending from similar but not identical basic parent magmas:

1. Gabbro (slightly undersaturated) → ferrogabbro (oversaturated) → ferrosyenite (oversaturated).

2. Olivine gabbro (highly undersaturated) → andesine essexite → oli-

---

[2] H. Korn and H. Martin, The Messum igneous complex in South-west Africa, *Geol. Soc. South Africa Trans.*, vol. 57, pp. 84–124, 1954; E. S. W. Simpson, The Okonjeje igneous complex, South-west Africa, *Geol. Soc. South Africa Trans.*, vol. 57, pp. 125–172, 1954.

TABLE 35. CHEMICAL COMPOSITIONS OF NEPHELINE SYENITES, IJOLITES, AND ASSOCIATED ROCKS

Constit- uent	1	2	3	4	5	6	7	8*	9
SiO_2	49.32	59.50	49.96	58.35	54.83	59.04	45.73	35.5	39.5
TiO_2	1.30	0.70	0.80	0.49	0.39	1.80	3.30	7.4	2.9
Al_2O_3	18.87	19.29	22.30	18.10	22.63	19.50	16.27	4.1	14.0
Fe_2O_3	1.90	1.65	2.54	3.27	1.56	1.44	2.93	9.4	7.6
FeO	6.39	2.32	2.25	3.51	3.45	2.10	8.05	6.9	2.8
MnO	0.15	0.03	0.03	0.04	Trace	Trace	0.55	0.1	0.2
MgO	3.15	1.10	0.68	0.50	Trace	0.24	3.63	11.9	4.3
CaO	7.43	2.53	3.86	2.23	1.94	2.46	9.99	19.7	18.3
Na_2O	5.53	5.97	8.11	6.79	10.63	6.78	4.68	0.6	5.2
K_2O	4.14	5.78	6.04	6.07	4.16	5.02	2.42	1.0	2.4
H_2O+	0.60	0.84	1.27	0.30	0.18	0.82	0.85 ⎫	1.3	1.2
H_2O-	0.03	0.12	0.10	0.04		0.13	0.25 ⎬		
P_2O_5	0.86	0.34	0.27	0.21		1.05	0.96	1.3	1.2
CO_2	0.25	0.28	1.88	0.17	0.06	Trace	0.38	0.2	0.5
Cl	0.28	Trace	0.40	0.30	0.82			Trace	
F	0.10	0.04	0.07	0.07				0.1	
Total	100.30	100.49	100.56	100.44	100.65	100.38	99.99	99.5	100.1

* BaO, .1; SrO, .1.

Explanation of Table 35

1. Andesine essexite, Okonjeje complex, Southwest Africa. (E. S. W. Simpson, *Geol. Soc. South Africa Trans.*, vol. 57, facing p. 172, Table IIa, No. 195, 1954.)
2. Pulaskite, Okonjeje complex. (Simpson, *op. cit.*, Table IIa, No. 143, 1954.)
3. Foyaite, Okonjeje complex. (Simpson, *op. cit.*, Table IIa, No. 47, 1954.)
4. Tinguaite, Okonjeje complex. (Simpson, *op. cit.*, Table IIa, No. 42, 1954.)
5. Nepheline syenite (ditroite), Ditro, Carpathians. (A. Streckeisen, *Das Nephe-linsyenit-Massif von Ditro, Schweizer min. petr. Mitt.*, Band 34, Heft 2, p. 371, No. 10, 1954.)
6. Alkali syenite (nordmarkite), Ditro. (Streckeisen, *op. cit.*, p. 369, No. 4, 1954.)
7. Essexite, Ditro. (Streckeisen, *op. cit.*, p. 371, No. 16, 1954.)
8. Average biotite pyroxenite, Iron Hill, Colorado. (E. S. Larsen, *U.S. Geol. Survey Prof. Paper, 197–A*, p. 36, 1942.)
9. Average ijolite, Iron Hill, Colorado. (Larsen, *op. cit.*, p. 36, 1942.)

goclase essexite → pulaskite (slightly undersaturated syenite). Highly feldspathoidal foyaites and tinguaites have developed in minor quantity, probably along a secondary lineage branching from the andesine-essexite magma. The minor elements show much the same pattern of distribution as in the olivine-basalt trachyte phonolite association. Ba and Sr rise to high concentrations in intermediate members (essexites and pulaskite) and fall sharply to low values in the end members of the series (foyaite and tinguaite). The latter are characterized by high Rb and Zr and by appreciable concentrations of Li, La, and Y.

In the Messum complex the following sequence of events has been reconstructed: [3]

1. Extrusion of tholeiitic basalts of the Stormberg lava series
2. Building of a huge volcanic cone—basalts and rhyolites and corresponding pyroclastics
3. Intrusion of a gabbro lopolith; differentiation of anorthosites
4. Cauldron subsidence of the whole volcano; upwelling of aplogranite on ring faults; marginal granitization
5. Intrusion of radial diabase dikes
6. Cauldron subsidence of the core; intrusion of syenite and then tinguaite
7. Intrusion of foyaite into the vent; nephelinization and syenitization of wall rocks
8. Intrusion of dikes of nepheline basalt and nephelinite

Field, petrographic, and chemical data regarding the Damaraland province are consistent with derivation of small amounts of syenite and nepheline-syenite magmas by differentiation of an undersaturated olivine-basalt magma.

The Ivigtut Province, Southwest Greenland.[4]   The mid-Paleozoic intrusive rocks of southwest Greenland include a number of separate plutons—some granitic, some composed mainly of nepheline syenites.   They probably belong to a single plutonic comagmatic series, which also includes the well-known cryolite pegmatite of Ivigtut.   One of the alkaline intrusive bodies near Ivigtut is the nepheline-syenite pluton of the Gronna Dal-Ika area, the exposed portion of which approximately 5 by 2 miles, invades and is completely surrounded by plagioclase-biotite gneisses, granitic gneisses, and amphibolites of the pre-Cambrian basement.   Mineralogically this pluton is a relatively simple type.   Rare elements and minerals are generally lacking, even in associated pegmatites.   The dominant rock is foyaite (microcline, albite, nepheline, cancrinite, biotite, aegirine-augite), which grades into pulaskite (microcline, altered nepheline and chloritized biotite), and near the borders of the intrusion into more mafic variants including the extreme type ijolite (nepheline, aegirine-augite).   Contacts with the country rock everywhere are sharp.   Presence of intrusive apophyses, and a general decrease in grain size of the syenite near the contacts, support the orthodox view that the pluton represents an intrusive body of magma.

[3] Korn and Martin, op. cit., p. 83, 1954.
[4] K. Callisten, Igneous rocks of the Ivigtut region, Greenland, Part I, *Meddelelser om Grönland*, vol. 131, no. 8, 74 pp., 1943.

The Khibine Pluton of Kola Peninsula, U.S.S.R.[5] The Khibine nepheline-syenite pluton of northwestern Russia, probably Devonian in age, is remarkable both for its great extent (over 500 square miles) and for its mineralogical and petrographic variety. Seven phases of intrusion, listed below in order of decreasing age, are recognized:

1. Syenites and nepheline syenites. Adjoining Archean gneisses show evidence of both injection and metasomatism by the syenitic magmas.

2. Coarse aegirine-nepheline syenites injected as a peripheral ring dike.

3. Trachytoid aegirine-nepheline syenites forming an imperfect ring within and adjoining the coarse rocks of phase 2.

4. An imperfect ring dike of poikilitic micaceous nepheline syenites of variable composition. The assemblage alkali feldspar, nepheline, mica, aegirine, astrophyllite is typical.

5. A massive cone sheet of silica-poor nepheline-rich rocks of the ijolite family.

6. Foyaites (alkali feldspars, nepheline, biotite, and hornblende or aegirine-augite) making the central core of the pluton.

7. Dikes of tinguaite, monchiquite, shonkinite, theralite, nepheline basalt, and leucite basalt.

Chemically the pluton is characterized by exceptional abundance of P, F, Ti, Zr, Nb, Sr, and rare earths (especially the Ce group). The great diversity of Zr-Nb-Ti silicates is unique. Rocks of extreme composition occur on an unprecedented scale and constitute valuable ores. Among these may be mentioned apatite ores, sphene ores, and aegirine pegmatites rich in rare earths. One body of apatite rock, consisting of apatite (45 to 65 per cent) nepheline, aegirine, and minor titanium minerals, is a flat lens $1\frac{1}{2}$ miles long and 500 ft. thick. By contrast with certain other nepheline-syenite provinces, calcite and other carbonates are rare.

The Iron Hill Stock of Colorado.[6] This is a composite alkaline stock of unknown age cutting pre-Cambrian gneisses. The principal component is biotite pyroxenite, which makes up 70 per cent of the exposed mass. The sequence of intrusion is as follows:

1. Melilite rock (uncompahgrite): melilite, diopside-hedenbergite, magnetite, phlogopite, perovskite, apatite.

2. Biotite pyroxenite (Table 35, analysis 8): an extremely variable rock, the most abundant variety of which consists of pyroxene (60 per cent), biotite (15 per cent), magnetite, perovskite, and accessories. Variants include biotite rock, magnetite-perovskite rock, apatite pyroxenite, nepheline pyroxenite, and others.

[5] The Kola Peninsula, 16th Internat. Geol. Cong. Guidebook, U.S.S.R., pp. 51–82, 1937.

[6] E. S. Larsen, Alkalic rocks of Iron Hill, Gunnison County, Colorado, U.S. Geol. Survey Prof. Paper, 197–A, 1942.

3. Ijolite (Table 35, analysis 9): nepheline, diopside-hedenbergite, melanite.

4. Soda syenite: sodic microperthite and aegirine-augite.

5. Nepheline syenite: sodic microperthite, nepheline, aegirine, biotite, and accessories.

6. Dikes of nepheline gabbro and quartz gabbro.

There is a large mass and several dikes of carbonatite (dolomite, calcite, ankerite), the position of which in the intrusive sequence is uncertain.

Nepheline Syenites Associated with Granite. In many alkaline plutonic complexes nepheline syenites are closely associated with quartz syenites and granites. In some such occurrences (*e.g.*, in the Damaraland province) field and time relationships indicate that the undersaturated and oversaturated rocks probably stem from independent sources. But more commonly a gradational relation between the two classes of rock strongly suggests community of origin. Tilley[7] has distinguished two general kinds of association:

1. Alkaline complexes closing with quartz syenite or granite

2. Alkaline complexes including granites but closing with nepheline syenite

Orthodox interpretation of the field, chemical, and petrographic data would ordinarily suggest that these are differentiation sequences—nepheline syenite to granite and vice versa. But such a relationship is opposed by experimental evidence in the system $NaAlSiO_4$-$KAlSiO_4$-SiO_2, The liquidus surface shows a high-temperature ridge coinciding with the alkali feldspar join (Fig. 14, page 120); and this ridge constitutes a thermal barrier separating undersaturated from oversaturated feldspathic liquids. It is possible that this barrier does not persist in the presence of mafic components (such as aegirine) or volatiles other than water. Further experimental evidence is needed.

Petrogenesis.[8] The course of differentiation of one of the most widely developed magmas—the alkaline olivine-basalt magma type—normally leads to end fractions of trachytic (syenite) and phonolitic (nepheline syenite) composition (cf. pages 196). The undersaturated condition of the differentiates appears to be inherited from the parent magma. There is no need to assume any other mode of origin for many bodies of nepheline syenite—especially where these are minor associates of more exten-

[7] Tilley, *op. cit.*, p. 332, 1958.

[8] C. H. Smyth, The genesis of alkaline rocks, *Am. Philos. Soc. Proc.*, vol. 66, pp. 535–580, 1927; N. L. Bowen, *The Evolution of the Igneous Rocks*, pp. 234–236, 240–257, Princeton University Press, Princeton, N.J., 1928; F. Chayes, Alkaline and carbonate intrusives near Bancroft, Ontario, *Geol. Soc. America Bull.*, vol. 53, pp. 449–512, 1942 (especially pp. 499–510); S. J. Shand, The present state of Daly's hypothesis of the alkaline rocks, *Am. Jour. Sci.*, vol. 243A, pp. 495–507, 1945; Tilley, *op. cit.*, pp. 334–337, 1958.

sive somewhat alkaline gabbro. Moreover if this view is accepted, concentration of volatile constituents and rare elements in nepheline syenites requires no special explanation, for these are the very components that tend to be concentrated in trachytes and phonolites. To explain on the above basis any relatively large alkaline complex that lacks significant alkaline gabbro members—the Kola Peninsula pluton, for example—would imply the existence of an immense subjacent mass of basic rock; but this is not an impossible supposition, for large alkaline complexes are rare, and extensive intrusions and outpourings of basic magma are common.

Close association of granitic with nepheline syenite members in certain complexes and provinces does not prove that the two have originated from a common parent magma. Elsewhere (page 287) we have discussed the mutual association of olivine basalt and rhyolite in some provinces, and concluded that they represent independent magmas rising along the same zone of crustal weakness or even alternately utilizing the same volcanic vent. Relations between granite and nepheline syenite in many complexes could be accounted for by assuming the existence of two magmas of independent origin. In other provinces, such as the Oslo district of Norway (page 345), where nepheline syenites are but minor associates of greatly predominant saturated and oversaturated rocks, it seems likely that—by some mechanism not precisely understood—the alkaline silica-deficient magma has differentiated from a parent syenite-monzonite magma saturated in silica.[9] Although we do not precisely comprehend the evolutionary process, we strongly suspect from alternate effusion of rhyolitic and trachytic lavas in some olivine-basalt provinces that in natural magmatic systems the experimentally determined thermal barrier between oversaturated and undersaturated feldspathic liquids is sometimes suppressed or overcome. Perhaps partial pressure of oxygen plays some significant role in determining the extent to which Fe and Ti enter into oxide or silicate phases, thus influencing the silica content of late liquids.

In literature on nepheline syenites the classic limestone assimilation hypothesis of Daly and Shand still is repeatedly invoked to explain derivation of feldspathoidal from saturated or oversaturated calc-alkalic magmas. Granitic or granodioritic magma is supposed to become desilicated by reaction with limestone to produce lime silicates such as make up skarns

[9] In the Oslo region Barth recognizes at least two parent magmas: alkaline olivine basalt (essexite) and syenite-monzonite (kjelsåsite and larvikite). The syenite-monzonite magma is slightly oversaturated or saturated. It gives rise to two lines of descent: an oversaturated line leading to quartz syenite, soda granite, and granite and an undersaturated line leading to nepheline syenites (e.g., lardalite). (T. F. W. Barth, The igneous rock complex of the Oslo region, *Norske Vidensk.-akad. Oslo Skr., I, Math.-Naturv. Kl. 9,* pp. 96–98, 1945; S. I. Tomkeieff, The Oslo petrographical province, *Sci. Progress,* vol. 45, pp. 429–446, 1957.)

and tactites. There are numerous records of nepheline syenite invading limestone and dolomite in regions where nearby granite intrusions could have supplied the parent magma.[10] However, in view of the widespread occurrence of calcareous sediments that is not surprising. Skarns are common enough at granite-limestone contacts; but marginal feldspathoidal variants of calc-alkaline plutonic rocks (mainly of basic and intermediate composition) are rare indeed, and are quantitatively insignificant.[11] Several decades of investigation have failed to confirm the efficacy of limestone assimilation as a significant factor in the development of nepheline-syenite magma.

Finally there is the possibility that primary nepheline-syenite magmas have developed repeatedly by differential fusion of appropriate materials deep within the crust. Tilley has tentatively suggested something of this kind [12]—relief of pressure resulting in decomposition of complex minerals (hornblende or biotite) or high-pressure phases such as jadeite or omphacite. This hypothesis remains to be tested by future experiments on the melting behavior of jadeite and other sodic pyroxenes at high water pressures. As the evidence now stands it seems unlikely that such different processes as fractional crystallization of olivine-basalt magma and partial decomposition of crustal rocks would yield identical magmas having the striking compositional characters typical of nepheline syenites. However, complete or partial fusion of subcrustal materials, possibly followed by reaction or differentiation deep within the crust, is perhaps the most likely mode of origin for the basic and ultrabasic alkaline magmas from which the rocks of the ijolite family and some nepheline syenites seem to have been derived.

In conclusion it should be noted that the composition of most nepheline syenites [13] lie in the low-melting region (adjacent to R in Fig. 14, page 120) of the silica-deficient sector of the system $NaAlSiO_4$-$KAlSiO_4$-SiO_2. Syenites closely approximate the point of minimum melting temperature on the alkali feldspar join ($Or_{30-35}Ab_{65-70}$). These facts confirm the magmatic origin of nepheline syenites and syenites; and they strongly suggest that nepheline syenite magmas come into existence in the early stages of differential fusion or in the last stages of fractional crystallization of natural systems undersaturated in silica. The authors' preference is for the latter alternative.

[10] E.g., J. F. McAllister, Quartz Spring Area, California Div. Mines Special Rept., no. 25, pp. 29–35, 1952.

[11] E.g., A. Knopf, The Boulder bathylith of Montana, Am. Jour. Sci., vol. 255, pp. 98, 99, 1957; Tilley, op. cit., p. 334, 1958.

[12] Tilley, op cit., p. 336, 1958.

[13] A. Streckeisen, Das Nephelinsyenit—Massif von Ditro (Siebenbürgen), Schweizer min. pet. Mitt., Band 34, Heft 2, p. 407, 1954; Tilley op. cit., p. 329, 1958.

METASOMATIC "NEPHELINE SYENITES" [14]

In the Haliburton-Bancroft area of southeastern Ontario [15] high-grade metamorphic rocks of the Grenville Series (dolomitic marbles, amphibolites, and paragneisses) have been extensively invaded by pre-Cambrian granites. The intrusions are mainly concordant and effects of granitization are widespread. Within this region there is a belt one hundred or more miles in length within which outcrop many small masses of nepheline- and corundum-bearing rocks of plutonic aspect, whose distribution seems to be related to that of the Grenville marbles. Though outcrops are numerous their total extent aggregates about one per cent of the exposed granites.[16]

Adams and Barlow envisaged marginal differentiation of granitic magma to give local feldspathoidal magmas; these were thought to have invaded, permeated, and partially replaced the Grenville marbles.[17] Other writers,[18] taking into account the small total volume of nephelinic rocks and their field association with the granites, have invoked the Daly-Shand hypothesis of local desilication of granitic magma by reaction with marble.

Recent detailed field work shows that the nepheline-bearing rocks consistently display certain field relationships that strongly suggest a metasomatic origin: [19] They are generally banded. They are structurally conformable with and similar to the surrounding metamorphic rocks. Mineralogically there is complete gradation from nephelinic rocks to nepheline-free metamorphics. Moreover there is no direct field evidence of nepheline syenites having been formed by reaction between limestone (marble) and granitic magma. Indeed the granitic rocks everywhere are younger than and intrusive into the nepheline-bearing rocks.

Detailed petrographic examination of 570 feet of core obtained by drilling at York River has revealed the extremely heterogeneous nature of the feldspathoidal rocks.[20] "Purer limestone bands are interdigitated with silicate assemblages which may be nepheline-free, nepheline-poor or nepheline-rich. The sections reveal the passage from relatively pure

[14] L. Moyd, Petrology of the nepheline and corundum-bearing rocks of southeastern Ontario, Am. Mineralogist, vol. 34, pp. 736–751, 1945; W. K. Gummer and S. V. Burr, Nephelinized paragneisses in the Bancroft area, Ontario, Jour. Geology, vol. 54, pp. 137–168, 1946; Tilley, op. cit., pp. 337–357, 1958.

[15] F. D. Adams and A. E. Barlow, Geology of the Haliburton and Bancroft areas, Ontario, Canada Geol. Survey Mem. 6, 1910.

[16] Chayes, op. cit., p. 494, 1942.

[17] Adams and Barlow cited by Tilley, op. cit., p. 338, 1958.

[18] E.g., Chayes, op. cit., 1942.

[19] Gummer and Burr, op. cit., p. 160, 1946; Tilley, op. cit., p. 339, 1958.

[20] Tilley, op. cit., pp. 344–349, 1958.

limestones containing dispersed biotite, amphibole or clinopyroxene to assemblages rich in plagioclase ± nepheline, with biotite or amphibole, often of hastingsite type." There is also widespread microscopic textural evidence of replacement of earlier minerals, notably plagioclase, by nepheline.[21]

There is a convincing case for origin of the Bancroft and York River nephelinic rocks by metasomatism. But there is less certainty regarding the nature of the replaced rock and the nature and source of solutions responsible for metasomatism. It has been suggested by Moyd that granitic emanations, first desilicated by reaction with limestones, then brought about nephelinization of "dark gneisses" of the Grenville group. Tilley, on the other hand, argues that the parent rock was pure dolomitic limestone and that this was progressively replaced by various silicates under the influence of undersaturated solutions (containing dissolved material equivalent to nepheline plus iron) emanating from nepheline-syenite magma.[22]

The following replacement series illustrate Tilley's conception of the course of metasomatism in the York River area:

1. Limestone → calcite-biotite-plagioclase → biotite-plagioclase-nepheline → biotite-nepheline

2. Limestone → hastingsite-plagioclase-calcite → hastingsite-plagioclase-nepheline → hastingsite-clinopyroxene-nepheline

CARBONATITES [23]

Occurrence. In the fourth of Brögger's great memoirs on the igneous rocks of southern Norway [24] he described an association of carbonatites (rocks composed mainly of carbonate) and alkaline plutonic rocks in the small intrusive complex of Telemark. Since then over thirty similar occurrences have been reported, many of them in Africa. Carbonatites invariably are associated with stocks, ring-dike complexes, or plugs of basic alkaline rocks in regions of crustal stability or rifting. Their field occurrence—compellingly suggestive of emplacement by intrusion—conforms to one or other of several patterns: [25]

---

[21] *Ibid.*, pp. 349–351.

[22] D. F. Hewitt (cited by Tilley, *op. cit.*, p. 356, 1958) has demonstrated the nearby existence of igneous-textured intrusive nepheline syenites that might well represent the source magma for the metasomatizing solutions invoked by Tilley.

[23] W. C. Smith, A review of some problems of African carbonatites, *Geol. Soc. London Quart. Jour.*, vol. 112, pp. 189–220, 1956; W. T. Pecora, Carbonatites, a review, *Geol. Soc. America Bull.*, vol. 67, pp. 1537–1556, 1956.

[24] W. C. Brögger, Die Eruptivgesteine des Kristianiagebietes—IV, *Kristiania Vidensk. Skr. 1, Math-Naturv. Kl. 9*, 1921.

[25] Pecora, *op. cit.*, Table 1, 1956.

1. Tabular dike-like bodies cross-cutting the host rocks. Two hundred such "veins," ranging from a few inches to 20 ft. in width, occur in the Mountain Pass complex of southern California,[26] where the main carbonatite mass is irregular in form, 2,400 ft. long and about 700 ft. in maximum width.

2. A central carbonatite core or plug enclosed in a ring complex of feldspathoidal rocks. This is the prevailing pattern in Africa.[27] In several African complexes the core itself is compound; concentric carbonatite rings, dipping steeply inward, differ in composition and in texture.[28] This pattern, and accompanying foliation in the carbonatites, strongly suggest emplacement by intrusion of carbonate, either as magma or as a plastic solid.

3. A swarm of outward-dipping carbonatite cone sheets concentric about a central alkaline complex. This is the structural pattern on Alnö and adjacent islands in southern Sweden,[29] where the distribution of still younger dikes, mainly of alnoite, is radial.

4. One or more irregularly shaped bodies within the igneous complex.[30]

Associated Igneous Rocks.[31]   The igneous rocks associated with carbonatites fall into two categories:

1. Rocks of the central stock or ring complex are mainly basic or ultrabasic alkaline types: ijolite, essexite, biotite pyroxenite, or even kimberlite (alkaline peridotite breccia). Nepheline syenites and alkali syenites, though commonly present, are subordinate.

2. Late dikes and other minor intrusions—typically postcarbonatite—include members of the nepheline-basalt (nephelinite), lamprophyre, or phonolite families.

Carbonates are minor, possibly primary, constituents of some of these igneous rocks, e.g., the alnoites of Sweden, the ijolites of Iron Hill,[32] and some kimberlites.[33]

Composition of Carbonatites.[34]   The main constituent of most carbonatites is calcite or dolomite. In some rocks the two minerals occur together. Ankerite, siderite, and manganiferous carbonates are much less

---

[26] J. C. Olson et al., Rare-earth mineral deposits of the Mountain Pass district, San Bernardino County, California, U. S. Geol. Survey Prof. Paper, 261, 1954.

[27] Smith, op. cit., pp. 198, 199, 1956.

[28] Smith, ibid., p. 201.

[29] H. von Eckermann, The alkaline district of Alnö Island, Sveriges geol. undersökning, Ser. Ca, no. 36, 1948.

[30] E.g., Larsen, op. cit., 1942.

[31] Smith, op. cit., pp. 205–208, 1956; Pecora, op. cit., Table 1, 1956.

[32] Larsen, op. cit., p. 22, 1942.

[33] J. Verhoogen, Les pipes de kimberlite du Katanga, Com. spécial Katanga, Annales, Service des Mines, Bruxelles, pp. 26, 46, 1940.

[34] Pecora, op. cit., pp. 1542–1547, 1956.

abundant.   Complex rare-earth-bearing carbonates such as bastnaesite are widely distributed as minor constituents, rarely making up as much as 10 per cent of the bulk composition of the carbonatite mass (as at Mountain Pass, California).   Silicate minerals include alkali feldspars, nepheline, pyroxenes, biotite, olivines, and others.   A great variety of other minerals may be present in variable amounts, among them apatite, monazite, barite, pyrochlore, perovskite, fluorite, iron and titanium oxides, and various sulfides.

Many minor elements attain high concentrations in carbonatites: Ba, Sr, Nb, Ti, P, rare earths, and Zr.   These elements are also abundant in alkaline igneous rocks associated with carbonatites.   The concentrations of Ba, Nb, and rare earths are higher in carbonatites than in associated alkaline rocks, and the concentration of Zr is lower.   The distribution of trace elements in carbonatites differs strikingly from patterns typical of sedimentary limestones.[35]

Fenitization.[36]   A kind of alkali metasomatism—fenitization—has commonly affected quartzo-feldspathic country rocks (granites, gneisses, sandstones) around carbonatite-alkaline complexes.   Quartz is replaced by alkali feldspars (orthoclase, albite), aegirine, or blue amphibole. Hornblende and biotite are transformed to aegirine.   Potash feldspars become clouded.   The end product is what Brögger termed fenite—a syenitic rock composed of alkali feldspars, aegirine, and various introduced minor minerals (blue amphibole, apatite, sphene).   It seems that fenitization, advancing ahead of the alkaline or carbonatite magma, developed immediately prior to rather than following emplacement of the igneous-carbonatite complex.

Petrogenesis.[37]   Brögger explained the Fen carbonatites as products of crystallization of intrusive carbonate magma, formed by melting of limestone by an essexitic magma which itself assumed an ijolitic composition in the process.   A completely different petrogenesis was reconstructed for the same rocks by Bowen: [38] intrusion of the nepheline syenite ijolite complex, hydrothermal replacement of some of these rocks by carbonates (in the order calcite, dolomite, siderite), and finally intrusion of mica peridotite.   He derived the carbonic solutions from the great masses of intrusive magma in the Oslo district immediately to the east.

As the mineralogical, chemical, and field characteristics of carbonatites have become more clearly defined and since their genetic connection with

---

[35] Smith, *op. cit.*, p. 193, 1956.

[36] *Ibid.*, pp. 208–212.

[37] *Ibid.*, pp. 212–216; Pecora, *op. cit.*, pp. 1549–1552, 1956.

[38] N. L. Bowen, The Fen area in Telemark, Norway, *Am. Jour. Sci.*, vol. 8, pp. 1–11, 1924; The carbonate rocks of the Fen area in Norway, *Am. Jour. Sci.*, vol. 12, pp. 499–503, 1926.

basic alkaline rocks has been established, opinion in favor of magmatic origin of carbonatites has grown. Many writers now advocate the existence of some kind of liquid carbonatite magma—either a primary magma rising from the depths and reacting with crustal rocks on the ascent,[39] or a carbonatite liquid differentiate from a parent peridotite, pyroxenite, ijolite, or shonkinite magma.[40] Deuteric and hydrothermal processes now tend to be assigned a minor or more doubtful role.

Until recently a formidable difficulty opposing the concept of the carbonatite magma was the high melting range of anhydrous carbonates.[41] At 1,000 bars pressure of CO_2, calcite melts at 1340°C. Griggs and Heard have found no trace of melting of calcite up to 800°C. at pressures as high as 5,000 bars of pure CO_2 or of pure water. Largely for this reason Pecora [42] concluded that "a carbonatite magma in the normal sense is less likely to exist than carbonate-rich solutions which at elevated temperatures and pressure can have a higher concentration of dissolved ingredients than normally believed for hydrothermal solutions." And Harker and Tuttle [43] stated that "it is unlikely that pure carbonatite magmas could occur in nature because the temperatures involved are probably above those necessary to cause some fusion in the country rocks."

A recent experiment of Paterson [44] resolved this difficulty. He reported rapid complete melting of calcite at 1000°C. and incipient melting at 900°C. at a combined CO_2 and water pressure of only 50 bars (respective partial pressures unknown). The powerful influence of water on lowering the melting point of calcite, first demonstrated by Paterson, has subsequently been confirmed by Wyllie and Tuttle,[45] who have shown that at pressures of 2,000 bars hydrous carbonatite magmas may remain completely liquid at temperatures of the order of 700°C.—such as Bowen considered to be the maximum consistent with the petrological requirements of the Fen paragenesis.

[39] Von Eckermann, op. cit., pp. 145–161, 1948.

[40] E.g., S. I. Tomkeieff, The role of carbon dioxide in igneous magmas, British Assn. Rept., 1938, Cambridge, Sect. C, pp. 417–418, 1938; B. C. King, The Napak area of southern Karamoja, Uganda, Uganda Geol. Surv. Mem. 5, pp. 46–47, 1949; C. A. Strauss and F. C. Truter, The alkali complex at Spitskop, Sekukuniland, eastern Transvaal, Geol. Soc. South Africa Trans., vol. 53, pp. 121–122, 1951.

[41] Bowen, op. cit., p. 500, 1926; Pecora, op. cit., pp. 1548, 1549, 1956.

[42] Pecora, op. cit., p. 1537, 1956.

[43] R. I. Harker and O. F. Tuttle, Studies in the system $CaO\text{-}MgO\text{-}CO_2$, Part 1, Am. Jour. Sci., vol. 253, p. 223, 1955.

[44] M. S. Paterson, The melting of calcite in the presence of water and carbon dioxide, Am. Mineralogist, vol. 43, pp. 603–606, 1958.

[45] P. J. Wyllie and O. F. Tuttle, Melting of calcite in the presence of water, Am. Mineralogist, vol. 44, pp. 453–459, 1959.

CHAPTER 14

Pegmatites

INTRODUCTION

It has already been noted in Chap. 3 that some magmas have a water content high enough to cause, under certain conditions, separation of a gas phase in equilibrium with the liquid and solid phases of the magma. The possible effects of this condition must be taken into account in discussing the nature and genesis of pegmatites.

The components of a magma which under favorable conditions enter mainly into a gas phase are generally (though somewhat loosely) referred to as the "volatile" components of the magma. They are, of course, also present in the liquid phase of the magma, though typically in much smaller concentrations. There is no way of telling what the relative proportions of volatile components may have been in the magma prior to consolidation. We do know that they make up only a small percentage of the total composition of common igneous rocks, though some volcanic glasses may contain as much as 8 or 10 per cent of water.

There is good reason to suppose that relatively small quantities of volatile components may appreciably modify the behavior of silicate melts. Since their molecular weights are low compared with those of most silicates or metallic oxides, their molar fractions may be large in spite of their small percentages by weight. The molar fraction of water in a water-albite melt containing 6 per cent H_2O by weight is 0.483. Thus the effect of a small amount (by weight) of water on the chemical potential of albite should be comparable with that of a much larger amount of some silicate of high molecular weight. Moreover, the molar volume of a volatile component, particularly when in the form of a gas phase at moderate pressure, is large, and the effect of changing pressure on equilibrium should thus be greater in a system containing minor volatiles than in the corresponding "dry" system. Elements such as Cl and F readily form chemical compounds (*e.g.*, $FeCl_3$, SnF_4) much more volatile and soluble in water than corresponding oxides and silicates. Finally, viscosity of silicate melts appears to be decreased notably by addition of water or fluorine,[1] this being ascribed by Buerger[2] to breaking down of Si—O—Si bonds in the presence of OH and F radicals. Although the magnitude of this effect has not been determined quantitatively, it is probably an important factor in controlling rates of processes which depend on the viscosity of the melt (crystallization, etc.). For all the above reasons, and especially for the first two, a closer study of systems with volatile components is necessary.

PHYSICAL CHEMISTRY OF SYSTEMS WITH VOLATILE COMPONENTS

Properties of a Gas. In Chap. 2 we wrote the chemical potential of any component i of a gas phase under the form

$$\mu_i = \mu_i^0 + RT \ln PN_i f_i \qquad (2\text{-}37)$$

In this equation μ_i^0 depends on the temperature only; P is the total pressure of the gas, N_i is the molar fraction of component i in this gas, and f_i is the "activity coefficient" of i. The quantity PN_i is the "partial pressure" of i and is often noted p_i. Since we always have $\sum_i N_i = 1$, we also have $\sum_i p_i = P$; that is, the total pressure of the gas is the sum of the partial pressures of its constituents. The quantity of $p_i f_i$ is the "fugacity" of i.

The activity coefficient f of a pure gas ($N = 1$) measures essentially the departure of this gas from "perfect" behavior. A perfect gas is defined as one for which $f = 1$. In such a case, we have also, by (2-37) and (2-29),

$$\frac{RT}{P} = v \qquad (14\text{-}1)$$

where v is the volume of one mole of the gas. Equation (14-1) is the form under which the perfect-gas law is usually written. All gases are found to obey this law at sufficiently low pressure and at sufficiently high temperature. For instance, water vapor behaves as a perfect gas above 1200°C. at pressures less than 100 bars.

All the thermodynamic properties of a perfect gas (free energy, entropy,

---

[1] H. Saucier, Quelques expériences sur la viscosité à haute température de verres ayant la composition d'un granite, *Bull. Soc. franç minéralogie*, vol. 75, pp. 1–45 and 246–284, 1952.

[2] M. J. Buerger, The structural nature of the mineralizer action of fluorine and hydroxyl, *Am. Mineralogist*, vol. 33, pp. 744–748, 1948.

etc.) are determined by the pressure and temperature; but for nonperfect gases it is necessary to know also the value of the activity coefficient, which depends on temperature and pressure in a manner that cannot be predicted from thermodynamic theory only. In gaseous mixtures of several components, the activity coefficient of any component depends also on the relative proportion of the other components. This dependence of the activity coefficient on composition is slight in gases at low pressure, but under high pressure, when the density of the gas is large and when the molecules of the various components are relatively close together, intermolecular forces, depending on the chemical nature of the molecules, come into play and may affect noticeably the activity coefficients. This would mean, for example, that the thermodynamic properties of water vapor would not be exactly the same in a dense gas consisting of water and CO_2 as they would in one consisting of water and HCl. Very little experimental work has been done so far in this connection, so that it is difficult to predict the behavior of dense gaseous mixtures within given ranges of temperature and pressure.

Liquid-Gas Equilibrium. Critical Point. Since a system of one component in two phases is univariant, a pure liquid may be in equilibrium with its pure vapor only at one temperature, at any given pressure. The relation between this temperature and the corresponding pressure is given by the Clausius-Clapeyron relation (2–40)

$$\frac{dP}{dT} = \frac{\Delta h}{T\,\Delta v} \tag{14-2}$$

where Δh is the heat of vaporization and Δv the change in volume when one mole passes from the liquid to the gas state.

In contradistinction to the behavior of a solid in equilibrium with its own liquid (melting), it is found that both Δh and Δv in Eq. (14–2) simultaneously tend to zero when the temperature and pressure are raised. When this happens (critical point) the liquid boils without absorption of heat or change in volume; dP/dT becomes indeterminate, and the boiling-point curve ceases. The two phases, gas and liquid, become indistinguishable. The pressure P_c and the temperature T_c at which this happens are respectively the critical pressure and the critical temperature. The values of P_c and T_c for certain gases are listed in Table 36.

At any pressure or temperature greater than P_c or T_c, respectively, the substance may be brought continuously, without any change of phase, from a state in which it behaves as a perfect gas to a state in which its density is comparable with that of an ordinary liquid and its activity coefficient is very different from 1. In such a state a "fluid" presumably resembles an ordinary liquid in many of its properties, e.g., viscosity and solvent power, but one would nevertheless hesitate to call it a "liquid," since it may be brought continuously into a state which would generally be described as gaseous. It is for this reason that we shall usually refer to supercritical phases as gases; but we must remember that if the pressure is sufficiently high and the temperature sufficiently low these "gases" may behave in many respects like ordinary liquids.

TABLE 36. CRITICAL TEMPERATURES AND PRESSURES AND DENSITIES
OF VARIOUS GASES

Gas	$T_c(^\circ C.)$	$P_c(kg./cm.^2)$	Density at critical point, gm./cm.^{3}
CO_2	31.1	75.5	0.46
HCl	51.4	84.5	0.42
SO_3	218.3	86.5	0.63
H_2O	374.0	224.9	0.4

The critical pressure of water is only 224.9 kg./cm.^{2} or 221 bars, a value which may be attained in the earth at a depth of less than 1 km.   Pure water in the crust at a depth greater than 1 km. is therefore normally in the super critical state, regardless of its temperature.   We may note also that the density of water "vapor" at 400°C. and 2,000 bars is 0.75 gm./cm.^{3} and is thus comparable with that of many liquids under normal conditions.

Two-component Systems. *Two-phase Equilibria (Liquid-gas).* The solubility of a gas in a liquid solvent in which it forms an ideal solution depends essentially on the partial pressure of the gas and is given by Henry's law

$$N_i = Kp_i \qquad (14\text{--}3)$$

where N_i is the molar fraction of gas i in its saturated solution, and K is a constant which depends on temperature only.   The solubility of i thus increases proportionally to its vapor pressure.   Conversely, the vapor pressure of any component of an ideal solution is proportional to its molar fraction in the solution.   If we consider more particularly the solvent, and if p_0 is the vapor pressure of this pure solvent at a given temperature, the vapor pressure p of the solvent in an ideal solution in which its molar fraction is N_0 will be given by Raoult's law

$$p = p_0 N_0 \qquad (14\text{--}4)$$

Since N_0 is always less than 1, the vapor pressure of the solvent is decreased by addition of a solute, and therefore its boiling point is raised. For instance, addition of 58 gm. of sodium chloride to 1,000 gm. of water would raise the boiling point by approximately 0.5°C.   It should be noted, however, that Henry's and Raoult's laws are valid only for ideal solutions; for nonideal solutions corrections must be made to the simple relations (14–3) and (14–4).

Binary systems consisting of a gas and a liquid phase may be studied by the method used for liquid-solid systems with solid solutions (see, for example, the system albite-anorthite discussed on page 100). It will be

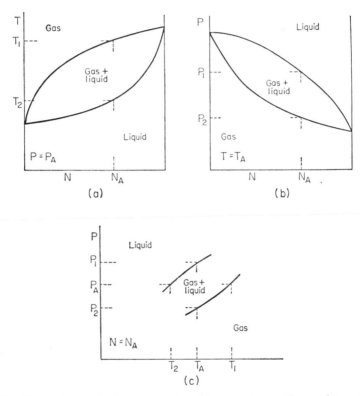

Fig. 56.   Gas-liquid equilibrium curves for a binary system with complete miscibility and no azeotropic mixture. (*a*) Temperature-concentration diagram for constant pressure; (*b*) pressure-concentration diagram for constant temperature; (*c*) pressure-temperature diagram for constant concentration.

found in the same manner that at a given temperature and pressure the compositions of the gas and liquid phases are usually different, the gas phase being richer in the component with the lowest boiling point. This is, of course, the basic principle of distillation. Exceptions to this rule may occur in certain systems for "azeotropic" mixtures, that is, for certain "azeotropic" compositions of the system in which the gas and liquid phases have exactly the same composition. Azeotropic mixtures boil or condense without any change in composition at a definite temperature.

In solid-liquid systems it is usually sufficient to represent phase diagrams in two coordinates T and N, because small changes in pressure do not alter the phase relationships appreciably. In liquid-gas systems, on the contrary, the effect of pressure, as already noted, is great; and it is usually necessary to consider phase diagrams in three coordinates, T, N, and P. Such three-dimensional diagrams are cumbersome, and one commonly prefers to consider only three two-dimensional diagrams drawn respectively for given constant values of P, T, and N. For instance, Fig. 56a represents a T-N phase diagram at a certain pressure P_A for a system with complete miscibility and no azeotropic mixtures. Figure 56b represents the corresponding P-N diagram for a certain temperature T_A, and Fig. 56c, drawn for a certain concentration N_A, gives for any pressure P_A the temperatures T_1 and T_2 at which condensation begins and is completed, respectively. Figure 56c shows also, for any temperature T_A, the respective pressures P_1 and P_2 at which vaporization begins and ends, as shown by referring back to Fig. 56b.

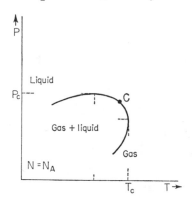

FIG. 57. Critical behavior in a binary system with given composition. The points P_c and T_c refer respectively to the highest pressure and the highest temperature at which the two phases may coexist in this system; C is the point at which the compositions of the two phases become identical.

Critical Phenomena in Two-component Systems. It may happen that the two branches of the curve drawn in Fig. 56c join in such a manner as to produce a single continuous curve, as represented in Fig. 57. In such a case, for any pressure or temperature greater than P_c and T_c, respectively, it is possible to bring the system continuously, *i.e.*, by gradual changes, from the liquid to the gaseous state, or vice versa. P_c and T_c are still referred to as the "critical pressure" and "critical temperature," but, in contradistinction to systems of one component, P_c and T_c no longer refer to the same point. The pressure corresponding to the temperature T_c is not the maximum pressure at which the two phases can coexist, nor is the temperature corresponding to P_c the maximum temperature of the two-phase system. The coordinates of the point C at which the compositions of the two phases become identical are neither T_c nor P_c. Furthermore, P_c and T_c correspond to one particular system of given composition N_A; for another composition N_B the critical points would be different, as shown in Fig. 58a.

It is customary to use the expression *critical point*, in connection with a two-component system, to designate the point at which the compositions of the two phases become identical, and the curve C of Fig. 58a which is tangent to all the separate curves for $N = N_A$, $N = N_B$, etc., at their respective critical points is the "critical curve" or "plait-point curve" (*Faltenspunktkurve*).

Figure 58*b* represents a T-N phase diagram drawn at the pressure P_1 indicated in Fig. 58*a*. Note the correspondence between temperatures T_1, T_2, T_3, T_4 in the two figures. All mixtures which do not lie in the range N_C to N_D may be brought without change of phase from a state of low temperature to a high one, *i.e.*, from a comparatively large density to a much smaller one; it is only for mixtures inside this range that boiling or condensation occurs. N_C and N_D are "critical" compositions, in the sense that they represent respectively the minimum and maximum concentrations at which two phases can coexist at the pressure P_1.

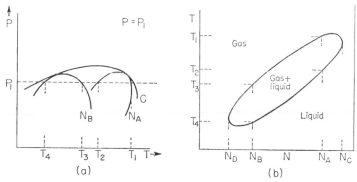

Fig. 58. Critical behavior in two-component systems. (*a*) Plait-point curve, *C*. (*After P. Niggli.*) (*b*) Range of composition in which gas and liquid phases may coexist at a given constant pressure.

The plait-point curves for different systems may differ as to shape. In one system the curve might slope continuously from the critical point of component 1 to that of 2, or it might show a maximum, as in Fig. 58*a*, or a minimum. Such a minimum is known to exist, for instance in the system ether—methyl alcohol. Very little is known as to the shape of the critical curves in systems of geologic interest. A summary of work in this field has been given by Booth and Bidwell.[3]

Three-phase Equilibria.[4] A system of two components distributed in three phases (solid, liquid, gas) is univariant, and the value of any one intensive variable, for instance temperature, will be sufficient to determine all the other intensive variables (pressure and molar fractions of each component in each phase).

[3] H. S. Booth and R. M. Bidwell, Solubility measurements in the critical region, *Chem. Rev.*, vol. 44, pp. 477–513, 1949.

[4] G. W. Morey, The application of thermodynamics to heterogeneous equilibria, *Franklin Inst. Jour.*, vol. 194, pp. 425–484, 1922.

It is particularly interesting to investigate such three-phase systems from the point of view of the relation which may exist between temperature and pressure, the pressure in the system being assumed to be equal to its vapor pressure. Suppose that, starting with a system in equilibrium, we change the temperature and pressure simultaneously in such a manner that equilibrium shall be maintained. We have then, by writing an equation of type (2–27) for each of the three phases,

$$\left.\begin{array}{l} -v^1\,dP + s^1\,dT + N^1_1\,d\mu_1 + N^1_2\,d\mu_2 = 0 \\ -v^2\,dP + s^2\,dT + N^2_1\,d\mu_1 + N^2_2\,d\mu_2 = 0 \\ -v^3\,dP + s^3\,dT + N^3_1\,d\mu_1 + N^3_2\,d\mu_2 = 0 \end{array}\right\} \qquad (14\text{--}5)$$

where, as usual, lower indices refer to a component and upper indices to a phase. If we eliminate $d\mu_1$ and $d\mu_2$ between these relations, we are left finally with an expression which involves only dP, dT, molar volumes and entropies and molar fractions, and which will show how the vapor pressure P varies as a function of temperature.

In particular, if component 1 is a volatile substance with very low melting point and 2 is a nonvolatile component, the vapor phase 1 will consist mainly, if not entirely, of component 1, while the solid phase 3 will consist of crystals of pure component 2. We have then $N^1_1 = 1$, $N^3_1 = 0$, $N^1_2 = 0$, $N^3_2 = 1$, and the solution of (14–5) is now found to be, omitting unnecessary indices,

$$\frac{dP}{dT} = \frac{s^1 N_2 - s^2 + s^3 N_2}{v^1 N_2 - v^2 + v^3 N_2} = \frac{N_1\,\Delta h_1 - N_2\,\Delta h_2}{T(N_1\,\Delta v_1 - N_2\,\Delta v_2)} \qquad (14\text{--}6)$$

where $\Delta h_1 = T(s^1 - \bar{s}_1)$ is the heat of evaporation from the melt of 1 mole of component 1, while $\Delta h_2 = T(\bar{s}_2 - s^3)$ is the heat of crystallization of 1 mole of component 2. Δv_1 is the change in volume of the volatile substance passing from melt to vapor phase; Δv_2 is the volume change when 1 mole of nonvolatile substance crystallizes from the melt.

Experimental investigations in a large number of binary systems involving water and various salts such as NaCl, KCl, $AgNO_3$, $K_2Si_2O_5$, etc., have shown that at high temperature the right-hand side of Eq. (14–6) is negative, meaning that the vapor pressure *rises* as the temperature falls.[5] If the system is subjected to an external pressure, it begins to "boil" when the value of the vapor pressure becomes equal to the external pressure. The temperature at which this happens is referred to as the "second boiling point"; like the ordinary, or first, boiling point, it depends on the external pressure. It has in addition the remarkable property that it results from cooling and that it corresponds to a liberation of heat, the heat given

[5] N. B. Keevil, Vapor pressures of aqueous solutions at high temperatures, *Am. Chem. Soc. Jour.*, vol. 64, pp. 841–850, 1942.

off by crystallization being greater than the heat required to evaporate the volatile component.

The second-boiling-point phenomenon results essentially from crystallization of the nonvolatile component, which causes the solution to become gradually enriched with respect to the other component, whose vapor pressure therefore increases. But as the temperature is lowered further, the effect of falling temperature on vapor pressure becomes noticeable; and the vapor pressure, having risen to a maximum, begins to decrease. The temperature at which the maximum pressure occurs will depend on the chemical nature of the system, and it is conceivable that certain systems might not show this phenomenon at all. For instance, the effect of temperature on the vapor pressure of the volatile component might be greater than the effect due to its increasing concentration; in such a case the vapor pressure would fall continuously throughout the whole range of temperature. On the other hand, the effect of increasing concentration might predominate, so that the vapor pressure would rise continuously until the whole amount of the nonvolatile component had crystallized. The albite-water system probably belongs to this latter type, although complications arise here because of limited solubility.

The rise in vapor pressure that has been observed experimentally in three-phase systems of two components is believed to occur also in magmas and has been advocated as a possible cause of volcanic explosions.[6] It should be noted, however, that this extrapolation is not always warranted. As may be seen from Eq. (14–6), the relation between vapor pressure and temperature depends on a number of variables: concentrations, molar volumes, and molar entropies of the various phases involved. Since these quantities will have different values in different magmas under different conditions, it cannot be assumed in all cases that dP/dT will be negative in a certain temperature range. Furthermore, it will readily be seen that the vapor pressure can be determined from the temperature alone only if one has a sufficient number of equations to eliminate all the undesired variables $d\mu_1$, $d\mu_2$, etc. In other words, the vapor pressure can be determined only if the system is univariant; and most magmas are probably not univariant systems, at least in the early stages of cooling where the number of components is much larger than the number of phases (see Eq. 2–50). It cannot be assumed therefore that the vapor pressure of a cooling magma must necessarily rise, and it is possible that in certain types of magma under certain conditions it never does.[7]

[6] G. W. Morey, The development of pressure in magmas as a result of crystallization, *Washington Acad. Sci. Jour.*, vol. 12, pp. 219–230, 1922.

[7] J. Verhoogen, Thermodynamics of a magmatic gas phase, *California Univ. Dept. Geol. Sci., Bull.*, vol. 28, pp. 91–136, 1949.

Critical Phenomena in Three-phase Systems.[8] Consider a binary system in which the vapor pressure happens to rise in a certain temperature range as a result of cooling. It is conceivable that at a certain stage of cooling the vapor pressure (that is, the pressure that will develop if the system is enclosed in a sufficiently strong container) may exceed the critical pressure corresponding to the composition of the system at this stage of the cooling. In other words, it may happen that the vapor-pressure curve intersects the plait-point curve, as shown in Fig. 59. In this figure, A and B represent respectively the critical points of the two pure components, and the curve AB is the plait-point curve. It is intersected at P and Q by the vapor-pressure curve, only a part of which has been drawn. A system in which this happens is said to be of "P-Q type." Note that Q may lie either to the right or to the left of M, the point representing the maximum vapor pressure.

The P-Q type of behavior is to be expected in systems in which the vapor pressure rises above the critical pressure of the more volatile component, provided that this critical pressure is not raised too strongly by addition of the less volatile component, that is, provided the solubility of this component in the vapor phase is not too great. For example, SiO_2 is only sparingly soluble in water, and the effect of dissolved SiO_2 on the critical pressure of water would be very small. Moreover, the vapor pressure in the system H_2O-SiO_2 probably attains high values, so that this system would presumably be of the P-Q type, although the points P and Q are probably very close together and close to A, the critical point of water. The system H_2O-$K_2Si_{12}O_5$ on the other hand, is probably not of the P-Q type because of the relatively high solubility of $K_2Si_{12}O_5$ in water.

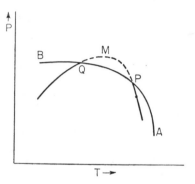

FIG. 59. Vapor pressure and plait-point curves in a system of the P-Q type. AB is plait-point curve.

Note that in the interval between P and Q the system ceases to be univariant, since only two phases are present, namely, the solid and the fluid, supercritical, phase. Thus the pressure in this interval cannot be determined from the temperature alone; it will be influenced for instance by the total volume of the container and the mass of each component; and it may change from one set of experimental conditions to another.

[8] See G. W. Morey and E. Ingerson, The pneumatolytic and hydrothermal alteration and synthesis of silicates, *Econ. Geology*, vol. 32, pp. 611ff., 1937; also P. Niggli, *Das Magma und seine Produkte*, pp. 235–302, Akademische Verlagsgesellschaft, Leipzig, 1937.

Consideration of the Water-Albite System.[9]   Let us consider the water-albite system, as its behavior may perhaps throw light on possible courses of crystallization of pegmatitic magma. We shall attempt to follow in detail the sequence of events when a liquid mixture of albite and water is allowed to cool slowly enough for equilibrium to be maintained throughout. As will be seen presently, the behavior of such a system depends largely on external conditions, which must therefore be carefully stated.

In the first instance, let us assume that the system is cooling under a given constant pressure less than the critical pressure of pure water. At a high temperature T_1 the system is assumed to consist of a saturated solution of water in albite, and of a vapor phase consisting of almost pure water. The behavior of these phases may be illustrated by hypothetical phase diagrams such as Fig. 60. The boiling point of albite is out of experimental reach, and as the solubility of albite in the vapor phase is very small at all temperatures within the magmatic range and at the relatively low pressure considered here, the vapor phase consists almost entirely of water. Hence the vapor and liquid curves have the general appearance of Fig. 60a. The solid-liquid equilibrium curves, on the other hand, will have the appearance of Fig. 60b. The right-hand part of this curve gives the lowering of the freezing point of albite by the addition of water, the slope of which may be determined either experimentally or (at least approximately) by assuming the solution water-albite to be ideal for low water concentrations. The left-hand part of the curve shows how the solubility of albite in water changes with temperature. The eutectic point for the mixture albite-water has not been represented because, if it exists, it occurs at some extreme dilution such that the point would be indistinguishable from the melting point of pure water.

If Figs. 60a and 60b are now combined into a single phase diagram, Fig. 60c, then, since the interval between the respective melting and boiling points of albite is much larger than that between the melting and boiling points of water, the solid-liquid and the liquid-gas equilibrium curves must intersect at A and B, and the solid-liquid equilibrium becomes metastable in the interval AB.

From Fig. 60c the behavior of a mixture of bulk composition X_1 during cooling from temperature T_1 is easily read. At this temperature the system consists of a gas phase of composition X_3 and a liquid of composition X_2. As the temperature falls the liquid becomes gradually enriched in water until, at temperature T_2, condensation of vapor is complete and the system now consists of a single liquid with composition X_1. This liquid remains stable during further cooling to temperature T_3, at which albite

[9] For experimental data on this system, see R. W. Goranson, Phase equilibria in the $NaAlSi_3O_8$-H_2O and $KAlSi_3O_8$-H_2O systems, *Am. Jour. Sci.*, vol. 35A, pp. 71–91, 1938.

begins to crystallize. The solution becomes gradually richer in water and cools down the curve *CA*. When the point *A* (temperature T_4) is reached, a gas phase of composition X_4 reappears, the system consisting

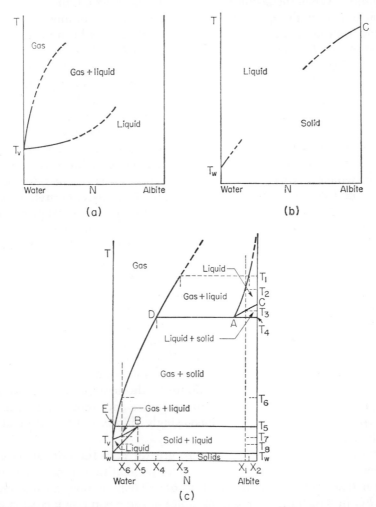

FIG. 60. Hypothetical phase diagram for the system water-albite at constant pressure. (*a*) Liquid-vapor equilibrium; (*b*) solid-liquid equilibrium; (*c*) three-phase equilibrium. (*After P. Niggli.*) *C* is the melting point of albite; T_w and T_v are respectively the melting point and the boiling point of water at the pressure considered. The albite content of the gas phase at *D* has been greatly exaggerated in order to make the diagram easier to read.

now of three phases (gas *D*, liquid *A*, and solid albite). But a three-phase, two-component system is univariant, and as the pressure has been maintained at a given, constant, value throughout, the three-phase equilibrium can exist only at one temperature (T_4); at any lower temperature only gas and solid albite can coexist. The temperature therefore remains constant at T_4 until the liquid phase A is converted to gas D and crystalline albite. When the last of the liquid has disappeared, the temperature falls once more with separation of very small amounts of albite as the gas changes from D to E (the albite content of gas D has been greatly exaggerated in Fig. 60c, so that the compositions of E and D are almost identical). When the temperature T_5 is reached, a three-phase equilibrium is again possible, and a liquid phase B reappears (composition X_5). This involves not only condensation of the vapor but simultaneous re-solution of a small amount of albite. Since B, like A, is an invariant point, this condensation and solution proceed at constant temperature until the gas phase is eliminated. At any temperature below T_5, the system consists of solid albite and a highly aqueous solution, the composition of which follows the curve BT_w. When T_w is reached, ice begins to form, and continues to do so until the mass consists entirely of solid albite and ice. It may be noted that at no stage did subequal proportions of water and albite develop in the melt. The only liquids which occurred were (1) the albite-rich liquid in the range T_1 to T_4, and (2) the water-rich solution in the interval T_5 to T_w.

A water-albite mixture of composition X_6 remains gaseous down to the temperature T_6, at which crystallization of albite begins. This continues with further cooling to temperature T_5, while the composition of the gas changes along the curve from D to E. At T_5 a liquid B forms and the solid albite is gradually resorbed, the temperature remaining constant until only gas E and liquid B remain. The temperature now falls again, and condensation of the gas proceeds to completion at temperature T_7. In the temperature interval from T_7 to T_8 the system remains entirely liquid, but at T_8 solid albite begins to crystallize once more, and continues to do so as the liquid becomes increasingly dilute, until albite and ice crystallize together at T_w

If the system cools under a pressure greater than the critical pressure of water, as would probably be the case for pegmatitic material, the phase diagram takes the appearance of Fig. 61. There is now an interesting range FG in which the solubility of albite in the gas phase increases with decreasing temperature, solid albite being redissolved as the temperature falls through this interval, only to crystallize again as the temperature falls from G to H. In this case the only liquids that can exist in the system are the albite-rich liquids in the high temperature range, for mixtures initially rich in albite.

Instead of assuming that the pressure is kept constant, let us suppose now that the system containing fixed amounts of water and albite is enclosed in a container of constant volume V, the walls of which are sufficiently strong to withstand any pressure that may develop as a result of cooling. The system is closed, in the sense that neither water nor albite is added to it or subtracted from it in the course of the experiment. Then, by Duhem's theorem (see page 27), the state of the system is completely determined, as regards intensive and extensive properties, by the values of two independent variables; but if the system consists of three phases it is univariant, and therefore only one of the independent variables can be an intensive one.   Let T and V be taken as the independent

variables; then as V remains constant for a given container, the pressure, mass, specific volume, and composition of each phase may be determined from the values of T and of the initial masses of the components. Thus in a series of experiments in containers of different volume and containing variable amounts of water and albite, it would be expected that at any temperature T the state of the system would not be the same in the different containers, these states differing either by the value of the pressure, or by the number of phases present, or by the composition of these phases, or by their relative masses.   Hence it is no longer possible to speak of "the" vapor pressure of the water-albite system at a given temperature, for

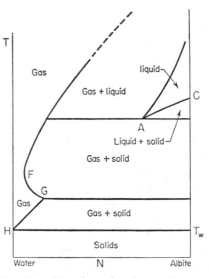

Fig. 61. Hypothetical phase diagram for the system water-albite at a pressure greater than the critical pressure of water. (*After P. Niggli.*)

this pressure will depend on the size of the container, its "degree of filling," and the proportion of water and albite initially present.

As a third case, it may be supposed that some method has been devised by which the system is allowed to cool under its own vapor pressure, the amount of water and albite being adjusted at each temperature in such a manner that three-phase equilibrium is maintained.   It is not clear how this could be done experimentally, but it is possible nevertheless to deduce how the system would behave from Goranson's curves.[10]   For instance, if the system initially contains 3 per cent of water, crystallization would begin at 1000°C., the pressure at this stage being 1,000 bars.   The concentration of water in the melt increases as cooling proceeds, and at 850°C. the vapor pressure has risen to 1,700 bars, the solution containing

[10] Goranson, *op. cit.*, 1938.

now 8 per cent of water. (Note that we have assumed that the quantities of water and albite present are precisely such that the melt could contain 8 per cent of water and that enough water would be left over to build up, in the available space, a pressure of exactly 1,700 bars.) From an extrapolation of Goranson's curves it appears that at 800°C. the pressure would have risen to 4,000 or 5,000 bars, the solution containing approximately 12 per cent water. Now this is the maximum amount of water that can be held in the melt under any pressure, so that with further cooling below this temperature the water content of the solution cannot increase. The vapor pressure thus reaches a maximum at this stage, and begins to drop with further cooling. As more albite crystallizes, more water goes into the vapor phase, and the amount of melt gradually decreases, until eventually the system consists entirely of solid albite and almost pure water vapor.

It is doubtful whether this behavior ever occurs in natural systems. As mentioned above, maintenance of the three-phase equilibrium in a system under its own vapor pressure implies at the same time (1) that the walls of the magma reservoir should be strong enough to withstand the pressure built up in the system, and (2) that the volume or the masses of the components should be adjusted at each stage of cooling in such a manner that the amount of water available in the gas phase should exert in the available space exactly the pressure required for the equilibrium. It is not likely that the two conditions could be fulfilled simultaneously. If the first condition were satisfied, the system could develop a pressure in excess of the weight of the overlying rocks, but the total volume of the system would presumably remain constant. If the walls were mechanically weak, the volume of the system might vary as crystallization proceeds, but the pressure presumably would then be constant and equal to the weight of the overlying column.

An alternative which may require consideration is that in which the gas phase on one hand, and the liquid and solid phases on the other hand, are at different pressures. This might occur if the wall rocks were permeable to the gas phase which could migrate outward through pores and fissures impervious to the more viscous melt. The system would then be enclosed in a "semipermeable membrane," permeable only to one component of the system, namely, water. Under these "osmotic" conditions the equilibrium would be noticeably changed. For a fuller discussion the reader may be referred to papers by Goranson and by Verhoogen.[11]

Solubility of Water in Silicate Melts. *The System Water*-$NaAlSi_3O_8$. Goranson [12] determined the solubility of water in a melt of albitic compo-

---

[11] R. W. Goranson, Silicate-water systems: the "osmotic pressure" of silicate melts, *Am. Mineralogist*, vol. 22, pp. 485–490, 1937; Verhoogen, *op. cit.*, pp. 125–128, 1949.

[12] Goranson, *op. cit.*, pp. 71–91, 1938.

sition at temperatures from 900° to 1200°C., and at pressures up to 4,000 bars; his results may be expressed in the form

$$x = \frac{P}{a + bP} \qquad (14\text{-}7)$$

where x is the solubility of water expressed in weight per cent, P is the pressure, and a and b are constants at any one temperature. When P is small, x is roughly proportional to P, as required by Henry's law (14-3), but this is no longer true at higher pressures. Equation (14-7) shows that when P is very large, x tends to the limiting value $1/b$; there is thus a maximum concentration which cannot be exceeded and which is about 11 per cent at 1200°C. and 11.82 per cent at 900°C. This behavior is probably common to most silicates and precludes the possibility of obtaining melts with, say, 50 per cent water.

Equation (14-7) is an empirical relation expressing experimental results; the solubility of water is governed by Eq. (2-47)

$$\frac{dN}{dP} = \frac{v^g - \bar{v}^s}{(\partial \mu / \partial N)_{P,T}}$$

where N is the molar fraction of water, v^g is the molar volume of water vapor, and $\bar{v}^s$ is the partial molar volume of water in the melt. Goranson finds that $\bar{v}^s$ is approximately 0.7 times the volume it would have if the solution were ideal.

Other Systems. There are but few determinations of the solubility of water in silicate melts. Goranson also made some measurements on the system $KAlSi_3O_8$ and on melts of granitic composition.[13] Tuttle and England [14] reported a solubility of 2.3 per cent water in SiO_2 melt at 1,200 bars and 1300°C. D. B. Stewart [15] found about 6 percent water in saturated melts in the ternary system $CaAl_2Si_2O_8\text{-}SiO_2\text{-}H_2O$ at 2,000 bars. Yoder, Stewart, and Smith [16] noted the large amount of water (about 11 per cent) in melts at 5,000 bars, 720°C., in the system $NaAlSi_3O_8\text{-}KAlSi_3O_8\text{-}CaAl_2Si_2O_6\text{-}H_2O$.

Influence of Water on the Melting Points of Silicates. It has already been mentioned that, because of its low molecular weight, a small amount of water should have a large effect on depressing the melting point of

[13] R. W. Goranson, The solubility of water in granite magmas, *Am. Jour. Sci.*, ser. 5, vol. 22, p. 481, 1931.

[14] O. F. Tuttle and J. L. England, A note on the system $SiO_2\text{-}H_2O$, *Geol. Soc. America Bull.*, vol. 66, p. 149, 1955.

[15] D. B. Stewart, in Annual report of the director of the Geophysical Laboratory, *Carnegie Inst. Washington Year Book*, no. 56, pp. 214–216, 1957.

[16] H. S. Yoder, D. B. Stewart, and J. R. Smith, in Annual report of the director of the Geophysical Laboratory, *Carnegie Inst. Washington Year Book*, no. 56, pp. 206–214, 1957.

anhydrous silicates. This is well shown by Goranson's experiments referred to above. Yoder [17] found that at 5,000 bars (water pressure), the melting point of anorthite drops from 1550° (for the anhydrous system) to 1234°C., and thus becomes lower than the melting point of diopside at the same pressure (1295°C.). The eutectic composition in the system $CaMgSi_2O_6$-$CaAl_2Si_3O_8$ is therefore notably shifted to the anorthite side (73 per cent by weight, as against 42 per cent in the anhydrous system), while its temperature is lowered from 1274° to 1095°C. At 5,000 bars of water pressure, orthoclase melts congruently at 876°C., albite at 748°C., the last liquid in the system Or-Ab-An-H_2O disappears at 695°C.

Nature of the Water-Silicate Melt. The lowering of the melting point of a component of a binary system is given by Eq. (5–3) (page 93) for the case of an ideal solution. If the solution is not ideal, an experimental determination of the composition of the saturated melt in equilibrium with the pure solid at a given temperature will allow a determination of the activity coefficient of the component. From available data, it appears that the introduction of water into a silicate melt does not appreciably change its nature. For instance, the activity coefficients of $CaAl_2Si_2O_8$ and $CaMgSi_2O_6$ are about the same, at comparable molar fractions, in the systems $CaAl_2Si_2O_8$-$CaMgSi_2O_6$ and $CaAl_2Si_2O_8$-$CaMgSi_2O_6$-H_2O.[18] The activity coefficient of $NaAlSi_3O_8$ at molar fraction 0.41 in the system $NaAlSi_3O_8$-H_2O is about 0.64. If Eq. (14–7) is extrapolated down to a pressure of 1 bar, the solubility of water in an albite melt at 1200°C. ($N_w = 10.8 \times 10^{-4}$) is of the same order of magnitude as the solubility of a nonpolar gas such as oxygen, nitrogen or argon, in a nonpolar solvent such as acetone, at 20°C. Wasserburg [19] has found that Goranson's data for the system $NaAlSi_3O_8$-H_2O may be explained in a very simple way. If the two components of a binary solution mix without any heat effects, the free energy of the solution differs from the sum of the free energies of the components by the term $T\Delta S$ where ΔS is the entropy change involved in mixing. This entropy change can easily be calculated by using the statistical definition of entropy $S = k \ln W$ (see page 10) and computing the number of ways W in which n_1 molecules of one type and n_2 molecules of the other types may be distributed in $(n_1 + n_2)$ positions; the expression for the chemical potential of a component of an ideal solution $\mu = \mu° + RT \ln N$ (see Eq. 2–38a) follows immediately therefrom. Wasserburg considers a simple model of a silicate-water solution

[17] H. S. Yoder, in Annual report of the director of the Geophysical Laboratory, *Carnegie Inst. Washington Year Book*, no. 53, pp. 106–107, 1954.

[18] Data were taken from Yoder, *op. cit.*, pp. 106–107, 1954.

[19] G. J. Wasserburg, The effects of H_2O in silicate systems, *Jour. Geology,* vol. 65, pp. 15–23, 1957.

containing n_2 molecules of silicate and n_1 molecules of water and he designates by r the number of shared oxygens per molecule of silicate, *i.e.*, the number of Si—O—Si bridges; this cannot be greater than 8 in albite. He then computes the entropy involved in an ideal mixture of rn_2 bridging oxygens and n_1 oxygens of the H_2O molecules and computes therefrom the chemical potentials of the components. Exact agreement with experimental data is obtained for $r = 8$. The result is not very sensitive to r, which could almost equally well be taken to be 7 or even 6; a value as low as 4 does, however, produce a notable discrepancy. Wasserburg's simple model is surprisingly good.

The Gas Phase of a Magma. *Complexity of the Problem.* From the preceding discussion it should be clear that the behavior of even a simple binary system involving a gas phase, such as water-albite, is too complex for brief description. This behavior depends, as shown above, on external circumstances, and even when those circumstances may be described simply (as where pressure is maintained at a constant value), the sequence of events is complex. It may involve a number of successive stages of condensation, boiling, crystallization, and resorption of early-formed crystals, and the sequence of these events may vary with variations in initial composition within a single system. Yet a binary system is simple compared with a multicomponent magma, and therefore the prediction of the detailed behavior of a magma must necessarily be a formidable, if not impossible, task. Behavior of magmas is still far from being understood, and a large proportion of what has been written about them is meaningless for want of sufficiently precise specification of the physical condition under which the magma is supposed to evolve. For instance, it is meaningless to state that at a given stage of cooling the vapor pressure of the magma must have risen to some specific value, for this pressure may be determined from the temperature alone only if the magma is univariant. Most magmas are not univariant in their early stage of cooling, and the value of the vapor pressure can be predicted only if the values of a sufficient number of variables (temperature, composition of certain phases, etc.) are known. It would be necessary also to know something of the physical and mechanical conditions under which the magma evolves (mechanical strength and permeability of the wall rock). It is thus very difficult to describe exactly the conditions that may prevail where magmas are known to exist, let alone those of the environments in which they may have formed. Treatment of the subject must therefore be very broad, and it is particularly important to avoid unwarranted generalizations drawn from a particular case.

Some thermodynamic properties of the gas phase in equilibrium with a liquid magma, and notably the conditions under which a gas phase may evolve, have been investigated in a publication to which the reader may be referred for further details.[20]   Only one important topic will be mentioned here, namely, solubility of rock-forming materials in a magmatic gas phase.

Solubility in a Magmatic Gas Phase.   The intensity of the metasomatic action displayed around certain igneous bodies and in areas of regional metamorphism (see page 562) apparently implies migration on a grand scale of substances, such as silicoaluminates, which in the solid form have only a negligibly small vapor pressure.   The manner in which this metasomatic action occurs, on the other hand, leaves little doubt that the replacing phase must have extreme mobility in order to penetrate and soak through rocks of low porosity and permeability, and it has generally been concluded that this mobile phase must have been essentially gaseous.   Such terms as *gaseous transfer* and *pneumatolysis* are often used in this connection, and the idea is widely held that under certain conditions gases must be able to dissolve large amounts of material having very low vapor pressure.

At the outset, however, it must be remembered that although there is evidence that large quantities of nonvolatile materials have been transported, sometimes over large distances, there is little evidence as to the concentration of these materials.   The same effect, as regards the total mass transported, could be obtained either from a small amount of the gas phase with high concentration of the transported substances, or from a large amount of a dilute gaseous solution.   For instance, regional feldspathization of quartzites does not necessarily imply the existence of high concentrations of K and Al in the permeating phase; there may have been only a small proportion of these elements in a large volume of gas.

A nonvolatile substance i will be said to be "soluble" in a gas phase when the mass of i in a given volume of the gas is greater than the mass of i in an equal volume of its own vapor at the same temperature.   Thus if addition of a gas j to a system consisting of solid i in equilibrium with its vapor causes further evaporation of i, j is said to be a "solvent" of i. In an example treated briefly in Chap. 2 (page 22), we noted that the vapor pressure of a solid increases as a function of the pressure applied to the solid.   It was also stated that a simple means of increasing this pressure is to add to the system solid-gas an inert gas which does not react chemically with either the solid or its vapor.   Addition of this inert gas thus increases the chemical potential of the solid, but leaves the chemical potential of the vapor unaffected, since the effect of the increase in pressure is compensated exactly by a decrease in its molar fraction.   Equilibrium may thus be restored only if the chemical potential of the substance of the solid in the gas phase is increased, and this requires further evapo-

---

[20] Verhoogen, *op. cit.*, 1949.

ration of the solid. The inert gas is thus a "solvent" of the solid, in the sense defined above.

In addition to this purely physical effect, there is perhaps a more important effect arising from the chemical nature of the gases present in the system. As explained earlier in this chapter, the activity coefficient f_i of component i of a dense gas phase depends on the chemical composition of the phase. It is thus conceivable that at a given temperature and total pressure the activity coefficient f_i might be large in a gas mixture containing a certain component j, and small in another mixture with component k. But if the vapor of i is in equilibrium with solid i, its chemical potential must remain constant at constant temperature and total pressure, and therefore, by (2–37), one must have

$$N_i f_i = \text{constant}$$

that is, the molar fraction of i in the gas phase is inversely proportional to its activity coefficient. Hence component k would be a solvent of i as compared to j.

What happens most commonly when a nonvolatile substance dissolves in a water vapor phase is probably quite different from the usual process of solution of an ionic compound in water. There is abundant evidence that solution of silica in steam actually represents a chemical reaction in which a volatile compound is formed, viz.,

$$SiO_2 + 2H_2O \rightarrow Si(OH)_4 \text{ gas}$$

the exact nature of the volatile compound still being somewhat uncertain.[21] Morey[22] gives the following figures referring to the solubility, in parts per million, in steam at 500°C. and 15,000 p.s.i. (1,020 bars):[23]

Al_2O_3	1.8	BeO	120
SnO_2	3.0	$CaCO_3$	120
$CaSO_4$	20	SiO_2	2,600
$BaSO_4$	40	GeO_2	8,700

It is interesting that, according to Morey, albite does not dissolve stoichiometrically at low pressure; in steam at 500°C. and 400 bars the SiO_2/Na_2O ratio is 8.65. At 2,000 bars, albite is carried over as such.

[21] See E. U. Franck, Zur Löslichkeit fester Stoffe in verdichteten Gasen, Zeitschr. physikal. Chemie, vol. 6, pp. 345–355, 1936; G. J. Wasserburg, The solubility of quartz in supercritical steam as a function of pressure, Jour. Geology, vol. 66, pp. 559–578, 1958.

[22] G. W. Morey, The solubility of solids in gases, Econ. Geology, vol. 52, pp. 225–251, 1957; The solubility of solids in water vapor, Am. Soc. Testing Materials Proc., vol. 42, pp. 980–988, 1942.

[23] G. W. Morey and J. M. Hesselgesser, The solubility of some minerals in superheated steam at high pressures, Econ. Geology, vol. 46, p. 821, 1951

A point which needs emphasis is that this "solubility" in a gas phase depends essentially on the density of the gas phase. Now water vapor acquires a notable density only under pressures exceeding the critical pressure, and therefore a water vapor phase of high density is necessarily "supercritical." There has been some confusion of thought in this connection, for it has also been assumed that any supercritical aqueous phase will have great solvent power. This is not necessarily true, for the density of a supercritical gas phase may vary continuously, according to the temperature and pressure, from that of an ordinary liquid to that of an extremely tenuous vapor obeying the perfect-gas law; in the latter case its solvent action will undoubtedly be very small. A high temperature would increase the vapor pressure of the solid substance, and therefore increase its concentration in the gas phase, but it appears probable that in some cases this increase might be partially offset by a corresponding decrease in density, and hence in solvent power, of the water vapor. Thus pressure is an important factor in determining the solvent action of water vapor.[24]

Conclusion. It is impossible to predict the precise conditions which may favor appearance of a gas phase, along with the liquid and crystalline phases of magma, under changing physical conditions. Relatively high concentration of volatile components in the system is, of course, one important factor. Where evolution of gas does occur, it may be expected to constitute one event in a complex sequence which may involve crystallization, melting, boiling, and condensation in any order and in some cases individually repeated during a single cycle of cooling. In rocks formed in the presence of magmatic gases, complex mineral assemblages showing evidence of incomplete adjustment to changing conditions are to be expected. It is not safe to assume that under plutonic conditions (high temperature and pressure) the efficiency of gases in transferring nonvolatile components in solution will necessarily be very great in every case and under all conditions within this range of temperature and pressure. The gas will indeed be in the supercritical state, but its solvent power, which increases with increasing pressure, might decrease with temperature, and will depend on composition. Clearly the possible complexity which may attend activity of gases in connection with such geological processes as reaction between magma and wall rock, plutonic metasomatism, and genesis of pegmatites, is very great and at present largely unpredictable.

[24] See, for example, G. C. Kennedy, A portion of the system silica-water, *Econ. Geology*, vol. 45, pp. 629–653, 1950.

NATURE AND OCCURRENCE OF PEGMATITES [25]

The great variability of pegmatites renders it difficult to define them concisely and adequately. They are always coarse (sometimes extremely so) and usually irregular in grain compared with plutonic rocks of similar composition. Presence of graphic intergrowths (most commonly microcline-quartz) and local development of crystal-lined cavities are typical but not ubiquitous features. By far the greater number of pegmatites are mineralogically and chemically similar to granite in that their main constituents are quartz, microcline, sodic plagioclase, and micas, associated with which, however, there may also occur any of a large number of rare minerals including some that are also found as accessory minerals in granitic rocks (tourmaline, apatite, sphene, monazite, zircon, fluorite, etc.). Pegmatitic equivalents of gabbro and diorite, consisting mainly of hornblende and plagioclase, are also known, but are much rarer than granitic pegmatites and will not be considered further. The common mafic minerals of pegmatites are hydrous—micas in acid pegmatites, amphiboles in more basic varieties. Presence of minerals containing phosphorus, fluorine, chlorine, sulfur, boron, etc., in mineralogically complex pegmatites also suggests that volatile substances have played an essential role in the origin of pegmatites in general. A further characteristic of the more complex pegmatites is comparatively high concentration of other rare elements, such as Li, Be, Mo, W, Th, Zr, Sn, Ta, Nb, etc., which do not fall within the category of "volatile constituents" but which form chlorides and fluorides with a considerably lower boiling point than the corresponding salts of metals such as Cu, Zn, Pb, etc., which occur only sparingly in pegmatites.

Acid pegmatites (*i.e.*, common pegmatites containing free quartz) fall into two mineralogical classes:

1. Simple pegmatites consisting of quartz, alkali feldspars (microcline and variable amounts of sodic plagioclase), and minor micas. The min-

[25] W. C. Brögger, Die Mineralen der Syenitpegmatitgänge der Südnorwegischen Augit und Nephelinsyenite, Zeitschr. Kristallographie, vol. 16, 1890; K. K. Landes, Origin and classification of pegmatites, Am. Mineralogist, vol. 18, pp. 35–56, 95–104, 1933; F. L. Hess, Pegmatites, Jour. Geology, vol. 28, pp. 447–462, 1933; W. T. Schaller, Pegmatites, Ore Deposits of the Western States, American Institute of Mining and Metallurgical Engineers, New York, pp. 144–151, 1933; E. Raguin, Géologie du Granite, pp. 119–131, Masson et Cie, Paris, 2nd ed., 1957; E. W. Heinrich, Pegmatites of Eight Mile Park, Fremont County, Colorado, Am. Mineralogist, vol. 33, pp. 420–448, 550–588, 1948; Pegmatites of Montana, Econ. Geology, vol. 44, pp. 307–335, 1949; E. N. Cameron, R. H. Jahns, A. H. McNair, and L. R. Page, Internal structure of granite pegmatites, Econ. Geology Mon. 2, 115 pp., 1949; R. H. Jahns, The genesis of pegmatites, Am. Mineralogist, vol. 38, pp. 563–598, 1078–1112, 1953; The study of pegmatites, Econ. Geology, 50th anniversary vol., pp. 1025–1130, 1955.

eral assemblage is essentially simple, and rare minerals are either absent or present as accessories (as in granites).

2. Complex pegmatites containing, in addition to quartz feldspars and micas, rare minerals in considerable abundance and variety—lepidolite, spodumene, tourmaline, topaz, cassiterite, beryl, tantalite, columbite, zircon, uraninite, thorite, apatite, amblygonite, etc. Individual crystals of some of these may be of immense size.

Pegmatites of the first class occur as swarms of dikes, veins, or flat lenses within or at the margins of batholiths and stocks of granite and granodiorite, or as constituents of regionally developed migmatite complexes (e.g., in the pre-Cambrian of Finland and eastern Canada). Complex pegmatites also may be associated with granitic intrusions, particularly in their marginal zones. Those connected with intrusions of syenite and nepheline syenite are likely to be particularly rich in rare minerals. The complex condition is apparently determined by factors of regional significance, for within one granite granodiorite province (e.g., that of southwestern New Zealand) pegmatites will be uniformly simple, while throughout another (e.g., the Appalachian province of North America) complex mineral assemblages will be the general rule.

Pegmatites consistently occur as minor rock bodies. Individual masses are usually tabular or flatly lenticular, and they range from a few inches to a few hundred yards in length. More massive bodies (dikes, lenses, or irregular pipes) are also known, in some cases several miles in length and locally 200 or 300 ft. in thickness.

ZONED PEGMATITES [26]

Many pegmatite bodies consist of concentric zones differing in mineralogy and texture, and usually showing gradational common boundaries. Apophyses from the inner may cut the outer zones; but the reverse is not true. A generalized sequence of zones—only partially developed in any one body—from the walls to the core is as follows:

1. Plagioclase-quartz-muscovite
2. Plagioclase-quartz
3. Quartz-plagioclase-perthite (± muscovite, biotite)
4. Perthite-quartz
5. Perthite-quartz-plagioclase-amblygonite-spodumene
6. Plagioclase-quartz-spodumene
7. Quartz-spodumene
8. Lepidolite-plagioclase-quartz
9. Quartz-microcline
10. Microcline-plagioclase-lithia micas-quartz
11. Quartz

[26] Cameron, Jahns, McNair, and Page, op. cit., pp. 13–70, 98–105, 1949.

The plagioclase of zones 1–3 (andesine to median albite) tends to be more calcic than that of the inner zones (nearly pure albite).

Zoned pegmatites are generally interpreted as products of crystallization of aqueous granitic magma in a closed system. Successively later fractions progressively enriched in water are concentrated inward, so that the core represents the final stage of crystallization in which an aqueous gas phase appears.

The highly variable texture and zoned structure of many of the economically important large pegmatite bodies make it difficult to estimate the average bulk composition of such rocks. A detailed survey of one of the large lithium pegmatite bodies of New Mexico [27] gave a mean composition ($SiO_2 = 74.5$, $Al_2O_3 = 14.8$, $CaO = 0.2$, $Na_2O = 3.3$, $K_2O = 5.4$) close to that of average granite; Li_2O (0.7) and F (0.9) were found to be unusually abundant.

PETROGENESIS

Pegmatites of regionally developed migmatites have been variously interpreted as products of magmatic injection (with or without concurrent metasomatism), as material sweated out of the host rock as a result of partial fusion (anatexis), or as concentrations of silica, alumina, and alkalies, formed by ionic diffusion through solid rocks. The authors' preference for the second of these modes of origin has already been noted (page 388), but it is likely, too, that some migmatite pegmatites are formed by magmatic injection and associated replacement processes, such as are discussed below.

If, as the authors believe, granitic batholiths and stocks are formed by intrusion of acid magmas, then associated pegmatites must also be of ultimately magmatic origin. They must have formed late in the magmatic cycle, for many of them cut the granite itself. Indeed concentration of volatile constituents, as a necessary condition for development of pegmatites, has long been attributed to development of a low-melting residual liquid fraction in the final stages of crystallization of acid magma. This raises the possibility that an aqueous gas phase as well as a water-saturated silicate melt may normally participate in the complex evolution of a mass of pegmatite.

There is strong mineralogical and structural evidence pointing in the same general direction. The general prevalence of graphic intergrowths of microcline and quartz, rapid variations in mineralogy and in grain, and a tendency for crystals of certain minerals (e.g., tourmaline) to develop radial orientation, all suggests that pegmatites do not crystallize directly from a melt of the same composition. Rather the picture presented is one of magmatic crystallization followed by a sequence of replacements,

[27] Jahns, op. cit., pp. 1078–1112, 1953.

such as are familiar enough in connection with growth of metalliferous veins from dilute aqueous solutions or from aqueous gases. The role of replacement in genesis of pegmatites, first propounded sixty years ago by Brögger, has been elaborated in papers (cited above) by F. L. Hess, Landes, Schaller, and others. It is also clear in some cases, from lack of offsetting of recognizable bands or streaks in the host rock where obliquely intersected by pegmatite veins, that the veins themselves may have formed by metasomatic replacement of the host rock.[28] Such replacements of host rock by pegmatite, or of one pegmatite mineral by another, must involve activity of highly mobile fluids—an aqueous gas phase of changing composition being the most likely medium adequate for this purpose.

It is readily understood that a late residual liquid (pegmatite magma) should be rich in quartz and alkali feldspars in proportions roughly corresponding to those in the low-melting range of the system orthoclase-albite-silica (cf. Fig. 53, page 350). That water and other volatile components—phosphorus, fluorine, chlorine, sulfur, etc., and volatile compounds of these such as $SnCl_4$, $FeCl_3$, etc.—should also become concentrated in the final residua of granitic magma is also to be expected under certain conditions. However, there are other rare elements characteristic of pegmatites which do not fit into either of these categories—e.g., beryllium, lithium. The main reason for concentration of this latter group is that these elements, at temperatures of magmatic crystallization, do not substitute for common elements in crystal lattices growing in the melt phase of the magma. They differ notably, as regards ionic radius, ionic charge, or some kinetic property of the ion,[29] from elements such as the alkali metals, calcium, magnesium, etc., which build common igneous minerals. Excluded from the stable crystalline phases—in which they would have high chemical potentials—they become stored up in the residual liquid phase and, under favorable conditions, perhaps in the gas phase when this ultimately appears.

The great mineralogical diversity of pegmatites is due partly to composition of the pegmatite magma, which is largely determined by that of the parent granitic magma. Probably of much greater importance is the wide range of possible variation in the sequence of phase changes that may occur during cooling of magmas relatively rich in water under various sets of conditions. Even within the simple two-component system albite-water, as has been seen (pages 412 to 416), there is a high degree

[28] Cf. G. E. Goodspeed, Dilation and replacement dikes, Jour. Geology, vol. 48, pp. 175–195, 1940; R. A. Chadwick, Mechanisms of pegmatite emplacement, Geol. Soc. America Bull., vol. 69, pp. 803–836 (especially 809–814), 1958.

[29] Cf. K. Rankama, On the geochemistry of tantalum, Comm. géol. Finlande Bull. 133, p. 58, 1944.

of complexity and variability as regards the order of appearance of gaseous and crystalline phases from liquids of different compositions cooling under various sets of conditions. The pegmatite magma, in the first place, can contain only a limited amount of water under given conditions. The solubility of water in a pegmatite melt of otherwise constant composition depends upon prevailing temperatures and pressures.[30] It is approximately 6 to 7 per cent at 750°C. and 1,000 bars; 10 per cent at 550°C. and 2,500 bars.[31] As feldspars and quartz crystallize from a pegmatitic melt still undersaturated in water, the water content gradually rises, but this effect may be partially or completely nullified by crystallization of amphiboles, micas, or other hydrous minerals from the melt. It is to be expected that under some, though not necessarily all, circumstances the liquid phase of a pegmatite magma from which quartz and microcline are separating will ultimately become saturated with water.[32] A second fluid phase, with a very high water content, now separates. At high temperatures this fluid will be gaseous (in the supercritical state) from the moment of separation. If, on the other hand, the temperature and pressure drop below the critical values for the aqueous fluid, this may separate in turn into two fluid phases, both very rich in water, i.e., into aqueous solution and water vapor. It is thus possible that, under some conditions within the crystallization range of pegmatites, crystalline phases, water-saturated silicate melt, dilute aqueous solution, and vapor phase rich in water may all coexist in equilibrium. Where crystallization of pegmatites takes place in the presence of two or three distinct fluid phases, the chance for mutual separation of different groups of elements according to their respective solubilities and "volatilities" is greatly increased. The ultimate differentiation of rare elements concentrated from a single parent magma into the manifold mineral assemblages of granites, pegmatites, hypothermal veins, epithermal veins, and even fumarole deposits probably depends upon the development of a succession of changing fluid phases as pictured above.

[30] The pressure on a plutonic magma is determined not only by depth, but by the strength of the rocks which wall the magma chamber, by their permeability to the fluid phases present, and in some cases by the composition (e.g., the water content) of the magma. If the wall rocks are mechanically weak and are highly permeable to a gas phase only, the pressure on the melt phase may be very much greater than that on the gas phase. Assuming completely impermeable wall rocks, which are nevertheless weak enough to transmit to the magma the whole load of the overlying rock column, the pressure range 1,000 to 2,500 bars corresponds to depths between 3½ and 10 km.

[31] Goranson, op. cit., pp. 481–502, 1931; F. G. Smith, Transport and deposition of non-sulphide vein minerals, Part III, Econ. Geology, vol. 43, pp. 535–546, 1948.

[32] Experimental work by F. G. Smith (op. cit., 1948) has shown that in the case of a pegmatitic melt left after 96 per cent crystallization of a magma initially containing 2 per cent water, crystallization at constant pressure of 2,500 bars would lead to saturation of the residual melt in water (10 per cent) at about 550°C.

Once a stage is reached at which crystals, water-saturated melt, and aqueous gas coexist, further crystallization is, as shown by the relatively simple albite-water system, a complex and highly flexible process involving exchange of material between all phases present as temperature falls. Where the permeability of the wall rock is high enough, the gas phase may be free to escape, while liquid and crystalline phases are retained in mutual contact. Respective pressures on liquid and gas may be markedly different at a given moment. Specially significant is the possibility that with falling temperature copious evolution of gas will continue as a result of the "second-boiling-point" phenomenon discussed on page 409. The essential condition for this to occur is that the vapor pressure of the liquid should increase with falling temperature as a result of increasing concentration of water in the silicate melt. This will not necessarily be so in every case. The particular behavior will depend upon a number of internal and external properties of the system—concentrations, volumes, entropies, etc., of the various phases, constancy of volume or of pressure for the system as a whole, and so on (cf. pages 412 to 416).

In spite of the complexity of the problem, various attempts have been made to generalize, from petrographic data, as to the possible sequences of crystallization in pegmatites. The most satisfactory scheme is that of Fersman,[33] which is too elaborate to present here. Several of his conclusions may be noted briefly, however. Fersman correlates the observed order in which pegmatite minerals appear (under favorable conditions) with four successive stages of crystallization governed by falling temperature: [34]

1. Magmatic stage, at which equilibrium is maintained between liquid and crystalline phases

2. Pegmatitic stage (roughly 800° to 600°C.), throughout much of which crystalline, liquid (silicate-melt), and gas phases coexist

3. Pneumatolytic stage (600° to 400°C., i.e., the range of temperature between the inversion point of quartz and the critical temperature of aqueous solutions), characterized by equilibrium between crystals and gas

4. Hydrothermal stage (400° to 100°C.) in which equilibrium is maintained between crystals, aqueous solution, and aqueous gas.

Microcline, quartz, and micas (in some cases lithium-bearing) are

[33] A. E. Fersman, Les pegmatites, leur importance scientifique et pratique, *Acad. Sci. U.S.S.R.*, Leningrad, 1931, translation in French by J. Thoreau, Louvain, 3 vols., 1951. A brief abstract, in French, has been published by N. Varlamoff (La répartition de la minéralisation d'après la clef géochimique de Fersman, *Soc. géol. Belgique Bull.*, tome 70, pp. 108–138, 1946).

[34] For application of Fersman's concept to a specific example of a complex pegmatite see P. Quensel, The paragenesis of the Varuträsk pegmatite, *Arkiv. Mineralogi och Geologi Stockholm*, Band 2, nr. 2, pp. 9–125, 1956 (especially 84–116).

referred to the pegmatitic and early pneumatolytic stages. Albite replaces microcline in the pneumatolytic stage, and may be followed by adularia or zeolites in the hydrothermal stage. Over the same range quartz continues to separate, and mica (represented first by muscovite and lepidolite) may ultimately give way to hydrothermal kaolin. A sequence of rare minerals from monazite and titanite (early pegmatitic) to phosphates, sulfates, and carbonates (hydrothermal), permits division of pegmatites into ten types. These are listed below (together with the minerals and elements in which they are characteristically enriched) in order of decreasing temperature:

a. Pegmatitic stage:
   Type 1. Allanite, monazite; Th, Y
   Type 2. Tantalite, columbite, uraninite; Ta, Nb, Ti, U, Y, etc.
b. Pneumatolytic stage:
   Type 3. Tourmaline, muscovite; B, F
   Type 4. Beryl, topaz; Be, F
   Type 5. Albite, lepidolite; Na, Li
   Type 6. Phosphates of Li, Mn and Fe; Li, P
c. Hydrothermal stage:
   Type 7. Cryolite; Al, F
   Type 8. Fluocarbonates of Ca, Mg, Fe, Mn
   Type 9. Sulfides of Fe, Cu, Zn
   Type 10. Zeolites, kaolin

Any individual mass of pegmatite may consist mainly of quartz, microcline, mica, and sodic plagioclase that have crystallized at the magmatic and pegmatitic stages. These may be the only minerals present, or they may be accompanied or partially replaced by one or more of the successive mineral assemblages listed above. For further detail the reader is referred to the tables which accompany Varlamoff's paper.

The zoned structure which is so characteristic of many pegmatites is generally attributed to crystallization within the pegmatitic stage.[35] To the pneumatolytic and hydrothermal stages belong the development of replacement bodies and fracture fillings, which cut across the zoned structure.[36] Where their total bulk is small compared with that of the pegmatite body as a whole, the replacement bodies may have formed from aqueous residua developed within the pegmatite itself. But external sources, e.g., residua developed from subjacent granitic magma, have been invoked to explain replacement bodies of large size.

In conclusion attention is drawn to a perplexing question: why are pegmatites uniformly simple in composition throughout one granite prov-

[35] E.g., Quensel, op. cit., p. 86, 1956.
[36] E.g., Cameron et al., p. 105, 1949.

ince, and predominantly complex in another?   Prevalence of simple pegmatites cannot be explained entirely by assuming general absence of *all*
minor elements in the parent granitic magma.   A partial answer has been
suggested by Rankama, who finds some chemical evidence indicating that
certain rare elements (Li, Be, Rb, Cs, Ba, Ta, Pb, rare earths) may have
been subject to secular concentration in the granitic crust of the continents.[37]   If this process were effective, complex pegmatites rich in these
elements should occur more frequently in association with young granites
than with those of the pre-Cambrian.   This is not obviously so.   The
present authors would correlate abundance of rare elements in complex
pegmatites partly with the physical conditions under which they formed.
A necessary condition must surely be development of a fluid phase with
high concentrations of the ions in question.   The most likely mechanism
seems to be copious and continued evolution of a vapor phase with high
solvent capacity under conditions (as described for the albite-water
system cooling under constant pressure) in which there will be a complex
sequence of crystallization, resorption, boiling, and condensation.   This
might account for the replacement phenomena observed in complex
pegmatites, and perhaps also for the splitting of the rare constituents of
the magma into those which occur in pegmatites and those (*e.g.*, sulfides
of base metals, carbonates, etc.) which are more commonly found in
hydrothermal veins.   A vapor solvent could well serve as a medium for
concentration of elements that are but sparingly present in the pegmatite
melt.   Favorable conditions for its development might be (1) initially
high water content in the silicate melt; (2) composition and environment
such that the water content of the silicate melt gradually increases, leading to rise in vapor pressure of the melt phase with falling temperature,
and thus to the "second-boiling-point" phenomenon; (3) high confining
pressure, which would increase the density, and hence the solvent capacity and chemical activity of the gas phase.   Under other conditions (*e.g.*,
low pressures, or conditions not conducive to formation of a dense gas
phase), the rare elements could be removed steadily from the pegmatite
system as components of dilute gases and ultimately of aqueous solutions.
Within the pegmatites they would then leave no trace of their former
presence, but some of them might ultimately be deposited in metalliferous
veins far from their granite source.   Under yet other conditions, rare
elements might fail to be concentrated in either a gas phase or an aqueous
solution; instead, they might occur in a dispersed condition throughout
the igneous rock itself (*e.g.*, boron in tourmaline granite).

[37] K. Rankama, On the geochemical differentiation in the earth's crust, *Comm. géol.
Finlande Bull.* 137, 1946.

CHAPTER 15

Environment, Origin, and Evolution of Magmas

PRIMARY MAGMAS

Definition and Criteria. A mass of magma of given composition conceivably could originate in one of three ways: (1) by tapping some body whose primitively liquid condition had been inherited from a remotely early stage in the earth's history; (2) by partial or complete fusion of preexisting solid rock; (3) by modification of preexisting magma, *e.g.*, by differentiation or contamination. The frequently used but seldom defined term *primary magma* should be restricted to magmas of the first two categories above, although in the authors' opinion no satisfactory evidence for existence of the first type has yet been brought forward. Magmas of the third group may be called *derivative magmas*.[1]

Criteria by which a primary magma may be recognized as such are somewhat vague. Probably the most satisfactory is a pronounced tendency for the magma to appear repeatedly throughout geologic time, in great quantities and in extensive individual bodies (lava floods, batholiths, lopolithic sheets, etc.), over large sectors of the earth's crust. A further criterion is predominance of corresponding rocks within one or more rock associations, the other members of which could have been derived from the primary magma by accepted modifying processes—differentiation, assimilative processes, etc.

Conversely there is a tendency to regard magmas as belonging to the derivative class when they occur habitually in small quantities, when they are constantly found in association with a magma conventionally considered as primary, and when derivation from the latter can be explained in terms of accepted modifying processes. Phonolitic (nepheline-syenite),

[1] Cf. N. L. Bowen, Magmas, *Geol. Soc. America Bull.*, vol. 58, p. 271, 1947.

trachytic, and dacitic magmas are usually considered to be derivative. Nevertheless it must be admitted that some magmas conventionally regarded as derivative could equally well be of primary origin.

Only two broad magma families—basaltic and granitic—satisfy completely the criteria of primary magmas defined above. Their claims to be regarded as primary magmas will now be considered in more detail.

Primary Basaltic Magma. There is general agreement among geologists that basaltic magma is primary. A statement by Bowen [2] sums up the situation thus:

The geologic record reveals that basaltic magma occupies a unique position in the igneous economy. It is accepted by many as the primary magma, or at least a primary magma. This type of molten matter has, in all ages, broken through the crust of the earth in the form of dikes of great extension and has poured out on the surface in great floods. It has insinuated itself as sills and other concordant intrusives between the beds of layered rocks and, in many of the larger masses of this kind, has, upon consolidation, itself developed a system of layers which are, in part at least, the result of crystallization differentiation in which gravity has apparently been a dominant control.

The primary status of basaltic magma is further supported by the prominent place it occupies in a number of different volcanic and plutonic igneous associations—e.g., the alkaline olivine-basalt, tholeiite, and spilitic kindreds, and the plutonic association of basic lopoliths. Moreover Bowen's synthesis of experimental and petrographic data has shown conclusively that the range of composition of all common associates of basalt is such as would be expected in the various crystalline and liquid fractions that could develop during fractional crystallization of basaltic magma. In many cases, e.g., trachytes of oceanic islands or peridotites of stratified basic lopoliths, the rocks associated with basalt or gabbro are habitually present in the rather small proportions demanded by such a mode of origin. The case for world-wide development of primary basaltic magmas is now satisfactorily established.

Over the past three or four decades opinion has been divided as to the ultimate source of basaltic magma. It has been variously attributed to fusion of basic rock in the deeper levels of the earth's crust [3] or to partial fusion of underlying feldspathic peridotite.[4] These possibilities will be examined more fully when we have considered the chemical and physical constitution of the outer part of the earth.

[2] *Ibid.*, p. 266.

[3] R. A. Daly, Volcanism and petrogenesis as illustrated in the Hawaiian Islands, *Geol. Soc. America Bull.*, vol. 55, pp. 1363–1400, 1944; Nature of the asthenosphere, *Geol. Soc. America Bull.*, vol. 57, pp. 707–726, 1946; W. Q. Kennedy, Crustal layers and the origin of magmas, *Bull. volcanologique*, sér. 2, vol. 3, pp. 24–41, 1938.

[4] N. L. Bowen, *The Evolution of the Igneous Rocks*, pp. 315–320, Princeton University Press, Princeton, N. J., 1928.

Primary Granitic Magma.   The question of origin of granitic rocks has been discussed in Chap. 12 and need not now be reviewed. The conclusion there reached is that there is a broad class of primary granitic magmas which develop in continental regions by differential fusion of varied rocks (including basic types) in the lower parts of the crust. Granitic magmas contrast sharply with basaltic magmas in several respects, other than with regard to average composition. They show a much wider range of chemical variation (quartz diorite to granodiorite to granite), even within a single petrographic province. They are represented mainly by plutonic rocks of batholiths, and these are apparently restricted to continental regions.

Granitic batholiths are developed on an immense scale in pre-Cambrian terranes all the world over. This could be explained by assuming that in pre-Cambrian times, under the influence of a temperature gradient steeper than that of today, the earth's crust was more subject to fusion than in subsequent times. On the other hand, it may be merely that the pre-Cambrian basements, as contrasted with sections eroded in younger rocks, expose material once situated relatively deep within the crust.

Post-Archean granites attain their greatest development in orogenic zones. The requirements of isostasy and the observed coincidence of extensive belts of negative gravity anomaly with young folds (e.g., in the East and West Indies)[5] strongly suggest that downward thickening of the sial as a bulge displacing underlying denser rocks is a necessary accompaniment of orogeny. Here are conditions favorable to development of magmas at the base of the crust—disturbance of temperature distribution, local concentration of rocks rich in radioactive elements, depression of great masses of fusible material (sial) into a zone of elevated temperature, and (perhaps most important of all) relatively rapid increase in temperature resulting from local convectional overturn in the mantle immediately beneath the orogenic zone. It is not surprising—though it is impossible to do more than speculate as to the sequence of events involved—that extrusion and intrusion of both granitic (rhyolitic) and basaltic magmas should so commonly be associated with large-scale folding.

Other Primary Magmas.   If it be admitted that primary basaltic and granitic magmas develop at different levels in the mantle and crust during orogeny, it is also reasonable to suppose that primary andesitic magma could likewise form as liquid squeezed away from partially fused basic rocks, or by more complete melting of sedimentary rocks in the lower

---

[5] Cf. H. H. Hess, Geological interpretation of data collected in the West Indies, *Am. Geophys. Union Trans.*, pp. 69–77, 1937; Major structural features of the Western North Pacific, *Geol. Soc. America Bull.*, vol. 59, pp. 438–440, 1948.

levels of the continental crust. This hypothesis of origin of andesitic lavas has already been discussed on pages 285 to 287.

There is also the possibility that peridotites and anorthosites are formed by uprise of corresponding primary magmas (not necessarily entirely or even largely liquid) generated in the lower levels of a basic crustal layer or in the mantle beneath (see pages 315 and 328). Existence of primary peridotite and anorthosite magmas is possible; indeed it is becoming increasingly difficult to explain the composition and great extent of very large bodies of these rocks in terms of any simple process of derivation from basaltic magma. An interesting aspect of modern petrology—as contrasted with orthodox thought 30 years ago—is a pronounced tendency to invoke a primary origin, or at least origin in the inaccessible depths, for magmas of many kinds—among them peridotite, anorthosite, biotite-pyroxenite, and even carbonatite. This trend reflects a prevalent difficulty in deriving these and other rocks from basaltic or granitic parent magmas by chemical processes completely consistent with experimental data. Perhaps the natural chemical systems involved are too complex to permit development of any scheme of petrogenic evolution explicable in every detail in terms of currently available laboratory data drawn from grossly simplified systems.

Conclusion. The main problem of the origin of magmas is twofold. In the first place, we must account for the world-wide occurrence of a primary basaltic magma of slightly variable composition (alkaline olivine-basalt to tholeiite), yet of remarkable uniformity in space and time. Then we must account for the occurrence, in continental areas only, of primary granitic magmas. Any solution of the problem must of course be based on what is known regarding the constitution of the earth, the possible sources of heat, and the temperature distribution therein. To this we now turn.

CONSTITUTION OF THE EARTH

The geophysical and geochemical evidence bearing on the constitution of the earth has been recently reviewed in a number of places [6] and need not be repeated here. The salient facts are that a thin crust covers a

[6] B. H. Mason, *Principles of Geochemistry*, 2nd ed., Wiley, New York, 1958; E. C. Bullard, The Interior of the Earth, in *The Earth as a Planet*, vol. 2, pp. 57–137, University of Chicago Press, 1954; K. E. Bullen, Seismology and the Broad Structure of the Earth's Interior, in *Physics and Chemistry of the Earth*, vol. 1, pp. 68–93, Pergamon Press, London, 1955; J. A. Jacobs, The Earth's Interior, in *Encyclopedia of Physics*, vol. 47, pp. 364–406, Springer, Berlin, 1956; P. Byerly, Subcontinental Structure in the Light of Seismological Evidence, in *Advances in Geophysics*, vol. 3, pp. 105–152, Academic Press, New York, 1956.

thick mantle which in turn surrounds the core. The thickness of the crust varies locally from about 6 to 8 km. below the oceans to 30 km. or more below the continents, where it seems to thicken in proportion to surface elevation. The boundary between crust and mantle, which is very sharp, is known as the Mohorovičić discontinuity. Above this discontinuity the average density is about 2.8 gm./cm.[3] It is doubtful whether there is any systematic "layering" in the crust, except locally, although it appears that the composition varies with depth. It is usually said that the upper part is dominantly granitic, the bottom being perhaps closer in composition to basalt or gabbro, but it should be remembered that such lithological identifications are based mostly on the velocity of propagation of elastic waves—a velocity that is not lithologically diagnostic; many different kinds of rocks have the same velocity, and continents could, for all we know, consist of metamorphic rather than of igneous rocks. The extreme thinness of the oceanic crust, the bottom of which lies some 10 to 12 km. below the surface of the oceans and 6 to 8 km. below their floor, makes it almost certain that magmas of the types erupted in the oceanic environment originate in the underlying mantle. As identical magmas also appear in continental areas, it may be inferred that basaltic magmas originate for the greatest part not in the crust but in the underlying mantle.

The mantle, in turn, is a very thick unit which extends without major break down to a depth of 2,900 km., almost halfway to the center of the earth. The velocity of longitudinal elastic waves just below the Mohorovičić discontinuity is very generally found to be close to 8.1 km./sec. in continental and oceanic areas alike. The density, obtained from considerations of isostatic balance between continents and oceans, must be close to 3.3 gm./cm.[3] Very few rock types have such characteristics, and the choice narrows down to either peridotite (or dunite) and eclogite. Of the two, the former seems preferable, because of the very widespread occurrence in basaltic volcanic ejecta, all over the world, of peridotite nodules of astoundingly uniform composition (olivine, enstatite, chromian diopside, spinel) which are also very similar to the widely distributed dunites and associated intrusive rocks; [7] eclogite, by contrast, is a fairly rare rock. It has been repeatedly suggested that eclogite is the main constituent of the upper mantle, the Mohorovičić discontinuity corresponding to a physical rather than to a chemical change, as eclogite is a high-pressure equivalent of the basalt or gabbro that may form the lower part of the crust. The pressure-induced transition from gabbro to eclogite would not, however, be a sharp one, as in a system of

[7] C. S. Ross, M. D. Forster, and A. T. Myers, Origin of dunites and of olivine-rich inclusions in basaltic rocks, Am. Mineralogist, vol. 39, pp. 693–737, 1954.

one component; in a multicomponent system it would spread over a certain pressure interval, in contradiction with the seismological observation that the Mohorovičić discontinuity is actually very sharp.

The composition of the main part of the mantle is still in doubt. Most authors agree that its composition should, on the whole, be close to that of peridotite or average stony meteorite. From the Mohorovičić discontinuity down to a depth of about 400 km. (zone B of Bullen), the velocity of seismic waves increases relatively slowly, although there may be, at depths around 80 to 100 km., a "low velocity" zone in which the velocity remains constant or even decreases slightly with depth; this, however, is still uncertain. From 400 to 900 km. (zone C of Bullen), the velocities increase more rapidly than would be expected from theoretical calculations of the effect of pressure on the elastic coefficients that control these velocities. There are also arguments derived from a determination of the earth's moment of inertia that show that the mantle cannot be physically or chemically homogeneous. Thus Birch [8] suggests that zone C corresponds either to a change in composition or to a phase change; and recent experimental work of Ringwood [9] brings out the possibility of a transition to a cubic (spinel) form of olivine, although it remains difficult to explain, on this hypothesis, the great spread (400 to 900 km.) of the transition zone. From 900 km. to the core boundary (zone D of Bullen), the velocities again increase slowly at the rate which Birch predicts from theory for a homogeneous material.

Propagation of S (shear) waves through the whole of the mantle leaves no doubt that it consists throughout of crystalline material; earlier suggestions that pressure could impart to a liquid (glass) above its melting point a rigidity comparable to that of the mantle have been disproved by experiments.[10] It is thus in crystalline material that the origin of basaltic liquid must be found.

TEMPERATURE DISTRIBUTION AND SOURCES OF HEAT IN THE EARTH

Temperature Gradient. Measurements in mines, tunnels, and bore holes indicate that temperature increases with depth at a rate which varies locally, but averages around 30°C. per km. (geothermal gradient).

[8] F. Birch, Elasticity and constitution of the earth's interior, *Jour. Geophysical Research,* vol. 57, pp. 227–286, 1952.

[9] E. A. Ringwood, The constitution of the mantle: I. Thermodynamics of the olivine-spinel transition, *Geochim. et Cosmochim. Acta,* vol. 13, pp. 303–321, 1958; II. Further data on the olivine-spinel transition; III. Consequences of the olivine-spinel transition, *ibid.,* vol. 15, pp. 18–29 and 195–212, 1958.

[10] F. Birch and D. Bancroft, The elasticity of glass at high temperatures and the vitreous basaltic stratum, *Am. Jour. Sci.,* vol. 240, pp. 457–490, 1942.

If the thermal conductivity K (cal./cm. sec. deg.) of the rocks in which the gradient is measured is known, the upward heat flux Q (cal./cm.^{2} sec.) is found from the simple relation

$$Q = \frac{K\,dT}{dh} \tag{15-1}$$

where h is depth measured in centimeters downward from the surface.

For most rocks, K falls in the range $3 - 8 \times 10^{-3}$ cal./cm. sec. deg. It decreases with increasing temperature, and is higher for basic and ultrabasic rocks than for granites and sediments. Its average surface value is usually around 4×10^{-3}, corresponding to a mean heat flow of about 1.2×10^{-6} cal./cm.^{2} sec. This heat, which is probably almost entirely of radioactive origin, is small when compared with, say, the heat received from the sun; it is nevertheless large when compared with the amount of heat required to account for volcanic activity. If all the heat flowing out of the earth could be used solely for the purpose of melting while none of it escaped, the normal heat flow would be sufficient to melt a layer of basalt one centimeter thick in about 30 years. The problem of the origin of basaltic magma is thus not so much that of finding adequate heat sources as that of explaining how relatively small amounts of heat may become locally concentrated to produce relatively small pockets of liquid in an otherwise crystalline mantle.

Radioactive Heat. In the steady state in which temperatures at all depths remain constant, the upward heat flow at depth h is obviously equal to the flow at depth $h + dh$ plus whatever amount of heat is generated, from radioactivity or other sources, between h and $h + dh$. Thus, if the rate of heat generation and the thermal conductivity were known at all depths, Eq. (15-1) could be integrated to give the temperature at any depth, knowing the temperature and heat flow at the surface. The manner in which radioactive matter is distributed in the earth is thus of paramount significance. The radioactive elements that need be considered are U, Th, and their disintegration products, and K^{40}. The heat liberated during disintegration is equivalent to the kinetic energy of the elementary particles emitted from the disintegrating nuclei, and is well known. It is then sufficient to know the amounts of these elements present in any rock to calculate the rate of heat generation in it. Formidable problems of sampling and analysis present themselves, as the radioactive elements do not appear to be homogeneously distributed; certain parts of exposed batholiths, for instance, are far more radioactive than others. Average figures indicate that the heat generation from all radioactive sources in granite is about 2.2×10^{-13} cal./gm. sec or 6×10^{-13} cal./cm.^{3} sec., taking the density of granite as 2.7 gm./cm.^{3} The concentration of radioactive elements decreases rapidly with decreasing silica content, so that the rate of heat generation in basalt is about one-seventh that for granite, and the rate for dunite or stony meteorites is of the order of 1×10^{-15} cal./gm. sec.[11] Very little is known of the radioactive content of metamorphic rocks. Yet these figures show that a column of granite 20 km. thick would liberate an amount of heat equivalent to the total heat flow at the surface, as $6 \times 10^{-13} \times 2 \times 10^6$ equals 1.2×10^{-6}. This result is a sufficient indication that continents do not consist entirely of granites of the type sampled at the surface, for a thickness of 35 or 40 km. of such granites would yield a heat flow greater than the measured one.

Birch [12] has given estimates of the temperature that could prevail at the bottom of the continental crust for various assumptions regarding the composition of the crust and its thermal conductivity; they range from about 300° to 700°C. This is much less than the figure that would be obtained merely by extrapolating the surface geothermal gradient ($30 \times 30 = 900$), precisely because of the gradual decrease in Q resulting from radiogenic heat sources within the continents, and because of a probable increase of K in the more basic rocks forming the lower part of the crust.

Because of the long accepted, but certainly erroneous view that continents consist dominantly of granite, and because of the high radioactive content of such rock, it had also been generally accepted that very little heat should be generated in the mantle; otherwise there would be an embarrassingly large amount of heat to dispose of in an unaccountable manner. This view has, however, been largely dispelled by measurements of heat flow on the ocean floor.[13] More measurements of heat flow have now been effected on the ocean floor than on land, and the average heat flow in oceanic areas turns out to be very close to that measured on land, the average for seven Atlantic stations being 0.93×10^{-6}, and the average for 25 Pacific stations 1.53×10^{-6} (all in cal./cm.² sec.). The extreme thinness of the oceanic crust precludes the possibility that all this heat would be generated in it; hence it must come from the underlying mantle. A remarkable feature of the oceanic heat flow is its great variability; certain areas (e.g., the Albatross Plateau) have an average heat flow many times greater than other areas (e.g., the Acapulco Trench).

The similarity in the observed values of heat flow in continental and oceanic areas is so remarkable, in spite of utterly different geologic setting, that it can hardly be accidental. Birch [14] has noted that this heat

[11] F. Birch, Physics of the Crust, in *The Crust of the Earth, Geol. Soc. Am. Special Paper 62*, p. 113, 1955.

[12] *Ibid.*, p. 114.

[13] E. C. Bullard, A. E. Maxwell, and R. Revelle, Heat Flow through the Deep Sea Floor, in *Advances in Geophysics*, vol. 3, pp. 153–181, Academic Press Inc., New York, 1956.

[14] F. Birch, Differentiation in the mantle, *Geol. Soc. Amer. Bull.*, vol. 69, pp. 483–485, 1958.

flow is about the same as would be generated in a uniform mantle with a content of radioactive matter comparable to that of average stony meteorite, the only difference between continents and oceans being perhaps that most of the radioactive matter in continental sectors is now concentrated in the crust as a result of differentiation, while in oceanic areas it remains in the mantle. It should be noted in this connection that any heat generated in the lower part of the mantle cannot contribute at present to the surface heat flow if transfer of heat occurs by conduction; as shown by Slichter,[15] the mantle is so thick, and its thermal conductivity is so low, that heat generated near the core boundary would take more than the age of the earth to reach the surface. Thus if Birch is correct, either transfer of heat in the mantle occurs by some process faster than conduction (radiative transfer or convection) or its radioactive content must have become largely concentrated in its upper reaches by some unknown process. In either case, the thermal gradient in the lower mantle is likely to be very low.

Temperature Distribution in the Mantle.[16]   *Temperature of Melting.* The seismological evidence, the evidence from the yielding of the earth to tidal deformation, and the period of the variation of latitude [17] leave no doubt regarding the high rigidity of the earth's mantle; as mentioned above, it is most unlikely that any material other than a crystalline solid could have such rigidity at the temperatures and pressures prevailing there. This implies that the temperature in the mantle is everywhere less than the melting point at that depth. Thus knowledge of the melting point sets an upper limit to the temperature at any depth.

As melting generally involves a change in volume, the melting point should depend on pressure, the dependence being given by the Clausius-Clapeyron equation (2–40). Exact calculations cannot be made, as we do not know the composition of the mantle. On the assumption that it consists dominantly of olivine, extrapolation of experimental data on the volume change and entropy of melting give an upper limit of about 6000°C. for the temperature at the bottom of the mantle, due allowance being made for probable changes with pressure of the parameters in the Clausius-Clapeyron equation. It also appears that the melting point of a solid is related to its elastic properties; in general hard substances, such as corundum, periclase, or diamond, also have high melting points. As

[15] L. B. Slichter, Cooling of the Earth, *Geol. Soc. America Bull.*, vol. 52, pp. 561–600, 1941.

[16] See J. A. Jacobs, The Earth's Interior, in *Encyclopedia of Physics*, vol. 47, pp. 364–406, Springer, Berlin, 1956; J. Verhoogen, Temperatures within the Earth, in *Physics and Chemistry of the Earth*, vol. 1, pp. 17–43, Pergamon Press, London, 1956; H. A. Lubimova, Thermal history of the earth with consideration of the variable conductivity of its mantle, *Geophysical Jour.*, vol. 2, pp. 115–134, 1958.

[17] See Bullard, *op. cit.*, pp. 95–101, 1954.

the elastic properties of the deep mantle are known from the velocity of propagation of seismic waves, it is possible to obtain an estimate of the change in melting point with depth.   This has been done by Uffen,[18] who finds a probable melting point of about 5000°C. at the core boundary. This may be taken as the maximum possible temperature at that depth.

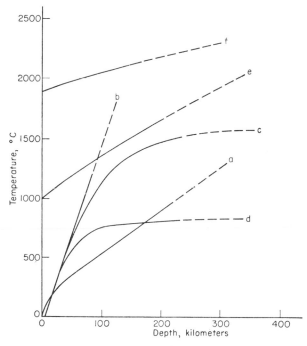

Fig. 62.   Possible temperature distributions in upper mantle.   Curve a is for a continental area, assuming that 75% of the surface heat flow originates in the upper 30 km. of the crust.   Curve b refers to oceanic areas, using the average oceanic heat flow and a rate of heat generation in the upper mantle equal to the average for stony meteorites.   Curves c and d, also for oceanic areas, assume progressively greater concentration of radioactive matter in the uppermost part of the mantle.   Curve e gives the probable minimum melting temperature of basalt as a function of depth, while curve f gives the melting point of forsterite.   Assumed conductivity of mantle: 8×10^{-3} cal./cm. sec. deg.

If one now uses Eq. (15-1) with the known value of the surface heat flow in oceanic areas, and the rate of heat generation for stony meteorites, one finds that the temperature must increase downward at such a rate that the melting point would soon be reached and exceeded (see curve b in Fig. 62); hence the mantle should be molten, which it is not.   The discrepancy could be corrected by assuming a strong concentration of

[18] R. J. Uffen, A method of estimating the melting point gradient in the earth's mantle, Am. Geophysical Union Trans., vol. 33, pp. 893–896, 1952.

radioactive matter in the upper mantle, as this would greatly reduce the gradient (curves c and d, Fig. 62). Geochemists, however, object that one cannot clearly see how such concentration could be effected within what appears seismologically to be a fairly uniform mass of peridotitic composition. Alternatively, it is possible to avoid wholesale melting in the mantle by assuming that the effective thermal conductivity in the mantle is greatly in excess of the laboratory values. This could happen in two ways. Silicate crystals are fairly permeable to infrared and visible radiation. When heated to temperatures of a thousand degrees or more they begin to glow, i.e., to give off radiation in the visible part of the spectrum. If they are sufficiently transparent, this radiation could be transmitted without appreciable absorption over distances of the order of a few millimeters, or a centimeter, or more, leading to a large increase in thermal conductivity.[19] Whether radiative heat transfer is effective in reducing the temperature gradient in the mantle is not definitely known, but it appears likely that it may play a part, at least at depths of the order of a few hundred kilometers.

Convection in the Mantle. There is yet another way in which the gradient may be reduced. Most solids expand on heating. If a liquid, originally of uniform density, is heated from below, the lower layers, being hotter, will expand more than the upper, colder ones, and instability sets in because of the inverted density stratification; the hot liquid rises to the surface, while the colder portions sink to the bottom. Convective motion of this sort results in a greatly increased rate of transfer of heat to the surface. The temperature gradient at which instability sets in, called the "adiabatic gradient," can be evaluated, and turns out to depend only on density, temperature, coefficient of thermal expansion, and specific heat. The instability gradient in a viscous fluid depends also somewhat on viscosity, and for solid materials such as may form the mantle, convection will start only when the stresses resulting from the unstable density stratification exceed the yield limit at which the material will begin to flow plastically. It may be shown, however, that these stresses increase proportionally to the linear dimensions of the system and may become sufficiently large where the dimensions of the convecting body are of the order of 1,000 km.

The adiabatic gradient in the mantle may be computed from seismic data and solid-state theory [20] and turns out to be very small, of the order of a fraction of a degree per kilometer, for temperatures less than the

[19] S. P. Clark, Jr., Radiative transfer in the earth's mantle, *Am. Geophysical Union Trans.*, vol. 38, pp. 931–938, 1957; A. W. Lawson and J. C. Jamieson, Energy transfer in the earth's mantle, *Jour. Geology*, vol. 66, pp. 540–551, 1958.

[20] J Verhoogen, The adiabatic gradient in the mantle, *Am. Geophysical Union Trans.*, vol. 32, pp. 41–43, 1951.

melting point.   Thus, if the adiabatic gradient prevailed throughout the
mantle, the temperature at its base could not much exceed 1500°C.   The
possibility of convection provides an additional reason why the temperature gradient in the mantle should be very low.

Other Estimates of Temperature in the Deep Mantle.   If it is assumed
that the outer core, which is liquid, consists dominantly of molten iron,
the temperature at the core-mantle boundary must be less than the melting
point of mantle material, but higher than the melting point of iron, which
has been evaluated by Simon,[21] and turns out to be somewhere between
3000° and 3400°C. at the corresponding pressure (1.3×10^6 bars).
Nickel and silicon dissolved in the core would probably somewhat lower
the melting point.   Verhoogen,[22] using solid-state theory and seismic
data, finds that the temperature at the core boundary probably does not
exceed 2500°C.   All these results appear fairly consistent, and suggest
that the temperature in the mantle rises slowly from a value of a few
hundred degrees at the Mohorovičić discontinuity (600°C. below continents?) to a maximum of 3000°C. or less at the core boundary.

ORIGIN OF BASALTIC MAGMA

Depth of Formation.   Where, then, in this relatively cold mass, the
mantle, do basaltic magmas form?   Some evidence bearing on their actual
depth of formation may perhaps be found in the fact that the rise in melting point with pressure is rather different for different minerals.[23]   It is
quite low (about 5°C./1,000 bars) for olivine and anorthite, about 13°C./
1,000 bars for diopside and enstatite, 26°C./1,000 bars for albite.   The
coefficient is generally greater for minerals with low melting points, so
that differences in melting point should tend to decrease as the pressure
rises.   Yoder[24] points out that anorthite and albite would melt at the
same temperature at 22,000 bars.   Similarly, the melting relationship
between olivine and pyroxene might be reversed, and the incongruent
melting relationship of enstatite and forsterite might vanish, as does that
of orthoclase and leucite at lower pressures.   It is not inconceivable that
a peridotite, which by fractional melting under normal conditions might
yield a basaltic liquid with relative concentration of alumina and alkalis,
would behave differently at pressures of 40,000 or 50,000 bars or more; at

---

[21] F. E. Simon, The melting of iron at high pressures, *Nature*, vol. 172, p. 746, 1953.
[22] J. Verhoogen, Thermal expansion of solids and the temperature at the boundary
of the earth's core, *Am. Geophysical Union Trans.*, vol. 36, pp. 866–874, 1955.
[23] J. Verhoogen, Petrological evidence on temperature distribution in the mantle of
the earth, *Am. Geophysical Union Trans.*, vol. 35, pp. 85–92, 1954.
[24] H. S. Yoder, Change in melting point of diopside with pressure, *Jour. Geology*,
vol. 60, pp. 364–374, 1952.

depths greater than, say, 200 km., it is likely that the first liquid to be produced would not be a normal basalt. Tholeiitic basalts could be produced by partial melting of peridotites at depths less than approximately 100 km., olivine basalts being possibly formed by partial melting at slightly greater depths.

The fact that lavas rarely, if ever, reach the surface at temperatures much higher than 1200°C. may also have a bearing on their depth of formation. We have noted that the melting temperature, or temperature range if we are dealing with a multicomponent system, increases with depth; the initial rate of increase for basalt has been estimated by Yoder [25] to be about 10°C./1,000 bars. At a depth of 500 km. the pressure is 175,000 bars, and the melting range of basalt at that pressure would be well over 2000°C. (see curve e, Fig. 62). If a basaltic melt could still form at that depth, it would thus rise to the surface at a temperature which would be only slightly less, the cooling due to expansion as it rises being very small. The fact that lavas of Kilauea are erupted at a temperature which is very close to the upper limit of their melting range at normal pressure speaks strongly for a shallow depth of formation.

The figures mentioned above refer to the melting of dry rock. It has been shown earlier (see page 417) that pressure increases the solubility of water in a silicate melt, thereby lowering its melting point. Assume that the original mantle material contains, say, 0.5 per cent water in hydrous minerals. Will this water notably reduce the temperature at which partial melting could occur?

The answer appears to be no. Partial melting of, say one-third of the initial mass would result in a water concentration in the melt of 1.5 per cent, assuming that all available water were dissolved. This estimate agrees with the fact that basaltic magmas probably do not contain much more than 1 per cent water when they reach the surface, judging from the gas/lava ratio in volcanic eruptions. Now Eq. (5–2) shows that the depression of the melting point $(T_m - T)$ of a pure substance is proportional to T_m, its melting temperature, and is inversely proportional to Δh, the heat of melting, and to the logarithm of its mole fraction in the melt. Pressure might change Δh somewhat, and would of course change T_m; but for a given mole fraction, the lowering of the melting point is likely to be nearly proportional to the melting point itself. Thus the lowering of the melting point by 1.5 per cent of water at a pressure such that the melting point is doubled is not likely to be much more than double what it is at ordinary pressure, namely, 50° to 100°.

A further estimate of the depth of formation of basaltic magma is obtained if one assumes that, once molten, the magma will rise to a height such that the pressure of the liquid column will balance the pressure

[25] *Ibid.*

existing at the depth where magma forms. Let h be this depth below the ocean floor, and H the total height of the liquid column; then equilibrium requires that $d_1h = d_2H$, where d_1 is the density of mantle material and d_2 that of the liquid. In the Hawaiian Islands, $H\text{-}h$ is about 10 km.; hence for $d_1 = 3.3$ and $d_2 = 2.8$, $h = 56$ km. The actual depth is probably somewhat greater than this, as volcanoes are known not to be in static equilibrium. We conclude that the most probable depth of formation of basaltic magma is in the range 50 to 100 km., the temperature required to produce such magmas at such depths being of the order of 1200° to 1500°C.

Possible Mechanisms. Basaltic magma, in copious amounts, has been generated throughout geologic time in continental and oceanic sectors alike. The heat carried to the surface, though large, is nevertheless small when compared to the normal heat flow: if the average rate of outpouring at the present time is taken to be 2 km.^{3}/year (which would, in 4 billion years, yield a total volume about equal to that of the whole crust, oceanic areas included) the total heat released at the surface, including the latent heat of crystallization and cooling, amounts to but 1 or 2 per cent of the total heat flow during the same interval. The problem therefore is not so much to find suitable sources of heat as to explain how melting can occur locally in an otherwise crystalline mantle everywhere below the temperature of incipient melting. The seismological evidence is indeed clear that there is no world-encircling layer of liquid basalt, and the possibility of the existence of small, local pockets of liquid material remaining from a time when the earth was molten may be dismissed on the grounds that (1) there is no convincing evidence to show that the earth ever was completely molten; (2) liquid basaltic melt, being much lighter than the surrounding crystalline material, would have risen to the surface long ago; and (3) such small masses located within a few hundred kilometers of the surface would, in a few billion years, lose enough heat to crystallize.

If it be assumed that the temperature, at the depth at which basaltic magma forms, is intermediate between the melting points of basalt at zero pressure and at the pressure prevailing at that depth, melting could occur if the pressure were released. This hypothesis of magma generation is possibly the most popular one, although it is unacceptable to the present writers. It is difficult indeed to see how pressure can effectively be reduced at such depths. The isostatic balance which generally exists points unequivocally to conditions under which the pressure, everywhere below the level of compensation, is hydrostatic; it becomes so whenever the yield point under stresses of long duration becomes negligible. The level of compensation is almost certainly less than 100 km., and most geologists would consider that hydrostatic conditions already prevail at depths of a few kilometers in the crust. If hydrostatic equilibrium prevails, there is no way, short of decreasing the mass of the earth, by which the pressure could be reduced.[26] For instance, formation of an arch transfers the weight of the arch from its center to its extremities, whence it is transmitted equally in all directions in the supporting medium. Loading or unloading a ship does not alter the hydrostatic pressure at a given depth in the water below the ship. Opening of tensional fractures extending all the way down to the magma source, if at all possible, would provide only transient relief, as the melt rising in the fracture would soon re-establish the orginal pressure at the base of the rising column. Furthermore, melting as a result of release of pressure would be essentially adiabatic, meaning that the latent heat of fusion (around 100 cal./gm. under ordinary conditions) would have to come from the mass itself which, in order to melt, would have to be initially at a temperature appreciably higher than the melting point at the reduced pressure; there is no evidence that such temperatures exist in the depth range in which basaltic magma is believed to form (curves a and c, Fig. 62). The hypothesis of magma generation by release of pressure thus seems unpromising.

Alternatively, one may consider possible mechanisms by which the temperature could locally be raised sufficiently to produce partial melting. The blanketing effect of thick series of sediments is negligible, and furthermore basaltic magma is produced in oceanic areas where no such blankets exist. Strain energy (= stress $\times$ strain) in a nonelastic medium is converted to heat, and it has been suggested that wherever intense deformation occurs in the mantle, enough heat might be generated to raise the temperature appreciably.[27] There is considerable doubt as to the magnitude of this effect and, additionally, large outpourings of basaltic magma have occurred in areas (e.g., Columbia Plateau, Hawaiian Islands) where there is no evidence of deformation. Neither does the North Atlantic volcanic area, including the active volcanoes of Iceland, seem to be connected with any major diastrophic process. On the other hand, basaltic eruptions are rare or absent in other areas (e.g., Himalayas) where deformation has been intense.

Partial melting could occur if mantle material could be transported upward (e.g., by convection) in a region where the original temperature gradient is greater than the adiabatic one, provided the initial temperature of the rising material is greater than the melting point at the place

[26] Existence of earthquake foci down to depths of the order of 700 km. may point to nonhydrostatic conditions in the mantle. The three principal stresses may indeed be unequal, leading to deformation. The maximum stress difference needed to produce deformation at any depth in the mantle is, however, presumably small as compared with the load (hydrostatic) pressure at that depth, and the average of the three principal stresses, which determines the melting point, cannot differ much from the load pressure.

[27] Bullard, op. cit., pp. 120–121, 1954.

where it comes to rest. Imagine, for instance, material rising from a depth where the temperature is originally 2000°C., to a level where the melting point is, say, 1500°C. The rising material cools by expansion as it moves upward; if this cooling is less than 500 deg., it will melt, at least partially. This is illustrated in Fig. 63, which shows the temperature distribution before and after a convectional overturn. The line a represents the original temperature gradient at which convection starts (adiabatic plus whatever is required to overcome fundamental strength + viscosity). The material that ends up in the upper half of the layer was originally in the lower half, and vice versa. Under appropriate conditions, curve b could intersect the melting-point curve, while a does not.

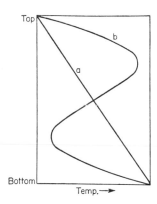

Fig. 63. Vertical temperature distribution (a) before, and (b) after, a convectional overturn. (*After H. Brooks.*)

Whether convection does occur in the mantle is not definitely known, nor is it known whether it could be effective in the upper part of the mantle where magmas are generated. Thus, although convection might lead to melting, it cannot be shown that it does, and the problem of the generation of magma remains as baffling as ever. Let it be repeated once more that the difficulty arises mainly from the intermittent and local character of magma generation. It would be relatively easy to find a possible distribution of radioactive material that would lead to permanent melting of a world-wide character (see curve b, Fig. 62), but this is denied by seismological evidence. It seems as if the temperature in the upper mantle were close to, although slightly less than, the minimum melting point, but were also subject to local and somewhat random fluctuations of unknown origin, possibly connected with convection in the deep mantle. These fluctuations presumably are reflected in the high variability of the heat flow in the Pacific basin.

ORIGIN OF GRANITIC MAGMA

In contradistinction to basalts, which occur everywhere, granitic rocks are restricted to continental areas, and more particularly to orogenic zones therein. This suggests that either the presence of the continents alters conditions in the underlying mantle in a manner such that granitic magmas could form therein, or that granitic magmas originate in the crust itself. The first mechanism has been suggested by Griggs [28] as an

[28] D. Griggs, Discussion, *Am. Geophysical Union Trans.*, vol. 35, p. 95, 1954.

autocatalytic process by which continents would continually grow. The second mechanism is consistent with the common association of granites with profoundly metamorphosed sediments and other crustal rocks.

It has been pointed out earlier (page 438) that the "normal" temperature at the bottom of the crust probably does not exceed 600°C. or so; the minimum melting point of granite, even at extreme water pressures, is at least 100 deg. higher. Thus, again, production of copious granitic liquid in the crust cannot be a "normal" phenomenon: it would not be sufficient merely to lower mechanically a mass of sediments to the deepest part of the crust to produce selective melting of a granitic fraction. On the contrary, formation of granite magma implies notable disturbances of temperature.

In the discussion of conditions leading to regional metamorphism, with which granites are intimately associated, it will be pointed out (Chap. 24) that regional metamorphism itself occurs only where heat flow has been notably increased. This increased heat flow, which apparently accompanies orogeny, may in fact be its most characteristic feature, deformation being but a subsidiary effect of a deep-seated thermal disturbance. What the nature of this disturbance is remains speculative, just as in the case of production of basaltic magma; again convection in the mantle may be the primary cause. Apparently the mantle must be the site of processes (transfer of radiogenic heat, convective motion, partial melting, and differentiation) far more complicated than is usually recognized—processes which lead to (1) the formation of basaltic magmas in the mantle itself; (2) increased heat flow, regional metamorphism, and formation of granitic magmas in the crust; and (3) diastrophism. Unsatisfactory as it is to leave such fundamental processes unexplained, it is doubtful that a valid interpretation can be reached until much more is known regarding the nature of the materials forming the mantle and the distribution of radioactive matter within it.

TENTATIVE SYNOPSIS OF MAGMATIC EVOLUTION

A tentative outline of courses of evolution leading from primary basaltic or granitic magma to common types of igneous rocks is given on page 448. It is open to revision and elaboration in the light of new data as these come to hand. Some links in the evolutionary chain are better established than others. In some cases linkage between two or more groups is obvious, but the nature of the link, *i.e.*, the mode of derivation of one group from the other, remains ambiguous or obscure. Nor is a possible mode of origin offered for every kind of igneous rock: the andesine-labradorite anorthosites and the alpine peridotites are notable omissions. Nevertheless the scheme as a whole is, in the authors' opin-

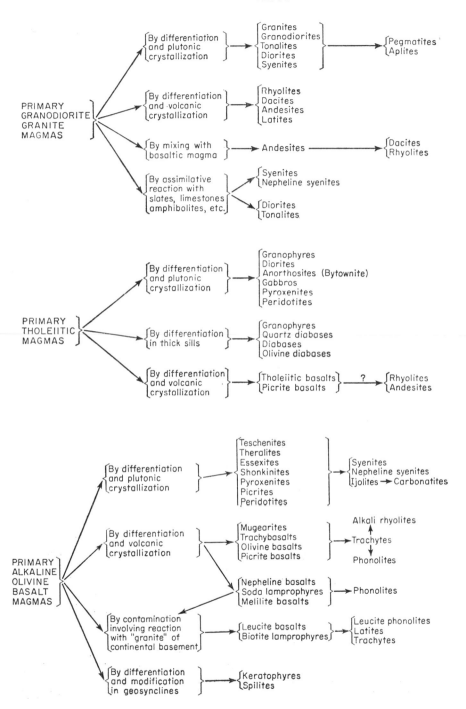

ion, broadly consistent with a vast quantity of data drawn from independent sources, including the chemistry, mineralogy, fabric, and field distribution of the rocks themselves, the experimentally determined behavior of silicate melts, the chemistry of meteorites, and the behavior of seismic waves as they penetrate the earth's interior.   In spite of its imperfect and provisional nature, some such synthesis of petrological knowledge is necessary as a background for research and as a basis for the more prosaic task of classifying igneous rocks.

CHAPTER 16

Scope of Metamorphism and Classification of Metamorphic Rocks

DEFINITION

Metamorphism is the mineralogical and structural adjustment of solid rocks to physical or chemical conditions which have been imposed at depths below the surface zones of weathering and cementation and which differ from the conditions under which the rocks in question originated.

Van Hise's [1] extension of the scope of metamorphism to include weathering and cementation has now been generally abandoned, since it so enlarged the field of metamorphism that the term became virtually synonymous with "rock alteration" and therefore had little practical value in petrology. Other writers have narrowed the field of metamorphism as defined above by excluding such rock transformations as involve addition or removal of material, i.e., changes into which metasomatism enters. Eskola has applied the term isochemical metamorphism to cover reactions involving no substantial change in composition. However, addition or removal of substances of high vapor pressure, notably water and carbon dioxide, may play an essential part in rock alterations universally classed as metamorphic; e.g., carbon dioxide is expelled in quantity during conversion of impure limestone to calc-silicate hornfels. Again, the metamorphism of basalt to green schists consisting of albite, chlorite, calcite, and epidote may involve not only extensive hydration and carbonation of the rock, but also removal of certain bases, especially lime. Where metamorphism is connected with intrusion of magma, widespread introduction of elements such as boron, fluorine, and chlorine is almost universal, as evidenced by the wide distribution of such minerals as tourmaline, scapolites, topaz, and apatite in the resultant rocks. It would, therefore, seem impossible to distinguish sharply between metamorphism of this

[1] C. R. Van Hise, A treatise on metamorphism, U.S. Geol. Survey Mon. 47, 1904.

type and that in which mineralogical and chemical results of metasomatism are more obvious. In accordance with the views of Eskola, Niggli, and Harker,[2] metamorphism may be considered as commonly approaching isochemical change; but it is also recognized that the content of the volatile constituents of the rock is affected and that with increasing metasomatic activity there is a transition from isochemical to allochemical metamorphism involving substantial addition or removal of material, as in albitization, serpentinization, or development of skarns.

When coming into contact with metamorphic phenomena for the first time, the student familiar with igneous petrology must bear in mind that metamorphic rocks have remained essentially solid throughout the processes of deformation and chemical reconstitution by which their present mineralogy and structure have evolved. Much of the chemical and mechanical reaction concerned has been effected, however, through the medium of aqueous or other pore fluids, the total amount of which at any instant never rises above a small fraction of the reacting mass. In migmatite zones, layers of igneous rock that have crystallized from magma may alternate with layers of metamorphic rock that have remained essentially solid during their reconstitution.

TYPES OF METAMORPHISM

Classic attempts to subdivide the field of metamorphism [3] have developed a terminology with a genetic flavor, implying that the causes and physical conditions of each kind of metamorphism are known. Here belong such terms as *injection, thermal, geothermal, pneumatolytic, hydrothermal, load, dynamic,* and *dislocation* metamorphism. Much of this terminology is ambiguous, either for lack of definitive criteria of petrogenesis, or because the implicit underlying assumptions (*e.g.,* with respect to the effects of load) are of doubtful validity.

The common metamorphic rocks occur in a few recognizable geological situations; and corresponding types of metamorphism may be defined in terms of field and associational criteria, without regard to still imperfectly understood causes and conditions of the phenomena in question. Three main types of metamorphism are recognized on this basis: [4]

[2] P. Eskola, *Die Entstehung der Gesteine* (Barth, Correns, Eskola), Springer, Berlin, 1939; P. Niggli, *Die Gesteinsmetamorphose* (Grubenmann, Niggli), Borntraeger, Berlin, 1924; A. Harker, *Metamorphism,* pp. 117, 135, Methuen, London, 1932.

[3] *E.g.,* R. A. Daly, Metamorphism and its phases, *Geol. Soc. Am. Bull.,* vol. 28, pp. 375–418, 1917; Niggli, *op. cit,* pp 179–181, 339, 1924; G W. Tyrrell, *The Principles of Petrology,* pp. 253–256, Methuen, London, 1926; Eskola, *op. cit.,* pp. 264–267, 1939.

[4] H. Williams, F. J. Turner, and C. M. Gilbert, *Petrography,* pp. 163–164, Freeman, San Francisco, 1954.

1. *Contact metamorphism.* This occurs in restricted zones (aureoles) adjacent to bodies of plutonic rocks.

2. *Regional metamorphism.* This is developed over large areas—hundreds or thousands of square miles in extent—not consistently or obviously related to igneous intrusion. It is best exemplified in root regions of fold mountains and in pre-Cambrian continental shields.

3. *Dislocation metamorphism.* This is restricted to zones of intense deformation such as major faults and "movement horizons."

Other kinds of metamorphism, usually rather restricted in their field occurrence, can be satisfactorily defined in petrogenic terms. For these, unambiguous field, mineralogical, or textural criteria of metamorphic conditions are available.

1. *Pyrometamorphism.* This is shown by xenoliths in flows and dikes of volcanic rocks—especially basalts. The field occurrence, as well as the partially fused condition of many such rocks, testifies to the extreme temperature of metamorphism implied by the name.

2. *Cataclastic metamorphism.* Mechanical deformation (plastic flow, rupture) of rocks without recrystallization or chemical reaction. The activity of such processes can be inferred without ambiguity, from the nature of the fabric which they imprint upon the rocks affected.

3. *Metasomatic metamorphism.* This involves substantial change in chemical composition, as deduced from chemical, mineralogical, and fabric criteria. Conversion of peridotite to antigorite schist or to soapstone and replacement of limestone by calc-silicate rocks are common examples.

4. *Retrogressive metamorphism (Diaphthoresis).* A high-temperature metamorphic mineral assemblage is converted to an assemblage (usually more hydrous) stable at lower temperature. The criteria are mineralogical and are interpreted in the light of experimental data.

Not every instance of metamorphism can be placed unequivocally in one or other of the types listed above. The glaucophane schists of California are not consistently related either to igneous contacts or to dislocation zones, though occurrences in both situations are known. Nor is their field occurrence continuous or extensive enough to warrant the term "regional." Metasomatism and retrogressive metamorphism (from eclogite) enter into the genesis of some but by no means the majority of Californian glaucophane schists.

OUTLINE CLASSIFICATION OF METAMORPHIC ROCKS

Metamorphic rocks are classified on the basis of several kinds of geological criteria selected for their genetic significance:

1. Mineralogical composition. This is readily determined by petro-

graphic means and gives information regarding the chemical composition of the parent rock, the nature and extent of metasomatism, and especially —in the light of experimental data—the physical conditions of metamorphism.

2. Structure and texture (together constituting fabric). Fabric elements inherited from the premetamorphic condition (such as relict bedding or porphyritic structure) give some clue to the nature of the parent rock. Metamorphic elements in the fabric (schistosity, lineation, and so on) collectively provide evidence of the physical conditions of metamorphism—especially the role and symmetry of deformation in reconstitution of the rock.

3. Chemical composition. Chemical analyses provide the most complete evidence of the nature of the parent rock and the extent to which it has been affected by metasomatism.

4. Field occurrence. This throws light upon the nature and ultimate causes of the pressure-temperature gradients controlling metamorphism.

The following classification [5] uses only well-established names. Definitions are based mainly on macroscopically visible structural criteria, but some classes (e.g., quartzite, marble) are defined on a mineralogical basis.

Hornfelses. Non-schistose rocks composed of a mosaic of equidimensional grains (granoblastic or hornfelsic texture). Schistosity is weak or lacking, and where present is inherited from a premetamorphic condition (e.g., bedding or slaty cleavage). Porphyroblastic texture is common. Hornfelses typically occur in contact aureoles, but regional occurrences are also known.

Buchites. Partially fused hornfelsic rocks occurring as xenoliths, usually in basalt or diabase; products of pyrometamorphism.

Slates. Fine-grained rocks with perfect planar schistosity (slaty cleavage) but lacking segregation banding. They are formed by regional metamorphism of fine-grained clastic sediments (mudstones, siltstones, tuffs), and are rich in colorless micas. *Spotted slates* are slates in which, as a result of incipient contact metamorphism, spots and porphyroblasts of new minerals (micas, cordierite, and andalusite) have begun to appear, while the slaty cleavage has been preserved or intensified by crystallization of parallel flakes of mica.

Phyllites. Rocks similar to slates but coarsening in grain as a result of somewhat more advanced metamorphism. New mica and chlorite impart a lustrous sheen to the schistosity (cleavage) surfaces.

Schists. Strongly schistose, commonly lineated, metamorphic rocks in which the grain is coarse enough to allow macroscopic identification of the component minerals. Segregation banding is usually prominent. Micaceous minerals are abundant; their high degree of preferred orienta-

[5] Williams, Turner, and Gilbert, *op. cit.*, pp. 173–176, 1954.

tion is reflected in the development of schistosity.   Schists are among the most widespread products of regional metamorphism.

Gneisses.   Coarse-grained irregularly banded rocks with discontinuous rather poorly defined schistosity.   The geneissic fabric reflects predominance of quartz and feldspars and a general lack of micaceous minerals.   Gneisses are products of high-grade regional metamorphism.

Granulites (= *leptites, leptynites* of some European petrologists). Even-grained metamorphic rocks poor in micas and rich in minerals such as quartz, feldspars, pyroxene, and garnet which lack a tabular or prismatic habit.   Some degree of segregation banding and especially alignment of flat lenses of quartz or feldspar typically impart a regular foliation to the rock.   Products of high-grade regional metamorphism.

Mylonites.   Fine-grained, flinty-looking, strongly coherent, banded or streaked rocks resulting from extreme granulation of coarse-grained rocks without noteworthy chemical reconstitution.   Eyes or lenses of the undestroyed parent rock may persist enclosed in a mylonitic groundmass. Products of extreme dislocation metamorphism.

Cataclasites.   Rocks formed by ruptural deformation (brecciation, partial granulation) of brittle parent rocks.   With decrease in grain size and development of banded structure cataclasites grade into mylonites.

Phyllonites.   Rocks macroscopically resembling phyllites but formed by mechanical degradation of initially coarser rocks (such as graywacke, granite, or gneiss).   Highly characteristic are silky films of newly crystalized mica or chlorite smeared out along the schistosity surfaces. Phyllonites are among the most characteristic and persistent products of dislocation metamorphism.

Quartzites.   Metamorphic rocks composed principally of recrystallized quartz.   Products of contact or regional metamorphism of sandstones (quartz arenites).

Marbles.   Metamorphic rocks composed principally of calcite or of dolomite.   Products of contact or regional metamorphism of calcareous sediments.

Amphibolites.   Metamorphic rocks composed principally of hornblende and plagioclase.   Foliation, due to alignment of amphibole prisms, is less conspicuous than in schists.   Segregation banding may or may not be developed.   Products of medium- to high-grade regional metamorphism of basic igneous rocks and of some impure calcareous sediments.

Serpentinites and Soapstones.   Rocks composed principally of serpentine minerals, talc, and chlorite, formed by metasomatic metamorphism of peridotites.

Each of the above classes may be subdivided on a chemical or mineralogical basis.   Most metamorphic rocks fall into one or other of six such divisions:

1. Pelitic; derivatives of pelitic (aluminous) sediments.

2. Quartzo-feldspathic; derivatives of initially quartzo-feldspathic rocks —sandstones, granites, and so on.

3. Calcareous; derivatives of limestones and dolomites.

4. Basic; derivatives of basic and semibasic igneous rocks and of some sediments (such as tuffaceous sands) rich in Ca, Al, Mg, and Fe.

5. Magnesian; derivatives of peridotites and of some magnesian (*e.g.*, montmorillonite-rich) sediments.

6. Ferruginous; derivatives of iron-rich sediments (cherts, ironstones).

CHAPTER 17

Chemical Principles of Metamorphism

EQUILIBRIUM IN METAMORPHIC ROCKS

The trend of chemical reaction in metamorphism is towards establishment of stable mineral assemblages—having minimal free energy under the physical conditions of metamorphism. There has long been a widely held belief, to which the authors subscribe, that the common mineral assemblages of metamorphic rocks closely approximate stability in relation to metamorphic conditions. A general tendency for compositional uniformity and lack of zoning in plagioclase and other minerals in metamorphic as contrasted with igneous rocks is evidence supporting this view. Moreover the simplicity and regularity of metamorphic paragenesis can scarcely be explained on any other basis (see page 503).

Since the field throughout which there may be complete chemical interaction between solid phases, even in the presence of pore solutions, is often restricted in metamorphism (as contrasted with magmatic crystallization in the presence of abundant liquid), so, too, the space field within which the minerals of a metamorphic rock may be considered as constituting an equilibrium assemblage is likewise restricted. For example, the assemblage hornblende plagioclase-biotite-epidote, which makes up streaks and narrow lenses intercalated in marbles of Doubtful Sound, New Zealand, and assemblages like calcite-epidote-diopside-plagioclase or calcite-zoisite-scapolite-quartz, which are typical of the enclosing marble, are to be regarded as distinct mineral associations each internally in a state of equilibrium. This conclusion applies to heterogeneous rocks even if the two assemblages occur side by side in a single thin section.

The mineral assemblage which makes up any metamorphic rock must usually be regarded as metastable under temperatures and pressures prevailing at the earth's surface. The fact that it has persisted unchanged through a long period of cooling and unloading prior to exposure by erosion shows that the velocity of adjustment of the assemblage in question to falling temperature was infinitely low (cf. pages 43 and 44).

There are also cases where one or more minerals of a metamorphic rock appear to be metastable with respect to temperature and pressure of metamorphism. This condition can be expected wherever chemical adjustment of a rock to metamorphic conditions has not proceeded to completion, i.e., where metamorphic conditions have been maintained only for short periods, or where the velocity of chemical reaction has been very low. Under such circumstances minerals known as relics tend to persist from an earlier paragenesis and so to be inherited by the metamorphic assemblage. The relation of relict minerals to members of the associated equilibrium assemblage has been discussed at some length by Eskola,[1] who states that relics are specially common in rocks that have been mineralogically adjusted to several sets of physical conditions in turn. When a mineral phase that is stable under one set of conditions is also stable under a later set of conditions, it persists as a stable relic in the new assemblage. Thus, the plagioclase of basalt may be retained without change as a stable relic in amphibolite resulting from metamorphic reconstitution at moderately high-temperature. Another common example is the persistence of quartz in most metamorphic rocks derived from arenaceous sediments. On the other hand, minerals that are metastable under the newly imposed conditions, but nevertheless persist as relics on account of low velocity of transformation, are termed unstable (metastable) relics. A familiar example of this latter condition is relict augite, occurring in residual crystals surrounded by rims of actinolite or chlorite, in greenschists derived from basic igneous rocks.

Rather exceptionally minerals formed by metamorphism are probably metastable from the moment of their first appearance. According to Ostwald's law (considered in greater detail on page 480), such phases tend to crystallize in preference to truly stable phases, when their appearance involves minimal disturbance of internal structure of the reacting system. So in the stable greenschist assemblage albite-epidote-chlorite-actinolite-sphene formed by regional metamorphism of basic rocks at low temperatures, it is by no means uncommon to find scattered grains of metastable green hornblende pseudomorphous after igneous augite or brown hornblende. Appearance of hornblende as a transitory phase involves relatively slight change in previously existing space-lattices of pyroxenes. The same thing applies to temporary crystallization of kyanite at the expense of andalusite during regional metamorphism of pelitic hornfels to almandine-mica schist.[2] In the above two examples, hornblende and kyanite, though metastable, are not relics.

[1] P. Eskola, *Die Entstehung der Gesteine* (Barth, Correns, Eskola), pp. 341, 343, Springer, Berlin, 1939.

[2] C. E. Tilley, The role of kyanite in the "hornfels zone" of the Carn Chuinneag granite (Ross-shire), *Mineralog. Mag.*, vol. 24, pp. 92–97, 1935.

FACTORS IN METAMORPHIC EQUILIBRIA:
THE VARIABLES IN METAMORPHISM [3]

Temperature. Many metamorphic reactions are the result of an increase in temperature, often accompanied by changes in other variables enumerated below. The increase in temperature may be brought on by injection of magma (contact metamorphism), or by deep burial in a geosyncline, or by an increase in the rate of heat flow (see Chap. 15). In the first case, the temperature reached in the country rock at some distance from the contact of the igneous body will always be less than the initial temperature of the intrusion, which is presumably always less than 1200°C., and will vary with time and distance from the contact. At depths of 10 to 20 km., corresponding to the maximum depth of burial in the thickest known geosynclines, temperatures of the order of 250°–450°C. might normally be expected. Higher temperatures may occur at greater depth or if there is a local increase in the rate of upward flow of heat coming from the mantle. Experimental evidence indicates that temperatures of the order of 700°–750°C. are not uncommonly attained in rocks undergoing metamorphism of the highest grade; this is also the temperature range at which a granitic melt would begin to form at sufficiently high water pressure.

Pressure. *Load Pressure P_l.* Another result of increasing depth of burial is an increase of pressure which is due to the superjacent load and is therefore referred to as the load pressure P_l. The rate of increase of P_l is about 250 to 300 bars per km., depending on the average density of the rock cover. It is usually assumed that pressure resulting from weight of overlying rocks is hydrostatic, as is the case within a liquid, but some departure from this rule is to be expected at shallow depths or wherever the rocks retain the ability to resist, without continuous deformation, stresses that are applied over a long period of time. No exact limit can be set to the depth at which the load pressure becomes truly hydrostatic, as this depends on the mechanical properties of the particular rocks considered; it is certainly less in highly deformable materials (*e.g.*, salt) than in hard, competent quartzite. The load pressures involved in metamorphism lie in the range 0 to 10,000 bars.

Stress Pressure P_s. In a static fluid the pressure is everywhere hydrostatic, and varies only with depth, $P = g\rho h$ where g is the acceleration of gravity, ρ is the fluid density, and h is depth below the free surface of the fluid. In a flowing non-viscous fluid, the pressure is still everywhere hydrostatic, but it varies from point to point and is not solely controlled by depth. Rocks, however, are not fluids, their most characteristic prop-

[3] W. S. Fyfe, F. J. Turner, and J. Verhoogen, Metamorphic Reactions and Metamorphic Facies, *Geol. Soc. America Mem. 73*, 1958.

erty being that they possess a "fundamental strength," *i.e.*, they deform continuously and nonelastically only when subjected to stress differences of sufficient magnitude. Thus any permanently deformed rock (*e.g.*, a fold) bears evidence of having been subjected to a nonhydrostatic state of stress, that is, a state in which the three principal components of stress are not equal.

The effect of a nonhydrostatic stress, with components X_1, X_2, X_3 on the chemical potential of a solid may be shown (see pages 474 to 476) to be approximately the same as that of a hydrostatic pressure $P_s = (X_1 + X_2 + X_3)/3$. Thus it may be necessary to consider a "stress pressure" P_s equal to the arithmetic mean of the 3 components of stress not due to load (as load is already fully accounted for by P_l). What the magnitude of these stresses X_1, X_2, X_3 may be in rocks undergoing deformation at the time of metamorphism is not exactly known, but experiments on deformation and fracturing of rocks under various conditions suggest that P_s cannot be greater than, say, 2,000 or 3,000 bars; at higher temperature, where rocks deform more easily, P_s is probably much less than this. Thus the general impression is that P_s may be disregarded, to a first approximation, when compared to P_l, and when dealing with deep-seated metamorphism; it may nevertheless be important in relatively shallow metamorphism.

Water Pressure P_{H_2O}. Chemical equilibrium in any reaction in which water appears as a reactant or a product must depend on the chemical potential of the water. This is so for any of the numerous metamorphic reactions involving dehydration, *e.g.*:

$$KAl_2(AlSi_3)O_{10}(OH)_2 \; + \; SiO_2 = KAlSi_3O_8 \; + \; Al_2SiO_5 \; + \; H_2O$$

| Muscovite | Quartz | K-feldspar | Sillimanite | Water |

At any given temperature, an increase in the chemical potential of water tends to drive the reaction to the left, while a decrease drives it to the right. At a given temperature, the chemical potential of water is conveniently measured by its fugacity [Eq. (2–35)], or, if its activity coefficient is known, by its partial pressure, which is equal to the total pressure in the gas phase if it should consist of pure water. If the reaction is carried out in a closed container, the pressure of the gas phase is also the pressure exerted on the solid phases, and only one variable is necessary to describe the pressure conditions. In nature, however, osmotic conditions may arise in which the water pressure P_{H_2O} (*i.e.*, the pressure in the pure water phase) is not equal to the pressure on the solid phases. For instance, in an open fracture extending vertically downward and filled with water, the water pressure at a depth h would be $P_{H_2O} = g\rho_w h$ where ρ_w is the density of water, while the load pressure exerted on the rock at the same depth would be $P_l = g\rho_r h$ where ρ_r is the rock density. Conversely, the water in the pores of a rock could be at a higher pressure than the rock itself; petroleum geologists are familiar with drill holes that encounter water at a higher pressure than the depth would suggest. This may also be the case in or near igneous bodies, as the water vapor pressure inside an exploding volcano, for instance, bears little relation to the load pressure on the magma itself. Thus, as long as rocks have strength the water pressure may be greater or less than the load pressure. Again, as with stress pressure, it would appear that in deep metamorphic zones the water pressure is approximately equal to the load pressure; this may not be true at shallow depths.

Other Pressure Variables. A further reason why the water pressure may differ from the load pressure is that the fluid phase may contain other components besides water. In a mixture of water and CO_2, the water pressure would be equal to the total fluid pressure times the molar frac- tion of water in the fluid. Similarly the carbon-dioxide pressure would differ from the total fluid pressure and the pressure on the solid phases. This would affect any chemical equilibrium in which CO_2 appears as a reactant or product. It may thus be necessary, in order to understand parageneses involving carbonates, to consider the carbon dioxide pressure as an additional variable. Similarly, the partial pressure of oxygen may control reactions involving a change in the state of oxidation of iron. Goldschmidt [4] has noted that a considerable reduction of ferric iron to ferrous iron commonly takes place in high-temperature metamorphism, either contact or regional; thus the partial pressure of oxygen may be an additional variable.

Note that the phase rule in its usual form (2–50) is valid only if two intensive variables (P and T) are sufficient to describe the state of the system. Should the pressures acting on the solid and gas phases be different, an additional variable is necessary to describe the state of the system, and the variance correspondingly increases by one unit.

Surface Tension Effects. It has been said that many transformations concerned in metamorphic rocks take place through the medium of a solution occupying intergranular pores of the rock in question. As we shall see presently, the fact that these pores are of very small size, as also are sometimes the grains of the rock itself, introduces additional com- plications in the equilibrium conditions; *e.g.*, it will be shown that, be- cause of surface tension, the solubility of a grain in a given solvent may depend on its size. More generally, metamorphic reactions are hetero- geneous reactions, *i.e.*, they involve two or more phases, and they occur therefore mainly at the interfaces between these phases, as at the contact between two different minerals or between a mineral and a solution. Under such conditions, the chemical properties of materials on their

[4] V. M. Goldschmidt, *Geochemistry*, p. 36, Clarendon Press, Oxford, 1954.

surfaces assume great importance. Conditions and processes occurring at the surface of a phase are somewhat different from those occurring within the phase. It is reported, for example, that melting begins at crystal boundaries of metals, at temperatures several degrees lower than the normally accepted melting points for the same metals in the mass.[5] Clearly, the study of surface phenomena has great importance in the study of heterogeneous reactions and hence in the whole field of petrogenesis. Growth and solution of crystals in contact with solvents, vaporization of liquids, and eutectic melting of two or more solids, may be cited as processes which essentially concern surfaces, rather than the internal structures of phases.

Clearly, matter constituting the boundary surface of a homogeneous phase cannot be physically identical with that situated within the phase. Any molecule, ion, or atom in a solid or liquid at rest is surrounded on all sides by other particles which exert upon it certain forces, the sum of which must average zero. But a particle at the boundary between two phases 1 and 2 is simultaneously subjected to forces exerted on the one side by particle of phase 1 and on the other side by particles of phase 2. If the kind or arrangement of particles is different in the two phases, these two forces will be unequal. To maintain equilibrium, it is therefore necessary that particles of both phases in the immediate vicinity of the boundary should become rearranged in such a manner as to balance the forces exerted between them. A "surface phase" may thus be envisaged as a thin layer having properties which may differ from those of the "volume phases" on either side; its thickness is of the order of the distances over which intermolecular and interionic forces are effective.

For the sake of convenience it is customary to treat a surface phase ideally as if it were a mathematical surface of vanishing thickness. But in order that the ideal system bounded by such a surface shall have properties identical with those of the actual system bounded by a surface phase of finite thickness, certain properties (e.g., amounts of energy, entropy, etc.) must be assigned to the mathematical surface. For instance, the surface energy so assigned must be such that the energy of the two adjacent phases (assumed to be homogeneous to the surface of contact) plus the surface energy shall be equal to the energy of the actual system. Again, the concentration of matter assigned to the surface must be such that the ideal and actual systems contain the same number of particles; this is shown, for instance, in Fig. 64. In order to make the actual system of Fig. 64a identical with the ideal system of Fig. 64b, we must imagine that the ideal surface of separation contains $(+2)$ particles of the type represented by small circles and $(+6)$ particles of the type

[5] W. I. Pumphrey and J. V. Lyons, Behaviour of crystal boundaries in aluminum and its alloys during melting, Nature, vol. 163, pp. 960–961, 1949.

represented by dots. If a is the area of the interface, the "adsorption" Γ_1 of component 1 at the interface is defined as $2/a$; that of 2 as $6/a$. More generally, if n_i^a is the number of moles of i that must be assigned to a surface of area a, the adsorption Γ_i^a is

$$\Gamma_i^a = \frac{n_i^a}{a} \qquad (17\text{-}1)$$

a quantity which may be positive, negative, or zero. It is true that Γ_i^a will depend upon the position of the surface, but this latter is not arbitrarily assumed, for it is determined by the condition that ideal and actual systems must be mechanically equivalent.

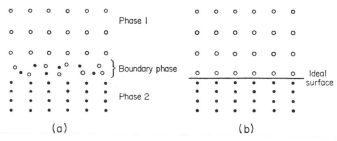

(a)   (b)

FIG. 64. Adsorption on surface phases. (*a*) Actual system with a boundary phase of finite thickness. (*b*) The same system with an ideal surface phase. In order that these two systems shall be equivalent with respect to total number of particles, it must be assumed that the ideal surface in (*b*) carries +2 particles of type 1 and +6 particles of type 2.

In the simple case where phase 1 is in contact with a vacuum, any particle on the surface of contact is attracted toward phase 1 but not toward the vacuum. Thus all surface particles, and hence the surface itself, are subject to a force acting toward the interior of the phase and tending therefore to reduce the area of the surface. This force acts normal to the surface, but since it produces the illusion of a contracting force acting within the face, it is convenient to describe is as "surface tension." Though it is possible to define surface tension precisely, it will be sufficient at this stage to illustrate it by analogy. The pressure P of familiar thermodynamics may be defined precisely by the relation

$$P = \left(\frac{\partial E}{\partial V}\right)_S$$

This means that if the volume of a system is to be increased by an amount δV without change of entropy S the amount of energy required is $P\,\delta V$. Similarly, the surface tension σ may be defined as the amount of work that must be done to stretch the surface by an amount ∂a at constant volume and entropy.

$$\sigma = \left(\frac{\partial E}{\partial a}\right)_{S,V} \tag{17-2}$$

"Surface energy" is defined as the energy which has to be expended to bring a particle from the interior of a phase to its surface. The difference between the two quantities surface tension and surface energy is small, and in most cases the two may be regarded as identical; both have the dimensions of energy per unit surface, or force per unit length.

The mechanical conditions for surface equilibrium are most important. It may be shown [6] that if two phases α and β are bounded by a curved surface in which there exists a surface tension $\sigma^{\alpha\beta}$, then between the two phases there must exist a difference in pressure, given by the relation

$$P^\alpha - P^\beta = \frac{2\sigma^{\alpha\beta}}{r^{\alpha\beta}} \tag{17-3}$$

where $r^{\alpha\beta}$ is the average radius of curvature of the interface $\alpha\beta$. The pressure is greatest on the concave side of the surface. This Eq. (17-3) carries a significant implication: if certain interfaces between phases are curved, the pressure will be different in different parts of the system, and the phase rule in its form (2-50) does not apply. It may be shown, by counting the number of variables and the number of relations between them, including relations of the type (17-3), that for a system of c components and φ phases with ($\varphi - 1$) interfaces, b of which are plane, the variance becomes

$$w = c + 1 - b$$

which reduces to the usual form ($w = c + 2 - \varphi$) if $b = \varphi - 1$ (all interfaces plane) but which, if all interfaces are curved ($b = 0$), becomes

$$w = c + 1$$

the variance now being independent of the number of phases. This example illustrates the fact, already pointed out, that the phase rule is valid only in the special case where only two physical variables are involved (for instance, temperature and a pressure which is uniform throughout the system).

The type of relation given by Eq. (17-3) is not limited to spheres. It is easily shown that for a cube of side a a similar relation holds, namely

$$P^\alpha - P^\beta = \frac{4\sigma^{\alpha\beta}}{a} \tag{17-4}$$

this, as Eq. (17-3), being a particular case of the more general relation

$$P^\alpha - P^\beta = \sigma\frac{dA}{dV}$$

[6] See for instance, E. A. Guggenheim, *Modern Thermodynamics by the Methods of J. W. Gibbs*, p. 168, Methuen, London, 1933.

equating the work obtained by expansion $[(P^\alpha - P^\beta)\, dV]$ to the work $\sigma\, dA$ done against surface tension by increasing the area by an amount dA.

It may be seen from (17–3) and (17–4) that the smaller the radius of curvature, or the smaller the size of the grain, the greater will be the pressure inside the phase. Since the chemical potential of any component of a phase depends on the pressure inside the phase, the chemical potentials will vary as the radius changes. For instance, the chemical potential of water in a small drop is greater than in a larger drop, and small drops consequently tend to coalesce to form larger ones. Similarly, small grains of a solid are unstable with respect to larger ones. All properties which are measured by the chemical potential (e.g., solubility) are liable to similar variation; [7] a small grain will be more soluble in a given solvent at a given temperature than a larger one, a prediction which has been verified by experiment. Similarly, the temperature at which certain minerals of the kaolinite group undergo thermal decomposition with liberation of water has been found to depend on the size of the particles, being somewhat lower for the smaller grains. A very small incipient crystal in the process of growth has a higher potential than a larger grain and can remain in equilibrium with a solution only if this is supersaturated with respect to larger grains. Thus crystallization may begin only if particles of sufficient size are present, and if these do not form spontaneously by statistical fluctuations, germs may have to be added from an external source.

If, instead of having a small grain surrounded by a solution, we consider a small amount of solution enclosed in a pore of the solid, the solubility relations are reversed; the solid tends to go into solution in the larger pores, while precipitation may take place simultaneously in the smaller ones. The tendency is to enlarge the larger cavities at the expense of the smaller ones, which tend to disappear. By this process the total area of the interface is reduced to a minimum.

The above theory has important applications in rock metamorphism. The observed tendencies for increase in average grain size during metamorphic recrystallization, and for growth of a few large porphyroblasts rather than many smaller grains, are presumably manifestations of the influence of surface tension existing at the boundaries of crystals. If a porous rock is impregnated with a multicomponent solution, resulting metasomatic effects will depend not only on the factors of pressure, temperature, and composition, but also on size and shape of the openings, surface tension of solution against various minerals present, and so on.

[7] J. Verhoogen, Geological significance of surface tension, Jour. Geology, vol. 56, pp. 210–217, 1948. Cf. also C. S. Smith, Grains phases and interfaces, Am. Inst. Min. Met. Eng. Trans., vol. 175, pp 15–51, 1948

Metasomatic effects will depend, therefore, to some extent, on the physical texture of the invaded rock, as well as upon its mineralogical and chemical composition. The same factors may be expected to influence the porosity and texture of cemented sediments—a topic which falls outside the scope of this book.

THE FREE ENERGY OF METAMORPHIC REACTIONS

Definition. Equation (2-29) shows that the free energy or chemical potential of a substance increases with increasing pressure at a rate proportional to its volume, and decreases with increasing temperature at a rate proportional to its entropy. As the stable phase, or assemblage of phases, is that which has least free energy for given P and T, a high pressure must necessarily favor the assemblage which has the smallest volume, while high temperature favors the assemblage with greatest entropy. Metamorphic reactions induced by increasing temperature run in the direction in which the entropy increases, $i.e.$, are endothermic. An exception to this rule occurs when the system is originally metastable, but cannot revert to the stable state because of a slow rate of reaction at low temperature. In such a case, an increase in temperature may, because of its kinetic effects, start an exothermic reaction. A probable instance of this kind is the conversion of pyroxene to hornblende during the metamorphism of diabase to amphibolite.

To know what phases in a system of given chemical composition would be stable at given P and T, it is necessary to know the free energies of all phases involved under the conditions considered. Free energies, as other forms of energy, are relative quantities measured with respect to an arbitrary zero. In chemical work it is invariably the practice to set the free energy of the elements as equal to zero when these elements are in their stable state at 298°K. and 1 bar; the "free energy" of a compound is then defined as its free energy of formation, $i.e.$, the free energy change of the reaction by which the compound forms from the elements. Such free energies are well known for common substances, but have been determined for only very few minerals. If they were known, it would be a simple matter of addition and subtraction to decide which assemblage has the lowest free energy under given conditions, and the equilibrium temperature and pressure would be easily found. Volume and entropy depend, however, on P and T, so that it is not sufficient merely to know their values under ordinary conditions. For instance, at 298°K., tridymite has a higher entropy than cristobalite, a fact which might lead to the erroneous conclusion that tridymite is a high-temperature modification of cristobalite. Actually, the entropy of cristobalite increases faster with increasing temperature than that of tridymite, and at the transition point, cristobalite has a higher entropy than tridymite. A similar situation exists for the pair andalusite-sillimanite.

Whatever their sign may be, entropy and volume differences are quite small for most metamorphic minerals, and so are the free energy differences involved in most metamorphic reactions. Kracek, Neuvonen, and Burley[8] find for the reaction albite = jadeite + quartz, a free energy change of only 1.6 Kcal. at 298°K. Such small free energies are difficult to measure experimentally. Extraneous effects, such as impurities, or different methods of preparing reagents, are likely to have a large effect, as illustrated by the same authors for the reaction nepheline + albite → 2 jadeite, for which they find free energy values ranging from -0.79 to -2.46 cal. (all $\pm$ 1.62 Kcal.), depending on the origin of the specimens. This smallness of the free energy of metamorphic reactions endlessly plagues the experimentalists, as it leads to unwanted appearance and persistence of metastable phases.

Entropy of Solids. *Pure Phases.* The molar volume of a solid phase is easily determined under ordinary conditions. To determine its volume at given P and T, one should know its coefficient of thermal expansion $\alpha = \dfrac{1}{V}\left(\dfrac{\partial V}{\partial T}\right)_P$ and isothermal compressibility $\beta = -\dfrac{1}{V}\left(\dfrac{\partial V}{\partial P}\right)_T$, and their variation with P and T. The coefficient α is usually of the order of $1 - 3 \times 10^{-5}$ per deg., while β is commonly of the order of $0.5 - 2 \times 10^{-6}$ per bar. Thus an increase in temperature of 1000°C. produces an expansion of a few per cent, whereas an increase in pressure of 10,000 bars produces a decrease in volume of the same order. Except in very precise work, the molar volume of a mineral under metamorphic conditions may be estimated without serious error from values of α and β under normal conditions.

Knowledge of the entropy is, however, much more difficult to obtain. To determine the entropy at any temperature T it is necessary to measure the specific heat $c_P = (\partial H/\partial T)_P$ at all temperatures from T down to absolute zero, the entropy being then computed by integration

$$S_T = \int_0^T c_P \, dT/T$$

Specific heat measurements are difficult and tedious. The effect of pressure on entropy must further be evaluated either experimentally or by means of the thermodynamic relation $(\partial S/\partial P)_T = -\alpha V$.

It turns out, very fortunately, that the entropy of many minerals can be calculated simply by adding the entropies of the oxides of which the mineral is composed. For instance, it is known experimentally that the en-

[8] F. C. Kracek, J. Neuvonen, and G. Burley, A thermodynamic study of the stability of jadeite, *Washington Acad. Sci. Jour.*, vol. 41, pp. 373–383, 1951.

tropy of MgO at 298°K., 1 bar, is 6.55 cal./deg. mole and that of quartz is 10.0 cal./deg. mole; that of clinoenstatite is 16.22, very close to the sum $10.0 + 6.55$. The entropy of forsterite Mg_2SiO_4 is 22.75, while that of the sum of the oxides is $6.55 \times 2 + 10.0 = 23.1$. The computed value for Al_2SiO_5 is $12.5 + 10.0 = 22.5$; the experimental values for sillimanite and andalusite are 22.97 and 22.3 respectively. Note that an error of 1 cal./ deg. in the reaction entropy entails an error of 1 Kcal. in the free energy of the reaction at 1000°K.

Existing small departures from this additivity rule may be accounted for in various ways.[9] It commonly turns out that when the entropy of a compound is less than the sum of the entropies of the constituent oxides, its volume is also less than the sum of the volumes of the oxides, the ratio of the entropy discrepancy to the volume discrepancy being very close to the ratio α/β. [This is suggested by the thermodynamic relation $(\partial S/\partial V)_T = \alpha/\beta$.][10] Other factors may also be taken into consideration. For instance, of two polymorphs, the harder one (in the mineralogical sense) is likely to have the lower entropy. The outcome is that, if experimental values are not available, entropy relations in solid phases may generally be estimated without very serious error.

Entropy of Mixing. The calculation of the entropy of a phase, or assemblage of phases, at high temperature is further complicated by the tendency to form phases of variable composition. If crystals of two species (phases) a and b are in contact, there is a tendency, just as in the case of liquids or gases, for the two phases to mix spontaneously. This is because mixing increases the internal disorder, and hence the entropy, of the system. At constant T and P the necessary condition for a process to occur spontaneously is that

$$\Delta G = \Delta H - T \, \Delta S < O$$

or, if we neglect changes in volume

$$\Delta E - T \, \Delta S < O$$

Thus if temperature is sufficiently high the entropy times the temperature may be larger than ΔE, and a process may occur spontaneously, although it increases the energy of the system. Two phases which do not mix at low temperature because the internal energy of the mixed phase is greater than the sum of those of the separate phases, may begin to mix appreciably when the temperature is sufficiently high. Departure from strict stoichiometric composition is a thermo-dynamic requirement for equilibrium in a system consisting of solid phases of different composition. The amount of these departures will increase with increasing temperature

[9] Fyfe, Turner, and Verhoogen, *op. cit.*, pp. 27–34, 1958.
[10] *Ibid.*, pp. 113–114.

and decrease with increasing energy of substitution for the system in question. The latter varies from case to case but usually is very high, except in isomorphous series where it may even be zero (as in the olivine series and in plagioclase). General lack of zoned structure in crystals of isomorphous series in metamorphic rocks testifies to the efficiency of spontaneous diffusion in crystal lattices at metamorphic temperatures if energy of ionic substitution is low. The wide range of ionic substitution in garnets and in pyroxenes of eclogites and granulites (cf. pages 555 and 558), as contrasted with those of amphibolites, could well be correlated with exceptionally high temperatures of metamorphism. In general, according to the above principle, high metamorphic temperatures should favor development of rocks of uniform composition—simple mineral assemblages consisting of few phases within which there is a wide range of ionic substitution. That this is so among eclogites and granulites is suggestive evidence that ionic diffusion between crystalline phases at high temperature and with little change in volume has played an important part in metamorphism of these two groups of rocks.

Entropy of Water. Many metamorphic reactions involve water, either as a separate gas phase or as a component of hydrous silicates. The entropy of water vapor is reasonably well known up to a pressure of 2,500 bars and a temperature of 800°C. The entropy of hydrous compounds may be estimated by the additivity rule, using for the entropy per mole of water under normal conditions a value of 9.6, which is close to the entropy of ice (9.1). By extrapolation the entropy of hydration, *i.e.*, the difference between the entropies of one mole of water in the gas phase and in the hydrous mineral, may be roughly estimated.[11] This difference is large at low pressure (about 30 cal./deg. mole) but decreases rapidly as the pressure increases. It would be less if the water were present as an adsorbed film rather than a separate gas phase, as the entropy of adsorbed water under normal conditions seems to be intermediate between those of liquid water and ice.[12]

Dependence of Equilibrium Temperature on Pressure. If the equilibrium temperature T_e for a given reaction is known at a certain pressure, its value at any other pressure may be estimated from the Clausius-Clapeyron equation (2–39)

$$\frac{dT_e}{dP} = \frac{\Delta V}{\Delta S}$$

For most metamorphic reactions, ΔV and ΔS have the same sign, so that generally the equilibrium temperature increases with increasing pressure. This is a direct consequence of the fact that dense minerals also have

[11] Fyfe, Turner, and Verhoogen, *op. cit.*, p. 118, 1958.
[12] *Ibid.*, p. 43.

small entropies. The coefficient dT_e/dP is usually of the order of 5×10^{-2} deg./bar for polymorphic transformations or for reactions between solid phases. Because of the dependence on temperature and pressure of the entropy of hydration (page 468), the coefficient dT_e/dP for reactions involving water is highly variable. It is large at low pressure but decreases rapidly with increasing pressure; at pressures of about 2,000 bars, it generally lies between 1×10^{-2} and 7×10^{-2} deg./bar, as for reactions without water. Note that for reactions involving water, the water pressure P_{H_2O} is relevant.

Coupled Reactions. When discussing metamorphic reactions, attention must be focused on the whole paragenesis rather than on individual minerals, as the stability range of a single mineral may be seriously affected by the presence or absence of other minerals. Consider the following separate reactions, with free energy changes ΔG_1 and ΔG_2

$$A = B + C \qquad \Delta G_1$$
$$C + E = D \qquad \Delta G_2$$

These two reactions may be combined into a single one

$$A + E = B + D \qquad \Delta G = \Delta G_1 + \Delta G_2$$

and if ΔG_2 is negative and sufficiently large, ΔG may be negative (*i.e.*, the reaction runs from right to left) although ΔG_1 is positive. Thus the breakdown of A depends on whether E is present or not. ΔG_2 is likely to be particularly large and negative if C is an oxide and E is silica. For instance, at 298°K., 1 bar

$$CaCO_3 = \quad CaO + CO_2 \qquad \Delta G = 31.2 \text{ Kcal.}, \Delta S = +38.4$$
$$CaCO_3 + SiO_2 = CaSiO_3 + CO_2 \qquad \Delta G = 10.6 \text{ Kcal.}, \Delta S = +28.9$$

showing that calcite will disappear at a lower temperature if silica is present. Similarly, the temperature at which muscovite breaks down

$$\text{Muscovite} = \text{orthoclase} + \text{corundum} + \text{water}$$

will be reduced by the coupling reaction

$$\text{Corundum} + \text{silica} = \text{sillimanite (or kyanite, or andalusite)}$$

which has a large negative free energy.

This fact, trivial as it may seem, is the thermodynamic basis for distinguishing sharply between metamorphic assemblages that do or do not contain quartz.

Relative Influences of Temperature and Pressure. As mentioned above, an increase in temperature favors the development of the phase assemblage with the largest entropy, while increase in pressure favors the assemblage with the smallest volume. For instance, rising temperature favors dehydration reactions, because of the relatively large entropy of water vapor.   There also exists a general relation that the density of characteristic minerals increases with the grade of metamorphism.   Thus, in the eclogites we find kyanite, pyrope-grossularite garnet, and diamond, all dense minerals.   In the glaucophane schists pyrope-almandine garnet is associated with jadeite and lawsonite, the latter being a dense hydrate of anorthite.   In the lowest grades, on the contrary, zeolites, clays, and other minerals of low density are present.   This strongly suggests that pressure is a controlling factor.   For instance, increasing pressure favors reaction from left to right in the following cases: [13]

   Olivine + anorthite → garnet
   Augite + anorthite → garnet + quartz
   Ilmenite + anorthite → sphene + hornblende
   Nepheline + albite → glaucophane + jadeite
   Anorthite + gehlenite + wollastonite → grossularite
   Andalusite → sillimanite → kyanite

However, even in the case of the reactions mentioned above, it cannot be assumed that any of these is actually a pressure-controlled metamorphic reaction, until the thermal relations of the chemically equivalent mineral assemblages have been taken into account, and the reaction products have been compared with known metamorphic parageneses.   Even where reduction in molar volume is obvious, temperature rather than pressure may have been the principal controlling factor.   The well-known tendency for garnet and quartz to be associated in what appear to be reaction borders separating augite from basic plagioclase in metamorphosed gabbros and allied rocks seems to confirm the validity of the second equation as representing an actual metamorphic transformation. Nevertheless, the reaction in question is probably not so simple.   The compositions of garnets so derived, from pure magnesian pyroxene on the one hand or pure diopside on the other, should theoretically be $(CaMg_2)Al_2Si_3O_{12}$ and $(Ca_2Mg)Al_2Si_3O_{12}$ respectively.   These, or any intermediate compositions, are virtually prohibited by the proved limited possibility for replacement of Ca by Mg (or Fe) in grossularite and pyrope-almandine.   The same objection holds against derivation of garnet from simple reaction between olivine and anorthite-rich plagioclase.[14] We are forced to conclude that if, as petrographic evidence suggests,

[13] Cf. P. Niggli, *Die Gesteinsmetamorphose* (Grubenmann-Niggli), p. 107, Borntraeger, Berlin, 1924.

[14] Garnets of the pyrope-almandine series containing as much as 35 to 40 per cent grossularite have been recorded in eclogites and could indeed have originated by simple reaction between hypersthene or olivine and anorthite.   However, in most eclogite garnets CaO is too low for such an origin (cf. Eskola, *op. cit.*, p. 364, 1939).

garnet does form by reconstitution of anorthite and either pyroxene or olivine, the process is not adequately represented by either of the two equations just cited, but usually involves participation of yet other associated minerals, or else removal of material from the reacting system. To evaluate the relative influence of pressure and temperature, the metamorphic paragenesis must be considered as a whole and compared with the full mineralogical composition of the parent rock. For instance, Alderman [15] represents the conversion of basic igneous rocks to eclogite by the following equation, in which the compositions of the minerals in question agree with the results of chemical analyses:

$$\begin{cases} Mg_2Fe_2Si_4O_{12} \\ Ca_2Mg_2, \ Fe^{++}Fe^{3+}Si_5AlO_{18} \end{cases} + \begin{cases} NaAlSi_3O_8 \\ Ca_2Al_4Si_4O_{16} \end{cases} + 2FeTiO_3 \rightarrow$$

$$\text{Augite} \qquad\qquad\qquad \text{Labradorite } Ab_1An_2 \qquad \text{Ilmenite}$$

$$(NaCa_2)MgFeAlSi_6O_{18} + (Fe_4Mg_3Ca_2)(Al_5Fe^{3+})Si_9O_{36} + SiO_2 + 2TiO_2$$

$$\text{Omphacite} \qquad\qquad \text{Almandine-pyrope} \qquad\qquad \text{Quartz} \qquad \text{Rutile}$$

Notable decrease in molar volume is entailed in this change, and since it is unlikely that the temperature of metamorphism departed markedly from the temperature range over which augite and labradorite normally crystallize from basic magma, pressure has almost certainly outweighed temperature as an influence determining the eclogite paragenesis. More difficult to assess are the relative parts played by temperature and pressure in retrogressive metamorphism of eclogite to amphibolite or to glaucophane schist.

It is again emphasized that stability of a mineral assemblage under given physical conditions depends upon the type of transformation by which equilibrium might be displaced. It was shown, for instance (page 40), that a system of pure phases may be stable with respect to a transformation at constant volume, but may be unstable with regard to the same reaction occurring at constant pressure and variable volume. Again, conditions for stability in thermally insulated systems are not the same as in systems from which heat may be dissipated as fast as it is produced. Care should therefore be taken to consider all such factors involved, in connection with problems referring to stability and metastability of mineral assemblages in metamorphic rocks.

It should be further noted that although the direction in which a certain metamorphic reaction runs may be controlled by pressure, it is doubtful whether the reaction would effectively be possible, because of kinetic considerations, without simultaneous increase in temperature. Eclogite would probably not form from basalt at the appropriate pressure but at ordinary temperature. It appears unlikely that metamorphism

[15] A. R. Alderman, Eclogites from the vicinity of Glenelg, Inverness-shire, Geol. Soc. London Quart. Jour., vol. 92, p. 511, 1936.

could take place within geologically reasonable lengths of time at temperatures much less than 300°C. In this sense, metamorphism appears invariably to be controlled by temperature.

Influence of Nonhydrostatic Stress. *Stress and Antistress Minerals.* An external force applied to the surface of a solid is transmitted and redistributed throughout the body, each portion of which exerts a certain force on neighboring portions. If we imagine any arbitrary surface cutting through the body, the force per unit area transmitted across this surface in the vicinity of any point is defined as the "stress" existing at this point; its magnitude depends on the orientation of the surface. By convention, a tension is considered positive, a compression negative. It can be shown that at any point in a stressed body there are three mutually perpendicular planes across which the stress is purely normal.[16] These three stresses X_1, X_2, X_3 ("principal stresses") are mutually perpendicular, and their values define completely the stress at this point. When the principal stresses are equal ($X_1 = X_2 = X_3$) the stress is said to be hydrostatic; the familiar "pressure" of thermodynamics belongs to this type. In all other cases, shearing stresses are present; that is, there are planes across which the action of one part of the body on the other has a component in the plane itself. It is shear which is responsible for flow of liquids; but solids, as they are usually defined, are capable of resisting such stresses over long periods of time without showing any appeciable permanent deformation. For the sake of brevity, we shall henceforth refer to a hydrostatic stress as a *pressure,* or a *confining pressure*; a nonhydrostatic state of stress will be referred to briefly as a *stress,* or a *shear.*

All naturally deformed rocks must at some stage have been subjected to high nonhydrostatic stresses, and it is therefore necessary to investigate the possible effects of such stresses upon chemical equilibrium in assemblages of metamorphic minerals. This rather controversial topic has been discussed especially by Johnston and Niggli, Harker, and Eskola.[17] Harker defined "stress minerals" as those whose fields of stability on a pressure-temperature diagram are extended by introduction of nonhydrostatic stress, while antistress minerals are those whose fields of stability are reduced under like conditions. A third class of neutral minerals is also recognized. Harker believes that in some cases stress reduces the field of stability to nil, while conversely there are other minerals that crystallize only under the influence of stress; such minerals are respec-

[16] For further consideration of stress, see, *e.g.*, A. E. H. Love, *A Mathematical Theory of Elasticity,* 4th ed., Cambridge University Press, 1927.

[17] J. Johnston and P. Niggli, The general principles underlying metamorphic processes, *Jour. Geology,* vol. 21, pp. 599–615, 1913; A. Harker, *Metamorphism,* pp. 147–151, Methuen, London, 1932; P. Eskola, A note on diffusion and reaction in solids, *Comm. géol. Finlande Bull. 104,* pp. 150–154, 1934.

tively termed *antistress* and *stress* minerals "in a very special sense." Evidence bearing upon the validity of this hypothesis accrues from three sources: (1) petrographic, (2) experimental, and (3) crystallographic.

1. Metamorphic rocks that have been deformed under stresses of high magnitude are readily distinguishable by their structural and petrofabric characters. The occurrence of certain minerals (stress minerals) is restricted to schists and other deformed rocks, while others (antistress minerals) are confined to undeformed rocks such as hornfelses and igneous rocks. The number of minerals belonging to the first category is very limited, chloritoid and possibly kyanite being the only common examples. Staurolite is typically a mineral of the schists but also occurs in pegmatites, while micas, chlorites, talc, the epidote minerals, and amphiboles, though highly characteristic of deformed rocks, are by no means confined to these. In contrast to such minerals, leucite, nepheline, sodalite, cancrinite, and scapolite are cited by Eskola as true antistress minerals incapable of withstanding high shearing stress and hence never found in deformed rocks. The incompatibility of andalusite, cordierite, and anorthite with stress is somewhat less sharply defined, though still distinct.

2. The simple laboratory experiments of Lea [18] showed that certain chemical compounds can be decomposed into their constituent elements under stress developed during grinding in a mortar. It would appear, therefore, that there is a class of mechanically unstable substances (to which antistress minerals would belong) whose crystalline space-lattices are incapable of withstanding high shearing stress and consequently are disrupted during deformation. Failure to synthesize in the laboratory such minerals as staurolite, epidote, or muscovite has been brought forward by Harker and others as evidence that these are essentially stress minerals incapable of crystallizing in systems not subjected to stress. On the other hand, Eskola draws attention to the synthesis of sericite in 1936 by Noll (who found incidentally that the velocity of reaction was extremely low) and concludes that the limited time available in laboratory experiments may well be responsible for failure to synthesize other minerals such as epidote and staurolite. Unsuccessful attempts have recently been made by Larsen and Bridgman to synthesize stress minerals by subjecting wet powdered samples of appropriate compounds to extreme shearing stress at very high confining pressures.[19] Thus mixtures of halloysite and orthoclase, in appropriate proportions to give sericite, remained unchanged. Chlorite-sericite and chlorite-silica-K_2CO_3 mixtures

[18] C. M. Lea, On endothermic reactions effected by mechanical force, *Am. Jour. Sci.*, vol. 146, pp. 241–244, 1893.

[19] E. S. Larsen and P. W. Bridgman, Shearing experiments on some selected mineral combinations, *Am. Jour. Sci.*, vol. 36, pp. 81–94, 1938.

likewise failed to yield biotite.   Andalusite showed no sign of incipient change to kyanite.   These experiments, though suggestive, are by no means conclusive, owing to the short interval of time (about 10 sec.) over which the high shearing stress could be maintained.

3. Eskola has also shown that the compatibility of a silicate mineral with stress is related both to the type of space-lattice structure and to density of ionic packing within the lattice.[20]   Silicates with sheet structure of the (Si_2O_5) type (micas, chlorites, talc) or with (Si_4O_{11}) band structure (amphiboles) are conspicuous among the constituents of deformed rocks, though orthosilicates with (SiO_4) groups (kyanite, staurolite, almandine, epidotes) and aluminosilicates with spongy frameworks (orthoclase, albite) are no less prominent.   On the other hand, the rock-forming silicates that show the greatest antipathy to stress belong exclusively to one or other of these last two classes (andalusite, cordierite, leucite, nepheline, cancrinite, scapolite, anorthite).   High density and close packing of ions within the space-lattice also appears to favor resistance to stress.[21]   Most of the silicate minerals typical of deformed rocks (kyanite, garnets, amphiboles, epidotes, staurolite, chloritoid, and probably micas) have a high index of packing for the crystal lattice, while antistress minerals (e.g. alkali feldspars, leucite, nepheline, cordierite, scapolite) mostly have a low packing index.

Thermodynamic Relations.[22]   A solid body is said to be elastic if (1) on application of stress it instantaneously becomes deformed (strained) by an amount depending solely on the magnitude of the stress; (2) on subsequent release of the stress it spontaneously reassumes its initial unstrained configuration.   Liquids are nonelastic with respect to shear.   They never reach a stable state of strain, for as long as stress is applied, they deform continuously by flow, and when stress is subsequently released, they fail to regain the initial configuration.   Between ideally elastic solids and nonelastic liquids, all gradations are possible.   Few solid bodies under stress exhibit ideally elastic behavior.   The solid may reach a stable condition of strain only after a definite interval of time, or it may never reach complete stability.   The velocity of deformation in such cases is usually a function of time and of magnitude of stress.   Strain is also liable to be influenced by the previous history of the specimen.

In the case of an ideally elastic body, the chemical potential of the strained solid must be uniform throughout; otherwise there would be a tendency to flow, the body would not be in internal equilibrium, and the

[20] Eskola, *op. cit.*, p. 333, 1939.

[21] H. W. Fairbairn, Packing in ionic minerals, *Geol. Soc. America Bull.*, vol. 54, pp. 1366, 1367, 1943.

[22] G. J. F. MacDonald, Thermodynamics of solids under nonhydrostatic stress, *Am. Jour. Sci.*, vol. 255, pp. 266–281, 1957.

process of stressing and straining would not be exactly reversible.   Thus, for any strained ideally elastic solid, the chemical properties such as solubility, vapor pressure, melting point, etc., remain respectively uniform at all points in the body, although these properties may be different from the corresponding properties of the unstrained body.   This difference in chemical potential between strained and unstrained body consists of the sum of two terms.   The first, which is the most important, may be written

$$d\mu = V \, dP \tag{17-5}$$

where P is the stress invariant defined by

$$- P = \frac{X_1 + X_2 + X_3}{3} \tag{17-6}$$

and V is the molar volume.   This relation (17-5) is identical in form with the usual relation giving dependence of chemical potential on hydrostatic pressure, as indeed is to be expected, since the quantity P defined by (17-6) is really a hydrostatic pressure.   The second term in the chemical potential will usually be much smaller than $d\mu$ and may be neglected in a first approximation.[23]   What remains is the very simple relation expressed in (17-5) above.   According to this, if the body is subjected to uniform compression $-C$ along a given axis (that is, X_3, acting parallel to the compression, $= -C$, and X_1 and X_2 both $= 0$), then, to a first approximation,

$$d\mu = \frac{V \, dC}{3} \tag{17-7}$$

It can be shown that any simple shear of magnitude S is equivalent to the combined effect of a tension S and a compression $-S$ acting across mutually perpendicular planes.   Thus the effect of a shear upon chemical potential will be the sum of the respective effects of a compression and a tension of equal magnitude and opposite sign.   Consequently, to a first approximation, a simple shear has no effect upon the chemical potential of an elastic body.

The problem is quite different in the case of nonelastic bodies, for as long as they have not attained a stable strained configuration, such bodies are not in a state of internal equilibrium, and the classic methods of thermodynamics cannot be applied.   The occurrence of flow and creep in stressed nonelastic bodies indicates variation in chemical potential throughout the body, for it is under the influence of such a condition that flow occurs.   It has been shown by one of the writers [23] that for nonelastic

[23] J. Verhoogen, The Chemical Potential of a Stressed Body, *Am. Geophysical Union Trans.*, vol. 32, pp. 251–258, 1951.

bodies, shear may exert a considerable influence on chemical potential in certain cases; it may lead, for instance, to a variation of the solubility from point to point over the surface of the sheared solid.

Apart from the complex case of nonelastic bodies, the result expressed above by Eq. (17–5) conflicts with those advanced by previous investigators.[24]   Riecke, for instance, calculated that the melting point of a body stressed by uniform longitudinal compression is depressed by a factor comparable with the second-order term which we neglected on the above approximation.   The effect is probably much larger than that predicted by Riecke's formula.   Riecke, however, found, in agreement with the present conclusion, that the melting point would be depressed uniformly throughout the stressed solid and on all its bounding faces.   His results appear to have been generally misinterpreted, for "Riecke's principle" has frequently been cited as stating that a stressed crystal tends to go into solution at faces where stress is applied and to elongate normal to the direction of stress as a result of reprecipitation of material on the free faces.   This is probably incorrect for the case of a homogeneously stressed elastic body.   However, if an elastic body is subjected to stresses which are non-homogeneous (i.e., which vary in magnitude from point to point), the quantity P as defined by Eq. (17–6) likewise varies, and with it the chemical potential.   Such a condition applied to a crystal in contact with its saturated solution could lead to solution at points of maximum stress and precipitation at points of least stress, as verified experimentally by Russell.[25]

It is likely that, in rocks composed of granular aggregates, stresses are not uniformly distributed, but are greatest at points of contact of hard grains.   Solution presumably tends to occur here, where compressive stresses are high, and under favorable conditions precipitation could occur elsewhere, so that the rock as a whole would recrystallize.   Moreover, even if stress is homogeneous and chemical potential uniform for each grain, both conditions will vary from grain to grain.   There is thus a tendency for recrystallization to a state such that stress throughout the rock is minimal.   Buerger and Washken[26] recrystallized powders of fluorite and periclase under stresses of the order of 40 tons/in.[2] at temperatures high enough to allow reaction between the solid grains to run with appreciable velocity.   They found that the critical temperature at

[24] E. Riecke, Ueber das Gleichgewicht zwischen einem festen homogen deformierten Koerper und einer flussigen Phase, Annalen der Physik, vol. 54, pp. 731–738, 1895; R. W. Goranson, Flow in stressed solids, an interpretation, Geol. Soc. America Bull., vol. 51, pp. 1023–1034, 1940.

[25] G. A. Russell, Crystal growth in solution under local stress, Am. Mineralogist, vol. 20, pp. 733–737, 1935.

[26] M. J. Buerger and E. Washken, Metamorphism of minerals, Am. Mineralogist, vol. 32, pp. 296–308, 1947.

which recrystallization of a given mineral will occur freely decreases with increasing stress.

Conclusion. It was stated in Eq. (17–4) above that the effect of stress on chemical potential is proportional to the molar volume, so that it is not surprising that minerals whose fields of stability seem to be restricted by nonhydrostatic stress (*i.e.*, antistress minerals) are those of large molar volume. However, these antistress minerals should also become unstable at high hydrostatic pressures, and there is no geological evidence to suggest that this is usually the case, although the transition andalusite → kyanite may be an example. Harker's concept of stress minerals "in a very special sense," capable of existence only under high shear, directly conflicts with the theoretically deduced principle that the effect of pure shear upon chemical potential of an elastic solid is very small.

The conflict between the petrographically established concept of stress and antistress minerals and the theoretical conclusions reached from thermodynamic considerations is perhaps less acute than might at first appear to be the case. Few solids are ideally elastic; indeed, there is good reason to believe that nonelastic behavior plays an important part in metamorphic deformation. For stressed nonelastic solids, the effect of simple shear on chemical potential may conceivably be large, and it is thus theoretically possible that minerals whose response to deforming forces normally departs notably from ideally elastic behavior might have their ranges of stability extended or limited by simple shear. Moreover, shear possibly has a marked "catalytic" effect in accelerating metamorphic reactions which otherwise would proceed extremely slowly, if at all. Deformation resulting from shear involves such processes as penetrative intergranular movement, agitation of reacting grains, reduction of grain size (and hence increase in chemical potential and in total area of reacting surfaces), repeated renewal of surfaces of contact, and storing of strain energy in deformed crystal lattices. These probably play an essential combined role in bringing about mineralogical transformations which, in the absence of shear, would be too slow to lead to observable results. Eskola, in particular, has emphasized the importance of catalytic effects of this kind in the crystallization of "stress minerals."

RATE OF METAMORPHIC REACTIONS [27]

Survival of Metamorphic Assemblages. Metamorphic rocks which have recrystallized at various temperatures, load pressures, and water pressures now coexist on the surface of the earth together with their unmetamorphosed equivalents. Their occurrence can be explained only if

[27] The matter is discussed at some length in: Fyfe, Turner, and Verhoogen, *op. cit.*, pp. 53–103, 1958.

they represent metastable assemblages that have survived cooling and unloading. The problem is different from that of the survival of metastable phases (*e.g.*, glass, zoned plagioclase feldspars) in chilled igneous rocks, for cooling and unloading of metamorphic rocks are essentially the result of slow geological processes such as erosion. Erosion cannot, on the whole, be very much faster than sedimentation, and the rate of unloading cannot be much greater than the rate at which load increased in the original geosyncline. Similarly, cooling cannot be much faster than heating. Why did reactions which occurred during slow heating or loading fail to reverse during equally slow cooling or unloading?

The matter is of some importance in the interpretation of metamorphic facies in terms of physical variables. At equilibrium, reactions occur with zero velocity (see page 36). A reaction runs at a finite velocity only when the reaction free energy, which is zero at equilibrium, has a finite value. A high-temperature phase forms only when the temperature exceeds the equilibrium value. The necessary "overstepping" into the field of stability of the forming phase may be very small (*e.g.*, melting), but it may also be large; Bridgman [28] notes, for instance, that for polymorphic transitions there may be a range of several hundred to a thousand bars or more on both sides of the equilibrium pressure, in which the reaction does not run. Thus the temperature or pressure at which a certain metamorphic assemblage developed could possibly be notably greater than the equilibrium temperature or pressure, and fall in the range of stability of still another, higher-temperature or higher-pressure assemblage. If this were commonly the case, the interpretation of metamorphic facies in terms of equilibrium temperatures would be hopelessly confused. It is thus of some interest to know precisely under what conditions metamorphic reactions will run at rates commensurate with geological time.

Theory of Reaction Rates. As mentioned on pages 48–49, the theory of absolute reaction rates predicts that the specific reaction rate K, measured in reciprocal time (sec^{-1}) will be of the form

$$K = \frac{kT}{h} e^{-\Delta G^{\ddagger}/RT} \qquad (17\text{-}8)$$

where k is Boltzmann's constant (1.38×10^{-16} ergs/deg.), h is Planck's constant (6.62×10^{-27} erg/sec.) and $\Delta G^{\ddagger}$ is the free energy of activation, *i.e.*, the difference between the free energies of the activated and initial states (see Fig. 65). Now

$$\Delta G^{\ddagger} = \Delta H^{\ddagger} - T \Delta S^{\ddagger} = \Delta E^{\ddagger} + P \Delta V^{\ddagger} - T \Delta S^{\ddagger} \qquad (17\text{-}9)$$

---

[28] P. W. Bridgman, *The Physics of High Pressure*, pp. 252–253, G. Bell, London, 1931.

so that at low pressure $(P \cong o)$

$$K = \frac{kT}{h} e^{\Delta S\ddagger/R} e^{-E/RT} \qquad (17\text{--}10)$$

where E is written instead of $\Delta E^{\ddagger}$ and is called the "activation energy." E is usually determined experimentally by measuring K at different temperatures and noting that

$$\ln K = C - \frac{E}{RT}$$

where C is a quantity which varies only slowly as a function of temperature; thus the slope of the curve obtained by plotting $\ln K$ versus $1/T$ is equal to $-E/R$.

Note the occurrence in $(17\text{--}10)$ of the term $\exp\ (\Delta S^{\ddagger}/R)$. A large energy of activation does not necessarily imply a slow reaction, as its effect may be largely offset by the entropy of activation, if it is positive. A negative entropy of activation, indicating a low probability of the activated state, also implies a slow reaction. Although not a necessary condition, it often happens that the entropy of activation has the same sign as the entropy of reaction itself; thus a reaction running in the direction of entropy increase, as would result from increasing temperature, is commonly faster than the opposite reaction resulting from cooling (*e.g.*, melting faster than crystallization, vaporization faster than condensation).

Any metamorphic reaction is a complicated process involving several steps, all of which proceed at different rates; the over-all rate is determined by the slowest of these. We turn to a consideration of these various steps.

Diffusion. The rate of a reaction between two solids A and B to form a third solid C may be controlled by the rate at which either A or B or both diffuse through the growing layer of C which separates them. Similarly, a metasomatic reaction between solid A and a solution depends on the rate at which the appropriate component of the solution diffuses through the reaction rim into A.

What little evidence there is on silicates seems to indicate that solid diffusion is probably effective at most over distances measured in centimeters in times measured in millions of years. Where ionic diffusion is involved, as would be the case for most silicates, diffusion rates decrease rapidly with increasing charge and radius of the diffusing ion. For instance, Li^{+} diffuses more rapidly through quartz than Na^{+}, which, in turn, has a greater diffusion coefficient than Mg^{++}. Highly charged cations, such as Al^{3+} or Si^{4+}, would hardly diffuse at all, in spite of their smaller size, at temperatures within the metamorphic range. Activation energies in silicates and oxides are highly variable, ranging from 17 Kcal.

for Li in quartz to 112 Kcal. for Fe in Fe_2O_3.[29] The properties, particularly the compressibility of the host crystal, have some bearing on diffusion rates, as diffusion will presumably be easier in easily deformable lattices. Diffusion rates are also sensitive to lattice defects; it appears, for instance, that diffusing ions in quartz avail themselves of vacant oxygen lattice positions. Impurities, as they bear on lattice defects, may have a large effect on diffusion rates.

Nucleation. Microscopic evidence shows that although reaction rims are not absent in metamorphic textures (cf. page 470), most metamorphic minerals do not develop simply by reaction between two adjacent pre-existing phases. Grains of the new phase form as nuclei which then start to grow. Nucleation—formation of the nucleus—may be in many cases the rate-controlling step.

Nucleation is usually slow, for the reason that a very small grain of a substance is essentially unstable with regard to larger gains (see page 463). For a given degree of supersaturation or overstepping, there is a critical radius or size below which the nucleus is unstable, and stable growth can start only when this critical radius has been reached, presumably as a result of fluctuations (page 34). An exact formulation of the problem is somewhat involved, but calculations tend to show [30] that the rate of nucleation is strongly temperature-dependent, and also depends on the reaction entropy: nucleation is fastest when the entropy change is smallest. Herein may lie the explanation of Ostwald's step rule, according to which the most likely phase to appear is not necessarily the most stable one; for instance, cristobalite may form (metastably) faster than quartz in a temperature range where quartz is stable.

Note that no nucleation is involved in second-order transitions. A second-order transition is characterized by the absence of a first-order discontinuity in thermodynamic properties (volume, heat content, entropy); rather, there is a gradual change over a temperature range. Order-disorder transitions are frequently of this type. The phase rule precludes the coexistence over a temperature range of two phases of the same composition, at the same pressure. A phase undergoing a second-order transition remains statistically homogeneous. Thus no surface energy is involved, as it would be at the contact of two distinct phases, and no nucleation problem arises. This may be one reason why second-order transitions (e.g., α quartz $\rightarrow$ β quartz) are usually fast.

The problem of nucleating crystals of complex composition is a formidable one. Should the nucleation of, say, talc, or cordierite, depend on chance collision of ions in solution, it would be extraordinarily slow, as

[29] See a summary of experimental data in Fyfe, Turner, and Verhoogen, *op. cit.*, p. 63, 1958.

[30] Fyfe, Turner, and Verhoogen, *op. cit.*, pp. 69–74, 1958.

the probability of simultaneous collision of such a large number of ions in the appropriate proportion would be vanishingly small. The facts rather suggest that complex minerals form by steps, either from simpler ones, or possibly, from complex ionic groups such as $(AlSiO_4)^-$, $(AlSi_2O_6)^-$, or $(Al_2AlSi_3O_{10}[OH]_2)^-$, or others yet more complicated, which may exist in solution. Winkler,[31] studying experimentally the metamorphism of clays, noted that the reacting phases (e.g., illite, muscovite, chlorite) disappeared within a relatively short time, to be followed only after some delay by the appearance of the new phases (e.g., biotite, cordierite), the formation of cordierite requiring about twice as long as that of biotite. It is interesting, though, that under Winkler's experimental conditions (2,000 bars water pressure, temperatures 400°–600°C.), reaction times for these important reactions were measured in days, not in thousands of years.

Influence of Water. Such rapid reactions, as compared with dry diffusion rates, leave no doubt as to the catalytic effect of water. This can readily be shown by experiments. For instance, Shaw[32] studied the reaction $SiO_2 + 2MgO \rightarrow Mg_2SiO_4$. In the dry state, the rate is controlled by diffusion, with SiO_2 migrating into MgO at an extremely small rate (diffusion coefficient about 10^{-16} cm.^{2}/sec. at 1200°C.). If the diffusion process is hastened by decreasing the grain size and increasing the area of contact of the reagents, the reaction rate increases notably. Yet the same mixture which, when dry, produced only 26 per cent of forsterite in four days at 1000°C., reacted completely in a matter of minutes at 450°C. in the presence of water. Very small rates in the absence of water have been found for other reactions,[33] such as MgO $+ MgSiO_3 \rightarrow Mg_2SiO_4$, where it would take 10^{70} years at room temperature to convert 1 mole of MgO if the surface of contact was 1 cm.^{2} For the reaction kyanite $\rightarrow$ mullite $+$ silica the amount (per cent) transformed at 700°K. in 10 million years would be immeasurably small.[34]

The influence of water is manifold. A crystal cannot grow faster than reagents are supplied at its surface. Diffusion to the growing surface would be much slower in the dry state than in an aqueous solution. The solvent power of water also helps to tear down the reacting phases, thus increasing the rate at which reagents are supplied. Shaw[35] found that in the presence of water the rate of formation of forsterite is probably

[31] H. G. F. Winkler, Hydrothermale Metamorphose karbonatfreier Tone, *Geochim. et Cosmochim. Acta*, vol. 13, pp. 42–69, 1957.

[32] C. M. Shaw, Doctoral dissertation (unpublished), University of California, Berkeley, 1955.

[33] R. Jagitsch, Geologische Diffusionen in Kristallierten Phasen, *Arkiv. f. Min. och Geol.*, vol. 1, pp. 65–84, 1949.

[34] Fyfe, Turner, and Verhoogen, *op. cit.*, pp. 77, 98, 1958.

[35] Shaw, *op. cit.*, 1955.

controlled by the rate of solution of quartz.   The solvent power of water
arises from its high dielectric constant which, however, notably decreases
with increasing temperature, particularly near the critical point, as also
does the solubility of quartz.   The effect is less marked at pressures well
above the critical pressure; in this range solubility seems to be strongly
dependent on the density of the solvent.   Thus in any environment where
water exists with an appreciable density one may expect a large kinetic
effect.

Note that it may not be necessary for the water to be present as a
separate phase: an adsorbed film may suffice.   This was shown in Shaw's
experiments mentioned above.   A fairly rapid reaction between SiO_2 and

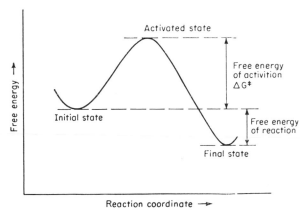

FIG. 65.   Diagram to illustrate free energy of activation.

MgO could still be obtained at low pressures and temperatures (425°C.,
50 bars) under conditions where the solubility of silica in the vapor phase
would be very small, but where an adsorbed film of water could still
exist.

Kinetics of Surface Reactions.   Any actual surface phase has finite
thickness, usually of the order of a few angstroms, and consists of particles
of both adjacent phases arranged in a manner which may differ from that
in either adjacent phase.   Because of this difference in physical structure,
a surface phase is sometimes referred to as constituting a "disorganized
layer"—though this does not necessarily imply structural analogy with
liquid or glass.   Clearly, the physical conditions in such surface phases
must influence the kinetics of metamorphic reactions to an important
degree.

Not only the arrangement but the very nature of the individual particles
in the surface phase may differ from that of adjacent phases.   For instance, the molecule H_2 dissociates to form two atoms of hydrogen when

it is adsorbed on the surface of palladium, the energy required for this dissociation being supplied by change in potential energy of hydrogen when brought into the attractional field of the palladium atoms. The phenomenon of catalysis, as is well known, is due primarily to this special state of the surface layer.   A catalyst is a substance whose mere presence increases the rate of a reaction in which it otherwise does not take part. This effect is due to the fact that the reacting molecules may be adsorbed at the surface of the catalyst in a state requiring a lower activation energy. In the same manner, diffusion may be faster over a boundary, i.e., within a surface phase, than through a volume phase, as illustrated by the following data for diffusion of thorium in tungsten: [36]

Type of diffusion	Free energy of activation, Kcal.
In solid solution	105.7
On grain boundary	76.9
On surface	55.1

It has been found, too, that the electrical conductivity of a glass, which depends on rate of diffusion in the glass of the ions which carry electrical charges, is usually greater over the surface than through the bulk of the sample, and may become many times greater (by a factor of 10^8 or so) if the glass is in contact with water-saturated air.[37]   This may be due to adsorption of water on the surface of the glass to form a film in which ions may diffuse readily.[38]   There is an obvious analogy between this adsorbed film and the "intergranular film" which figures so prominently in recent petrological literature as the medium through which metasomatism and metamorphism are supposed to be effected.

There is thus much experimental evidence to indicate that, because of their peculiar physical state, surface phases may have properties that are not shared by adjacent phases but which have a particularly important influence upon the kinetics of various reactions.   It is not yet possible to predict quantitatively this kinetic effect of surface phases, but certainly it may play an essential role in the processes of rock metamorphism.

Effect of Pressure on Reaction Rates.   Equations (17–8) and (17–9) show that the reaction rate K depends on pressure through the term $\exp{(-P\,\Delta V^{\ddagger})}$ where $\Delta V^{\ddagger}$ is the activation volume.   If the activated state has a greater volume than the initial state, the rate will decrease

[36] S. Glasstone, K. J. Laidler, and H. Eyring, *The Theory of Rate Processes*, p. 541, McGraw-Hill, New York, 1941.

[37] G. W. Morey, *The Properties of Glass*, p. 450, Reinhold, New York, 1938.

[38] See also, in this connection, the effect of water on recrystallization of amorphous silica, in G. Cohn and I. M. Kolthoff, Thermal aging of silica, *Festskrift J. Arvid Hedvall*, p. 97–116, Göteborg, 1948.

with increasing pressure. There is evidence that for simple processes in homogeneous phases, such as diffusion, $\Delta V^\ddagger$ and $\Delta S^\ddagger$ are simply related: [39] $\Delta S^\ddagger/\Delta V^\ddagger \cong \alpha/\beta$ (see page 467). Thus $\Delta V^\ddagger$ and $\Delta S^\ddagger$ would have the same sign. For polymorphic transformations, the activated state probably corresponds to a disorganized "vapor-like" condition with larger entropy and larger volume. The positive volume of activation would then indicate a decrease in rate with increasing pressure. On the other hand, it has been noticed that in aqueous systems rates tend to increase at constant temperature with increasing pressure. This may be due in part to increased solubility of the reagents resulting from increasing density of the vapor phase, or to the increase in the dielectric constant of the water, or to the effect of pressure on the rate of solution. Pressure also affects nucleation rates, particularly in polymorphic transformations in which the new phase has a larger volume than the initial phase; thus reaction resulting from an increase in pressure should run faster, other things being equal, than reactions resulting from decreasing pressure.

Influence of Deformation. It has long been known that mechanical deformation, such as intense grinding, considerably hastens reactions between solid phases (see page 473). The effect of such deformation is probably mainly to increase diffusion rates and possibly rates of crystal growth. This may be accomplished by grinding away the reaction zone between adjacent phases so as to provide continuously fresh surfaces on which the reaction can readily take place. Deformation also promotes intracrystalline diffusion by creating dislocations, i.e., areas of misfit within the crystalline lattice. These dislocations are also probably largely responsible for growth of crystals, as they may circumvent the requirements of nucleation of new layers on the surface of the growing crystal.[40] The effect of deformation on diffusion rates can be studied in ionic crystals in which the diffusing particles are ions which, as they move, produce an electrical current. The electrical conductivity of such crystals is closely related to the ionic diffusion rates. Experiments show that the compression of powders usually increases diffusion rates by a factor of 10 or 100. Thus deformation certainly does have a catalytic effect, although it may be difficult to assess its magnitude in rocks.

[39] A. W. Lawson, Correlation of $\Delta S^\ddagger$ and $\Delta V^\ddagger$ in simple activated processes in solids, *Phys. and Chem. of Solids*, vol. 3, pp. 250–252, 1957.

[40] On the very important subject of dislocations, diffusion, and crystal growth, see W. Dekeyser and S. Amelinckx, *Les dislocations et la croissance des cristaux*, Masson, Paris, 1955; W. T. Read, Jr., *Dislocations in Crystals*, McGraw-Hill, New York, 1953; A. R. Verma, *Crystal Growth and Dislocations*, Academic Press, New York, 1953; F. Seitz, Fundamental Aspects of Diffusion of Solids, in *Phase Transformations in Solids*, pp. 77–148, Wiley, New York, 1951; *Imperfections in Nearly Perfect Crystals*, Wiley, New York, 1952.

Kinetic Irreversibility of Metamorphism. Earlier it was concluded from general petrographic evidence that most metamorphic rocks consist essentially of assemblages of solid mineral phases which coexisted in equilibrium at the time of metamorphism. This conclusion is qualified by recognition, in some rocks, of minerals which are presumed to have been metastable with reference to metamorphic conditions. It was also stated that most metamorphic assemblages—like most igneous rocks— must be metastable at surface temperatures and pressures. Acceptance of these statements raises two fundamental problems of metamorphic petrogenesis. The first concerns the range of temperature and pressure represented by a given metamorphic assemblage. The second refers to the possibility that the rate of metamorphic reaction is notably greater in response to rising than to falling temperature.

Mostly metamorphism has been effected during a considerable but limited period of high temperature, within a much longer period of changing temperature and pressure which closed with exposure of the metamorphic product at the earth's surface. With what part of this sequence of changing physical conditions should metamorphism be corre- lated? It has been tacitly assumed by many writers that the maximum temperature to which the rock has been subjected during its metamorphic history is that which is recorded in the mineral assemblage ultimately exposed at the surface. But this is not wholly compatible with the gen- erally accepted view that mineralogical variation in metamorphic rocks of the same composition is largely due to difference in physical conditions, especially temperature and water pressure, of metamorphism. For ex- ample, if basic igneous rocks are normally metamorphosed to the assem- blage andesine-hornblende at high temperatures and to albite-epidote- actinolite-chlorite at lower temperatures, how is it that the former assemblage so commonly cools unchanged through temperatures at which the second is presumed to be stable? The common failure of reactions induced by increasing temperature or increasing pressure to reverse when temperature or pressure are subsequently decreased indicates a kind of kinetic irreversibility which requires explanation.

Consider particularly the effect of temperature. We have previously noted that reaction rates generally increase with increasing temperature, and that these rates become appreciable only at temperatures somewhat removed from the equilibrium value. Consider a high-temperature assemblage A and an equivalent low-temperature assemblage B, the equi- librium temperature being T_e. Assemblage A will form from B at a temperature $T_e + \Delta T$, say. Consider now the ratio m of the forward rate B $\rightarrow$ A at $T_e + \Delta T$ to the backward rate A $\rightarrow$ B at $T_e - \Delta T$. Using the theory of absolute reaction rates as expressed by Eq. (17–8), it may be shown [41] that m depends essentially on the ratio $\Delta T/T_e$, on $\Delta H^{\ddagger}$, the enthalpy of activation, and on ΔS, the reaction entropy (not the entropy of activation). For a typical polymorphic transition in a dry system, where $\Delta H^{\ddagger}$ would be large and ΔS small, the overstepping ΔT may be large, and m can be very large: the forward reaction could, for instance, run 10^6 times faster at a temperature $100°$ above equilibrium than the backward reaction would at $100°$ below equilibrium. Persistence of the high-temperature form is thus very likely. On the contrary, for a typical metamorphic reaction involving hydration or dehydration, $\Delta H^{\ddagger}$ may be small, ΔS large, and m would be small. Probability of survival would be low, and would decrease with increasing pressure. The necessary overstepping, which is inversely proportional to the entropy of reaction, would also be small, meaning that during slow heating the mineralogical composition of a rock would remain at all times nearly adjusted to the prevailing temperature. If so, a high-grade rock, for instance a kyanite-staurolite pelite schist, should have gone in succession through all intermediate stages corresponding to the zones of chlorite, biotite, and almandine. Field evidence is not definite on this point, and possibly in some instances rocks reached their high-temperature form without going through any intermediate stages. This would indicate very rapid heating.

In brief, the irreversibility of metamorphic reactions involving sluggish phase transformations in the absence of water is understandable; the irreversibility of reactions involving loss or gain of water is not, on purely thermodynamic grounds. Most probably such irreversibility is due to the loss of water itself, the system being open and the water released by dehydration having been driven out during the high-temperature stage. Absence of water during cooling of the now anhydrous system is a possible reason why pyroxenes do not revert to amphiboles, or why micas and chlorites fail to develop from anhydrous minerals such as cordierite, almandine, orthoclase, or sillimanite. The low permeability of completely recrystallized metamorphic rocks is a factor operating against late hydration or carbonation.

Retrogressive metamorphism (diaphtoresis) [42]—the mineralogical readjustment of high-temperature metamorphic assemblages to a lower temperature—does occur locally; it is generally attributed to strong deformation or to hydrothermal activity subsequent to the main stage of metamorphism. Conversion of amphibolite (hornblende-andesine-biotite) to greenschist (chlorite-actinolite-albite-epidote-sericite) in a zone

[41] Fyfe, Turner, and Verhoogen, op. cit., pp. 97–98, 1958.

[42] For discussion of retrogressive metamorphism and reference to individual cases, see F. J. Turner, Mineralogical and structural evolution of the metamorphic rocks, Geol. Soc. America Mem. 30, pp. 299–304, 1948.

of strong differential movement, and hydrothermal replacement of serpentine rocks by talc-magnesite schists are well-known illustrations. This interpretation of retrogressive phenomena conforms to the principles just discussed. It is based on the assumption that, over the lower range of metamorphic temperatures, or below 300°C., metamorphic reactions proceed at an appreciable rate only when accelerated by strong deformation or by passage of a wave of aqueous or other pore fluids. Recognition of zones of retrogressive metamorphism may thus be a significant step in the solution of structural problems in metamorphic terranes. It should be emphasized that retrogressive metamorphism can be diagnosed with certainty only where metastable or partially destroyed relicts of the high-temperature assemblage persist in an association of low-temperature minerals. Relict garnet or staurolite, mantled with chlorite, in schists composed of muscovite-chlorite-quartz may be cited as an illustration. This raises the possibility that some rocks, lacking obvious metastable relict—especially strongly deformed rocks of phyllitic aspect—may really be products of retrogressive metamorphism that has proceeded to completion.

CHAPTER 18

Metamorphic Zones and Metamorphic Facies

METAMORPHIC ZONES

Evolution of the Becke-Grubenmann Concept of Depth Zones. It has already been stated that the mineral assemblage in any metamorphic rock may be regarded as the product of partial or complete chemical adjustment of a preexisting association of minerals to a particular physical and chemical environment. Since temperature and pressure, the main physical conditions that govern metamorphism, are subject to variation with increasing depth below the earth's surface, some general connection between depth and metamorphic paragenesis is to be expected. Such a relationship was first envisaged by Sederholm in 1891, and upon it is based Becke's concept of depth zones in metamorphism, which was elaborated by Grubenmann and subsequently by Niggli.[1]

Becke, in framing his twofold division of the field of metamorphism according to depth, in 1903, emphasized the influence of a "volume law" according to which crystallization of dense minerals is caused by the increased pressure that almost invariably accompanies metamorphism.[2] On the other hand, he also appreciated the influence of high temperature, which acts in the opposite direction to pressure in most metamorphic reactions. Recrystallization in the upper zone of relatively low temperature was, therefore, pictured as dominated by pressure, and favorable to the formation of dense, often hydroxyl-bearing, silicates (such as micas, chlorites, epidote, chloritoid, and actinolite). In the lower zone, rising temperature due to geothermal heat was thought to overshadow high

[1] P. Niggli, *Die Gesteinsmetamorphose,* Borntraeger, Berlin, 1924, *e.g.,* pp. 374, 375, 398–400; A. Knopf, in *Geol. Soc. America 50th Anniversary Vol.,* pp. 352–354, 1941.

[2] For a summary of Becke's ideas see: W. S. Fyfe, F. J. Turner and J. Verhoogen, Metamorphic reactions and metamorphic facies, *Geol. Soc. America Mem. 73,* pp. 4–5, 1958.

pressure, thus prohibiting crystallization of the hydroxyl-bearing minerals, the place of which was taken by silicates comparable with those occurring in igneous rocks: pyroxenes, feldspars, olivine, garnets, etc.

In 1904 appeared the first edition of Grubenmann's great work *The Crystalline Schists*, which in 1924 was completely rewritten and in many respects modified by Niggli. Grubenmann's classification of metamorphic rocks, adopted in its essentials by Niggli, rests on chemical composition and on depth of metamorphism. Three depth zones are defined:

1. The upper or *epi zone:* mechanical and chemical metamorphism giving rise mainly to hydrous silicates; temperature relatively low or moderate, hydrostatic pressure for the most part low, nonhydrostatic stress often high but sometimes lacking. Typical minerals include sericite, chlorite, chloritoid, stilpnomelane, antigorite, talc, brucite, actinolite, epidote, zoisite, albite, glaucophane, manganiferous garnet, calcite, dolomite, magnesite.

2. The middle or *meso zone:* principally chemical metamorphism; temperature and pressure higher than in the epi zone, nonhydrostatic stress often very high though sometimes lacking. Typical minerals include biotite, muscovite, staurolite, kyanite, anthophyllite, cummingtonite-grunerite, actinolite, common hornblende, alkali hornblende, epidote, zoisite, sodic plagioclase, almandine, calcite, brucite. Not infrequently the meso zone is characterized by association of epi-zone minerals with other species typical of the kata zone.

3. The lower or *kata zone:* long-continued chemical recrystallization, often unaccompanied by deformation; temperature high, hydrostatic pressure usually very high though sometimes low, nonhydrostatic stress weaker than in the other two zones and in many cases quite lacking. The mineral composition approaches that which would develop by direct crystallization from an appropriate igneous melt. Typical minerals include biotite, potash feldspars, sillimanite, andalusite, enstatite-hypersthene, olivine, diopside-hedenbergite, omphacite, common hornblende, alkali hornblende, aegirine-augite, jadeite, cordierite, garnets (almandine, pyrope, grossularite, andradite), spinel, plagioclase (often highly calcic), anorthite, vesuvianite, scapolite, humites, monticellite, calcite, periclase.

Grubenmann's original scheme had no place for the products of purely contact metamorphism but was extended by Niggli to remedy this defect. High-temperature contact metamorphism such as occurs in the innermost zone of an aureole, as well as accompanying perimagmatic (pneumatolytic) metasomatism and injection metamorphism, are all included by Niggli as processes of the kata zone, regardless of the actual depth (commonly shallow), and hence pressure, at which they may have operated. In the same way the products of low-temperature contact metamorphism

near the outer boundary of an aureole, and rocks that have been affected
by low-temperature metasomatism (often hydrothermal) involving introduction of magmatically derived material, are classed by Niggli among
the rocks of the epi zone. As thus amended, the three "zones" are no
longer defined by depth but by temperature.[3]

The influence of depth upon conditions of metamorphism must be
admitted. But in any particular instance, other factors such as proximity
to injected bodies of hot magma, or situation with respect to active
orogenic zones where temperature gradients are abnormally steep, may
outweigh absolute depth in this connection. Thus, while a term such as
epimetamorphism may be of some value in suggesting a particular broad
range of metamorphic conditions normally encountered in the shallower
rather than in the deeper levels of the outer continental crust, it is too
vaguely defined and too closely associated with the idea of universal depth
control to be used in framing a basis for classification of metamorphic
rocks. Moreover, Grubenmann's threefold classification of metamorphic
rocks (into "epi," "meso," and "kata," divisions), though retained by
Niggli, is too simple to represent the wide range of physical conditions
reflected in the varied mineralogical assemblages developed in rocks of
this general class.

Zones of Progressive Regional Metamorphism. *Zones in the Dalradian
Schists of Scotland.* A decade before the depth-zone hypothesis was
fomulated by Becke and Grubenmann, Barrow,[4] working on the Dalradian schists of Scotland, mapped in the field a series of zones of progressive regional metamorphism, based on mineralogical transformations
in derivatives of pelitic (argillaceous) sediments and correlated in a
general way with increasing temperature and pressure. This was a most
important advance in the evolution of the concept of regional metamorphism. Barrow's conclusions, as later extended, substantiated, and to
some extent modified by Tilley and others,[5] may be summed up as follows: The rocks of the Scottish Dalradian are regionally metamorphosed
geosynclinal sediments and associated pyroclastics ("green beds"). Mineralogical variation in pelitic members is correlated with variation in
temperature, and to a less extent pressure, at the time of metamorphism.

[3] Cf. P. Niggli, Some hornfelses from Saxony and the problem of metamorphic
facies, *Am. Mineralogist,* vol. 35, p. 869, 1950.

[4] G. Barrow, On an intrusion of muscovite-biotite gneiss in the south-east Highlands
of Scotland, *Geol. Soc. London Quart. Jour.,* vol. 49, pp. 330–358, 1893; On the
geology of lower Dee-side and the southern Highland border, *Geol. Assoc. Proc.,* vol.
23, pp. 268–284, 1912.

[5] C. E. Tilley, Metamorphic zones in the southern Highlands of Scotland, *Geol. Soc.
London Quart. Jour.,* vol. 81, pp. 100–112, 1925; G. L. Elles and C. E. Tilley, Meta-
morphism in relation to structure in the Scottish Highlands, *Royal Soc. Edinburgh
Trans.,* vol. 56, pt. 3, pp. 621–646, 1930.

This variation reflects the continuous change in degree or grade of metamorphism corresponding to a temperature-pressure gradient. Each zone of progressive metamorphism is defined by an index mineral, the first appearance of which (in passing from low to higher grades) marks the isograd, or line of equal grade, defining the outer limit of the zone in question. Since the stability of any mineral cannot be considered separately from that of the mineral association of which it is an integral member, it follows that rocks of constant chemical composition must be selected for zonal mapping in terms of index minerals. Barrow chose pelitic rocks, both on account of their wide distribution in the Dalradian and because of their sensitivity to variation in temperature and pressure. The sequence of index minerals, in order of increasing metamorphic grade, is chlorite, brown biotite, almandine, staurolite, kyanite, sillimanite; isograds have been drawn to mark the first appearance of each of these. In the zone of chlorite (Barrow's "zone of digested clastic mica") the indefinite colorless mica (e.g. illite) and chlorite (celadonite) of the original sediment have recrystallized as muscovite and an aluminous chlorite, which, as the metamorphic grade advances, themselves react to give red-brown biotite, the first appearance of which defines the outer boundary of the biotite zone. At this stage the typical rock is a biotite schist containing quartz, either chlorite or muscovite, and less commonly albite. A manganese-rich garnet of the spessartite-almandine series may occur as a minor constituent of low-grade rocks, even within the zone of chlorite; but the entry, in abundance, of porphyroblastic nonmanganiferous almandine is a sharply defined stage of metamorphism and marks the edge of the almandine zone, which is characterized by coarse, garnetiferous, mica schists containing biotite, muscovite, or both micas. The remaining three zones occupy more limited areas on the map. Staurolite, like chloritoid in rocks of lower grade, is restricted to rocks of high iron content, and its value as an index mineral is on this account somewhat doubtful, since in normal pelitic rocks lower in iron there is direct passage from the zone of almandine to the kyanite zone, without intermediate crystallization of staurolite. In the sillimanite zone pelitic rocks are represented by coarse gneisses containing biotite, garnet, and sillimanite, together with potash feldspar which now largely takes the place of muscovite. Cordierite may also be present. The presence of extensive bodies of granite within the sillimanite zone led Barrow to the conclusion, still held by some workers in this field, that the whole series of metamorphic zones mapped by him had developed under the influence of temperatures induced by the intrusive granite mass. Tilley's view is that the temperature in any zone was determined largely by depth of burial, modified in the deeper levels by proximity to masses of injected granitic magma.

Zones in Pelitic Schists Elsewhere. Where metamorphic zones have been mapped in other countries, they have been based, more often than not, upon pelitic assemblages. The sequence established for the Scottish Highlands has been generally confirmed, though there may be minor departures, probably due to local peculiarities in rock composition or in metamorphic conditions. Thus in the Sulitelma region of Norway, Vogt [6] found that oligoclase or more calcic plagioclase is the typical index mineral (instead of staurolite and kyanite) for the highest grades of metamorphism, and correlated this feature with the somewhat calcic and feldspathic nature of the pelitic sediments in this region. In the Littelton-Moosilauke district of New Hampshire [7] a low-grade (chlorite) zone is succeeded by a middle-grade zone, the typical rocks of which contain quartz, biotite, muscovite, garnet, and staurolite, and this in turn passes into a high-grade zone with the entry of sillimanite into the pelitic paragenesis. Though biotite tends to crystallize at a grade somewhat below that requisite for the formation of garnet and staurolite, there is some doubt whether temperature or rock composition is responsible for the distribution of staurolite within the middle-grade zone. Kyanite is present in pelitic rocks throughout the whole of the middle-grade zone in an adjacent area.

The modern French school recognizes a scheme of zones based on both the fabric and the mineralogy of pelitic schists and gneisses.[8] These zones, first mapped in the Massif Central of France, are broadly correlated with depth estimated from field observations in that region:

1. Non-metamorphic zone: shales. Depth, 0 to 4,000 m.

2. Zone of upper mica schists: sericite schists. Depth, 4,000 to 7,000 m.

3. Zone of lower mica schists: muscovite-biotite schists; garnet, staurolite, kyanite may be present. Depth, 7,000 to 10,000 m.

4. Zone of upper gneiss: mineralogy is similar to that of zone 3, but the structure is gneissic. Depth, 10,000 to 14,000 m.

5. Zone of lower gneiss: biotite-sillimanite-garnet gneisses without muscovite. Depth, 14,000 to 20,000 m.

Zones in Nonpelitic Rocks. Zones of progressive regional metamorphism have been mapped for other rocks than pelitic sediments, notably for derivatives of basic igneous rocks, calcareous sediments, and gray-

[6] T. Vogt, Sulitelmafeltets Geologi og Petrografi (with English summary), *Norges geol. undersökelse 121*, 1927.

[7] M. P. Billings, Regional metamorphism of the Littleton-Moosilauke area, New Hampshire, *Geol. Soc. America Bull.*, vol. 48, pp. 463–566, 1937; C. A. Chapman, Geology of the Mascoma quadrangle, New Hampshire, *Geol. Soc. America Bull.*, vol. 50, pp. 127–180, 1939.

[8] J. Jung and M. Roques, Introduction à l'étude zonéographique des formations cristallophylliennes, *Services Carte géol. France Bull.*, no. 235, t. 50, 1952 (especially pp. 12–19).

wackes. Correlation of these zones with isogradic zones in pelitic schists is illustrated in Table 37.[9]

Interpretation of Zones. Most workers in the field of regional metamorphism correlate the sequence of zones of progressive metamorphism with a temperature gradient, induced either by regional uprise of granitic magma, or by depth of burial, or by some complex combination of factors (perhaps including both of those just mentioned) developed in zones of active orogeny.[10] This is the interpretation tentatively accepted by the present authors. It should be noted, however, that petrologists who on the contrary attribute granite batholiths to metasomatism *in situ* would correlate mineralogically distinct zones developed in rocks of common parentage, with different degrees and kinds of metasomatism induced by a series of advancing "fronts," the innermost of which corresponds to the granite (completely granitized rock) itself.[11] In spite of current differences in opinion as to the origin and metamorphic function of granite— whether it represents the principal cause or the ultimate product of regional metamorphism—the authors still believe that existence of zones of regional metamorphism reflects the varied mineralogical response of chemically similar rocks to different physical conditions—especially temperatures. This does not mean that rocks in zones of high metamorphic grade must necessarily have passed previously through the full sequence of changes corresponding to lower grades in the same region. Nor must metamorphism necessarily have covered precisely the same period of time

[9] Vogt, *op. cit.*, 1927; Elles and Tilley, *op. cit.*, 1930; J. D. Wiseman, The central and south-west Highland epidiorites, *Geol. Soc. London Quart. Jour.*, vol. 90, pp. 354–417, 1934; Billings, *op. cit.*, 1937; F. J. Turner, Progressive regional metamorphism in southern New Zealand, *Geol Mag.*, vol. 75, pp. 160–174, 1938; W. Q. Kennedy, Zones of progressive regional metamorphism in the Moine schists of the western Highlands of Scotland, *Geol. Mag.*, vol. 86, pp. 43–56, 1949. For other examples of zones of regional metamorphism see F. J. Turner, Mineralogical and structural evolution of metamorphic rocks, *Geol. Soc. America Mem. 30*, pp. 37–39, 1948; H. L. James, Zones of regional metamorphism in the pre-Cambrian of northern Michigan, *Geol. Soc. America Bull.*, vol. 66, pp. 1455–1488, 1955; J. J. Reed, Regional metamorphism in south-east Nelson, *New Zealand Geol. Surv. Bull.*, no. 60, 1958.

[10] For various opinions in this connection, see Barrow, *op. cit.*, 1893; V. M. Goldschmidt, Die Kalksilikatgneisse und Kalksilikatglimmerschiefer des Trondhjem-Gebiets, *Kristiania Vidensk. Skr., Math.-Naturv. Kl. 10*, 1915; Elles and Tilley, *op. cit.*, 1930; J. W. Ambrose, Progressive kinetic metamorphism of the Missi series near Flinflon, Manitoba, *Am. Jour. Sci.*, vol. 32, pp. 257–286, 1936; Billings, *op. cit.*, 1937; H. H. Read, Metamorphism and igneous action, *Nature*, vol. 144, pp. 729–731, 772–774, 1939; W. Q. Kennedy, On the significance of thermal structure in the Scottish Highlands, *Geol. Mag.*, vol. 85, pp. 229–234, 1948; James, *op. cit.*, pp. 1483–1485, 1955.

[11] Cf. H. H. Read's discussions of regional metamorphism in relation to regional granitization, in Meditations on granite, Part II (*Geol. Assoc. Proc.*, vol. 55, pp. 69, 70, 1944); and This subject of granite (*Sci. Progress*, vol. 34, p. 668, 1946).

TABLE 37. CORRELATION OF ZONES OF REGIONAL METAMORPHISM DESIGNATED BY FIRST APPEARANCE OF INDEX MINERALS

Parent rock:	Pelitic sediments		Somewhat feldspathic pelitic sediments	Graywackes	Basic igneous rocks	Calcareous sandstones	Ferruginous cherts
Example:	Scotland: Dalradian schists	New Hampshire: Ordovician and Silurian schists	Norway: Sulitelma region	New Zealand: Otago and Westland schists	Scotland: Dalradian epidiorites	Scotland: Calc-silicate granulites of Moine series	Michigan: Iron formation
Index minerals of zones: Chlorite	Chlorite	Chlorite	Chlorite (albite)	Chlorite (albite, stilpnomelane)	Chlorite (actinolite, albite)	Chlorite and albite	Stilpnomelane
Biotite	Biotite		Biotite	Biotite			
Almandine	Almandine	Almandine (biotite, kyanite, staurolite)	Almandine		Almandine (hornblende)	Almandine (zoisite, biotite)	
			Oligoclase	Oligoclase	Oligoclase to labradorite	Hornblende	Grunerite Garnet
Kyanite (staurolite)	Kyanite (staurolite)					Bytownite	
Sillimanite	Sillimanite	Sillimanite				Augite	

in every zone. The mineralogical assemblage of any zone has formed during that last incident of petrogenesis when reaction was rapid enough to permit a state of equilibrium to develop between the solid phases of the system.

Conditions favoring rapid reaction include high temperature, contemporary or previous deformation, and (for common reactions involving water and carbon dioxide) permeability of the rock with respect to a fluid phase. Hydration reactions in a relatively anhydrous system (*e.g.*, conversion of basalt to greenschist) can proceed with appreciable velocity only when there is a continuous flow of aqueous fluid into the reacting system; this implies not only high permeability of the system but a continuously maintained gradient in water pressure between the system and its immediate environment. The optimum combination of all these conditions need not occur simultaneously and need not be of the same duration in zones differently situated with respect to foci of orogeny and magmatic activity. It is perhaps to variations in such relations, involving both space and time, that the observed differences in detail between described zonal series may be attributed.

The great extent of the chlorite zone of southern New Zealand (province of Otago) perhaps implies the existence of a very gentle horizontal temperature gradient outward from a central axis of orogeny. Alternatively the rocks everywhere may have remained permeable to water and CO_2 streaming continuously upward from the depths, so that equilibrium was established everywhere during the final low-temperature stage—not necessarily attained simultaneously throughout the province as a whole. Long-continued and repeated deformation, evident from the fabric of the Otago schists, may have been the principal factor in maintaining permeability, and may have contributed to the activation energy of metamorphic reactions.

Contrast the Otago province with the more usual situation where the various zones are more nearly equal in width and extent. It is not unlikely that the mineral assemblage of each zone represents equilibrium near the local maximum temperature—again not necessarily reached simultaneously in all zones. The high-grade water-poor mineral assemblages of the inner zones have retained their identity during subsequent cooling largely for lack of water. Either they have become impermeable through metamorphic crystallization, or there was no external source of water (such as a crystallizing body of magma) during the cooling episode.

Zones of Progressive Contact Metamorphism. In some contact aureoles there are concentric zones marked by mineralogical or textural differences in rocks of the same general composition. These represent gradients in temperature, and perhaps in water pressure, developed consequently upon intrusion of hot magma into relatively cool rocks.

The character of the gradient, and hence the width of the aureole and of each zone, depends upon several factors: temperature of intrusive magma; regional temperature and pressure of host rock; water pressures developed at different stages after intrusion (*e.g.*, as a result of boiling of magma); chemical nature and permeability of rocks in the aureole.

The Comrie aureole of the Perthshire Highlands, Scotland, illustrates progressive contact metamorphism round an intrusive diorite complex.[12] Outside the aureole the slate, which lies within the chlorite zone of regional metamorphism, is a well-cleaved rock consisting of muscovite, chlorite, quartz, iron ores, and many accessory constituents. Three zones are recognizable within the aureole:

1. Zone of spotted slates: Minute ill-defined ovoid spots, notably richer in muscovite and chlorite than is the surrounding matrix, are the only indications of contact metamorphism in this outermost zone. Similar spotting is characteristic of incipient contact metamorphism in many other aureoles. According to Harker[13] the spots in some cases are composed largely of glass. This presumably is rather a chemically precipitated gel, more or less equivalent to the chlorite-muscovite spots of the Comrie rocks.

2. Zone of biotite: Biotite, as in regional metamorphism, is the first new mineral to crystallize, probably as a result of reaction between muscovite and either iron ore or chlorite. Chemical reconstitution may still be far from complete, and the original schistosity therefore persists.

3. Zone of cordierite: With the entry of cordierite the rock rapidly becomes completely reconstituted, so that the biotite schists of the previous zone merge into cordierite hornfelses in this, the innermost zone. The principal rock types are quartz-bearing hornfelses rich in cordierite, hypersthene, and biotite, but silica-poor rocks with spinel or corundum and no quartz are also represented. Andalusite is rather rare. Epidiorites in the outer part of the aureole have been converted to hornblende-plagioclase-biotite hornfelses, which pass into pyroxene-bearing types in the innermost zone.

Grout[14] has compared the contact effects adjacent to intrusive bodies of granite and gabbro, and finds that, in Minnesota, pelitic slates are converted into chemically identical biotite schists for distances of 220 yards to 5 or 10 miles from visible contacts with granite intrusions. Regular zones of progressive metamorphism based upon mineral composition are lacking. Where slates have been metamorphosed by the Duluth gabbro, on the other hand, three zones may be recognized:

---

[12] C. E. Tilley, Contact metamorphism in the Comrie area of the Perthshire Highlands, *Geol. Soc. London Quart. Jour.*, vol. 80, pp. 22–71, 1924.

[13] A. Harker, *Metamorphism,* pp. 15, 16, Methuen, London, 1932.

[14] F. F. Grout, Contact metamorphism of the slates of Minnesota by granite and by gabbro magmas, *Geol. Soc. America Bull.*, vol. 44, pp. 989–1040, 1933.

1. Between 500 ft. and 50 ft. from the visible contact, spotted rocks have developed, consisting of quartz, feldspars, and biotite, and in some cases irregular porphyroblasts of cordierite.

2. A transition zone.

3. Within 5 ft. of the contact, a zone of hornfelses consisting of quartz, labradorite (more calcic than the plagioclase of the outer zone), biotite, pyroxenes (diopside, hypersthene or both), and, in many rocks, cordierite. Chemically the rocks of this innermost zone are said to have been affected by addition of CaO, MgO, FeO, and TiO_2 from the magma, and removal of K_2O, SiO_2, and water.

In general, schists rather than hornfelses are considered by Grout to be the normal products of contact metamorphism in granite aureoles, whereas hornfels is the characteristic rock of contact zones adjacent to intrusions of gabbro.

THE MINERALOGICAL PHASE RULE

Definition. In his classic study of contact metamorphism in the Oslo region of Norway, Goldschmidt [15] found some remarkably simple and consistent relationships in the mineral paragenesis. In spite of their wide diversity in composition, the common quartz-bearing hornfelses of this district are simple mineral assemblages, each consisting of four or five out of ten common minerals: quartz, potash feldspar, plagioclase, andalusite, cordierite, biotite, hypersthene, diopside, grossularite, wollastonite. Certain pairs of these minerals are consistently developed to the exclusion of other chemically equivalent pairs; thus anorthite-hypersthene appears instead of andalusite-diopside, which is unknown at Oslo. Moreover there is constant correlation between chemical and mineralogical composition; for a given range of chemical composition, the mineral assemblage is invariably the same.

This regularity in paragenesis was recognized by Goldschmidt as very strong evidence pointing to prevalence of equilibrium in the Oslo hornfelses. He argued that if this generalization applies to most common metamorphic mineral assemblages, they must conform to the requirements of the phase rule [16]

$$w = c + 2 - \varphi \qquad (2\text{-}50)$$

where w is the variance (number of "degrees of freedom") of the system, c is the number of components, and φ is the number of phases existing

[15] V. M. Goldschmidt, Die Kontaktmetamorphose im Kristianiagebiet, *Kristiania Vidensk. Skr., I, Math.-Naturv. Kl. 11*, 1911. For discussion of Goldschmidt's observations and conclusions, see Fyfe, Turner, and Verhoogen, *op. cit.*, pp. 5–8, 1958.

[16] Goldschmidt, *op. cit.*, p. 123, 1911.

together in equilibrium. If metamorphism occurred with zero variance—
i.e., if the mineral assemblage could crystallize only at one temperature
and one pressure—the chance that these exact conditions would be realized and maintained under natural conditions is so exceedingly remote as
to preclude the possibility of its occurrence in rocks. This same limitation applies even where the variance is one. For a paragenesis such as
albite-chlorite-epidote-sphene to appear commonly among metamorphic
rocks (as is actually the case), it must be stable over a wide range of
both temperature and pressure, so that the variance of the system must
be 2 or more. If, then, $w \geqq 2$, the equation

$$w = c + 2 - \varphi$$

becomes

$$\varphi \leqq c$$

This may be stated as Goldschmidt's "mineralogical phase rule": The
maximum number of crystalline minerals that can coexist in stable equilibrium is equal to the number of components in the rock in question (cf.
page 31).

Application. Application of the mineralogical phase rule tends to be
qualitative rather than strictly quantitative, for metamorphic rock groups
are multicomponent systems in which it is difficult to fix the number of
components or to evaluate the possibility of reaction between the phases
represented. The principal use of the phase rule in physical chemistry is
to determine the variance or the number of coexisting phases in limited
systems. In metamorphic petrology the problem is a different one. We
are concerned rather with comparing the possible assemblages of mineral
phases that can develop in all common rocks (*i.e.*, within a very broad
multicomponent system) over a certain range of pressures and temperatures. We are also concerned with the question of whether equilibrium
has or has not been reached in the development of a particular mineral
assemblage.

The procedure is as follows: First the value of c is taken as the minimum number of independent components, in terms of which may be
defined all mineral phases observed to occur within the broad system
comprised by common rocks, under a given set of metamorphic conditions. The composition of silicate minerals is conventionally stated in
terms of 10 main constituent oxides, namely: SiO_2, TiO_2, Al_2O_3, Fe_2O_3,
FeO, MgO, CaO, Na_2O, K_2O, H_2O. Allowing for the fact that Na_2O
enters into plagioclase, that Fe_2O_3 is capable of being accommodated as
a substitute for Al_2O_3 in minerals such as cordierite, and that FeO substitutes for MgO, and assuming that $(Si, Ti)O_2$ is one component, it is
possible to write the number of independent components as six. It does
not matter precisely which six are selected, but one possible grouping

(cf. page 504) would be as follows: (Si, Ti)O_2; $K_2OAl_2O_3$; (Ca, Na$_2$)OAl_2O_3; (Al, Fe)$_2O_3$ (in excess of the Al_2O_3 already grouped with Na_2O, CaO, and K_2O); (Mg, Fe)O; H_2O. The observed mineral assemblages are now tabulated and compared in the light of the prediction, by the mineralogical phase rule, that the number of associated mineral phases in any equilibrium assemblage should not exceed six—provided the number of components has been correctly estimated as six. In the Oslo hornfelses this requirement is met; individual rocks consist of assemblages of two to six main minerals. This suggests, but does not prove, general attainment of equilibrium in the rocks in question, and correct diagnoses of the number of components in the whole broad system.

It is impossible to represent graphically all possible variations in mineralogy within a broad system of more than four components, and for clear representation on a planar phase diagram, the maximum number of components that can be treated is three. For some metamorphic provinces it is possible to treat a broad range of rocks in terms of three components, the relative proportions of which determine the observed variation in mineralogy. This is true of the Oslo hornfelses. In these Na_2O is present in but a single phase (plagioclase), the content of SiO_2 is so high that quartz is invariably present, and orthoclase is a possible member of every assemblage. The nature and relative amounts of the remaining mineral phases, never more than three in number, are determined by the relative proportions of the three components (Al, Fe)$_2O_3$, CaO, and (Mg, Fe)O. On the corresponding phase diagram (Fig. 75, page 521) is plotted the composition of each of the observed mineral phases (with the exception of quartz, albite, and orthoclase). If the components have been correctly diagnosed, and if equilibrium is general in the Oslo hornfelses, it will be possible to divide the three-component diagram (Fig. 75) into triangular fields such that any rock whose chemical composition falls within a given triangle will be found to consist of the three mineral phases located at the apices of that triangle. This is the main application of the mineralogical phase rule. It is found to hold good for the hornfelses of the Oslo aureole and for high-temperature hornfelses in certain other regions as well. Similarly, if there should be four independent components—expressed on a tetrahedral model—it should be possible to group the observed mineral assemblages into pyramidal fields of four phases each.[17]

Some exceptions to the general rules outlined above have been noted. For example, in some hornfelses which otherwise resemble those of the Oslo aureole, almandine garnet has been found as a common associate

[17] T. F. W. Barth, Structural and petrologic studies in Dutchess County, New York, *Geol. Soc. America Bull.*, vol. 47, pp. 819–822, 1936; J. B. Thompson, The graphical analysis of mineral assemblages in pelitic schists, *Am. Mineralogist*, vol. 42, pp. 842–858, 1957.

of the assemblage andalusite-cordierite-plagioclase; again, in the Oslo hornfelses, biotite often appears as a fourth phase in assemblages such as andalusite-cordierite-hypersthene. This could be interpreted as evidence of disequilibrium—though this is most unlikely in view of the regularity with which such associations recur in different metamorphic regions. The most likely interpretation is that the number of components that has been deduced is too few. In almandine, the FeO/MgO ratio is consistently high; in cordierite of hornfelses, the same ratio is invariably low. (Fe, Mg)O cannot therefore be considered as a single component in cordierite-bearing rocks relatively rich in iron. The excess iron over that which can be taken up in cordierite appears as an additional phase— garnet or, if there is sufficient K_2O, biotite.[18]   There are other conditions of metamorphism for which it is impossible to treat the full range of common rocks in terms of three or even four components. For example, in greenschists formed by low-temperature regional metamorphism, the compositions of such common mineral phases as chlorites, stilpnomelanes, spessartite, actinolite, magnetite, and chloritoid can be correlated only by assuming that MgO, FeO, Fe_2O_3, MnO, and Mn_2O_3 may all play the part of independent components. In metamorphic rocks of all kinds the various degrees to which different elements may substitute for one another in the various stable mineral phases is a most important factor in determining the number of phases in mineral assemblages.

Mobile Components.   Experimental studies on petrologically significant phase equilibria have largely been carried out in closed systems. Metamorphic systems on the other hand are essentially closed for some components but open for others, notably water and carbon dioxide. On this basis Korzhinsky [19] distinguished between inert and mobile components in applying the mineralogical phase rule to metamorphic rocks (cf. page 31). He concluded that the maximum number of phases in equilibrium in a metamorphic system with a variance of 2 is equal to the number of inert components, and is not affected by ideally mobile components.

A similar conclusion has been reached by Thompson [20] from thermodynamic reasoning. If a component such as water or H_2S is free to enter or leave the system, its chemical potential is presumably fixed by conditions outside the system (e.g., conditions in the magma from which H_2O

[18] P. Eskola, *Die Entstehung der Gesteine* (Barth, Correns, Eskola), p. 317, Springer, Berlin, 1939.

[19] D. S. Korzhinsky, Mobility and inertness of components in metasomatosis, *Acad. Sci. U.S.S.R.*, Sér. geol., no. 1, pp. 35–60 (English summary, pp. 58–60), 1936; *Physicochemical Basis of the Analysis of the Paragenesis of Minerals* (Russian edition, 1957), English translation, pp. 61–64, Consultants Bureau, New York, 1959.

[20] J B. Thompson, The thermodynamic basis for the mineral facies concept, *Am. Jour. Sci.*, vol. 253, pp. 79–81, 1955.

issues), and is therefore no longer a variable to be determined from conditions inside the system. The statement that the component is mobile is equivalent to the statement that its chemical potential has a fixed, constant value that is independent of what happens in the system. As the variance of a system is the difference between the number of intensive variables and the number of relations to determine them, any additional condition reduces the variance by one unit. Thus, if m is the number of mobile components, the variance is reduced by m units and

$$w = c - m + 2 - \varphi$$

In particular, for a system with variance 2, the maximum number of phases is equal to p, where $p = c - m$ is the number of inert components. This is essentially Korzhinsky's conclusion.

As pointed out in the discussion of Goldschmidt's mineralogical phase rule, there is only a very small probability of ever finding rocks that formed under invariant or univariant conditions; for the probability of coexistence of precisely fixed temperatures, pressures, and concentrations at any point in the earth would be very small. Any effect that reduces the variance of a system also reduces the probability of its occurrence in nature. The reduction in variance resulting from the mobility of one or several of the components of a geological system must therefore be compensated by a corresponding increase in variance brought about by a reduction in the number of phases. We thus expect the number of phases to be reduced by 1 for each component that is, or becomes, mobile. Consider for example a system with components CaO, MgO, SiO_2, CO_2, H_2O (siliceous dolomite metamorphosed in the presence of water). If both CO_2 and H_2O are mobile, the maximum number of phases we could expect to find would be 3, although a great many more minerals individually could be stable (calcite, dolomite, forsterite, enstatite, diopside, serpentine, quartz, brucite). Four-phase assemblages could be stable if the variance were increased by allowing load pressure to differ from fluid pressure, or if either CO_2 or H_2O were not mobile. According to Goldschmidt's rule, 5 phases (including a pore fluid) could coexist if all components were inert.

Note that the phase rule, in its form (2.50), applies to open and closed systems as well; the special conditions just discussed refer to the case where the mobile component i is assumed to have the same chemical potential in both the system and its immediate surroundings, so that there is no tendency for i to move in or out of the system once the latter has reached internal equilibrium. How commonly such conditions exist is not known, a fact which militates against rigorous application of the phase rule to metamorphic paragenesis. An effect capable of qualitative assessment is reduction in the number of phases in the metamorphic assemblage; indeed many metasomatic rocks consist of but one or two minerals. Little is known as to the relative velocities of diffusion of ions of different kinds in the metamorphic environment. Clearly, these are liable to great variation, with variation in such factors as composition of the pore fluid, prevailing temperature and pressure, mineralogical environment of the fluids, and pore size. It is important, however, to note that the bulk composition of the rock may become modified and the resultant mineral assemblage tends to be simplified by outward diffusion of one or more of the original components. For example, low-grade metamorphism of basalt, which normally yields the assemblage albite-epidote-actinolite-chlorite-sphene, may lead to such contrasted assemblages as albite-epidote-chlorite-sphene or epidote-quartz if the system is open with regard to appropriate components. Conversely, free entry of fluids containing new components, or containing a high concentration of some component already present in the system, may lead to development of "abnormal" phases in the ultimate assemblage. For example, introduced boron may become fixed as tourmaline or axinite; fluorine may cause chondrodite to appear instead of forsterite, or topaz in place of andalusite.

METAMORPHIC FACIES

Definition.[21]   Shortly after Goldschmidt's account of metamorphism in the Oslo region appeared, Eskola published an important monograph on the relation between chemical and mineral composition of metamorphic rocks of the Orijärvi region of Finland. The same broad range of chemical composition was represented in the two areas. In Finland, as in Norway, there was a constant correlation between mineral and chemical composition. But Eskola recognized certain striking mineralogical differences between the Oslo paragenesis and that of Orijärvi, as set out below:

Oslo	*Orijärvi*
Potash feldspar + andalusite	= Muscovite
Potash feldspar + cordierite	= Biotite + muscovite
Potash feldspar + hypersthene + anorthite	= Biotite + hornblende
Hypersthene	= Anthophyllite

[21] P. Eskola, On the relation between chemical and mineralogical composition in the metamorphic rocks of the Orijärvi region, *Comm. géol. Finlande Bull. 44*, pp. 114–117, 1915; The mineral facies of rocks, *Norsk geol. tidsskr.*, vol. 6, pp. 143–194, 1920; C. E. Tilley, The facies classification of metamorphic rocks, *Geol. Mag.*, vol. 61, pp. 167–171, 1924; N. L. Bowen, Progressive metamorphism of siliceous limestone and dolomite, *Jour. Geology*, vol. 48, pp. 272–274, 1940; Fyfe, Turner, and Verhoogen, *op. cit.*, pp. 3–20, 1958; Korzhinsky, *op. cit.*, pp. 61–79, 1959.

Eskola framed his concept of metamorphic facies to explain such differences in metamorphic paragenesis, which he attributed to differences in the physical conditions (notably temperature and pressure) of metamorphism. In statements by Eskola and others regarding the nature of metamorphic facies, there has been some confusion between geological criteria and genetic (physical) interpretation of these.[22] Here we define a metamorphic facies, in purely geological terms, as a series of metamorphic mineral assemblages conforming to the following specifications:

1. Some or all of the assemblages are commonly associated in space and time (e.g., in an aureole or in a mappable metamorphic zone). The whole association tends to recur in other regions and in rocks of different ages.

2. The mineral composition of each assemblage is strictly a function of the bulk chemical composition as it now exists—regardless of possible effects of metasomatism.

3. The total number of essential mineral phases in common rocks of the facies as a whole is relatively small—perhaps a dozen. Each assemblage consists of a limited number (usually 2 to 6) of these phases.

4. There is no textural or other evidence of mutual replacement by minerals of the same facies.

It is the whole association of mineral assemblages that defines a metamorphic facies. Many of these individually are common to two or more facies; e.g., diopside-plagioclase-grossularite and quartz-plagioclase-cordierite-andalusite are common to the Oslo and Orijärvi parageneses. But a few assemblages are peculiar to each facies—and after one of these the facies customarily is named. We now recognize a pyroxene-hornfels facies (named after the diagnostic assemblage diopside-hypersthene-plagioclase) to cover the Oslo hornfelses and mineralogically similar rocks from other contact aureoles. The whole rock association of Orijärvi, and similar assemblages from elsewhere, are grouped in a hornblende-hornfels facies named after the characteristic assemblage hornblende-plagioclase.

Interpretation. In interpreting the physicochemical significance of facies the authors tentatively accept two assumptions made by Eskola: (1) each facies corresponds to metamorphism under some particular range of physical conditions; (2) the ideal mineral assemblages constituting a facies represent systems which reached equilibrium during metamorphism.

There is overwhelmingly strong geological evidence in support of these assumptions, for it is difficult to explain on any other basis the regularity of paragenesis and the consistent relation of mineral to chemical composition which are the essence of metamorphic facies as defined above. So

[22] This has been discussed more fully by Fyfe, Turner, and Verhoogen, *op. cit.*, pp. 9, 10, 18, 1958.

we envisage the course of metamorphic reaction as being determined essentially by the relative values of the free energies of the participating assemblages of phases. These, in turn, are controlled by physical variables, notably temperature and pressure. In recent years it has become increasingly apparent that there are several partially independent pressure variables: load pressure of grain on grain P_l; fluid pressure P_f; partial pressures of the principal components of the pore fluid—water pressure P_{H_2O}, carbon dioxide pressure P_{CO_2}, oxygen pressure P_{O_2}, and so on.

Graphic Representation of Facies Characteristics. The consistent correlation between mineralogical and chemical composition within any facies or subfacies can be demonstrated by phase diagrams drawn for three or four components. The simplest type is a triangular three-component diagram. The apices represent the three components, and any point within the triangle represents a unique chemical composition. Except in very simple systems—such as siliceous dolomitic limestones—it is impossible to treat the natural mineral assemblages in terms of only three components. Tetrahedral four-component diagrams,[23] and various ingenious projections of these onto a plane,[24] can be used for more rigorous treatment of phase equilibria; but they are more difficult to construct and will not be considered further here.

The three components upon which Eskola bases his triangular ACF diagrams are Al_2O_3, CaO, and (Mg, Fe)O. The effect of silica upon the possible mineral assemblages can be illustrated by comparing two ACF diagrams, one for rocks with excess SiO_2 crystallizing as free quartz (the most widely applicable case), and the other for rocks deficient in SiO_2. In the same way special diagrams could, where necessary, be constructed for rocks with excess K_2O, deficient H_2O, and so on. The following rules for calculating the percentages of the components A, C, and F from a chemical analysis of the rock are given by Eskola.[25]

To correct for the accessories ilmenite, sphene, and magnetite, the percentages (by weight) of these minerals are first estimated by micrometric analysis as i, s, and m respectively. Then from the percentage of FeO (by weight) subtract 50% i and 30% m; likewise subtract 30% s from CaO and 70% m from Fe_2O_3. The percentages of the various oxides, corrected as above, are now calculated as molecular percentages, and CaO is then further corrected for calcite and apatite by subtracting molecular amounts equivalent to 3 P_2O_5 + CO_2. A, C, and F can now be reckoned as follows and are finally recalculated so that A + C + F = 100: A = Al_2O_3 + Fe_2O_3 − (Na_2O + K_2O); C = CaO; F = MgO + FeO + MnO. Note that A is given by the total alumina

[23] Cf. Barth, *op. cit.*, pp. 819–822, 1936.
[24] For an elegant treatment of pelitic rocks in terms of four components, see Thompson, *op. cit.*, 1957.
[25] Eskola, *op. cit.*, p. 347, 1939.

and ferric iron, less the quantity that would be required to combine with total alkali as feldspar.

Comparison of the composition diagrams prepared for rocks of the same general facies, as developed in different regions, has revealed the existence of minor, but consistently recurring, departures from the standard typical metamorphic parageneses. The alternative assemblages thus recognized, provided they represent true equilibria, may be used to define subfacies of the facies in question. For instance, Eskola's amphibolite facies is founded on the critical assemblage hornblende-plagioclase, which is stable, in rocks of appropriate composition, over a wide range of temperature and pressure. Within a small part of this range, corresponding to conditions in the staurolite zone of regional metamorphism, the pair staurolite-almandine is stable in rather ferruginous pelitic rocks; under other conditions, represented by the kyanite zone, the pair kyanite-almandine appears in rocks of the same composition. Both pairs of minerals are associated in the field with hornblende-plagioclase rocks. They may therefore be used to define subfacies of the amphibolite facies. In this book Eskola's amphibolite facies has been divided into two independent facies one of which has several subfacies.

Disequilibrium Assemblages. Metamorphic facies are based upon petrographically recognizable mineral assemblages which we tentatively interpret as equilibria. This is not tantamount to assuming general prevalence of equilibrium in metamorphic rocks. Rather we interpret a particular widely developed assemblage of essential minerals as being in equilibrium e.g., quartz-muscovite-biotite-almandine in pelitic schists of the garnet zone. Other minerals, locally developed in shears, veins, and pseudomorphs can be recognized by petrographic criteria as products of incipient reaction leading toward an assemblage belonging to another facies. For example, chlorite may partially replace garnet or biotite in the pelitic assemblage just mentioned. Where assemblages belonging to two facies are associated in one rock or in one small area, their mutual relations in space and time—which may be determined by petrographic and field observations—afford valuable data on which to reconstruct the course of metamorphism.

High-grade metamorphic assemblages composed mainly of anhydrous minerals commonly show local or partial replacement by low-grade assemblages, e.g., pyroxene → amphibole; garnet → chlorite; plagioclase → albite + clinozoisite. Such reactions probably occur during cooling following the main high-temperature phase of metamorphism; and they have been limited by lack of water, which could only become available by diffusion inward from some external source. Low-grade metamorphic assemblages on the other hand commonly include "unstable relics" which have survived from the premetamorphic condition, probably because of low reaction velocities in the lower range of metamorphic temperatures. Thus in low-grade schists of the chlorite zone in New Zealand, unstable relics of augite and hornblende (constituents of the parent tuffaceous sand) survive in a metamorphic assemblage albite-epidote-chlorite-actinolite, typical of the greenschist facies.

Physicochemical Significance of the Isograd.[26] The physical variables that could affect the first appearance of an index mineral in rocks of a given composition and thus determine the configuration of the corresponding isograd are T, P_l, P_f, P_{H_2O} and perhaps others like P_{O_2}, P_{CO_2} and P_F. The geologically possible combinations of these are so diverse that one might almost expect each metamorphic paragenesis unique. That facies do recur t they are few in number .nan a dozen) can only mean that the physical variables enumerated above are not wholly independent in the geological environment. The number of common patterns of physical conditions in metamorphism must be rather small.

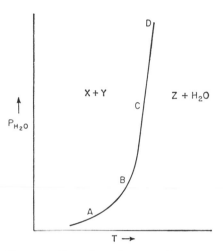

Fig. 66. Hypothetical curve of univariant equilibrium for reaction $X + Y = Z + H_2O$. AB = conditions of contact metamorphism; CD = conditions of regional metamorphism.

Although the geothermal gradient is not everywhere the same, temperature and load pressure generally increase simultaneously in regional metamorphism. And at depths of more than a few miles fluid pressure probably is close to load pressure. As a first approximation it seems reasonable to assume that for deep-seated metamorphism $P_f = P_l$. A further approximation is that for metamorphic reactions of hydration and dehydration $P_{H_2O} = P_f$; and that for metamorphic reactions involving decarbonation of limestone and dolomite, $P_{CO_2} = P_f$.

The isograds drawn for progressive regional metamorphism of pelitic rocks—as in the Dalradian schists of Scotland—mostly represent dehydration reactions; e.g., appearance of almandine at the expense of chlorite, reaction between muscovite and quartz to give sillimanite and potash feldspar. The physical conditions to which such reactions are most sensitive are T and P_{H_2O}. The corresponding curves of univariant equilibrium, which also depict the possible range of physical conditions represented

[26] Fyfe, Turner, and Verhoogen, *op. cit.*, pp. 181–185, 237–239, 1958.

by the corresponding isograds will have the general shape of Fig. 66.   At high pressures of regional metamorphism (C, D in Fig. 66), where P_{H_2O} approximates P_l, reactions are relatively insensitive to pressure, and isograds are approximately isothermals.   For contact metamorphism at low pressures the situation is quite different: reactions are equally sensitive to P_{H_2O} and to T, and P_{H_2O} and P_l are to a large degree independent variables.   The isograds of contact aureoles and the facies of contact metamorphism cover a diversity of physical conditions which can be only partially represented on a diagram such as Fig. 66 (A, B).

In the light of a rapidly growing volume of experimental data relating to simple metamorphic systems, some facies boundaries and isograds can now be evaluated in terms of T, P_l, P_{H_2O}, and P_{CO_2}.[27]   The inferences drawn from experiment do not always agree with those based on the voluminous data of petrology and field geology.   Characteristic of modern metamorphic petrology is a general reassessment and refinement of geological and experimental observations with a view to resolving mutual conflicts.

[27] For a survey of the situation as it existed at the end of 1955, see Fyfe, Turner, and Verhoogen, *op. cit.*, 1958.

CHAPTER 19

Facies of Contact Metamorphism

THE EIGHT METAMORPHIC FACIES OF ESKOLA

Five metamorphic facies were originally defined by Eskola; the sanidinite (low pressure), hornfels (moderate pressure), and eclogite facies (high pressure), all representing high-temperature metamorphism; the amphibolite (moderate temperature) and greenschist facies (low tempeature), both corresponding to conditions of moderate pressure. In Eskola's latest statement, which takes into account the opinions of the various writers cited in the previous chapter, three additional facies are incorporated into the scheme, a number of subfacies are defined, and the original hornfels facies is renamed the pyroxene-hornfels facies.[1] Table 38 indicates Eskola's tentative views concerning the relative positions of

TABLE 38. RELATION OF METAMORPHIC FACIES AND IGNEOUS FACIES (GIVEN IN ITALICS) TO TEMPERATURE AND PRESSURE ACCORDING TO P. ESKOLA, 1939

Temperature increasing ⟶

	Development of zeolites in igneous rocks			Sanidinite facies (*diabase facies*)
Pressure increasing ↓	Greenschist facies	Epidote-amphibolite facies	Amphibolite facies (*hornblende-gabbro facies*)	Pyroxene-hornfels facies (*gabbro facies*)
				Granulite facies
		Glaucophane-schist facies		Eclogite facies (*eclogite facies*)

[1] P. Eskola, *Die Entstehung der Gesteine* (Barth, Correns, Eskola), p. 344, Springer, Berlin, 1939.

the various facies in relation to physical conditions; the igneous equivalents, where such exist, are shown in italics for comparison.

Eskola's scheme was expanded and modified first by Turner [2] and more recently by Fyfe, Turner, and Verhoogen,[3] whose classification is followed in this book. Eskola's terminology of 1939 is retained as nearly as possible.

THE FACIES OF CONTACT METAMORPHISM

On the basis of petrographic and field criteria most metamorphic facies fall into one or other of two classes—those of contact and of regional metamorphism. This twofold division reflects a real difference between two common environments of metamorphism. The one typically develops locally, at relatively shallow depths, late in orogeny. The other is regional, deep-seated, and more or less synchronous with the climax of orogeny. There is a range of metamorphic temperatures and pressures that spans the gap between these two environments; and corresponding transitional facies will ultimately be established when the mineralogy of rocks so metamorphosed has been more fully described.

The physical conditions of contact metamorphism are believed to encompass a broad range of temperature (limited above by fusion and below by kinetics of reaction) and relatively low pressures consistent with shallow depth. We envisage temperatures ranging from perhaps 300° to 800°C. (rarely 1000°C.), and load pressures of the order of 100 to 3,000 bars. Water pressures vary considerably. In permeable rocks at shallow depths they could approximate the hydrostatic head of a column of water reaching to the surface; but near a boiling mass of water-saturated magma the water pressure could build up to values exceeding the load pressure and limited only by the breaking strength of the rock.

The facies of contact metamorphism are listed below in order of increasing temperature (at constant water pressure) or of decreasing water pressure (at constant temperature):

1. Albite-epidote-hornfels (formerly albite-epidote-amphibolite facies; actinolite-epidote-hornfels subfacies)

2. Hornblende-hornfels (formerly amphibolite facies; cordierite-anthophyllite subfacies)

3. Pyroxene hornfels

4. Sanidinite

[2] F. J. Turner, Mineralogical and structural evolution of the metamorphic rocks, *Geol. Soc. America Mem. 30*, pp. 61–107, 1948.

[3] W. S. Fyfe, F. J. Turner, and J. Verhoogen, Metamorphic reactions and metamorphic facies, *Geol. Soc. America Mem. 72*, pp. 201–202, 1958.

ALBITE-EPIDOTE-HORNFELS FACIES [4]

The albite-epidote-hornfels facies typically occurs on the outer margins of zoned aureoles and passes inward into rocks of the hornblende-hornfels facies. Rarely it may appear in a weakly developed aureole consisting of only the one facies.[5]   Because of the low temperature of metamorphism recrystallization tends to be imperfect, and the paragenesis is obscured by persistence of unstable relics and by the fine grain of the metamorphic products.   Mineral assemblages have much in common with those of the greenschist facies.   Indeed it is only the difference in field

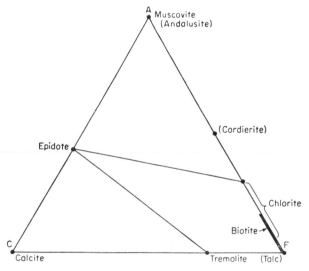

Fig. 67.   Albite-epidote-hornfels facies: ACF diagram for assemblages with excess SiO_2. Quartz and albite are possible members of each assemblage. Minerals in brackets are stable only in K_2O-deficient assemblages (lacking potash feldspar).

occurrence that warrants erection of a separate albite-epidote-hornfels facies.

Some typical assemblages for rocks with excess SiO_2 are as follows (Fig. 67):

A. Pelitic hornfelses and spotted slates:

Quartz-albite-muscovite-biotite (with cordierite or andalusite)

Quartz-albite-epidote-micas

[4] J. S. Flett, The geology of the country round Bodmin and St. Austell, *England Geol. Survey Mem.*, pp. 99–100, 1909; Fyfe, Turner, and Verhoogen, *op. cit.*, pp. 203–205, 1958.

[5] T. DeBooy, Géologie de la région de Francardo (Corse), Doctoral thesis, University of Amsterdam, pp. 26–29, 1954.

B. Basic hornfelses: [6]
   Albite-epidote-actinolite-chlorite-quartz
C. Marbles (Fig. 68):
   Calcite-talc-quartz [7]
   Calcite-tremolite-quartz
   Calcite-spinel-diaspore

In silica-deficient marbles dolomite may appear in association with calcite and either talc or tremolite (cf. Fig. 68). The incoming of

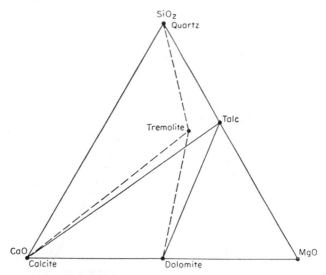

FIG. 68.   Albite-epidote-hornfels facies: assemblages in derivatives of siliceous limestones and dolomites.   Full lines = low temperatures or high P_{CO_2}; broken lines = higher temperatures or lower P_{CO_2}.   For intermediate conditions calcite-talc-tremolite is a possible assemblage.

diopside in such assemblages as calcite-tremolite-diopside and calcite-diopside-quartz marks a transition to the hornblende-hornfels facies in which other diopside-bearing assemblages are typical.

HORNBLENDE-HORNFELS FACIES [8]

Field Occurrence.   Rocks of the hornblende-hornfels facies occur extensively in contact aureoles.   Where these are zoned, as at Oslo, an inner zone of higher grade (pyroxene-hornfels facies) may separate the horn-

[6] *Ibid.*, pp. 27, 28.

[7] C. E. Tilley, Earlier stages in the metamorphism of siliceous dolomites, *Mineralog. Mag.*, vol. 28, pp. 272–276, 1948.

[8] Fyfe, Turner, and Verhoogen, *op. cit.*, pp. 205–211, 1958.

blende-hornfelses and their associates from the igneous contact. Xenoliths in granite and granodiorite commonly consist of assemblages of the
hornblende-hornfels facies. Rarely, as in Aberdeenshire,[9] rocks of this
facies occur on a regional scale and are not obviously related to intrusive
igneous bodies.

Diagnostic Characteristics. A very wide range of paragenesis, characterized by development of the assemblage hornblende-plagioclase in
basic rocks, was embraced by Eskola's amphibolite facies. One subfacies,
formerly named after the characteristic hornfels assemblage cordierite-
anthophyllite, is here elevated to the status of an independent facies now

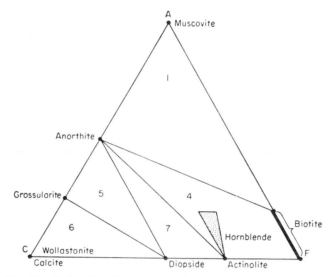

Fig. 69. Hornblende-hornfels facies: ACF diagram for rocks with excess SiO_2 and
K_2O. Quartz and microcline are possible members of each assemblage.

called the hornblende-hornfels facies. The others are products of regional metamorphism and are here grouped collectively in an almandine-
amphibolite facies. The classic description of the hornblende-hornfels
facies is Eskola's account of contact metamorphism in the Orijärvi region
of Finland.[10]

From the albite-epidote-hornfels facies the hornblende-hornfels facies

[9] A. Harker, *Metamorphism*, pp. 230–235, Methuen, London, 1932; H. H. Read,
Metamorphism and migmatization in the Ythan Valley, Aberdeenshire, *Edinburgh
Geol. Soc. Trans.*, vol. 15, pp. 265–279, 1952.

[10] P. Eskola, On the petrology of the Orijärvi region in southwestern Finland,
Comm. géol. Finlande Bull. 40, 1914; On the relation between chemical and min-
eralogical composition in the metamorphic rocks of the Orijärvi region, *Comm. géol.
Finlande Bull. 44*, 1915.

differs in the development of hornblende-plagioclase rather than albite-epidote-actinolite, and in the presence of diopside-, forsterite-, and grossu-larite-bearing assemblages in calcareous rocks. From the pyroxene-hornfels facies it differs in the widespread development of hornblende (instead of diopside-hypersthene) and of micas (instead of andalusite or cordierite associated with potash feldspar), and in rarity or absence of wollastonite. Presence of andalusite, cordierite, or anthophyllite in potash-deficient rocks, absence of epidote and kyanite, and rarity of almandine and staurolite, distinguish the hornblende-hornfels from the almandine-amphibolite facies.

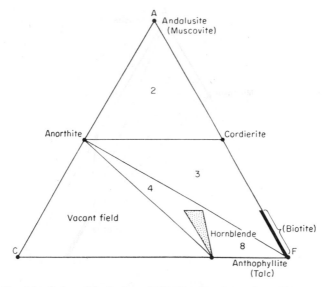

Fig. 70.   Hornblende-hornfels facies: ACF diagram for rocks with excess SiO₂ and deficient in K₂O.

Mineral Assemblages.[11]   *Assemblages with Excess* SiO₂.   Figures 69 to 71 illustrate the principal mineral assemblages with excess SiO₂. Some of the commonest, numbered to correspond with triangular fields of Figs. 69 to 71, are:

A. Pelitic
  2. Quartz-muscovite-andalusite-cordierite (-plagioclase)
  3. Quartz-muscovite-biotite-cordierite (-plagioclase)

[11] Eskola, *op. cit.*, 1914, 1915; G. A. Joplin, A comparison of the Rydal and Hartley exogenous contact zones, *Linnean Soc. New South Wales Proc.*, vol. 61, pp. 151–154, 1936; C. E. Tilley, Anthophyllite-cordierite granulites of the Lizard, *Geol. Mag.*, vol. 74, pp. 300–309, 1937; Y. Seki, Petrological study of hornfelses in the median zone of Kitakami mountainland, Iwate Prefecture (Japan), *Saitama University Sci. Repts.*, Ser. B, vol. 2, no. 3, pp. 307–361, 1957.

B. Quartzo-feldspathic
   1. Quartz-plagioclase-microcline-muscovite-biotite
   4. Quartz-plagioclase-microcline-biotite-hornblende
C. Calcareous
   5. Diopside-grossularite-plagioclase-quartz
   6a. Diopside-grossularite-wollastonite-quartz
   6b. Diopside-grossularite-calcite-quartz
D. Basic
   4. Plagioclase-hornblende (-quartz-biotite)
   7. Plagioclase-hornblende-diopside (-quartz-biotite)

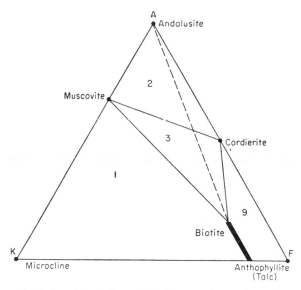

Fig. 71.   Hornblende-hornfels facies: AKF diagram for rocks with excess SiO_2 and Al_2O_3 (triangle 1, Fig. 69; triangles 2, 3, Fig. 70).   Full lines = Orijärvi assemblages; broken lines denote stability of andalusite-biotite elsewhere. $A = Al_2O_3 - (CaO + Na_2O + K_2O)$; $K = K_2O$; $F = (FeO + MgO)$.

E. Magnesian
   8. Talc-tremolite (-quartz)
   9. Anthophyllite-cordierite (-quartz)

Neither cordierite nor talc can accommodate more than a limited amount of FeO substituting for MgO.   This accounts for the presence of an additional iron-bearing phase—almandine [12] or anthophyllite—in some cordierite-bearing assemblages.   Spessartite-almandine appears in rocks containing some manganese.[12]

Andalusite and cordierite are restricted to rocks deficient in potash,

[12] C. E. Tilley, On garnet in pelitic contact zones, *Mineralog. Mag.*, vol. 21, pp. 47–50, 1926.

i.e., those containing no potash feldspar.   Their occurrence is thus more limited than in the pyroxene-hornfels facies where both minerals may co-exist with orthoclase or microcline.   Sillimanite may take the place of andalusite in the immediate vicinity of granite contacts.   Wollastonite—a common mineral in the pyroxene-hornfels facies—typically is not found in the hornblende-hornfels facies.   Its occurrence is confined to products of metamorphism at high temperatures or low pressures of CO_2 transi-tional to those of the pyroxene-hornfels facies.

Assemblages Deficient in SiO_2.   The most widely developed silica-

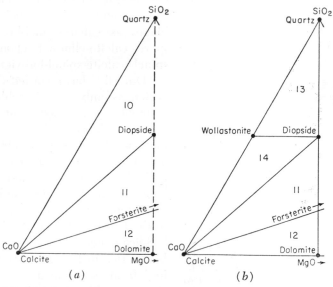

Fig. 72.   Hornblende-hornfels facies: assemblages in derivatives of siliceous lime-stones and dolomites; water absent.   (*a*) At low temperature, high P_{CO_2}; (*b*) At high temperature, low P_{CO_2}.

deficient assemblages are derivatives of slightly siliceous or marly lime-stones and dolomites.   Some magnesian hornfelses also fall in this cate-gory.   Some typical assemblages (Figs. 72, 73) are as follows:

A. Calcareous
  11. Calcite-diopside-forsterite
  12. Calcite-dolomite-forsterite
  14. Calcite-diopside-wollastonite

B. Magnesian
  17a. Forsterite-talc-tremolite-clinochlore
  17b. Anthophyllite-clinochlore
  18. Forsterite-tremolite-diopside
  19. Forsterite-brucite-spinel (-diopside)

Assemblage 14 is formed near the upper temperature limit of the facies and overlaps into the pyroxene-hornfels facies, of which it is more typical. Other calcareous assemblages formed from dolomitic limestones are 18 and 19, with calcite as an additional phase. They can form only at moderate partial pressures of both water and CO_2, and so correspond to pressure conditions intermediate between those of Fig. 72 ($P_{H_2O} = 0$) and 73 ($P_{CO_2} = 0$). Further variety is introduced by the presence of Al_2O_3 in the form of clay impurity, and of fluorine or boron as mobile components in the fluid phase. To this category belong such assemblages as calcite-spinel-brucite-diaspore, calcite-clintonite-chondrodite-spinel, calcite-spinel-ludwigite.

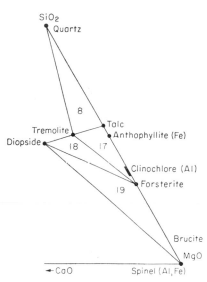

FIG. 73. Hornblende-hornfels facies: magnesian assemblages stable at 500° to 600°C. and $P_{H_2O} = 1,000$ to 2,000 bars. (*After H. S. Yoder.*)

Durrell [13] has recorded the magnesian assemblages clinochlore-talc-actinolite and forsterite-talc-tremolite-clinochlore (17a) in contact-metamorphosed serpentinites of the Sierra Nevada. Associated basic rocks have been converted to hornblende-diopside-plagioclase hornfels; so all three assemblages must belong to the hornblende-hornfels facies.

Silica-deficient magnesian assemblages rich in FeO and Al_2O_3 resulting from metasomatism of basic igneous rocks in contact aureoles include the following: [14]

Anthophyllite-cordierite-spinel-plagioclase
Anthophyllite-cordierite-spinel-diaspore
Hornblende-plagioclase-spinel-diaspore

It is probable that diaspore, and likewise clinochlore, which sometimes appears as an additional phase, are products of incomplete reaction at relatively high water pressures during cooling, for the number of phases in some assemblages containing these minerals seems too great for divariant equilibrium. Moreover the assemblage diaspore-spinel is stable only at temperatures below about 430°C. (see page 520).

[13] C. Durrell, Metamorphism in the southern Sierra Nevada northeast of Visalia, California, *California Univ. Dept. Geol. Sci. California,* vol. 25, pp. 65–88, 1940.

[14] C. E. Tilley, Metasomatism associated with the greenstone hornfelses of Kenidjack and Botallack, Cornwall, *Mineralog. Mag.,* vol. 24, pp. 181–202, 1935; cf. also J. A. W. Bugge, Geological and petrological observations in the Kongsberg-Bamble formation, *Norges geol. undersökelse,* vol. 160, pp. 82–103, 1943.

Physical Conditions of Metamorphism.   On the basis of mineralogical characteristics and field occurrence Eskola attributed the hornblende-hornfels facies of Orijärvi to lower temperatures and higher pressures of metamorphism than the pyroxene-hornfels facies of Oslo.   This general conclusion remains valid today.   A more accurate picture of the physical conditions of metamorphism in the hornblende hornfels facies emerges from experimental data on carbonates, micas, and clays.

If siliceous dolomites and limestones of various compositions are subjected to increasing temperatures at some constant value of CO_2 pressure (say 300 bars), a series of reactions involving progressive elimination of CO_2 from the assemblage of solid phases takes place.   To account for natural assemblages in rocks of such composition, and the order of their appearance in zoned aureoles, Bowen [15] drew up a series of 13 reactions of this kind.   Corresponding reaction curves (curves of univariant equilibrium) on a pressure-temperature diagram were assumed to be of the same general shape as that thermodynamically constructed by Goldschmidt [16] for the reaction

$$CaCO_3 + SiO_2 \rightleftarrows CaSiO_3 + CO_2$$

Calcite   Wollastonite

Each of Bowen's 13 reactions or steps of metamorphism is marked by disappearance of some mineral or assemblage which is stable only below the temperature of the reaction in question:

Below step 1: dolomite-quartz
Below step 2: dolomite-tremolite
Below step 3: calcite-tremolite-quartz
Below step 4: calcite-tremolite
Below step 5: dolomite
Below step 6: calcite-quartz
Below step 7: calcite-forsterite-diopside
Below step 8: calcite-diopside
Below step 9: calcite-forsterite
Below step 10: calcite-wollastonite
Below step 11: calcite-akermanite
Below step 12: spurrite-wollastonite
Below step 13: spurrite-akermanite

This list does not include certain minerals that remain stable at all temperatures above that of the step at which they first appear.   To complete

[15] N. L. Bowen, Progressive metamorphism of siliceous limestone and dolomite, *Jour. Geology*, vol. 48, pp. 225–274, 1940.

[16] V. M. Goldschmidt, Die Gesetze der Gesteinsmetamorphose mit Beispielen aus der Geologie des Südlichen Norwegens, *Kristiania Vidensk. Skr., I Math.-Naturv. Kl.* 22, 1912.   This curve has since been revised by A. Danielsson, Das Calcit-Wollastonit-gleichgewichte, *Geochim. et Cosmochim. Acta*, vol. 1, pp. 55–69, 1950.

the picture therefore, the following list of minerals formed at successive steps in order of increasing temperature is added:

At step 1: Tremolite $[Ca_2Mg_5Si_8O_{22}(OH)_2]$

At step 2: Forsterite (Mg_2SiO_4)

At step 3: Diopside $(CaMgSi_2O_6)$

At step 5: Periclase (MgO)

At step 6: Wollastonite $(CaSiO_3)$

At step 7: Monticellite $(CaMgSiO_4)$

At step 8: Akermanite $(Ca_2MgSi_2O_7)$

At step 10: Spurrite $(CaCO_3 \cdot 2Ca_2SiO_4)$

At step 11: Merwinite $[Ca_3Mg(SiO_4)_2]$

At step 12: Larnite (Ca_2SiO_4)

Reactions corresponding to the first six steps are:

1. Dolomite + quartz → calcite + tremolite + CO_2
2. Dolomite + tremolite → forsterite +calcite + CO_2
3. Calcite + tremolite + quartz → diopside + CO_2
4. Calcite + tremolite → diopside + forsterite + CO_2
5. Dolomite → periclase + calcite + CO_2
6. Calcite + quartz → wollastonite + CO_2

Hydroxyl-bearing minerals—tremolite, talc, brucite—can form only if the partial pressure of water in the gas phase exceeds some limiting value. For tremolite this must be low, for it is one of the commonest minerals of contact metamorphism. This is why Bowen has treated tremolite as if it were an anhydrous phase. Talc and brucite are much more restricted in their occurrence and so are not included above. We note however that if partial pressures of water are high enough talc appears even earlier than tremolite (step 1).[17]

Curves of univariant equilibrium based on thermodynamic data or on experimental synthesis are now available for several of Bowen's 13 steps. Three of these are shown in Fig. 74:

$$(2)^{18} \text{ Dolomite} + \text{silica} \rightleftarrows \text{calcite} + \text{forsterite} + CO_2$$

This reaction in complete absence of water presumably occurs under conditions close to those of Bowen's step 2 which is possible only in a hydrous system. Both mark the first appearance of forsterite. They must be close to the low-temperature boundary of the hornblende-hornfels facies.

$$(6)^{19} \text{ Calcite} + \text{silica} \rightleftarrows \text{wollastonite} + CO_2 \quad (\text{Bowen's step 6})$$

[17] Tilley, op. cit., 1948.

[18] W. F. Weeks, A thermochemical study of equilibrium relations during metamorphism of siliceous carbonate rocks, Jour. Geology, vol. 64, pp. 259–261, 1956.

[19] Danielsson, op. cit., 1950.

The fact that wollastonite marbles are sometimes associated with horn-blende hornfelses shows that the high-temperature boundary lies some-where to the right of step 6.

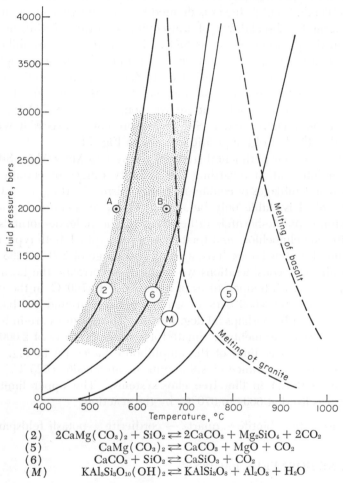

(2) $2CaMg(CO_3)_2 + SiO_2 \rightleftarrows 2CaCO_3 + Mg_2SiO_4 + 2CO_2$
(5) $CaMg(CO_3)_2 \rightleftarrows CaCO_3 + MgO + CO_2$
(6) $CaCO_3 + SiO_2 \rightleftarrows CaSiO_3 + CO_2$
(M) $KAl_3Si_3O_{10}(OH)_2 \rightleftarrows KAlSi_3O_8 + Al_2O_3 + H_2O$

FIG. 74. Curves of univariant equilibrium for four reactions relating to the horn-blende-hornfels facies. For 2, 5, and 6, $P_{CO_2} = P_{fluid}$; for M and the two melting curves $P_{H_2O} = P_{fluid}$. Field of the hornblende-hornfels facies (boundaries not defined) is stippled.

(5)[20] Dolomite $\rightleftarrows$ periclase + calcite + CO_2 (Bowen's step 5)

Experimental studies of dissociation of dolomite have shown that this reaction occurs at temperatures considerably greater than those necessary

[20] R. I. Harker and O. F. Tuttle, Studies in the system CaO-MgO-CO₂, Part I, *Am. Jour. Sci.*, vol. 253, pp. 209–224, 1955.

to give wollastonite in siliceous limestone. The order of Bowen's steps 5 and 6 thus is reversed. Periclase, except perhaps at very low pressures of CO_2, seems to be a mineral of the pyroxene-hornfels facies.

The data of petrography show that the high-temperature boundary of the hornblende-hornfels facies is defined by three nearly coincident curves representing the breakdown of muscovite, iron-rich biotite, and hornblende in the presence of excess SiO_2. Pure magnesian amphiboles and micas remain stable at temperatures several hundred degrees higher; but we have little precise information as to the limits of stability of metamorphic ferruginous biotites and hornblendes. The curve for destruction of muscovite in a silica-deficient environment (M in Fig. 74)[21] is close to the wollastonite curve; and if SiO_2 were present in excess it would be displaced to the left of the position shown in Fig. 74.

Finally Yoder's experimental study of the system $MgO-Al_2O_3-SiO_2-H_2O$ has given information regarding the stability ranges of certain natural magnesian assemblages of contact metamorphism.[22] At $P_{H_2O} = 1,070$ bars diaspore-spinel is stable only below 430°C.—a temperature appropriate to the albite-epidote-hornfels rather than the hornblende-hornfels facies. Forsterite-talc-clinochlore and forsterite-brucite-spinel, both typical of the hornblende-hornfels facies, have a combined range of 500° to 665°C.

From all these considerations we tentatively correlate the hornblende-hornfels facies with temperatures of about 550° to 700°C. in the pressure range $P_{H_2O} = 1,000–3,000$ bars; at water pressures around 500 bars, temperatures would be perhaps 50 deg. lower. These figures are in harmony with Winkler's experimental estimate for a water pressure of 2,000 bars.[23] He placed the lower limit of the amphibolite facies (A in Fig. 74) between the first appearance of sillimanite or andalusite (525°C.) and of cordierite (560°C.) in lime-free clay systems. The upper limit (B in Fig. 74) was fixed at 660° to 670°C. by the reaction

$$\text{Muscovite} + \text{biotite} + \text{quartz} \rightarrow \text{cordierite} + \text{potash feldspar.}$$

PYROXENE-HORNFELS FACIES [24]

Field Occurrence and Diagnostic Characteristics. Rocks of the pyroxene-hornfels facies are found in the innermost zones of some contact

[21] H. S. Yoder and H. P. Eugster, Synthetic and natural muscovites, *Geochim. et Cosmochim. Acta*, vol. 8, pp. 225–280, 1955.

[22] H. S. Yoder, The $MgO-Al_2O_3-SiO_2-H_2O$ system and the related metamorphic facies, *Am. Jour. Sci.*, Bowen vol., pp. 569–627, 1952 (especially pp. 600–601, Figs. 12–14).

[23] H. G. F. Winkler, Experimentelle Gesteinsmetamorphose, I, *Geochim. et Cosmochim. Acta*, vol. 13, pp. 42–69, 1957.

[24] Fyfe, Turner, and Verhoogen, *op. cit.*, pp. 211–213, 1958.

aureoles. Their development is favored especially by shallow depths (where pressures are low) and proximity to intrusions of basic magma (where temperatures are high).

Mineralogical characteristics are as follows: association of andalusite (sillimanite) and cordierite with potash feldspar in pelitic rocks; presence of hypersthene in magnesian rocks and of wollastonite in marbles and skarns; restricted occurrence of biotite and almandine; absence of hornblende and muscovite. These collectively distinguish the pyroxene-hornfels from the hornblende-hornfels facies.

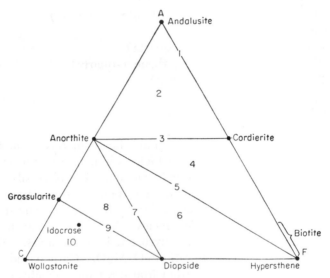

FIG. 75. Pyroxene-hornfels facies: ACF diagram for rocks with excess SiO_2. Quartz and orthoclase are possible members of each assemblage. (*After V. M. Goldschmidt.*)

Mineral Assemblages.[25] *Assemblages with Excess* SiO_2. Goldschmidt recognized ten classes of quartz-bearing hornfelses in the Oslo region (1 to 10, Fig. 75), and to these Tilley added two noncalcareous assemblages (1A, 1B) from the Comrie aureole of Scotland. The principal

[25] V. M. Goldschmidt, Die Kontaktmetamorphose im Kristianiagebiet, *Kristiania Vidensk. Skr., I. Math.-Naturv. Kl. 11*, 1911; C. E. Tilley, Contact metamorphism in the Comrie area of the Perthshire Highlands, *Geol. Soc. London Quart. Jour.*, vol. 80, pp. 22–71, 1924; Contact-metamorphic assemblages in the system CaO-MgO-Al_2O_3-SiO_2, *Geol. Mag.*, vol. 62, pp. 363–367, 1925; G. D. Osborne, The contact metamorphism and related phenomena in the neighborhood of Marulan, New South Wales, *Geol. Mag.*, vol. 68, pp. 289–314, 1931; T. Watanabe, Geology and mineralization of the Suian district, Tyôsen, Korea, *Hokkaido Imp. Univ. Fac. Sci. Jour.*, 4th ser., vol. 6, nos. 3–4, pp. 236–246, 1943; R. W. Chapman, Contact-metamorphic effects of Triassic diabase at Safe Harbor, Pennsylvania, *Geol. Soc. America Bull.*, vol. 61, pp. 191–220, 1950.

assemblages, any of which may contain both quartz and potash feldspar, are as follows:

A. Pelitic and quartzo-feldspathic
1. Quartz-orthoclase-andalusite-cordierite (-biotite)
2. Quartz-orthoclase-plagioclase-andalusite-cordierite (-biotite)
3. Quartz-orthoclase-plagioclase-cordierite (-biotite)
4. Quartz-orthoclase-plagioclase-cordierite-hypersthene
B. Calcareous
8. Plagioclase-diopside-grossularite
9. Diopside-grossularite (-idocrase)
10. Diopside-grossularite-wollastonite (-idocrase)
C. Basic
5. Plagioclase-hypersthene (-quartz)
6. Plagioclase-diopside-hypersthene (-quartz)
7. Plagioclase-diopside (-quartz)
D. Magnesian
Talc-quartz [26]
Hypersthene-quartz

Andalusite is the common aluminum silicate in this as in the hornblende-hornfels facies: but in both facies sillimanite may occur instead in the immediate vicinity of a plutonic contact. The physical significance of this behavior is not understood. The pelitic assemblage cordierite-orthoclase (-hypersthene) is chemically equivalent to quartz-muscovite-biotite of the hornblende-hornfels facies. Yet some biotite not uncommonly occurs in pelitic rocks of the pyroxene-hornfels facies. Almandine occurs only in rocks with a high ratio FeO/MgO. Garnets and pyroxenes of skarns and tactites resulting from iron-silica metasomatism of marble approach andradite and hedenbergite. Appearance of idocrase rather than grossularite and of scapolite instead of anorthite have been attributed to appreciable partial pressures of fluorine and chlorine respectively.

Assemblages Deficient in SiO_2. Figures 76 to 78 illustrate the paragenesis of rocks deficient in silica.

1. Derivatives of dolomitic limestones with silica as the sole impurity fall within the left-hand half of Fig. 76. In order of decreasing silica and increasing magnesia content, the silica-deficient assemblages are calcite-wollastonite-diopside, calcite-diopside-forsterite, calcite-forsterite-periclase. None is critical of the pyroxene-hornfels facies. The first two are found in the hornblende-hornfels facies; the first and third overlap into the sanidinite facies.

[26] Yoder (*op. cit.*, p. 601, 1952) records the assemblage talc-quartz as stable up to 795°C. at $P_{H_2O} = 1,070$ bars, *i.e.*, through a good part of the pyroxene-hornfels facies (cf. Fig. 79). We know of no record of talc-quartz as a natural assemblage in this facies.

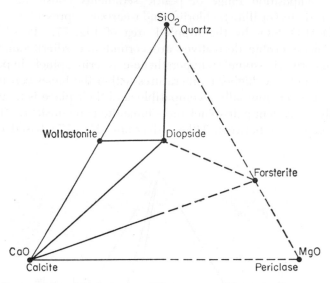

Fig. 76. Pyroxene-hornfels facies: equilibrium assemblages in derivatives of silica-bearing dolomitic limestones, at considerable P_{CO_2}.   (*After N. L. Bowen.*)

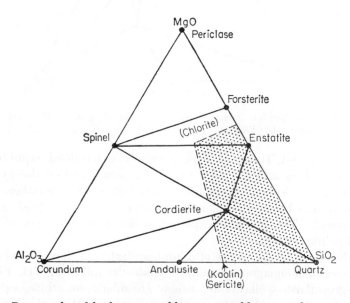

Fig. 77. Pyroxene-hornfels facies: equilibrium assemblages in the system MgO-Al₂O₃-SiO₂. In natural assemblages hypersthene appears instead of enstatite. Stippled area = field of pelitic sediments.   (*After C. E. Tilley.*)

2. The composition range of pelitic sediments consisting of kaolin minerals, sericite (or illite), chlorite, and quartz is represented in the system $MgO-Al_2O_3-SiO_2$ by the stippled area of Fig. 77. Possible silica-deficient metamorphic derivatives are corundum-cordierite-andalusite,[27] corundum-cordierite-spinel,[27] hypersthene-cordierite-spinel, hypersthene-forsterite-spinel. At higher temperatures within the facies cordierite and corundum become mutually incompatible, and their place is taken by the chemically equivalent pair spinel and sillimanite (andalusite). There are some discrepancies between Tilley's diagram (Fig. 77) based on petro-

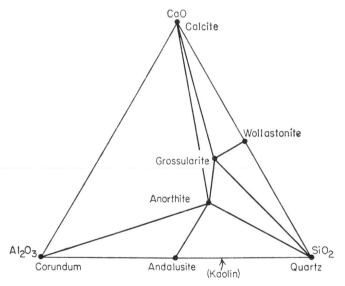

Fig. 78. Pyroxene-hornfels facies: equilibrium assemblages in the system CaO-$Al_2O_3SiO_2$, at considerable P_{CO_2}.

graphic data and Yoder's[28] experimentally determined equilibria for $P_{H_2O} = 1,070$ bars and $T = 655°$ to $830°C$.—conditions of the pyroxene-hornfels facies. Yoder records talc-quartz instead of hypersthene-quartz at temperatures below $795°C$., and forsterite-cordierite instead of hypersthene-spinel. The probable explanation lies in the presence of iron in natural pelitic systems.

3. Silica-poor assemblages of aluminosilicates of lime developed in derivatives of nonmagnesian marls include the following (cf. Fig. 78); calcite-grossularite-wollastonite, calcite-grossularite-anorthite, calcite-anorthite-corundum, andalusite-anorthite-corundum. If marls are consid-

[27] Biotite and orthoclase are additional phases in natural pelitic assemblages (Tilley, *op. cit.*, pp. 42–49, 1924).

[28] Yoder, *op. cit.*, p. 601, 1952.

ered as consisting essentially of calcite, kaolin, and quartz, the commonest assemblage of corresponding calsilicate hornfelses without free quartz must be calcite-grossularite-wollastonite.

4. Either calcite or spinel is a constituent of all but the most aluminous of the silica-poor assemblages in the above three systems. By analogy one or both of the same two minerals is present in almost all hornfelses without quartz that can be referred to the system $CaO\text{-}MgO\text{-}Al_2O_3\text{-}SiO_2$.

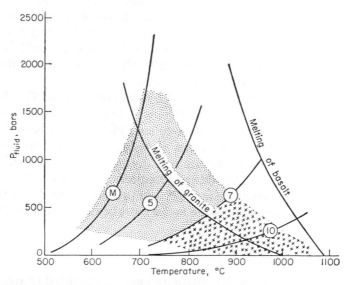

(5) $CaMg(CO_3)_2 \rightleftarrows CaCO_3 + MgO + CO_2$
(7) $2CaCO_3 + Mg_2SiO_4 + CaMgSi_2O_6 \rightleftarrows 3CaMgSiO_4 + 2CO_2$
(10) $3CaCO_3 + 2CaSiO_3 \rightleftarrows 2Ca_2SiO_4 \cdot CaCO_3 + 2CO_2$
(M) $KAl_3Si_3O_{10}(OH)_2 \rightleftarrows KAlSi_3O_8 + Al_2O_3 + H_2O$

Fig. 79.   Physical conditions of the pyroxene-hornfels (stippled) and sanidinite facies (crosses).   Curves of univariant equilibrium are:
For 5, 7, and 10, $P_{CO_2} = P_{fluid}$; for M and the two melting curves $P_{H_2O} = P_{fluid}$.

Possible four-phase assemblages are as follows (analogies with Gold-schmidt's classes being noted in parentheses): [29]
A. Derivatives of silica-poor chloritic and aluminous sediments:
    Corundum-andalusite-cordierite-anorthite (cf. Class 2)
    Corundum-spinel-cordierite-anorthite (cf. Class 2)
    Spinel-cordierite-anorthite-hypersthene (cf. Class 4)
B. Derivatives of basic igneous rocks:
    Spinel-anorthite-hypersthene-olivine
    Spinel-anorthite-diopside-olivine
    Anorthite-hypersthene-diopside-olivine (cf. Class 6)

[29] Cf. Tilley, op. cit., 1925.

C. Impure calcareous dolomite and marly sediments:

Spinel-diopside-anorthite-grossularite (cf. Class 8)
Calcite-spinel-diopside-grossularite
Calcite-spinel-anorthite-grossularite
Calcite-spinel-diopside-olivine
Calcite-spinel-periclase-olivine
Calcite-spinel-corundum-anorthite
Calcite-diopside-grossularite-wollastonite (cf. Class 10)

Physical Conditions of Metamorphism. From stratigraphic evidence Goldschmidt concluded that metamorphism in the Oslo region occurred at shallow depths. The corresponding load pressures would be about 400 bars, locally increasing to perhaps 1,000 bars. Pressures of this order seem typical of the pyroxene-hornfels facies.

The probable temperature range—for values of P_{H_2O} and P_{CO_2} between a few hundred and 1,500 bars—is shown in Fig. 79. The nature of the common boundary with the hornblende-hornfels facies has been discussed already (page 520). At pressures of a few hundred bars the high-temperature boundary is determined by step 7 of Bowen's series (Calcite + Forsterite + Diopside $\rightleftharpoons$ Monticellite + CO_2). This marks the incoming of the sanidinite facies.[30] At pressures exceeding 1,000 bars the upper temperature limit of the facies must be determined by the onset of fusion.

SANIDINITE FACIES [31]

The field occurrence of the sanidinite facies is restricted to xenoliths in lavas and dike rocks, fragments in breccias, and local developments of contact rocks in the immediate vicinity of near-surface intrusions. As a consequence of the low pressure of metamorphism, water and other substances of high vapor pressure, which normally play an important part in facilitating reaction and crystallization, escape readily from the system and may not be available for this function. Chemical reconstitution of the rock tends to be retarded and approaches completion in the short time available only if extreme temperatures are reached (pyrometamorphism). The mineralogical consequences of this combination of conditions are as follows:

[30] The corresponding curve of univariant equilibrium (7 in Fig. 79) is taken from R. I. Harker and O. F. Tuttle, The lower limit of stability of akermanite, *Am. Jour. Sci.*, vol. 254, pp. 468–478, 1956 (especially Fig. 2, p. 474).

[31] H. H. Thomas, On certain xenolithic Tertiary minor intrusions in the island of Mull, *Geol. Soc. London Quart. Jour.*, vol. 78, pp. 229–259, 1922; C. E. Tilley, On larnite and its associated minerals from the contact-zone of Scawt Hill, Co. Antrim, *Mineralog. Mag.*, vol. 22, pp. 77–86, 1929; Eskola, *op. cit.*, pp. 347–349, 1939; S. O. Agrell and J. M. Langley, The dolerite plug at Tievebulliagh near Cushendale, Co. Antrim, *Royal Irish Acad. Proc.*, vol. 69, sect. B, no. 7, pp. 93–127, 1958.

1. Chemical and thermal equilibrium are rarely attained. The number of associated minerals therefore is likely to exceed that demanded by the mineralogical phase rule, and unstable relics from other metamorphic facies are common.

2. High-temperature minerals, analogous with the products of crystallization of dry melts in the laboratory, appear in the mineral assemblages of the sanidinite facies; tridymite, mullite, monticellite, melilite, and larnite are examples.

3. Sanidine, often with a high content of soda, is a critical mineral of this facies. Whether stable or metastable at the time of crystallization, its presence in a mineral assemblage indicates rapid cooling from an unusually high temperature of metamorphism.

4. As a result of partial or complete fusion, glass is sometimes present in rocks of the sanidinite facies.

Although it is true that high temperature is a characteristic condition of metamorphism in the sanidinite facies, the range of temperature involved is probably considerable. The extent and type of metasomatism also may vary greatly. Moreover some of the characteristic minerals are stable over narrow ranges of conditions. Consequently, it is impossible to represent adequately all possible mineral assemblages upon a simple ACF diagram, though an attempt in this direction has been made by Eskola in the two diagrams shown here, with slight modifications, as Figs. 80 and 81. In these, mullite is shown as the stable aluminous phase for systems with excess silica. This accords with experimental observations upon crystallization of dry silicate melts, and is borne out by the occurrence of mullite in aluminous buchites belonging to the sanidinite facies. Also in agreement with laboratory observations, the possibility of considerable replacement of Ca by Mg in high-temperature wollastonite is indicated in the bottom left corner of Fig. 80. Magnesia-bearing wollastonites are still unrecorded in metamorphic rocks, but Tilley has described an occurrence of ferruginous wollastonite in a contaminated dolerite intimately associated with contact-altered limestones of the sanidinite facies at Scawt Hill in Antrim.

The sanidinites of Laacher See occur as fragments enclosed in trachyte and trachyte tuff and have been derived from pelitic schists by pyrometamorphism involving pneumatolytic introduction of soda from the trachyte magma.[32] Typical mineral assemblages include as members such minerals as sodic sanidine, cordierite, spinel, corundum, hypersthene, and sillimanite, often accompanied by unstable relics of other facies, e.g., garnet and staurolite. Glass is sometimes present.

The aluminous xenoliths described by Thomas, from minor intrusions

[32] R. Brauns, *Die kristallinen Schiefer des Laacher See-Gebietes und ihre Umwandlung zur Sanidinit,* Stuttgart, 1911.

of tholeiite in the island of Mull, present a somewhat different para-genesis from which sanidine is absent. Typical associations are anorthite-corundum-spinel, cordierite-spinel-mullite, cordierite-glass, mullite-glass. Associated, partially fused quartzo-feldspathic xenoliths contain tridymite which has crystallized around nuclei of unfused quartz enclosed in glass, but there is no means of determining whether the tridymite formed originally as a stable or as a metastable phase.

The possible mineral assemblages that may develop in calcareous rocks within the sanidinite facies are very varied, and some care is necessary in interpreting the petrographic data, especially where disequilibrium is

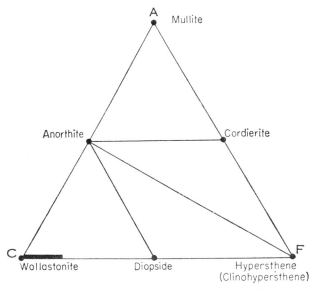

Fig. 80.   Sanidinite facies: ACF diagram for rocks with excess SiO_2 (*After P. Eskola.*)

suspected. Eskola states that garnets of all types are unstable in the sanidinite facies, so that in place of grossularite (a characteristic mineral of the pyroxene-hornfels facies) anorthite and wollastonite constitute a stable association. This accords with the observed association of pseudo-wollastonite and anorthite as products of crystallization from dry melts of appropriate composition in the system $CaO\text{-}Al_2O_3\text{-}SiO_2$, and with the occurrence of wollastonite and calcic plagioclase in metamorphosed lime-stone zenoliths in some volcanic rocks.[33]   The assemblage wollastonite-augite-labradorite occurs in the contaminated diabase of Scawt Hill.

In silica-deficient calcareous assemblages of the sanidinite facies there are a number of highly characteristic minerals formed at step 7 and sub-

[33] A. Lacroix, *Les enclaves des roches volcaniques,* pp. 146, 147, 210, Protat, Mâcon, 1893.

sequent steps in Bowen's series of metamorphic reactions. In probable order of appearance (cf. page 518) these are: monticellite, melilite, tilleyite,[34] spurrite, rankinite,[34] merwinite, and larnite.

The formation of monticellite at the expense of diopside and forsterite at step 7 marks the transition from the pyroxene-hornfels facies to the sanidinite facies. The corresponding equation is

$$CaMgSi_2O_6 + Mg_2SiO_4 + 2CaCO_3 \rightleftarrows 3CaMgSiO_4 + 2CO_2$$

$$\underset{\text{Diopside}}{\quad} \underset{\text{Forsterite}}{\quad} \underset{\text{Calcite}}{\quad} \underset{\text{Monticellite}}{\quad}$$

The reactions at higher temperatures, steps 8 to 13, could be used, wherever convenient, to define a number of subfacies within the sanidinite

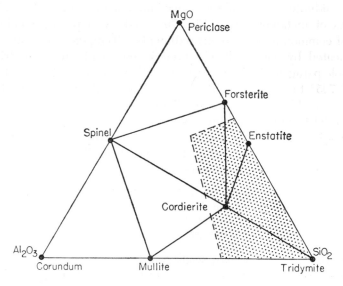

Fig. 81. Sanidinite facies: equilibrium assemblages in the system $MgO\text{-}Al_2O_3\text{-}SiO_2$. (*After P. Eskola.*)   Stippled area = field of pelitic sediments.

facies. It would seem sufficient, however, to recognize two subfacies:

1. A high-temperature, low-pressure, larnite-merwinite-spurrite subfacies, characterized by assemblages containing at least one of the minerals larnite, merwinite, rankinite, spurrite, or tilleyite.[35]

---

[34] Added to Bowen's list by C. E. Tilley, A note on progressive metamorphism of siliceous limestones and dolomites, *Geol. Mag.*, vol. 88, pp. 175–178, 1951.

[35] Tilley, *op. cit.*, 1929; The gabbro-limestone contact zone of Camas Mòr, Muck, Inverness-shire, *Comm. géol. Finlande Bull. 140,* pp. 97–105, 1947; J. H. Taylor, A contact metamorphic zone from the Little Belt Mountains, Montana, *Am. Mineralogist*, vol. 20, pp. 120–128, 1935; A. O. Woodford, R. A. Crippen, and K. B. Garner, Section across Commercial quarry, Crestmore, California, *Am. Mineralogist*, vol. 26, pp. 351–391, 1941; C. W. Burnham, Contact metamorphism of magnesian limestones at Crestmore, California, *Geol. Soc. America Bull.*, vol. 70, pp. 879–920, 1959.

2. A monticellite-melilite subfacies corresponding to somewhat lower temperatures or higher pressures than subfacies 1.[36]   Presence of either or both monticellite or melilite is the only certain criterion of this subfacies.

Occurrence of critical silica-deficient lime-silicate assemblages of both groups is relatively rare.[37]   Moreover, they tend to be complicated by the presence of any of a large number of hydrous lime silicates (afwillite, custerite, crestmoreite, riversideite, etc.), which are best interpreted as products of incipient retrogressive metamorphism activated by circulating solutions as temperature and pressure of CO_2 dropped subsequently to the main metamorphism.   These minerals, therefore, do not belong to the sanidinite facies.

The sanidinite facies embraces maximum temperatures and minimum pressures of metamorphism (Fig. 79), and overlaps the conditions of fusion of common noncalcareous sediments.   Temperatures above 870°C. are indicated by commonly observed inversion of quartz to tridymite. The whole paragenesis of Fig. 81 was synthesized by Yoder [38] at temperatures of 795° to 990°C. and $P_{H_2O} = 1,070$ bars.   The silica-deficient calcareous assemblages that are so diagnostic of the facies crystallize at 800° to 1000°C. at CO_2 pressures below 500 bars.   The larnite-merwinite-spurrite subfacies must be stable only at very low pressures of CO_2; its boundary doubtless is very close to curve 10 of Fig. 79,[39] which marks the incoming of spurrite at Bowen's step 10.

[36] Harker, op. cit., p. 96, fig. 36, 1932.   W. T. Schaller, Monticellite from San Bernardino County, California, Am. Mineralogist, vol. 20, pp. 815–827, 1935.

[37] For details of phase assemblages at each step of metamorphism see Bowen, op. cit., 1940; Fyfe, Turner, and Verhoogen, op. cit., pp. 214–215, 1958.

[38] Yoder, op. cit., p. 601, 1952.

[39] O. F. Tuttle and R. I. Harker, Synthesis of spurrite and the reaction wollastonite + calcite ⇌ spurrite + carbon dioxide, Am. Jour. Sci., vol. 255, pp. 226–234, 1957.

CHAPTER 20

Facies of Regional Metamorphism

THE FACIES OF REGIONAL METAMORPHISM

The facies of regional metamorphism are listed below approximately in order of increasing temperature, as indicated by field relations in zones of progressive metamorphism.

1. Zeolitic (newly defined) [1]
2. Greenschist. Three subfacies are:
    a. Quartz-albite-muscovite-chlorite (formerly muscovite-chlorite)
    b. Quartz-albite-epidote-biotite (formerly biotite-muscovite)
    c. Quartz-albite-epidote-almandine (formerly albite-epidote-amphibolite facies, chloritoid-almandine subfacies)
3. Glaucophane-schist (previously equated with the greenschist facies by the authors). This facies probably corresponds to the same temperature range as that of the greenschist facies but higher pressures.
4. Almandine-amphibolite. Four subfacies are:
    a. Staurolite-almandine [2]
    b. Kyanite-almandine-muscovite [2] } (formerly staurolite-kyanite)
    c. Sillimanite-almandine-muscovite } (formerly sillimanite-
    d. Sillimanite-almandine-orthoclase } almandine)
5. Granulite. Two subfacies are:
    a. Hornblende-granulite
    b. Pyroxene-granulite
6. Eclogite

[1] W. S. Fyfe, F. J. Turner, and J. Verhoogen, Metamorphic reactions and metamorphic facies, *Geol. Soc. America Mem.*, p. 215, 1958.
[2] G. H. Francis, Facies boundaries in pelites in the middle grades of regional metamorphism, *Geol. Mag.*, vol. 93, pp. 353–368, 1956.

ZEOLITIC FACIES [3]

There must be a transition, with increasing depth of burial, between diagenesis and regional metamorphism. Many of the changes involved —*e.g.*, reconstitution of clays, crystallization of quartz and alkali feldspars, destruction of high-temperature minerals, and precipitation of carbonates—are common to both. Where the bulk of the rock, including even coarse particles of sand grade, is substantially affected, the process may properly be called metamorphic. Commonly a criterion of incipient metamorphism is schistosity, for this is the result of ruptural deformation which reduces grain size and accelerates reaction even at relatively low temperatures.

Exceptionally, even without the aid of deformation, chemically unstable rocks may become completely converted to low-temperature assemblages rich in zeolites and duplicating the products of diagenesis. These assemblages constitute the zeolitic facies. This is based upon Coombs's description of metamorphism of andesitic volcanic sands buried to depths of between 20,000 and 30,000 ft.[4] Corresponding conditions are load pressures of 2,000 to 3,000 bars and temperatures of perhaps 200° to 300°C. The unstable nature of the parent material and its high content of combined water (in glass) are probably responsible for the unusually complete response of the rocks to such low temperatures.

Metamorphism was preceded by diagenetic reactions, the products of which are still preserved in the upper part of the section: crystallization of heulandite from glass; reaction between glass and trapped sea water to give analcite; incipient albitization of plagioclase. In the lower levels, where conditions of the zeolitic facies were fully realized, the following characteristic assemblages developed at the expense of plagioclase, ferromagnesian minerals, and diagenetic zeolites:

Laumontite-albite-quartz (-sphene-celadonite)

Quartz-albite-pumpellyite

Quartz-adularia-pumpellyite

Ferruginous epidote, prehnite, and calcite are accessory minerals in some rocks. Prehnite seems to be a characteristic mineral in rocks of this facies elsewhere.[5] Presence of laumontite instead of clinozoisite distinguishes the zeolitic from the greenschist facies. Upward transition to the zone of diagenesis (depth, 10,000 to 20,000 ft.; temperatures, 100° to 200°C.) is marked by prevalence of heulandite rather than laumontite.

---

[3] Fyfe, Turner, and Verhoogen, *op. cit.*, pp. 173, 215–217, 1958.

[4] D. S. Coombs, The nature and alteration of some Triassic sediments from Southland, New Zealand, *Royal Soc. New Zealand Trans.*, vol. 82, pt. 1, pp. 65–109, 1953.

[5] R. N. Brothers, The structure and petrography of graywackes near Auckland, New Zealand, *Royal Soc. New Zealand Trans.*, vol. 83, pt. 3, p. 478, 1956.

GREENSCHIST FACIES [6]

Definition. The greenschist facies includes the common products of low-grade regional and dislocation metamorphism. Greenschists are characterized by abundance of the green minerals chlorite, epidote, and actinolite. The diagnostic assemblage that distinguishes them from amphibolites of higher grade (almandine-amphibolite facies) is quartz-albite-epidote, which is prominent, too, in associated quartzo-feldspathic and pelitic schists. As the grade of metamorphism advances within the zone of almandine as defined for pelitic rocks, there is a sudden change in composition of plagioclase associated with epidote, from albite An_{0-7} to oligoclase or andesine An_{15-30}. This is microscopically recognizable and makes a convenient point at which to draw the high-temperature boundary of the greenschist facies. Our definition extends the facies to include assemblages of somewhat higher grade than those covered by Eskola's original greenschist facies—notably albite-epidote-hornblende, which occurs in the low-grade portion of the almandine zone and which Eskola assigned to a separate epidote-amphibolite facies.

Characteristic Minerals. The common chlorites of the greenschist facies are aluminous—prochlorites with FeO/MgO about 0.7 to 0.8.[7] Magnesian chlorites low in Al_2O_3, especially antigorite, occur mainly in metamorphosed serpentinites. Chloritoid, found only in rocks high in FeO and Al_2O_3 and low in K_2O, is restricted to this facies.

The principal white mica [8] is muscovite (identified by X-ray technique as the 2M polymorph). With it, in highly aluminous rocks, may be associated paragonite. Records of this mineral in the greenschist facies are rare,[9] partly because X-ray techniques must be used to distinguish it from muscovite, and partly because all the excess Al_2O_3 of most pelitic rocks is taken up by muscovite, chlorite, and chloritoid. However, experimental data in the silica-deficient system [10] orthoclase-albite-

[6] Fyfe, Turner, and Verhoogen, *op. cit.*, pp. 217–224, 1958.

[7] C. O. Hutton, Metamorphism in the Lake Wakatipu region, Western Otago, New Zealand, *New Zealand Dept. Sci. and Ind. Research, Geol. Mem. 5*, pp. 17–19, 1940; A. Miyashiro, Chlorite of crystalline schists, *Geol. Soc. Japan Jour.*, vol. 63, no. 736, pp. 1–8, 1957.

[8] Cf. H. S. Yoder and H. P. Eugster, Synthetic and natural muscovites, *Geochim. et Cosmochim. Acta*, vol. 8, pp. 225–280, 1955; H. S. Yoder, in Annual report of the director of the Geophysical Laboratory, *Carnegie Inst. Washington Year Book*, no. 56, pp. 232–237, 1957.

[9] J. L. Rosenfeld, Paragonite in the schist of Glebe Mountain, Southern Vermont, *Am. Mineralogist*, vol. 41, pp. 144–147, 1956; H. Harder, Untersuchungen an Paragoniten und an natriumhaltigen Muscoviten, *Heidelberg Beitr. Min. u. Petrog.*, Band 5, pp. 227–271, 1956 (especially pp. 249–252).

[10] H. P. Eugster and H. S. Yoder, in Annual report of the director of the Geophysical Laboratory, *Carnegie Inst. Washington Year Book*, no. 54, pp. 124–127, 1955.

corundum suggest that at water pressures above 2,000 bars the reaction

$$\text{Albite} + \text{aluminum silicate} + H_2O \rightleftarrows \text{paragonite} + SiO_2$$

runs from left to right at temperatures up to about 600°C. It is also clear from experiment that the stable aluminum silicate phase at low grades in the greenschist facies is pyrophyllite,[11] the rarity of which in petrographic descriptions may also be partly due to its optical similarity to muscovite. The characteristic amphiboles of the greenschist facies, except in the almandine zone, are tremolite and actinolite. The latter may be a deep bluish green variety indistinguishable optically from hornblende which occurs only beyond the almandine isograd; but it is low in Al_2O_3. A mineral of the clinozoisite-epidote series is almost ubiquitous. Stilpnomelane is confined to this and the glaucophane-schist facies, as also is pumpellyite which takes the place of epidote in incompletely metamorphosed basic rocks of the chlorite zone. Carbonates and sphene are widely distributed.

Subfacies and Conditions of Metamorphism. Mineralogical changes in pelitic schists at the classic biotite and almandine isograds are used to divide the greenschist facies into three subfacies defined respectively by the assemblages

　　a. Quartz-albite-muscovite-chlorite

　　b. Quartz-albite-epidote-biotite

　　c. Quartz-albite-epidote-almandine

Estimates of temperatures and pressures of low-grade regional metamorphism are little better than a guess. A possible range [12] compatible with experimental data on the stability of greenschist minerals, and taking into account the general lack of metamorphism in many deeply filled geosynclines, is 300° to 500°C. and $P_{H_2O} = 3,000$ to 8,000 bars.

Quartz-Albite-Muscovite-Chlorite Subfacies.[13]   The low-temperature

[11] R. Roy and E. F. Osborn, The system Al_2O_3-SiO_2-H_2O, *Am. Mineralogist,* vol. 39, pp. 853–855, 1954.

[12] Fyfe, Turner, and Verhoogen, *op. cit.,* pp. 166–173, 1958.

[13] For a detailed account of the petrography of rocks of this subfacies, with chemical analyses of rocks and of individual minerals, see Hutton, *op. cit.,* 1940. For other data see C. E. Tilley, On some mineralogical transformations in crystalline schists, *Mineralog. Mag.,* vol. 21, pp. 34–46, 1926; A. Harker, *Metamorphism,* pp. 209–214, Methuen, London, 1932; J. D. H. Wiseman, The central and south-west Highland epidiorites, *Geol. Soc. London Quart. Jour.,* vol. 90, pp. 357–378, 1934; C. O. Hutton and F. J. Turner, Metamorphic zones in northwest Otago, *Royal Soc. New Zealand Trans.,* vol. 65, pp. 405–406, 1936; F. J. Turner and C. O. Hutton, Some porphyroblastic albite-schists from Waikouaiti River, Otago, *Royal Soc. New Zealand Trans.,* vol. 71, pt. 3, pp. 223–240, 1941; F. J. Turner, Origin of piedmontite-bearing quartz-muscovite schists of northwest Otago, *Royal Soc. New Zealand Trans.,* vol. 76, pt. 2, pp. 246–249, 1946; N. J. Snelling, Note on the petrology and mineralogy of the Barrovian metamorphic zones, *Geol. Mag.,* vol. 94, pp. 297–304, 1957.

limit of the greenschist facies is determined largely by reaction kinetics. Under the accelerating stimulus of deformation and pore fluids, at some range of temperature which may be in the vicinity of 300°C., the velocities of a number of reactions become appreciable and metamorphism sets in. The clays, micas (1Md and 1M polymorphs such as illite and "sericite"), and "chlorites" (celadonite and glauconite) of sediments become reorganized to muscovite (the 2M polymorph), paragonite, and aluminous chlorite; lime zeolites give way to clinozoisite and epidote, and

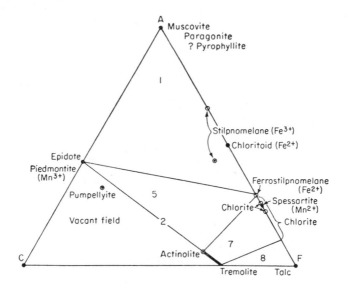

FIG. 82. Greenschist facies, quartz-albite-muscovite-chlorite subfacies: ACF diagram for rocks with excess SiO₂, at low P_{CO_2}. Quartz and albite are possible members of each assemblage. Circled points are analyzed minerals from the chlorite zone of Otago, New Zealand. (See C. O. Hutton, *Mineralog. Mag.*, vol. 25, pp. 172–206, 1938; *ibid.*, pp. 207–211, 1938; D. S. Coombs, *Mineralog. Mag.*, vol. 30, pp. 113–135, 1953.)

soda zeolites to albite; high-temperature minerals such as plagioclase, augite, and hornblende are replaced by appropriate combinations of albite, epidote, chlorite, actinolite, sphene, and calcite.

The minerals so formed make up the assemblages that define the quartz-albite-muscovite-chlorite subfacies. They remain stable over a range of temperature whose upper limit (at the biotite isograd) is marked by several reactions which seem to take place over much the same interval of temperature and pressure:

Muscovite + chlorite $\rightleftarrows$ biotite + chlorite

Chlorite + calcite + quartz $\rightleftarrows$ actinolite + H_2O + CO_2

Dolomite + quartz + H_2O $\rightleftarrows$ tremolite + calcite + CO_2

Since these reactions are affected differently by P_{H_2O} and P_{CO_2}, there is some overlap between the assemblages on opposite sides of the biotite isograd.

Mineral assemblages typical of the subfacies (Figs. 82 to 84) are:

A. Pelitic schists
   (1) Quartz-muscovite-chlorite-albite (-epidote-tourmaline); with chloritoid in rocks low in K_2O and high in Al_2O_3.

B. Quartzo-feldspathic schists
   (1) Quartz-albite-muscovite (-epidote); potash-feldspar is a possible but uncommon member.

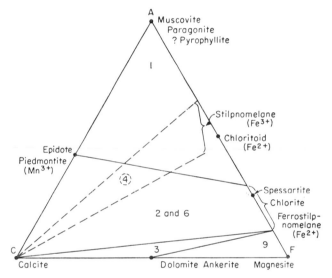

Fig. 83. Greenschist facies, quartz-albite-muscovite-chlorite subfacies; ACF diagram for rocks with excess SiO_2, at considerable P_{CO_2}. Quartz and albite are possible members of each assemblage.

C. Calcareous schists
   (2) Calcite-epidote-tremolite-quartz (-chlorite)
   (3) Calcite-dolomite-chlorite-tremolite
   (4) Calcite-stilpnomelane (-sphene)

D. Basic schists
   (5) Albite-epidote-chlorite-actinolite-sphene (-stilpnomelane-quartz)
   (5) Albite-epidote-chlorite-sphene (-stilpnomelane-quartz)
   (6) Albite-epidote-chlorite-calcite-sphene (-stilpnomelane-quartz)

E. Magnesian schists
   (7) Chlorite-tremolite

(8) Talc-tremolite (-chlorite-quartz)

(8) Talc-serpentine-tremolite

(9) Talc-magnesite (-dolomite)

F. Iron- and manganese-rich [14]

(1) Quartz-muscovite-piedmontite-spessartite (-tourmaline-barite)

(10) Quartz-spessartite-stilpnomelane

(10) Magnetite-spessartite-stilpnomelane

(10) Magnetite-epidote-chlorite-stilpnomelane

Rhodonite-rhodochrosite-spessartite

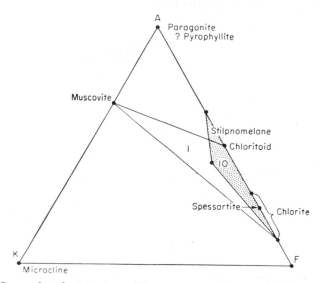

Fig. 84. Greenschist facies, quartz-albite-muscovite-chlorite subfacies; AKF diagram for rocks with excess SiO_2 and Al_2O_3 (Field 1, Figs. 82, 83). Stippled area = non-micaceous schists.

Quartz-Albite-Epidote-Biotite Subfacies. To this facies belong rocks of the biotite zone as defined for pelitic schists. Mineral assemblages for rocks with excess silica (Figs. 85, 86) are as follows:

A. Pelitic schists [15]

(1) Biotite-muscovite-quartz (-albite-epidote)

(2) Muscovite-chloritoid-quartz (-albite-epidote)

(3) Muscovite-chloritoid-chlorite-quartz (-albite-epidote)

(4) Biotite-muscovite-chlorite-quartz (-albite-epidote)

[14] Hutton, op. cit., pp. 38–47, 1940; C. O. Hutton, Contributions to the mineralogy of New Zealand, Part IV, Royal Soc. New Zealand Trans., vol. 84, pt. 4, pp. 791–803, 1957.

[15] Tilley, op. cit., 1926; Harker, op. cit., pp. 214–217, 1932; J. A. Noble and J. O. Harder, Stratigraphy and metamorphism in a part of the northern Black Hills, South Dakota, Geol. Soc. America Bull., vol. 59, p. 956, 1948; Snelling, op. cit., 1957.

B. Quartzo-feldspathic schists [16]
   (5) Quartz-albite-microcline (-biotite-muscovite-epidote)
   (6) Quartz-albite-microcline-biotite (-epidote)
C. Basic schists [17]
   (7) Actinolite-epidote-albite-chlorite-sphene (-quartz-biotite)
   (8) Epidote-albite-chlorite (-quartz-biotite-sphene)
   (11) Albite-epidote-actinolite (-quartz-microline-biotite)
D. Magnesian schists [18]
   (9) Talc-actinolite-chlorite (-quartz)
   (9) Serpentine (-talc-actinolite)

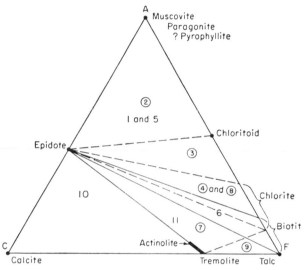

FIG. 85. Greenschist facies, quartz-albite-epidote-biotite subfacies; ACF diagram for rocks with excess SiO_2. Broken lines and circled numbers refer to assemblages deficient in K_2O (*i.e.*, lacking potash feldspar). Quartz and albite are possible members of each assemblage.

E. Calcareous schists [19]
   (10) Calcite-epidote-tremolite (-quartz)
   (11) Albite-epidote-actinolite (-quartz-microline-biotite)

[16] J. W. Ambrose, Progressive kinetic metamorphism of the Missi series near Flinflon, Manitoba, *Am. Jour. Sci.*, vol. 32, pp. 263–269, 1936.
[17] F. C. Phillips, Some mineralogical and chemical changes induced by progressive metamorphism in the green bed group of the Scottish Dalradian, *Mineralog. Mag.*, vol. 22, pp. 239–256, 1930; Wiseman, *op. cit.*, pp. 357–378, 1934; B. Mason and S. R. Taylor, The petrology of the Arahura and Pounamu series, *Royal Soc. New Zealand Trans.*, vol. 82, pt. 5, pp. 1065–1070, 1955.
[18] H. H. Hess, Hydrothermal metamorphism of an ultrabasic intrusion at Schuyler, Virginia, *Am. Jour. Sci.*, vol. 26, pp. 377–408, 1933.
[19] Harker, *op. cit.*, pp. 256, 260, 1932.

Quartz-Albite-Epidote-Almandine Subfacies.[20]   The quartz-albite-epidote-almandine subfacies is distinguished from subfacies of lower grade by the presenec of almandine (in place of iron-bearing chlorite) in pelitic rocks and of hornblende instead of nonaluminous actinolite in basic rocks. Magnesian chlorite may accompany the essentially ferrous phases almandine and chloritoid in rocks too low in potash for magnesium to be completely accommodated in biotite.   Kyanite though by no means a common

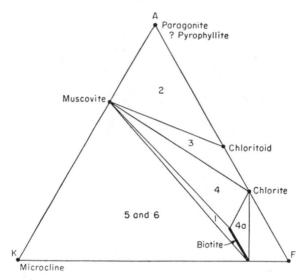

FIG. 86.   Greenschist facies, quartz-albite-epidote-biotite subfacies; AKF diagram for rocks with excess SiO_2 and Al_2O_3.   (*After N. J. Snelling.*)

mineral has been recorded sufficiently often in highly aluminous rocks to suggest that it, rather than pyrophyllite, is now the stable aluminum silicate.   Chloritoid appears in association with muscovite and/or almandine (but not with biotite) in rocks rich in both aluminum and iron.[21] Widespread retrogressive replacement of almandine and biotite by chlori-

[20] P. Eskola, On the petrology of eastern Fennoscandia, I, *Fennia 45*, no. 19, pp. 1–93, 1925; J. Suzuki, Petrological study of the crystalline schist system of Shikoku, Japan, *Hokkaido Imp. Univ. Fac. Sci. Jour.*, 4th ser., vol. 1, no. 1, pp. 27–111, 1930; Wiseman, *op. cit.*, pp 378–392, 1934; Ambrose, *op cit.*, pp. 257–286, 1936; Noble and Harder, *op. cit.*, pp. 956–959, 1948; W. Q. Kennedy, Zones of progressive regional metamorphism in the Moine schists of the western Highlands of Scotland, *Geol. Mag.*, vol. 86, pp. 43–56, 1949; Snelling, *op. cit.*, 1957; Fyfe, Turner, and Verhoogen, *op. cit.*, p. 224, 1958.

[21] Cf. D. H. Williamson, Petrology of chloritoid and staurolite rocks north of Stonehaven, Kincardineshire, *Geol. Mag.*, vol. 90, pp. 353–354, 1953; Snelling, *op. cit.*, 1957.

toid in Japan has been cited by Seki [22] as evidence that chloritoid is unstable in the present subfacies and is stable only in the chlorite zone. Characteristic mineral assemblages (Figs. 87, 88) are as follows:

A. Pelitic schists

    (1) Biotite-muscovite-almandine-quartz-albite (-epidote)

    (7) Biotite-muscovite-quartz-albite (-epidote-microcline)

    (2) Muscovite-chloritoid-almandine-quartz (-albite-epidote-chlorite)

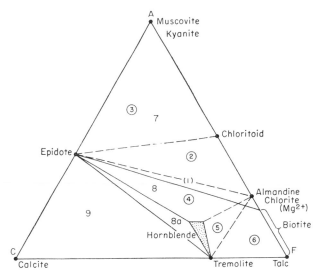

Fig. 87.   Greenschist facies, quartz-albite-epidote-almandine subfacies; ACF diagram for rocks with excess SiO_2.   Broken lines and circled numbers refer to assemblages deficient in K_2O.   Quartz and albite are additional phases in all assemblages; microcline only in assemblages lacking kyanite, chloritoid, almandine, and talc.   Presence of calcite implies considerable P_{CO_2}.

    (3) Muscovite-chloritoid-quartz (-chlorite)

    (3a) Muscovite-chloritoid-kyanite-quartz (-chlorite) [23]

B. Quartzo-feldspathic schists

    (7) Quartz-albite-microcline (-epidote-almandine-muscovite-biotite)

C. Amphibolites (Basic schists)

    (4) Hornblende-albite-epidote-almandine (-biotite-quartz)

    (8) Hornblende-albite-epidote (-biotite-quartz)

[22] Y. Seki, On chloritoid rocks in the Kitakami median metamorphic zone, northeastern Japan, *Saitama Univ. Sci. Rept.*, Ser. B, vol. 2, pp. 223–263, 1954 (reprinted) in the *Tsuboi Commemorative Volume*, Tokyo, 1955).

[23] P. Bearth, Geologie und Petrographie des Monte Rosa, *Geol. Karte Schweiz Beitr.*, 96, pp. 68, 69, 85, 1952.

D. Magnesian schists
   (5) Hornblende (tremolite) -chlorite-almandine
   (6) Talc-tremolite-chlorite
E. Calcareous schists
   (9) Calcite-epidote-tremolite-quartz

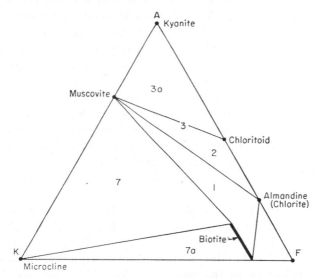

Fig. 88. Greenschist facies, quartz-albite-epidote-almandine subfacies; AKF diagram for rocks with excess SiO_2 and Al_2O_3 (Fig. 87, triangle 7).

GLAUCOPHANE-SCHIST FACIES [24]

Field Occurrence. Glaucophane schists and associated rocks containing lawsonite or jadeite are virtually restricted to post-Paleozoic geosyn-

[24] H. S. Washington, A chemical study of the glaucophane schists, *Am. Jour. Sci.*, vol. 21, pp. 35–59, 1901; J. Suzuki, On some soda pyroxene and amphibole bearing quartz schists from Hokkaido, *Hokkaido Imp. Univ. Fac. Sci. Jour.*, 4th ser., vol. 2, no. 4, pp. 339–353, 1934; H. W. Quitzow, Diabasporphyrite und Glaucophangesteine in der Trias von Nordkalabrien, *Gesell. Wiss. Göttingen Nachr., Math.-Phys. Kl.*, IV, N. F., vol. 1, pp. 83–118, 1935; N. L. Taliaferro, Franciscan-Knoxville problem, *Am. Assoc. Petroleum Geologists Bull.*, vol. 27, pp. 159–182, 1943; R. N. Brothers, Glaucophane schists from the North Berkeley Hills, California, *Am. Jour. Sci.*, vol. 252, pp. 614–626, 1954; W. P. de Roever, Genesis of jadeite by low-temperature metamorphism, *Am. Jour. Sci.*, vol. 253, pp. 283–298, 1955; W. T. Bloxam, Jadeite-bearing metagraywackes in California, *Am. Mineralogist*, vol. 41, pp. 488–496, 1956; I. Y. Borg, Glaucophane schists and eclogites near Healdsburg, California, *Geol. Soc. America Bull.*, vol. 67, pp. 1563–1584, 1956; A. Miyashiro and S. Banno, Nature of glaucophanitic metamorphism, *Am. Jour. Sci.*, vol. 256, pp. 97–110, 1957; Fyfe, Turner and Verhoogen, *op. cit.*, pp. 224–228, 1958; T. W. Bloxam, Glaucophane schists near Valley Ford, California, *Am. Jour. Sci.*, vol. 257, pp. 95–112, 1959.

clines.[25]   They are metamorphosed basalts, tuffs, graywackes, and cherts. Because of their geosynclinal location they tend to be associated broadly with serpentinites; but there seems to be no general causal connection between intrusion of ultramafic rocks and metamorphism of the geosynclinical filling to glaucophane schists.

In the Franciscan formation of the Californian Coast Ranges glaucophane-, lawsonite-, and jadeite-schists occur locally at hundreds of localities in an otherwise unmetamorphosed terrane.   There can be no doubt that conditions of metamorphism are sharply localized; and there is no obvious consistent connection between metamorphism and igneous intrusion.   Elsewhere, e.g., in Corsica and the Swiss Alps, glaucophane- and lawsonite-schists are closely associated with rocks of the lowest grade (quartz-albite-muscovite-chlorite subfacies) of the greenschist facies. There is also a marked tendency for some glaucophane schists to be associated with eclogites, but in many if not in all such occurrences there is field and textural evidence showing that the eclogite is in process of being converted by retrogressive metamorphism to assemblages of the glaucophane-schist facies.

Mineral Paragenesis.   Several characteristic minerals are confined to the glaucophane-schist facies: lawsonite, jadeite, acmite-jadeite, glaucophane, and crossite.   Stilpnomelane and pumpellyite occur only in the glaucophane-schist and in the greenschist facies (quartz-albite-muscovite-chlorite subfacies) or in the zeolitic facies.   Many greenschist assemblages, e.g., chlorite-epidote-sphene, muscovite-chlorite-sphene, calcite-chlorite, are widely distributed also in the glaucophane-schist facies. Biotite, the universal index of advancing metamorphic grade, is totally lacking.   Other high-temperature minerals—hornblende, diopside-jadeite, almandine—are commonly if not invariably unstable relics inherited from an earlier phase of metamorphism.[26]

In glaucophane-schists and their associates, MgO, FeO, MnO, Fe_2O_3, and Al_2O_3 behave as independent components; so it is impossible to construct three-component diagrams adequately representing the great variety of mineral assemblages in this facies.   The following are selected assemblages drawn especially from the California paragenesis:

A. Pelitic
   Muscovite-chlorite-quartz (-glaucophane)
B. Quartzo-feldspathic
   Quartz-jadeite (-muscovite-glaucophane)
   Quartz-jadeite-lawsonite-glaucophane
   Quartz-lawsonite-glaucophane

[25] H. M. E. Schürmann, Beiträge zur Glaucophanfrage, *Neues Jahrb. Mineral. Monatsh.*, p. 63, 1951.
[26] Cf. Borg, *op. cit.*, 1956.

Quartz-muscovite-stilpnomelane-glaucophane

Quartz-albite-crossite

C. Basic

Lawsonite-glaucophane (-sphene)

Lawsonite-pumpellyite-glaucophane (-sphene)

Almandine-lawsonite-glaucophane-muscovite

Epidote-pumpellyite-glaucophane (-sphene)

Lawsonite-jadeite-glaucophane (-sphene)

Albite-epidote-chlorite-muscovite (-glaucophane)

D. Calcareous

Calcite-epidote-chlorite-glaucophane

E. Ferruginous (derivatives of chert)

Quartz-spessartite-stilpnomelane-glaucophane

Quartz-crossite-aegirine-spessartite-stilpnomelane

Definition of the Facies. The significance of glaucophane as a facies index—probably because of its spectacular optical properties—has perhaps been overrated. There are extensive metamorphic terranes where the whole paragenesis is that of the greenschist facies (quartz-albite-muscovite-chlorite subfacies) except for local presence of glaucophane as an additional phase in a greenschist assemblage. Such are the glaucophane-epidote-albite schists of Queensland and of Anglesey, Wales, and the calcite-epidote-chlorite-glaucophane schists of the Swiss Alps. The authors assign this paragenesis to the greenschist facies, noting that the presence of glaucophane indicates conditions transitional to those of the glaucophane-schist facies.

The glaucophane-schist facies is here restricted to the paragenesis in which glaucophane schists are associated with assemblages containing lawsonite, jadeite-quartz, aegirine, or pumpellyite—typically all four. This is the facies represented in California, Celebes, parts of Japan, and elsewhere.

Physical Conditions.[27] The glaucophane-schist facies has such close mineralogical analogies with the quartz-albite-muscovite-chlorite subfacies of the greenschist facies, that the physical conditions governing the two cannot be greatly different. Two mineralogical differences appear to be highly significant from the physiochemical viewpoint:

1. Although iron-rich epidote is rather common in the glaucophane schists, lawsonite is the characteristic Ca-Al-silicate phase instead of clinozoisite. It has been shown experimentally and by thermodynamic reasoning that high water pressure and low temperature favor crystallization of lawsonite.

2. Albite, ubiquitous in the greenschist facies, is rare in glaucophane schists. Its place is taken by the chemically equivalent pair quartz-

[27] Fyfe, Turner, and Verhoogen, *op. cit.*, pp. 174–178, 226, 1958.

jadeite,[28] which has been shown to be stable, even at low temperatures, at high load pressures.

We tentatively conclude that the physical conditions of the glaucophane-schist facies are low temperatures (perhaps 300° to 400°C.) overlapping those of the greenschist facies, and very high pressures of water and load. Such conditions might develop in deeply buried, water-saturated sediments in regions of exceptionally low thermal gradient connected in some way with geosynclinal conditions.

Metasomatic introduction of soda and iron seems to have accompanied the development of some—though by no means all—glaucophane schists.[29] Moreover rocks of this facies commonly show conspicuous effects of metamorphic differentiation, notably monomineralic segregation veins of such diverse phases as glaucophane, lawsonite, pumpellyite, jadeite, chlorite, actinolite, and so on. We conclude that the fluid phase is unusually mobile and active in glaucophane-schist metamorphism. In spite of this there are many rocks of this paragenesis whose compositions do not differ appreciably from those of the parent basalts, cherts, and graywackes from which they were derived.

ALMANDINE-AMPHIBOLITE FACIES

Definition and Field Occurrence.[30] The amphibolite facies of Eskola [31] included all metamorphic assemblages associated with the diagnostic basic assemblage hornblende-plagioclase (oligoclase or more calcic types). Fyfe, Turner, and Verhoogen have erected separate facies for the amphibolitic assemblages of contact and those of regional metamorphism. The classic occurrence of the latter is in the Barrovian zones of medium to high grade from the middle of the almandine zone onward.[32] The diagnostic mineral assemblages, restricted to the facies of regional metamorphism, are hornblende-plagioclase-almandine and hornblende-plagioclase-epidote; the facies is named accordingly the almandine-amphibolite facies.

[28] Jadeites of this paragenesis commonly contain an appreciable amount of acmite —which accounts for association of jadeite quartz and albite, without textural evidence of disequilibrium, in some rocks.

[29] E.g., R. Michel (Les schistes cristallins des massifs du Grand Paradis et de Sesia-Lanzo, *Sciences de la Terre*, tome 1, nos. 3–4, pp. 211–232, 238–239, 272–278, 1953) describes glaucophanization of amphibolitic mica schists and albitization of pelitic mica schists as cognate effects of a younger metamorphism superposed upon assemblages of the greenschist and almandine-amphibolite facies formed during an earlier metamorphism.

[30] Fyfe, Turner, and Verhoogen, *op. cit.*, p. 228, 1958.

[31] P. Eskola, *Die Entstehung der Gesteine* (Barth, Correns, Eskola), pp. 351–355, Springer, Berlin, 1939.

[32] Cf. Wiseman, *op. cit.*, pp. 378–385, 1934.

By no means all amphibolites of this paragenesis, however, contain almandine. Abundance of almandine, staurolite, or kyanite, and absence of cordierite and andalusite, distinguish pelitic assemblages of the almandine-amphibolite facies from those of the hornblende-hornfels facies of contact metamorphism.

Subfacies. Francis [33] recognized three subfacies in pelitic rocks of higher grade, corresponding respectively to the staurolite, kyanite, and sillimanite zones of the Scottish Highlands. These are here retained, with slight modification of nomenclature, and a fourth subfacies (No. 3, below) is added to accommodate somewhat different pelitic assemblages widely represented in the United States and believed to have formed at lower pressures. These are listed below in order of increasing metamorphic grade (temperature):

1. Staurolite-almandine
2. Kyanite-almandine-muscovite
3. Sillimanite-almandine-muscovite
4. Sillimanite-almandine-orthoclase

Staurolite-Almandine Subfacies.[34] The paragenesis of the staurolite-almandine subfacies is illustrated in Figs. 89, 90a.

In pelitic rocks [35] staurolite, almandine, or (rarely) kyanite is associated with micas but not with potash feldspar; in highly potassic rocks micas and potash feldspar may occur together. The common micas are red-brown biotite [36] and muscovite. Paragonite [37] occurs only in highly aluminous schists, in association with garnet, staurolite, or kyanite. Kyanite is restricted to rocks very rich in Al_2O_3 and low in alkali—a combination rare in pelites. Staurolite,[38] which is chemically equivalent to chloritoid of the greenschist facies, occurs in rocks rich in Al_2O_3 and FeO and deficient in K_2O. Typical pelitic assemblages are:

(2) Quartz-staurolite-almandine-muscovite-plagioclase (-biotite)

(3) Quartz-almandine-muscovite-biotite-plagioclase (-epidote)

Chloritoid may accompany or take the place of staurolite in the almandine zone. In highly aluminous rocks the paragenesis is:

---

[33] G. H. Francis, *op. cit.*, 1956.

[34] Fyfe, Turner, and Verhoogen, *op. cit.*, pp. 229, 230, 1958.

[35] *E.g.*, Harker, *op. cit.*, pp. 224–226, 1932; H. H. Read, On the geology of Central Sutherland, *Scotland Geol. Survey Mem.*, pp. 38–40, 1931; T. F. W. Barth, Structural and petrologic studies in Dutchess County, New York, *Geol. Soc. America Bull.*, vol. 47, pp. 775–850, 1936; Noble and Harder, *op. cit.*, pp. 941–976, 1948 (especially pp. 956–960); D. Wyckoff, Metamorphic facies in the Wissahickon schist near Philadelphia, Pennsylvania, *Geol. Soc. America Bull.*, vol. 63, pp. 25–58, 1952.

[36] Snelling, *op. cit.*, pp. 300–304, 1957.

[37] H. Harder, *op. cit.*, pp. 242–263, 1956.

[38] D. H. Williamson, *op. cit.*, pp. 353–361, 1953.

(1) Quartz-kyanite [39]-staurolite-muscovite-plagioclase (-biotite-paragonite)

The paragenesis in quartzo-feldspathic rocks is:

(7) Quartz-microcline-plagioclase-biotite (-muscovite-epidote)

Derivatives of basic igneous rocks are amphibolites,[40] many of them garnetiferous. Coexistence of a medium plagioclase (An_{25} to An_{45}) with epidote is common in these rocks and indicates high load pressure, sometimes perhaps locally augmented by stress. Rocks rich in MgO may contain cummingtonite or anthophyllite. Typical assemblages are:

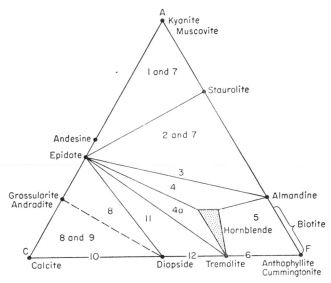

Fig. 89. Almandine-amphibolite facies, staurolite-almandine subfacies; ACF diagram for rocks with excess SiO_2. Quartz and plagioclase are possible additional phases; microcline only in assemblages lacking kyanite, staurolite, or almandine.

(4) Hornblende-plagioclase-almandine-epidote (-quartz-biotite)

(4a) Hornblende-plagioclase-epidote (-quartz-biotite)

(11) Hornblende-plagioclase (-diopside)

In magnesian assemblages anthophyllite, its aluminous counterpart

[39] In some regions such as west-central New Hampshire (J. B. Thompson, The graphical analysis of mineral assemblages in pelitic schists, Am. Mineralogist, vol. 42, p. 851, Fig. 5, 1957) sillimanite appears instead of kyanite—probably reflecting lower load pressures.

[40] Wiseman, op. cit., pp. 354–417, 1934; T. A. Dodge, Amphibolites of the Lead area, northern Black Hills, South Dakota, Geol. Soc. America Bull., vol. 53, pp. 561–584, 1942; S. Matthes and H. Krämer, Die Amphibolite und Hornblendgestcinc im mittleren kristalliner Vor Spessart, Neues Jahrb. Mineral. Abh., Band 88, Heft 2, pp. 225–272, 1955.

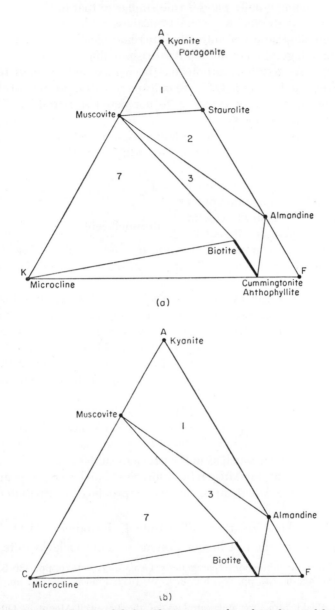

Fig. 90. (a) Almandine-amphibolite facies, staurolite-almandine subfacies; AKF diagram for rocks with excess SiO₂ and Al₂O₃. Quartz, plagioclase, and epidote are possible additional phases. (b) Almandine-amphibolite facies, kyanite-almandine-muscovite subfacies; AKF diagram for rocks with excess SiO₂ and Al₂O₃. Quartz and plagioclase are possible additional phases.

gedrite, or cummingtonite plays a role similar to that of the chlorites and talc in the greenschist facies. Some typical assemblages are:

(5) Cummingtonite (or anthophyllite)-hornblende-almandine
(6) Cummingtonite (or anthophyllite)-tremolite

The paragenesis gedrite-staurolite has been recorded in rocks relatively rich in Al_2O_3 and lacking K_2O.[41] In calcareous rocks, presumably because of high pressures, neither wollastonite nor periclase is stable. Common assemblages are as follows:

(8) Calcite-diopside-epidote (-plagioclase-quartz)
(9) Calcite-diopside-grossularite (-quartz)
(10) Calcite-diopside-phlogopite
(10), (12) Calcite-diopside-tremolite
(11) Calcite-hornblende-epidote

Diopside-forsterite-calcite ⎱
Forsterite-dolomite-calcite ⎰ silica-deficient

Epidote may be accompanied by sodic or medium plagioclase (An_{25} to An_{45} in most cases). Scapolite appears in place of plagioclase in some rocks.

Kyanite-Almandine-Muscovite Subfacies. At the kyanite isograd in the Barrovian sequence staurolite becomes unstable and the pair kyanite-almandine takes its place in pelitic rocks (Fig. 90*b*). Otherwise the rocks of the kyanite zone resemble those of the staurolite zone (staurolite-almandine subfacies). However, pelitic schists are so widespread and the kyanite zone so clearly recognizable in the field that erection of a distinct subfacies is warranted.[42]

The characteristic pelitic assemblages (cf. Fig. 90*b*) are:

(1) Quartz-kyanite-muscovite-almandine-plagioclase (-biotite)
(2) Quartz-almandine-muscovite-biotite-plagioclase (-epidote)

Other assemblages are much as in the previous subfacies.

Sillimanite-Almandine-Muscovite Subfacies.[43] The first appearance of sillimanite in some American metamorphic provinces is attributed to the reaction

$$\text{Staurolite} + \text{quartz} \rightleftarrows \text{sillimanite} + \text{almandine} + H_2O$$

The equivalent reaction at higher pressures would yield kyanite instead

[41] C. E. Tilley, Kyanite-gedrite paragenesis, *Geol. Mag.*, vol. 76, pp. 326–330, 1939.

[42] Cf. Wyckoff, *op. cit.*, pp. 28 (Fig. 3c), 49, 1952; Francis, *op. cit.*, pp. 355, 356, 1956.

[43] M. T. Heald, Structure and petrology of the Lovewell Mountain Quadrangle, New Hampshire, *Geol. Soc. America Bull.*, vol. 61, pp. 46–50, 74, 75, 1950; Wyckoff, *op. cit.*, pp. 28 (Fig. 3c), 30, 31, 48, 49, 1952; C. A. Chapman, Structure and petrology of the Sunapee Quadrangle, New Hampshire, *Geol. Soc. America Bull.*, vol. 63, pp. 384–389, 420–422, 1952; H. L. James, Zones of regional metamorphism in the pre-Cambrian of northern Michigan, *Geol. Soc. America Bull.*, vol. 66, pp. 1455–1488, 1955 (especially pp. 1462, 1466–1467, 1472, 1477).

of sillimanite, as at the kyanite isograd of the Scottish Highlands. To accommodate the sillimanite-bearing assemblages developed at this stage, *i.e.*, at the "first sillimanite isograd," a separate subfacies, named after the diagnostic assemblage sillimanite-almandine-muscovite is here erected. Typical pelitic assemblages are:

(1) Quartz-sillimanite-muscovite-almandine-plagioclase (-biotite)

(2) Quartz-almandine-muscovite-biotite-plagioclase

Plagioclase is oligoclase-andesine in pelitic schists, andesine to labradorite

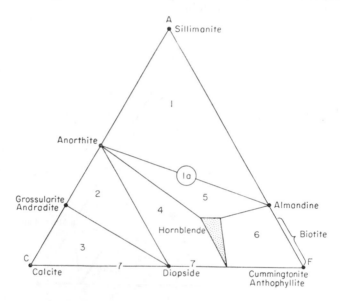

Fig. 91.   Almandine-amphibolite facies, sillimanite-almandine-orthoclase subfacies; ACF diagram for rocks with excess SiO₂ and K₂O.   Quartz and potash feldspar are possible additional phases.

in amphibolites, and anorthite in calc-granulites.   Epidote is absent or negligible.   Except for this and for the presence of sillimanite in place of kyanite, the subfacies is generally similar to the kyanite-almandine-muscovite subfacies.

Sillimanite-Almandine-Orthoclase Subfacies.[44]   The paragenesis of the sillimanite-almandine-orthoclase subfacies is illustrated in Figs. 91 and 92. Some typical assemblages are listed below.

[44] W. R. Browne and F. L. Stillwell, *Appendices* to E. C. Andrews, The geology of the Broken Hill district, *New South Wales Geol. Surv. Mem. 8*, 1922; Wiseman, *op. cit.*, pp. 394–396, 1934; F. F. Osborne, Petrology of the Shawinigan Falls district, *Geol. Soc. America Bull.*, vol. 47, pp. 197–228, 1936; Wyckoff, *op. cit.*, pp. 26–35, 48, 49, 1952; Chapman, *op. cit.*, p. 420, 1952; Francis, *op. cit.*, pp. 357–359, 1956; Fyfe, Turner, and Verhoogen, *op. cit.*, pp. 230–232, 1958.

A. Pelitic and quartzo-feldspathic:
  (1) Quartz-sillimanite-almandine-orthoclase (-plagioclase-biotite)
  (1a) Quartz-orthoclase-plagioclase-almandine-biotite
  (8) Quartz-orthoclase-plagioclase-biotite
B. Calcareous:
  (2) Anorthite-diopside-garnet-quartz
  (3) Calcite-diopside-garnet-quartz
  (4) Anorthite-diopside-hornblende-quartz
  (7) Calcite-diopside-tremolite

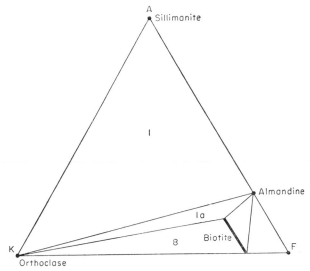

FIG. 92. Almandine-amphibolite facies, sillimanite-almandine-orthoclase subfacies; AKF diagram for rocks with excess SiO_2 and Al_2O_3. Quartz and plagioclase are possible additional phases.

C. Basic:
  (4) Hornblende-plagioclase (-diopside-quartz)
  (5) Hornblende-plagioclase-almandine (-quartz)
D. Magnesian:
  (6) Cummingtonite (or gedrite)-hornblende-almandine (-plagioclase)
  (7) Cummingtonite-tremolite
      Olivine-hornblende-pleonaste

Mutual Relations of the Four Subfacies.[45]   Everywhere the staurolite-almandine subfacies corresponds to the lowest grade of metamorphism within the almandine-amphibolite facies. Some of the reactions that

[45] Wyckoff, *op. cit.,* 1952; Chapman, *op. cit.,* 1952; Francis, *op. cit.,* 1956.

mark progressive metamorphism of pelitic rocks beyond the staurolite zone are:

(1) Staurolite + quartz $\rightleftarrows$ kyanite + almandine + H_2O
(2) Staurolite + quartz $\rightleftarrows$ sillimanite + almandine + H_2O
(3) Kyanite $\rightleftarrows$ sillimanite
(4) Muscovite + quartz $\rightleftarrows$ sillimanite + orthoclase + H_2O

Each has its unique curve of univariant equilibrium on a pressure-temperature diagram. The physical significance of the sillimanite isograd in

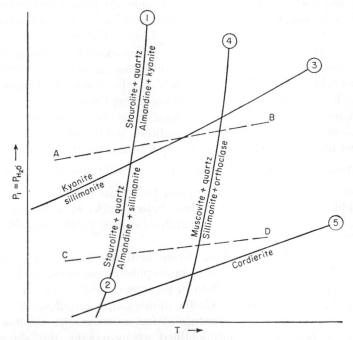

FIG. 93. Almandine-amphibolite facies; hypothetical curves of univariant equilibrium for reactions leading to appearance of sillimanite. AB = inferred temperature gradient in Scottish Highlands; CD = inferred gradient in New Hampshire.

any particular province will depend upon which of the curves corresponding to reactions (3) to (4) is first encountered along the prevailing temperature-pressure gradient of metamorphism. In the Barrovian sequence of Scotland, reaction (1) marks the kyanite isograd. The sillimanite isograd [46] is defined by reaction (3) closely followed by reaction (4); so the sequence of progressive metamorphism is from the staurolite-almandine to the kyanite-almandine-muscovite to the sillimanite-almandine-orthoclase subfacies (A to B in Fig. 93). In New Hampshire reaction (2) is followed by (4), so that here the zonal sequence of sub-

[46] Cf. Harker, *op. cit.*, pp. 228, 229, 1932; Francis, *op. cit.*, p. 357, 1956.

facies is (1) staurolite-almandine, (2) sillimanite-almandine-muscovite, (3) sillimanite-almandine-orthoclase (C to D, Fig. 93).

Mineral assemblages of the sillimanite zone—in fact assemblages of high-grade facies in general—tend to be complicated and partially obscured by reactions taking place in response to (1) falling temperature (during unloading), and (2) fluctuations in water pressure connected with melting and freezing of a granitic magmatic component in migmatite areas. So sillimanite, muscovite, and potash feldspar may be associated in one rock; and textural evidence may indicate partial replacement of early mica by sillimanite or late "sericitization" of kyanite or sillimanite after the peak temperature of metamorphism. This mixing of facies in no way invalidates recognition of the sillimanite-almandine-muscovite and sillimanite-almandine-orthoclase subfacies as separate entities in provinces where the high-grade assemblages have remained unmodified during unloading following the peak of metamorphism.

The two sillimanite-bearing subfacies must be stable over a wide range of pressure, their respective stability fields (cf. Fig. 93) being separated by the curve of univariant equilibrium for the reaction

$$\text{Muscovite} + \text{quartz} \rightleftarrows \text{sillimanite} + \text{potash feldspar} + \text{water}$$

In the lower pressure range there must be a field transitional to that of the hornblende-hornfels facies. Here belong rocks in which cordierite appears in the sillimanite- and almandine-bearing pelitic assemblages of regional metamorphism.[47]

Conditions of Metamorphism. From field relationships and mineralogical evidence, temperatures of metamorphism must generally be higher in the almandine-amphibolite than in the greenschist facies. The upper limit may be near 700° or 750°C.; for in migmatite complexes the higher grades must overlap the temperature of fusion of granite. Unless pyrophyllite has been widely misidentified as muscovite, the almandine-amphibolite facies must be largely within the stability field of one of the polymorphs of Al_2SiO_5. According to Kennedy's experimental data,[48] pyrophyllite remains stable up to 600° to 660°C. at water pressures of 5,000 to 10,000 bars; thus, unless water pressures generally are much less than load pressures, 550°C. is about the minimum permissible temperature for the facies over this range of pressure.

High water pressures are indicated by prevalence of amphiboles and micas—especially muscovite and hornblende. Widespread occurrence of

[47] H. H. Read, Metamorphism and migmatization in the Ythan Valley, Aberdeenshire, *Edinburgh Geol. Soc. Trans.*, vol. 15, pp. 265–279, 1952 (especially pp. 276, 277). Thompson, *op. cit.*, p. 856, Fig. 9, 1957; Fyfe, Turner, and Verhoogen, *op. cit.*, p. 211, 1958.

[48] G. C. Kennedy, Pyrophyllite-sillimanite-mullite equilibrium relations to 20,000 bars and 800°C., *Geol. Soc. America Bull.*, vol. 66, p. 1584, 1955.

kyanite and of epidote associated with medium plagioclase shows that the almandine-amphibolite facies covers pressures far exceeding those of the hornblende-hornfels facies. General absence of cordierite and andalusite points to the same conclusion. An anomaly (not yet resolved) is presented by Kennedy's experimental data on the kyanite-sillimanite transition.[49] At 700°C. this is said to take place at a load pressure of 14,000 bars, consistent with depths far below the bottom of the crust. In nature kyanite certainly forms at temperatures of this order well within the crust.

We conclude that metamorphism in the almandine-amphibolite facies covers a temperature range of perhaps 550° to 750°C., and pressures normally between 4,000 and 8,000 bars. Assemblages transitional to the hornblende-hornfels facies perhaps correspond to pressures as low as 3,000 bars.

GRANULITE FACIES

Definition. Among the products of high-grade, deep-seated regional metamorphism is a group of gneissic rocks (termed *granulites* by German and Finnish petrologists) which are marked by a highly distinctive fabric and by mineralogical peculiarities sufficiently outstanding to form the basis of a distinct metamorphic facies—the granulite facies of Eskola.[50] The granulites [51] of Saxony, Ceylon, and Scandinavia are, in the main, quartzo-feldspathic garnet or pyroxene gneisses with little or no mica, in which quartz occurs in flattened lenticles oriented (by grain form) parallel to the foliation.[52] Alternation of light "acid" and dark "basic"

[49] *Ibid.*, p. 1584.

[50] Eskola, *op. cit.*, p. 360, 1939.

[51] H. Rosenbusch and A. Osann, *Elemente der Gesteinslehre*, pp. 676, 682, 686, Erwin Nägele, Stuttgart, 1922; K. H. Scheumann, Ueber eine Gruppe bisher wenig beachteter Orthogneise des Granulitgebirges und deren Einschlichtung, *Min. pet. Mitt.*, vol. 47, pp. 403–469, 1936; F. D. Adams, The geology of Ceylon, *Canadian Jour. Research*, vol. 1, pp. 444–498, 1929; T. Sahama, Die Regelung von Quarz und Glimmer in den Gesteinen der Finnisch-Lappländischen Granulitformation, *Comm. géol. Finlande Bull. 113*, pp. 1–110, 1936; J. A. W. Bugge, Geological and petrographical investigations in the Kongsberg-Bamble formation, *Norges geol. undersökelse 160*, pp. 15–18, 1943; P. Eskola, On the granulites of Lapland, *Am. Jour. Sci.*, Bowen vol., pp. 133–171, 1952; On the mineral facies of charnockites, *Jour. Madras University*, B, vol. 27, no. 1, pp. 101–119, 1957; R. A. Howie, The geochemistry of the charnockite series of Madras, India, *Royal Soc. Edinburgh Trans.*, vol. 62, pt. 3, pp. 725–768, 1955; Fyfe, Turner, and Verhoogen, *op. cit.*, pp. 159–161, 232–234, 1958.

[52] The term *granulite* has also been used in quite different senses by British and French petrologists. According to British usage a granulite is any rather fine-grained, quartzo-feldspathic metamorphic rock without conspicuous schistosity (*e.g.*, the Moine "granulites"), while French workers have applied the same term to the muscovite granites.

bands is typical but not essential.   Highly characteristic is the association
of hypersthene- and diopside-bearing assemblages with rocks containing
pyrope-almandine and perthitic alkali feldspar.

Two subfacies are now recognized:

(1) The hornblende-granulite subfacies, in which hornblende and/or
biotite are present in the garnetiferous and pyroxenic assemblages.

(2) The pyroxene-granulite facies, lacking hornblende and biotite.
For a given water pressure the second subfacies represents the higher
range of temperatures; but because it most clearly typifies the granulite
as contrasted with the almandine-amphibolite facies it will be treated
first.

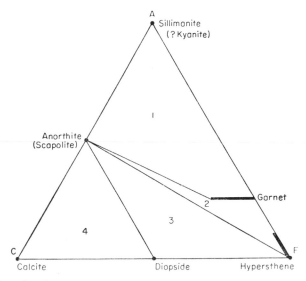

FIG. 94.   Granulite facies, pyroxene-granulite subfacies; ACF diagram for rocks with
excess SiO₂ and K₂O.   Quartz and perthite are possible additional phases.

Pyroxene-Granulite Subfacies.   The paragenesis of silica-saturated
rocks of the pyroxene-granulite facies is illustrated in Fig. 94.   The following assemblages—also occurring individually, with minor variations,
in other high-temperature facies—collectively constitute a unique and
characteristic association:

(1) Quartz-perthite-garnet (-plagioclase-kyanite or sillimanite).   Here
belong the khondalites of India and Ceylon—metamorphosed aluminous
sediments consisting mainly of sillimanite, garnet, perthite, and quartz.

(2) Quartz-perthite-hypersthene (-garnet-plagioclase).   This is the
normal assemblage of charnockites—hypersthene-bearing granitic gneisses
some of which are of metamorphic, others of direct magmatic origin (cf.
page 346).

(2) Plagioclase-hypersthene-garnet (-quartz-perthite).

(3) Plagioclase-hypersthene-diopside (-quartz-orthoclase). These last two assemblages are developed in the pyroxene granulites (*"Trapp-granulite"*) and in their mineralogical equvialents the noritic members of the charnockite series of igneous rocks.

(3a) Plagioclase-hypersthene-diopside-garnet—a silica-deficient basic assemblage.

(4) Diopside - plagioclase - calcite - quartz; diopside - scapolite - calcite - quartz. Pressures are high enough to prohibit reaction between calcite and quartz to give wollastonite. This question has been elaborated by Korzhinsky in connection with his approximately equivalent Aldan facies, named after the Aldan massif of Siberia.[53] It is characterized by absence of wollastonite, grossularite, vesuvianite, and periclase, none of which is stable at extreme pressure. In calcareous rocks deficient in silica, the stable assemblages (as in the almandine-amphibolite facies) are diopside-calcite-forsterite and forsterite-calcite-dolomite, the reigning pressures being too high to allow dissociation of dolomite in the absence of silica.

The individual minerals in rocks of the granulite facies display a number of distinctive and persistent peculiarities. Hypersthene is strongly pleochroic from green to deep pink to yellow, the intensity of the absorption tints apparently having little relation to the iron content; analyses of hypersthenes [54] show a high content of Al_2O_3 (6 to 9 per cent by weight). Garnets have a much wider range of composition than do the almandines of the amphibolite facies; they are pyrope-almandines similar to those of the eclogites, and sometimes contain as much as 55 per cent pyrope and up to 20 per cent grossularite. Microperthitic structure is characteristic of the alkali feldspar, and the plagioclase similarly tends to be antiperthitic. Scapolite commonly accompanies or substitutes for plagioclase. Rutile is common in acid, and ilmenite in basic assemblages; sphene is invariably absent. Rocks undersaturated in silica contain olivines, corundum, and green spinel. The last-named mineral (like magnetite in rocks of other metamorphic facies) may even appear in rocks containing free quartz.

Hornblende-Granulite Subfacies. In some granulite provinces hornblende and/or biotite are persistently associated with what is otherwise the typical granulite paragenesis. There is no textural evidence of disequilibrium between associated phases, and recognition of a distinct subfacies seems warranted. This is the hornblende-granulite subfacies.

Pelitic and quartzo-feldspathic assemblages resemble those of the

---

[53] D. S. Korzhinsky, Dependence of mineral stability on depth, *Soc. russe minéralogie Mem.*, vol. 66, no. 2, pp. 369–396, 1937.

[54] A. W. Groves, The charnockite series of Uganda, British East Africa, *Geol. Soc. London Quart. Jour.*, vol. 91, pp. 150–207, 1935; Eskola, *op. cit.*, p. 152, 1952.

almandine-amphibolite facies (sillimanite-almandine-orthoclase sub-facies). In calcareous assemblages the pair andesine-clinozoisite commonly occurs in place of the calcic plagioclase characteristic of pyroxene granulites. Typical basic assemblages are: plagioclase-hornblende-diopside; plagioclase-hornblende-diopside-hypersthene; plagioclase-hornblende-diopside-almandine. Diopside-almandine-hypersthene is a magnesian assemblage sometimes miscalled "eclogite."

Cordierite is an essential constituent of some granulites—especially biotite-bearing types.[55] Much more rarely highly magnesian granulites contain sapphirine [56] in such assemblages as sapphirine-gedrite-hypersthene, sapphirine-hypersthene-cordierite-anthophyllite, sapphirine-cordierite-enstatite, and sapphirine-enstatite-spinel. Petrographic records are inadequate to show whether cordierite and sapphirine occur in the hornblende-granulite facies alone, or in both subfacies. It is even possible that they might be used to define a third subfacies.[57]

Conditions of Metamorphism. In some regions successive zones characterized respectively by the almandine-amphibolite facies, the hornblende-granulite subfacies, and the pyroxene-granulite facies constitute a progressive sequence of advancing grade.[58] Elsewhere the pyroxene-granulite and the hornblende-granulite assemblages seem, from the rather imperfect records available, to be intermingled in the field. Elsewhere again there is textural evidence, in the form of coronas and other replacement structures, that pyroxene-granulite assemblages are in process of retrogressive metamorphism to hornblende-granulite or almandine-amphibolite assemblages. These field and textural relationships are consistent with the widely prevalent view that the granulite facies represents high temperatures of metamorphism.

The pyroxene-granulite assemblages are essentially anhydrous. This does not necessarily imply either low water pressure or lack of water in the parent rocks; high temperature by itself could account for development of anhydrous assemblages even at high water pressures. Indeed it is difficult to imagine any means, other than high-temperature metamorphism, by which pelitic rocks of normal water content could become converted to anhydrous assemblages of the pyroxene-granulite facies.

[55] Eskola, *op. cit.,* pp. 140, 149, 162–164, 1952.

[56] T. Vogt, Mineral assemblages with sapphirine and kornerupine, *Comm. géol. Finlande Bull. 140,* pp. 15–24, 1947; T. N. Muthuswami, Sapphirine (Madura), *Indian Acad. Sci. Proc.,* vol. 30, pp. 295–301, 1949; E. R. Segnit, Sapphirine-bearing rocks from MacRobertson Land, Antarctica, *Mineralog Mag.,* vol. 31, pp. 690–697, 1957.

[57] Cf. Eskola, *op. cit.,* p. 164, 1952.

[58] A. F. Buddington, Adirondack igneous rocks and their metamorphism, *Geol. Soc. America Mem. 7,* pp. 267–282, 1939; Fyfe, Turner, and Verhoogen, *op. cit.,* p. 232, 1958.

It seems likely that the granulite facies spans a considerable range of pressure. Presence of kyanite and rutile in typical granulites and the high aluminum content of granulite pyroxenes suggest very high pressures. But sillimanite appears instead of kyanite in many granulites; and the presence of cordierite in others cannot be reconciled with pressures of more than a few thousand bars. When the distribution and paragenesis of cordierite granulites is more completely investigated it may be necessary to erect a cordierite-granulite subfacies corresponding to relatively low pressures in the granulite facies.

We conclude that the granulite facies represents maximum temperatures of regional metamorphism, perhaps of the order of 700° to 800°C. Pressures normally are high, but may in some instances extend down to only a few thousand bars. Falling temperature, with sustained or increased water pressure, would cause transition via the hornblende-granulite subfacies to the almandine-amphibolite facies. The hornblende-granulite subfacies is correlated with crystallization on the curves of univariant equilibrium for the reactions

Hornblende $\rightleftarrows$ pyroxenes + water
Biotite $\rightleftarrows$ almandine + orthoclase + hypersthene + water

A system rendered anhydrous by crystallization in the proxene-granulite facies could cool down these curves if water were continually supplied from crystallizing magma or some other source—provided the rock were sufficiently permeable to the diffusing aqueous fluid. The complex interplay of metamorphism, fusion, injection, and magmatic crystallization in the deep levels of the crust could cause fluctuations of temperature and water pressure on a scale sufficient to account for the observed field relations between the pyroxene granulites, the hornblende granulites, and rocks of the almandine-amphibolite facies.

ECLOGITE FACIES

The eclogite facies [59] is based upon the highly distinctive critical association omphacite-garnet, developed in rocks of gabbroid composition. Many eclogites are composed entirely of the two minerals mentioned, and feldspar is completely lacking in most. The pyroxene omphacite differs

[59] P. Eskola, On the eclogites of Norway, *Kristiania Vidensk, Skr., 1. Math.-Naturv. Kl.*, 8, 1921; *op. cit.*, pp. 363–367, 1939; A. R. Alderman, Eclogites from the vicinity of Glenelg, Inverness-shire, *Geol. Soc. London Quart. Jour.*, vol. 92, pp. 488–530, 1936; H. G. Backlund, Zur genetischen Deutung der Eclogite, *Geol. Rundschau*, vol. 27, pt. 1, pp. 47–61, 1936; C. E. Tilley, The paragenesis of kyanite-eclogites, *Mineralog. Mag.*, vol. 24, pp. 422–432, 1936; H. S. Yoder, The jadeite problem, *Am. Jour. Sci.*, vol. 48, pp. 24–48, 1950; Borg, *op cit.*, 1956; Fyfe, Turner and Verhoogen, *op. cit.*, pp. 154–158, 235–237, 1958.

strikingly from pyroxenes of normal igneous rocks in its high content of Na_2O and Al_2O_3 and negligible TiO_2; it can be regarded as a diopside-jadeite solid solution with some additional Al replacing Ca. Equally characteristic is the composition of the garnets, lime-bearing almandine-pyropes with 25 to 70 per cent pyrope and 12 to 40 per cent grossularite. Chemical equivalence of olivine gabbro and bimineralic eclogite may be illustrated by the simplified equation:

$$3CaAl_2Si_2O_8 + 2NaAlSi_3O_8 + 3Mg_2SiO_4 + n\,CaMg(SiO_3)_2 =$$

Labradorite   Olivine   Diopside

$$3CaMg_2Al_2(SiO_4)_3 + 2NaAl(SiO_3)_2 + n\,CaMg(SiO_3)_2 + 2SiO_2$$

Garnet   Omphacite   Quartz

Kyanite on the one hand, or enstatite on the other, may rank as essential constituents of eclogites containing excess of Al_2O_3 or $(Mg, Fe)O$ respectively, over the amount that can be accommodated in omphacite and garnet. Rutile is a characteristic accessory, but ilmenite is also known. The typical eclogite assemblages, therefore, are omphacite-garnet (-rutile), omphacite-garnet-kyanite (-rutile), omphacite-garnet-enstatite (-rutile). The presence of primary hornblende and zoisite in some eclo-gites (hornblende eclogites) suggests transition between the eclogite and the almandine-amphibolite facies.

Secondary crystallization of hornblende, plagioclase, and other minerals at the expense of omphacite and garnet is very common in eclogites and may be attributed to incipient chemical adjustment of the rock to falling pressure and temperature—an instance of retrogressive metamorphism (diaphthoresis). In the field complete gradation may be traced from un-altered eclogite, through eclogite amphibolites containing relict garnet and omphacite together with newly generated plagioclase and horn-blende, to amphibolites of normal composition. In some cases myrme-kite-like intergrowths of diopside and plagioclase first replace omphacite, and then in turn pass over into the amphibole-plagioclase association of the almandine-amphibolite facies. Elsewhere, e.g., in the Franciscan of California, eclogites show every stage of retrogressive metamorphism to glaucophane schists.

Eskola [60] lists four principal modes of occurrence of eclogites:

1. Inclusions in kimberlites, basalts, and basalt breccias, where they are habitually accompanied by "olivine nodules" (having the composi-tion and structure of peridotite), sillimanite schists, garnet-bearing ultra-basics, and other rocks whose mineralogy indicates either igneous or metamorphic origin at great depths.

2. Streaks and bands enclosed in dunites and their serpentinized equivalents. These merge into rocks which depart from the strict defini-

[60] Eskola, op. cit., p. 366, 1939.

tion of eclogite in that their pyroxene is chrome diopside containing little or no soda.

3. Lensoid masses enclosed in migmatite gneiss. These have been interpreted by Eskola as fragments of larger masses originating in depth and brought upward into geosynclines with the rising granitic intrusions. Whatever their mode of origin may have been, eclogites of this type characteristically grade into amphibolite as contact with granite is approached, and amphibolitization (retrogressive metamorphism) of the eclogite may be correlated with injection of granite and development of the surrounding migmatites.

4. Bands associated with amphibolites, mica schists, and the like, in regions of alpine folding and deformation. The associated rocks in many such occurrences clearly belong to facies very different from that of eclogites. Such is the case with small masses of eclogite locally associated with glaucophane, lawsonite, and jadeite schists and serpentinites in the Franciscan formation of California.

The great density of eclogites (3.35 to 3.6) compared with chemically equivalent gabbros (2.9 to 3.1) and the high specific gravity of even minor constituents (kyanite, rutile, diamond) suggest the influence of very high pressure. High temperature is inferred from the presence of pyroxenes, lack of silicates containing (OH), and a wide range of isomorphous substitution of cations in pyroxenes and garnets of eclogites. So there is rather general agreement that eclogites normally crystallize deep within or even beneath the crust. Conditions prevailing at a depth of 40 km. ($T = 700°C$. or more; $P = 13,000$ bars), even if water pressures approach load pressure, are consistent with available data regarding the stability range of eclogite minerals. It is true, however, that in the complete absence of water (an unlikely condition in a natural metamorphic environment) the eclogite assemblage omphacite-garnet might be stable at much shallower depths, e.g., 15 km., $T = 400°C.$, $P = 5,000$ bars. Backlund concluded, from studies in the Caledonian system of Norway, that great depth is not necessary for the development of eclogite. He advocated high temperature and local stress of great magnitude, which would have the effect of raising the equivalent pressure of values normally attained only at deep levels in the crust. Korzhinsky [61] has even advanced the unorthodox view that eclogites normally are formed at only moderate depth, and so are foreign to the abyssal pre-Cambrian complexes in which the prevailing metamorphic facies is that of the granulites.

The present authors would emphasize four characteristic features of eclogites which seem to reflect peculiar conditions of origin. First is their distinctive mineralogy and exceptional density. Second is their restricted range of chemical composition corresponding approximately with that of

[61] Korzhinsky, *op. cit.*, pp. 392, 393, 1937.

basic igneous rocks.   Third is the high susceptibility of eclogites to retrogressive metamorphism, underlining a marked difference in metamorphic facies of eclogites and the rocks with which they are commonly associated. Finally, there is a pronounced tendency for eclogites to occur in a number of situations which, while differing radically among themselves, are alike in bearing the unmistakable imprint of strong intrusive or tectonic transport of rock from the depths.   This applies to kimberlite pipes, to peridotite and serpentinite intrusions, to deformed migmatite complexes, and to zones of dislocation in deformed geosynclines.   These peculiarities of eclogites are compatible with the hypothesis, foreshadow by Eskola,[62] that eclogites normally form at such depths within the crust, or in the outer part of the mantle, that only basic and ultrabasic material is present; that eclogite, whether properly regarded as metamorphic, as igneous, or as primary material of an eclogite shell of the earth, finds its way into the upper continental crust only where an adequate mechanism or vehicle of transport is active; and finally, that eclogite is so susceptible to retrogressive metamorphism to amphibolite or glaucophane schist at lower temperatures and pressures, that it survives in partially altered condition only where reaction velocities have been too low for its complete destruction in the time available.   This is the hypothesis tentatively adopted in this book.

[62] Eskola, *op. cit.*, p. 367, 1939.

CHAPTER 21

Chemical Changes Accompanying
Metamorphism

INTRODUCTORY STATEMENT

As contrasted with the products of weathering or with ore deposits, metamorphic rocks are, with certain exceptions, generally similar in chemical composition to the rocks from which they have been derived. Not infrequently this similarity amounts almost to identity, especially if the water content of the rocks in question is disregarded. However, although it is convenient in discussing certain aspects of metamorphism to assume ideal chemical reconstitution in a closed system without addition or removal of material, this assumption is at best only an approximation, and commonly departs from reality to such an extent as to be most misleading. In applying the facies principle we have made no such assumption, but have considered metamorphic rocks as end products of chemical reaction, without regard to the exact composition of the parent rock. The equilibrium assemblage of minerals in any metamorphic rock has been determined by (1) the physical conditions of metamorphism, and (2) the ultimate chemical composition attained by the rock in question. Although minor components such as B_2O_3, P_2O_5, and Cl, which are often introduced into the system from an external source, have for the most part been neglected in the discussion on metamorphic facies, it is of course recognized that their presence can lead to the appearance of appropriate additional mineral phases in any of the assemblages considered.

It now remains to consider the direction and extent to which the chemical composition of common rocks may change during metamorphism, and the means by which such changes are effected. Two broad processes, often proceeding simultaneously, are concerned, namely, metasomatism and metamorphic differentiation.

561

GENERAL CHARACTER OF METASOMATISM

By Lindgren [1] metasomatism has been defined as "the process of practically simultaneous capillary solution and deposition by which a new mineral of partly or wholly differing chemical composition may grow in the body of an old mineral or mineral aggregate." As defined by Goldschmidt,[2] metasomatism is more limited in its scope and is conceived as "a process of alteration which involves enrichment of the rock by new substances brought in from the outside," such enrichment taking place "by definite chemical reactions between the original minerals and the enriching substances." In this chapter we are concerned primarily with that part of the field of metasomatism which was considered by Goldschmidt, *i.e.*, with metamorphism involving introduction and removal of certain substances, with corresponding change not only in the mineralogy but in the chemical composition of the rocks affected. It is nevertheless important to remember that even when metamorphism takes place in a closed system without any change in the bulk composition of the mass, replacement of the original mineral assemblage by a metamorphic assemblage through the medium of pore fluids is governed by the laws of metasomatism discussed below and would be included within the broad domain of metasomatism as originally defined by Lindgren.

Metasomatism of silicate rocks is effected by reaction between the constituent minerals and chemically active solutions or gases streaming through the rock pores under the influence of a pressure gradient. The effects of metasomatism are especially conspicuous in the immediate vicinity of bodies of granite and granodiorite. Whether the fluids concerned are regarded as the last aqueous residues of crystallizing magma, or as a wave extending in advance of a "front" of granitization, there can be no doubt as to the general connection between emplacement of granitic masses and metasomatism of the adjacent country rocks. The complexity and variety of metasomatism of this kind reflects the complex possibilities of phase changes (*e.g.*, gas $\rightleftarrows$ liquid) and compositional variation in the fluid itself as it migrates along strongly developed pressure and temperature gradients. Metasomatism is also likely to accompany contact metamorphism in the vicinity of undoubtedly intrusive masses of more basic plutonic rocks such as diorite and gabbro. Moreover, in some cases the gases or solutions active in metasomatism may be largely of nonmagmatic origin, in that they are derived directly from rocks in process of metamorphism. For example, transformation of limestones or dolomites into

[1] W. Lindgren, *Mineral Deposits*, p. 91, 4th ed., McGraw-Hill, New York, 1933.
[2] V. M. Goldschmidt, On the metasomatic processes in silicate rocks, *Econ. Geology*, vol. 17, pp. 105–123, 1922.

silicate rocks by high-grade metamorphism yields an abundance of carbon dioxide, which, migrating upward into zones of lower temperature, may there play an important part in metasomatism.

When the agents of metasomatism are of magmatic origin, the elements introduced into the adjacent rocks are often those in which the parent magma itself is strikingly deficient. This phenomenon has been referred to by Lodochnikow [3] as the principle of polarity. It is illustrated by such well-known types of alteration as lime and soda metasomatism in connection with peridotites, albitization by diabases (not necessarily spilitic), and development of iron and magnesian silicate skarns at contacts between granite and limestone. The phenomenon of polarity reflects contrasted compositions of the dilute aqueous solutions (or gases) and the silicate melts (magmas) from which they separate.

In most rocks affected by metasomatism the textural details of the parent rock are faithfully preserved. The sharp outlines of crystals of enstatite persist unchanged in serpentinites; corals and plant stems retain their internal structure in perfect detail though now replaced by silica; the ophitic texture of diabase survives replacement of the rock by kaolin minerals. From such evidence Lindgren [4] concluded that in rigid rocks metasomatism takes place without change of volume of the rock affected, and that space for precipitation of the new mineral is provided by simultaneous solution of that which it replaces. General petrographic experience tends to confirm the wide though not universal applicability of Lindgren's "law of constancy of volume." On thermodynamic grounds it is easy to see that increase in volume is unlikely when metasomatism occurs at depth, because of the work that would have to be done against the external pressure. But there seems to be no obvious thermodynamic barrier to metasomatic reactions involving reduction in volume. Changes in volume certainly accompany metamorphism associated with rock deformation.

Where there is textural evidence that metasomatism has occurred without change of volume, the course of the corresponding chemical reaction cannot be represented quantitatively by the conventional simple equation which merely balances equal weights of material on the right and left sides respectively, without taking volume changes into account. It was shown earlier in this book (pages 318 and 319) that if serpentinization of peridotite occurs without noteworthy change in volume, some such reaction as the following must be assumed:

[3] W. Lodochnikow, Serpentine und Serpentinite der Iltschir-lagerstätte, *U.S.S.R. Central Geol. Prosp. Inst. Trans.*, fasc. 38, p. 743 (with German summary, pp. 728–770), 1936.

[4] W. Lindgren, Volume changes in metamorphism, *Jour. Geology*, vol. 26, pp. 542–554, 1918.

$$5Mg_2SiO_4 + 4H_2O \rightarrow 2H_4Mg_3Si_2O_9 + 4MgO + SiO_2$$

Olivine	Serpentine	(Removed in
(700 gm.; 218 cc.)	(552 gm.; 220 cc.)	solution)

By similar reasoning, Lindgren deduced that the conventional equation for sericitization of orthoclase,

$$3KAlSi_3O_8 + H_2O + CO_2 \rightarrow KAl_3Si_3O_{10}(OH)_2 + K_2CO_3 + 6SiO_2$$

Orthoclase   Muscovite

should be discarded, if metasomatism has been governed by the law of equal volumes, for an expression approximating

$$6KAlSi_3O_8 + 4H_2O + CO_2 + 3Al_2O_3 \rightarrow$$

Orthoclase

$$4KAl_3Si_3O_{10}(OH)_2 + K_2CO_3 + 6SiO_2$$

Muscovite

From the examples cited, it is obviously unsafe to interpret metamorphic reactions quantitatively, in terms of simple equations that actually correspond to reactions involving marked change of volume in closed systems. Such equations are useful enough in demonstrating chemical equivalence (by weight) of two assemblages of minerals stable under different physical conditions and for indicating qualitatively the general direction in which chemical reaction may have proceeded. Thus, to show that muscovite and quartz, an association in the hornblende-hornfels facies, are chemically equivalent to the assemblage orthoclase-andalusite in the pyroxene-hornfels facies, it is permissible to write the equation

$$KAl_3Si_3O_{10}(OH)_2 + SiO_2 \rightleftarrows KAlSi_3O_8 + Al_2SiO_5 + H_2O$$

Muscovite   Orthoclase   Andalusite

However, it cannot be inferred that this equation necessarily represents completely the chemical reconstitution of muscovite slate to andalusite hornfels, even though it is clear from petrographic evidence that destruction of muscovite and complementary crystallization of orthoclase and andalusite take place during the process. There is a tendency to treat the chemical side of metamorphism in terms of relatively simple equations based upon the evidence supplied by pseudomorphs, but it is in just such cases that the law of equal volumes most clearly applies. For example, although pseudomorphs of almandine replacing chlorite are sometimes formed during progressive metamorphism of pelitic rocks, and although the two minerals are roughly comparable in chemical composition, we are not justified in assuming that metamorphism involves a simple reaction of the type:

$$3H_4(Mg, Fe)_2Al_2SiO_9 + H_4(Mg, Fe)_3Si_2O_9 + 4SiO_2 \rightarrow$$

Prochlorite

$$3(Mg, Fe)_3Al_2(SiO_4)_3 + 8H_2O$$

Garnet

Instead, replacement of chlorite by the same volume of almandine, with addition and removal of various oxides in appropriate amounts, and certainly with some substitution of FeO for MgO, is part of a complex chemical process in which all the minerals present take some part.

During the course of contact metamorphism, metasomatism is particularly effective, and under such conditions there is usually no accompanying deformation nor change in volume of the rock mass as a whole. Consequently, a fairly accurate estimate of the extent to which various substances have been added or removed during metamorphism may be made by comparing chemical analyses of the metamorphosed and of the unmetamorphosed rock. A number of illustrations are given in Lindgren's classic study of contact metamorphism of siliceous limestone by monzonite porphyry at Bingham, Utah,[5] where it was shown that silica, iron sulfide, and magnesia were added, and lime and carbon dioxide were removed during metamorphism of the Yampa limestone.

THERMODYNAMIC CONSIDERATIONS

Statement of the Problem. The course of any metasomatic reaction is determined by the chemical potentials of the various components in the pore fluid and in the mineral phases concerned (cf. Chap. 2, pages 15 and 17). A necessary condition for conversion of orthoclase to muscovite by exchange of alumina for potash, as in the third equation on page 564, is that the sum of the chemical potentials of the reactants should exceed the sum of the chemical potentials of the products [each chemical potential being multiplied by the appropriate stoichiometric coefficient; cf. Eq. (2-21)]. Now, for a given temperature and pressure, the chemical potential of each of the crystalline phases remains constant. But that of each component of the pore solution increases with concentration. Hence, for muscovitization of orthoclase to proceed, it is necessary that the concentrations of ions brought in by the solutions (Al^{3+}, OH^-, and so on) should exceed certain limiting values, and concentrations of ions to be removed, notably K^+, should be maintained below some other limiting values. These values will of course depend on temperature and pressure. The important point is that the course of metasomatism is controlled largely by concentrations of participating ions in the active fluids.

Formidable difficulties arise, however, when one attempts to trace changes in concentration in space and time, in order to interpret correctly the succession of events attendant upon metasomatism. To the complexity arising from our usual ignorance of initial conditions (tempera-

[5] W. Lindgren, Contact metamorphism at Bingham, Utah, *Geol. Soc. America Bull.*, vol. 35, pp. 507–534, 1924.

ture, pressure, concentrations in fluid) must be added that accruing from additional factors such as temperature and pressure gradients, the velocity of the percolating fluid, reaction velocities, the size and shape of capillary openings through which the fluid percolates and their effect on solubilities, etc.

Percolation Velocity and Reaction Rate. To illustrate the problem, consider first the relatively simple case of a fluid penetrating a rock consisting of a single mineral soluble in the fluid (*e.g.*: water $+ CO_2$ and limestone). Note that the solubility of $CaCO_3$ in a carbonated solution depends on temperature, total pressure, partial pressure of CO_2, and radius of the pores (see page 464) in such a way that solution might occur in large pores or fissures while precipitation takes place in smaller pores. Assume uniform pore size and an average percolation rate v. Clearly, if v is much greater than the rate of solution, the fluid will travel a great distance through the rock before equilibrium is reached; conversely, if v is small, saturation, or near-saturation, will be reached at a short penetration distance. The problem has been carefully examined by Weyl [6] who concludes that the rate of solution of calcite is controlled by the rate of diffusion of solute away from the crystal-fluid boundary. Let the corresponding diffusion coefficient be D, while a is the radius of the straight circular capillaries through which the fluid is assumed to flow in laminar fashion. Weyl shows that the penetration, *i.e.*, the distance from the boundary at which the fluid reaches 90 per cent of saturation would be $0.57 \ v \ a^2/D$. For values of v corresponding to flow below the water table, this penetration depth would be measured in centimeters or millimeters; the thickness of the zone in which solution is actually taking place at any time would thus be very small. It would be larger in the case of slow reactions.

Consider next the more complicated case, discussed by Ramberg,[7] of a fluid carrying Na^+ ions percolating through a rock consisting of orthoclase, muscovite, and quartz. An exchange reaction might occur by which an Na-K feldspar would be formed at the expense of orthoclase, the final Na concentration in the feldspar depending on the Na^+ concentration in the fluid. At a certain distance from the boundary, the Na^+ concentration would have dropped below the level at which exchange is possible, while the K^+ concentration would have increased. In this manner, a zone of Na-K feldspar, with variable Na/K ratio, would develop. Next, it might happen that the equilibrium K^+ concentration of the fluid, as it leaves this zone, would be sufficient to cause the reaction

[6] P. K. Weyl, The solution kinetics of calcite, *Jour. Geology*, vol. 66, pp. 163–176, 1958.

[7] H. Ramberg, *The Origin of Metamorphic and Metasomatic Rocks*, pp. 199–200, University of Chicago Press, 1952.

$$2K^+ + KAl_3Si_3O_{10}(OH_2) + 6SiO_2 = 3KAlSi_3O_8 + 2H^+$$

Muscovite   Orthoclase

to proceed from left to right. Thus a second zone would develop, consisting of orthoclase + muscovite, or orthoclase + quartz, depending on the initial amount of quartz in the rock. The H^+ concentration would build up in turn, while the K^+ concentration dropped to the level at which reaction ceases; thus a third zone might form in which orthoclase is transformed to kaolin. The boundaries between these zones would be sharp (assuming slow percolation and rapid reactions), but would be moving forward in the direction of flow; eventually, if an unlimited supply of fluid were available, the whole rock would be metasomatosed to the assemblage Na-K feldspar + quartz. If the volume of fluid were limited, zoning might remain visible.

Korzhinsky's Metasomatic Zoning. Korzhinsky has made gallant attempts to define the rules of "infiltration zoning," [8] as distinct from "diffusion zoning" that arises when the fluid ascends through open fissures and reacts with the enclosing rock by diffusion through immobile pore-solutions in the walls of the fissure. Considering more particularly infiltration zoning, Korzhinsky notes that continuity conditions must be satisfied: the amount of any component of the fluid leaving any zone must be equal to the amount that entered it, plus or minus the amount added or subtracted by reaction within the zone. Thus the concentrations in successive zones are related to each other, and to the rate at which the zone boundaries move forward. On this basis, Korzhinsky arrives at the following rules: [9]

1. As the result of the action of percolating solutions on a multicomponent rock, a number of sharply delimited metasomatic zones must be formed, and these will increase simultaneously, the lower ones advancing upon the upper ones (for upward-moving fluid).

2. Replacement occurs only at the zone boundaries. Within the zones, only change in composition of variable minerals (solid solution) is possible.

3. In each zone, the number of minerals is equal to that of the inert components (see page 32).

4. From the upper to the lower zone, the number of inert components and the number of minerals gradually decrease, while the number of mobile components correspondingly increases.

[8] D. S. Korzhinsky, Differential mobility of components and metasomatic zoning in metamorphism, *18th Internat. Geol. Cong. Repts.,* Part III, pp. 65–72, London, 1950. A complete bibliography of Korzhinsky's work up to 1956 appears in J. B. Thompson, Local equilibrium in metasomatic processes (in *Researches in Geochemistry,* P. H. Abelson, ed., pp. 427–457, Wiley, New York, 1959). This paper also contains an excellent discussion of the phase rule and its application to metasomatism.

[9] Korzhinsky, *op. cit.,* p. 69, 1950.

5. The higher the solubility of a component and its ability to percolate with the solvent, the earlier its passage to the mobile state; inertness is maintained even in the lowest zones by the least soluble, *i.e.*, the least mobile components.

To illustrate Korzhinsky's view by a simple example, consider a rock originally consisting of three minerals A, B, C, while the percolating fluid is mainly oversaturated with mineral F. Assume that the rate of solution [10] is C > B > A. The succession of zones would then be as follows

$$\overset{\rightarrow}{F \mid F + A \mid F + A + B \mid A + B + C}$$

the final product, as zone boundaries move forward, being a monomineralic rock F. The tendency of metasomatic processes to decrease the number of phases is, according to Korzhinsky, in conflict with the suggested metasomatic origin of granites, which are polymineralic, with nearly constant ratios of the constituent phases.

The problem of metasomatic zoning is clearly a very difficult one. The petrologist observes spatial relations of minerals, from which he must infer also a time sequence of events in an essentially irreversible process. None of the many reactions involved is likely to be as simple as solution of some components with simultaneous deposition of others; it is more likely that a complicated sequence of reactions will take place between pore fluid and solid phases of variable composition. The initial composition of the pore fluid is, of course, unknown and must be deduced from the observations. Kinetic variables (rates of percolation, of solution, and of reaction) are undoubtedly important as are also effects arising from the curvature of interfaces occurring between solids and pore solutions (see page 464).

"Acid" Metasomatism.[11]   Examples will be given later (*e.g.*, "Kaolinization," page 577) of metasomatic reactions that depend on the concentration of hydrogen ions, or *p*H, of the metasomatic fluid. It is useful to remember in this connection that a neutral solution (*i.e.*, a solution in which the concentrations of H^+ and OH^- are equal) has a *p*H of 7 only at ordinary temperatures, as the dissociation constant of water varies with temperature; at 300°C. a neutral solution has a *p*H of about 5.8. In addition, the dissociation constants of acids (HCl, HF, H_2CO_3, etc.) also vary with temperature, with the result that at high temperature, water becomes a stronger acid than most common ones. For instance the reactions

[10] Note that Korzhinsky now mentions rate of solution rather than solubility [D. S. Korzhinsky, *Theory of infiltration metasomatic zoning*, Acad. sci. U.S.S.R., 1954 (English translation on pp. 19–32)].

[11] W. S. Fyfe, F. J. Turner, and J. Verhoogen, Metamorphic reactions and metamorphic facies, *Geol. Soc. America Mem. 73*, pp. 142–147, 1958.

$$NaCl + HOH = NaOH + HCl \text{ (gas)}$$
$$NaF + HOH = NaOH + HF \text{ (gas)}$$
$$CaF_2 + HOH = CaO + 2HF \text{ (gas)}$$

will run from left to right at sufficiently high temperatures. Although solutions escaping from an intrusive body may well carry very large amounts of acids that are strong at 25°C., there would actually be very little reaction between such acids and wall rock until the temperature has appreciably dropped. It is thus unlikely that much halogen fixation would occur at temperatures in the magmatic or high-grade metamorphic range. Acid or neutral conditions should be rare, and very localized, in natural high-temperature environments.

PRINCIPAL TYPES OF METASOMATISM

Outline. Four principal types of metasomatism were recognized by Goldschmidt: (1) metasomatism of silicate rocks and quartz rocks; (2) metasomatism of carbonate rocks; (3) metasomatism of salt deposits; (4) metasomatism of sulfide deposits. In the field of metamorphism we encounter chemical changes that fall within the first, and to a less extent the second, of these categories. The more detailed treatment of these that now follows is based upon that of Eskola,[12] who adopted a simplified version of Goldschmidt's elaborate chemical classification of metasomatic processes in silicate rocks, at the same time taking into account their geological environments. Eskola discusses metasomatism under five major headings: (1) alkali metasomatism, (2) lime metasomatism, (3) iron-magnesia-silicate metasomatism, (4) metasomatism with introduction of Si, Sn, B, Li, F, Cl, S, (5) carbon dioxide metasomatism.

Alkali Metasomatism. The simple compounds of the alkali metals, including silicates and carbonates, are relatively soluble in water, and in consequence alkali metasomatism is common in metamorphism of silicate rocks as well as in the latest stages of evolution of igneous rocks (zeolitization, development of spilites, crystallization of pegmatites).

On a chemical basis Goldschmidt classifies alkali metasomatism as follows:

1. Metasomatic exchange of alkalis (*e.g.*, formation of myrmekite; replacement of microcline by albite in pegmatites)

2. Fixation of alkali by excess Al_2O_3 in the precipitating rock (*e.g.*, introduction of albite or potash feldspar into the metamorphic derivatives of shales)

3. Fixation of alkali by silicates of iron and magnesium (*e.g.*, replacement of hornblende by biotite in granitization of amphibolite)

[12] P. Eskola, *Die Entstehung der Gesteine* (Barth, Correns, Eskola), pp 375–392, Springer, Berlin, 1939.

4. Reaction between alkaline aluminates (in solution) and quartz to give alkali feldspar in the contact zones of intrusions of nepheline syenite and other highly alkaline rocks (*e.g.*, contact conversion of granite to a metasomatic rock of syenitic composition—fenite)

In migmatite zones bordering deep-seated plutonic intrusions, there is complete gradation between metasomatic metamorphism and reactive assimilation, in which metasomatism also plays a part but which falls into the category of igneous rather than metamorphic phenomena. Eskola conveniently draws an arbitrary line between igneous and metamorphic reactions, according to whether the agent of metasomatism is on the one hand a truly magmatic water-rich silicate melt, or on the other the much more dilute aqueous solution of silica and other oxides (or its gaseous equivalent) which according to the experiments of Goranson is, from the moment of its origin, sharply differentiated from the residual magma proper. In the province of metamorphism, so limited, the most important instances of alkali metasomatism are those that involve fixation of alkali by excess alumina in pelitic rocks (group 2 above).

According to the frequently cited view of Goldschmidt, alkaline solutions expelled from crystallizing granite penetrate outward into pelitic rocks where, as the temperature diminishes, they now react with minerals such as micas and chlorites that contain excess alumina, with generation of alkali feldspar. A rock metasomatically enriched in alkali, in the form of potash feldspar or albite, results. A classic example is furnished by Goldschmidt's account of gradual transition, in the Stavanger area of Norway,[13] from quartz-muscovite-chlorite schists, resulting from normal regional metamorphism of pelitic sediments, through porphyroblastic albite schists of much more sodic composition [14] to augen gneisses in which potash also has been augmented and a granitic composition is approached.

The precise nature of the mineralogical changes in any given case of alkali metasomatism will be determined by the $(K_2O + Na_2O)/Al_2O_3$ ratios in the parent rock and in the introduced fluid, and by temperature and pressure. Other effects than growth of alkali-feldspar porphyroblasts may be expected in pelitic rocks. During the period of maximum temperature following immediately upon injection of the liquid magma, the adjacent pelitic rocks have been converted to hornfelses and schists containing such assemblages as

[13] V. M. Goldschmidt, Die Injektionsmetamorphose im Stavanger-Gebiete, *Kristiania Vidensk. Skr., Math.-Naturv. Kl. 10,* 1921.

[14] It should be noted that the presence of porphyroblastic albite in schists in general is not necessarily an indication that soda has been introduced into the rock. Nonmetasomatic metamorphic derivatives of basic igneous rocks and of graywackes commonly contain albite porphyroblasts (cf. F. J. Turner, Mineralogical and structural evolution of metamorphic rocks, *Geol. Soc. America Mem. 30,* p. 116, 1948).

Orthoclase-andalusite (pyroxene-hornfels facies)

or

Muscovite-andalusite ⎱
Muscovite-microcline ⎰ (hornblende-hornfels facies)

Later, alkali metasomatism may occur at lower temperatures or higher water pressures, well within the range of the hornblende-hornfels facies, and the composition of the metamorphic rock becomes progressively altered in a direction leading toward the microcline corner of the AKF diagram (Fig. 71, p. 514). Andalusite may now become replaced by muscovite (cf. muscovitization of sillimanite schist as described by Billings in New Hampshire [15]). In rocks originally containing no andalusite or sillimanite, alkali metasomatism takes the form of crystallization of microcline (or albite) at the expense of muscovite. The high-grade association orthoclase-andalusite becomes unstable even in the absence of alkali-bearing solutions and, provided adequate water pressure is maintained, is replaced by muscovite without addition of further potash. If the molecular ratio $(Na_2O + K_2O)/Al_2O_3$ approximates unity in the original rock, as in quartzo-feldspathic hornfelses without mica, alkali metasomatism can occur only by crystallization of nonaluminous alkaline minerals such as glaucophane, riebeckite, or aegirine—"fixation of alkali by silicates of iron and magnesium," in Goldschmidt's classification.

Alkali metasomatism is by no means confined to contact zones connected with intrusions of granitic and syenitic rocks. The residual solutions expelled from basic and even ultrabasic magmas are often sufficiently sodic to bring about notable enrichment of the neighboring rocks in soda. In the vicinity of sills of albite diabase or sometimes normal diabases, shales may be converted to fine-grained rocks, adinoles, composed largely of albite and quartz.

Lime Metasomatism. There are several ways in which lime metasomatism may enter into metamorphism:

1. Calc-silicates such as diopside, grossularite, idocrase, and scapolite sometimes form as constituents of reaction skarns at contacts between marble and silicate rocks (amphibolite and so on) during regional metamorphism. This is an instance of metamorphic diffusion with transfer of lime across the original limestone boundary. In central Sweden, ore bodies consisting of hematite or ferruginous quartz are bordered by reaction skarns of magnetite, diopside, andradite, and tremolite, formed by reaction with adjoining dolomite rocks during regional metamorphism.[16]

[15] M. P. Billings, Introduction of potash during regional metamorphism in western New Hampshire, *Geol. Soc. America Bull.*, vol. 49, pp. 289–302, 1938.

[16] N. H. Magnusson, The evolution of the lower Archaean rocks in central Sweden and their iron manganese and sulphide ores, *Geol. Soc. London Quart. Jour.*, vol. 92, pp. 332–359, 1936.

2. Metasomatic introduction of CaO into recently emplaced minor intrusions of granite, pegmatite, syenite, and related rocks may result in endometamorphic development of such minerals as hornblende, diopside, or scapolite, replacing primary igneous minerals. Granitic rocks containing diopside more commonly originate, however, by direct crystallization of the calc-silicates from magmas that have absorbed lime by assimilative reaction with limestone—a process outside the scope of metamorphism.

3. Metamorphism of initially noncalcareous sediments to calc-silicate rocks at contacts with intrusions of peridotite, and metasomatic replacement of plagioclase by grossularite, hydrogrossular, prehnite, or idocrase in diorite or gabbro (rodingite) dikes cutting serpentinite, are well-known instances of lime metasomatism connected with ultrabasic intrusions.[17] The added lime here is probably magmatic. Some writers regard it as having been concentrated in residual solutions left after crystallization of the ultrabasic "magma," while others envisage it as having been set free during serpentinization of diopsidic pyroxene in the peridotite.

4. In low-grade metamorphism of basic igneous rocks, removal of CaO and Al_2O_3 in aqueous solution, with complementary introduction of epidote or prehnite into adjacent rocks, is a common phenomenon. Epidotization and prehnitization of basic lavas of the spilitic kindred [18] have already been discussed (page 268).

Iron-Magnesia-Silicate Metasomatism. For descriptive purposes it is convenient to distinguish two principal types of iron-magnesia-silicate metasomatism, namely, introduction of iron and magnesium into limestone, with crystallization of calc-silicates rich in these metals, and metasomatic development of noncalcic ferromagnesian minerals in silicate rocks and in quartzites.

To the first category belong skarns [19] formed in contact zones by reaction between limestone and iron- and silica-bearing solutions or gases

[17] F. J. Turner, The metamorphic and intrusive rocks of southern Westland, *New Zealand Inst. Trans.*, vol. 63, pp. 269–276, 1933; J. Suzuki, On the rodingitic rocks within the serpentinite masses of Hokkaido, *Hokkaido Univ. Fac. Sci. Jour.*, ser. 4, vol. 8, pp. 419–430, 1953 (reprinted in *Tsuboi Commemorative Volume*, Tokyo, 1955).

[18] For an account of the interplay of soda and lime metasomatism in connection with the evolution of spilitic rocks (a case of autometasomatism) see Turner, *op. cit.*, pp. 120–124, 1948.

[19] P. Eskola, On the petrology of the Orijärvi region in south-western Finland, *Comm. géol. Finlande Bull. 40*, pp. 225–233, 241, 1914; A. F. Buddington, Adirondack igneous rocks and their metamorphism, *Geol. Soc. America Mem. 7*, pp. 44, 168, etc., 1939; W. T. Holser, Metamorphism and associated mineralization in the Philipsburg region, Montana, *Geol. Soc. America Bull.*, vol. 61, pp. 1053–1090, 1950; C. E. Tilley, The zoned contact skarns of the Broadford area, Skye, *Mineralog. Mag.*, vol. 29, pp. 621–666, 1951; G. C. Kennedy, Geology and mineral deposits of Jumbo Basin, Southeastern Alaska, *U.S. Geol. Surv. Prof. Paper 251*, pp. 13–35, 1953.

from intrusive masses of granite, granodiorite, etc. Andradite and heden-bergite-rich pyroxene are the most prominent constituents of skarns formed from pure limestones, but diopside, tremolite, and phlogopite may appear when the parent rock was dolomitic. Also conspicuous in skarn assemblages are hematite, quartz, hornblende, wollastonite, and various calc-silicates, while wide distribution of fluorite, scapolite, and various metallic sulfides points to possible importance of such elements as fluorine, chlorine, and sulfur as "carriers" of iron. In some cases metasomatism has been pneumatolytic, the iron having been introduced in some volatile form such as FeF_3 or $FeCl_3$; but more generally hydrothermal reactions following upon, and sometimes much later than, the main phase of thermal metamorphism, and governed by somewhat lower temperatures, seem to have been responsible for development of skarns. At times it is difficult to distinguish between skarns of the type just described, reaction skarns formed by lime metasomatism of silicate rocks and iron ores adjacent to limestones, and calc-silicate rocks formed by normal contact metamor-phism of impure limestones without accession of material from extrane-ous sources.

The classic examples of magnesia-silicate metasomatism in silicate rocks and quartzites occur in the pre-Cambrian of Finland and Sweden.[20] In the Orijärvi district of Finland, introduction of magnesium, iron, and silicon from the Orijärvi granodiorite led not only to development of tremolite, hedenbergite, hornblende, and andradite-hedenbergite skarns in adjacent limestones, but also to metasomatic transformation of quartzo-feldspathic leptites into cordierite-anthophyllite gneisses with comple-mentary removal of calcium and sodium.[21] At the same time amphibo-lites were converted to cummingtonite amphibolites by exchange of mag-nesium for calcium in the original hornblende. Contemporaneously with the magnesia-iron metasomatism local introduction of sulfides gave rise to copper sulfide ores. In central Sweden a similar sequence of events was complicated by premetamorphic concentration of iron and manganese in certain members of the leptite formation. These gave rise to quartz-hematite ores, quartz-magnetite ores, and ferruginous skarn ores during the main phase of regional metamorphism and granite intru-

---

[20] Eskola, op. cit., pp. 167–222, 252–264, 1914. P. Geijer, Recent work on the Archaean sulphide ores of Fenno-Scandia, Econ. Geology, vol. 16, pp. 279–288, 1921; Magnusson, op. cit., pp. 343–344, 1936.

[21] Recently H. V. Tuominen and T. Mikkola (Soc. géol. Finlande Compte rendu, no. 23, pp. 67–92, 1950) have shown that metasomatism at Orijärvi has been localized along limbs and crests of folds. But their view that the cordierite and anthophyllite rocks are products of kinematically controlled metamorphic differentiation has been criticized by Eskola (ibid., pp. 93–102) who still assigns an important role in the origin of these rocks and associated sulfide ores to fluids derived from the Orijärvi granodiorite.

sion, but a somewhat later magnesia metasomatism (with accompanying introduction of sulfides) caused replacement of ferruginous calc-silicates by noncalcic ferromagnesian silicates such as gedrite and cummingtonite in the skarns, and metasomatic development of cordierite and gedrite schists from originally quartzo-feldspathic leptites. The exact nature of the agent of metasomatism is not clear in either case. There are many recorded instances where iron metasomatism has almost certainly been made possible through the activity of chlorine or fluorine as "carriers" of iron, even though these elements have, in the course of metasomatism, formed highly soluble compounds, and so have passed completely from the system, without leaving any imprint upon the mineral assemblages of the resulting skarns. It is therefore possible that some unknown "carrier" of magnesium has functioned similarly in the magnesia metasomatism just discussed. It is also possible, and on the whole more likely, that dilute aqueous solutions of magnesium salts emanating from the cooling granite have caused insoluble silicates of magnesium to crystallize in place of silicates containing calcium or sodium, the corresponding calcic or sodic compounds having been more soluble than the magnesian salts under prevailing conditions of metasomatism. Petrologists who regard granites as products of metasomatism in place interpret basic (desilicated) contact zones of cordierite and anthophyllite rocks as "basic fronts" (magnesium fronts) enriched in magnesium expelled from the adjacent zone of granitization.[22]

Metasomatism with Introduction of Si, Sn, B, Li, F, Cl, or S. *Special Mineral Assemblages.* The complex effects of pneumatolytic and hydrothermal introduction of nonmetallic elements into rocks adjacent to plutonic intrusions and the economically important processes of ore genesis connected therewith cannot be considered here in any detail. Special mineral assemblages resulting from this kind of metasomatism commonly conform to one or other of the types listed below.[23]

Boron Metasomatism. Boron metasomatism near intrusions of granite or granodiorite may cause a borosilicate (tourmaline, axinite, or rarely datolite and danburite) or even a borate (ludwigite, kotoite) to appear in the metamorphic assemblage. A classic example is provided by the rocks in the contact aureoles of the granites of Cornwall and Devon,[24] where

---

[22] Details and illustrations of this interpretation of alkali and iron-magnesia-silicate metasomatism are given by D. L. Reynolds (The sequence of geochemical changes leading to granitization, *Geol. Soc. London Quart. Jour.*, vol. 102, pp. 389–446, 1946).

[23] Cf. Goldschmidt, *op. cit.*, 1922; Eskola, *op. cit.*, pp. 387–390, 1939.

[24] J. S. Flett, The geology of the country around Bodmin and St. Austell, *England Geol. Survey Mem.*, 1909, pp. 65–67, 101–104; A. A. Fitch, Contact metamorphism in southeastern Dartmoor, *Geol. Soc. London Quart. Jour.*, vol. 88, pp. 576–609, 1932; S. O. Agrell, Dravite-bearing rocks from Dinas Head, Cornwall, *Mineralog. Mag.*, vol. 26, pp. 81–93, 1941.

tourmalinization of aluminous rocks (slates and pelitic hornfelses), and local complete autometasomatism of the granite itself, have yielded the same ultimate assemblage, namely, tourmaline-quartz—another instance of mineralogical convergence in metasomatism. Clearly, removal of potash and other bases is involved in extensive tourmalinization of this kind. So too, pneumatolytic development of axinite in calcareous rocks within the same aureoles is part of a complex metasomatism, since association of axinite with andradite and hedenbergitic pyroxene points to simultaneous introduction of boron and iron.

As contrasted with local crystallization of abundant tourmaline which is clearly referable to emanations from adjacent granites, the widespread dissemination of small amounts of tourmaline in regionally metamorphosed schists presents a special problem. Goldschmidt and Peters [25] have shown that argillaceous marine sediments and some sedimentary iron ores may contain sufficient boron (0.1 per cent B_2O_3) to account for the presence of tourmaline as a minor constituent of their metamorphic derivatives; and therefore much of the tourmaline of pelitic or ferruginous schists can probably be attributed to the boron present in the parent sediment. This explanation is inadequate, however, when the tourmaline-bearing rocks are of other than pelitic composition, especially in the cases of greenschists derived from basic lavas and of marbles formed from limestones and dolomites, where the parent rocks are consistently poor in boron.

Fluorine Metasomatism. Introduction of fluorine, as of boron, tends to take place in the vicinity of granitic intrusions. Autometasomatism of granite [26] involves replacement of potash feldspar by muscovite and topaz, with complementary removal of potash, and leads to local development of quartz-muscovite-topaz rock (greisen), with cassiterite and lepidolite as possible additional constituents where volatile compounds of tin and lithium have been present in the magmatic gases. Special fluorine-bearing phases resulting from fluorine metasomatism of calcareous and dolomitic rocks include fluorite, fluor-apatite, members of the chondrodite-humite group, and rarely custerite; but phlogopite, idocrase, and pargasite may all contain appreciable amounts of fluorine, and their presence in calc-silicate rocks may often be due to fluorine metasomatism in conjunction with contact metamorphism. Simultaneous introduction of fluorine, boron, and iron into dolomitic marble is not uncommon.[27]

Scapolitization. Appearance of scapolite in a metamorphic assemblage is possible when chlorine or carbon dioxide are participating components

[25] V. M. Goldschmidt and C. Peters, Zur Geochemie des Bors, *Gesell. Wiss. Göttingen Nachr., Math.-phys. Kl.,* pp. 403–407, 528–545, 1932.

[26] Flett, *op. cit.,* pp. 67–68, 1909.

[27] Tilley, *op. cit.,* 1951.

in the metamorphic system.[28]   In some rocks scapolite takes the place of plagioclase, while in others the two minerals are associated.   Sundius, in a detailed discussion of the occurrence and origin of scapolites in rocks of the Kiruna region of Sweden, pointed out that, while scapolitization is usually a metasomatic (pneumatolytic) process, lime scapolites may also develop in limestones during normal regional or contact metamorphism, the necessary volatile components (CO_2, H_2O, SO_3) being supplied by materials already present in these rocks.   Pneumatolytic scapolites are generally of more sodic composition, and correspondingly richer in chlorine, than scapolites formed by normal metamorphism.   They occur not only in calcareous rocks, where their presence implies addition of soda as well as of chlorine, but also in a variety of silicate rocks such as amphibolites and metagabbros.   In northern Queensland [29] soda-chlorine metasomatism of calcareous shales has produced large bodies (10 miles by 2,000 to 5,000 feet) of scapolite-pyroxene granulite containing 2 lbs. of chlorine per cubic foot.   Pneumatolytic origin of scapolite rocks may be clearly demonstrable when they occur associated with skarns in contact aureoles; but there are also instances of widespread metasomatic scapolitization, remote from igneous intrusions, to which the term "regional-pneumatolytic metamorphism" has been applied.   Thus, throughout much of the Kiruna region, development of scapolite in amphibolites and other feldspathic rocks is in no way related to proximity of outcrops of igneous intrusions but has nevertheless been shown by Sundius to be due to introduction of Cl and CO_2 from magmatic gases or solutions, which have also supplied small amounts of such elements as phosphorus, sulfur, titanium, iron, and copper now present in associated apatite, sphene, and sulfide minerals.   In the Kiruna region, as also in the Norwegian apatite veins, scapolitization is genetically connected with intrusion of basic magma (gabbro, norite, pyroxenite).   Sometimes, however, scapolite-bearing rocks occur in aureoles surrounding intrusions of acid composition, as exemplified by the granite, soda granite, and nordmarkite intrusions of the Oslo district.   Again the widespread presence of scapolite in metamorphosed gabbros in the northwestern Adirondacks is attributed by Buddington to the influence of chlorine-bearing emanations from intrusive bodies of granitic magma.   Alternatively, the general phenomenon of scapolitization of gabbroid rocks has been explained by some writers as a process of autometasomatism analogous to saussuritization.

Sulfur Metasomatism.   Dissemination of pyrite and pyrrhotite through skarns and magnesium-iron-silicate rocks formed by high-temperature

[28] N. Sundius, Geologie des Kirunagebiets, 4, *Vitensk. Prakt. Unders. Lappland,* pp. 195–224, 1915; Buddington, *op. cit.,* pp. 260–267, 1939.

[29] A. B. Edwards and B. Baker, Scapolitization in the Cloncurry district of northwestern Queensland, *Geol. Soc. Australia Jour.,* vol. 1, pp 1–33, 1954.

metasomatism, or through sericite quartzites produced at lower temperatures by alkali metasomatism of quartzo-feldspathic rocks, has been attributed to reaction between iron-bearing silicates (*e.g.*, chlorite or biotite) and magmatically derived hydrogen sulfide. It has been suggested by Goldschmidt [30] that the commonly observed association of graphite and sulfides in rocks of this kind is due to reaction between ferruginous silicates and volatile compounds of carbon and sulfur such as CS_2 and COS. At still lower temperatures and pressures, andesitic rocks are hydrothermally altered by alkaline solutions of H_2S to propylites—mixtures of quartz, chlorite, epidote, alkali feldspars, and zeolites, with disseminated pyrite. Under similar physical conditions feldspars and other aluminous silicates are converted to alunite by acid solutions of sulfur compounds (SO_3, SO_2, sulfates, or free sulfuric acid). Local deposits of metallic sulfide ores are commonly associated with rocks that have been affected on a broader scale by any of the above-mentioned types of sulfur metasomatism.

Kaolinization, Sericitization. Eskola has summarized in the following words the important conclusions reached experimentally by Noll as to the physical conditions governing the alternative development of kaolin, montmorillonite, sericite, analcite, and pyrophyllite, by hydrothermal metasomatism of feldspars, feldspathoids, and so on: [31]

Kaolin originates from SiO_2- or Al_2O_3-gels in neutral solutions free of alkali metals or in acid solutions containing alkali metals, at temperatures below 400°C.; on the other hand, montmorillonite forms in alkaline solutions of the alkali metals, and at higher concentrations of potash, sericite appears. At very high concentrations of alkali, zeolites, especially analcite, appear. Pyrophyllite originates in silica-rich systems under conditions otherwise the same as for kaolin, only, in contrast to this latter at temperatures upward from 400°C. Geological experience confirms the conclusion that pyrophyllite is a high-temperature hydrothermal form. Also sericite has often been formed at the expense of feldspars. At lower temperatures (below 400°C.) it depends upon the alkali concentration of the active solutions whether zeolites, sericite, or kaolin are formed. From the province of hydrothermal metasomatism of feldspars these processes pass over to the hydrolytic breaking up of silicate minerals in weathering.

The system has recently been reinvestigated by Hemley,[32] who determined experimentally the stability fields of some of the important minerals as a function of temperature and of the ratio of potassium to hydrogen ion concentration. Hemley finds that, at a given temperature and pres-

[30] Goldschmidt, *op. cit.*, p. 117, 1922.

[31] Eskola, *op. cit.*, p. 390, 1939.

[32] J. J. Hemley, Some mineralogical equilibria in the system K_2O-Al_2O_3-SiO_2-H_2O, *Am. Jour. Sci.*, vol. 257, pp. 241–270, 1959.

sure, with increasing K^+/H^+ ratio the fields of kaolinite, mica, and potash feldspar are successively traversed; the same sequence is observed when the temperature is increased at constant K^+/H^+ ratio. Above about 350°C., at a water pressure of about 1,000 bars, the assemblage mica-kaolinite changes to mica-pyrophyllite-boehmite. As the temperature is raised, the equilibrium mica kaolinite, or mica K-feldspar, requires greater concentration of acid, in keeping with the general trend mentioned on page 568.

Carbon Dioxide Metasomatism. Whereas silica readily displaces carbon dioxide from carbonates at moderate and high temperatures, many silicates are converted with equal ease to carbonates by hydrothermal reaction with solutions containing carbon dioxide or soluble carbonates, at low temperature. Autometasomatic replacement of such minerals as feldspars, augite, and olivine by carbonates is a common deuteric process illustrated by igneous rocks of widely different composition. Magnesian silicates seem particularly susceptible to this type of alteration. Hydrothermal metamorphism of serpentinites and actinolite rocks to talc-carbonate rocks will be discussed further in the next section. A fine example of carbon-dioxide metasomatism in connection with gold deposition is afforded by A. Knopf's account of hydrothermal alteration of rocks adjacent to the auriferous quartz veins of the Mother Lode system of California.[33] Immense quantities of carbon dioxide have been introduced into the wall rocks (greenstones, slates, quartzites, schists, serpentinites), which, regardless of their original nature, show every stage of replacement by ankerite, with sericitization, albitization, and widespread dissemination of minor amounts of pyrite, arsenopyrite, and gold, as cognate results of metasomatism. The quartz of the lode system represents part of the silica that was displaced from the wall rocks by carbon dioxide and so passed into the circulating solutions during the process of ankeritization.

The gold, sulphur, carbon dioxide and certain other constituents were probably supplied by exhalations that issued at a high temperature from a deep-seated consolidating granitic magma, as was also a part of the thermal energy of the ore-forming solutions. After these exhalations had condensed to water that carried the other constituents dissolved in it, the motive power that caused this "magmatic water" to rise was doubtless the gravity potential of a meteoric circulation whose paths were determined by the fissure systems.[34]

Steatitization of ultrabasic rocks, defined by Hess [35] as "that process of hydrothermal alteration of an ultrabasic which in its final stages results

[33] A. Knopf, The Mother Lode system of California, *U.S. Geol. Survey Prof. Paper* 157, 1929.

[34] Knopf, *ibid.*, p. viii.

[35] H. H. Hess, The problem of serpentinization, and the origin of certain chrysotile asbestos talc and soapstone deposits, *Econ. Geology*, vol. 28, pp. 634–657, 1933.

in the formation of a talcose rock," may be accomplished simply by addition of silica, and in some cases water, to serpentinized peridotites. More commonly, carbon dioxide metasomatism is involved, and dolomite or magnesite then appear as constituent phases of the end product. The opinion is widely held that steatitization, unlike serpentinization, which often precedes it, is a hydrothermal process connected with intrusion of granitic magma. But there are cases where the only available sources of the active solutions are the ultrabasic body itself or the geosynclinal sediments in which it is enclosed. Here it would seem that steatitization is a local aftermath of serpentinization, caused by prolonged activity, at lower temperatures, of waters similar in origin to those which had been responsible for earlier serpentinization (cf. pages 320 and 321). A number of reactions leading to development of talc-carbonate schists are possible.[36]

1. By simple addition of CO_2, serpentine may be converted to talc-magnesite rock without appreciable change in volume.

$$2H_4Mg_3Si_2O_9 + 3CO_2 \rightarrow H_2Mg_3Si_4O_{12} + 3MgCO_3 + 3H_2O$$

Serpentine	Talc	Magnesite
(220 cc.)	(140 cc.)	(84 cc.)

In the presence of lime-bearing solutions, dolomite may form instead of magnesite by a reaction involving exchange of CaO for MgO, a possible equation being

$$2H_4Mg_3Si_2O_9 + 1.23CaO + 2.46CO_2 \rightarrow$$

Serpentine   (68 gm.)   (108 gm.)
(552 gm.; 220 cc.)

$$H_2Mg_3Si_4O_{12} + 1.23CaMg(CO_3)_2 + 1.77MgO + 3H_2O$$

Talc   Dolomite   (71 gm.)   (54 gm.)
(378 gm.; 140 cc.)   (226 gm.; 80 cc.)

2. At higher temperatures, ultrabasic rocks that originally contain CaO, or are exposed to lime-bearing solutions, tend to give actinolite rock or (if Al_2O_3 is present) actinolite-chlorite rock instead of serpentine, as products of hydrothermal metamorphism. At lower temperatures in the greenschist facies, partial substitution of CO_2 for SiO_2 subsequently gives the assemblage talc-dolomite in place of actinolite, while chlorite likewise tends to be replaced by talc, provided Al_2O_3 and some MgO can be removed in solution from the system. Equations for reaction without change of volume are

[36] H. H. Hess, Hydrothermal metamorphism of an ultrabasic intrusion at Schuyler, Virginia, *Am. Jour. Sci.*, vol. 26, pp. 377–408, 1933; H. H. Read, The metamorphic geology of Unst in the Shetland Islands, *Geol. Soc. London Quart. Jour.*, vol. 90, pp. 662–666, 1934; P. Haapla, On the serpentine rocks in northern Karelia, *Comm. géol. Finlande Bull. 114*, 1936.

(1)

$$Ca_2Mg_5Si_8O_{22}(OH)_2 + 4CO_2 \rightarrow 2CaMg(CO_3)_2 + H_2Mg_3Si_4O_{12} + 4SiO_2$$

Tremolite	Dolomite	Talc	(Removed
(810 gm.; 270 cc.)	(368 gm.; 130 cc.)	(378 gm.; 140 cc.)	in solution)

and (2)

$$4H_4Mg_2Al_2SiO_9 + 6H_4Mg_3Si_2O_9 + 13.2SiO_2 \rightarrow$$

Pennine
(2,768 gm.; 1,025 cc.)

$$7.3H_2Mg_3Si_4O_{12} + 4.1MgO + 4Al_2O_3 + 12.7H_2O$$

Talc   (Removed in solution)
(2,759 gm.; 1,022 cc.)

The chemical data recorded by Hess for an occurrence of soapstone at Schuyler, Virginia, show that the course of hydrothermal metamorphism has there been governed by just such reactions.[37]

3. At low metamorphic temperatures and sufficiently high pressures of CO_2, even talc becomes unstable when exposed to solutions rich in CaO and undergoes progressive replacement by dolomite, giving dolomite rock as the ultimate product of metasomatism. An equation for dolomitization of talc without change in volume is

$$H_2Mg_3Si_4O_{12} + 2.15CaO + 4.3CO_2 \rightarrow$$

Talc
(378 gm.; 140 cc.)

$$2.15CaMg(CO_3)_2 + 0.85MgO + 4SiO_2 + H_2O$$

Dolomite   (Removed in solution)
(396 gm.; 141 cc.)

4. Alternatively dolomite may form by direct replacement of serpentine without the intervening appearance of talc. The composition of the active solutions, especially as regards concentration of lime and carbon dioxide, is probably the factor determining whether talc and dolomite, or dolomite alone, are end products of metasomatism.

5. At sufficiently high pressures of CO_2 under conditions allowing removal of silica from the system, talc may be converted to magnesite-quartz rock at the lowest temperatures of hydrothermal metamorphism. The equation for reaction without change of volume is

$$H_2Mg_3Si_4O_{12} + 3CO_2 \rightarrow$$

Talc
(378 gm.; 140 cc.)

$$3MgCO_3 \quad + \quad 2.51SiO_2 \quad + \quad H_2O \quad + \quad 1.49SiO_2$$

Magnesite	Quartz		Silica, removed in solution
(252 gm.; 84 cc.)	(151 gm.; 56 cc.)		(89 gm.)

Direct replacement of serpentine by an equal volume of magnesite and quartz under similar conditions could be expressed as follows:

[37] Turner, *op. cit.*, pp. 133–135, 1948.

$H_4Mg_3Si_2O_9 + 3CO_2 \rightarrow$
Serpentine
(276 gm.; 110 cc.)

$$3MgCO_3 \quad + \quad 1.17SiO_2 \quad + \quad 2H_2O \quad + \quad 0.83SiO_2$$

Magnesite	Quartz		Silica, removed in solution
(252 gm.; 84 cc.)	(70 gm.; 26 cc.)		(50 gm.)

The end product of this second reaction would consist of 78 per cent magnesite plus 22 per cent quartz. Both equations given above are based upon the assumption that the material removed in solution is silica. If, however, constancy of volume during metasomatism were maintained by removal of magnesia in solution, then the end product of replacement of serpentine would consist of 62 per cent magnesite plus 38 per cent quartz. Wellman has described, from the Cobb River district of New Zealand, some magnesite-quartz rocks formed on an extensive scale by hydrothermal replacement of serpentine, through an intermediate talc-magnesite stage.[38] The average composition of the completely altered rocks is 76 per cent magnesite plus 24 per cent quartz, and all the analyzed rocks representing various stages in replacement of talc-magnesite by magnesite-quartz have an excess of magnesite over the amount (68 per cent) demanded by the equation in a closed system:

$H_4Mg_3Si_2O_9 + 3CO_2 \rightarrow$
Serpentine

$$1\tfrac{1}{2}MgCO_3 + \tfrac{1}{2}H_2Mg_3Si_4O_{12} + 1\tfrac{1}{2}CO_2 + 1\tfrac{1}{2}H_2O \rightarrow$$
Magnesite   Talc

$$3MgCO_3 + 2SiO_2 + 2H_2O$$
Magnesite   Quartz

Metasomatism of the Cobb River rocks has therefore been accompanied by removal of silica, and the composition of the end product agrees closely with that demanded by reaction without volume change.

METAMORPHIC DIFFERENTIATION

The term *metamorphic differentiation* was introduced into geological literature by Stillwell to cover collectively the various processes by which contrasted mineral assemblages develop from an initially uniform parent rock during metamorphism.[39] Segregations of biotite-hornblende and of epidote-labradorite enclosed in amphibolite, complementary hornblendic and feldspathic layers in amphibolite, and garnet porphyroblasts in otherwise fine-grained schists were cited as typical products of metamorphic differentiation. To the more general process of migration of rock com-

[38] H. W. Wellman, Talc-magnesite and quartz-magnesite rock, Cobb-Takaka district, *New Zealand Jour. Sci. Technology*, vol. 24, no. 3B, pp. 103B–127B, 1942.

[39] F. L. Stillwell, The metamorphic rocks of Adelie Land, *Australasian Antarctic Exped., 1911–1914, Sci. Repts.*, ser. A, vol. 3, pt. 1, 1918.

ponents during metamorphism, whether or not segregation of the migrating materials led to development of definite metamorphic differentiates, the term *metamorphic diffusion* was applied by the same writer. The importance of metamorphic diffusion in obliterating or blurring some premetamorphic lithological boundaries was also emphasized. The fact that certain materials migrate and are capable of segregation during metamorphism had, of course, been recognized by various writers prior to Stillwell's enunciation of the problem. For over a decade the concept of metamorphic differentiation was neglected except in the writings of Stillwell himself and of workers who had at some time been associated with him.[40] British opinion over this period was largely influenced by the views of Harker, who, arguing from his experience that delicate features of rock fabric such as fine bedding frequently survive complete metamorphism (especially in undeformed rocks), concluded that metamorphic diffusion is effective only over very short distances, and that metamorphic differentiation has a strictly limited if not negligible scope. In 1932 Eskola,[41] at that time unaware of Stillwell's contributions to the subject, independently drew attention to the petrogenic importance of metamorphic differentiation and discussed in some detail the chemical principles involved. Under this stimulus several British and American petrologists have since published accounts of phenomena of this kind, and metamorphic differentiation has now assumed an established status among the recognized processes of rock metamorphism.[42]

Stillwell discussed metamorphic differentiation in terms of elemental chemical processes—solution, "solid diffusion," and "force of crystallization" of growing porphyroblasts. Eskola envisaged the process as dominated by three more or less independent "principles" which he termed the concretion principle (illustrated by growing porphyroblasts) and the principles of solution (of locally unstable materials) and of enrichment in the most stable constituents (by local precipitation). The value of such analysis of a long-continued complex physicochemical mechanism is doubtful. The writers would rather regard metamorphic differentiation as a result of differential migration of the component ions of the metamorphic system through short distances, under the influence of local

[40] *E.g.*, J. A. Dunn, The geology of North Singhbhum including parts of Ranchi and Manbhum districts, *India Geol. Survey Mem.*, vol. 54, 1929.

[41] P. Eskola, On the principles of metamorphic differentiation, *Comm. géol. Finlande Bull.* 97, pp. 68–77, 1932.

[42] *E.g.*, F. J. Turner, The development of pseudostratification by metamorphic differentiation in the schists of Otago, New Zealand, *Am. Jour. Sci.*, vol. 239, pp. 1–16, 1941; S. Gavelin, Lime metasomatism and metamorphic differentiation in the Adak area, *Sveriges geol. Undersökning*, Arb. 45 (1951), no. 2, pp. 1–52, 1952; W. A. Roberts, Metamorphic differentiates in the Blackbird mining district, Idaho, *Econ. Geology*, vol. 48, pp. 446–456, 1953.

gradients in chemical potential. These gradients may be determined by such factors as (1) differences in pressure or nonhydrostatic stress, as between rock and open fissures; (2) differences in size and shape of grains (*i.e.*, in surface energy); (3) exsolution of forcign matter from crystals in process of inversion. Moreover, different types of ions tend to have different rates of migration under given conditions of temperature and pressure, since the rate of ionic migration is influenced by such factors as size of ions, their relative concentrations in pore solutions (influenced in turn by the relative solubilities of the various minerals involved), and rate of change of their chemical potentials in response to any of the variable factors just mentioned. Finally, it must be remembered that ionic migration and localized recrystallization need not necessarily lead to metamorphic differentiation. On the contrary, they not infrequently obliterate or blur originally sharp surfaces of mineralogical discontinuity in initially heterogeneous rocks.

One of the most familiar results of metamorphic differentiation is growth of porphyroblasts of such minerals as garnet, albite, and cordierite. Equally common is development of veins or laminae of simple mineral composition in initially homogeneous rocks. Though the laminae so produced are individually thin (usually only a few millimeters, seldom more than a few centimeters thick), the process whereby they form tends to be effective over large areas. Common instances are quartz-albite veins in low-grade pelitic and quartzo-feldspathic schists, veins of epidosite (quartz-epidote) in amphibolites and greenschists, and quartz-calcite veins in many kinds of low-grade schists. Read [43] has discussed in some detail the origin of kyanite-bearing quartz veins occurring in high-grade pelitic schists on Unst in the Shetland Islands. The normal pelitic schist is a highly aluminous rock consisting mainly of muscovite, quartz, kyanite, chloritoid, staurolite, and iron ore, but the rock immediately adjacent to any of the quartz-kyanite veins contains no quartz and is very much richer in kyanite, which here typically makes up between 65 per cent and 75 per cent of the total composition. The quartz-kyanite veins are regarded as products "of endogenous secretion during metamorphism"—a process of solution and recrystallization of the two component minerals through the medium of an aqueous pore fluid, the total volume of which was at no stage more than a small fraction of the volume occupied by the crystalline phases of the system. The veins and the adjoining kyanite-rich rocks are complementary products of metamorphic differentiation.

A typical instance where metamorphic differentiation of this type has led to regional development of laminated structure in schists is afforded by the quartzo-feldspathic schists of the greenschist facies that outcrop

[43] H. H. Read, On quartz-kyanite rocks in Unst, Shetland Islands, *Mineralog. Mag.*, vol. 23, pp. 317–328, 1933.

over an area of 500 to 700 square miles in the province of Otago, New Zealand.[44] These are coarse-grained blastophyllonites, which have evolved from parent massive graywackes by intense mechanical granulation, accompanied in the later stages by mineralogical reconstitution under the influence of slowly rising and finally long-sustained temperature. The schistosity is in the main subhorizontal or gently dipping and has developed parallel to the principal shear surfaces of the deformed rocks. It is accentuated by the laminated structure now under discussion—"foliation" in the sense used by Harker. Alternate laminae, a few millimeters in thickness, are composed principally of quartz and albite on the one hand, and of chlorite, epidote, and muscovite on the other, but each of these assemblages contains in small quantities the minerals that dominate the other. Furthermore, in a given rock specimen the same variety of a variable mineral series (as chlorite, epidote, stilpnomelane, and amphibole) is usually common to both types of laminae, a fact which implies attainment of chemical equilibrium throughout the rock as a whole, so that free diffusion of materials, not only within the limits of a single layer but between adjacent layers, too, must have been possible during metamorphism. Comparison of chemical analyses of laminated schists with those of slightly metamorphosed or unaltered graywackes from adjoining districts shows that the bulk composition has remained approximately constant during chemical reconstitution of the schists. Nor does the composition of the leucocratic layers (quartz-albite or quartz-albite-calcite) persistently present in all types of schist in this region suggest derivation from a magmatic source. Rather the minerals in question are those that are most readily dissolved and redeposited under the prevailing conditions of low-grade metamorphism, so that they tend in consequence to become mutually associated in any rock of suitable chemical composition (pelitic and quartzo-feldspathic schists or greenschists). This conclusion accords with the general thesis of Read that the quartz-rich veins and stringers which are so widely distributed in regionally metamorphosed rocks of all grades are mainly products of metamorphic differentiation, derived from the rocks in which they occur. Laminated structure of the type just described is very general in metamorphic rocks, especially those of the greenschist facies, and has often been mistaken for bedding inherited from the parent rock. Its true origin can only be demonstrated beyond doubt in regions such as southern New Zealand where there is complete transition from laminated schists into massive parent graywackes in which bedding is seldom recognizable.

Since deformation plays a conspicuous part in metamorphism leading to development of laminated schists in general, and of other rocks (such as albite-chlorite-epidote schists) whose chemical compositions have been

[44] Turner, op. cit., 1941.

appreciably changed by metamorphic differentiation, it is appropriate to consider the extent to which purely mechanical processes may contribute to metamorphic differentiation. Deformation synchronous with crystallization is visualized by Sander [45] as the combined result of two contrasted types of movements that act simultaneously: direct componental movements involving gliding or rotation or other differential movements of crystals, and indirect componental movements of a chemical nature, which include transport of material by solution, diffusion, and redeposition, or by migration of the pore fluids themselves, in so far as such movements are related to the process of deformation. The part played by indirect componental movements in metamorphic differentiation has already been discussed. The auxiliary effects of direct componental movements (the mechanical factor in deformation) may be illustrated by referring to a particular case that has just been described. In the province of Otago, southern New Zealand, the earlier stages of transition from massive graywacke to quartz-albite-epidote-chlorite-muscovite schist are characterized by deformation in which direct componental movement (rupture and displacement of grains) greatly outweighs indirect movement which is expressed in incipient recrystallization and neomineralization. This essentially cataclastic metamorphism, by reducing the grains to slate-like fineness and by initiating schistosity (shear surfaces) which facilitates movement of pore solutions, greatly increases the effectiveness of solution, diffusion, and chemical reaction, which in any case are accelerated as the temperature rises. Ultimately, as described in a previous section, rocks that still lie well within the chlorite zone become completely reconstituted and develop a conspicuously laminated structure parallel to the main schistosity as a result of metamorphic differentiation.

Sander and Eskola [46] emphasize the interplay of mechanical and chemical activity in the process of metamorphic differentiation, but they assign an auxiliary role to the mechanical factor. In Eskola's opinion the relatively high solubility of quartz, albite, and calcite is the essential property that facilitates segregation of these minerals into veins and layers during low-temperature metamorphism. Schmidt,[47] on the other hand, developed an ingenious hypothesis of metamorphic differentiation by purely mechanical means. He pointed out that a comparable structure is produced mechanically in rolled wrought iron, which consists of alternating layers of ferrite and perlite. By analogy he suggested that major slip surfaces which first form in the early stages of deformation of a heterogenous rock are irregular and discon-

[45] B. Sander, *Gefügekunde der Gesteine,* Springer, Vienna, pp. 269–275, 1930.

[46] Eskola, *op. cit.,* pp. 406, 407, 1939.

[47] W. Schmidt, *Tektonik und Verformungslehre,* pp. 183–187, Borntraeger, Berlin, 1932.

tinuous and tend to be located as far as possible within crystals, or aggregates of crystals, of those minerals (quartz, calcite, albite) that can most readily adjust their lattices to the imposed stresses, either by translation gliding or by twin gliding or by recrystallization. These early-formed slip surfaces in the fabric are interrupted at first by projecting crystals and crystalline aggregates of minerals such as micas, chlorites, and epidotes, which are less susceptible to deformation by gliding. As deformation proceeds, these obstructing minerals are mechanically eliminated and segregated, as penetrative movement in the major slip surfaces rolls out the plastic quartz and calcite into more and more regular and continuous bands. Within the micaceous or actinolitic bands, the platy or prismatic crystals of the dominant constituents are drawn into subparallel position (orientation according to crystal form) by flow movement involved in deformation. Schmidt's hypothesis explains how two physically dissimilar sets of minerals could be separated from one another; but to what extent this mechanism actually operates in metamorphism remains dubious in view of the undoubted fact that solution and crystallization usually play an important role in shaping the fabric of deformed rocks, and in the absence of experimental proof that under metamorphic conditions quartz and albite are more ductile than mica, chlorite and epidote. It would seem, then, that development of laminated structure in regionally metamorphosed rocks is a complex process involving solution, crystallization, and mechanical deformation, acting together or in alternating combinations as the selective agents of metamorphic differentiation.

CHAPTER 22

The Fabric of Metamorphic Rocks

FABRIC RELICS IN METAMORPHIC ROCKS

General Statement. Under the physical conditions imposed in metamorphism, rocks undergo internal adjustments that trend toward a state of structural as well as chemical equilibrium. When metamorphism is ideally complete a new assemblage of minerals and a new fabric have completely replaced the minerals and fabric of the parent rock. It frequently happens, however, that fabric relics (palimpsest structures), like mineral relics, survive metamorphism and provide valuable indications of the parentage of the metamorphic rock. In this connection, as will be apparent from the following discussion, difficulty is often experienced in distinguishing between what is truly relict and what is metamorphic in the rock fabric.

Fabric Relics Inherited from Sedimentary Rocks. Bedding, the most characteristic structure of sedimentary rocks, commonly persists in metamorphosed sediments as a relict banded structure, which may even be accentuated by metamorphic differentiation. Varved schists have been described from several localities in the pre-Cambrian of Finland.[1] The alternating sandy and argillaceous layers of the parent rocks still retain their identity in the schists, as clear-cut, light-colored quartzose, and dark micaceous (sometimes staurolite-bearing) bands, respectively. The parentage of these rocks, and hence the relict nature of the banding, has been verified by comparing chemical analyses of the light and dark portions of the banded schists with those of corresponding layers of a Pleistocene varve rock. Again in the schists and "granulites" of the Moine series of Scotland, in spite of the coarse grain and high grade of metamorphism that has been reached, relict sedimentary structures such as color-banding

[1] P. Eskola, Conditions during the earliest geological times, *Acad. Sci. Fenn. Annales,* ser. A, vol. 36, no. 4, pp. 9–19, 1932.

(due to variation in biotite content), graded bedding, cross-bedding, and films of iron ore and other heavy minerals still persist.[2]   Not to be confused with relict bedding, however, is the laminated structure, due to metamorphic differentiation of alternating quartzo-feldspathic and micaceous layers, that so often develops parallel to schistosity, and hence to bedding as well, in regionally metamorphosed rocks.   The low-grade schists of Otago, New Zealand, to which reference was made in the previous chapter, exemplify this purely metamorphic banding, which probably is much commoner than is generally suspected.

The term *helicitic structure* is applied to curved or contorted lines of inclusions (graphite, iron ore, mica, and so on) preserved within coarse crystals (porphyroblasts) of minerals such as albite, biotite, staurolite, and chloritoid (Fig. 101a, page 613).   The strings of inclusions are relict structures inherited from a parallel fabric, either sedimentary or metamorphic, that existed prior to growth of the porphyroblasts.   In true helicitic structure corrugation of the strings of inclusions preceded static crystallization of the enclosing crystals, so that not only the linear arrangement, but its curved or contorted pattern too, is relict.   More usually the trend of the strings of epidotes and other minerals enclosed in porphyroblasts of albite is S-shaped, as a result of partial rotation of the growing crystals of albite, in which case the direction of rotation is given by the attitude of the S (Fig. 101c, page 613).   A normal S corresponds to counterclockwise rotation of the porphyroblast, while the reverse pattern Ƨ indicates a clockwise movement.[3]   By contrast there are also instances where lines of inclusions develop during the later stages of metamorphism and constitute an element in the truly metamorphic fabric.   Thus Ingerson found that the linear trend of inclusions of mica in porphyroblastic albites of schists in parts of Pennsylvania and Maryland is in no way relict, but is directly related to the crystal lattice of the enclosing albite.[4] Here the mica is younger than the albite and has been formed during late hydrothermal activity by solutions penetrating the host crystals along planes of greatest permeability in the lattice.

In derivatives of coarse-grained sediments such as conglomerates, grits, and sandstones, the outlines of individual grains or pebbles often survive metamorphism.   The resultant relict fabric is termed *blastopsammitic* or *blastopsephitic* according to whether the parent rock was sandstone or conglomerate respectively.   In metamorphosed conglomerates, pebbles

[2] *E.g.*, G. Wilson, J. Watson and J. Sutton, Current bedding in the Moine series of north-western Scotland, *Geol. Mag.*, vol. 90, pp. 377–387, 1953.
[3] *E.g.*, E. B. Bailey, The metamorphism of the south-west Highlands, *Geol. Mag.*, vol. 60, pp. 317–331, 1923.
[4] E. Ingerson, Albite trends in some rocks of the Piedmont, *Am. Jour. Sci.*, vol. 25, pp. 127–141, 1933.

retain their identity with great persistence even when subjected to metamorphism of high grade involving deformation of sufficient intensity to cause marked change of external form.   It has usually been assumed that flattened or ellipsoidal pebbles in such rocks afford clear evidence as to the nature and extent of deformation that has accompanied metamorphism, but the possibility that the shape of the pebbles is purely relict must not be dismissed, even when the individual pebbles are aligned with their greatest dimensions parallel to the schistosity of the surrounding matrix, until it is found that their internal fabric yields evidence of deformation.   Actually, if the pebbles have indeed been mechanically flattened or otherwise distorted during metamorphism, the imprint of deformation upon the internal fabric will be clearly demonstrable by modern methods of structural investigation.[5]

Ripple mark, cross-bedding, concretions, pisolites, and outlines of fossils may all on occasion survive as fabric relics in metamorphic rocks.   Fossils are seldom recognizable in rocks of the higher grades of metamorphism, but in weakly metamorphosed rocks they may give reliable indications as to the extent to which the rocks in which they occur have been deformed.   Ripple mark and cross-bedding closely simulate purely deformational structures, such as drag folds and other small-scale corrugations which occur far more frequently in schistose rocks, and it may be necessary to investigate the microscopic fabric in minute detail to identify ripple mark with confidence.

Fabric Relics Inherited from Igneous Rocks.   Relict structures distinctive of igneous rocks (*e.g.*, porphyritic, ophitic, granitoid) are often locally preserved in rocks affected by contact or hydrothermal metamorphism and may even survive considerable deformation.   Terms such as *blastoporphyritic* and *blastophitic* are applied to such structures.   Blastoporphyritic structure is particularly persistent.   In the aureole that surrounds the Bluff norite intrusion in southern New Zealand [6] basic lavas and dike rocks, including spilitic types, have been converted to hornfelses of the hornblend-hornfels facies, consisting essentially of hornblende and labradorite or andesine (albite in derivatives of spilite).   While the groundmass of the typical hornfels has been completely reconstituted to a hornblende-plagioclase mosaic, the original phenocrysts of augite and plagioclase retain their crystallographic form, even though the former are now represented by pseudomorphs of hornblende.   In associated quartz

[5] B. W. D. Elwell, The lithology and structure of a boulder-bed in the Dalradian of Mayo, Ireland, *Geol. Soc. London Quart. Jour.*, vol. III, pp. 71–84, 1955; D. Flinn, On the deformation of the Funzie conglomerate, Fetlar, Shetland, *Jour Geology*, vol. 64, pp. 480–505, 1956.

[6] H. Service, An intrusion of norite and its accompanying contact metamorphism at Bluff, New Zealand, *Royal Soc. New Zealand Trans.*, vol. 67, pt. 2, pp. 185–217, 1937.

keratophyres, which likewise have reached internal chemical equilibrium during metamorphism, the structure is consistently blastoporphyritic, and in some rocks relict orthophyric structure is still clearly recognizable. The mineral assemblage is albite (or oligoclase)-orthoclase-quartz-biotite-hornblende, with manganiferous garnet in some rocks.   In the Scottish Highlands,[7] blastophitic and blastoporphyritic structures are commonly recognizable in epidiorites of the biotite zone of regional metamorphism, and locally in rocks of higher metamorphic grade where deformation has been less intense.   Blastoporphyritic structure is likewise widely prevalent among the leptites of Sweden.   Even the granulites of Saxony, in spite of metamorphism at high temperature and great depth, include rocks in which relict porphyritic and granitic structures can be identified in confirmation of the ultimate igneous origin of these rocks.   Generally speaking it is unsafe to assume that the large feldspars and hornblendes of high-grade metamorphic rocks are survivals from an originally porphyritic igneous fabric, since there is a strong tendency for these minerals, like garnet, staurolite, biotite, and others, to form large porphyroblasts of purely metamorphic origin.   Sharply idiomorphic outline (especially in feldspars), poverty in inclusions, and prevalence of Carlsbad, Carlsbad-albite, Manebach, or Ala twinning in plagioclase are indications, though by no means infallible, of relict origin.

Metamorphic derivatives of plutonic rocks characteristically possess a banded or streaky structure (gneissic structure).   Dark and light minerals alternately predominate in rather coarse discontinuous subparallel streaks, within which tabular and prismatic crystals also show a marked dimensional parallelism.   In some the structure is purely metamorphic (the result of deformation of the solid rock), and in others it is mainly a relict igneous structure that has been imprinted during magmatic flow prior to complete solidification of the rock.   It is usually difficult to differentiate precisely between the igneous and the metamorphic elements in the gneissic fabrics of such rocks, or even to distinguish between unmetamorphosed plutonic rocks whose gneissic structure is of purely igneous origin and rocks that have been deformed in the solid state and hence possess a fabric which is, in part at least, metamorphic.   In the absence of granulation, undulose extinction, or marked preferred orientation in the crystals of quartz—the mineral most sensitive to deformation—any parallel arrangement of mica flakes or of prisms of hornblende may safely be interpreted as due to magmatic flow.   Such cases are relatively rare, however, since the forces which bring about sustained flow of a highly viscous, largely crystalline magma are usually sufficiently powerful to cause rupture and incipient granulation of some of the grains of quartz and feld-

---

[7] J. D. H. Wiseman, The central and south-west Highlands epidiorites, *Geol. Soc. London Quart. Jour.*, vol. 90, pp. 357, 358, 405–407, 1934.

spar in the last stages of magmatic flow. The result is a protoclastic (nonmetamorphic) structure.   More often than not, the deforming forces continue to operate after crystallization is complete and consequently then imprint a truly metamorphic fabric upon the quartz and perhaps upon other minerals as well.   This metamorphic element in the fabric is susceptible to identification by modern methods involving measurement of crystal orientation, but it may still remain uncertain whether the whole, or only the last stage, of the structural evolution of the rock was subsequent to complete consolidation.

In mylonites and other rocks resulting from cataclastic metamorphism (*i.e.,* deformation, rupture, granulation, and differential movement of grains without appreciable chemical reconstitution), fabric relics are usually conspicuous and readily identified.   Particularly common are islands or augen of quartz and feldspar, in which a relict hypidiomorphic granular fabric inherited from a parent granitic rock can at times be recognized.   The term *porphyroclastic* has been applied to the fabric of mylonites with isolated large grains or aggregates of quartz or of feldspar, enclosed in a cataclastic fine-grained matrix.

Fabric Relics Inherited from Metamorphic Rocks.   The possibility that strings of inclusions enclosed in porphyroblasts, whether the structure is truly helicitic or not, represent relics of a parallel fabric of metamorphic origin has already been noted.   Similarly, schistose rocks, especially at low grades of metamorphism, often show traces of an earlier parallel fabric (*s*-surfaces in the terminology of Sander) which may be either metamorphic or sedimentary and which is in process of obliteration by the prevailing schistosity.   Even when these early *s*-surfaces can be attributed with some confidence to metamorphism, it is still uncertain whether there have been two distinct periods of metamorphism, or whether an earlier and a later phase of a single metamorphism have been recorded in the fabric.   The former alternative is the more likely when there is evidence that the later phase of metamorphism was retrogressive, *i.e.,* of lower grade than the earlier phase, as when relict *s*-surfaces are enclosed within large staurolites or biotites (unstable relics of the almandine-amphibolite facies) in muscovite-chlorite schists of the greenschist facies.   In blastomylonites and blastophyllonites, cataclastic degradation of a rock has been followed in a subsequent phase of metamorphism by chemical reconstitution, with the result that fabric relics referable to the mylonitic or phyllonitic stage may still be recognizable in the completely reconstituted rock.

Recognizable pseudomorphs also may be classed as fabric relics and at times provide the sole evidence of polymetamorphism.   Thus, aggregates of white mica, pseudomorphous after andalusite or cordierite in pelitic mica schists, point to a period of thermal metamorphism preceding the hydrothermal or regional metamorphism in which the rock reached its present state of chemical equilibrium. In much the same way Hess has interpreted pseudomorphs of talc after actinolite in certain soapstones as evidence that hydrothermal alteration of the parent serpentine rock has proceeded in two stages—a high-temperature (actinolite-forming) stage, followed by a low-temperature (talc-forming) stage which determined the present mineral composition of the rock.

CRYSTALLOBLASTIC FABRIC AND CRYSTALLOBLASTIC SERIES

Characteristics of the Crystalloblastic Fabric. The fabric of any metamorphic rock is determined broadly by one or both of two contrasted processes, *viz.*, mechanical strain, rupture, and differential movement of preexisting mineral grains, and chemical reconstitution involving growth of new crystals in an essentially solid medium. Fabrics resulting from this latter process are called *crystalloblastic,*[8] a term introduced by Becke to cover structures of the crystalline schists but nowadays generally extended to include also fabrics of like origin and character in rocks resulting from contact or metasomatic metamorphism. The peculiarities of the crystalloblastic fabric, as contrasted with fabrics of igneous rocks, can be correlated directly with a mode of origin wherein every individual crystal exerts its own force of crystallization against a resistance offered by the enclosing solid medium and its constituent competing crystals. These peculiarities are worth discussing in some detail since they provide satisfactory criteria by which rocks may be distinguished as of metamorphic rather than magmatic origin.

The following characteristics of crystalloblastic fabric were recognized by Becke [9] and by Grubenmann:

1. Crystalloblastic fabric was attributed to simultaneous growth of all the component crystals. [While there is certainly a marked tendency for simultaneous development of the constituent minerals of a metamorphic rock (since the goal of metamorphism is a state of chemical equilibrium between all the associated minerals), it is nevertheless possible, using petrofabric analysis, to trace definite sequences of mineral development in many rocks which have been affected to a notable degree by deformation or by metasomatism during their metamorphism.]

2. As compared with constituents of igneous rocks, the majority of the

[8] F. Becke, Ueber Mineralbestand und Struktur der Kristallinen Schiefer, *Akad. Wiss. Wien Denkschr.*, vol. 75, p. 35, 1913; H. Rosenbusch and A. Osann, *Elemente der Gesteinslehre*, Erwin Nägele, Stuttgart, pp. 638–642, 1923; U. Grubenmann and P. Niggli, *Die Gesteinsmetamorphose*, pp. 417–444 (and accompanying illustrations), Borntraeger, Berlin, 1924; A. Harker, *Metamorphism*, pp. 30–44, Methuen, London, 1932.

[9] Becke, *op. cit.*, pp. 35, 36, 1913.

grains in metamorphic rocks are irregular in outline (xenoblastic). Some minerals nevertheless persistently tend to occur in sharply bounded (idioblastic) crystals. The faces which develop strongly always belong to simple crystallographic forms and commonly are limited to a single zone or to such faces as lie parallel to planes of good cleavage. Unterminated prisms of amphibole and andalusite, micas and chlorites with well-defined basal planes but ragged edges, and unit rhombohedrons of magnesite and dolomite are all typical instances. Note the contrast between the typical xenoblastic outline of feldspars of metamorphic rocks and the tendency toward partial idiomorphism displayed by feldspars of granitic rocks of magmatic origin.

3. The larger crystals tend to be packed with small inclusions (often idioblastic crystals) of other minerals, giving rise to what is termed "sieve structure" (diablastic structure). This type of fabric is favored by relatively rapid crystallization of the host mineral about sparsely scattered nuclei and hence is typical of minerals such as garnet, cordierite, chloritoid, and staurolite, which are purely metamorphic and have not developed by enlargement of seed crystals present in the parent rock. Under other conditions, grains of such minerals (perhaps quartz, graphite, mica) as are making no contribution to a growing porphyroblast become crowded aside as growth proceeds. The regular geometric patterns produced by symmetrical localized concentration of inclusions in porphyroblasts, e.g., in chiastolite and in chloritoid, depend upon variation in force of crystallization in different directions within the crystal. Harker [10] describes the evolution of cruciform patterns in prisms of chiastolite thus:

The dark area in the centre represents the nucleus of the crystal, unable at that stage to free itself of inclusions. Subsequent growth has been effected by a thrusting outward, which was most effective in the directions perpendicular to the prism-faces. Much of the foreign matter brushed aside in this growth accumulated on the edges of the prism and was enveloped by the growing crystal. The arms of the dark cross represent thus the traces of the prism-edges as the crystal grew. Finally re-entrant angles may be left, but more usually these are filled in by the latest growth of the crystal, remaining full of inclusions. . . .

4. Zonary structure is seldom shown in individual crystals belonging to isomorphous series, and when present, it is said not to obey the rules which govern zoning of the same minerals in igneous rocks. It is doubtful if most modern petrographers would agree with the second part of this generalization, but the first part suggests a criterion (abundance of zoned plagioclases) by which diorites and granodiorites may be distinguished as magmatic in origin.

[10] Harker, *op. cit.*, p. 42, 1932.

5. In crystalline schists there is a strong tendency for crystals of pronounced prismatic or tabular habit, as in amphiboles and micas, to develop a preferred orientation with their greatest dimensions in subparallel position. This is partly the result of rotation and deformation of crystals and partly due to growth of new crystals with appropriate orientation. Schistosity arising from this latter process is termed "crystallization schistosity."

6. The minerals of metamorphic rocks may be arranged in a crystalloblastic series in order of decreasing "force of crystallization."

Interpretation of Crystalloblastic Series. Becke's concept of the crystalloblastic series (crystalloblastic order, idioblastic series) marks an important advance in interpretation of metamorphic fabric. It is an arrangement of metamorphic minerals in order of decreasing "force of crystallization," so that crystals of any of the listed minerals tend to assume idioblastic outlines at surfaces of contact with simultaneously developed crystals of all minerals occupying lower positions in the series. The series recognized by Becke [11] is as follows:

Sphene, rutile, magnetite, hematite, ilmenite, garnet, tourmaline, staurolite, kyanite

Epidote, zoisite

Pyroxene, hornblende

Breunnerite, dolomite, albite

Mica, chlorite

Calcite

Quartz, plagioclase

Orthoclase, microcline

Becke's crystalloblastic series includes only the minerals that occur typically in the crystalline schists—products of regional metamorphism—but it is obvious that the characteristic minerals of the hornfelses likewise may be classed respectively as high (e.g., andalusite, forsterite, idocrase) or lower (e.g., cordierite, scapolite) in the crystalloblastic series. Since the "force of crystallization" of a mineral presumably varies with such variable physical conditions as temperature, and since it must be affected too by the chemical environment in which crystallization proceeds, exceptions to Becke's general scheme are to be expected. In low-grade albitic schists from New Zealand and elsewhere, as also in the leptites of Sweden and Finland, albite occupies a much lower position in the crystalloblastic series than that assigned to it by Becke. Its normal position would seem to be with quartz and the other feldspars. Sphene is one of the highest-ranking minerals in the crystalloblastic series; yet in many amphibolites and chlorite schists, where it has developed at the expense of ilmenite, it persistently takes the form of rounded, drop-like grains with no trace of crystallographic boundary surfaces.

In discussing growth of crystals in a solid medium, it is possible to

[11] Becke, *op. cit.*, pp. 42, 43, 1913.

visualize "force of crystallization" in two opposite senses. In connection with the crystalloblastic series the term is customarily used to denote that property by virtue of which a grain tends to develop its own crystal form against the resistance of the surrounding solid mass. In this sense the prism faces of actinolite exert a greater "force of crystallization" than do the basal plane and other terminal faces, so that a crystal of actinolite habitually shows well-formed prisms with ragged terminations. On the other hand, the universally prismatic or acicular habit of actinolite indicates a pronounced tendency for crystals of this mineral to grow more rapidly parallel to the c axis than in directions normal to c, and from this point of view actinolite might be said to exert a maximum "force of crystallization" parallel to the c axes. Similarly, crystals of mica, chloritoid, andalusite, and staurolite show sharply defined crystal faces parallel to the direction of elongation or the plane of flattening. Eskola's term "form energy" is perhaps preferable to "force of crystallization," to denote potentiality to develop crystal form in mineral grains growing within a solid medium.

Becke pointed out that the crystalloblastic series corresponds closely to an arrangement in order of decreasing specific gravity, and he concluded in consequence that dense "molecular" packing is the factor mainly responsible for form energy of high magnitude. This view has been elaborated by Eskola [12] in terms of the modern conception of crystal structure. Of the silicate minerals, those that stand high in the series are orthosilicates whose crystal lattices are built up of independent $[SiO_4]$ tetrahedra, viz., sphene, garnet, staurolite, kyanite, epidote minerals, andalusite, sillimanite, idocrase, zircon, forsterite. Further down come the metasilicates with chain and band structures (pyroxenes and amphiboles), then silicates with sheet structure (micas, chlorites, talc, chloritoid), and finally tectosilicates with open, three-dimensional frameworks, notably the feldspars, quartz, and probably cordierite. Some such relation between density of ionic packing and "form" energy can be expected from what was stated in Chap. 17 regarding surface energy. Surface energy —the energy required to form a new unit of surface—was shown to depend on the amount of work done against the forces exerted by neighboring particles, in bringing an ion or an atom from within the phase in question to a surface position. The magnitude of such forces must be related to the strength of interionic bonds and to the number of bonds per unit of surface. All other factors remaining constant, surface energy will therefore increase with density of ionic packing and, like the latter, will vary according to the crystallographic orientation of the surface concerned. It

---

[12] P. Eskola, *Die Entstehung der Gesteine* (Barth, Correns, Eskola), pp. 278, 279, Springer, Berlin, 1939.

has been computed,[13] for example, that in lattices of the halite type the ratio of respective surface energies on {110} and on {100} planes is 315: 116. It is because of such variation in surface energy as a function of crystallographic orientation that crystals develop faces. Eighty years ago Gibbs [14] showed that if a_i is the area of any facet and σ_i the corresponding surface energy, the condition for equilibrium is that, for a given volume,

$$\sum a_i \sigma_i = \text{minimum}$$

the sum being extended to all the facets of the crystal. It is easily seen that if the surface energy were the same for all orientations, this condition is that the total area must be minimum for a given volume; in other words, the crystal would necessarily be a sphere.

Various attempts have been made to demonstrate experimentally that a growing crystal face exerts a force against its surroundings and to measure this "force of crystallization" (believed to be responsible notably for the widening of fissures in which vein minerals form). Becker and Day,[15] for instance, succeeded in growing crystals of alum which would raise, by a few tenths of a millimeter, a weight of 1 kg. applied to their upper faces. They concluded that the force of crystallization per unit area is of the same order of magnitude as the resistance which crystals offer to crushing. However, the problem of evaluating exactly this force, and of determining thereby which of two crystals competing for the same space will assert its faces against the other, is extremely difficult. Indeed, if a growing crystal exerts against the surrounding medium a force which varies with direction, then the medium exerts on the crystal an equivalent reaction which also must vary according to direction. Thus the crystal as it grows is subjected to nonhydrostatic stress and is affected by the complicated relations discussed on pages 474 to 477. The only generalization that may safely be stated is that the work done by the growing crystal against external forces cannot exceed the free energy of crystallization from the pore fluid. This latter may be expressed by the difference between the respective chemical potentials of the crystal and an equivalent amount of the same substance in the fluid from which the crystal grows. This difference depends on temperature and on concentration. Thus the force of crystallization of a growing crystal will depend, among other things, on temperature and on the degree of supersaturation as well as the nature of the pore fluid with which it is in contact. Moreover the

[13] *E.g.*, see F. Seitz, *The Modern Theory of Solids*, p. 97, McGraw-Hill, New York, 1940.

[14] *The Collected Works of J. W. Gibbs*, p. 332, vol. I, Longmans, New York, 1928.

[15] G. F. Becker and A. L. Day, The linear force of growing crystals, *Washington Acad. Sci. Proc.*, vol. 7, pp. 283–288, 1905. See also S. Taber, The growth of crystals under external pressure, *Am Jour. Sci.*, vol. 41, pp. 532–556, 1916.

mechanical properties of the enclosing medium, as determined by the rock fabric and the properties of individual constituent minerals, help to determine the magnitude of the work done by any growing crystal.

Kinetics of crystal growth are probably just as important as surface energy relations in determining the habit and size of crystals. In a given crystal, difference in status of similar ions on different faces presumably gives rise to corresponding differences in energy of activation for the process of growth on these faces. Hence different faces of one crystal tend to grow at different rates. Growth rates in general also depend markedly upon rates of diffusion of ions to the growing faces. A face in continued contact with a supersaturated pore solution will grow much more rapidly than will a face constrained by close contact with another crystal. The mechanism is complicated by a tendency for particles to migrate some distance over a crystal surface before becoming finally attached to the growing lattice. Rates of growth, and hence habit, may also be materially affected by the presence, in the pore solution, of compounds capable of being adsorbed selectively on certain faces; e.g., elongated gypsum crystals can be caused to grow into short tabular prisms by adding sodium citrate to the solution from which they are growing.[16] Herein may lie an explanation of the tendency for crystals of a given mineral at one locality to assume a common habit, which differs from that shown by crystals of the same mineral from some other occurrence. All these habits may actually be metastable with respect to the habit obeying Gibbs' law.

Cleavage, like external crystal faces, depends on surface energy.[17] It is probably for this reason that the crystallographic forms most commonly developed on partially idioblastic crystals in metamorphic rocks are those parallel to which there is good cleavage.

The size reached by crystals of a particular mineral under given conditions of metamorphism also appears to be related in some way to the "force of crystallization" of the mineral, for there is a general, though by no means universal, tendency for the coarser constituents of a rock to occur in idioblastic crystals. A notable exception is cordierite, which shows a strong tendency to build large porphyroblasts which sometimes attain giant size, as in the Orijärvi region of Finland; yet cordierite of metamorphic rocks is never idioblastic. Albite is another mineral which tends to develop large porphyroblasts, though it is low in the crystalloblastic series. Coarse grain is favored by high temperature of metamorphism and probably also by protraction of the period over which metamorphic temperatures are maintained, even when the latter are rela-

---

[16] E. K. Rideal, How crystals grow, *Nature,* vol. 164, pp. 303–305, 1949.

[17] M. D. Shappell, Cleavage in ionic minerals, *Am. Mineralogist,* vol. 21, pp. 75–102, 1936.

tively low. Though the grain size of a rock on theoretical grounds should increase without limit during prolonged metamorphism, growth of very large crystals is actually rare. Even in purely thermal metamorphism in the absence of deforming movements, recrystallization not infrequently results in diminution of grain size, as when a phenocryst of feldspar, while retaining its original external form, is converted to an aggregate of smaller grains. Indeed Joplin,[18] as a result of studies on thermal metamorphism of gabbro xenoliths in granites from Australia and elsewhere, concludes that reduction of coarse gabbro to a fine-grained aggregate of pyroxene and plagioclase is a characteristic preliminary to assimilative reaction and hybridization. The presence of finely divided, chemically inert substances, notably graphite, which collect upon the surfaces of growing crystals, may so impede chemical reaction and growth of crystals during metamorphism as to impose upon the fabric of the resultant rock a conspicuously fine grain.[19] Experimental recrystallization (annealing) of cold-strained coarse marble yields a relatively fine-grained aggregate of unstrained grains.

Although the crystalloblastic fabric persistently shows a number of highly characteristic features, it must be realized in conclusion that the physical processes leading to its development are extremely complex and as yet are not amenable to quantitative treatment. The factors involved in the development of faces on a crystal growing in a solid medium are numerous: surface energy, concentration of the solution, change in solubility under nonhydrostatic stress, mechanical properties of the resisting medium, strain energy in the parent material, and all the factors involved in the kinetics of crystal growth (nucleation, rate of diffusion to the growing faces, free energies of activation, impurities, etc.).

PREFERRED ORIENTATION OF CRYSTALS IN METAMORPHIC ROCKS

One of the most distinctive features of the fabric of crystalline schists is preferred orientation of crystals of constituent minerals, i.e., the tendency for crystals of the same mineral to assume parallel, or partially parallel, crystallographic orientation. When, for example, the majority of the muscovite flakes in a mica schist lie with {001} parallel or inclined at low angles to the plane of schistosity, the crystals of mica are said to possess a preferred orientation. The fact that the majority of the crystals depart

[18] G. A. Joplin, Note on the origin of basic xenoliths in plutonic rocks, Geol. Mag., vol. 72, pp. 227–234, 1935.

[19] P. Eskola, op. cit., pp. 24–27, 1932; T. W. Gevers, Comparative notes on the Pre-Cambrian of Fennoscandia and South Africa, Comm. géol. Finlande Bull. 119, pp. 50, 51, 1937.

slightly, and a few diverge strongly, from the ideal orientation is immaterial, provided the tendency toward parallel orientation can be discerned. Preferred orientation of idioblastic tabular, prismatic, or fibrous crystals has long been recognized as characteristic of schists and gneisses. The terms *lepidoblastic* and *nematoblastic* are widely used to denote crystalloblastic fabrics marked respectively by predominance of subparallel crystals of platy and of prismatic or fibrous habit, as contrasted with the *granoblastic* fabric of rocks such as hornfels, in which the mineral grains are equidimensional and lack obvious preferred orientation.

It is convenient to distinguish descriptively between preferred orientation of inequidimensional crystals according to their external crystal form (*Regelung nach Korngestalt; Formregelung*) and preferred orientation of equidimensional grains according to their space-lattice structure (*Regelung nach Kornbau; Gitterregelung*). The obvious tendency toward parallelism of the *c* axes in prisms of amphibole, and the equally conspicuous subparallel orientation of {001} in flakes of mica, are instances of preferred orientation according to crystal form. It is of course accompanied by a preferred orientation of the crystal lattices, but whether this is consequent upon form orientation (as was at one time universally assumed), or whether the reverse relationship holds good, is not obvious. It is possible, for example, that the flakes of muscovite in a mica schist are crystallographically parallel only because the habit of the crystals is tabular; but it is also conceivable that during crystallization of the schist the mica crystals have become arranged with {001} of the space-lattice in subparallel position, the resultant dimensional parallelism of the tabular crystals being purely incidental. Each of these alternatives has been shown to be valid in particular cases.

A much less obvious, but highly important, type of preferred orientation is that governed purely by space-lattice structure of the mineral grains concerned. It is displayed by minerals such as quartz, calcite, and feldspar, which rank low in the crystalloblastic series and therefore occur in metamorphic rocks in xenoblastic, usually equidimensional, grains. A special universal-stage technique is necessary to demonstrate in detail the preferred orientation of the mineral grains (cf. pages 625 to 627). If preferred orientation is strongly developed and simple in pattern (as where there is a strong tendency for the *c* axes of quartz to be aligned in one direction in the schistosity), this condition may be rendered obvious in the course of routine examination with an ordinary microscope, by similarity in interference tint, extinction position, and compensation behavior of the majority of the grains. Not uncommonly the grains of calcite or quartz in a deformed rock, though xenoblastic, are lensoid in outline, and show a marked tendency toward arrangement with their greatest dimensions in parallel position. Nevertheless there can still be no doubt that any preferred orientation of crystallographic vectors, such as optic axes or twin lamellae, is controlled entirely by the space-lattice structure of the crystals, since the outlines of the grains have no constant crystallographic significance.

Preferred orientation of crystalline grains is a common feature of the fabrics of igneous and some sedimentary rocks, as well as of most metamorphic rocks. It has also been studied extensively in connection with industrial materials such as metals and ceramic bodies. In most cases preferred orientation may be correlated with growth or accumulation of crystals in a continuous contemporaneously flowing [20] medium. To this category belong the fabrics of many igneous rocks, some water- and wind-laid sediments, rolled metals, worked pottery bodies, and particularly rocks affected by deformation during metamorphism. All such fabrics have been classed by Sander as tectonite fabrics.[21] The fundamental assumption underlying interpretation of preferred orientation phenomena in tectonite fabrics, and thus in deformed rocks in particular, is that the geometric symmetry of the preferred orientation pattern is closely related to the symmetry of the flow movement which accompanied evolution of the fabric in question. In just such a way we are used to judging the motion of wind by preferred orientation of wheat stalks bent harmoniously in response to that motion, or that of water by the orientation of floating logs or weed. And in just such a way the direction of a prevailing wind can be inferred from the bedded fabric and external form of sand dunes. There are other rock fabrics (nontectonites) which cannot be related to motion of the medium in which they have developed. Some of these show preferred orientation patterns usually classified as growth fabrics. The fundamental assumption in connection with their interpretation is that crystal growth has been conditioned by an anisotropic medium offering greater resistance in some directions than in others. Thus parallel alignment of quartz prisms normal to a vein wall, and of hornblende prisms in the schistosity planes of *Garbenschiefer*, are both conventionally interpreted as due to elongation of growing crystals along paths of minimum resistance to growth.

Preferred orientation of crystals in metamorphic, and especially in deformed, rocks has been studied, and its implications debated, for more than a century. These studies have received fresh impetus since the technique of petrofabric analysis was developed by Sander, Schmidt, and coworkers in Austria and Germany during the period following the First

---

[20] The term flow is here used broadly to cover relative movement of particles within a medium that retains its continuity and cohesion; it includes plastic flow of crystals and crystalline aggregates as well as flow of liquids.

[21] B. Sander, *Gefügekunde der Gesteine*, Springer, Vienna, 1930.

World War. The literature on this topic is now too extensive and too specialized to be reviewed adequately in a book such as this, but in the chapter which follows some of the results achieved and some of the speculations advanced will be summarized.[22]

[22] For reviews, in English, of the methods and results of petrofabric analysis (with full reference to original European literature), the reader is referred to the following: E. B. Knopf and E. Ingerson, Structural petrology, *Geol. Soc. America Mem.* 6, 1938; H. W. Fairbairn, *Structural Petrology of Deformed Rocks*, Addison-Wesley, Cambridge, Mass., 1949; F. J. Turner, Mineralogical and structural evolution of metamorphic rocks, *Geol. Soc. America Mem. 30*, pp. 161–282, 1948.

CHAPTER 23

Special Features of Fabric of Deformed Rocks

STRESS AND STRAIN IN DEFORMATION OF ROCKS

Definitions. A body subjected to stress becomes temporarily or permanently deformed as compared with the same body in an unstressed condition. The changes in volume and shape induced by application of stress are included together as *strain*. Provided the stress is applied for a relatively short time and does not exceed a critical value—the *elastic limit*—the strained body instantaneously reverts to its unstrained state when the stress is released. Strain of this type is called *elastic strain*. The elastic limit of a given material is more or less constant for the standard conditions of a laboratory test—relatively rapid application of stress at atmospheric confining pressure. It tends to have much higher values at high confining pressures.

If the stresses developed within the body exceed the elastic limit of the material, permanent deformation (*plastic strain*) ensues, and the accompanying differential movement of particles within the strained mass is termed *plastic flow*. The permanently deformed body at any stage of plastic strain retains its coherence, but does not revert to its initial form on release of the external force—*e.g.*, a flattened disk of clay, a sheet of rolled metal, a bent iron bar. Plastic strain at given temperature and confining pressure cannot proceed indefinitely, for if it is carried to certain limits, which vary with the rate of deformation, rupture of the test specimen occurs. The strength of the material at the time of rupture is termed the *ultimate strength;* it may be either greater in magnitude or less than the elastic limit. Substances such as quartz and feldspar, which are incapable of plastic deformation under ordinary temperatures and pressures, are termed *brittle,* in contrast with *ductile* materials such as gold, copper, and the metals in general, which can undergo very extensive plastic deformation before failing by rupture. It must be emphasized, however,

602

that the values of elastic limit and ultimate strength of a given material are liable to great variation with variation in temperature, confining pressure, and even duration of deformation. Substances which are brittle under ordinary conditions may therefore be ductile at high temperature or at great confining pressures, and substances which can sustain a high stress when this is rapidly applied (as in an ordinary laboratory strength test) may rupture at much lower stresses when these are slowly applied over many days.

Experimental Observations. Experimentally determined stress-strain relations for calcite and marble [1] under geologically possible conditions of temperature and pressure are illustrated in Fig. 95. Some points to note are as follows:

1. In each experiment a small amount of elastic strain (steep part of curve) is followed, after the yield point, by major plastic strain (flatter part of curve).

2. The stress sustained by the specimen rises steadily with increasing plastic strain—a condition known as strain hardening. For marble this effect is reduced by lowering the strain rate (Fig. 95c).

3. The strength of a single crystal varies according to its orientation in the stress field (Fig. 95a). This reflects differences in the mechanism of gliding and in the resolved shear stress on the active glide system (cf. page 615).

4. For a given temperature and degree of strain (say 3 per cent) the strength of the aggregate (marble) is greater than the mean strength of individual crystals of different orientations.

5. High temperature lowers the strength of calcite (Fig. 95b) and marble (Fig. 95c).

At room pressure marble is brittle, and calcite is ductile (to a limited degree) only for orientations that favor twin gliding (e.g., extension perpendicular to $\{01\bar{1}2\}$, e, or compression perpendicular to $\{10\bar{1}0\}$, m). Confining pressures of a few kilobars greatly increase strength and ductility of both crystals and marble. Ductility is also increased by high temperature. Thus at 5,000 bars confining pressure and 800°C. (still more than 500°C. below the melting point of calcite) local elongation of 1,000 per cent has been achieved in marble without rupture; the specimen behaves like ductile metal.

Marble is considerably weaker and more ductile than other monomineralic rocks that have been tested over the same range of conditions

[1] D. T. Griggs, F. J. Turner, and co-authors, Experimental deformation of Yule marble, *Geol. Soc. America Bull.*, vol. 62, pp. 853–862, 863–886, 887–906, 1951; vol. 62, pp. 1385–1406, 1952; vol. 64, pp. 1327–1342, 1343–1352, 1953; vol. 67, pp. 1259–1294, 1956; F. J. Turner, D. T. Griggs, and H. Heard, Experimental deformation of calcite crystals, *Geol. Soc. America Bull.*, vol. 65, pp. 883–934, 1954.

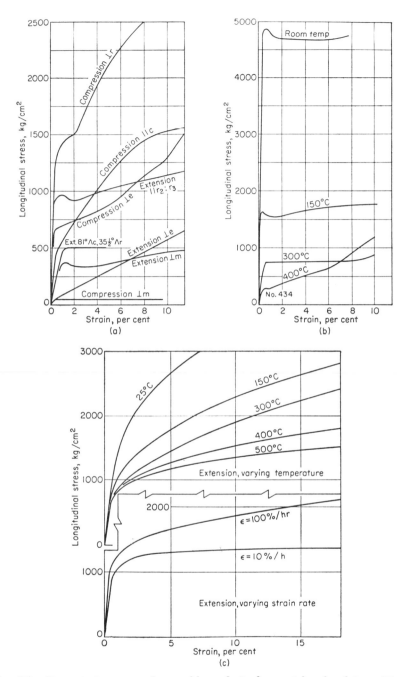

Fig. 95. Stress-strain curves for marble and single crystals of calcite. (*Experiments of D. T. Griggs and H. Heard.*) (*a*) Calcite crystals: 300°C., confining pressure (c.p.) 5,000 bars. (*b*) Calcite crystals: compression at 30° to *c* and 75° to *r*; c.p. 5,000 bars. (*c*) Yule marble: c.p. 5,000 bars, except the 400°C. experiment (3,000 bars); upper set of curves are for cylinders cut parallel to mean trend of *c* axes of calcite in marble; the two lower curves are for cylinders cut normal to this direction. $c = \{0001\}$; $r = \{10\bar{1}1\}$; $e = \{01\bar{1}2\}$; $m = \{10\bar{1}0\}$.

(20° to 800°C.; 3,000 to 10,000 bars).[2]   Dolomite rock,[3] though moderately ductile at 380°C. and 3,000 bars, is much less so, and three times stronger, than marble under the same conditions.   Enstatite pyroxenite and dunite are somewhat ductile at 500° to 800°C.; but quartz and quartzite remain persistently brittle and immensely strong.

The calcite lattice, like that of any common rock-forming mineral, differs fundamentally from the lattice of a metal in that its structure is ionic. Yet calcite and marble, when deformed under conditions conducive to ductility, have been found to behave in precisely the same way as metals. It is probable, then, that much of what is known regarding the behavior of metals during deformation [4] may be applied to metamorphic deformation of rocks—especially monomineralic rocks such as marble, dolomite, quartzite, and dunite.

Strain Ellipsoid.   The geometry of homogeneous strain in a given body may be described ideally in terms of a strain ellipsoid developed from an initially spherical sample of the unstrained body.   In its most general form this is a triaxial ellipsoid, whose three mutually perpendicular axes of greatest, mean, and shortest length are respectively designated A, B, and C.   On the basis of observed fabric data, such as the shapes of distorted fossils and pebbles and the directions of slip surfaces and tension joints in deformed rocks, some writers have attempted to construct strain ellipsoids representing the deformation involved in specific instances of metamorphism.   The degree of correspondence between a reconstructed strain ellipsoid of this kind and the actual deformation depends upon the nature of fabric data employed.   Distorted or ruptured fossils, pisolites, pebbles, or sedimentary strata sometimes provide reliable information as to deformation, at least within limited fields.   But more than one dynamic interpretation can be placed upon the disposition in space of such structures as tension joints, schistosity, strain-slip cleavage, and lineations; for rocks are nonhomogeneous bodies, and the observed or deduced strain is the cumulative result of elastic deformation, subsequent plastic flow, and even rupture of the component crystals or crystal aggregates, with or without chemical reconstitution of the rock—whereas the mathematical relationships between stress, strain, and movement in the strain ellipsoid (as conceived, for example, by Becker) presuppose homogeneous elastic strain of homogeneous material.   Moreover geological strain commonly is

[2] D. T. Griggs, F. J. Turner, and H. Heard, Deformation of rocks at 500° to 800°C., *Geol. Soc. America Mem.* (in press), 1959.

[3] F. J. Turner, D. T. Griggs, H. Heard, and L. Weiss, Plastic deformation of dolomite rock at 380°C., *Am. Jour. Sci.*, vol. 252, pp. 477–488, 1954; J. Handin and H. W. Fairbairn, Experimental deformation of Hasmark dolomite, *Geol. Soc. America Bull.*, vol. 66, pp. 1257–1273, 1955.   D. Higgs and J. Handin, Experimental deformation of dolomite single crystals, *Geol. Soc. America Bull.*, vol. 70, pp. 245–278, 1959.

[4] C. S. Barrett, *Structure of Metals*, McGraw-Hill, New York, 1952.

far from being strictly homogeneous.   Clearly it is impossible to represent
a nonhomogeneous folded structure in terms of a strain ellipsoid.

KINEMATICS OF STRAIN IN ROCKS

The Movement Plan.   Evidence of differential movement of the component parts of deformed rocks has long been recognized in the field, in
hand specimens, and in microsections.   To this category belong folded
bedding, certain types of schistosity (*e.g.*, strain-slip cleavage), slickenside surfaces and associated lineations, rolled porphyroblasts with spiral
trains of inclusions, bent crystals, and certain types of twin lamellae
(notably those parallel to $\{01\bar{1}2\}$ in calcite and to $\{02\bar{2}1\}$ in dolomite).
It was first shown by Sander that in a given rock unit (a hand specimen,
an outcrop, or a map area) this tangible direct evidence of movement
conforms to a uniform over-all pattern distinguished above all by its
geometric symmetry.   This pattern he called the movement plan (*Bewegungsbild*) of deformation.   Sander also correlated the symmetry of the
deformed fabric as a whole—especially the patterns of preferred orientation of constituent minerals—with that of the movement plan (cf. page
628).   Thus Sander and his school interpret kinematically the geometry of strain and of the whole fabric of strained rocks; and this is the
approach that is emphasized in this chapter.   So strain and fabric of
deformed rocks are considered as products of integrated internal movements without reference to the forces that initiated and controlled the
motion.   Dynamic interpretation of the movement plan so synthesized is
a further step, usually obscured by some degree of ambiguity.

Some Ideal Movement Plans.[5]   The movement plan deduced from
fabric data may show analogies—though seldom identity—with one or
other of a number of ideal types that have figured in geological literature
since the end of the last century.   While emphasizing that their strict
geological application is limited, we outline several of these briefly below.

1. *Affine deformation by movement on one set of parallel slip planes.*
This type of deformation (sometimes termed simple shear) may be
illustrated by equal relative displacement (in a constant direction of
slip) of all adjacent cards in a deck.   Figure 96 shows sections parallel
to the deformation plane (the plane containing the slip direction and
normal to the slip plane) of the strain ellipsoid at successive stages of

---

[5] G. Becker, Experiments on schistosity and slaty cleavage, *U.S. Geol. Surv. Bull.*,
no. 241, 1904; Current theories of slaty cleavage, *Am. Jour. Sci.*, vol. 24, pp. 1–17,
1907; B. Sander, *Einführung in die Gefügekunde der geologischen Körper*, Pt. I, pp.
33–66, Springer, Vienna, 1948; L. E. Weiss, Structural analysis of the basement system at Turoka, Kenya, *Overseas Geology and Mineral Resources*, vol. 7, pp. 3–35,
123–153, Her Majesty's Stationery Office, London, 1959.

deformation. The ellipsoid is triaxial. Its mean axis B (normal to the plane of Fig. 96) maintains a constant length, equal to the diameter of the original sphere, throughout strain. The slip plane SS, at every stage, is one of the two circular sections of the strain ellipsoid. As deformation proceeds, the second circular section $S'S'$ rotates both in space and with reference to coordinates in the strained mass, so that it coincides with different sets of particles at successive stages of strain. The movement plan has monoclinic symmetry, with AC as plane and B as axis of symmetry. Crystal gliding on a single glide system (cf. page 615) is its most significant geological counterpart.

2. *Nonaffine deformation by movement on one set of parallel slip planes.* Here the relative displacement of adjacent layers varies, and a straight line drawn upon the deformation plane prior to deformation assumes a

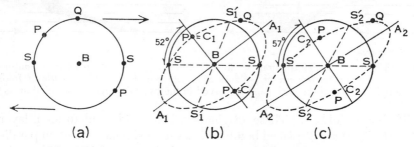

(a)   (b)   (c)

FIG. 96. Affine, deformation by slip on one set of s-planes SS. (a) Section through sphere prior to deformation. (b) and (c) Sections through strain ellipsoids at successive stages of deformation. AA, B, and CC are principal axes of ellipsoids; SS and $S'S'$ are circular sections. Identical particles of matter in all these figures are represented by solid circles identically lettered.

curved or even folded form (Fig. 97a), although no flexure is involved in the movement. Such folded patterns are called shear or slip folds (*Scherfalten*). They are sometimes recognizable in the contorted color bands of slates and phyllites and are characterized by the constant thickness of any individual band as measured parallel to the slip direction in any part of a fold. The fold axis is the intersection of the slip plane and the folded surface (bedding or banding) and may be inclined at any angle to the slip direction. The slip planes may be clearly recognizable as a schistosity, or "cleavage," cutting and displacing bedding, and itself oriented subparallel to the axial planes of the slip folds. This axial-plane relation affords useful field criteria for identifying overturned limbs of folds in outcrops of deformed rocks whose bedding dips at varying angles in some constant direction.[6] Where the cleavage dips more gently

[6] G. Wilson, The relationship of slaty cleavage and kindred structures to tectonics, *Proc. Geol. Assn.*, vol. 62, pp. 263–302, 1946.

than the bedding which it intersects, the bed in question is overturned. Steeply dipping cleavage cutting gently dipping bedding on the other hand shows that the beds are "right way up." Considerable importance is attached by Schmidt [7] to what he terms *Gleitbrett* folds (literally,

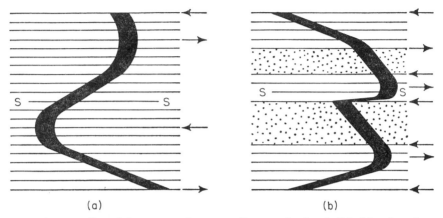

(a)   (b)

Fig. 97.   Nonaffine deformation of an initially straight band (black) of uniform thickness, by slip on one set of s-planes SS. (a) Slip folds. (b) *Gleitbrett* folds, with original trend of the dark band preserved in the two *Gleitbretter* (stippled areas).

"glide-board folds"). These originate when rigid, undeformed layers (*Gleitbretter*) are displaced bodily as a result of slip movement on parallel surfaces in intervening mechanically weak layers (cf. Fig. 97b). Another geologically important instance is nonaffine translation gliding in a single crystal.

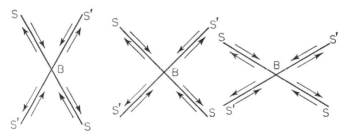

Fig. 98.   Movement pictures for three successive stages of affine deformation ("flattening") by simultaneous symmetrical slip on two intersecting sets of s-planes, SS and S'S'.

3. *Affine deformation by symmetrical movement on two sets of slip planes.*   This is the type of deformation which Sander terms flattening (*Plättung*); it is also known as pure shear. The straight lines of Fig. 98

[7] W. Schmidt, *Tektonik und Verformungslehre*, pp. 81–87, Borntraeger, Berlin, 1932.

represent two sets of slip planes intersecting in an axis B, normal to the plane of the figure. Then the latter is the deformation plane, and an initial sphere is transformed into an ellipsoid, the principal axes of which maintain a fixed position (both in space and within the strained mass) throughout deformation. The maximum and minimum axes A and C bisect the angles between the slip planes, while the length of the mean axis B remains equal to the diameter of the original sphere. The slip planes rotate about the B axis of the ellipsoid, but coincide with the same material particles at every stage of deformation and are circular sections (*i.e.*, planes of no distortion) of the ellipsoid (cf. Fig. 99). In cases of extreme strain they ultimately approach the AB plane of the ellipsoid,

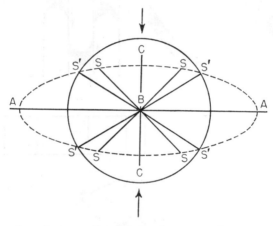

Fig. 99. AC section of strain ellipsoid developed by "flattening" under a simple compressive force (shown by arrows). $S'S'$ are early slip planes now rotating toward a "dead" position; SS are newly developed active slip planes.

where the value of shearing stress is nil. Consequently the first slip planes to develop are likely to approach and ultimately reach a "dead" position where internal frictional resistance to movement outweighs the motivating shearing stress. Slip upon these particular surfaces may now cease, and any further deformation requires initiation of new symmetrical movements upon sets of planes (new slip surfaces) parallel to which the shearing stress is high enough to break down internal resistance. These also rotate toward the "dead" position as strain proceeds. The symmetry of the movement plan is orthorhombic in that there are three mutually perpendicular planes of symmetry parallel to AB, BC, and CA of the strain ellipsoid. There is a rather common type of schistosity marked by parellel alignment of lensoid grains of quartz or calcite and orthorhombic symmetry of fabric, and this has been correlated by Sander with the type of movement plan just described—flattening (*Plättung*) under a compressive force normal to the schistosity. This correlation is consistent with available geological data; but correlation with any other geologically possible orthorhombic movement plan would be equally plausible.

4. *Composite and nonplane deformations.* The movement plan of metamorphic deformation commonly must be more complex and less

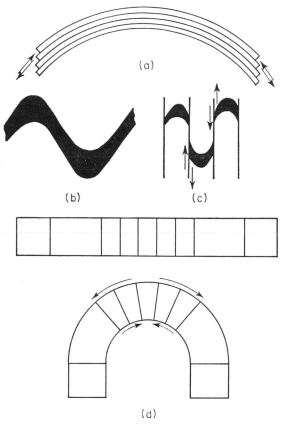

FIG. 100. Folding involving flexure. (*a*) Flexural-slip folding of competent layers. (*b*) Flexural-slip folding of incompetent layer. (*c*) Shearing out of fold limbs in incompetent material during flexural slip. (*d*) Folding by simple flexure (below) of originally straight bar (above).

regular than any of the three types of plane deformation just described. For example slip may proceed simultaneously or alternately, but with different velocities, on two or more intersecting sets of slip planes, which therefore rotate at different rates with reference to fixed coordinates. The movement plan has monoclinic symmetry. Again, in strongly deformed rocks of orogenic zones, there is commonly evidence that the rock has elongated at right angles as well as parallel to the main direction of tectonic transport (*i.e.*, parallel to B as well as to A of the strain ellipsoid). This is no longer a plane deformation. In both the above cases the strain may be considered as a composite of two appropriate simple plane deformations.

5. *Deformation involving flexure.* Two fundamentally distinct types of nonaffine deformation lead to development of folded patterns in metamorphic rocks. In slip folding (already considered under 2) there is "internal rotation" of lines and surfaces, *e.g.*, the dark bands of Fig. 97, within the strained body, but there is no rotation of the mass as a whole or of any unstrained portions of it (*e.g.*, *Gleitbretter*). On the other hand, in folding that involves bending (flexure), undeformed portions of the mass, *e.g.*, the limbs of a fold, rotate bodily in space ("external rotation"), while the material located along the fold axes is simultaneously affected by strain involving internal rotation (Figs. 100*a* and 100*d*). Most rocks that show conspicuous effects of folding involving flexure are not mechanically isotropic, but have a laminated structure, so that bending of individual bands is accompanied by slip movements upon the intervening surfaces of discontinuity. The result is a flexural-slip fold. If the rock is made up entirely of mechanically resistant (competent) layers, these tend to deform by pure flexure, with simultaneous slip movements upon the intervening surfaces (Fig. 100*a*). Less resistant (incompetent) laminated material tends to be thrown into folds with thickened crests and attenuated limbs, the movement being mainly within rather than between individual bands (Fig. 100*b*). With extreme deformation the limbs are sheared out (Fig. 100*c*), and movement is then limited to slip upon the surfaces of rupture so produced. If a fold is cylindroidal, the movement plan and the folded fabric are monoclinic: there is one symmetry plane normal to the fold axis.

MECHANISM OF DEVELOPMENT OF PREFERRED ORIENTATION IN TECTONITE FABRICS

Componental Movements in Deformation of Rocks.[8] In the previous section strain was pictured as the result of differential movement (slip) on surfaces within the strained body. The essential feature of plastic strain, as contrasted with rupture, is that the strained body at every stage remains coherent and strong. This is exemplified by plastic strain of a single crystal by gliding on planes of weakness in the space-lattice. Plastic strain of rocks, which are composed of many crystals often belonging to several mineral species, cannot be defined so precisely, but the term nevertheless has been widely applied by geologists to any permanent

[8] B. Sander, *Gefügekunde der Gesteine*, pp. 262, 263, Springer, Vienna, 1930; E. B. Knopf, Petrotectonics, *Am. Jour. Sci.*, vol. 25, pp. 460–462, 1933.

deformation throughout which the rock maintains essential cohesion (and hence, too, its strength), regardless of the extent to which local microfracturing and displacement of individual grains may have entered into the process. This maintenance of spatial continuity of the rock mass as strain develops is a condition essential to the evolution of tectonite fabrics in metamorphism, for the characteristic symmetry of the fabric is determined by symmetry of movement in a continuous medium undergoing strain. The strain itself is the integrated product of many movements affecting fabric elements of all sizes (ions, twin lamellae, crystals, aggregates of crystals, and portions of the intergranular solution) and conforming to a general movement plan. All such are termed componental movements (*Teilbewegungen*) by Sander, who further distinguishes between direct componental movements, involving crystal gliding or relative displacement of grains, and indirect movements, which include transport of atoms and ions by such means as solution and redeposition, diffusion through pore solutions, or movement of pore solutions by convection, in so far as all these movements are related to deformation.

The time relation between direct and indirect componental movements in metamorphism [9] is brought out by recognizing that crystallization (*i.e.*, indirect componental movement) may fall into any of three categories:

1. Pretectonic: crystallization prior to deformation. Common criteria are partial granulation of crystals, bending of cleavages and twin lamellae, presence of undulose extinction, and development of visible glide lamellae (*e.g.*, $\{01\bar{1}2\}$ twin lamellae of calcite).

2. Paratectonic (syntectonic): crystallization broadly synchronous with deformation. Some crystals show criteria of pretectonic crystallization, while others (of the same or of different minerals) appear to have crystallized after deformation ceased. One of the clearest criteria of paratectonic crystallization in coarse-grained schists is the presence of porphyroblasts of such minerals as albite, garnet, or staurolite, enclosing S-shaped lines of inclusions the symmetrically curved trend of which, merging without break into the trend of similar lines in the surrounding matrix, bears witness to rotation of the porphyroblasts during their growth. Sander [10] cites instances of finely corrugated schists, in which the flakes of mica lying within the protected areas on the concave sides of the arches of microfolds are sharply crystallized, while those aligned along the convex sides have been twisted and bent by deformation which locally has outlasted crystallization. Paracrystalline folding may also result in the close association, within a single microsection, of folds with sharply

---

[9] B. Sander, *Einführung in die Gefügekunde der Geologischen Körper*, II, pp. 295–306, Springer, Vienna, 1950.

[10] Sander, *op. cit.*, pp. 245, 246, 1930.

crystallized micas and others in which the mica flakes are obviously deformed.

3. Post-tectonic: crystallization subsequent to deformation. This tends to obliterate some effects of mechanical deformation, e.g., undulose extinction, distorted crystal outlines, microjoints, and so on. The major slip surfaces in the fabric, although healed by such processes, offer directions of minimum resistance to crystal growth, and hence tend to be rendered more conspicuous by crystallization of tabular micas or prismatic horn-

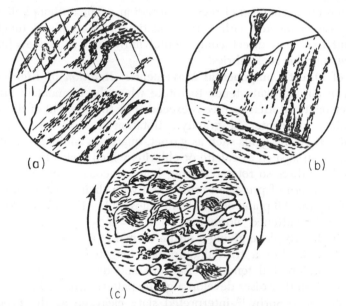

Fig. 101.   Lines of inclusions (si) in porphyroblasts of albite in albite schists.   (a) Helicitic structure (precrystalline deformation).   (After F. J. Turner and C. O. Hutton.)   (b) Postcrystalline deformation indicated by abrupt change in direction of si at crystal boundaries.   (After F. J. Turner and C. O. Hutton.)   (c) Paracrystalline deformation involving clockwise rotation of growing porphyroblasts.   (After E. B. Bailey.)

blendes with strong dimensionally controlled preferred orientation. Minerals low in the crystalloblastic series (e.g., quartz, calcite) crystallize as a mosaic of equant grains.   (Grains formed by pretectonic crystallization tend to be lensoid; those formed by paratectonic crystallization may be either equant or lensoid.)   It is possible also to distinguish three types of deformation in relation to crystallization, viz.: postcrystalline, paracrystalline, and precrystalline deformation. These are illustrated by different types of albite porphyroblasts in Fig. 101.

The fabric characters of a metamorphic rock, including the patterns and degrees of preferred orientation displayed by its component minerals, are end products of some particular combination or sequence of direct and indirect componental movements. It is appropriate, therefore, to review some of the types of componental movement that may be involved in rock deformation, and to note in what ways they are likely to influence the character of the fabric and particularly the state of preferred orientation of constituent minerals. At the outset it should be noted that since direct componental movements are more readily treated, both mathematically and experimentally, than are indirect componental movements, the latter have been somewhat neglected or have been oversimplified in discussions on mechanism of rock deformation. The authors believe that the processes connected with recrystallization probably play a much more important role than direct componental movement in the evolution of most metamorphic rock fabrics.

Role of Direct Componental Movements. *Rotation of Rigid Grains.* A grain in an aggregate tends to rotate when the moment of frictional resistance on its external surface is exceeded by the moment of the applied forces about its center of gravity. Rotation is most effective where mechanically strong gains such as garnet, quartz, or albite are embedded in a weak matrix, such as calcite or chlorite, which yields by plastic flow. Rigid rods or discs so rotated during homogeneous strain of the matrix develop a pattern of preferred orientation with their greatest dimensions approaching the AB plane of the strain ellipsoid.[11] For a biaxial ellipsoid symmetrically flattened normal to C $(A = B > C)$ rods tend to lie radially in the AB plane giving a girdle normal to C, and discs tend to be parallel to AB. For a triaxial ellipsoid elongated in B $(A > B > C)$ there is a preferential tendency for rods and discs to align themselves parallel to B; so that discs develop a girdle pattern around B.

A century ago Sorby[12] interpreted slaty cleavage as the product of rotation of mica flakes under a compressive force acting normal to the cleavage; and ever since then this mechanism has been accepted by some petrologists as an explanation of the origin of some types of schistosity. However, parallel alignment of coarse unstrained interlocking flakes of mica in mica schists can scarcely be due to rotation of the crystals as they now exist; and the meager experimental evidence available suggests that under geological conditions mica is mechanically weaker than its commonest associates quartz and feldspar. Although mica of metamorphic rocks commonly shows patterns of preferred orientation consistent with those predicted for orienting by rotation, it is improbable, in the authors'

---

[11] A. March, Mathematische Theorie der Regelung nach Korngestalt bei affine Deformation, *Zeitschr. Kristallographie*, Band 81, Hefte 3-4, pp. 285-297, 1932; B. Sander, *op. cit.*, pp. 103-113, 1950.

[12] H. C. Sorby, On the theory of slaty cleavage, *Phil. Mag.*, vol. 12, pp. 127-129, 1856.

opinion, that this is a significant process in the evolution of mica fabrics of schists and slates.   On the other hand it could be highly effective in orienting clay minerals and sedimentary micas during premetamorphic compaction of water-saturated silts and shales.   Here the flowing matrix would be interstitial water.

Quartz is a mineral that tends to crystallize in equant grains in metamorphic rocks.   However, it has been found that at high temperature and confining pressure quartz tends to fracture into needlelike fragments whose long dimensions are parallel to simple crystal directions such as the c axis and inter-rhombohedral edges.   Rotation and alignment of such fragments, followed by post-tectonic crystallization from the nuclei so oriented, could account for most of the known patterns of preferred orientation of quartz in metamorphic rocks.[13]   This is no proof that the process is geologically significant.   It can scarcely apply to the development of such patterns in coarsely recrystallized quartz of metacherts or in quartzite so slightly deformed as to retain much of its sedimentary texture.[14]   Nevertheless, no other hypothesis consistent with experimental data has yet been advanced in explanation of the known patterns of preferred orientation of quartz in deformed rocks.

Some of the clearest textural evidence of rotation is afforded by mechanically strong minerals of equant habit such as garnet and albite.   These, as indicated by spiral trains of inclusions, are committed to continual rotation in the plastic matrix; and because they lack a tabular or prismatic habit they never attain significant preferred orientation.

Plastic Deformation of Grains.   In petrology the closest approximation to homogeneous strain of a homogeneous body is afforded by plastic strain (gliding) in a single crystal.[15]   Gliding occurs parallel to some plane and direction of weakness in the crystal lattice, determined by closest packing of similar ions.   In translation gliding, though the external shape of the grain changes, its lattice remains unstrained and intact, for gliding of one layer over another is limited to some whole number of inter-ionic spacings.   In twin gliding ($e.g.$, $\{01\bar{1}2\}$ gliding in calcite) each layer moves through a certain fraction of an inter-ionic spacing relative to the layer beneath, and a new lattice, twinned in relation to the initial lattice, is formed.   Consequently greater strain can result from translation than from twin gliding, though the latter is the more potent orienting mechanism.   Twin gliding on $\{01\bar{1}2\}$ in calcite reorients the lattice through an angle of 52°.

[13] D. T. Griggs and J. F. Bell, Experiments bearing on the orientation of quartz in deformed rocks, *Geol. Soc. America Bull.*, vol. 49, pp. 1723–1746, 1938.

[14] Cf. H. W. Fairbairn, The stress-sensitivity of quartz in tectonites, *Tschermaks min. pet. Mitt.*, Band 4, Hefte 1–4, pp. 75–80, 1954.

[15] M. Buerger, Translation gliding in crystals, *Am. Mineralogist*, vol. 15, pp. 1–20, 1930; Barrett, *op. cit.*, pp. 336–352, 1952.

In metallic aggregates and in marble, plastic strain tends to be homogeneous: each grain, by gliding on one or more suitably stressed glide systems, becomes strained to approximately the same degree as its neighbors and as the aggregate.[16]   In each grain gliding begins on that glide system for which the resolved shear stress (determined by its orientation in the stress field) is maximal.   Under the constraint imposed by its neighbors the grain, as strain progresses, rotates [17] bodily in space in the sense opposite to that of internal gliding.   This changes the orientation of the grain lattice in relation to neighboring lattices and with respect to the applied force.   In consequence the values of resolved shear stress on the potential glide systems within the grain change continually, and in this way a second or even a third glide system may become activated.   Ultimately a stable orientation may be achieved when two or more glide systems, symmetrically oriented with respect to the applied force, operate simultaneously, and corresponding rotational effects are mutually cancelled.

The role of plastic gliding as an orienting mechanism in an aggregate of ionic crystals is illustrated by the behavior of marble in the course of experimental compression and extension at 300° to 600°C. and 5,000 bars:

1. There are two principal glide mechanisms for calcite: twin gliding on $\{01\bar{1}2\} = e$ (three glide systems); translation gliding on $\{10\bar{1}1\} = r$ (three systems).   The first leaves visible traces in the form of twin lamellae, except that a lattice twinned to completion is indistinguishable from an untwinned lattice.   The second leaves no visible microscopic trace; yet it is the most effective mechanism of strain.   This shows that caution should be exercised in assessing the tectonic significance of microscopically obvious lamellar structure such as Boehm lamellae in quartz and pinacoidal "deformation lamellae" in olivine, which have been very generally interpreted as manifestations of active glide systems responsible for the orientation of the grains in which they occur.   There is no foundation whatever for this belief.   Deformation bands in metal crystals and in calcite are commonly transverse to the active glide system. Visible lamellae in quartz and olivine are probably late structures super-

[16] G. I. Taylor, Plastic strain in metals, *Inst. Metals. Jour.*, vol. 62, pp. 307–324, 1938; F. J. Turner, D. T. Griggs, R. H. Clark, and R. H. Dixon, Deformation of Yule marble, Part VII, *Geol. Soc. America Bull.*, vol. 67, pp. 1273, 1274, 1956.

[17] This is called external rotation since it can be described only with reference to externally situated coordinates.   Gliding also can cause rotation of internal surfaces of discontinuity such as cleavages or twin lamellae with reference to internal coordinates (lattice directions).   This is internal rotation.   If it is directed clockwise, the sense of accompanying external rotation of the grain is counterclockwise.   External rotation brings the active glide system progressively more nearly normal to a compressive force or parallel to a tensile force (Turner, Griggs, and Heard, *op. cit.*, pp. 898–901, 1954; Turner, Griggs, Clark, and Dixon, *op. cit.*, pp. 1278–1284, 1956).

posed upon an already oriented fabric; they give no clue to the orienting mechanism.

2. High strains (>40 per cent shortening; >80 per cent elongation) induce a high degree of orientation of the calcite lattices in what seem to be stable orientations. The c axis then is inclined at between 10° and 30° to the axis of compression, or between 60° and 80° to the axis of extension.

3. Individual grains assume lensoid outlines tending to conform to the strain ellipsoid of the aggregate.

4. In calcite twinning on $\{01\bar{1}2\}$ occurs much more readily than translation on $\{10\bar{1}1\}$ (although this difference in strength nearly vanishes at 800°C.). In slightly strained aggregates incipient twinning—here the main mechanism of deformation—is shown by glide systems for which the coefficient of resolved shear stress is high. Assuming that the same relation holds good for visible twin lamellae in slightly deformed natural calcite fabrics, it is possible to calculate a statistical mean direction for the maximum compressive stress σ_1 responsible for the deformation.[18]

The Role of Indirect Componental Movements.   Growth of competing crystals in an antisotropic solid medium involves so many factors, and the experimentally determined data bearing on these are so inadequate, that the effects of indirect componental movements in rock deformation are impossible to predict except in a most generalized manner.[19]   Four such generalized conclusions are stated below:

1. During recrystallization of a stressed monomineralic aggregate—*e.g.*, quartzite or marble—solution of highly stressed grains or portions of grains presumably is accompanied by crystallization of new material either as outgrowths from less stressed portions of grains or as new crystals in intergranular cavities. Possible effects are reduction of pore space, development of an even-grained aggregate somewhat coarser than the parent aggregate, and general elongation of grains in the plane normal to the maximum compressive stress.   This last factor has long been appealed to—as "Riecke's principle" (see page 476)—as an important, or even the dominant, mechanism in the development of schistosity. It could scarcely apply, however, to the very numerous schists, especially those of the pelitic class, which are many times coarser in grain than were the parent rocks prior to metamorphism. Moreover, any preferred orientation of space-lattice in such rocks should be symmetrical about the normal

[18] D. B. McIntyre and F. J. Turner, Petrofabric analysis of marbles from Mid-Strathspey and Strathavon, *Geol. Mag.*, vol. 90, pp. 225–240, 1953.

[19] It is for this reason that some writers, *e.g.*, Schmidt (*op. cit.*, p. 171, 1932), while admitting the effectiveness of crystallization, confine their discussion of fabric evolution to the influence of direct componental movements.

to the plane of grain elongation (*i.e.*, of schistosity). This is never the case in that most important class of deformed rocks—defined later in this chapter as B-tectonites—in which a single axis of symmetry coincides with the lineation within the schistosity plane. Preferred orientation according to space-lattice cannot develop by the so-called "Riecke mechanism" in rocks composed of such minerals as quartz and calcite whose elongated or lensoid form is not correlated with crystallographic habit. This is confirmed by the experimental work of Fairbairn,[20] who produced a "synthetic quartzite" by compressing quartz sand in weak sodium carbonate solution under high confining pressures. The individual grains assumed an elongate form (maximum elongation, 2:1) and distinct dimensional orientation, but failed to develop any preferred orientation of space-lattice. It seems reasonable to conclude that the principle discussed in this paragraph is not an important factor in the development of schistose fabrics where associated preferred orientation patterns of crystal lattices are well developed.

2. In crystals of many metamorphic minerals, certain crystallographic directions are axes of most rapid crystal growth (for example, the *c* axis in amphiboles; directions in {001} in micas and chlorites). Crystal nuclei that happen to be oriented with such directions in or near to fabric planes (for example, *s*-planes of most kinds) offering minimum resistance to crystal growth, tend to survive and to develop at the expense of less favorably oriented grains. Paratectonic or post-tectonic crystallization governed by this principle is probably a very important factor in development of coarse tabular micas and prismatic amphiboles with their longer dimensions in planes of schistosity. Here dimensional orientation must be accompanied by orientation of space-lattice.

3. Crystals of many minerals (*e.g.*, calcite, quartz) are markedly anisotropic with respect to such properties as compressibility and other elastic moduli. The relative stabilities of different crystals of such a mineral in a stressed rock will be affected by their crystallographic orientation in relation to directions of maximum and minimum stress. Recrystallization involves elimination of unstable and growth of stable grains. Some type of preferred orientation of the space-lattice in relation to stress axes is therefore to be expected in monomineralic fabrics formed by paratectonic crystallization. Bain[21] attributed strong simple patterns of preferred orientation in recrystallized marbles of Vermont to alignment of *c* axes of calcite (the direction of maximum compressibility) parallel to the maximum compressive stress. Just such an effect has been observed in

[20] H. W. Fairbairn, Synthetic quartzite, *Am. Mineralogist*, vol. 35, pp. 735–748, 1950.

[21] G. Bain, The central Vermont marble belt, *Guidebook, New England Intercollegiate Geological Association, 1938.*

calcite aggregates formed by paratectonic crystallization in Yule marble deformed experimentally at 400° to 600°C. and 5,000 bars.[22]

4. Strain hardening (cf. Fig. 95c) in cold-strained crystalline aggregates is generally attributed to the development and storing of lattice defects—especially dislocations—within the strained crystals. The strain energy so stored increases the free energy of the aggregate compared with that of unstrained material. Subsequent heating may bring about substantial recovery from strain hardening, with reduction in the free energy of the system. One effect, well known to metallurgists, is *polygonization,* a process whereby a single strained grain becomes divided into several homogeneous, strain-free subgrains of slightly different orientation. This is believed to involve migration of dislocations from an initially random distribution in the strained grain into planar arrays at the subgrain boundaries. These are approximately normal to the previously active glide planes of the crystal lattice. The development of sharply bounded extinction bands in bent grains of quartz and of olivine seems to be an effect analogous to polygonization in metals. At higher temperatures, about halfway between the absolute zero and the melting point, the cold-worked fabric becomes much more strongly modified by *recrystallization* (*annealing*).[23] Unstrained crystals, developing from new nuclei, completely replace strained grains of the cold-worked fabric. At the same time stored strain energy is released, and recovery from strain hardening is complete. With prolonged annealing at high temperature recrystallization enters a second phase sometimes termed *grain growth.* Certain grains in the strain-free aggregate now slowly grow at the expense of less favored grains, and a general coarsening of fabric results. Annealing may, and commonly does, cause profound change in the pattern of preferred orientation of the cold-worked aggregate—one pronounced tendency being toward random orientation. Annealing recrystallization occurs readily in calcite and marble shortened 20 per cent at room temperature and 5,000 bars and subsequently heated (the pressure being maintained) to between 600° and 800°C. It is probably a geologically important process responsible for unoriented fabrics of monomineralic rocks—notably marbles and quartzites—that nevertheless show conspicuous effects of deformation (lineation, schistosity, and so on).

Concluding Statement. Before it is possible to evaluate the relative roles of direct and indirect componental movements of various kinds in rock deformation, it will be necessary to have much more information than is at present available upon the behavior of crystals and aggregates

[22] Turner, Griggs, Clark, and Dixon, *op. cit.,* p. 1272, 1956; Griggs, Turner, and Heard, *op. cit.,* 1959.

[23] Cf. Barrett, *op. cit.,* pp. 485–509, 1952; P. A. Beck, Annealing of cold-worked metals, *Phil. Mag. Suppl.* (*Adv. in Physics*), vol. 3, no. 11, pp. 245–324, 1954.

of various minerals under controlled conditions. A few experimental results on the behavior of calcite, marble, dolomite, and quartz deformed at high confining pressures are available. Only when these are greatly augmented, and when it is possible to match different fabric patterns of metamorphic rocks with fabrics artificially developed during experimental deformation, will it be possible to correlate natural fabrics with particular orienting mechanisms. At present this can be done only for postcrystalline deformation of marble and possibly dolomite.

Nevertheless, from the complex and as yet inadequately elucidated processes of rock deformation there emerge fabrics whose symmetry and patterns of preferred orientation consistently conform to one or other of several clearly recognizable types. Before we discuss these types we must refer to the general procedure of petrofabric analysis by means of which they may be recognized.

PETROFABRIC ANALYSIS OF DEFORMED ROCKS

Fabric Axes. The fabric of a deformed rock is described with reference to three mutually perpendicular axes named *a, b,* and *c* as in crystallography. These are defined descriptively and are selected according to conventional procedure designed to bring out as simply as possible the symmetry of the fabric:

1. For orthorhombic fabrics with three mutually perpendicular planes of symmetry, that symmetry plane which coincides with the most prominent foliation (schistosity) is selected as *ab;* the intersection of a second symmetry plane with *ab,* especially if marked by a lineation in the fabric, is selected as *b.*

2. In monoclinic fabrics [24] there is a single symmetry plane, and the normal to this is *b.* The *ab* plane is any prominent surface of schistosity or cleavage normal to the symmetry plane (cf. Fig. 102).

3. In triclinic fabrics there is no plane of symmetry. Any prominent schistosity or *s*-surface may be selected as *ab,* and any lineation within it as *b.*

The procedure outlined above is based upon criteria susceptible to direct observation and measurement. The further step of kinematic or dynamic interpretation of fabric—as in Sander's correlation between symmetry of fabric and symmetry of movement—is more subjective and open to controversy, but in the present authors' opinion is most valuable.

Megascopic Surfaces and Directions in the Fabric of Deformed Rocks. *S-surfaces.* The term *s-surfaces* (*s-planes*) is used by Sander to denote sets of parallel planes of mechanical inhomogeneity in deformed rocks.

[24] F. J. Turner, Lineation, symmetry and internal movement in monoclinic tectonite fabrics, *Geol. Soc. America Bull.,* vol. 68, pp.1–13, 1957.

It is a purely descriptive, nongenetic term mnemonically signifying possible identity of s-planes with planes of stratification, slip, or schistosity of any kind, but premetamorphic parallel fabrics originating in other ways (*e.g.*, primary flow banding in igneous rocks) also fall within the category of s-surfaces, provided the rock is rendered mechanically anisotropic by their presence.   These s-surfaces are surfaces of potential it not of actual yielding in connection with deformation.   They include visible surfaces of differential movement, either continuous or discontinuous in space, as well as statistical surfaces which are defined purely by preferred orientation of crystallographic planes or axes of one or more minerals and which need not pass continuously across the whole fabric.   Either they may originate during metamorphism, or alternatively they may be relics of a premetamorphic anistropic condition that has been emphasized by

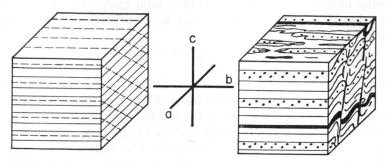

Fig. 102.   Fabric axes (*a*, *b*, *c*) in relation to plane of most prominent schistosity (*ab*) and lineation (*b*) in two monoclinic fabrics.

movement or crystallization during metamorphism.   Any set of s-surfaces of metamorphic origin constitutes a schistosity or foliation in the rock. To avoid confusion that is likely to result from attempts to classify particular sets of s-surface as "fracture" or as "flow cleavage," general use of the nongenetic term *s-surface*, now thoroughly established in the literature of structural petrology, is recommended.   It may be found possible in the course of fabric analysis to classify a given set of s-planes as products of slip movements or of deformation by "flattening," or as relict stratification, in which cases use of genetic terms such as s-planes of slip (*Scherungs-s*), s-planes of "flattening" (*Plättungs-s*), s-planes of stratification, and so on is justified.   Alternatively, especially when there is some doubt concerning their mode of origin, the various sets of s-surfaces that can be identified in a rock may simply be designated numerically, preferably in order of development if such is decipherable: for example, S_1 (stratification), S_2 (plane of "flattening"), S_3, S_4 (plane of slip).   This nomenclature has the advantage that it is independent of any assumption as to the relative parts played respectively by chemical reconstitution (indirect componental movements) and by mechanical processes such as slip and rotation (direct componental movements) in bringing about preferred orientation of crystals with reference to the various s-planes. In practice it may, in specific instances, be difficult if not impossible to evaluate the relative importance of these two contrasted processes, which nevertheless have frequently been cited as the essential criteria for distinguishing between "fracture cleavage" and "flow cleavage."

Lineation.[25] Careful examination of the most prominent s-surfaces in almost any deformed rock reveals the presence of a parallel rectilinear element in the fabric, termed *lineation,* which typically is determined by one or any combination of the following characters:

1. Intersection of two or more sets of s-surfaces

2. Dimensional parallelism of prismatic or platy crystals (amphiboles, chlorites, micas) or of stretched pebbles, sand grains, fossils, vesicles, or other relict elements in the fabric that have suffered elongation during metamorphic deformation

3. Linear grooves, or microcorrugations (drag folds) developed in the s-surfaces

Very commonly lineation is parallel to the *b* axis of a monoclinic fabric and to the axis of major folds. Lineation of this type is termed *Striemung* by Sander. Usually, though not invariably, identification of lineation with the *b* fabric axis can be confirmed by evidence of rotational movements around it, obtained either from direct observation of swirls, crumpling, and the like, on polished hand specimens, or from petrofabric diagrams depicting preferred orientation of constituent minerals. Dimensional orientation of elongated fabric elements with their long axes parallel to *b* may be the result of any or all of three processes, *viz.:* rotation of prismatic crystals, stretched pebbles, and other elongated elements about *b* during deformational flow at right angles to *b*; elongation of pre-existing fabric elements parallel to *b* and recrystallization of prismatic and tabular minerals, synchronous with deformation; postdeformational growth of prismatic and tabular crystals with their longest axes aligned in the direction of minimum resistance to growth in an anisotropic fabric resulting from deformation.

In some monoclinic fabrics, especially in mylonites and in slickenside films, the most conspicuous lineation lies in the symmetry plane, and so coincides with the *a* fabric axis. There are other schists in which *a* and *b* lineations are both present. Finally there are triclinic fabrics [26] in which the most pronounced lineation is simply the intersection of two nonsyn-

[25] E. Cloos, Lineation, *Geol. Soc. America Mem. 18,* 1946; Weiss, *op. cit.* pp. 19–21, 1959.

[26] Cf. L. E. Weiss, Fabric analysis of a triclinic tectonite, *Am. Jour. Sci.,* vol. 253, pp. 225–236, 1955.

chronous surfaces and is not simply related to other elements in the fabric—such as the pattern of preferred orientation of quartz or calcite.

Joints. In a deformed rock mass there is usually a well-defined joint system composed of several differently oriented sets of parallel joints, which may have a simple relation to the rest of the fabric.[27] Joints are produced by rupture of the rock under stress exceeding the breaking strength, and may therefore be classified on dynamic grounds as shear joints and tension joints respectively, according to whether rupture has taken place on surfaces of maximum shearing stress or at right angles to the direction of maximum tension within the strained mass.

Since it can scarcely be imagined that open megascopic joints could survive plastic flow of the rocks they traverse, any such joint system in a mass of rock whose fabric bears the imprint of plastic deformation must have originated in the last stages of deformation, or alternatively should be attributed to postdeformational rupture. The former alternative is perhaps favored by the general and striking tendency for the main sets of joints to be oriented simply in relation to the metamorphic fabric. It is also possible, however, that this relationship is an indirect one. For example, joints might develop by contraction of a deformed mass in process of cooling from the relatively high temperatures of metamorphism, and the anisotropic deformational fabric of the strained mass might then determine the orientation of the joints so formed. A more likely explanation is based upon the experimentally proven fact that the elastic limit of a rock at high confining pressures, comparable with those that prevail in metamorphism, is much higher than at atmospheric pressure. At the close of deformational metamorphism the rock is left in a state of elastic strain, but at some stage in subsequent unloading, the internal stresses within the strained mass exceed the elastic limit (as the latter is lowered with reduction of pressure) and are relieved by rupture. Whether directly or indirectly related to metamorphic strain, however, the joint pattern encountered in deformed rocks usually contributes to an important degree to the fabric pattern. This applies particularly to two classes of joints, respectively termed *cross joints* and *longitudinal joints*.

Most significant in this respect, and almost universal in deformed rocks, are tension joints, customarily termed *cross joints* or *ac* joints, which are in most instances inclined at about 10° to the *ac* plane of the fabric. They are nearly perpendicular to all visible *s*-surfaces, to the lineation *b*, and to the strike of folds formed during the deformation. They may take the form of microscopic hair-like cracks, which, when closely spaced and healed with a filling of parallel platelets of mica, occasionally impart a

[27] Cf. E. Cloos, The application of recent structural methods in the interpretation of the crystalline rocks of Maryland, *Maryland Geol. Survey*, pp. 73–78, 1937; E. S. Hills, *Outlines of Structural Geology*, Nordeman, New York, pp. 91–97, 1940.

marked fissility or cleavage to the rock; or they may appear as extensive master joints conspicuously cutting across the strike of folded rocks. Longitudinal (strike) joints, parallel to the b axis of the fabric, also occur widely in metamorphic and folded terranes. Their position in relation to the fabric axes varies, and they are classed together as ($h0l$) joints. They include both tension joints (for example, bc joints) and shear joints.

In many deformed rocks, joints which fall into neither of the above categories also occur, e.g., shear joints parallel to the a fabric axis—($0kl$) joints.

Outline of Procedure in Petrofabric Analysis.[28]   The field work which forms an integral part of any investigation of this kind involves mapping, in as great detail as possible, of the available macroscopic elements in the fabric, such as s-surfaces, lineations, and joints. Geographically oriented specimens of representative rocks are collected from selected outcrops. From every such specimen may be cut microsections, oriented where possible with reference to tentatively selected axes a, b, and c, or, if conspicuous s-surfaces and lineation are lacking, with reference to geographic coordinates. It is then possible to compare data determined microscopically and in the field.

The following are typical steps in petrofabric analysis of a metamorphic rock:

1. The orientations of all sets of s-surfaces, fractures, and so on, that can be recognized on polished surfaces of the specimen or in microsections cut from it are measured and plotted either upon a clinographic projection (block diagram) of the specimen, or upon a suitable circular projection (cf. Fig. 103). In the latter case an equal-area projection is preferable to the stereogram of crystallography, since it may be compared directly with diagrams showing preferred orientation of minerals, for which, following the procedure adopted by Schmidt, a Lambert equal-area projection is universally employed. The corresponding projection net is usually termed a *Schmidt net.* In contrast with the custom in crystallographic projection, it is usual in petrofabric analysis to project the lower hemisphere of the sphere of reference. When the plane of projection is the geographic horizon, this convention facilitates visualization of downward-dipping surfaces and directions within the rock mass, as contrasted with a projection of the upper hemisphere which shows the same features produced upward above the land surface.

2. For each of the principal minerals, such details as the following are recorded: grain size, habit, inclusions, cracks and twinning, evidence of rotation or deformation, relative abundance, relations to other minerals. It may be advisable to separate grains of one mineral into several cate-

---

[28] Knopf and Ingerson, *op. cit.,* pp. 209–262, 1938; J. C. Haff, Preparation of petrofabric diagrams, *Am. Mineralogist,* vol. 23, pp. 543–574, 1938; Weiss, *op. cit.,* 1959.

gories for subsequent measurement of orientation.[29] For example, in quartz-albite-muscovite-chlorite schists from Otago, New Zealand, it was necessary to distinguish between three types of quartz grains: small rounded grains enclosed in large porphyroblasts of albite; coarse irregular grains in narrow pencils trending parallel to the lineation b; and similar grains locally building up prominent rods of nearly pure quartz several centimeters in diameter, also lying parallel to b.

3. The degree of preferred orientation of crystals of appropriate minerals is next estimated statistically. For each mineral the orientation of a large number (usually 100 to 500) of representative grains in an oriented section is determined by measuring in each grain a selected crystallo-

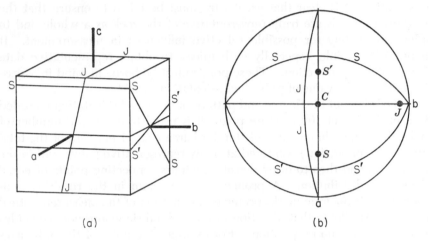

(a)   (b)

Fig. 103. (a) Clinographic projection of three sets of ($h0l$) schistosity planes, ab, SS, and $S'S'$, and one set of ($0kl$) joints JJ. (b) Equal-area projection (lower hemisphere) of the same planes; corresponding poles are lettered C, S, S', and J.

graphic or optical direction, which is then plotted upon an equal-area projection by means of a Schmidt net. Each grain is thus represented by one point (a pole) upon the projection. Some minerals are more suitable for this type of investigation than others. Until now attention has been focused mainly upon quartz [optic axis], calcite [optic axis; {01$\bar{1}$2} lamellae; {10$\bar{1}$1} cleavage], and micas [{001} cleavage], and to a less extent feldspars [{001} and {010} cleavages; X, Y, Z], hornblende [{110} cleavages; c crystal axis], and olivine [X, Y, Z]. The facility with which cleavages and twin lamellae can be measured depends upon the orienta-

[29] E.g., F. J. Turner and C. O. Hutton, Some porphyroblastic albite schists from Waikouaiti River, Otago, Royal Soc. New Zealand Trans., vol. 71, pt. 3, pp. 223–240, 1941; C. S. Ch'ih, Structural petrology of the Wissahickon schist near Bryn Mawr, Pennsylvania, Geol. Soc. America Bull., vol. 61, pp. 923–956, 1950.

tion of the particular grain within the rock section; planes nearly normal to the section are readily measured, but those subparallel to the section are invisible and cannot be located.   In the case of the {001} cleavage of mica, for example the central portion of the projection constitutes a "blind spot" and must necessarily remain unfilled, even if the grains completely lack preferred orientation.   The tabular habit of mica, parallel to {001}, intensifies this anomaly, for a microsection tends to encounter a much greater number of crystals oriented with {001} nearly perpendicular to the plane of the section than of grains with {001} subparallel to the section.   For optically biaxial minerals, it is of course possible to determine the three principal directions X, Y, and Z in any crystal.   From these observations it is clear that great care must be taken to ensure that the measured crystals are truly representative of the rock as a whole and to allow adequately for possible selective influences in measurement.   It may be impossible, especially with micas, to obtain representative data from a single thin section.   These points must be borne in mind in assessing critically the value of published orientation data.[30]

4. To bring out the degree and pattern of preferred orientation recorded in a projection of 100 or more poles, it is usual to count the number of poles that fall within a circle of standard area (e.g., 1 per cent of the total area of the projection), centered at many representative points within the projection, and then to draw density contours connecting points of equal density of distribution.   A pronounced maximum in the resulting contoured fabric diagram marks preferred orientation of the measured optical direction parallel to that direction in the rock fabric whose pole coincides with the maximum in question.   For example, the poles of the optic axes of 150 grains of quartz in a slightly gneissic granite are shown in Fig. 104a, and the corresponding contoured diagram is reproduced in Fig. 104b. The latter brings out a marked tendency for the optic axes of the grains of quartz to lie within a plane (the plane of the projection) nearly perpendicular to the megascopic lineation (b) and hence to the plane of schistosity (ab).   A further tendency for the quartz axes to be concentrated in two sectors (XX and YY in Fig. 104b) symmetrically inclined to the schistosity at angles of about 30° is also obvious.

5. A fabric diagram such as Fig. 104b is a graphic representation of the preferred orientation of crystals of a selected mineral within a microsection.   If there is any doubt as to whether certain maxima or minima shown on the diagram are real or are due to chance, a second diagram based on a new set of measurements may be prepared for comparison with the first.   It is preferable to use a second microsection of similar

---

[30] Cf. Turner and Hutton, op. cit., pp. 231, 232, 1941; E. B. Knopf, Fabric changes in Yule marble after deformation in compression, Am. Jour. Sci., vol. 247, pp. 433–461, 1949.

orientation for this purpose.   Alternatively, elemental diagrams for each 100 or 200 grains may be prepared in the first place, and subsequently all the measurements may be replotted together as the basis of a single, collective diagram.   Maxima or minima that appear in only one such diagram of a homogeneous fabric are fortuitous.[31]   It is also advisable to measure sufficient grains of the same mineral in each of three mutually perpendicular microsections (usually cut at right angles to *a*, *b*, and *c* respectively), to permit construction of three independent diagrams, which then give a picture of the preferred orientation as observed from three different directions.   The three diagrams can be used in conjunction, either to test

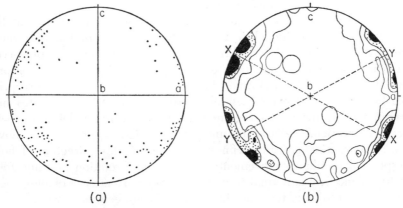

(a)   (b)

Fig. 104.   Orientation diagrams for quartz grains in a foliated granite, New Zealand. (*a*) Poles of optic axes of 150 grains projected upon the *ac* plane of the fabric.   (*b*) Contours drawn upon the same diagram to represent respective concentrations of 6%, 4%, 2%, and 0.7%, per 1% area of projection sphere.

the homogeneity of the fabric or to fill in the "blind spot" of one diagram from data recorded in the other two.

   Interpretation of Results. *The Problem.*   The orientation data recorded in fabric diagrams are used, in conjunction with megascopic features such as *s*-surfaces, lineations, joints, and fold axes, to reconstruct the movement plan of deformation.   Into this reconstruction hypothesis and personal judgment both enter to a considerable extent; but there is considerable agreement on some aspects of the problem, notably those relating to symmetry and homogeneity of movement and of deformation. Interpretation of specific details of fabric such as maxima in orientation diagrams is much more speculative.   Much of it indeed is valueless.

---

[31] If the pattern of preferred orientation is very weak, it may be advisable to test its significance by statistical means as described by F. Chayes, in H. W. Fairbairn, *Structural Petrology of Deformed Rocks*, pp. 297–326, 1950).

Significance of Fabric Symmetry.[32]   In connection with natural or experimental deformation there are four independent sets of geometric data each of which integrates to some simple plan of symmetry.   These relate to (1) the system of applied forces, (2) the geometry of strain (as shown for example by a strain ellipsoid), (3) the movement plan of deformation, and (4) the deformed fabric.   In experimental strain, (1), (2), and (4) can be directly measured; and something of the nature of (3) can in some instances be observed as well (*e.g.*, where deformation is accomplished by twin gliding or by diagonal shear of an aggregate[33]).   In natural strain only the deformed fabric is susceptible to complete measurement; some unambiguous effects of movement (rolled crystals, folded structures) are visible in some but not all fabrics.

Sander and Schmidt formulated the fundamental hypothesis that the symmetry of a tectonite fabric reflects the symmetry of the movement plan of deformation.   Where direct evidence of movement is preserved in a fabric, the movements (folding, rolling, or slip) do indeed conform to the symmetry of the deformed fabric.   Moreover the symmetry principle holds good for experimentally deformed metallic aggregates and ceramic bodies, for which the movement plan can be determined by direct observation.[34]   Three mutually perpendicular axes of the movement plan are kinematically defined: a is the principal direction of flow, transport, or rectilinear movement; b is an axis of rotation (rolling, folding, turbulence); c is normal to a and b.   Several patterns of movement plan, believed to be of tectonic significance, have been recognized:

1. Orthorhombic, with three symmetry planes, ab, bc, and ac.   The direction of maximum extension (transport) is a; planes or curved surfaces of relative displacement of particles intersect in b; movement is essentially normal to b, though some flow (extension) parallel to b is also possible.

2. Monoclinic, with one symmetry plane, ac.   The movement plan is characterized by flow normal to and rotation around b.   Any conspicuous direction of movement normal to b may be selected as a.

3. Triclinic, with no symmetry plane.   Selection of a, b, and c is arbitrary; but a or b is usually chosen so as to bring out some prominent element of transport or rotation in the complex movement plan.   Triclinic symmetry of movement may arise in several ways.   Two nonsynchronous deformations lacking a common plane of symmetry may be superposed to give a triclinic compound system.   Movement in shear folding is triclinic if, as commonly is the case, the direction of shear is oblique to the inter-

[32] Turner, *op. cit.*, 1957; Weiss, *op. cit.*, pp. 10–12, 1959.

[33] Turner, Griggs, and Heard, *op. cit.*, pp. 901–909, 1954; Turner, Griggs, Clark, and Dixon, *op. cit.*, pp. 1285–1289, 1956.

[34] Barrett, *op. cit.*, pp. 459–480, Figs. 19–40, 1952; W. O. Williamson, Lineations in three artificial tectonites, *Geol. Mag.*, vol. 92, pp. 53–62, 1955; Turner, *op. cit.*, pp. 11, 12, 1957.

section of the shear plane with the folded structure (Fig. 105). The latter direction (*e.g.*, the bedding-cleavage intersection in some rocks) is the axis of folding and so has the attributes of a *b* kinematic axis; but the direction of shear has the qualities of an *a* axis. Either, but not both, may be selected as an axis of the movement plan.

Kinematic interpretation of fabric symmetry as advocated by Sander and Schmidt has proved most fruitful. The principle is further elaborated below in the light of experimental data, especially those drawn from Griggs's experiments on marble:

1. Symmetry of tectonite fabrics can be correlated with symmetry of the movement plan of deformation. Thus *a*, *b*, and *c* of the fabric can be equated with *a*, *b*, and *c* of the movement plan; for in both the axes are defined in terms of symmetry.

2. The symmetry elements of the movement plan are reproduced in the geometry of strain. Symmetry of strain, however, may be of a higher order than symmetry of movement. Thus simple shear with monoclinic symmetry (cf. Fig. 96) produces a strain ellipsoid of orthorhombic symmetry; but the symmetry plane of movement is a symmetry plane of the ellipsoid.

3. Symmetry of strain is compounded of symmetry of the applied force superposed on symmetry of the initial fabric.[35]

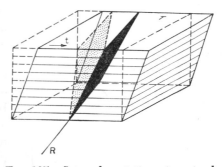

Fig. 105. Internal rotation in simple shear. (*After I. Borg and F. J. Turner.*) The body outlined in broken lines is deformed to the shape outlined by full lines by shear on planes *T* parallel to the direction *t*. Simultaneously the stippled plane rotates about *R* to the position shown in solid black. *R* is oblique to *t*.

In Griggs's experiments the test material, Yule marble, has a fabric characterized by marked preferred orientation of *c* axes of calcite in one direction; this is an axis of axial symmetry with a principal plane of symmetry perpendicular thereto. The force systems, axial compression or extension, also have this symmetry. Where the principal symmetry axis of the applied force coincides with that of the initial fabric (Fig. 106*a*), both the strain form and the new pattern of preferred orientation of calcite axes retain the same axis; the symmetry of both remains axial (Fig. 106*c*). Where the symmetry axis of the force system is normal to that of the initial fabric, the two are superposed to give orthorhombic symmetry of strain and of the newly oriented fabric (Fig. 106*b*, *d*). Even in a specimen elongated

[35] Cf. Turner, Griggs, Clark, and Dixon, *op. cit.*, pp. 1269, 1274, 1956; Turner, *op. cit.*, pp. 12–15, 1957.

(in the neck region) to ten times its initial length, the influence of the original fabric upon the ultimate fabric and upon the geometry of strain is still strongly impressed.

Homogeneity. A fabric is said to be homogeneous when it shows identical reproducible features in all samples of equal extent within a given field. Individual fields may be thin sections or parts of thin sections, hand specimens, outcrops, or even areas measured in square miles. Clearly, a given rock formation may have a fabric which is homogeneous within the field of a hand specimen but nonhomogeneous within a field of 100 m.², or vice versa. A homogeneous fabric reflects deforming movements which have conformed to a uniform plan within the field in question. Nonhomogeneous fabrics are commonly shown by rocks affected by flexural-slip or by shear folding, and by rocks in which grains of one mineral (e.g., quartz) enclosed in porphyroblasts (e.g., of albite) have a different orientation from that of the same mineral in the rock matrix. Although the fabric of a single flexural-slip fold is nonhomogeneous, a complex of many folds having a common axis may be statistically homogeneous for large fields of sampling many square miles in extent.

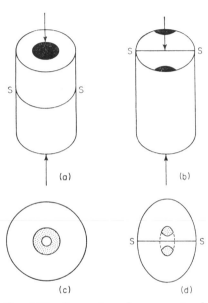

FIG. 106. Symmetry of compressional strain in Yule marble. Full black = mean orientation of *c* axes in initial fabric, projected on end of cylinder; SS = initial foliation. (*a*) Compression normal to foliation; (*b*) compression parallel to foliation. (*c*) Cross section of cylinder *A* after deformation; (*d*) cross section of cylinder *B* after deformation. Stippled areas show orientation of *c* axes in the deformed fabric.

Significance of Fabric Details. Interpretation of individual maxima on orientation diagrams depends to a large extent upon hypotheses (of which there may be many) relating to orienting mechanisms of the minerals concerned. For example, in the case of quartz, there are at least five, and possibly a dozen, different observed relations between *c*-axis maxima in orientation diagrams and schistosity of the corresponding rocks. Different writers interpret these in terms of various assumed orienting mechanisms: internal translation gliding of individual grains on basal plane, prisms, or rhombohedrons of the lattice; fracture of grains into elongated splinters and rotation of these into parallel alignment; recrystallization under stress, without either gliding or fracture. Preferred orientation patterns in micas are on the whole more simply related to schistosity of the rock in which they occur, but little is positively known as to the processes by which these minerals become oriented. Experiments on plastic strain of calcite revealed a mechanism of gliding and orienting unsuspected from petrofabric observations on natural fabrics. Until more experimental data on development of fabric under controlled conditions become available, this remains the most speculative aspect of petrofabric analysis. The all too common practice of invoking new orienting mechanisms (on purely petrographic evidence) for every newly observed pattern of preferred orientation has done much to obscure the results of petrofabric research.

In spite of such ambiguities, there is now a large mass of consistent information as to preferred orientation patterns in relation to schistosity and lineation. On the basis of such information it is possible to define descriptively a number of standard structural types of fabric among schistose (foliated) deformed rocks. These will now be reviewed.

COMMON TYPES OF TECTONITE FABRIC

Fabrics of S-tectonites. *Definition.*[36] Sander recognizes two contrasted, though not sharply separable, patterns of tectonite fabric, and accordingly divides tectonites into two classes termed *S-tectonites* and *B-tectonites* respectively.

S-tectonites are rocks the fabrics of which are dominated by one set of visible s-planes (expressed megascopically as a single well-defined planar schistosity), linear parallelism of fabric elements being inconspicuous or absent. Corresponding to this lack of lineation is an equally characteristic lack of girdle patterns in orientation diagrams and of any evidence of external rotation (*e.g.*, flexure) in connection with deformation. As lineation and girdles become obvious, S-tectonites merge into B-tectonites.

S-tectonites with One Set of s-planes. Mylonites and slickensides are mutually similar in mode of origin and in their typically localized occurrence as thin layers or films of comminuted rock material smeared upon, or milled between, surfaces of strong differential movement. Consequently they resemble each other also in fabric. The nature, and at times even the precise direction, of movement connected with their development can be reconstructed more surely than is the case with most other deformed rocks. They provide perhaps the clearest illustrations of those rock fabrics of the S-tectonite class that have evolved under the influence of penetrative movement upon one set of parallel or subparallel s-surfaces. Certainly mylonites and slickensided rocks are products of

[36] Sander, *op. cit.*, pp. 58, 220, 221, 1930; Knopf and Ingerson, *op. cit.*, pp. 68–70, 1938.

metamorphic deformation of great intensity, for the notable relative displacements with which they commonly are associated have been achieved by penetrative movements distributed through but small thicknesses of rock. For discussion of the petrography, field occurrence, and classification of mylonites and allied rocks, the reader is referred to other works.[37]

As originally defined by Lapworth, mylonites are strongly coherent, fine-grained, conspicuously laminated rocks, formed by extreme microbrecciation and milling of rocks during movement on fault surfaces. Metamorphism is dominantly cataclastic, *i.e.*, is achieved mainly by direct componental movements, with little or no growth of new crystals; but mylonites, by virtue of the high confining pressures under which they originate, always retain a strongly coherent condition and characteristically present an aphanitic or even flinty appearance in hand specimen.[38] Most mylonites, as seen under the microscope, consist of an intensely granulated streaked matrix through which are scattered porphyroclasts of the more resistant minerals in various stages of mechanical degradation. The characteristic fluxional lamination of many mylonites reflects a sliding movement in a single set of parallel *s* planes, which, with some simultaneous rotation of the larger surviving grains about *b*, is no doubt the dominant movement concerned in the formation of mylonites. There are, however, mylonite fabrics [39] that show evidence of more complex movements such as folding by flexural slip, or movement upon symmetrically developed *s*-planes intersecting in *b*. Some of these should be classed as B-tectonites.

The term *slickenside* is restricted by Sander to individual surfaces of slipping without limitation as regards number, spacing, or degree of accompanying recrystallization, but in all cases characterized by lineation that coincides with the direction of movement, *i.e.*, with the *a* axis of the fabric.[40] Some slickensides show also an additional lineation, parallel to

---

[37] P. Quensel, Zur Kenntnis der Mylonbildung, *Upsala Univ., Geol. Inst., Bull.*, vol. 15, pp. 91–116, 1916; E. B. Knopf, Retrogressive metamorphism and phyllonitization, *Am. Jour. Sci.*, vol. 21, pp. 1–27, 1931; A. C. Waters and C. D. Campbell, Mylonites from the San Andreas fault zone, *Am Jour. Sci.*, vol. 29, pp. 473–503, 1935; C. E. Tilley, The dunite-mylonites of St. Paul's Rocks (Atlantic), *Am. Jour. Sci.*, vol. 245, pp. 483–491, 1947; H. H. Read, Mylonitization and cataclasis in acidic dikes in the Insch (Aberdeenshire) gabbro and its aureole, *Geologists' Assoc. Proc.*, vol. 62, pp. 237–247, 1951; J. S. Scott and H. I. Drever, Frictional fusion along a Himalayan thrust, *Royal Soc. Edinburgh Proc.*, sect. B, vol. 65, pt. 2, pp. 121–142, 1953.

[38] Brittle rocks such as granite, failing by diagonal shear under confining pressures of a few thousand bars, maintain their strength while developing mylonitic structure in the shear zone (Griggs, Turner, and Heard, *op. cit.*, 1959).

[39] Waters and Campbell, *op. cit.*, p. 492, 1935; J. Christie, D. B. McIntyre, and L. E. Weiss, Appendix to D. B. McIntyre, The Moine thrust, *Geologists' Assoc. Proc.*, vol. 65, pp. 219–223, 1954.

[40] Sander, *op. cit.*, p. 227, 1930; *op. cit.*, II, pp. 268, 269, 1950.

the *b* axis. Other surfaces, in which the only linear structure is a lineation parallel to *b*, are placed in a distinct category as pseudo-slickensides. Sander suggests that a relatively high velocity of movement favors building up of a lineation parallel to *a* rather than to *b*. In mylonites, too, the direction of lineation may coincide with either *a* or *b*, or both types may be represented in the same rock. The term *slickenside mylonite* is applied to the layer of mylonitized rock—often of microscopic thickness—that forms a veneer on most slickensides.

Preferred orientation of minerals in mylonites and slickenside films is mostly difficult to evaluate on account of the fine grain of such rocks. In fine-grained quartzose mylonites, a strong tendency for quartz grains to lie with their optic axes subparallel to *a* may be demonstrated by simultaneous compensatory behavior of all grains, when a gypsum plate is inserted parallel to *a* while the section is viewed between crossed nicols. On the other hand, extreme granulation in dunite mylonites from Milford Sound, New Zealand, has resulted in strong dimensional parallelism of the elongated fragments so formed, but in no perceptible preferred orientation of the olivine space-lattice.

Sander has described from Melibokus, Odenwald, the fabric of a slickenside mylonite developed on shear surfaces traversing granite; [41] post-tectonic crystallization has rendered individual grains of quartz and biotite large enough for satisfactory measurement. The essential features of the fabric of this rock are as follows:

1. The slickenside surfaces S show a lineation R, marked by fine corrugation, coupled with a parallel alignment of small prisms of hornblende.

2. Crystals of biotite show a marked tendency to lie with {001} parallel to R, giving a girdle pattern in the fabric diagram (Fig. 107*a*), which identifies R as the *b* axis of the monoclinic fabric. The strong maximum in Fig. 107*a* is evidence of a further tendency for the majority of the crystals of biotite to be oriented with {001} subparallel to S ($= ab$).

3. Grains of quartz are elongated at right angles to R, thus giving a microscopic second lineation, and show very pronounced preferred orientation of the space-lattice with the optic axis [0001] parallel to S and normal to R, that is, parallel to *a* of the fabric (Fig. 107*b*). This illustrates one of the common orientation rules for quartz in deformed rocks. How it originates is unknown.

4. Microsections cut parallel to S show numerous closely spaced rectilinear hair cracks normal to R. These are the microscopic equivalents of *ac* tension joints.

Mylonites proper are products of almost purely cataclastic metamorphism. As chemical processes—recrystallization and neomineralization—enter more and more into deformation, there is continuous transition

[41] Sander, *op. cit.*, pp. 228–231, D22–27, 1930; *op. cit.*, II, pp. 145–147, 1950.

from mylonites to a group of rocks variously referred to as augen schists, mylonite gneisses, flaser rocks, and blastomylonites. These pass in turn into completely crystalloblastic schists which tend toward regional development over board zones of deformational metamorphism, as contrasted with the localized occurrence of mylonitic rocks in restricted zones of major dislocation. The Laurel gneiss of Maryland [42] appears to be such a rock. It is a roughly foliated rock of igneous aspect, without recognizable lineation, and consisting of quartz, feldspars, biotite, and muscovite. It is thought to have been formed from the associated Wissahickon oligoclase-mica schist by plastic deformation involving granulation and recrystallization under high temperature and stress and in the presence of

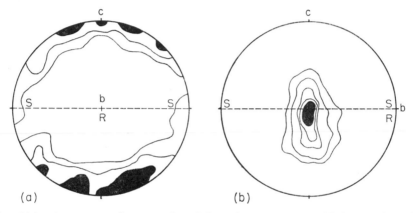

FIG. 107. Orientation diagrams for slickenside mylonite, Melibokus, Odenwald. (*After B. Sander.*) (*a*) Biotite: poles of {001} cleavage in 140 crystals in section normal to lineation, *b*, and to slickenside plane SS. Contours at 4%, 2%, 1%, per 1% area. (*b*) Quartz: optic axes of 138 grains in section normal to SS and parallel to *b*. Contours at 16%, 12%, 4%, 0.7%, per 1% area.

abundant water. Orientation diagrams for {001} in micas (Fig. 108*a*) show strong maxima coinciding approximately with the pole of the foliation plane S. The arcuate form of the area occupied by muscovite poles, when compared with a small circle of the projection sphere (broken ellipse) drawn around the pole of S, is seen to result from distortion accompanying projection and not from incipient development of a real girdle in the plane of projection. Quartz (Fig. 108*b*) has developed preferred orientation with the *c* axes of the crystals parallel to a particular direction in S, as indicated by a point maximum on the orientation diagram. There is also a decided tendency for quartz grains to be elongated in this same direction, which can thus be identified as either

[42] R. W. Chapman, Pseudomigmatite in the Piedmont of Maryland, *Geol. Soc. America Bull.*, vol. 53, pp. 1299–1330, 1942.

the a or the b axis of the orthorhombic fabric. The former is preferred since alignment of c axes of quartz in a of the fabric is common in tectonites.

S-tectonite Fabrics of Granulites and Quartzites.   Granulites, according to current usage of the term in Europe, are quartzo-feldspathic rocks of high metamorphic grade, poor or lacking in mica, and typically characterized structurally by a single conspicuous megascopic plane schistosity, S_1, of great regularity, which is determined mainly by parallel dimensional orientation of flat lenses of coarse-grained quartz ("granulitic" quartz) set in a quartzose matrix of smaller equidimensional grains.   In many granulites, $(h0l)$ planes S_2, S_3, and so on, which may be either

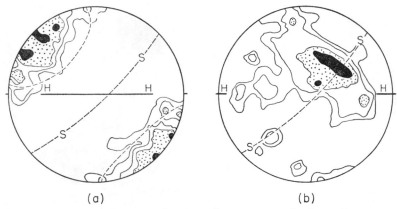

(a)   (b)

Fig. 108.   Orientation diagrams for Laurel greiss, Maryland. HH = geographic horizontal, SS = megascopic foliation.   (*After R. W. Chapman.*)   (*a*) Muscovite: poles of {001} cleavage in 100 crystals.   Contours at 8%, 5%, 3%, 1%, per 1% area. Broken ellipse is a small circle drawn at 40° angular distance from the pole of the foliation SS.   (*b*) Quartz: optic axes of 200 large grains in the same section.   Contours at 6%, 4%, 2%, 1%, per 1% area; maximum concentration, 8%.

visible or merely statistically defined by preferred orientation of quartz axes (as suggested by the maxima of Fig. 110*a*), are paired symmetrically about the megascopically conspicuous foliation S_1.   The symmetry of such fabrics approximates the orthorhombic.   Normally the crystals of muscovite and biotite show a high degree of preferred orientation with {001} parallel to S_1, so that, as regards mica patterns and megascopic structural characters, the fabrics conform to the requirements of S-tectonites.   Nevertheless on the basis of quartz fabric considered alone, many granulites would be classed as B-tectonites, for ac girdles and paired concentrations of [0001] suggesting $(h0l)$ planes are common (Fig. 110*a*). Other granulite fabrics, in which two mutually perpendicular B axes can be detected in the quartz fabric, are treated in a separate section as $B \perp B'$-tectonites.

The patterns of preferred orientation for quartz and mica typical of granulites proper are also commonly encountered in deformed quartzites and in quartzo-feldspathic schists. Two examples are noted briefly as follows:

In the pre-Cambrian of Finland strongly metamorphosed quartzites described by Hietanen [43] are steeply dipping rocks, the foliation of which is generally interpreted as having developed normal to a compressive force. Some are B-tectonites with well-developed girdles, others are B ⊥ B'-tectonites (see page 646), while still others are S-tectonites with strong point maxima and poorly defined girdles in the quartz diagrams. Two examples of the latter class are reproduced in Fig. 109. In one

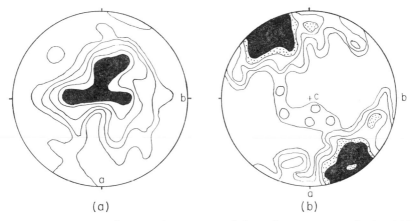

(a)   (b)

Fig. 109. Orientation diagrams for quartz in deformed quartzites, Finland. (*After A. Hietanen.*) (*a*) Simsiö. Optic axes of 202 grains. Contours at 4%, 3%, 2%, 1%, 0.5%, per 1% area. (*b*) Rautakero. Optic axes of 104 grains. Contours at 4%, 3%, 2%, 1%, 0.5%, per 1% area.

(Fig. 109*a*), the quartz axes are concentrated approximately at right angles to the foliation plane *ab*. Although this was the first type of preferred orientation to be recognized in quartz of deformed rocks, it has subsequently proved to be rare except in quartzites and granulites of the pre-Cambrian. The second diagram (Fig. 109*b*) illustrates a feature characteristic of many granulites and quartzites, namely, a triclinic symmetry of total fabric, derived from lack of coincidence between the symmetry plane of the quartz fabric and that of the megascopic fabric. The principal maximum represents an alignment of quartz axes within the foliation plane *ab* but oblique to *a* and *b* of the fabric. This quartz pattern possibly records the imprint of a strain later than the principal

[43] A. Hietanen, On the petrology of Finnish quartzites, *Comm. géol. Finlande Bull.* 122, 1938.

deformation, which is now expressed by the schistosity and by the mica fabric; but other kinematic interpretations are possible (cf. pages 652, 653).

The Moine series of northwest Scotland consists largely of quartzo-feldspathic schists ("granulites" of Scottish Geological Survey Memoirs) derived from arenaceous sediments by medium- or high-grade regional metamorphism. Semipelitic and pelitic schists are also represented. The megascopic fabric of some quartzose schists is dominated by a single plane foliation, with or without lineation parallel to b, but microscopically the constituent grains of quartz and feldspar tend to be equidimensional, and strongly lenticular quartz comparable with that of European granu-

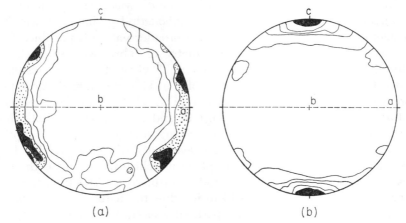

(a)   (b)

Fig. 110.   Orientation diagrams for quartz-muscovite schist of the Moine series, Scotland.   (*After F. C. Phillips.*)   (*a*) Quartz: optic axes of 250 grains.   Contours at 5%, 4%, 3%, 2%, 1%, per 1% area.   (*b*) Muscovite: poles of {001} cleavage in 250 crystals.   Contours at 20%, 15%, 10%, 5%, 1%, per 1% area.

lites is lacking.   A detailed petrofabric investigation of these rocks by Phillips [44] has yielded remarkably consistent results.   In any of the mica diagrams (*e.g.*, Fig. 110*b*) the most conspicuous feature is a strong maximum indicating parallelism of {001} with S (*ab*), but there is also a tendency for incipient (rarely complete) girdles to develop parallel to *ac* of the fabric.   Quartz axes likewise tend to be oriented in the *ac* plane, yielding partial or complete *ac* girdles in fabric diagrams; but typically there is a sector of strong concentration extending about 45° on either side of *a*, with prominent maxima near the sector boundaries (Fig. 110*a*).

The orthorhombic pattern of quartz orientation illustrated in Fig. 110*a* has been recorded in many quartz-rich S-tectonites and has been variously

[44] F. C. Phillips, A fabric study of some Moine schists and associated rocks, *Geol. Soc. London Quart. Jour.*, vol. 93, pp. 581–620, 1937; The microfabric of Moine schists, *Geol. Mag.*, vol. 82, pp. 205–220, 1945.

interpreted.[45]   Some writers have assumed alignment of crystallographic glide planes (e.g., {10$\bar{1}$1} or {0001}) in (h0l) slip planes of the fabric, symmetrically paired about the schistosity S (ab of the fabric).   Others have suggested alignment of assumed rhombohedral glide planes parallel to the schistosity plane, which on this interpretation would be a plane of slip in the rock fabric.   Another hypothesis appeals to alignment of fractured splinters with their long axes (the edges {10$\bar{1}$1} : {22$\bar{4}$3}) parallel to the a fabric axis, and the face {10$\bar{1}$1} parallel to ab (the schistosity). Actually there is no evidence, other than that supplied by the quartz pattern itself, that symmetrically developed (h0l) slip planes exist and have functioned in the deformation of these rocks.   Nor is there direct evidence as to whether the schistosity plane ab was ever a plane of slip, as is assumed on the hypothesis of rhombohedral gliding.   However, a satisfactory tentative interpretation of a broader nature is possible, if we are content merely to correlate symmetry of fabric with symmetry of deforming movement.   From this point of view there is an essential similarity between the quartz pattern of Fig. 110a, considered in conjunction with the associated mica pattern (Fig. 110b), and the movement picture of a simple compression (Figs. 98 and 99, page 608).   Both show orthorhombic symmetry.   There is a strong suggestion that the type of S-tectonite fabric shown in Fig. 110 evolved during compression by a force acting normal to the present plane of schistosity.   This is the conclusion tentatively accepted here, and accordingly schistosity of this type is provisionally classified as schistosity due to "flattening" (Plättung). It is not necessary to accept any particular orienting mechanism, nor to correlate individual maxima of the quartz diagram with particular assumed slip planes of the fabric.   Such may later prove possible when the gliding mechanism of quartz has been investigated in the laboratory, and when the relative roles of direct and indirect componental movement in rock deformation are better understood.

Fabrics of B-tectonites. *Definition.*[46]   B-tectonites are tectonites in whose fabrics a linear parallelism of elements with reference to the b (=B) axis of the fabric is the outstanding structural feature.   There are some B-tectonites in which the B axis is defined solely by intersecting sets of (h0l) planes, which typically are conspicuous in hand specimen as intersecting planes of schistosity.   Where one of these planes of schistosity becomes notably more conspicuous than the others and the lineation becomes correspondingly faint, rocks of this class merge into S-tectonites

[45] Knopf and Ingerson, *op. cit.*, pp. 143–149, 1938; F. J. Turner, Mineralogical and structural evolution of the metamorphic rocks, *Geol. Soc. America Mem. 30,* pp. 255–266, 1948.

[46] Sander, *op. cit.*, pp. 220–222, 1930; Knopf and Ingerson, *op. cit.*, pp. 70, 71, 153–156, 1938.

of the "flattened" type. The *B* axis of most B-tectonites is an axis of rotation of individual crystals, layers, and *s*-surfaces (*e.g.*, in flexural folds); *ac* girdles are then conspicuous in the orientation diagrams for constituent minerals.   B-tectonites of this type have been termed *R-tectonites*, to emphasize their rotational character.   They are not sharply distinguished from other B-tectonites, for the *B* axis in most cases is an axis both of rotation and of intersection of *s*-surfaces.   Nor are they sharply separable from S-tectonites having incipient girdles about the *b* axis.   Most deformed rocks belong to the B-tectonite class.

Example 1: Wissahickon Schist of Pennsylvania.[47]   The Wissahickon

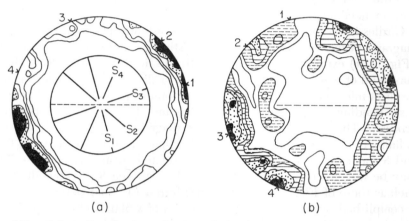

(a)   (b)

Fɪɢ. 111.   Orientation diagrams for Wissahickon schist, Pennsylvania.   (*After E. Cloos and A. Hietanen.*)   (*a*) Muscovite: poles of {001} cleavage in 150 coarse crystals.   Contours at 8%, 6%, 4%, 2.5%, 0.7%, per 1% area.   (*b*) Quartz: optic axes of 200 grains.   Contours at 5%, 4%, 3%, 2%, 1%, 0.5%, per 1% area.   Broken line is the geographic horizontal plane.   Maxima are numbered 1, 2, 3, 4 to correspond respectively to *s*-planes S_1, S_2, S_3, S_4.

schist, as exposed in the general vicinity of Bryn Mawr, Pennsylvania, consists essentially of quartz, muscovite, biotite, and plagioclase in varying proportions, with garnet and staurolite as possible additional constituents.   The most conspicuous element in the megascopic structure is a primary stratification S_1 marked by alternating quartzo-feldspathic and micaceous layers and accentuated by metamorphism.   A second plane of schistosity S_2, described as a "flow cleavage" and defined by parallel alignment of some of the recrystallized mica, intersects the earlier S_1 surfaces at angles of about 20° in the schists whose fabrics are illustrated in the accompanying fabric diagrams (Fig. 111).   The *s*-surfaces of the

[47] E. Cloos and A. Hietanen, Geology of the "Martic Overthrust" and the Glenarm series in Pennsylvania and Maryland, *Geol. Soc. America Special Paper*, 35, pp. 139–185, 1941; C. S. Ch'ih, *op. cit.*, 1950.

S_1 series show widespread evidence of megascopic folding and of accompanying corrugation on a microscopic scale. Where this minute crumpling is most conspicuous, it is associated with demonstrable slip on a third set of s-planes, S_3, which lie subparallel to the axial planes of the corrugations and constitute a "fracture cleavage"—the youngest of the visible s-structures. All three sets of s-surfaces intersect in a single b ($=B$) axis, which is also the axis normal to the girdles that dominate preferred orientation diagrams for minerals of the Wissahickon schist. It would seem that, although the various recognizable sets of s-surfaces have developed in a definite sequence, $viz.$, S_1, S_2, S_3, each has been active at a different stage of a single major deformation; for all conform to the same monoclinic pattern of symmetry.

Girdles in the plane normal to B are consistently present in orientation diagrams of both micas and quartz. Distinct maxima in the mica girdles (Fig. 111a) express a tendency for {001} to lie in S_1 and in S_2. The S_1 maximum of many diagrams is drawn out or divided into submaxima as a result of microcorrugation (external rotation of S_1 about B), and in some cases bending of individual mica crystals may be attributed to the same process. The maxima in the quartz diagrams (Fig. 111b) are less definite, but many correspond to an alignment of optic axes parallel to S_2 and S_3; a fourth maximum, not simply related to any visible s-surface, may be thought of as defining a statistical surface S_4. From diagrams such as those of Fig. 111, it has been inferred (1) that deformation was accomplished by successive slip on a series of s-planes (S_1, S_2, etc.) progressively rotating through the stress field, and (2) that mica became oriented in response to early slip on S_1 and S_2, while orientation of quartz developed later under the influence of S_3 and S_4. There is no obvious inconsistency between this detailed movement plan and the petrofabric data. But other interpretations are possible. The authors prefer to invoke slip only for surfaces (such as S_3) which show microscopic evidence of such. It is safer to reconstruct the movement plan only in broad outline—a plan with monoclinic symmetry, involving rotation around and shear transverse to the b axis of the fabric. Regarding the number of active slip planes and the manner in which mica and quartz became oriented, very little is known.

Example 2: Calcareous Phyllonite, Brenner, Tyrol.[48] Phyllonites are intensely deformed rocks superficially resembling phyllites; but whereas evolution of phyllite from slate involves progressive increase of grain, the fine-grained structure of a phyllonite is the result of reduction of grain during deformation of originally coarser rocks.[49] Nevertheless, phyllo-

---

[48] Sander, *op. cit.*, p. 233, D139–150, 1930.
[49] H. Williams, F. J. Turner and C. M. Gilbert, *Petrography*, pp. 206–208, Freeman, San Francisco, 1955.

nites are by no means purely cataclastic rocks, for recrystallization and neomineralization have usually been active in their development. During phyllonitization, earlier s-structures become closely folded by flexural slip on a megascopic or a microscopic scale, in the course of which the arches thicken and the limbs are stretched until they rupture. Deformation in the final stages is achieved purely by slip upon the resultant ruptural s-surfaces. These are essentially the same s-surfaces as existed in the rock prior to deformation, but they now have been transposed by folding so that they may even lie at right angles to the trend of equivalent s-surfaces in adjacent undeformed rocks. Most phyllonites belong to the B-tectonite class. The calcareous phyllonite from Brenner, which Sander described, is a rock with pronounced pencil structure in which the most conspicuous element of the megascopic fabric is a very strong lineation b ($= B$), parallel to which the individual pencils break out under a blow from the hammer. Measurement of $\{001\}$ in 222 flakes of muscovite (unselected) gives a diagram (Fig. 112c) in which are five distinct maxima. Optic axes of quartz and of calcite, and poles of visible $\{01\bar{1}2\}$ lamellae in calcite, also show strong girdles concentric with the axis b of the mica girdle. Monoclinic symmetry of fabric is very strong. Sander correlated $\{001\}$ maxima for mica, c-axis minima for quartz and calcite, and $\{01\bar{1}2\}$ maxima for calcite as indicated by numbers in Fig. 112; and from this correlation he deduced the existence of five ($h0l$) surfaces parallel to the mica concentrations of Fig. 112c. In the absence of any knowledge as to the orienting mechanisms of quartz and mica, the value of detailed correlation of this kind is doubtful. Development of $\{01\bar{1}2\}$ lamellae in calcite is known to be a relatively trivial feature dating from the final stage of deformation. What is most significant in the fabric of the Brenner phyllonite is its homogeneity, and its pronounced monoclinic symmetry indicating movements around and transverse to the b fabric axis (lineation). By comparison with experimentally developed fabrics,[50] the cleft-girdle pattern of Fig. 112a is consistent with plastic deformation of calcite involving elongation parallel to the girdle axis b, and all-round compression normal to b.

Example 3: Rocks Affected by Flexure.[51] The fabric of any rock that has been deformed by flexure or by flexural slip belongs to the R-tectonite class. The patterns assumed by the folded s-surfaces may be distinguished from shear folds by such criteria as the following (Fig. 113):

1. The thickness of any competent layer, measured radially, is constant (Fig. 113a).

[50] Turner, Griggs, Clark, and Dixon, *op. cit.*, p. 1276, Fig. 10D, 1956; Turner, *op. cit.*, p. 14, Fig. 6C, 1957.

[51] Sander, *op. cit.*, pp. 243–262, 1930; Knopf and Ingerson, *op. cit.*, pp. 79–83, 154–161, 1938; Turner, *op. cit.*, pp. 217–220, 1948; Sander, *op. cit.*, II, pp. 286–289, 1950.

2. In the fold arches spaces tend to be formed between the outer sur-faces of competent layers and the inner surfaces of overriding compe-tent layers. Such spaces are filled by incompetent material squeezed

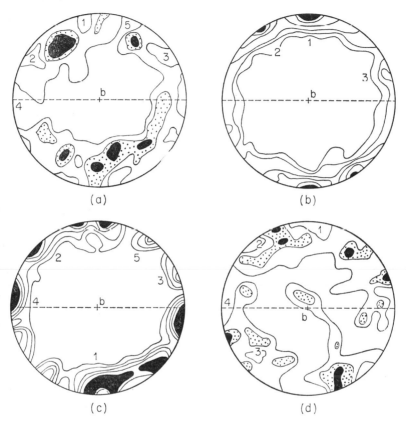

(a)   (b)   (c)   (d)

Fig. 112. Orientation diagrams for calcareous pencil phyllonite, Brenner, Tyrol. (*After B. Sander.*) (*a*) Calcite: optic axes of 263 grains. Contours at 3%, 2%, 1%, per 1% area. Minimum concentrations are numbered to correspond with maxima of accompanying diagrams, (*b*) and (*c*). (*b*) Calcite: 117 poles of $\{01\bar{1}2\}$ lamellae (one per grain). Contours at 6%, 5%, 4%, 3%, 2%, 1%, per 1% area. (*c*) Mus-covite: poles of $\{001\}$ cleavage in 222 crystals. Contours at 5%, 4%, 3%, 2%, 1%, per 1% area. (*d*) Quartz: optic axes of 146 grains. Contours at 3%, 2%, 1%, per 1% area. Minimum concentrations are numbered to correspond with maxima of (*b*) and (*c*).

between the competent bands, or by material deposited from solution (Fig. 113*a*).

3. In a fold arch the direction of curvature of one layer may locally be reversed with respect to that of adjacent layers (Fig. 113*b*), a condition totally incompatible with shear folding.

4. Folds formed by flexural slip are higher and wider in the thicker (and therefore more competent) bands than in thinner bands which offer less resistance to deformation by buckling (Fig. 113c).

Other distinctive characteristics are revealed in the orientation patterns which evolve in rocks that have been deformed by flexure. In this connection comparison of partial diagrams, each of which presents phe-

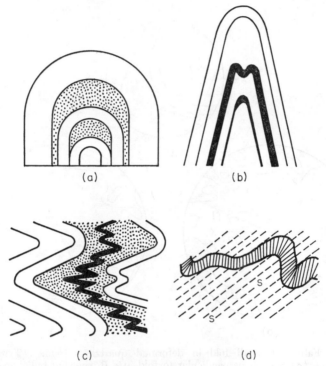

(a)   (b)

(c)   (d)

FIG. 113. Patterns of folds involving flexure. (a) Ideal case of thickening of arches in incompetent material (stippled) while competent bands (blank) maintain even thickness. (b) Reversal of direction of curvature in a fold arch. (c) Relation of thickness of bands to size of folds (diagrammatic). Stippled area represents sheared incompetent material. (d) Flexure of competent layer accompanied by shearing of adjacent incompetent layers on slip planes SS. (*After B. Sander, Fig. 128, 1930.*)

nomena of preferred orientation for a limited field of the fold, is especially instructive, since the fabric of a flexural or flexural-slip fold is characteristically inhomogeneous as contrasted with that of many shear folds. Two examples are summarized briefly:

1. The fabric of a competent layer differs notably from that of adjacent incompetent layers, the orientation patterns of which may have been determined by simple shear on subparallel s-planes, or by close, small-scale

folding, followed by rupture and shearing out of fold limbs leading ultimately to complete transposition of the old s-surfaces (Fig. 113d).

2. Where external rotation about B has been the dominant process in flexure of a competent layer, the internal fabric itself may have been rotated bodily, with little or no modification by internal movements within the folded layer. By an imaginary process of "unrolling," the homogeneous predeformational fabric may be deduced from an inhomo-

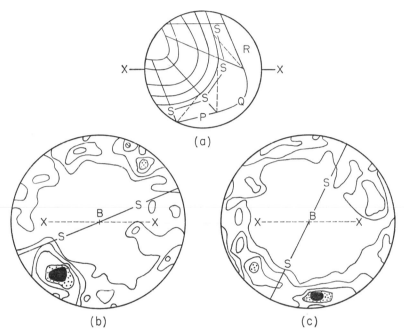

FIG. 114. Fabric of small fold in deformed quartzite, Brenner, Tyrol. (*After B. Sander.*) (a) Section perpendicular to fold axis B, showing radial sectors P, Q, and R, in which quartz axes were measured separately. S = mean trend of folded s-surface in each sector; broken lines show directions of concentration of quartz axes in each sector. (b) Quartz: optic axes of 300 grains in sector P. Contours at 5%, 4%, 3%, 2%, 1%, per 1% area. (c) Quartz: optic axes of 300 grains of sector Q. Contours at 5%, 4%, 3%, 2%, 1%, per 1% area.

geneous externally rotated fabric of this kind. A classic instance is provided by a folded quartzite from near Brenner, Tyrol, investigated by Sander.[52] Three hundred quartz axes were measured and plotted for each of three sectors (P, Q, R of Fig. 114a), together making up the arch of a fold in a field about 4 cm. square. In all three diagrams (e.g., Fig. 114b, and 114c) there is a well-defined B girdle, within which a strong maximum is situated at an angular distance of about 25° from the mean

[52] Sander, *op. cit.*, pp. 257, 258, D162–166, 1930.

trend (S) of the visible s-surfaces within each of the measured sectors. The 900 quartz axes plotted together upon a collective diagram yield a B girdle with maxima distributed over an arc of about 90°. The fabric within the field of the fold is inhomogeneous, but it has been built up by external rotation (folding) of a homogeneous, predeformational fabric in which the optic axes of the quartz grains tended to be concentrated at about 25° to the visible (now folded) s-planes. The symmetry of the fabric is monoclinic.

Fabrics of B $\wedge$ B'-tectonites and B $\perp$ B'-tectonites. Two girdles with mutually inclined axes can be recognized in some orientation diagrams, especially those for quartz; and corresponding fabrics, on the assumption that the girdles have developed about distinct B-axes, have been termed $B \wedge B'$-*tectonites*.[53] Usually the two girdles intersect at right angles or nearly so, and the term $B \perp B'$-*tectonite* is then applicable. Though the majority of B $\wedge$ B'-tectonites are recognizable as such only through the pattern of preferred orientation exhibited by quartz, comparable two-girdle diagrams have also been recorded in some cases for calcite and for mica.

One way in which a B $\wedge$ B'-tectonite fabric may develop is by partial overprinting of one B-tectonite fabric upon another during the course of two independent deformations governed by unrelated systems of stress. Larsson[54] has investigated the fabric of a Swedish granite mylonite in which quartz conforms to a B $\wedge$ B' pattern that has evolved in this way. Quartz and mica diagrams prepared from the undeformed parent granite show only traces of preferred orientation, but a schistose phase of the same rock within a zone of deformation proves to be a tectonite with a distinct foliation S perpendicular to which a well-defined B girdle appears in the quartz diagram. This same girdle G_1 is clearly recognizable in the quartz fabric (Fig. 115a) of a granite mylonite from the same deformation zone, but here it is accompanied by a second girdle, G_2, of later origin, which intersects G_1 at 40°. Veinlets of quartz, locally developed within the mylonite, yield simple orientation patterns in which the late girdle, G_2, occurs alone. The independent origin and the order of appearance of the two girdles of Fig. 115a are thus clearly established. The prin-

---

[53] Sander, *op. cit.*, pp. 239–243, 1930; *op. cit.*, II, pp. 272–277, 354, D146–156, 1950. Schmidt, *op. cit.*, pp. 176–178, 1932; T. G. Sahama, Die Regelung von Quartz und Glimmer in den Gesteinen der Finnisch-Lappländischen Granulitformation, *Comm. géol. Finlande Bull.*, no. 113, pp. 1–110, 1936; F. F. Osborne and G. K. Lowther, Petrotectonics at Shawinigan Falls, Quebec, *Geol. Soc. America Bull.*, vol. 47, pp. 1343–1370, 1936; Knopf and Ingerson, *op. cit.*, pp. 155–156, 197, 198, 1938; Turner, *op. cit.*, pp. 220–223, 256, 257, 1948.

[54] W. Larsson, Die Svinesund-Kosterfjord Ueberschiebung, *Sveriges geol. undersökning*, vol. 32, no. 1, pp. 3–32, 1938.

cipal slip plane in each phase of deformation was the megascopic foliation S.

In contrast with the case just described, in many, perhaps most, tectonite fabrics with intersecting girdles, evidence such as the following

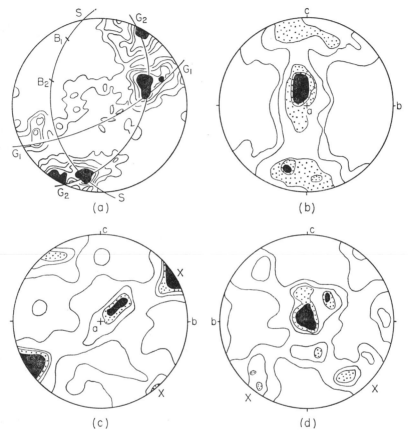

FIG. 115.   Orientation diagrams for quartz (optic axes) in $B \wedge B'$ and $B \perp B'$-tecton-
ites.   (a) 200 grains in granite-mylonite, Svartskär, Sweden.   Contours at 5%, 4%,
3%, 2%, 1%, per 1% area.   (After W. Larsson.)   (b) Collective diagram for quartz
in gneisses, Mugl series, Obersteiermark.   (After W. Schmidt.)   (c) 151 grains in
granulite, Finland.   Contours at 4%, 3%, 2%, 0.7%, per 1% area.   (After T. G.
Sahama.)   (d) 115 grains in granulite, Saxony.   Contours at 4%, 3%, 2%, 1%, per
1% area.   (After B. Sander.)

strongly suggests that the two-girdle pattern is the product of a single
deformation having orthorhombic symmetry:

1. Within a given tectonic province, e.g., the granulite areas of Finnish
Lappland described by Sahama, the two girdles typically appear side by
side and developed to the same degree in any particular fabric.   This

uniformity of pattern seems to preclude independent development of the two girdles.

2. The angle of intersection of the girdles usually approximates to 90°, and $B \perp B'$-tectonites therefore constitute the most widely distributed class of $B \wedge B'$-tectonites. The $B \perp B'$ pattern therefore seems to be a stable pattern rather than an unstable transitional type of fabric.

3. The two-girdle fabric typically is related symmetrically both to the macroscopic fabric and to any single-girdle pattern that may be recognized in the preferred orientation of other constituents of the same rock. This suggests a close genetic relationship between all the elements that make up the fabric of the rock as a whole.

4. The symmetry of the whole fabric usually is orthorhombic.

There are two particular types of $B \perp B'$-tectonite patterns of quartz orientation which are widely distributed and to which the above remarks specially apply. In one, illustrated in Fig. 115b, the usual ac girdle is accompanied by a partially developed girdle in the bc plane, so that $B = b$ and $B' = a$. The second type, shown in Figs. 115c and 115d, is commonly encountered in granulites and related rocks. There are two mutually perpendicular $(0kl)$ girdles within which optic-axis maxima are developed at the point of emergence of the a fabric axis and at the two points (X) where the girdles cut the bc plane of the fabric. Such patterns have been interpreted in several ways. It has been claimed that some if not all features of the fabric reflect slip movements in the ab plane (the principal schistosity). This conclusion is still an unsubstantiated speculation and is here rejected.

NONTECTONITE FABRICS IN DEFORMED ROCKS

Preferred orientation of minerals in many metamorphic rocks can be correlated, at least in part, with chemical recrystallization and neomineralization that bear no direct relation to deformation.[55] When, for example, a slate is converted by contact metamorphism to a micaceous schist, or when chemical reaction and crystallization outlast the deformational stage of regional metamorphism, any new preferred orientation that develops is controlled entirely by the anisotropic nature of the medium in which the crystals have grown. Fabrics that originate in this manner have been termed *belteroporic*. Their evolution and their genetic interpretation are governed by a number of recognized principles, summarized briefly below:

1. In an anisotropic medium there are certain directions of maximum

[55] Sander, *op. cit.*, pp. 156–162, 283, 1930; F. J. Turner, Structural petrology of quartzose veins in the schists of eastern Otago, *Royal Soc. New Zealand Trans.*, vol. 71, pt. 4, pp. 307–324, 1942.

ease of growth (*Wegsamkeit*), *e.g.*, a direction of lineation or any direction in an existing *s*-plane in a deformed rock. Any surface to which a direction of greatest ease of growth is perpendicular is termed a *blastetrix*. If the blastetrix directly influences the space-lattice orientation of the growing crystals, it is said to be anisotropic; if there is no such influence, it is isotropic. For example, in an R-tectonite undergoing postdeformational recrystallization, the plane normal to *B* may be a blastetrix in that the direction *B* offers least resistance to growing crystals of quartz, which consequently tend to elongate parallel to *B*. In the resulting quartz fabric it may be possible also to detect a space-lattice pattern of preferred orientation (*e.g.*, a girdle of optic axes normal to *B*), inherited from the deformational stage of metamorphism. The blastetrix would thereby be shown to have been anisotropic. Though preferred orientation patterns of deformational origin may be partially preserved during recrystallization subsequent to deformation, our attention is confined in this section to the development of nontectonite growth fabrics controlled by one or more isotropic blastetrices.

2. The most obvious effect of an isotropic blastetrix is the development of a dimensional orientation of grains with their long axes perpendicular to the blastetrix.

3. For minerals such as garnet which usually show no pronounced tabular or prismatic crystallographic habit, the dimensional orientation mentioned above is not accompanied by preferred orientation of the crystal lattices. Crystals of mica, chlorite, or amphibole, on the other hand, persistently tend to develop a pronounced tabular or prismatic habit, *i.e.*, they tend to grow more rapidly in some crystallographic directions than in others. The same is true for quartz and calcite of veins. Any such grain, whose crystallographic direction of most rapid growth happens to coincide with a direction of minimum resistance to growth in the anisotropic fabric, has a greater chance of survival than have less favorably oriented grains, which tend to be suppressed as crystallization proceeds. For minerals of this kind, a dimensional orientation in the belteroporic fabric may be accompanied by preferred orientation of the crystal lattices. However, lattice orientation unaccompanied by cognate dimensional orientation can never be attributed to growth from an isotropic blastetrix.

4. A fabric diagram depicting preferred orientation of a mineral controlled purely by growth from a blastetrix is symmetrical about the normal to the blastetrix. It may show either (1) a point maximum coinciding with the pole of the blastetrix, or (2) a girdle with the pole of the blastetrix as center, or (3) no significant preferred orientation of the space-lattice. For example, a hornblende fabric controlled by growth of acicular crystals parallel to a lineation *B* inherited from a previous deformation will give orientation diagrams showing (1) a concentration of *c* crystal

axes at the point of emergency of B, and (2) even distribution of {110} cleavage poles through a girdle in the plane normal to B. Crystals of quartz and calcite, growing in a vein with the vein wall as blastetrix, tend to be elongated either parallel or at right angles to the c crystal axis. In diagrams based on measurements of optic axes, preferred orientation of the first type gives a point maximum coinciding with the pole of the vein wall, while the second type gives a girdle pattern in the plane of the wall.

5. The commonest type of growth fabric found in metamorphic rocks is that which is controlled by a plane of maximum ease of growth afforded by a preexisting schistosity (foliation), as for example in *Garbenschiefer*. Any plane standing at right angles to the schistosity has the properties of a blastetrix. Hornblende assumes a preferred orientation pattern in which the c crystal axes lie in a girdle, the plane of which coincides with the schistosity. In the case of micas, the poles of {001} are concentrated in a point maximum at the pole of the schistosity plane, just as in some S-tectonite fabrics. If, however, the normal (Y) to the optic axial plane is measured and plotted, it will usually show in S-tectonites a monoclinic pattern with a tendency toward concentration either parallel or at right angles to the b fabric axis, but in a true growth fabric of mica there should be no significant orientation of Y within the schistosity plane, since all directions in this plane have equal influence upon the growing tabular crystals.

TECTONIC SIGNIFICANCE OF SCHISTOSITY AND LINEATION IN DEFORMED ROCKS

Résumé of the Nature of Schistosity and Lineation. A linear parallelism of fabric elements, to which we have referred as lineation, and one or more sets of subparallel s-surfaces can be recognized in hand specimens and microsections of most deformed rocks. When any parallel fabric of this kind, originating through penetrative movement that has intimately affected the component fabric elements, imparts to the hand specimen a recognizable fissility, the rock is said to be schistose. Terms such as *schistosity, foliation, slaty cleavage,* and *flow cleavage* have been variously employed by different writers to describe parallel fabrics in metamorphic rocks, and considerable ambiguity attends their current usage. Not only are genetic terms like *flow cleavage* widely applied to fabrics whose mode of origin has not been demonstrated, but also there is no uniformity of definition of such fundamental nongenetic terms as *foliation*. Synonymous use of *foliation* and *schistosity*, to cover all those megascopically conspicuous parallel fabrics of metamorphic origin which impart a definite fissility to the rocks in which they occur, perhaps accords best with current terminology and has therefore been adopted here. Thus defined, *schistosity* (foliation) embraces most classes of s-surfaces encountered in metamorphic rocks. While schistosity is related to the presence of parallel surfaces in the fabric, the term *lineation* is reserved for that component in the fabric which is determined by parallelism of linear fabric elements, such as axes of prismatic crystals, intersections of s-surfaces, or axes of microfolds.

Structural characters that are very generally, but in no case universally, associated with schistosity and lineation include the following: a widely prevalent tendency for platy or prismatic crystals of minerals such as mica, amphibole, and chlorite, to show dimensional parallelism (preferred orientation according to crystal form); a tendency for parallel alignment of lensoid grains of quartz, calcite, or feldspar; a tendency for mineral cleavages to lie parallel to the schistosity surfaces (as in micas and chlorites) or to the lineation (as in hornblende); a very strong tendency for preferred orientation of all component minerals according to space-lattice structure, which is manifest even for minerals such as quartz, calcite, and feldspars which usually occur in equidimensional grains lacking dimensional orientation; more or less plane surfaces of rupture—often capable of identification as slip surfaces—parallel to one or more sets of schistosity planes; laminated structure resulting from segregation of contrasted mineral assemblages into alternating layers parallel to the schistosity (complementary products of metamorphic differentiation).

Schistosity (Foliation) in Relation to Deforming Forces and Movements. *Rival Hypotheses.* Schistosity or foliation is the most conspicuous structural feature of deformed rocks and is comparatively easily measured in the field and recorded upon maps. Furthermore it commonly preserves a regular trend over considerable areas, and it clearly is related in some manner to the stress plan and the movement plan of metamorphism. Correlations between schistosity and deforming forces and movements therefore figure prominently in tectonic syntheses.[56]

From what has already been stated in this chapter it must be obvious that schistosity is no simple phenomenon, and that a number of different processes and kinds of componental movement, in various combinations, may be active in its development, so that schistosity of one type or another may arise in a number of alternative situations with regard to each of several distinct types of movement plan. Several schistosities of similar or different kinematic significance commonly may be recognized in a single rock specimen. It is therefore impossible to formulate any simple

[56] *E.g.*, Sander, *op. cit.*, pp. 97–103, 1930; Fortschritte der Gefügekunde der Gesteine, *Fortschr. Min. Krist. Petr.*, vol. 18, 1934, pp. 142–144; Schmidt, *op. cit.*, pp. 199–203, 1932; F. Bonorino, El origen mecánico de la esquistosidad, *Fac. Ciencias Universidad Buenos Aires Contrib. Cient.*, vol. 2, no. 2, pp. 29–94, 1958.

and universally applicable hypothesis of the genetic and tectonic significance of schistosity. From this condition arises much of the conflict between accepted theories of schistosity, in each of which some mode of origin, legitimately applicable to particular instances, has been emphasized to the exclusion of others which in actual fact have comparable or greater importance. Even today, when the manifold possibilities as to the meaning of schistosity have been convincingly demonstrated by the structural petrologist, there is a general reluctance among many geologists to abandon the convenient assumption that schistosity is a simple structure capable of simple interpretation by some universally applicable theory.

It is not proposed here to review historically the growth of ideas on schistosity, but it may be appropriate to mention three widely different views, each of which has been accepted by one group or another of metamorphic geologists as being universally applicable in the tectonic interpretation of schistosity:

1. By Sharpe and Sorby, schistosity was attributed to a compressive force acting at right angles to the plane of schistosity, a view still reflected in accounts of rock deformation in metamorphism.[57]

2. The Wisconsin geologists, notably Van Hise, Leith, and Mead, have interpreted "flow cleavage," i.e., all schistosity involving preferred dimensional orientation of prismatic, tabular, or flattened crystals, as invariably lying parallel to the AB plane of the strain ellipsoid.[58] The axis of maximum shortening in the strained mass is therefore assumed to be perpendicular to the plane of "flow cleavage" in all cases. This view agrees with 1 above only in the special case (possibly a common one) of "flattening," i.e., of deformation under a simple compression.

3. According to Becker,[59] schistosity surfaces are planes of minimum cohesion determined by slip movements parallel to surfaces of high resolved shear stress. Among modern European writers, Sander and Schmidt in particular recognize the importance of this type of schistosity.

Most writers who have investigated the schistosity problem from the standpoint of fabric analysis believe that none of the above hypotheses should be either universally applied as an explanation of schistosity nor yet totally rejected. On the contrary, they tend to recognize several distinct types of schistosity, each capable of subdivision on the basis of fabric criteria.

[57] D. Sharpe, On slaty cleavage, Geol. Soc. London Quart Jour., vol. 5, pp. 111–115, 1849; Sorby, op. cit., 1856; A. Harker, Metamorphism, pp. 153–155, 193–195, Methuen, London, 1932.

[58] E.g., W. Mead, Studies for students: folding, rock flowage and foliated structures, Jour. Geology, vol. 48, p. 1010, 1940.

[59] E.g., G. Becker, Current theories of slaty cleavage, Am. Jour. Sci., vol. 24, pp. 1–17, 1907.

Shear Schistosity.   In regions such as the European Alps, where alpine deformation and regional metamorphism have been active together, rock deformation has involved differential movement on closely spaced shear surfaces which now survive as schistosity.   In some rocks, bedding provided the surfaces of minimum strength along which slip was initiated, so that schistosity parallels bedding and conforms to the configuration of contemporary flexural folds.   More commonly, shear oblique to the bedding is concentrated in incompetent beds; and, as deformation progresses, remnants of competent beds become aligned parallel to the local trend of the shear schistosity in immediately adjacent incompetent rock.[60]   In regions of recumbent folding and low-angle thrusting, "bedding schistosity" of this class may maintain a subhorizontal attitude over wide areas. Such an attitude is thus no criterion of "load metamorphism."

In some rocks of this class, notably in mylonites and phyllonites of the S-tectonite group, a single schistosity, identified as the *ab* plane of the fabric and of the movement plan, predominates.   More commonly, several sets of schistosity surfaces intersect in a lineation (*b* fabric and kinematic axis) which is also parallel to axes of contemporary folding, and the fabric belongs to the B-tectonite class.   Lineation may even be so strong that the rock breaks into pencils rather than slabs.

The symmetry of the fabric typically is monoclinic; *i.e.*, there is a single plane of symmetry (*ac*) which is normal to the lineation (*b*).   But lineation parallel to *a*, and thus lying in the symmetry plane, may also be present and, in slickensided and mylonitized rocks, may completely overshadow the *b* lineation.   Orientation diagrams for *c* axes of quartz and calcite and for {001} in mica all show girdle patterns around the *b* axis. The more conspicuous the *b* lineation, the more sharply defined are the girdles of orientation diagrams.   Tension joints nearly normal to *b* may help to identify the lineation.

In rocks affected by shear folding, shear schistosity develops parallel to the axial planes of folds so formed.   This is one type of "axial-plane" cleavage common in slates and phyllites.[61]   The symmetry of the macroscopic fabric (bedding, cleavage, lineation) is monoclinic, the symmetry plane being normal to the lineation *b* defined by intersection of the cleavage with the folded *s*-surface (usually bedding).   While the axis of the mica girdle commonly coincides with *b*, that of the quartz girdle may

[60] Cf. A. E. J. Engel, Studies of cleavage in metasedimentary rocks of the northwest Adirondack Mountains, New York, *Amer. Geophys. Union Trans.*, vol. 30, pp. 767–784, 1949.

[61] Cf. E. S. Hills, *Outlines of Structural Geology*, pp. 104–112, Wiley, New York, 1953.   For evolution of shear schistosity by secondary deformation of *s*-surfaces previously affected by flexural slip, see L. E. Weiss and D. B. McIntyre, Structural geometry of Dalradian rocks at Loch Levin, Scottish Highlands, *Jour. Geology*, vol. 65, pp. 582, 583, 1957.

be oblique to *b*. The over-all symmetry of the fabric is then triclinic. A likely though still unverified explanation is that the direction of shear is oblique to the bedding-cleavage intersection and normal to the axis of the quartz girdle; so that the triclinic fabric reflects a triclinic movement plan compounded of two synchronous monoclinic movements—shear on the plane of cleavage, and folding (rotation) of bedding about the bedding-cleavage intersection.[62]

Compression Schistosity.   Some deformed fabrics are dominated by a single planar cleavage or schistosity which is more readily correlated with compression normal to than with shear parallel to the schistosity plane.

1. Many pre-Cambrian granulites have an orthorhombic fabric whose most conspicuous megascopic element is a steeply dipping planar foliation, marked by parallel alignment of flat lenses of coarse quartz, calcite, or feldspar. The foliation *ab* dominates the preferred orientation patterns of micas and quartz. Girdle patterns tend to be weak, incomplete, or absent; and lineation tends to be correspondingly inconspicuous. Whether the lineation should be identified as the *a* or the *b* fabric axis is a matter of opinion. The over-all symmetry of the fabric and the lensoid outlines of grains and aggregates of quartz and feldspar suggest, although without certain proof, that deformation has been achieved by compression normal to the foliation. This has been termed "flattening" (*Plättung*) by Sander.

2. Close folding of fine-grained sediments by flexural slip is commonly accompanied or followed by development of a type of slaty cleavage subparallel to the axial planes of the folds.[63] If, as seems likely from the regularity and widespread occurrence of such structures, folds and cleavage have formed successively in response to the same stress system, the cleavage must be approximately normal to a simple compressive force. Axial-plane schistosity (cleavage) of this kind must represent flow of mechanically weak material in a plane of low shear stress nearly normal to the compression. It thus approximates AB of the strain ellipsoid. Lineation resulting from extension (*e.g.*, elongation of oolites, stretching of pebbles) within AB, may be parallel to the axis of folding (*b* of the movement plan and fabric); or alternatively it may be normal to the fold axes and parallel to *a* of the fabric. The symmetry of the fabric is usually monoclinic.

AB Schistosity.   Where flattened bodies such as pebbles have assumed

[62] Weiss, *op. cit.*, 1955.

[63] For various views on the geometrical and time relations of cleavage to folds, and the depth of cover necessary for development of cleavage, see: M. P. Billings, Field and laboratory methods in the study of metamorphic rocks, *New York Acad. Sci. Trans.*, Ser. 2, vol. 13, pp. 47–50, 1950; Hills, *op. cit.*, 1953; P. Fourmarier, Remarques au sujet de la schistosité, *Geol. en Mijnbouw*, n. ser., no. 2, pp. 47–56 (with discussion by M. G. Rutten, pp. 57, 58, and L. U. De Sitter, pp. 58, 59), 1956.

parallel orientation as a result of bodily rotation during plastic deformation of the enclosing matrix, they may impart a foliation to the rock mass as a whole. This foliation is parallel to the AB plane of the strain ellipsoid, but it need not lie normal to a compressive force. Indeed, if plastic deformation of the matrix is itself the result of slip on one set of s-planes, there will be two planes of schistosity: (1) the s-planes of the matrix, and (2) the plane in which the rotated pebbles are now aligned. Neither of these can be interpreted as having formed normal to a compressive force. Moreover, the foliation imparted by dimensionally oriented pebbles cannot be interpreted as slip-plane foliation. It constitutes a third type in our classification.

Cases such as the above may not be common. But some students of rock fabric assume that a similar mechanism of passive rotation of mica tables in a plastically deforming matrix of quartz may be responsible for the strong dimensional parallelism of mica flakes in many tectonites. This is still an untested hypothesis. However, it raises the possibility that planar schistosity marked by parallel micas in some rocks may represent the AB plane of a strain ellipsoid resulting from slip on an obliquely intersecting set of shear surfaces, which may also be recognized in some rocks. Where deformation of this type had reached an advanced stage, the s-planes of slip would intersect the AB plane of the ellipsoid at so low an angle that the two would merge to give a compromise schistosity approximating to the type described above as compression schistosity.

Conclusion. Until more experimental data are available regarding development of oriented fabrics under controlled conditions, some degree of ambiguity must attend tectonic and kinematic interpretation of schistosity. In the meantime, we recognize that schistosity may be correlated alternatively with shear planes or with the AB plane of the strain ellipsoid. In most cases fabric criteria are much more compatible with one of these alternatives than with the other. Where schistosity is identified with the AB plane of the strain ellipsoid, it does not necessarily follow that strain was induced by compression normal to AB, but a close approach to this condition seems probable in "flattened" rocks (usually S-tectonites) with orthorhombic symmetry of fabric, and in rocks where "axial-plane cleavage" is associated with flexural-slip folds.

CHAPTER 24

Metamorphism in Relation to Magma and to Orogeny

STATEMENT OF PROBLEM

Metamorphism is a phenomenon of high temperature. The lower limit of temperatures at which metamorphic reactions commonly occur in silicate rocks is uncertain, but general lack of metamorphic effects in water-saturated sedimentary rocks that have been buried for millions of years at depths where temperatures of the order of 150°C. or more are prevalent [1] justifies arbitrary assumption of a figure of the order of 200°C. The upper limit of the range of metamorphic temperatures may be drawn where a silicate-melt phase begins to appear. As the proportion of melt to crystals increases, the mass becomes mobile and assumes the character of magma in the broad sense of the term as employed in this book. Metamorphic and magmatic phenomena thus merge into each other. The range of transition temperatures, depending on situation and nature of the rocks concerned, may perhaps be of the order of 700° to 900°C.

Field observations have clearly established a tendency for close association of metamorphic with plutonic rocks, such as might be expected from theoretical considerations alone. It is true that metamorphic rocks have been recorded far from the nearest outcropping igneous rocks, and there are numerous instances where large bodies of magma have invaded rocks that still remain almost unmetamorphosed. But the strong tendency for mutual association of metamorphic and plutonic rocks in the field is abundantly clear, and in many cases an equally close connection, in time, between metamorphic and magmatic activity has been established. This applies especially if we consider only rocks customarily regarded as being of high metamorphic grade.

[1] F. J. Turner, Mineralogical and structural evolution of the metamorphic rocks, *Geol. Soc. America Mem. 30*, pp. 287, 288, 1948.

Several explanations have been advanced to explain the generalization stated in the preceding paragraph.[2] Some of these are as follows:

1. Metamorphic temperatures may be determined by proximity to bodies of hot magma (which also supply water and other active fluids that play an important role in metamorphism).

2. Alternatively, metamorphic temperatures may be due to heat coming from some nonmagmatic source (e.g., the earth's interior; mechanical work done in deformation; radioactive disintegrations). Here, too, several correlations are possible.

  a. Associated igneous rocks (e.g., the granitic element of migmatites) may be products of melting at temperatures exceeding the upper limit of the metamorphic range.

  b. Alternatively, the associated plutonic (e.g., granitic) rocks may be not truly igneous, but rather of metasomatic origin; metamorphic and plutonic activity would then be cognate processes activated by an upward flow of "emanations" from the deep levels of the crust or the mantle.

  c. It is also possible that metamorphism and injection of magma may be independent processes which tend to be associated broadly in place and time because each is favored by high temperature and by orogenic deformation.

Each of these views has found favor with one or another student of the petrology of deep-seated rocks. Probably none need be discarded completely.

Many metamorphic rocks bear the unmistakable imprint of severe deformation. The banded or foliated fabrics of others have been variously interpreted as the result of deformation or as relics inherited from a premetamorphic condition. Thus arises this question: is deformation the principal factor in bringing about regional metamorphism, or is it an incidental effect of subordinate importance?[3] Another important question concerns the role of compositional change in metamorphism. Most petrologists agree that many instances of metamorphism involve some degree of change in bulk composition of the rocks affected—even if this relates principally to volatile and minor constituents. Are the progressive mineralogical variations developed in rocks of similar parentage, within any series of metamorphic zones, primarily due to differential metasomatism or to temperature gradients?

Questions such as these are interrelated. Radically different answers are offered by petrologists of the "transformationist" school, on the one hand, and by those who, on the other hand, consider granites to be products of crystallization of initially homogeneous melts. We have given

[2] Cf. H. H. Read, A commentary on place in plutonism, Geol. Soc. London Quart. Jour., vol. 104, pp. 155–206, 1948.

[3] Cf Read, op. cit., pp. 187–196, 1948.

reasons in Chap. 12 for our tentative acceptance of the view that most granites are igneous rocks, in the sense that they have crystallized from partly crystalline magmas containing a silicate-melt phase in sufficient quantity to impart general mobility. This view colors the opinions expressed below as to relations between metamorphism and the emplacement of plutonic bodies. The hypothesis which comprises the sum of these opinions is admittedly tentative; it seems compatible with much of the evidence of igneous and metamorphic petrology. Since metamorphism embraces phenomena which have occurred over a very broad range of environmental and physical conditions, a number of typical cases will now be considered individually.

CONTACT METAMORPHISM

Contact Metamorphism Related to Basic Intrusions. There are instances where contact metamorphism has undoubtedly been caused by injection of large bodies of essentially liquid magma and where metamorphic temperatures must have been controlled by temperature gradients connected with outward flow of heat from the injected magma. Such are the contact zones developed along the margins, and especially near the lower surfaces, of great sheets of diabase and gabbro exemplified by the "Karroo dolerites" of South Africa and by the "Duluth gabbro" of Minnesota.[4] In favorable situations, effects of metamorphism in sensitive rocks such as slate may be recognizable throughout an aureole extending 500 to 1,000 ft. from the visible contacts. But on the whole aureoles tend to be narrower than those which surround comparable bodies of granite, and there are many instances where diabase sills several hundred feet in thickness have scarcely affected rocks within a foot or two of contacts.

Broad aureoles margining basic intrusions may show obvious progressive zoning, e.g., from slate, to biotite schist, to cordierite or hypersthene hornfelses. Mineral assemblages so developed—e.g., cordierite-hypersthene-plagioclase-biotite, quartz-biotite-orthoclase-plagioclase—are in most cases identical with assemblages recorded for rocks of similar parentage in some granite aureoles. Metasomatic changes, such as introduction of CaO and FeO and removal of K_2O, and in some cases partial fusion, have been demonstrated for pelitic xenoliths enclosed in the gabbro itself and for hornfelses within a few feet of contacts. But throughout the

[4] F. F. Grout, Contact metamorphism of the slates of Minnesota by granite and by gabbro magmas, *Geol. Soc. America Bull.*, vol. 44, pp. 989–1040, 1933; D. L. Sholtz, The magmatic nickeliferous ore deposits of East Griqualand and Pondoland, *Geol. Soc. South Africa Trans.*, vol. 39, pp. 92–126, 1936; F. Walker and A. Poldervaart, Karroo dolerites of the Union of South Africa, *Geol. Soc. America Bull.*, vol. 60, pp. 613, 614, 625–627, 1949.

main extent of broad aureoles, chemical change is on the whole insignificant. The very striking mineralogical and textural differences between hornfels, schist, spotted slate, and parent slate can only be correlated with differences in temperature and in duration of metamorphism.

We conclude that the occurrence of metamorphic aureoles margining basic intrusions is satisfactorily explained by the classic theory of contact (essentially thermal) metamorphism caused by intrusion of hot magma. The far-reaching chemical changes noted in xenoliths and in wall rock in immediate contact with the basic magma are such as would be expected from Bowen's theory of assimilative reaction: (1) conversion of solid rock to those phases—pyroxenes, plagioclase—with which the reacting magma was saturated; (2) partial or complete fusion of material rich in quartz and alkali feldspar, leaving a residue of such minerals as cordierite, spinel, sillimanite in initially aluminous xenoliths.

Contact Metamorphism Related to Granitic Bodies. The varied relations between granites and granodiorites and the rocks with which they are in contact have been reviewed rather fully in Chap. 12 (pages 358 to 366). Only a few of the points there considered need be recapitulated here.

Conditions of metamorphism in contact zones bordering granitic bodies can be reconstructed with least ambiguity where isolated moderate-sized masses of granitic rocks occur in terranes little affected by regional metamorphism. The Dartmoor granites of Cornwall, the Comrie diorite complex of Scotland, and some of the granitic batholiths and stocks of Minnesota illustrate this condition. Aureoles varying in width from a few hundred feet (*e.g.,* at Comrie) to hundreds or thousands of yards (in Minnesota) show zones of progressive metamorphism in which the various mineral assemblages are closely similar to those developing from corresponding parent rocks in contact aureoles adjoining gabbro intrusions elsewhere. Except for marked chemical modification shown by xenoliths or by rocks within a foot or two of contacts, the metamorphic rocks of many of these zoned aureoles are almost identical chemically with the rocks from which they originated. Here, as in gabbro aureoles, we see the varied response of chemically similar rocks to a thermal gradient which prevailed during the period of metamorphism.

There are also many cases where metamorphism at granite contacts has been accompanied by notable introduction of alkali (*e.g.,* the Stavanger region of Norway) or of magnesia and iron (*e.g.,* the Orijärvi district of Finland), extending for long distances beyond visible contacts. Here is the combined influence of chemical and thermal gradients. But it should be noted that variation in chemical composition, as now observed from the granite contact outward, can readily be explained as the result of waves of fluids—rich in appropriate ions—expelled from crystallizing granite magma at the time of metamorphism. Likewise, the many recorded complex and gradational contacts between granites and schists, and the migmatite zones observed at the borders of some granitic bodies, can all be interpreted as zones of intense and prolonged reaction between intrusive granitic magma and host rocks already heated (*e.g.*, by burial) to metamorphic temperatures.

The alternative hypothesis that granites, like the hornfelses and contact schists which surround them, are products of metasomatism of solid rocks in place was criticized in Chap. 12 on the basis of the data of igneous petrology. It is unsatisfactory, too, as a part of a theory of metamorphism. The "doctrine of fronts," as it is sometimes termed, fails to account for the chemical and mineralogical analogies between rocks of granite and of gabbro aureoles respectively. It is difficult, too, to reconcile with the doctrine of fronts the discrepancy in size between very large granitic batholiths and the narrow aureoles which may border them. Moreover, the steep thermal gradient falling sharply outward from a granite contact, and now expressed by petrographic variation across a zoned aureole, can scarcely be explained on a hypothesis of pure metasomatism. It is readily understood, however, if it be assumed that the central granite was emplaced as an intrusive mass of magma notably hotter than the invaded rocks.

REGIONAL METAMORPHISM

Statement of Problem. In current geologic literature, regional metamorphism is variously attributed to one or another, or to some combination, of the following causes:

1. Regional invasion of rocks by granitic magma
2. Regional deformation under tangentially directed pressure in zones of alpine folding (regional metamorphism equals dynamic, kinetic, or dislocation metamorphism)
3. Deformation under vertically directed load resulting from deep burial (regional metamorphism equals load metamorphism)
4. Static recrystallization at high temperatures resulting from deep burial (regional metamorphism equals geothermal metamorphism)
5. Regional metasomatism effected by waves of chemically active fluids, expelled from a zone of rocks undergoing granitization, and proceeding in a succession of slowly advancing "fronts" now reflected in corresponding zones of progressive metamorphism

Characteristically regional metamorphism leads to complete mineralogical reconstitution and to development of a conspicuously schistose (foliated) fabric in the rocks concerned. A satisfactory hypothesis of regional metamorphism must therefore account for the high temperatures that may be inferred from the observed assemblages of metamorphic minerals, and at the same time must explain the building up of a stress system or a movement plan compatible with the observed metamorphic fabrics. Actually temperature and stress have operated simultaneously as individual factors in a set of metamorphic conditions, but it is perhaps desirable first to review separately the possible sources of heat and the alternative means by which the stress systems and movement plans, deduced from fabric data, may have originated. The role of deformation in regional metamorphism will be considered first.

Deformation as a Factor in Regional Metamorphism. *General Statement.* The widespread occurrence of regionally metamorphosed rocks in orogenic zones and the almost universal evidence of deformation preserved in the fabrics of such rocks (*e.g.*, distorted pebbles and fossils; crumpled s-surfaces; rotated porphyroblasts; tectonite patterns of preferred orientation) show beyond doubt that regional metamorphism typically involves rock deformation; but opinion differs as to whether deformation is essential or merely incidental to metamorphism. If the former alternative is correct, it still remains to be decided whether vertically directed load and horizontally directed compression are equally effective deforming forces, or again what particular factor in deformation actually initiates and controls the chemical changes in question. Such questions may be discussed in conjunction, if for the purpose of argument two general cases are distinguished, *viz.*, (1) metamorphism clearly associated with orogeny, and (2) regional metamorphism (usually in pre-Cambrian terranes) not so obviously related to major crustal disturbance.

Regional Metamorphism in Proved Orogenic Zones. When rocks are deformed, heat may be generated in two ways:

1. The work of nonelastic deformation is dissipated as heat. It can be shown, however, that even the most drastic deformation in rocks under stresses reaching their breaking strength is accompanied by a rise in temperature of no more than 10° or so. Similarly, heat generated by viscous flow under stresses such as are likely to develop within rock masses is negligible. Goguel [5] has computed the heat equivalent of the mechanical energy involved in the deformation of some sections of the Jura Mountains and of the Alps, and finds it to be of the order of a few calories per gram, quite insufficient to produce any metamorphic effect. Nor are such effects visible in some of the intensely deformed sediments of these mountains. There is, in general, very little detailed correlation between grade of metamorphism and degree of folding, or between times of deformation and of recrystallization.

2. When rupture occurs and is followed by differential movement on

[5] J. Goguel, Introduction à l'étude mécanique des déformations de l'écorce terrestre, *Services Carte Géol. France Mém.*, 1948.

the surface of rupture, a certain amount of heat may be generated by friction. This heat cannot exceed the elastically stored energy of deformation prior to rupture, and much of this in any case tends to be released not as heat but as clastic waves, as is well known in the case of seismic waves set up by sudden failure of rock masses along fault zones. The problem has been studied by Jeffreys [6] who gives numerical examples in which the temperature on a fault surface could reach the melting points of common rocks.

The well-known association of black, possibly glassy, pseudo-tachylite veins with mylonites suggests that frictional heat, when rapidly generated, can indeed bring about local fusion of the rocks affected. But even here, high temperatures have not been maintained sufficiently long to permit chemical reconstitution of the deformed rocks, for evidence of recrystallization and neomineralization is conspicuously lacking in mylonites. If frictional heat is seldom effective in promoting metamorphism in restricted zones of intense deformation, it must be of even less significance as a factor in regional metamorphism in which deformation is distributed through much greater thickness of rocks. By Niggli, Harker, and others it has been argued that dissipation of mechanically generated heat by conduction, though a slow process, generally is efficient enough to nullify the influence of heat from such sources in regional metamorphism.[7]

The contrary opinion has been expressed by some writers. For instance, De Lury has suggested that magmas may be developed on a large scale under deep-seated conditions by frictional heat.[8] His treatment of the problem is far from quantitative and must be regarded as unsubstantiated speculation. It has been argued, too, that in some cases the mutual relations in the field of metamorphic zones, stratigraphic sequences, and zones of intense deformation are more satisfactorily explained on the assumption that metamorphic temperatures were controlled by mechanically generated heat than by any other hypothesis.[9] In the absence of any knowledge as to possible existence and distribution of unexposed subjacent granitic or other igneous bodies, such opinions must be treated with caution.

A definite conclusion on the whole problem cannot be reached until such time as quantitative data may become available. In the meantime, the opinion that mechanically generated heat plays but a subordinate role

[6] H. Jeffreys, On the mechanics of faulting, *Geol. Mag.*, vol. 79, pp. 291–295, 1942.

[7] P. Niggli, *Die Gesteinsmetamorphose*, p. 204, Borntraeger, Berlin, 1924; A. Harker, *Metamorphism*, pp. 330, 331, Methuen, London, 1932.

[8] J. S. De Lury, Generation of magma by frictional heat, *Am. Jour. Sci.*, vol. 242, pp. 113–129, 1944.

[9] J. W. Ambrose, Progressive kinetic metamorphism of the Missi series near Flinflon, Manitoba, *Am. Jour. Sci.*, vol. 32, pp. 257–286, 1936. Cf. also A. Harker, *op. cit.*, pp. 331–338, 1932.

in maintaining temperatures of regional metamorphism seems well justified. Since it is thus unlikely that frictional heat contributes notably to
metamorphic temperatures, and in so far as high nonhydrostatic stresses
incidental to deformation are not essential to the crystallization of most
metamorphic minerals, deformation cannot be regarded as the principal
direct cause of regional metamorphism. In other words, deformation has
not, in most cases, played the principal role in bringing about the combination of physical conditions that has determined the nature of the
assemblage of metamorphic minerals. In particular instances stress incidental to deformation may locally raise the pressure to values permitting
the crystallization of a high-pressure phase such as kyanite or jadeite.

It is recognized that many mineral assemblages, which in reality are
unstable at relatively low temperatures within the metamorphic range,
may be heated to such temperatures for long periods of time without appreciable internal readjustment (metamorphism). If this were not so,
high-grade metamorphic assemblages, and high-temperature parageneses
found in plutonic rocks such as gabbro, would always be replaced by low-
temperature assemblages during slow cooling and unloading preparatory
to exposure of such rocks at the surface. It is equally well established
that when similar rocks are strongly deformed under otherwise identical
conditions, especially in the presence of aqueous solutions (often of magmatic origin), complete metamorphism takes place. So deformation may
exercise an all-important "catalytic" influence, initiating and promoting
the chemical adjustment which is the essence of metamorphism. This
applies especially to low-grade metamorphism (e.g., in development of
slates and phyllites), for chemical reaction in silicate assemblages is so
sluggish at low temperatures that chemical equilibrium may never be
reached except under the accelerating influence of contemporaneous
deformation.

Regional Metamorphism Resulting in Subhorizontal Schistosity. In
many parts of the world, and especially in pre-Cambrian terranes, there
are extensive areas of regionally metamorphosed rocks, throughout which
the schistosity (foliation) dips at low angles. It was suggested by Milch,
Daly, and others [10] that these rocks are products of "load metamorphism"
—chemical reconstitution under vertically directed load [11] at temperatures conditioned mainly by depth. Daly drew attention particularly to
the relatively undisturbed condition of bedding, and common coincidence

[10] L. Milch, Beiträge zur Lehre des Regionalmetamorphismus, *Neues. Jahrb.* Beilage-
Band 9, pp. 101–128, 1894; R. A. Daly, A geological reconnaissance between Golden
and Kamploops, B. C., *Canada Geol. Survey Mem. 68*, pp. 40–53, 110, 1915; Metamorphism and its phases, *Geol. Soc. America Bull.*, vol. 28, pp. 400–406, 1917.
[11] Actually the stress system set up by crustal loading would be much more complex than the simple vertical compressive stress assumed by such writers.

of schistosity with bedding, in pre-Cambrian schists of sedimentary origin in the Shuswap terrane of southern British Columbia. He noted that where the schists stand at high angles there is good evidence that dislocation was postmetamorphic. Further, he recorded that there is no obvious relation between grade of metamorphism and "amount of crustal deformation" as deduced from degree of folding.

The problem of progressive load metamorphism, with special reference to pelitic rocks, was more recently discussed in some detail by Born,[12] who concluded that sediments of deeply sinking geosynclines normally undergo epi-zonal load metamorphism prior to folding and that, in deformational metamorphism which normally accompanies folding, the imprint of the earlier load metamorphism may become obliterated or rendered too obscure to be recognized. Born gave the following ideal downward sequence for progressive load metamorphism of pelitic sediments in areas not affected by orogeny:

1. Zone of plastic clay (thickness about 600 m.)
2. Zone of nonplastic friable shales
3. Zone of slaty or shaly rocks with horizontal schistosity determined by vertically directed load (lower limit about 8,000 m., according to Daly)
4. Epi zone of regional metamorphism, with fully crystallized low-grade schists having horizontal schistosity

The interzonal boundaries would of course be indefinite. Passage from the zone of plastic clay to that of shale was said to be marked by conversion of colloidal clay minerals, with adsorbed potash, magnesia, lime, and soda, to finely crystalline sericitic mica and presumably to chlorite as well. As these become more coarsely crystalline in the third zone, horizontal schistosity develops, and there is gradual transition into the zone of fully reconstituted low-grade schists—the epi zone of regional metamorphism. It was suggested by Born that more precise definition and subdivision of the various zones might some day be founded upon the type and coarseness of mica as revealed by X-ray investigation.

The hypothesis of load metamorphism as advocated by Milch, Daly, and Born is open to serious criticism. The present writers agree that temperatures well within the metamorphic range may result solely from deep burial of sediments in sinking geosynclines, and further that sensitive mineral assemblages in salt beds and coals, and to a less extent in tuffs and clays, may become chemically reconstituted in response to the temperatures so imposed upon them. Suppose, in addition, that metamorphism of even the less sensitive silicate rocks does actually take place under physical conditions controlled solely by deep burial. The rocks in question, provided they maintain a high rigidity, admittedly are subject to

[12] A. Born, Ueber zonare Gliederung in höheren Bereich der Regionalmetamorphose, *Geol. Rundschau,* vol. 21, pp. 1–14, 1930.

a vertically directed force of the overlying load.   Surely, however, there is little justification for assuming that regionally subhorizontal schistosity must in general have developed under the direct influence of this vertical force, merely because we admit that such a force exists and because it is conceded that in one of the recognized general types of metamorphic deformation (that which we have termed "flattening") schistosity may develop in the plane normal to compression.   On the contrary, several independent lines of evidence point to a totally different interpretation of subhorizontal "undisturbed" schistosity:

1. There are many recorded instances where silicate rocks have certainly failed to respond chemically to long-sustained conditions of temperature and pressure induced by burial to great depth.

2. Schistosity typically is completely lacking in hornfelses, the normal products of contact metamorphism.   In hornfels fabrics there is thus no imprint of the vertically directed load that must have acted upon the rocks during their metamorphism.   If horizontal schistosity in regionally metamorphosed rocks is to be attributed to load, then a similar fabric should commonly be encountered in the rocks of deep-seated contact aureoles, and nonschistose hornfelses should be confined to aureoles formed at shallow depths.

3. Detailed mapping in some of the classic fields of "load metamorphism" has shown that other interpretations of the field data are at least equally satisfactory.   Brock's analysis of metamorphism in the Shuswap terrane of British Columbia—Daly's type area for load metamorphism—is significant in this connection.[13]   In this region the "Shuswap series," consisting of high-grade schists and gneisses intimately injected with granite, is overlain by a great thickness of metasediments and greenstones with subhorizontal or gently dipping schistosity.   Their metamorphism is attributed by Brock to high temperature resulting partly from deep burial and partly from copious injection of granitic magma into the lower members ("Shuswap series"), probably in Mesozoic time.   The invaded rocks are believed to have become domed up above the slowly rising intrusions of granitic magma, and at the same time to have yielded by "flattening" so that the resultant schistosity now conforms to the gently curving upper surfaces of the granite stocks.   It must be admitted that this hypothesis at least offers a possible alternative to that of load metamorphism.

4. According to Daly's hypothesis of load metamorphism, only minor deformation, incidental to recrystallization under load, accompanies development of schistosity.   In modern terminology, this type of schistosity could result from slight "flattening" in the plane normal to the applied load, or from mimetic crystallization of mica and amphiboles, with their

[13] B. B. Brock, The metamorphism of the Shuswap terrane of British Columbia, *Jour. Geology*, vol. 42, pp. 673–699, 1934.

long axes in the surfaces of maximum ease of growth afforded by sedimentary bedding. Petrofabric analysis of representative rock specimens from provinces of supposed load metamorphism should assist in elucidating this aspect of the problem. The available petrofabric data, though still meager, collectively are opposed to the hypothesis of load metamorphism. The petrofabric characters of two selected rocks from the Shuswap terrane, examined by Gilluly,[14] were found to resemble those of typical B-tectonites from the Alpine region of Europe, and point to strong penetrative movement during metamorphism. A strong lineation parallel to B, girdle patterns in the mica and quartz diagrams, and in one case the presence of several intersecting planes of schistosity constitute telling evidence against static crystallization under load. Osborne and Lowther[15] have arrived at similar conclusions from a much more comprehensive study of gently dipping, high-grade metasediments of the Grenville series in eastern Canada. These rocks bear the unmistakable imprint of strong deformation. The schistosity is "directly connected with the regional stresses that have produced the folding," for orientation diagrams of quartz and biotite "show a definite arrangement of the quartz axes, and biotite poles of {001}, about the tectonic axes of the fold." Most important of all, the tectonic axis b ($= B$), visibly expressed in the fabric by lineation parallel to b and by ac joints normal to b, maintains a constant trend throughout the whole area. A fabric resulting from load metamorphism should show axial symmetry about the axis of loading; or in the event of a faint lineation developing during flattening in the plane normal to the axis of loading, the trend of lineation should vary at random over any large area. The constant trend of lineation in the area described by Osborne and Lowther shows that lineation and schistosity are products of lateral flow in a single, well-defined direction. The movement plan is in fact incompatible with simple load metamorphism. Finally, brief reference may be made to the low-grade schists of eastern and central Otago in southern New Zealand. Here, too, the schistosity is substantially horizontal over wide areas. But petrofabric analysis has brought out the clearest evidence of strong penetrative movement, involving external rotation and microfolding about the direction of lineation b ($= B$) and slip upon intersecting sets of s-planes (including the visible surfaces of schistosity). Here again, the lineation conforms to a regionally constant direction.

For the reasons outlined above, vertically directed load is considered to be ineffective as a factor in the development of regionally subhorizontal

[14] J. Gilluly, Mineral orientation in some rocks of the Shuswap terrane as a clue to their metamorphism, Am. Jour. Sci., vol. 28, pp. 182–201, 1934.

[15] F. F. Osborne and G. K. Lowther, Petrotectonics at Shawinigan Falls, Quebec, Geol. Soc. America Bull., vol. 47, pp. 1343–1370, 1936 (especially p. 1355).

schistosity, except in so far as it may assist lateral spreading of deeply buried rocks already in process of plastic flow, in some definite direction determined by tangential forces. It is on the whole probable, though difficult to demonstrate, that, partly under the influence of superincumbent load, incipient mimetic crystallization of mica in deeply buried sediments of sinking geosynclines may at times intensity the existing s-planes of sedimentary stratification, which are the potential slip planes of subsequent metamorphic deformation, and thus may contribute slightly to the ultimate development of schistosity.

Sources of Heat for Regional Metamorphism. Regional metamorphism unquestionably reveals the existence, at the time of recrystallization, of temperatures several hundred degrees higher than those normally prevailing on the surface. Most, if not all, metamorphic reactions resulting from an increase in temperature are endothermic. The problem thus arises as to the source of the enormous amounts of heat required to convert, say, a shale to a high-grade gneiss. Heat developed during mechanical deformation may be disregarded as quantitatively very small (see page 660). The obvious remaining sources are

1. The normal outflow of radiogenic heat from the crust and mantle
2. Heat brought into the metamorphic rocks by upward migration of
    a. magma
    b. juvenile fluids, mainly water

With regard to 2a, we note that regional metamorphism typically does not occur at the time when basic magma, sometimes in copious amounts, is intruded into geosynclinal masses or erupted on the sea floor. Later magmas, mostly granitic or granodioritic, although they do act as heat sources (cf. their contact-metamorphic effects) are presumably in very large part palingenetic. Whatever the source of such magmas may be, they are not normal constituents of the earth's crust or mantle; to consider them as the main heat source of regional metamorphism merely displaces, without solving, the problem of the ultimate origin of the heat. The same applies to 2b; although it may be difficult to prove or disprove the existence, during metamorphism, of an upward flow of juvenile water rising from the mantle and transferring heat into the rocks undergoing metamorphism, we are left with the problem of finding where this water comes from, and why.

With regard to 1, it seems to be generally accepted that deep burial alone may account for metamorphic temperatures. An ordinary assumption is that sediments buried at the bottom of a geosynclinal mass 20 km. thick will automatically become heated to the temperature normally prevailing at that depth, as deduced from the geothermal gradient; and the distribution with depth of various metamorphic assemblages or facies is deduced from diagrams on which is plotted a "normal" depth-temperature curve. Without further emphasizing inherent uncertainties regarding the "normal" temperature distribution in the crust (see Chap. 15) it may be useful to point to difficulties arising from the assumption that regional metamorphism is just an effect of burial combined with a "normal" rate of heat flow from the interior to the surface of the earth.[16]

Consider first the depth at which various metamorphic facies would develop under the effect of a "normal gradient" of about 30°/km. near the surface gradually decreasing to about 10°/km. at the bottom of the crust, with an average of about 20°/km. Consider particularly the almandine-amphibolite facies, the low-temperature limit of which may be taken, for the sake of the argument, to correspond to a temperature of about 550°C. at a water pressure of 4,000 bars. Assume further that the water pressure is everywhere equal to the load pressure $P = \rho g h$. The equilibrium temperature for metamorphic reactions, as noted in Chap. 17, rises with increasing pressure at a rate which is generally, at high pressure, in the range $1 - 7 \times 10^{-2}$ deg./bar. Let β be this rate, while γ is the average geothermal gradient. If T_0 is the equilibrium temperature at pressure P_0, the equilibrium temperature T at pressure P is

$$T = T_0 + \beta(P - P_0) = T_0 + \beta(g_\rho h - P_0)$$

while the temperature at depth h (where the pressure is P) is γh. Equating the two yields

$$h = \frac{T_0 - \beta P_0}{\gamma - \beta g \rho}$$

Numerical substitution will convince the reader that if γ is anywhere near the "normal" value suggested above, the amphibolite facies could not develop at depths less than 30 or 35 km., while the granulite facies could exist only at depths greater than 80 km. or so. This is a strong suggestion that the geothermal gradient during metamorphism must be appreciably higher than its "normal" value.

A further consideration is pertinent. Imagine first a geosynclinal series gradually sinking as it thickens, and assume that the heat flow rising from the basement into the sediments retains at all times its usual surface value 1.2×10^{-6} cal./cm.^{2} sec. It is a simple matter to show that the sediments will heat up to the normal temperature corresponding to their depth of burial almost as fast as they are deposited; by the time 10 km. of sediments have accumulated, the bottom layers would be nearly at a temperature of 300°C. But remember now that when temperatures reach the metamorphic range, reactions start which will most generally

[16] See W. S. Fyfe, F. J. Turner, and J. Verhoogen, Metamorphic reactions and metamorphic facies, *Geol. Soc. America Mem. 73*, chap. 6, 1958.

be endothermic: heat is absorbed in the reaction, and the temperature cannot rise beyond the equilibrium point until the reaction is completed. Thus the temperature distribution at any time in the geosynclinal mass will depend on the nature and rate of the metamorphic reactions as well as on the rate at which heat flows into it from below, the general effect being that as long as reactions proceed, the temperature at any depth will be *less* than "normal." The effect is appreciable,[17] and cannot be dismissed on the grounds that it is transitory and will disappear when the reactions become complete; for the mineralogy recorded in metamorphic rocks is that corresponding to the temperature prevailing at the time of the last reaction, not that prevailing some time later. So the depth at which temperatures correspond to the almandine-amphibolite facies would be larger than that deduced in the preceding paragraph.

Thus, if one wishes to avoid the conclusion that 50 km. or more of material has been removed by erosion wherever rocks of the almandine-amphibolite facies are exposed, it is necessary to conclude that during regional metamorphism both the temperature gradient and the upward heat flow are considerably higher than, and perhaps twice or three times as great as, their normal values. Regional metamorphism is not a mere consequence of deep burial; it occurs only when and where heat, and possibly water, enter the rocks undergoing metamorphism in abnormal amounts. This conclusion has been reached many times before.[18]

As pointed out, regional metamorphism is typically associated, in space and time, with intense deformation leading to the formation of fold-mountain ranges. Mountain ranges of this type are believed, from geophysical evidence, to have "roots," *i.e.*, the crust is thicker below a mountain range than elsewhere. Thickening of the crust must thus occur at some stage of orogenesis. As the crust is presumably the source of a notable fraction of the radiogenic heat that reaches the surface, thickening of the crust also implies an increase in heat flow. It can be shown that if the crust is sufficiently radioactive, a notable thickening (*e.g.*, doubling the original thickness) would also be conducive to the development, at its base, of temperatures well within the range of melting of granitic rocks, which, as we noted in Chap. 15, is not true in the "normal" case for a "normal" thickness of the crust. This, however, does not imply that abundant granitic magma could form at the time of thickening; for the rate of heating would be extremely small, and melting could not occur earlier than some tens of millions of years after thickening (recall that a perfectly insulated piece of granite would require some 30 million years to melt itself). Thus thickening of the "root" may have something to do

---

[17] Fyfe, Turner, and Verhoogen, *op. cit.*, pp. 194, 195.

[18] W. H. Bucher, Fossils in metamorphic rocks; a review, *Geol. Soc. America Bull.*, vol. 64, pp. 275–300, 1953.

with postmetamorphic and postorogenic generation of andesitic magmas, but presumably has little to do with metamorphic processes themselves.

The source of metamorphic heat is thus, like that of magmatic heat, essentially unknown; both probably are connected with deep-seated (convective?) disturbances in the mantle. It may be worth pointing out that, although regional metamorphism is sometimes considered as an aftereffect of mountain-making, it is, from the standpoint of energy, a much more impressive phenomenon. (Compare the amount of energy involved in deformation of Alpine type with the amount of heat required to heat rocks to metamorphic temperatures, including heats of reaction and, locally, heats of fusion.) It might be more correct to describe mountain-making as a secondary mechanical effect of the thermal disturbances, whatever their nature may be, which also cause regional metamorphism.

Conclusion. Large-scale crustal deformation (alpine folding), batholithic intrusion, and regional metamorphism are geographically associated, broadly synchronous but partially independent manifestations of thermal (and mechanical?) disturbances in the mantle, not in the crust. Typically each of these processes increases in intensity from the margins toward the center of the orogenic zone, while intrusive activity and grade of metamorphism also increase (though not indefinitely) with depth. Highly simplified and conjectural relationships of these processes, as pictured by recent writers, are shown diagrammatically in Figs. 116 and 117. None of these schemes is entirely satisfactory.

In the upper zones, regional metamorphism is essentially a chemical adjustment of rocks to temperatures and pressures imposed by depth of burial in areas of abnormally high heat flow where, in addition, simultaneous or nearly simultaneous mechanical deformation is responsible for folding and development of schistosity. Deformation is almost essential to low-grade regional metamorphism of all but specially sensitive rocks (e.g., salts, vitric tuffs, and coals), in that it accelerates and renders effective chemical reactions that otherwise would proceed infinitely slowly. At still shallower depths and lower temperatures, on the margins of orogenic belts, regional metamorphism may pass laterally into almost purely cataclastic (kinetic) metamorphism, without noteworthy chemical reconstitution of the rocks affected.

In the lower levels, where regional metamorphism reaches medium to high grades, heat transported upward by invading bodies of granitic magma makes an important contribution to metamorphic temperatures. Magmatically derived fluids assist deformation in accelerating metamorphic reactions and may be active in bringing about metasomatism on a large scale. At still lower levels, high-grade regional metamorphism, ultimately culminating in fusion of rocks to give primary magmas, occurs

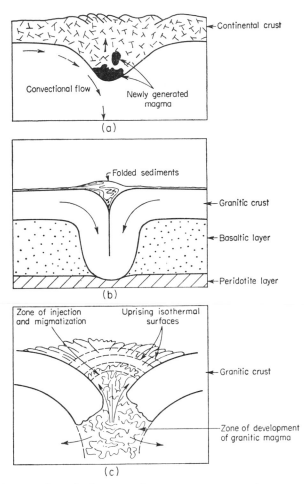

FIG. 116. Sections through the roots of orogenic zones according to various convectional hypotheses of orogeny. (*a*) Down-buckling of the continental crust due to convection (shown by arrows) in the mantle. (*As pictured by D. Griggs, Am. Jour. Sci., vol. 237, p. 644, 1939*). (*b*) Down-buckling of granitic layer (blank) through basaltic layer (stippled) to impinge on peridotite shell of the mantle. A generalized section through the Alps has been superposed to scale. (*After H. H. Hess, Am. Philos. Soc. Proc., vol. 79, pp. 75, 79, 1938.*) Note that deformation and flow in the "peridotite," layer obviously are inadequately represented. (*c*) Melting of down-buckled granitic layer, with plutonic intrusion, migmatization, and regional metamorphism in the overlying folded sedimentary cover. (*As pictured by W. Q. Kennedy, Geol. Mag., vol. 85, p. 233, 1948.*)

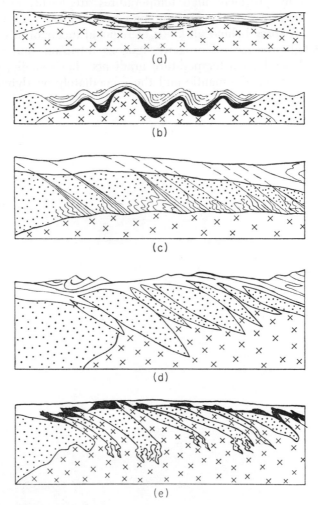

Fig. 117.   C. E. Wegmann's conception of regional metamorphism in relation to granitization and migmatite development in orogenic zones. (*After E. Raguin, Géologie du Granite, pp. 122, 123, 1946.*) (*a*) Section across a sediment-filled geosyncline prior to folding. The "infrastructure" already is encroaching on the lower part of the sedimentary filling. (*b*) Section across the same geosyncline after moderate folding. Note the advance of the granitized "infrastructure" into cores of the folded sedimentary "superstructure." (*c*) and (*d*) Sections through another orogenic zone showing successive stages of breaking down of the basement rocks of the "superstructure" by large-scale thrust faulting. (*e*) Section through the same zone at an advanced stage of orogeny. The "infrastructure" has invaded dislocation zones in the "superstructure" and has encroached considerably upon the "superstructure" as a whole. Covering sediments of "superstructure," ruled. Basement rocks of "superstructure," stippled. "Infrastructure," crossed. Rocks affected by regional metamorphism, solid black.

where generally prevalent high temperatures are locally elevated still further by concentration of energy by some means unknown.   The fundamental driving force behind all kinds of regional metamorphism, as well as the related processes of orogeny, igneous intrusion, and volcanism, must itself depend upon temperature gradients—horizontally, vertically, and in time—within the mantle and the immediately overlying portions of the earth's crust.

Index

Arizona potassic province, 189
Arran, western Scotland, 332–334
Artificial melts, crystallization from (see Phase diagrams)
Ascension Island volcanic province, 189
Assimilation, magmatic, 85–87, 148–154, 156–160
(See also Basaltic magma; Granitic magma)
Associations of igneous rocks, 79–80
Atlantic Ocean petrographic province, 187–189
Atlantic suite, 80
Atlas, L., 60, 131
Auckland volcanic province, New Zealand, 175
Aureoles, contact, 495–497, 657–659
Australia, charnockites, 346–347
granites, 346–347, 375
southeastern, alkaline basalt association, 190
Western, West Kimberley potassic province, 240–242
Autointrusion, 84
Autometasomatism, 578
Azeotropic mixtures, 406

B-tectonites, 638–645
fold pattern in, 643–645
B ∧ B′-tectonites, 645–647
B ⊥ B′-tectonites, 645–647
Backlund, H. G., 368, 369, 557
Baddley, E., 352
Bailey, E. B., 221, 588
Bain, G., 618
Baker, B., 576
Balk, R., 323, 351, 354–357
Ballachulish granodiorite, Scotland, 158–159
Bancroft, D., 346
Bancroft, Ontario, alkaline plutonic rocks, 397–398
Banks Peninsula, New Zealand, 174
Banno, S., 541
BaO, in basic rocks, 198, 253
content of potassic lavas, 237, 238, 241, 242, 248, 249
Barlow, A. E., 361, 397
Barrer, R. M., 44
Barrett, C. S., 605, 615, 619, 628
Barrow, G., 490–491
Barth, T. F. W., 90, 99, 105, 108, 113, 145, 191, 199, 200, 209, 215, 217, 232, 322, 325, 330, 344, 345, 359, 375, 384, 395, 499, 504, 545
Bartrum, J. A., 262, 266

Basalt, fusion, 154–155, 442–443
in volcanic associations, 164–202, 203–234, 236–250, 272–288
Basaltic magma, assimilation by, 148–154
of granite, 148–150
of limestone, 152–154, 200
of shale, 150–151
of siliceous rocks, 151–152
crystallization, 146–148
differentiation, 146–148, 167–168, 172–173, 175–184, 189, 196–200, 211–216, 225–226, 229–230, 232–234, 306
iron enrichment in, 142–143, 147, 306
origin, 227–233, 442–446
primary, 164, 227–233, 432
Basaltic magma types, 203–209, 227–223
Basement rocks, influence on volcanic associations, 232, 249–250, 287, 306–307
Basic fronts, 366
Basic rocks, contact metamorphism (see Hornfelses, basic)
definition, 57
reaction with granitic magma, 157–158
regional metamorphism (see Schists, basic)
Basification of granite, 156–159
Batholiths (see Granite batholiths)
Battey, M. H., 266–272
Bearth, P., 540
Beck, P. A., 619
Becke, F., 488, 592–595
Becker, C. F., 596, 606, 651
Bederke, E., 251, 255
Beger, P. J., 254
Bell, J. F., 615
Benson, W. N., 75, 76, 112, 151, 164–172, 174, 183, 184, 191, 192, 262, 263, 307–310, 316–320
Bewegungsbild, 606
Bidwell, R. M., 408
Billings, M. P., 336, 353, 492, 493, 571, 653
Bingham, Utah, 565
Biotite, differential fusion, 254
of igneous rocks, 135–139
resorption, 138–139
stability, 135–139
Birch, F., 100, 436, 438
Birunga volcanic area, Africa, 236–240
Black, G. P., 221
Blackwelder, E., 352
Bloxam, T. W., 541
Bohemian Mittelgebirge, alkaline lavas, 191, 193
Boiling point, second, 409, 428